AGROECOSYSTEM DIVERSITY

AGROECOSYSTEM DIVERSITY

RECONCILING CONTEMPORARY AGRICULTURE AND ENVIRONMENTAL QUALITY

Edited by

GILLES LEMAIRE
Honorary Director of Research, INRA Lusignan, France

PAULO CÉSAR DE FACCIO CARVALHO
UFRGS, Federal University of Rio Grande do Sul, Porto Alegre, Brazil

SCOTT KRONBERG
USDA - Agricultural Research Service, Northern Great Plains Research Laboratory, Mandan, North Dakota, USA

SYLVIE RECOUS
Director of Research, INRA, FARE laboratory, Reims, France

Academic Press is an imprint of Elsevier
125 London Wall, London EC2Y 5AS, United Kingdom
525 B Street, Suite 1650, San Diego, CA 92101, United States
50 Hampshire Street, 5th Floor, Cambridge, MA 02139, United States
The Boulevard, Langford Lane, Kidlington, Oxford OX5 1GB, United Kingdom

Library of Congress Cataloging-in-Publication Data
A catalog record for this book is available from the Library of Congress

British Library Cataloguing-in-Publication Data
A catalogue record for this book is available from the British Library

ISBN: 978-0-12-811050-8

For information on all Academic Publications visit our website at
https://www.elsevier.com/books-and-journals

Publisher: Andre Gerhard Wolff
Acquisition Editor: Nancy Maragioglio
Editorial Project Manager: Amy Clark
Production Project Manager: Omer Mukthar
Designer: Matthew Limbert

Typeset by TNQ Technologies

In Memoriam

Martino Nieddu

On the 11th of June 2018, Martino Nieddu, Professor of Economics at University of Reims Champagne-Ardenne, passed away. Since the beginning of his PhD in the early 1990s, he focused on agricultural dynamics. More precisely, he worked on the long-period trajectories of French agriculture. To understand it, he used a regulation theory approach in order to depict the endogeneous coevolution of agricultural accumulation regimes, agricultural practices, and agroindustry institutions. At the end of the 1990s, he started to highlight the emerging interest of agroindustries in non—food valorization. It brought him to be one of the first and main French economists to study green chemistry and biorefinery and to introduce a transition to sustainability studies in France. His colleagues and Ph-D students will remember him for his great ability to manage the constant back and forth between deep empirical knowledge and a clear view of theoretical advances, especially in the field of Agricultural and Sustainability Transitions.

His influence on these research topics is certainly related to his strong involvement in the organization at the head of the laboratory REGARDS at the University of Reims Champagne-Ardenne. Notably, he opened this research unit to pluridisciplinary investigations with economics and other social sciences, but also with natural sciences.

Contents

III

HETEROGENEITY WITHIN AND AMONG AGROECOSYSTEMS AND DYNAMICS OF BIODIVERSITY

11. Can Increased Within-Field Diversity Boost Ecosystem Services and Crop Adaptability to Climatic Uncertainty?

ISABELLE LITRICO AND CHRISTIAN HUYGHE

12. The Future of Sustainable Crop Protection Relies on Increased Diversity of Cropping Systems and Landscapes

JONATHAN STORKEY, TOBY J.A. BRUCE, VANESSA E. MCMILLAN, AND PAUL NEVE

13. Grassland Functional Diversity and Management for Enhancing Ecosystem Services and Reducing Environmental Impacts: A Cross-Scale Analysis

MICHEL DURU, L.D.A.S. PONTES, J. SCHELLBERG, J.P. THEAU, AND O. THEROND

14. Local and Landscape Scale Effects of Heterogeneity in Shaping Bird Communities and Population Dynamics: Crop-Grassland Interactions

VINCENT BRETAGNOLLE, GAVIN SIRIWARDENA, PAUL MIGUET, LAURA HENCKEL, AND DAVID KLEIJN

IV

DIVERSIFIED AGROECOSYSTEMS AT FARM LEVELS FOR MORE SUSTAINABLE AGRICULTURE PRODUCTION?

15. Integration of Crop and Livestock Production in Temperate Regions to Improve Agroecosystem Functioning, Ecosystem Services, and Human Nutrition and Health

SCOTT L. KRONBERG AND JULIE RYSCHAWY

16. Integrated Crop-Livestock Systems as a Solution Facing the Destruction of Pampa and Cerrado Biomes in South America by Intensive Monoculture Systems

ANIBAL DE MORAES, PAULO CÉSAR DE F. CARVALHO, CARLOS ALEXANDRE COSTA CRUSCIOL, CLAUDETE REISDORFER LANG, CRISTIANO MAGALHÃES PARIZ, LEONARDO DEISS, AND R. MARK SULC

17. Toward Integrated Crop-Livestock Systems in West Africa: A Project for Dairy Production Along Senegal River

GILLES LEMAIRE, BERNARD GIROUD, BAGORÉ BATHILY, PHILIPPE LECOMTE, AND CHRISTIAN CORNIAUX

18. Silvopastoral Systems in Latin America for Biodiversity, Environmental, and Socioeconomic Improvements

ROGERIO MARTINS MAURICIO, RAFAEL SANDIN RIBEIRO, DOMINGOS SÁVIO CAMPOS PACIULLO, MAURONI ALVES CANGUSSÚ, ENRIQUE MURGUEITIO, JULIAN CHARÁ, AND MARTHA XOCHITL FLORES ESTRADA

V

SOCIO-ECONOMIC OPPORTUNITIES FOR AND LOCKING EFFECTS AGAINST DIVERSIFICATION OF AGROECOSYSTEMS AT FARM AND BEYOND FARM LEVELS

19. The Economic Drivers and Consequences of Agricultural Specialization

DAVID J. ABSON

20. Practices of Sustainable Intensification Farming Models: An Analysis of the Factors Conditioning Their Functioning, Expansion, and Transformative Potential

GIANLUCA BRUNORI, SIMONA D'AMICO, AND ADANELLA ROSSI

21. Environmental Benefits of Farm- and District-Scale Crop-Livestock Integration: A European Perspective

PHILIPPE LETERME, THOMAS NESME, JOHN REGAN, AND HEIN KOREVAAR

22. Payment for Unmarketed Agroecosystem Services as a Means to Promote Agricultural Diversity: An Examination of Agricultural Policies and Issues

CLEMENT A. TISDELL AND CLEVO WILSON

23. Agricultural Policies and the Reduction of Uncertainties in Promoting Diversification of Agricultural Productions: Insights From Europe

AUDE RIDIER AND PIERRE LABARTHE

24. Technological Lock-In and Pathways for Crop Diversification in the Bio-Economy

MARIE-BENOIT MAGRINI, NICOLAS BÉFORT, AND MARTINO NIEDDU

VI GLOBAL ASPECTS

25. Opening to Distant Markets or Local Reconnection of Agro-Food Systems? Environmental Consequences at Regional and Global Scales

GILLES BILLEN, LUIS LASSALETTA, JOSETTE GARNIER, JULIA LE NOË, EDUARDO AGUILERA, AND ALBERTO SANZ-COBEÑA

26. Scaling From Local to Global for Environmental Impacts From Agriculture

MARTY D. MATLOCK

27. An Evolutionary Perspective on Industrial and Sustainable Agriculture

JOHN GOWDY AND PHILIPPE BAVEYE

28. Current and Potential Contributions of Organic Agriculture to Diversification of the Food Production System

VERENA SEUFERT, ZIA MEHRABI, DOREEN GABRIEL, AND TIM G. BENTON

Contributors

David J. Abson Leuphana University, Lüneburg, Germany

Eduardo Aguilera Universidad Pablo de Olavide, Sevilla, Spain

Bagoré Bathily Laiterie du Berger, Richard Toll, Sénégal

Philippe Baveye UMR ECOSYS, AgroParisTech, Université Paris-Saclay, Thiverval-Grignon, France

Nicolas Béfort Chair in Industrial Bioeconomy, Neoma Business School; European Center in Biotechnology and Bioeconomy, Reims, France

Tim G. Benton School of Biology, University of Leeds, Leeds, United Kingdom

Isabelle Bertrand Eco&Sols, INRA, Univ Montpellier, CIRAD, IRD, Montpellier SupAgro, Montpellier, France

Gilles Billen Sorbonne Université, CNRS, EPHE, UMR 7619 METIS, Paris, France

Juliette Bloor INRA-UREP, Clermont-Ferrand, France

Timothy M. Bowles University of California Berkeley, Department of Environmental Science, Policy, and Management, Berkeley, California, USA

Vincent Bretagnolle CEBC-CNRS, Beauvoir-sur-Niort, France

Toby J.A. Bruce School of Life Sciences, Keele University, Staffordshire, United Kingdom

Gianluca Brunori University of Pisa - Department of Agriculture, Food and Environment (DAFE) - ITALY

Mauroni Alves Cangussú Centro Brasileiro de Pecuária Sustentável, Imperatriz, Brazil

Paulo César de F. Carvalho University Federal of Rio Grande do Sul (UFRGS), Porto Alegre, Brazil

Abad Chabbi Institut National de la Recherche Agronomique (INRA), URP3F, Lusignan, France; INRA, Ecosys, Thiverval-Grignon, France

Julian Chará CIPAV – Centro para la Investigación en Sistemas Sostenibles de Producción Agropecuaria, Cali, Colombia

Juan Cruz Colazo EEA San Luis, INTA & National University of San Luis, Villa Mercedes, Argentina

Christian Corniaux CIRAD, Unité SELMET, Dakar, Sénégal

Carlos Alexandre Costa Crusciol São Paulo State University (UNESP), College of Agricultural Science, Botucatu, Brazil

Simona D'Amico University of Pisa - Department of Agriculture, Food and Environment (DAFE) - ITALY; Union for Ethical BioTrade, Amsterdam - The NETHERLANDS

William Deen University of Guelph, Department of Plant Agriculture, Guelph, Ontario, Canada

Leonardo Deiss Doutorando, University Federal of Paraná (UFPR), Curitiba, Brazil

Luc Delaby INRA-PEGASE, Agrocampus Ouest, Rennes, France

Christian Dupraz INRA, UMR System, University of Montpellier, France

Michel Duru AGIR, INRA, Université de Toulouse, Auzeville, France

Martha Xochitl Flores Estrada Fundación Produce Michoacán, Morelia, Mexico

Alan J. Franzluebbers USDA – Agricultural Research Service, Raleigh, NC, United States

Doreen Gabriel Institute for Crop and Soil Science, Julius Kühn-Institut, Braunschweig, Germany

Josette Garnier Sorbonne Université, CNRS, EPHE, UMR 7619 METIS, Paris, France

Francois Gastal FERLUS, INRA, Lusignan, France

Amélie C.M. Gaudin University of California Davis, Department of Plant Sciences, Davis, California, USA

Bernard Giroud SAFE Nutrition, 8998 Sacré Coeur 3, Dakar, Sénégal

John Gowdy Rensselaer Polytechnic Institute, Troy, NY, United States

Henrik Hauggaard-Nielsen Department of People and Technology, Roskilde University, Roskilde, Denmark

Laura Henckel CEBC-CNRS, Beauvoir-sur-Niort, France

John Hendrickson Northern Great Plains Research Laboratory, USDA-Agricultural Research Service, Mandan, North Dakota, United States

Olivier Huguenin-Elie Agroscope, Forage Production and Grassland Systems, Zurich, Switzerland

Christian Huyghe Paris Siège - INRA - 147 rue de l'Université, F-75338, Paris, France

Erik Steen Jensen Swedish University of Agricultural Sciences, Department of Biosystems and Technology, Alnarp, Sweden

Eric Justes Formerly INRA, Joint Research Unit AGIR (Agroecology, Innovations, Territories), INRA-INPT-University of Toulouse, Castanet Tolosan, France; Currently CIRAD, Joint Research Unit SYSTEM, CIRAD - INRA SupAgro, Montpellier, France

David Kleijn Plant Ecology and Nature Conservation, Wageningen University, Wageningen, The Netherlands

Katia Klumpp INRA-UREP, Clermont-Ferrand, France

Hein Korevaar Plant Research International, Wageningen, The Netherlands

Scott L. Kronberg USDA – Agricultural Research Service, Northern Great Plains Research Laboratory, Mandan, ND, United States

Pierre Labarthe INRA-SAD, UMR AGIR, Castanet-Tolosan Cedex, France

Claudete Reisdorfer Lang University Federal of Paraná (UFPR), Curitiba, Brazil

Gwenaëlle Lashermes FARE laboratory, INRA, Université Reims Champagne-Ardenne, Reims, France

Luis Lassaletta CEIGRAM, Agricultural Production Universidad Politécnica de Madrid, Madrid, Spain

Gerry Lawson Centre for Ecology and Hydrology, Edinburgh, Scotland

Philippe Lecomte CIRAD, Unité SELMET, Dakar, Sénégal

Gilles Lemaire INRA, Centre de Recherche Poitou-Charentes, 86600, Lusignan, France

Julia Le Noë Sorbonne Université, CNRS, EPHE, UMR 7619 METIS, Paris, France

Philippe Leterme Agrocampus Ouest, Rennes, France; UMR 1069 SAS INRA/Agrocampus Ouest, Rennes, France

Isabelle Litrico P3F UR 004 - INRA - Le Chêne RD150, F-86600, Lusignan, France

Marie-Benoit Magrini AGIR, Université de Toulouse, INRA, Castanet-Tolosan, France

Marty D. Matlock University of Arkansas, Fayetteville, AR, United States

Rogerio Martins Mauricio Bioengineering Department, Universidade Federal de São João del-Rei (UFSJ), São João del-Rei, Brazil

Vanessa E. McMillan Rothamsted Research, Harpenden, United Kingdom

Zia Mehrabi Institute for Resources, Environment and Sustainability (IRES), University of British Columbia, Vancouver, BC, Canada

Paul Miguet CEBC-CNRS, Beauvoir-sur-Niort, France

Anibal de Moraes University Federal of Paraná (UFPR), Curitiba, Brazil

Enrique Murgueitio CIPAV – Centro para la Investigación en Sistemas Sostenibles de Producción Agropecuaria, Cali, Colombia

Thomas Nesme Bordeaux Sciences Agro, University of Bordeaux, Gradignan, France; UMR 1391 ISPA, Villenave-d'Ornon, France

Paul Neve Rothamsted Research, Harpenden, United Kingdom

Martino Nieddu REGARDS, Université de Reims Champagne-Ardenne, Reims, France

Domingos Sávio Campos Paciullo Embrapa Dairy Cattle, Rua Eugênio do Nascimento, Juiz de Fora, Brazil

Cristiano Magalhães Pariz UNESP, School of Veterinary Medicine and Animal Science, Botucatu, Brazil

Sylvain Pellerin INRA, Bordeaux Sciences Agro, Univ. Bordeaux, ISPA, Villenave-d'Ornon, France

Mark B. Peoples CSIRO Agriculture and Food, Canberra, ACT, Australia

L.D.A.S. Pontes IAPAR - Agronomic Institute of Paraná, Ponta Grossa-PR, Brazil

Sylvie Recous FARE laboratory, INRA, Université Reims Champagne-Ardenne, Reims, France

John Regan UMR 1391 ISPA, Villenave-d'Ornon, France

Leah L.R. Renwick University of California Davis, Department of Plant Sciences, Davis, California, USA

Rafael Sandin Ribeiro Bioengineering Department, Universidade Federal de São João del-Rei (UFSJ), São João del-Rei, Brazil

Aude Ridier Agrocampus Ouest, UMR SMART LERECO, Rennes Cedex, France

Adanella Rossi University of Pisa - Department of Agriculture, Food and Environment (DAFE) - ITALY

Cornelia Rumpel CNRS, Institute of Ecology and Environmental Sciences Paris, (IEES), Thiverval-Grignon, France

Julie Ryschawy Université de Toulouse, AGIR UMR 1248, INRA, INPT-ENSAT, Auzeville, France

Alberto Sanz-Cobeña ETSI Agronómica, Alimentaria y Biosistemas. Universidad Politécnica de Madrid, Madrid, Spain

J. Schellberg Institute of Crop Science and Resource Conservation, University of Bonn, Bonn, Germany

Verena Seufert Institute for Resources, Environment and Sustainability (IRES), University of British Columbia, Vancouver, BC, Canada; Institute of Meteorology and Climate Research - Atmospheric Environmental Research (IMK-IFU), Karlsruhe Institute of Technology (KIT), Garmisch-Partenkirchen, Germany

Gavin Siriwardena Terrestrial Ecology, British Trust for Ornithology, The Nunnery, Norfolk, United Kingdom

Jonathan Storkey Rothamsted Research, Harpenden, United Kingdom

R. Mark Sulc The Ohio State University, Columbus, OH, United States

J.P. Theau AGIR, INRA, Université de Toulouse, Auzeville, France

O. Therond UMR LAE, INRA, Université de Lorraine, Colmar, France

Clement A. Tisdell School of Economics, The University of Queensland, Brisbane, Australia

Cairistiona F.E. Topp Crop and Soil Systems, SRUC, Edinburgh, United Kingdom

Françoise Vertès INRA-SAS, Agrocampus Ouest, Rennes, France

Christine A. Watson Crop and Soil Systems, SRUC, Aberdeen, United Kingdom

Jeroen Watté Werkgroep voor Rechtvaardige en Verantwoorde Landbouw, Brussels, Belgium

Michael Williams Department of Botany, School of Natural Sciences, Trinity College Dublin, Dublin, Ireland

Clevo Wilson QUT Business School, Economics and Finance, Queensland University of Technology, Brisbane, Australia

Foreword

TRADEOFFS BETWEEN DIVERSITY AND INTENSIFICATION AS THE BASIS FOR SUSTAINABLE AGRICULTURE SYSTEMS

Most of the negative environmental impacts of modern agriculture are in general attributed to a too-high level of use of energy and chemical inputs for achieving high levels of food production necessary for feeding a very large human population. So intensification of agriculture seems to have reached its limit in most industrialized countries, and some people are calling for a limitation or even a decrease of the intensification per unit land area to protect or restore the environment. But other people claim that if such a solution were to be generalized across the whole planet, it would inevitably create food security problems in the near future. Is this dilemma absolutely irremediable? Or is there any degree of freedom for resolving this contradiction? It is the question that this book proposes to deal with.

To answer this question, a working hypothesis has been formulated and is developed and analyzed through the different chapters: the negative impacts of modern agriculture on the environment we can observe today could be linked more to a too-high degree of simplification and/or homogeneity of agriculture systems at fields, farm, landscape, and region levels than to a too-high level of intensification of production. As historically, in all industrialized countries, intensification of agriculture production has been strictly linked with field and farm size increase, simplification of cropping systems, reduction in the range of production, disconnection between crop and livestock production systems, etc. These processes being the result of the paradigm of economy of scale, leading to the strong uniformity of landscapes and regions, it is difficult to distinguish and to analyze the causality chain of processes responsible for this deterioration of the environment. If the cause of these negative impacts would be more directly linked to the loss in diversity of agriculture production systems, then it should be possible, by restoring this diversity, to maintain possibilities for increasing agriculture production while minimizing environmental impacts.

Environmental impacts linked to emissions to the atmosphere and hydrosphere are mainly due to an imbalance between C–N–P decoupling and coupling processes within the agroecosystem, which has been highly aggravated by the historical link between intensification, simplification, and homogeneity of agriculture production systems. Resilience of agroecosystems is achieved by intimately linking C–N–P decoupling processes for providing nutrients resources to organisms with recoupling processes for recycling these nutriments and for reducing losses. Intensification of agriculture production tends to increase decoupling and to reduce recoupling for increasing nutrient availability. Hence, a high diversity within agroecosystems can be viewed as a means to restore a better balance between decoupling and recoupling through spatial and temporal interactions among components of the system. Thus, a biogeochemical functional analysis of agroecosystems must be performed for developing biogeochemical engineering of agroecosystems at the different levels of field, farm, or landscape. These aspects are developed in Sections I and II.

Loss of biodiversity is also highly linked to the decrease in diversity within agroecosystems and among them at the landscape level as the consequence of the reductions in trophic networks and habitats. In the same way, the genetic diversity within agroecosystems has to be reconsidered in light of this integrated vision. These aspects are developed in Section III.

By restoring and increasing "diversity" at all levels of organization, field, farm, landscape, and region, we can postulate that it should be possible to maintain a high level of productivity of agroecosystems while a satisfactory level of environment quality and biodiversity could be maintained or restored. Grassland-arable cropping integration and agroforestry are two important ways for diversifying agriculture systems owing to their high degree for coupling C–N–P. Examples and performances of these integrated systems in different regions of the planet are analyzed. This aspect is developed in Section IV.

Diversification of agriculture systems must be analyzed not only at farm but also at landscape, regional, and continental levels for matching environmental and biodiversity issues with socioeconomic drivers. If diversity of agroecosystems cannot be maximized at farm level owing to several socioeconomic constraints, then it should be necessary to analyze at which conditions some of the necessary "recoupling" processes could be achieved beyond the farm gate. Socioeconomic analyses are necessary for studying the possible ways for disconnecting specialization/homogeneity from intensification (economy of scale) and for promoting the links between intensification and diversity (economy of scope). This analysis has to be performed at farm but also at regional, national, and international levels for identifying locking and lever effects for optimizing agriculture policies. These aspects are analyzed in Section V.

Global analysis on equilibrium between food/nonfood production of agriculture (agroforestry) and animal/plant in human diets (integrated crop-livestock systems) is necessary for optimizing the tradeoff between agriculture production and environment quality at the level of the planet. How do we balance between local objectives such as water quality or biodiversity with global objectives such as greenhouse gas emission and climate change? Which diversity for which objectives? These aspects are developed in Section VI.

Sustainable agroecosystems should be viewed as engineering systems conceived as connected networks of decoupling-recoupling spots that required a high temporal and spatial diversity. In such a system, energy and nutrients should circulate and recycle efficiently, making high overall productivity attainable without overly high environmental impacts. This general principle is very simple. Nevertheless, there exists a wide variety of solutions for sustainable agroecosystems through local combinations of elementary and highly diverse components. In conclusion, each solution must be conceived, developed, calibrated, and evaluated according to local constraints and should vary from one place to another. But they should all abide by the same general principles of increasing the level of diversity.

Gilles Lemaire (INRA, France)
Paulo Carvalho (UFRGS, Brazil)
Scott Kronberg (USDA, USA)
Sylvie Recous (INRA, France)
September 20, 2018

SECTION I

C–N–P CYCLES IN AGROECOSYSTEMS AND IMPACTS ON ENVIRONMENT

CHAPTER

1

Plant—Soil Interactions Control CNP Coupling and Decoupling Processes in Agroecosystems With Perennial Vegetation

Cornelia Rumpel[1], *Abad Chabbi*[2,3]

[1]CNRS, Institute of Ecology and Environmental Sciences Paris, (IEES), Thiverval-Grignon, France; [2]Institut National de la Recherche Agronomique (INRA), URP3F, Lusignan, France; [3]INRA, Ecosys, Thiverval-Grignon, France

INTRODUCTION

Plants are living organisms that are of crucial importance for biogeochemical cycling and soil quality. As autotrophic organisms, they play an important role in coupling of CNP and other elements through biomass formation with CO_2 fixation by photosynthesis and mineral nutrient uptake as key processes (Fig. 1.1). By the production of above- and belowground litter, plants are also the most important source of organic matter as C and an energy source for various ecosystem processes, including soil organic matter (SOM) formation (Kuzyakov and Domanski, 2000). Moreover, their impact on biogeochemical cycling of C, N, and P may control C sequestration and nutrient release. Most plant litter input into soils occurs belowground, the way that soil carbon is mostly root carbon (Rasse et al., 2005), its distribution being controlled by root systems of the vegetation (Jogbbagy and Jackson, 2000). Roots are particularly efficient in promoting aggregation of soil mineral particles and soil structural stability, but they may also contribute to SOM formation through their chemical composition, which is more chemically recalcitrant, compared to the one from aboveground biomass. Roots are located in close proximity to soil minerals, and the carbon they release is thus prone to stabilization through interactions with the mineral phase (Rasse et al., 2005). Moreover, root litter is deposited across different soil layers, even if preferentially within shallow horizons, thus contributing to SOM enrichment of deep horizons (Rumpel and Kögel-Knabner, 2011).

Growing plants have nutrient requirements, which they are able to meet through mobilization

Agroecosystem Diversity
https://doi.org/10.1016/B978-0-12-811050-8.00001-7

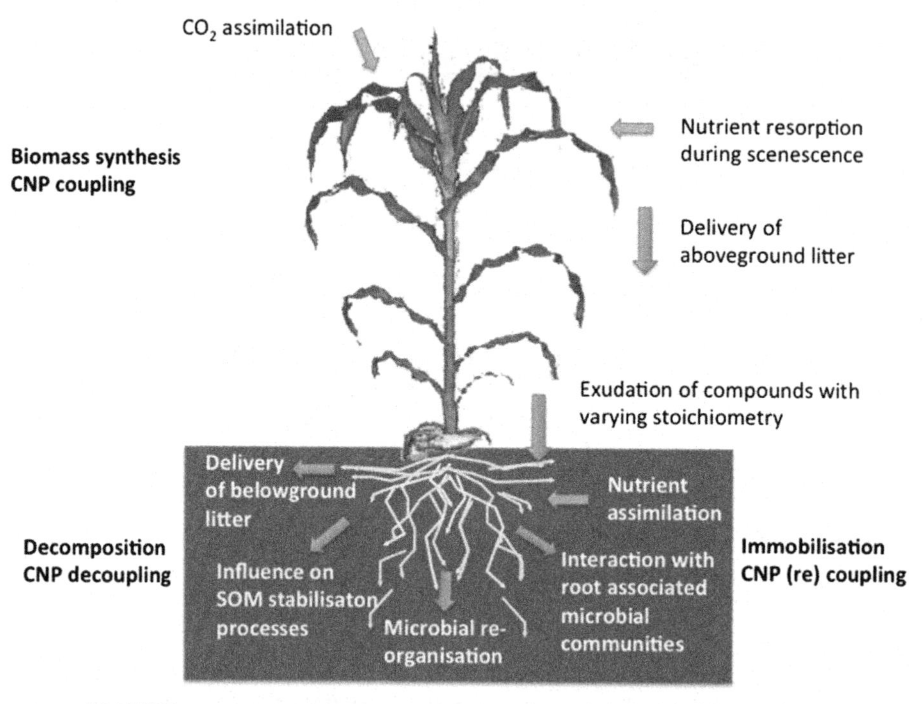

FIGURE 1.1 Plant influences elemental coupling and belowground processes.

from the soils' mineral phase and its organic matter pool, which is a large nutrient reservoir. However, N and P present in SOM in organic form need to be transformed into mineral form (NH_4^+, NO_3^-, PO_4^-) before plant uptake. This process requires CNP decoupling by heterotrophic microorganisms. Through rhizosphere processes, plants are able to control microbial activity leading to SOM decomposition and mineral nutrient release. They also control nutrient release from soil minerals through their influence on sorption/desorption processes and through symbiosis with mycorrhizal fungi. These fungi receive C assimilates from the plant and, in turn, provide access to soil nutrients that are not directly available to plants. Symbiosis with mycorrhizal fungi may not only be beneficial in terms of nutrient acquisition for plants, but it may also, through the competition with decomposers, control soil C sequestration (Averill et al., 2014). Moreover, plants are associated with other beneficial organisms such as rhizobia and growth-promoting bacteria, which may determine C, N, and P cycling at the root's surface (Nuccio et al., 2013).

In agricultural systems with perennial vegetation, plant activity may be influenced by management practices. For example in grassland, mowing may induce plant reactions and increase root exudation, thereby stimulating the release of SOM-bound nitrogen, which will be used for regrowth after defoliation (Hamilton et al., 2008). In grasslands, plant community composition is of crucial importance not only in terms of aboveground productivity, but also in terms of belowground processes. It was found that high biodiversity increases soil aggregation (Peres et al., 2013) as well as carbon storage, mainly through the properties of the associated root systems (Lange et al., 2015). Moreover, plants react strongly to environmental disturbance, such as increasing atmospheric CO_2 concentrations, increasing temperatures, and drought. These reactions need to be understood for single plants as well as for plant communities to be used to attenuate environmental changes. We hypothesized that the management of plant–soil interactions is of crucial importance for element cycling and the sustainability of agroecosystems with perennial vegetation. The aim of this chapter is to review the mechanisms by which single plants and plant communities affect C, N, and P cycles and thereby ecosystem services. In agroecosystems with perennial vegetation, these effects may then be managed to optimize the provision of ecosystem services, in particular, nutrient availability, soil protection against erosion, and C storage.

COUPLING AND DECOUPLING OF C, N, AND P CYCLES: A PREREQUISITE FOR PROVISION OF ECOSYSTEM SERVICES FROM AGROECOSYSTEMS

Agroecosystems, in particular through sustainable use of soils, may provide important regulating, as well as provisioning services including climate change mitigation and food production. However, agroecosystems are also prone to increased environmental pollution and release of greenhouse gas emissions, with poor management. Therefore it is important to understand how C, N, and P cycles interact to avoid uncontrolled decoupling of theses cycles and loss of molecules, such as CO_2, NO_3^-, or PO_4^-, which are harmful for our environment if present in excess. As outlined before, CNP cycles are coupled and decoupled through processes involving plants and microorganisms essentially by three processes: synthesis of plant biomass, decomposition and synthesis of microbial biomass or immobilization (Figs. 1.1 and 1.2). Carbon compounds generated by plants are used for the formation of different tissue types by including other elements, i.e., N and P. These elements accumulate in organic forms in the various plant organs above ground as well as below ground. In modern agroecosystems, crop breeding tends to maximize yields and therefore to decrease C allocation to roots, which are, however, most important for soil C storage (Kell, 2011). After the plant's death, its constituting organic matter containing N and P tightly bound to C through chemical bonding is returned to soil as litter and will undergo microbial decay, leading ultimately to decoupling of C, N, and P (Fig. 1.2). In general, microbial processing narrows C, N, and P stoichiometric ratios during litter decomposition, leading to immobilization and reorganization of N and P by decomposers (Mooshammer et al., 2014), until a critical value, when decomposers switch to net nutrient mineralization (Berg and McClauherty, 2003). The extent and nature of the decomposition and stabilization processes operating in soil will determine whether nutrients are released in mineral form or accumulate in organic form as SOM. Both processes are required for optimal provision of ecosystem services: C accumulation to mitigate climate change, for ensuring aggregation, air and water supply, and N and P release for securing soil fertility. The nature of organic matter is dynamic (Waksman, 1936), so organic matter may be most useful when it is degrading (Janzen, 2006). This is not only related to its capacity to provide nutrients for plants but also to its ability to provide structural stability through improving aggregation by delivering labile compounds to microorganisms (Six et al., 2004).

When deposited on the soils' surface, decomposition of organic matter is controlled by its chemical composition, i.e., its lignin and nutrient content, as well as pedoclimatic conditions. When the organic matter is incorporated in the mineral soil by bioturbation or directly deposited within the mineral soil as root litter, the controls on its fate are less clear. The amount of organic matter remaining in soil is the balance of input and output. Output generally occurs

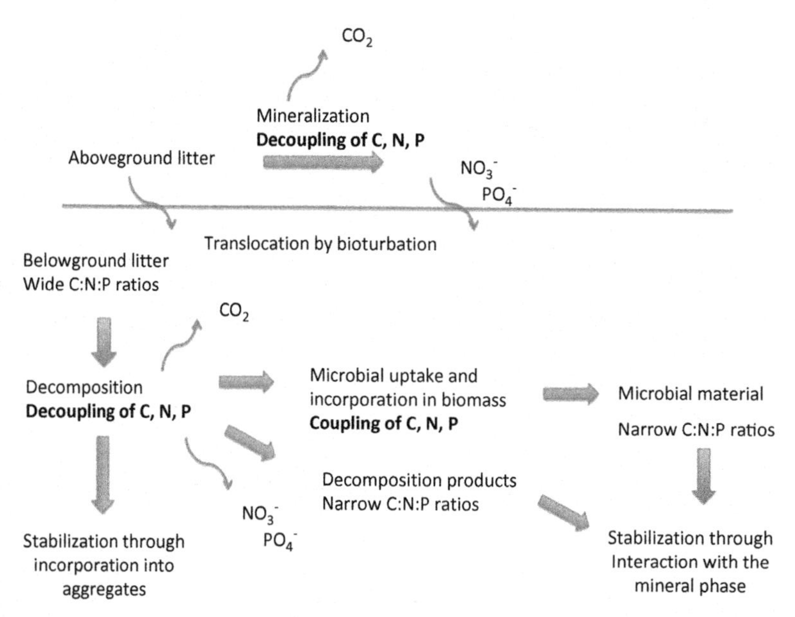

FIGURE 1.2 Processes involved in the coupling and decoupling of C, N, and P during decomposition and stabilization of organic matter from plant litter.

after complete mineralization of plant litter. In the recent literature, two processes have been identified to be responsible for slowing down microbial degradation, thereby increasing the residence time of SOM to decades or centuries. These are incorporation into soil aggregates and physicochemical protection due to adsorption on the soil's mineral phase (Lehmann and Kleber, 2015). The nature of SOM protected by these two processes is very different (Fig. 1.2). While aggregate occluded SOM consists of partly decomposed plant litter, OM in association with the mineral phase may be composed of small molecules containing high amounts of nitrogen (Kleber et al., 2007). Aggregate formation is strongly dependent on the production of microbial sugars following the degradation of fresh plant material (Six et al., 2004). SOM stored in soils has similar C:N:O:P ratios throughout the world's ecosystems (Kirkby et al., 2011), which is in the range of microbial material. This corroborates recent observations that microbial residues, rather than intact plant material, make an important contribution to SOM persisting within soil (Miltner et al., 2012). Stoichiometric ratios of SOM are much lower than those of plant material, suggesting tight coupling of C, N, and P in organic molecules, such as proteins, chitin, and DNA. Microbial decay leading to accumulation of gluing polysaccharides as well as narrowing of stoichiometric ratios may therefore be a prerequisite for organic matter stabilization. It has been shown that C storage in soil can be enhanced by nutrient addition, most probably by enhancing microbial activity. Therefore, SOM storage has an associated nutrient cost (Richardson et al., 2014), and microbial use efficiency may be its controlling factor (Cotrufo et al., 2013). Recently, it was suggested that agricultural systems, which facilitate the transformation of plant C into microbial biomass, may effectively build SOM (Kallenbach et al., 2015). Thus, while decoupling of C, N, and P following decomposition of plant litter may be necessary for ecosystem services such as protection from erosion, soil fertility, and water holding capacity, its tight coupling in microbial products may be crucial for soil C sequestration and nutrient retention. The CNP decoupling step during the initial phase of degradation, while leading to massive loss of C in form of CO_2, in perennial systems may not lead to N and P loss because of plant and microbial uptake.

PLANT EFFECTS ON C, N, AND P CYCLING

Plants are the most important source of carbon input into soils. Through photosynthetic activity, they fix CO_2 from the atmosphere, which is afterward transferred to different plant organs and incorporated in various organic molecules. These molecules are of various types, they have different stoichiometric ratios, and they may be more or less easily decomposable by soil microorganisms (Kögel-Knabner, 2002; Zechmeister-Boltenstern et al., 2015). Plants control to a certain extent the C:N:P stoichiometric ratios of their aboveground litter by resorption of N and P during leaf senescence (Fig. 1.1). In general, the amount of N and P resorbed depends on the nutrient status of the ecosystem, with greater proportions of N and P being resorbed at nutrient-poor sites (Richardson et al., 2005). This internal NP reuse is a specific characteristic of perennial vegetation systems, thus preventing N and P losses through conservative cycling. P tends to be resorbed more than N on a global scale, leading to higher N:P ratios of dead litter compared to living leaves (Mulder et al., 2013). However, the major energy source and nutrient flow pathway in terrestrial ecosystems are fine roots (Yuan and Chen, 2010), which contribute between 40% and 80% of total litter input into soils. Roots in general have higher C:N and C:P compared to leaves, and their capacity for nutrient resorbtion is much lower (Freschet et al., 2010). Therefore the C:N and C:P ratios of belowground litter are generally much higher than those of aboveground litter. Despite these large differences, the N:P ratios of these materials are rather similar, ranging between 40:1 and 43:1 globally (Zechmeister Boltenstern et al., 2015), indicating a common functional stoichiometry of the living plants (Yuan et al., 2011).

During the plant's life and after its death, molecular constituents of aboveground and belowground plant tissues are introduced into soil (Figs. 1.1 and 1.2), where they trigger microbial activity and induce the so-called priming effect, leading in most cases to acceleration of microbial activity and degradation of native SOM (Bingeman et al., 1953; Jenkinson et al., 1985; Kuzyakov et al., 2000). Plant-derived compounds with varying C chemistry and stoichiometric ratios induce contrasting priming effects (Hamer and Marschner, 2005) because they stimulate the activity of different microbial populations (Fontaine et al., 2003). The nature of plant input is contrasted with root-derived C input through exudates, litter, and sloughing cells having different effects compared to aboveground litter input. Thus, two specific spheres can be differentiated. The rhizosphere, below ground in the vicinity of roots, is a space with intense microbial activity and organic matter turnover, due to the high availability of labile compounds following exudation and rhizodeposition. Conditions in the rhizosphere are very different from those of the detritusphere around dead litter deposits. The detritusphere is a temporal hotspot of microbial activity, and element fluxes peak within a few hours to days after litter deposal (Kuzyakov and Blagodatskaya, 2015). In the following days to weeks, a succession of microbial population occurs, and activity and organic matter turnover slow down as more recalcitrant material accumulates. Priming effects and therefore C mineralization as well as nutrient release through the decoupling of C, N, and P cycles in these two spheres are contrasting (Kuzyakov, 2010).

Uncoupling of C, N, and P cycles is also dependent on the placement of plant litter (Fig. 1.2). Aboveground litter is deposited at the soil surface, where most of its decomposition may occur, unless it is transported into soil by bioturbation or tillage. Decomposition of aboveground litter may be more intense, leading to complete mineralization of organic matter, while belowground litter may be protected from microbial decay, thus conserving CNP coupling without transformation and release of mineral nutrients (Fig. 1.2). However, placement of litter (above ground or within the mineral soil) is not as important as litter type (root or shoot) for controlling C decomposition (Hatton et al., 2015). Litter degradation is controlled by its chemical composition, in particular, its lignin to N ratio (Sanaullah et al., 2010). Therefore, management practices in agroecosystems with perennial plants may control to some extent C sequestration and nutrient release through their impact on species choice with more or less root biomass, contrasting root:shoot ratios, and plant litter quality (more or less decomposable). Another option may be influencing plant resorption of nutrients, i.e., by choosing the time of harvest (before or after senescence). Moreover, C storage is affected by priming effects, which could be controlled by management of harvesting residues, introducing more or less fresh plant material with contrasting C:N:P ratios into soil, thereby stimulating different microbial populations and inducing contrasting priming effects (Kuzyakov, 2010). The quantitative effects and underlying biogeochemical mechanisms are still under investigation. Experiments using dual stable isotope labeling will probably allow revelation of the role of different litter and rhizosphere inputs on priming organic matter mineralization.

C, N, AND P CYCLING IN THE RHIZOSPHERE

The greatest impact of plant activity on biogeochemical cycling of elements is noted in the rhizosphere, which is defined as the soil around living roots (Hiltner, 1904). Plants control physical, chemical, and biologic processes within this space through exudation of a wide range of compounds, including organic acids, sugars, and other secondary metabolites, among which are signalling molecules. Through their activity, plant roots may affect microbial communities involved in the decomposition as well as the stabilization of organic matter. Plant roots may affect microbial decomposition through (1) decreasing mineral nutrient availability to soil microorganisms due to plant uptake (Schimel et al., 1989), (2) changing the physical and chemical environment in the rhizosphere (Shields and Paul, 1973), (3) increasing organic substrate supply, and (4) enhancing microbial turnover due to fungal grazing. However, interactions at the rhizosphere level between soil microorganisms and roots of different plant communities are complex and remain poorly understood (Cheng and Kuzyakov, 2005). The rhizosphere effects may include pH alterations, a process that may lead to the release of plant nutrients through exchange processes. Moreover, roots provide energy to beneficial microorganisms, such as rhizobia, mycorrhizae, and growth-promoting microorganisms, which in turn improve the plants' acquisition of N and P. Growth-promoting microorganisms trigger nutrient availability through the production and release of hydrolytic enzymes for organic N (Ollivier et al., 2011) and organic P (Rodriguez et al., 2006). The acquisition of N may be enhanced through biologic N_2 fixation through diazotrophs, which can be free living or associated with plants (Galloway et al., 2008). Phosphorus acquisition by plants is also largely supported by growth-promoting microorganisms able to solubilize inorganic P strongly bound to the mineral phase through the exudation of organic acids (Bhattacharyya and Jha, 2012). Moreover, growth-promoting microorganisms may be able to stimulate nutrient uptake by plants through influencing transmembrane transport (Bertrand et al., 2000).

Symbiosis with mycorrhiza fungi greatly improves plants' P nutrition, especially in P-deficient soils (Barea et al., 2008). However, N nutrition was also shown to be improved following mycorrhiza colonization by several mechanisms (Bücking and Kafle, 2015). Mycorrhiza colonization changes greatly the ecosystem's C, N, and P cycling, mainly through their effect on plant physiology, including tissue elemental composition, hormone balance, and C flow (Richardson et al., 2009). In response to changing environmental conditions, such as drought stress, plants adapt through increasing exudation, and this adaptation may depend on the community composition (Sanaullah et al., 2012) and the extent of mycorrhization. Thus, agricultural management may influence resistance of plants to environmental stresses by species choice and inoculation. C, N, and P cycling in the rhizosphere may be in particular influenced by the use of contrasting plant types, i.e., gramineous species requiring mineral N fertilization and leguminous species, able to fix atmospheric N, but requiring high amounts of P. The introduction of leguminous plants in grasslands with gramineous species was found to change the soil P forms and the biochemical composition of SOM (Crème et al., 2017, 2016), most probably due to rhizospheric processes. Inoculation with arbuscular mycorrhiza fungi may be a suitable strategy for enhancing N and P availability for plants as well as their stress resistance (Bücking and Kafle, 2015). There is an evident research need concerning the mechanisms by which above- and belowground vegetation exerts a synergistic control on biogeochemical cycling and how these mechanisms can be influenced by human activity to improve productivity and stress resistance.

PLANT–MICROBIAL CROSSTALK AT THE SOIL–ROOT INTERFACE

Foraging for nutrient "hot spots" is a key strategy by which some plants maximize nutrient gain from their carbon investment in root and mycorrhizal hyphae. Foraging strategies may depend on costs of root construction, with thick roots generally costing more per unit length than thin roots. To maximize cost-effective resource use, plants are not always investing in root biomass or rigidity. Investment in mycorrhizal associations or other plant growth-promoting microorganisms may represent an alternative strategy for cost-effective nutrient foraging. Plants are able to interact with root-associated beneficial microorganisms to improve their defense as well as nutrition by rhizodeposition of various substances originating from sloughed-off root cells, mucilages, volatiles, and exudates that are released from damaged and intact cells (Jones et al., 2009). These compounds may shape rhizosphere microbial communities within a small spatiotemporal window related to root apices (Dennis et al., 2010). Plants communicate with microbial populations in the rhizosphere in various ways. They mediate positive as well as negative interactions through exudation of high and low molecular weight compounds (Badri and Vivanco, 2009). For example, plants excrete antimicrobial compounds to cope with pathogens, and they are able to initiate positive interactions, such as rhizobia colonization by exudation of flavonoids, which regulate nodule development through auxin transport inhibition (Haichar et al., 2014). The first step of mycorrhizal colonization of plant roots is induced by excreting strigolactone as signalling molecules. The concentration and structure of strigolactones determines arbuscular mycorrhizal development (Ruyter-Spira et al., 2013). Moreover, they also induce hormonal excretion by fungi, which trigger symbiose development in plants (Gutjahr, 2014). Plants are able to regulate arbuscle development and lifetime through hormones having negative effects on fungal colonization and a number of other interactions with growth-promoting bacteria. As a possible control mechanism, it was suggested that they might have the possibility to manipulate gene expression and behavior in associated bacteria

TABLE 1.1 Functions of Plant Hormones

Plant Hormone	Function
Abscisic acid	Stress resistance
Auxin	Plant growth
Cytokinins	Regulators of plant growth and development
Ethylene	Plant development, defense
Gibberellins	Growth stimulators
Brassinosteroids	Growth and development
Jasmonates	Plant development, survival, and reproduction
Salicylic acid	Plant growth and defense
Strigolactones	Root plasticity, interaction with microbes

(Pii et al., 2015). It seems that hormonal signalling molecules are exchanged by plants and microorganisms to be able to achieve adequate response to environmental conditions. The complexity of these interactions is far from being understood, but it may be the clue for explaining plant responses to a changing environment (Pozo et al., 2015). Plant hormones and the hormonal crosstalk may play a pivotal role in resistance to abiotic stresses (Table 1.1). Recently, it has been suggested that phytohormone engineering represents an important platform for stress tolerance and that developing the technology could be an important step forward toward stress-resistant crops (Wania et al., 2016).

CONCLUSION

Plants influence biogeochemical cycling of C, N, and P and associated ecosystem services in various ways, mainly through their impact on SOM turnover. They provide aboveground and belowground litter for decomposition. The quality and placement of this material may determine its fate in soil and its contribution to ecosystem services. Labile plant compounds determine the formation of soil aggregates and therefore soil structure being related to aeration and water-holding capacity. Moreover, input of plant-derived labile compounds supports microbial decomposition of organic matter, thereby promoting nutrient release and the possibility of soil C sequestration in the form of microbial products. All these plant effects are linked with the plants' community composition, which may have different effects as compared to monocultures. In general, more diverse systems are beneficial in terms of nutrient acquisition and carbon storage.

Other important plant effects on C, N, and P cycles are related to rhizosphere processes, which are controlled by plants' secretion of root exudates including low molecular compounds. These compounds may be antimicrobial or favorable for microbial colonization. They are regulating plant–microbial interactions at the plant–soil interface, via the rhizosphere priming effect and hormonal crosstalk. The rhizosphere-priming effects remain unclear regarding the effect of different substrates on SOM fractions with contrasting stability. Experiments with continuous $^{13}CO_2$ labeling may help elucidate the quantitative effect of rhizosphere priming on biogeochemical cycling of elements. Other knowledge gaps exist with regard to hormonal crosstalk between plants and microorganisms. These interactions need to be understood to be able to exploit the intrinsic biologic potential of rhizosphere processes to increase crop nutrient use efficiency and C sequestration. Furthermore, evidence is accumulating that plant traits related to mycorrhizal symbiosis, i.e., mycorrhizal type and the degree of plant root colonization by mycorrhizal fungi, have important consequences for carbon, nitrogen, and phosphorus cycling in soil. The question of how plant and soil biogeochemical pools vary among vegetation types and plants with different mycorrhizal types is a new and exciting research challenge that needs further investigation.

The stimulation of plant productivity in response to global change (e.g., rising atmospheric CO_2 concentrations) can potentially compensate climate change feedbacks. However, this will depend on the allocation of C resources within vegetation, nutrient availability, and plant feedback with soil microorganisms. These dynamic adjustments within single plants will result in changes in above- and belowground stoichiometric relations via biomass production and root exudation. They will have an impact on the community level for different vegetation types, which will ultimately control the response of agroecosystems to global change.

Acknowledgments

This work was supported and benefited from the European Commission through the FP7 projects ExpeER (Experimentation in Ecosystem Research, Grant Agreement Number 262060) and AnaEE (Analysis and Experimentation in Ecosystems, Grant Agreement Number 312690). The authors also acknowledge the ANR (AnaEE Service ANR-11-INBS-0001; Mosaik ANR-12-AGRO-0005), AEGES (ADEME), INRA, Allenvi, and CNRS-INSU for financial support of the SOERE-ACBB. Any opinions, findings, and conclusions or recommendations expressed in this chapter are those of the authors and do not necessarily reflect the views of our sponsors.

References

Averill, C., Turner, B.L., Frinzi, A.C., 2014. Mycorrhiza-mediated competition between plantsand decomposers drives soil carbon storage. Nature 505, 543–545.

Badri, D.V., Vivanco, J.M., 2009. Regulation and function of root exudates. Plant Cell Environment 32, 666–681.

Barea, J.M., Ferrol, N., Azcón-Aguilar, C., Azcón, R., 2008. Mycorrhizal symbioses. In: White, P.J., Hammond, J.P. (Eds.), The Ecophysiology of Plant-phosphorus Interactions. Springer, Dordrecht, pp. 143–163.

Berg, B., McClaugherty, C., 2003. Plant Litter: Decomposition, Humus Formation, Carbon Sequestration. Springer-Verlag, Berlin, Germany.

Bertrand, H., Plassard, C., Pinochet, X., Touraine, B., Normand, P., Cleyet-Marel, J.C., 2000. Stimulation of the ionic transport system in Brassica napus by a plant growth-promoting rhizobacterium (Achomobacter sp.). Canadian Journal of Microbilogy 46, 229–236.

Bhattacharyya, P.N., Jha, D.K., 2012. Plant growth-promoting rhizobacteria (PGPR): emergence in agriculture. World Journal of Microbiological Biotechnology 28, 1327–1350.

Bingeman, C.W., Varner, J.E., Martin, W.P., 1953. The effect of addition of organic materials on the decomposition of an organic soil. Soil Science Society of America Proceedings 17, 34–38.

Bücking, H., Kafle, A., 2015. Role of arbuscular mycorrhizal fungi in the nitrogen uptake of plants: current knowledge and research gaps. Agronomy 5, 587–612.

Cheng, W., Kuzyakov, Y., 2005. Root effect on soil organic matter decomposition. In: Wright, S., Zobel, R. (Eds.), Roots and Soil Management: Interactions between Roots and the Soil. Agronomy Monographs No. 48, American Society of Agronomy, Crop Science Society of America, Soil Science Society of America, Madison, Wisconsin, USA, pp. 119–143.

Cotrufo, M.F., Wallenstein, M.D., Boot, C., Denef, K., Paul, E., 2013. The microbial efficiency-matrix stabilisation (MEMS) framework integrates plant litter decomposition with soil organic matter stabilization: do labile plant inputs form stable organic matter? Global Change Biology 19, 988–995.

Crème, A., Rumpel, C., Gastal, F., Mora, M.-L., Chabbi, A., 2016. Effect of grasses and a legume grown in monoculture or mixture on soil organic matter and phosphorus forms. Plant and Soil 402, 117–128.

Crème, A., Chabbi, A., Gastal, F., Rumpel, C., 2017. Biogeochemical nature of grassland soil organic matter under plant communities with two nitrogen sources. Plant and Soil 415, 189–201.

Dennis, P.G., Miller, A.J., Hirsch, P.R., 2010. Are root exudates more important than other sources of rhizodeposits in structuring rhizosphere bacterial communities? FEMS Microbiol Ecol 72 (3), 313–327. Jun.

Fontaine, S., Mariotti, A., Abbadie, L., 2003. The priming effect of organic matter: a question of microbial competition? Soil Biology and Biochemistry 35, 837–843.

Freschet, G.T., Cornelissen, J.H.C., van Logtestijn, R.S.P., Aerts, R., 2010. Substantial nutrient resorption from leaves, stems and roots in a subarctic flora: what is the link with other resource economics traits? New Phytologist 186, 879–889.

Galloway, J.N., Townsend, A.R., Erisman, J.W., Bekunda, M., Cai, Z.C., Freney, J.R., Martinelli, L.A., Seitzinger, S.P., Sutton, M.A., 2008. Transformation of the nitrogen cycle: recent trends, questions and potential solutions. Science 320, 889–892.

Gutjahr, C., 2014. Phytohormone signaling in arboscular mycorrhiza development. Current Opinion in Plant Biology 20, 26–34.

Haichar, F.Z., Santaella, C., Heulin, T., Achouak, W., 2014. Root exudates mediated interactions belowground. Soil Biology and Biochemistry 77, 69–80.

Hamer, U., Marschner, B., 2005. Priming effects in soils after combined and repeated substrate additions. Geoderma 128, 38–51.

Hamilton, E.W., Frank, D.A., Hinchey, P.M., Murray, T.R., 2008. Defoliation induces root exudation and triggers positive rhizospheric feedbacks in a temperate grassland. Soil Biology and Biochemistry 40, 2865–2873.

Hatton, P.-J., Castanha, C., Torn, M.S., Bird, J.A., 2015. Litter type control on soil C and N stabilization dynamics in a temperate forest. Global Change Biology 21, 1358–1367.

Hiltner, L., 1904. Uber neuer Erfahrungen und Probleme auf dem Gebiet der Bodenbakteriologie unter besonderer Berücksichtigung der Gründüngung und Brache. Arbeiten der DLG 98, 59–78.

Janzen, H.H., 2006. The soil carbon dilemma: shall we hoard it or use it? Soil Biology and Biochemistry 38, 419–424.

Jenkinson, D.S., Fox, R.H., Rayner, J.H., 1985. Interactions between fertilizer nitrogen and soil nitrogen—the so-called 'priming' effect. Journal of Soil Science 36, 425–444.

Jobbagy, E.G., Jackson, R.B., 2000. The vertical distribution of soil organic carbon and its relation to climate and vegetation. Ecological Applications 10, 423–436.

Jones, D., Nguyen, C., Finlay, D.R., 2009. Carbon flow in the rhizosphere: carbon trading at the soil–root interface. Plant Soil 321, 5–33.

Kallenbach, C., Grandy, A.S., Frey, S.D., Diefendorf, A.F., 2015. Microbial physiology and necromass regulate agricultural soil carbon accumulation. Soil Biology and Biochemistry 91, 279–290.

Kell, D.B., 2011. Breeding crop plants with deep roots: their role in sustainable carbon, nutrient and water sequestration. Annals of B 108 (3), 407–418.

Kirkby, C.A., Kirkegaard, J.A., Richardson, A.E., Wade, L.J., Blanchard, C., Batten, G., 2011. Stable soil organic matter: a comparison of C: N:P: S ratios in Australian and other world soils. Geoderma 163, 197–208.

Kleber, M., Sollins, P., Sutton, R., 2007. A conceptual model of organo-mineral interactions in soils: self assembly of organic molecular fragments into zonal structures on mineral surfaces. Biogeochemistry 85, 9–24.

Kögel-Knabner, I., 2002. The macromolecular organic composition of plant and microbial residues as inputs to soil organic matter. Soil Biology and Biochemistry 34, 139–162.

Kuzyakov, Y., 2010. Priming effects: interactions between living and dead organic matter. Soil Biology and Biochemistry 42, 1363–1371.

Kuzyakov, Y., Domanski, G., 2000. Carbon input by plants into the soil. Review. Journal of Plant Nutrition and Soil Science 163, 421–431.

Kuzyakov, Y., Friedel, J.K., Stahr, K., 2000. Review of mechanisms and quantification of priming effects. Soil Biology and Biochemistry 32, 1485–1498.

Lange, M., Eisnhauer, N., Sierra, C.A., Bessler, H., Engels, C., Griffiths, R.I., Mellado-Vazquez, P.G., Malik, A.A., Roy, J., Scheu, S., Steinbeiss, S., Thomson, B.C., Trumbore, S.E., Gleixner, G., 2015. Plant diversity increases soil microbial activity and carbon storage. Nature Communications 6, 6707.

Lehmann, J., Kleber, M., 2015. The contentious nature of soil organic matter. Nature 528 (7580), 60–68.

Miltner, A., Bombach, P., Schmidt-Brücken, B., Kaestner, M., 2012. SOM genesis: microbial biomass as a significant source. Biogeochemistry 111, 41–55.

Mooshammer, M., Wanek, W., Hammerle, I., Fuchslueger, L., Hofhansl, F., Knoltsch, A., Schnecker, J., Takriti, M., Watzka, M., Wild, B., Keiblinger, K.M., Zechmeister-Boltenstern, S., Richter, A., 2014. Adjustment of microbial nitrogen use efficiency to carbon: nitrogen imbalances regulates soil nitrogen cycling. Nature Communications 5, 3694,. https://doi.org/10.1038/ncomms4694.

Mulder, C., Ahrestani, F.S., Bahn, M., Bohan, D.A., Bonkowski, M., Griffiths, B.S., Guicharnaud, R.A., Kattge, J., Krogh, P.H., Lavorel, S., Lewis, O.T., Mancinelli, G., Naeem, S., Penuelas, J., Poorter, H., Reich, P.B., Rossi, L., Rusch, G.M., Sardans, J., Wright, I.J., 2013. Connecting the green and Brown worlds: allometric and stoichiometric predictability of above- and below-ground networks. In: Woodward, G., Bohan, D.A. (Eds.), Advances in Ecological Research, vol. 49, pp. 69–175.

Nuccio, E.E., Hodge, A., Pett-Ridge, J., Herman, D.J., Weber, P.K., Firestone, M.K., 2013. An arbuscular mycorrhizal fungus significantly modifies the soil bacterial community and nitrogen cycling during litter decomposition. Environmental Microbiology 15, 1870–1881.

Ollivier, J., Töwe, S., Bannert, A., Hai, B., Kastl, E.M., Meyer, A., Su, M.X., Kleineidam, K., Schloter, M., 2011. Nitrogen turnover in soil and global change. FEMS Microbiology Ecology 78 (1), 3–16.

Pérès, G., Cluzeau, D., Menasseri, S., et al., 2013. Mechanisms linking plant community properties to soil aggregate stability in an experimental grassland plant diversity gradient. Plant Soil 373, 285–299.

Pii, Y., Mimmo, T., Tomasi, N., Terzano, R., Cesco, S., Crecchio, C., 2015. Microbial interactions in the rhizosphere: beneficial influences of plant growth-promoting rhizobacteria on nutrient acquisition process. A review. Biology and Fertility of Soils 51, 403–415.

Pozo, M.J., López-Ráez, J.A., Azcón-Aguilar, C., García-Garrido, J.M., 2015. Phytohormones as integrators of environmental signals in the regulation of mycorrhizal symbioses. New Phytologist 205, 1431–1436.

Rasse, D.P., Rumpel, C., Dignac, M.-F., 2005. Is soil carbon mostly root carbon? Mechanisms for a specific stabilisation. Plant and Soil 269, 341–356.

Richardson, S.J., Peltzer, D.A., Allen, R.B., McGlone, M.S., 2005. Resorption proficiency along a chronosequence: responses among communities and within species. Ecology 86, 20–25.

Richardson, A.E., Barea, J.M., McNeill, A.M., Prigent-Combaret, C., 2009. Acquisition of phosphorus and nitrogen in the rhizosphere and plant growth promotion by microorganisms. Plant and Soil 321, 305–339.

Richardson, A.E., Kirkby, C.A., Banerjee, S., Kirkegaard, J.A., 2014. The inorganic nutrient cost of building soil carbon. Carbon Management 5, 265–268.

Rodriguez, H., Farag, R., Gonzalez, T., Bashan, Y., 2006. Genetics of phosphate solubilisation and its potential applications for improving plant growth-promoting bacteria. Plant and Soil 287, 15–21.

Rumpel, C., Kögel-Knabner, I., 2011. Deep soil organic matter—a key but poorly understood component of terrestrial C cycle. Plant and Soil 338, 143–158.

Ruyter-Spira, C., Al-Babili, S., van der Krol, S., Bouwmeester, H., 2013. The biology of strigolactones. Trends in Plant Science 18, 72–83.

Sanaullah, M., Chabbi, A., Lemaire, G., Charrier, X., Rumpel, C., 2010. How does plant leaf senescence of grassland species influence decomposition kinetics and litter compounds dynamics? Nutrient Cycling in Agroecosystems 88, 159–171.

Sanaullah, M., Chabbi, A., Rumpel, C., Kuzyakov, Y., 2012. Carbon allocation in grassland communities under drought stress followed by ^{14}C pulse labelling. Soil Biology and Biochemistry 55, 132–139.

Schimel, J.P., Jackson, L.E., Firestone, M.K., 1989. Spatial and temporal effects on plant microbial competition for inorganic nitrogen in a California annual grassland. Soil Biology and Biochemistry 21, 1059–1066.

Shields, J.A., Paul, E.A., 1973. Decomposition of 14C labelled plant material under field conditions. Canadian Journal of Soil Science 53, 297–306.

Six, J., Bossuyt, H., Degryze, S., Denef, K., 2004. A history of research on the link between (micro)aggregates, soil biota, and soil organic matter dynamics. Soil Tillage Research 79, 7–31.

Waksman, S.A., 1936. Humus: Origin, Chemical Composition, and Importance in Nature. The Williams & Wilkins Company, Baltimore.

Wania, S.H., Kumarb, V., Shriramc, V., Sah, S.K., 2016. Phytohormones and their metabolic engineering for abiotic stress tolerance in crop plants. The Crop Journal 4, 162–176.

Yuan, Z.Y., Chen, H.Y.H., 2010. Fine root biomass, production, turnover rates, and nutrient contents in boreal forest ecosystems in relation to species, climate, fertility, and stand age: literature review and meta-analyses. Critical Reviews in Plant Sciences 29, 204–221.

Yuan, Z.Y., Chen, H.Y.H., Reich, P.B., 2011. Global-scale latitudinal patterns of plant fine-root nitrogen and phosphorus. Nature Communications 2, 344. https://doi.org/10.1038/ncomms1346.

Zechmeister-Boltenstern, S., Keiblinger, K.M., Mooshammer, M., Penuelas, J., Richter, A., Sardans, J., Wanek, W., 2015. The application of ecological stoichiometry to plant-microbial-soil organic matter transformations. Ecological Monographs 85, 133–155.

CHAPTER

2

C–N–P Uncoupling in Grazed Grasslands and Environmental Implications of Management Intensification

Françoise Vertès[1], *Luc Delaby*[2], *Katia Klumpp*[3], *Juliette Bloor*[3]

[1]INRA-SAS, Agrocampus Ouest, Rennes, France; [2]INRA-PEGASE, Agrocampus Ouest, Rennes, France; [3]INRA-UREP, Clermont-Ferrand, France

INTRODUCTION

Grasslands provide key environmental services, as well as important economic and societal services, via grazing-based livestock production systems. In Europe, grasslands are characterized by a large diversity in grassland types and management practices, varying in type and intensity. Most common grassland types comprise short-duration grass and/or legume-based leys (i.e., alfalfa), temporary sown grasslands (short < 5 years to long 5–12 years) (Peeters et al., 2014), and permanent seminatural grasslands (>10 years) (e.g., Soussana et al., 2004; Huyghe et al., 2014). Over the last 50 years, significant areas of European grassland have been converted to arable crops and to sown pastures (15 M ha) as a result of an intensification of biomass production for human food and animal feeding via the use of concentrates and soybean in the animal diet (FAOSTAT, 2011). Increasing reliance on fertilizer inputs has also led to an overall intensification of systems and a specialization of livestock production system at the farm (e.g., Lemaire et al., 2015; Moraine et al., 2014), regional (e.g., Thieu et al., 2011), or higher scales (e.g., Galloway et al., 2008; Billen et al., 2009; Peyraud et al., 2014; Godinot et al., 2016). Agricultural intensification and increased inputs (N, P, K fertilizers and/or imported animal feed) affects the speed and magnitude of transformations of elements in the main nutrient cycles, with implications for the balance between coupling and uncoupling of carbon (C) with other nutrient elements (Faverdin and Peyraud, 2010). Intensification may also modify the spatial distribution of nutrient transformations and losses at larger scales: where livestock production is

Agroecosystem Diversity
https://doi.org/10.1016/B978-0-12-811050-8.00002-9

geographically separated from feed production areas, this may promote nutrient transfers and incomplete recycling of nutrients.

In the present chapter, we focus on nutrient cycling and uncoupling of nutrient cycles in grazed grasslands in space and time, describing C, N, P, and K fluxes and their heterogeneity and variability at the field scale. In particular, we examine element fluxes associated with cattle. We highlight environmental concerns associated with grazing intensification and identify the main levers to improve system efficiency and reduce polluting emissions at the local and larger scale.

BIOGEOCHEMICAL CYCLING IN GRAZED GRASSLANDS

Grazed grasslands are dynamic, complex ecosystems characterized by fast-growing perennial vegetation and the presence of domestic herbivores (Parsons et al., 2000). Fluxes of C, N, P and K are thus usually higher in grassland versus crop ecosystems. Unlike annual crops, grassland production implies a trade-off between leaf removal (by grazing or cutting) and leaving sufficient plant material to photosynthesize and replace tissues throughout the year (Parsons et al., 2011). Botanical composition as well as management practices are variable; sown grasslands are generally poor in plant species composition, while permanent grasslands present a large botanical diversity as a result of interactions between soil, climate, and management practices. This complexity in vegetation cover (species composition, plant functional groups, and traits) and management practices has significant consequences for biomass production but also for the biogeochemical cycling of nutrients.

C,N,P,K Cycling and (Un)coupling at the Field Scale

Grassland ecosystems are characterized by substantial stocks of C located largely below ground in roots and soil (Jones and Donnelly, 2004). This C sequestered by grasslands is the difference between C inputs, via fixation of C from the atmosphere by plants (photosynthesis), and heterotrophic respiration, biomass removal (harvest, grazing), and changes in soil C stocks (i.e., losses through lixiviation, runoff, etc.). C gain via photosynthesis is mainly controlled by environmental abiotic conditions (radiation, temperature, and water and nutrient availability), whereas C losses through respiration and lixiviation (Kindler et al., 2011) are largely influenced by management and climate factors (soil temperature, humidity; e.g., Bahn et al., 2008). Indeed, the nature, frequency, and intensity of biomass exports play a key role in the C cycling and balance of grasslands. In grazed grasslands, much of the primary production is ingested by animals and returned to the soil in the form of feces (nondigestible carbon; 25%–40% of the intake, depending on the digestibility of diet); the remainder is returned to the soil in the form of plant litter or root exudates. This fresh organic matter inputs, generally rich in energy and readily decomposed by microorganisms, contributes to heterotrophic respiration and exchange of C with the atmosphere in the form of CO_2. Soil C inputs, as senescent above- and belowground biomass or rhizodeposition, can reach more than 7 t C per ha per year in grazed grasslands (e.g., Whitehead et al., 1990; Vertès and Mary, 2014). Part of C exports may be compensated by imports of organic C fertilizers through farm manure and slurry application.

Nitrogen is an essential element for organisms to grow, maintain, and reproduce, and it is the main limiting nutrient in most terrestrial ecosystems. Nitrogen exists in a wide variety of mineral and organic forms, more or less mobile, which are linked in four main processes: N fixation, mineralization (conversion of organic N to ammonia), nitrification (oxidation of ammonia to nitrates/nitrite), and denitrification (reduction of nitrates to gaseous N). N absorption and/or fixation by plants is highly dependent on abiotic factors (N availability, soil humidity, temperature, etc.) but also on microbial activity in soil and plant growth (e.g., Lemaire et al., 1997; Butterbach-Bahl et al., 2011; Schwinning and Parsons, 1996). Nitrogen fixation by legumes can reach up to 400 kg per ha per year in temperate grasslands (Voisin and Gastal, 2015; Anglade et al., 2015). N losses from grassland ecosystems occur via leaching (mainly NO_3^- and dissolved organic N DON, Tilman et al., 1996), via volatilization (NH_3) and as a by-product of soil microbial processes (denitrification/nitrification in the form of N_2O, N_2, NOx; e.g., Butterbach-Bahl et al., 2013). As gross N fluxes are higher in grasslands compared to crops (Murphy et al., 2003a), N emissions leading to environmental impacts, i.e., ground water pollution (Ryden et al., 1984; Di and Cameron, 2002; Velthof et al., 1998) and greenhouse gas emissions (e.g., Luo et al., 2010; Stehfest and Bouwman, 2006; Flechard et al., 2007), can be high when N availability in soil exceeds N needs of plants (Ledgard et al., 1999; Ball and Ryden, 1984), or where soil cover (i.e., patches of bare soil, fallow) is low.

Most plants contain only about 0.2% of phosphorus (P) in biomass. Nevertheless, this small amount is critically important, as P is an essential component of adenosine triphosphate (ATP), cell development, and DNA formation. Insufficient soil P can result in delayed plant maturity, reduced flower development, low seed quality, and decreased yield. Phosphorus exists in many different forms in soil, as plant-available inorganic P or three forms that are not plant available (organic P, adsorbed P, and primary mineral P). Those forms can become available for plants mainly though weathering, mineralization, and desorption, where immobilization, precipitation, and adsorption decrease plant-available P (Hinsinger, 2001; Morel et al., 2000). While the use of N fertilizers seems often necessary to maintain grassland productivity in most managed grassland systems, soil P stocks of most temperate agricultural soils seem to be sufficient, in particular in breeding areas where large amounts of manure are spread (e.g., intensive dairy regions in Europe). Nonetheless, N_2 fixation by legumes can be limited by phosphorus (e.g., Soussana et al., 2010; Hogh-Jensen and Schjoerring et al., 2010), resulting in reduced yields (Liebisch et al., 2013). Runoff of particulate (soil-bound) P in eroded sediment, as well as mineral fertilizers and dissolved organic P from applied manure and animal dejections, are together with leaching the major cause of P losses from agricultural systems (Bouraoui et al., 2009; Schoumans et al., 2014).

A nutrient closely related to nitrogen and needed in large amounts is potassium (K). Potassium is taken up by plants in amounts only second to N, and the offtake with harvested plant material under intensive production can even exceed that of N. In contrast to phosphorus, which has very low mobility and can either originate from soil organic matter or recycled organic phosphorus, potassium is extremely mobile (Whitehead, 2000) and is derived exclusively from the soil mineral fraction (Murphy et al., 2003b). The nutrient cycling and the availability of K differ substantially between soils, and K is often a limiting factor in sandy soils. Grassland management also has a large influence on K requirements, with greater requirements in mown compared to grazed grasslands (Kayser and Isselstein, 2005). Leaching of K from grassland is usually low, but high levels of available soil K, high K input from soil improvers (in particular slurry, after the solid-liquid separation process), or urine patches may lead to increased losses.

In grasslands, as in all ecosystems, C and nutrient elements are coupled (1) during plant growth via photosynthesis and nutrient assimilation, and (2) in soils via microbial immobilization and soil organic matter dynamics (Soussana and Lemaire, 2014; Rumpel et al., 2015; Fig. 2.1). Uncoupling of C and nutrient elements arises during decomposition of organic matter (temporal uncoupling), and it is promoted by the

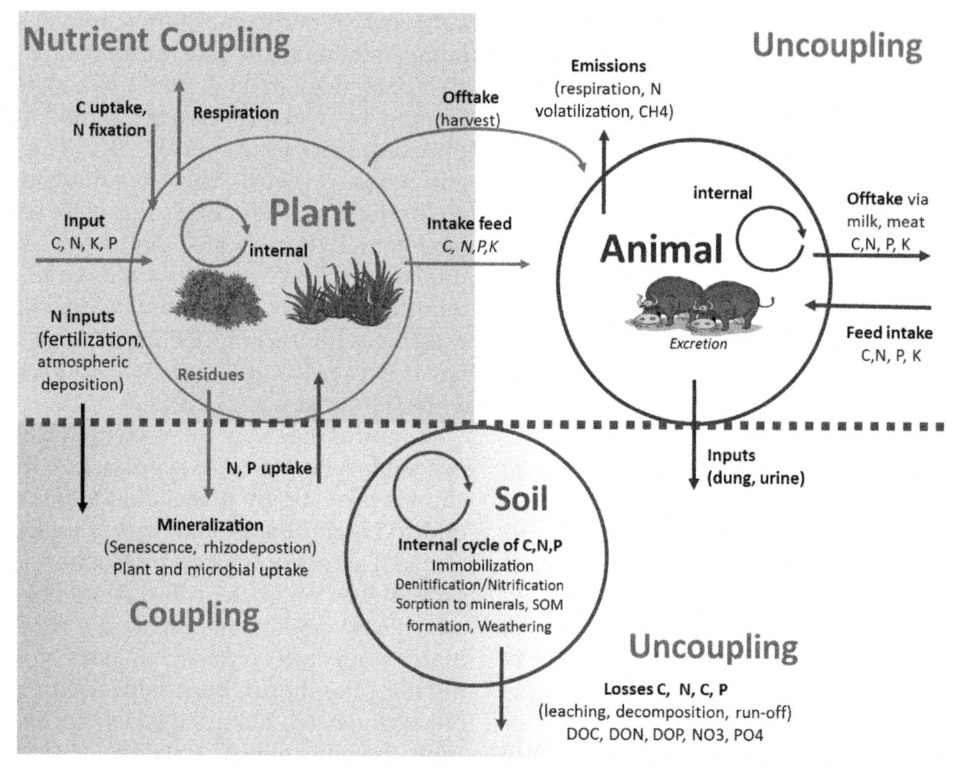

FIGURE 2.1 CNP cycling in managed grasslands showing coupling and uncoupling of nutrients and related ecosystem fluxes and processes. *Adapted from Rumpel, C., Crème, A., Ngo, N.T., Velásquez, G., Mora, M.L., Chabbi, A., 2015. The impact of grassland management on biogeochemical cycles involving carbon, nitrogen and phosphorus. Journal of Soil Science and Plant Nutrition, 15, 353–371.*

activities of grazing animals (temporal and spatial uncoupling) (Fig. 2.1).

Effects of Grazing on C,N,P,K Cycling

Grazing domestic herbivores have a large impact on grassland productivity, plant community structure, and biogeochemical cycling (Haynes and Williams, 1993). Effects of grazing are driven by plant tissue removal (defoliation), excretion (urine and dung deposits), and trampling, which exerts mechanical pressure and causes physical damage to the vegetation where animals pass repeatedly. In general, grazing results in a reduction in aboveground standing biomass and litter production, as well as changes in plant nutrient status (Bakker et al., 2006). Direct effects of grazing on plant growth rates and nutrient use may be further enhanced by indirect grazing effects and plant–soil feedbacks that modify nutrient availability and longer term biogeochemical cycling (Bardgett and Wardle, 2003; Hutchings et al., 2007). However, the magnitude of ecosystem responses may vary depending on the timing, frequency, and intensity of grazing, as well as pedoclimatic factors (Franck et al., 2000). Trampling effects may results in plant damage and soil compaction, affecting the soil structure, the earthworm populations (Cluzeau et al., 1992), and the vegetation dynamics, impacting water and nutrient flows (Blouin et al., 2013).

Grazing animals consume a variable proportion of shoot biomass, ingesting up to 60% of aboveground dry matter production in intensively grazed pastures (Lemaire and Chapman, 1996). Herbivores tend to prefer short grass in a vegetative, N-rich condition, although the degree of selectivity varies on animal type (Adler et al., 2001; Dumont et al., 2011). At the individual level, defoliation affects plant photosynthetic capacity and carbon allocation, with implications for plant–plant interactions and community-level productivity (Diaz et al., 2007b). Defoliation may also modify plant–soil interactions via changes in the quality and quantity of root exudates and litter inputs to the soil (Bardgett et al., 2003; Wardle et al., 2004). In the short-term, defoliation-induced increases in root exudation promote microbial activity and availability of P and N through positive feedback mechanisms (Hamilton and Frank, 2001). In addition, defoliation can increase leaf nutrient concentrations over short time scales as a result of nutrient reallocation or increases in N mineralization and N uptake (Hamilton and Frank, 2001; Shahzad et al., 2015). In the longer term, litter quality and quantity may be driven by changes in plant community composition; prolonged defoliation tends to promote fast-growing, defoliation-tolerant species or unpalatable species with physicochemical plant defenses (Milchunas et al., 1988). Fast growing and palatable species, typical of nutrient-rich and managed grasslands have higher quality (lower C:N) litter that promotes positive plant–soil feedbacks, increased nutrient cycling, and in fine plant productivity (Bardgett and Wardle, 2003). The C:N ratio of plant litter also determines the mean residence time of soil organic matter carbon pools (Klumpp et al., 2007, 2009); high-quality litter (low C:N ratio) of fast-growing species leads to rapid degradation by bacteria and short residence time of C. In contrast, grasslands adapted to low grazing levels are generally characterized by slow-growing plant species and lower aboveground net primary productivity, a microbial community dominated by fungi, as well as greater N retention and C storage (Bardgett et al., 2001, 2005).

Grazing animals promote spatial heterogeneity in CNP pools and fluxes via uneven patterns of defoliation and animal returns (Orwin et al., 2009; Bloor and Pottier, 2014). Herbivore foraging leads to patches of short and tall stands of vegetation within the field that may persist over time due to selective animal foraging and social behavior (Parsons and Dumont, 2003). In addition, large herbivores tend to avoid foraging in the immediate vicinity of dung pats, leading to the development of taller and more phenologically advanced vegetation close to dung (Gillet et al., 2010). Animal returns represent high-concentration nutrient "hotspots" with relatively limited lateral diffusion to the surrounding area. Consequently, grazed grasslands can be considered a mosaic of patches of variable vegetation height, with or without the presence of urine and dung. This variability of plant and soil properties across space has the potential for significant effects on biotic interactions, plant community dynamics, and ecosystem function (Bloor and Pottier, 2014). In particular, spatial heterogeneity may modify nutrient pools and fluxes via changes in plant traits, plant resource uptake, and nutrient use, which in turn impact the soil and plant-soil feedbacks.

NUTRIENT FLUXES BETWEEN PLANTS AND ANIMALS

Soil and vegetation compartments provide a strong coupling at the small scales between C and nutrient cycles in grassland ecosystems, promoting belowground C sequestration and reducing N losses to the atmosphere and hydrosphere (Soussana & Lemaire, 2014). Ingestion of biomass by animals (and subsequent digestion and excretion) drives substantial uncoupling (small to medium scales) of C, N, and P, as well as spatial redistribution of nutrients. In this section, we examine transfers from the plant to the animal and consider the fate of nutrients from animal excreta.

Element Intake by Animals

The ability of animals to grow, reproduce, and produce outputs such as milk or wool relies on the intake of essential nutrients. For ruminants, ingested biomass generally consists of grass-based products (grazing, hay, silage) and crop-based products (maize, fodder beet, sorgo), and increasingly cereal-legume mixtures (silage as rough forage or grain as concentrates). The quantity of ingested biomass differs according to animal category (i.e., sheep, dairy, beef, etc.) and live weight, and it can be estimated in terms of gross energy (e.g., megajoules per day) or dry matter (e.g., kilograms per day).

Forage nutrient content or quality of grass-based animal feed varies depending on the botanical composition of the grassland and plant phenological stage, as well as underlying soil fertility (Table 2.1). Temporary sown grassland (<5 years, e.g., grass-ley) are usually fertilized and sown with high productive grasses species (e.g., in Central EU *Lolium* sp., *Dactylis, Festuca arundinacea*) in species mixtures with legumes (*Trifolium repens or T. pratense, Medicago sativa*) in variable proportions (about 30% legumes, ensuring N self-sufficiency; e.g., Lüscher et al., 2014). Accordingly, forage biomass can account for N contents between 2% and 3.5% for green grass and 0.1%–0.5% of P. Permanent grasslands (>5 years long term or never ploughed) are more variable, including low to high productive grasslands depending on climate zone, soil fertility, and agricultural management (fertilization, grazing, etc.). Depending on availability of mineral nitrogen in soil, N content is negatively linked to biomass (and thus C) according to the N-dilution curve (Lemaire and Salette, 1984), while N:P ratios are more constant (N:P of 6–8, Jouany et al., 2004; N:K of 1.2–1.75, Hoosbeek et al., 2002).

When grass biomass is removed via mowing, all harvestable vegetation is exported from the field, whereas for grazing, biomass removal via intake varies with stocking density (up to 90% of the available biomass under severe grazing). The remaining vegetation parts, i.e., ungrazed grass turfs and biomass below 5 cm (stubble, roots, stolons, etc.), contribute to nutrient returns to soil (i.e., fresh litter).

Element Excretion by Animals

Grazing animals retain a relatively small proportion of the nutrients that they ingest in their feed; 65%–90% of nutrients ingested by domestic herbivores are excreted as urine or dung (Bloor et al., 2012). The chemical composition of animal excreta varies depending on the form of excretion and the quantity and quality of forage ingested, as well as on the species of animal and their physiologic state (Haynes and Williams, 1993; Vérité and Delaby, 2000). In general, dung represents a substantial input of C and P, with lower N and K content, whereas urine represents a substantial input of N and K. N fecal excretion depends on dry mass intake (8 g N/kg DM intake on average). For C, a large

TABLE 2.1 Harvestable Biomass and Nutrient Content (Range of Values, From INRA, 2010)

	Biomass t DM/ha	C (kg/ha)	N (kg/ha)	P (kg/ha)	K (kg/ha)
Temporary grassland	7–14	3000–6000	150–400	20–50	180–420
Permanent grassland	4–10	1700–4250	90–300	10–36	100–300
Maize silage	10–18	4500–8100	110–220	18–32	100–180

part is also eliminated by respiration (CO_2 is 40% of intake, Faverdin et al., 2007), with small losses by ruminant eructation (CH_4: 5%–7%, Doreau et al., 2011).

In dairy production, the nutrients exported in milk represent between 20% and 30% for N, 35%–40% for P, and less than 10%–20% for K of the total intake (Table 2.2). This proportion largely depends on the balance between diet supplies and animal needs (the higher the animal needs, the lower the proportion of nutrients exported in animal products). Nonetheless, domestic animals have a low conversion efficiency of ingested plant proteins into meat and dairy products: it takes more than 3 kg of plant protein to produce 1 kg of milk protein and between 5 and 10 kg of plant protein to produce 1 kg of bovine protein (protein conversion: 10% dairy heifers to 25% lactating cow and 5.2% for sheep, Peyraud and Peeters, 2016). For beef and sheep, as the growth and fattening functions are less efficient than milk production, the nutrients fixed in the animal products represent a lower proportion of intake. Nutrient decoupling is thus the consequence of different pathways associated with animal metabolism that eliminate the excess of nutrient intake. The degree of nutrient decoupling can be estimated from animal diet and production level using animal-based models (Maxin and Faverdin, 2006; Faverdin et al., 2007).

Daily nutrient intake rates for individual animals can be scaled up to the field level, assuming that production of maize (12 t DM/ha) and grass (8 t DM/ha) provides 675 and 400 rations, respectively (Table 2.3). Protein concentrate, mainly based on soya or rapeseed bean meal, must be added to dairy cow feed to improve maize-based animal diets. These concentrates modify the balance between animal ingestion of grass in the field (uptake of nutrients) and return of nutrients to the field in the form of excreta by increasing the input of nutrients at the field level over time from external sources (rather than a simple redistribution of nutrients within the field). Consequently, a grass-based grazing regime promotes a more balanced distribution of nutrients at the large spatial scale as well as synchrony of nutrient coupling/uncoupling at the local scale.

TABLE 2.2 Daily C, N, P, and K Fluxes for a Dairy Cow With Respect to Diet

Ration		C (g/day)[a]	N (g/day)	P (g/day)	K (g/day)	C/N	N/P
Maize-based	Intake	8566	429	65	232	20.0	6.6
	Milk	1905 (22%)	132 (31%)	25 (39%)	42 (18%)	14.4	5.2
	Feces	2518 (29%)	158 (37%)	38 (59%)	35 (15%)	15.9	4.1
	Urine	75 (1%)	113 (26%)	0.3 (0%)	155 (67%)	0.7	377
Grass-based	Intake	7303	490	61	510	14.9	8.0
	Milk	1632 (22%)	113 (23%)	22 (35%)	36 (7%)	14.4	5.2
	Feces	2313 (32%)	139 (28%)	38 (62%)	77 (15%)	16.6	3.6
	Urine	115 (2%)	208 (42%)	0.3 (0%)	397 (78%)	0.6	693

Values are presented for a maize-based ration (average of 16 kg DM maize silage + 3 kg soya bean meal) and a grass-based diet with a daily intake of 17 kg DM. Nutrient fluxes as a proportion of intake are given between brackets.

[a] *C–CH_4 and C–CO_2 (g/day) are, respectively, 321 and 3662 (42% of intake) and 285 and 2884 (39% of intake) for the two rations.*

TABLE 2.3 Field-Level C, N, P, and K Fluxes for Dairy Cows per Unit Area (ha) for a Maize- and Grass-Based Diet, Where a Hectare of Maize- and Grass-Based Diet Corresponds to 675 and 400 Daily Rations, Respectively

Diet		C (kg/ha)	N (kg/ha)	P (kg/ha)	K (kg/ha)
Maize[a]	**Ingested[b] biomass**	**4872**	**121**	**19**	**108**
	Milk	1286	89	17	28
	Feces	1700	107	26	23
	Urine	51	76	0.2	105
	Exports	***3037***	***272***	***43.2***	***156***
Grass	**Ingested biomass**	**2921**	**196**	**24**	**204**
	Milk	653	45	9	14
	Feces	925	56	15	31
	Urine	46	83	0.1	159
	Exports	***1624***	***184***	***24.1***	***204***

[a] *To be well composed, a maize silage–based ration should be supplemented with external protein concentrate.*
[b] *Local ingested biomass.*

Where Do the Nutrients Go? The Fate of Animal Excreta

In most livestock systems, animal excreta either returns directly to the field or is collected and stored for subsequent agricultural use as compost, manure, or slurry (sometimes treated or used for methane production). For grazed grasslands, excretal return to the field promotes nutrient cycling and redistribution, which in turn influences the structure and functioning of grassland systems (Haynes and Williams, 1993). Once deposited on the pasture, the constituent elements of urine and dung are either incorporated into the soil prior to being taken up by plants and microorganisms (before or after nutrient transformations) or "lost" from the grassland system due to leaching and gaseous emissions. The degree of leaching losses and gaseous emissions partly reflects the ambient conditions (soil moisture content, rainfall, temperature, vegetation cover) (Haynes and Williams, 1993; Leterme et al., 2003). In the case of dung, the majority of C is transformed into CO_2, and less than 20% of C persists in the soil (Bol et al., 2000). Dung decomposition is associated with small losses of NH_3 and slow mineralization of organic N, which generates a build-up of nitrate under the dung patch (Bloor et al., 2012). Dung addition has also been shown to increase extractable P and K in soil, although water-soluble K may be subject to rapid leaching (Aarons et al., 2004). Inputs of dung-derived C, N, and P often increase microbial biomass and induce microbial priming effects, promoting plant N content and/or growth in the longer term (Bloor, 2015). However, large dung pats from cattle may also have transient negative effects on plant growth by blocking plant access to light and increasing plant senescence at the local scale.

In the case of urine, the majority of N is in the form of urea, which shows rapid hydrolysis and subsequent mineralization; high levels of urine-derived mineral N exceed plant N demand and increase the risks of significant N losses (NH_3 volatilization, NO_3^- leaching, N_2O losses) (Haynes and Williams, 1993, Van Groenigen et al., 2005) and sometimes NH_4^+ losses (Leterme et al., 2003). Fate of urine N depends on the time of year: recycling in plant and soil is more important in spring compared to autumn (Decau et al., 2003; Vertès et al., 2008), and reducing stocking rates from late summer to winter is proposed as a mitigation action to decrease leaching risks (Jarvis et al., 2011; Schoumans et al., 2014). Nevertheless, urine patches are usually associated with a significant increase in plant growth and plant N content, and they can induce a shift toward grasses rather than legumes at the local scale (Vertès et al., 1997; Lemaire et al., 2000). In some cases, plant scorching, which can lead to bare soil, is observed in urine patches; this phenomenon is generally attributed to root death due to NH_3 (aq) toxicity.

In the case of "off-pasture" animal excreta (manure), solid manure is mainly feces, possibly with some bedding material (such as straw), whereas liquid manure or slurry is a mixture of urine and feces. In general, manure is a complex organic matrix in which bacteria and other microorganisms digest carbohydrates, proteins, and fats into smaller organic molecules and finally into different chemical compounds and inorganic molecules that can be taken up by plant roots or that can be lost in gaseous form to the atmosphere or in dissolved form to ground and surface water. The nutrients in manure are directly related to animal feed and vary between different fodders and feed ingredients (Table 2.4). Manure, if applied correctly and in the right quantity, is an excellent natural fertilizer (i.e., nitrogen, phosphorus, and potassium), containing also organic matter, numerous plant nutrients (magnesium, calcium, sulphur), and trace elements (zinc, copper, molybdenum, manganese, cobalt, and selenium). The challenge is thus to couple manure nutrients into the food chain instead of decoupling and losing them to the environment (Leip et al., 2015).

The majority of recovered animal waste is stored in livestock and outdoor buildings or in the field (Box 2.1), in which aerobic and/or anaerobic conditions of manure storage and management determine the type of degradation of the organic matter and therefore the associated gaseous emissions. As a result, CH_4 emissions are generally higher for liquid products (slurry) than for solid products (manure) and lower after aerobic application in the field. CH_4 emissions during manure storage also depend on the storage conditions (e.g., emissions increase with temperature), duration, and the animal species. N_2O is emitted under both aerobic and anaerobic conditions and is therefore more important for manure than for liquid dejections. Although rational manure management can reduce the undesired emissions of non-CO_2 greenhouse gases (CH_4, N_2O, NH_3), gaseous emissions and loss of particulate matter and phosphorus (P or phosphate P_2O_5) remain an issue for animal housing, manure storage, transport, and during and after field application (Chadwick et al., 2011).

TABLE 2.4 The Organic Matter, N, P, and K Contents for the Main Cattle Manure Types Expressed per kg Fresh Product

	DM %	OM %	Ntot (g/kg)	P (g P_2O_5/kg)	K (g K_2O/kg)	Nmineral (% N tot) (%)	C/N*
Cattle manure (straw ++)	22–25	18	58	2.3	9.6	5–20	10
Cattle manure (straw +)	19–25	16	51	2.3	6.2	20–40	13
Cattle compost	33–35	21	80	5	14	2–10	5
Cattle slurry (mean dilution)	6–7.5	6	26	1	3.1	40–60	9
Cattle brown waters	0.35	0.2	0.35	0.12	0.4	60–70	–
Pig slurry[a]	6.8	4.6	58	3.2	4.8	60–70	3

[a] Can be spread on grasslands.

*From Quideau, P. 2010. Les effluents d'élevage, les co-produits de traitement et leur incidences environnementales. In: Espagnol, S., Leterme, P. (Eds.), Elevage et Environnement, Educagri-Quae Editions, 119–186 and * for C/N from Chadwick, D., Sommer, S., Thorman, R., Fangueiro, D., Cardenas, L., Amon, B., Misselbrook, T. 2011. Manure management: implications for greenhouse gas emissions. Animal Feed Science and Technology 167, 514–531.*

BOX 2.1

MANURE MANAGEMENT AND CNP FLUXES

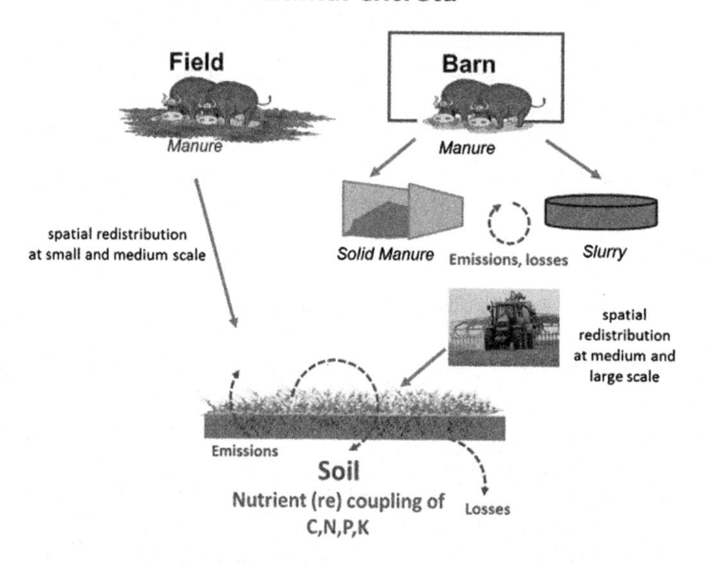

On average, animal manure represents 27, 14, and 37 kg per ha of fertilized agricultural area for N, P, and K respectively (mean values for France, Unifa, 2009). Due to the uneven geographic distribution of breeding farms, these values may be higher per ha for intensive livestock production areas (the EU-N Directive limit is 170 kg organic N/ha in these areas).

Annual N, P, and K excretion can be calculated for animal herds based on their diet and housing. Two examples are presented in Fig. 2.3, comparing N emissions during grazing, housing, and manure storage for two cattle systems. The balance between grazing, housing, and manure management (storage then spreading) is critical for reducing gaseous losses, in particular ammonia, while feed and temperature are the major factors influencing ammonia emissions in cattle stables. N_2O and N_2 emissions remain poorly understood.

GRAZING INTENSIFICATION: ENVIRONMENTAL CONCERNS AND MITIGATION STRATEGIES

Grazing and grasslands can reduce the carbon footprint of milk and beef production for pasture-based livestock systems by between 33% and 50% (Crosson et al., 2011; Peyraud et al., 2010), as well as markedly reducing emissions related to housing, manure management, and spreading (Pellerin et al., 2017). However, nutrient losses increase with intensification of herbage use efficiency (the proportion of herbage production harvested or consumed by animals) and animal stocking densities due to a greater number of excretal "hotspots" that promote leaching and gaseous emissions (Box 2.2). Animal loading also has a direct impact on the soil through soil compaction and soil degradation (Schils et al., 2013), increasing frequency of grassland renovation and associated high C and N losses (Vertès et al., 2007). Key tradeoffs between grazing intensification and environmental impacts have to be considered (Soussana and Lemaire, 2014): maximization of herbage use by animals versus nutrient returns to soil, improved forage quality to reduce enteric CH_4 versus decomposability of herbage to increase mean residence time of soil organic C, and maximization of animal stocking density versus reduction of CH_4 and N_2O emissions.

Mitigation Options at the Field and Landscape Scale

As we have seen in the previous sections, grazing leads to changes in C and nutrients in grassland systems via multiple, interacting processes. Consequently, strategies for sustainable grassland management that minimize nutrient losses and greenhouse gas emissions can only be achieved using a holistic approach, considering interactions and tradeoffs between the main element cycles.

A number of grassland management options have been suggested to mitigate climate change and reduce the negative environmental effects of intensification in grasslands (e.g., Dorioz et al., 2011). For widespread adoption by farmers, mitigation option and adaption measures need to be technically effective and economically efficient (cost beneficial, i.e., low production losses, labor) (Moran et al., 2011). The main mitigation options involve (1) optimization of grazing intensity and grazing season at the system level, (2) spatial (re)distribution and coupling of C:N cycles across field and barn systems, (3) restoration of degraded pasture, (4) use of biologic fixation to provide N inputs, improve forage quality, and enhance digestibility and subsequent nutrient cycling rates, and (5) optimization of N fertilization practices (i.e., timing and quantity) in line with vegetation requirements (Table 2.5).

When upscaling from the field or farm scale to the regional scale, mitigation options need to consider how transfers of nutrients such as N and P interact with landscape features. The magnitude of N and P fluxes (annual inputs, soil stocks, transfers to water), the preferential flow paths for these transfers, and the effects of landscape buffers differ in both intensity and spatial and temporal distribution (Fig. 2.4; Gascuel et al., 2009). For example, nitrate is very mobile and easily transferred in the landscape. Consequently a buffer interface for nitrate is only efficient if it combines a low flow velocity, a high biologic activity, and a source of carbon, as is the case in wetlands, hedges, and ponds. Riparian areas and wetlands are particularly efficient at nitrogen removal. Topographic indices can be used to identify effective buffer areas and their influence at the catchment level (Gascuel et al., 2009; Vertès et al., 2009).

Transfers of nutrients at the landscape scale are exacerbated by the geographic separation of livestock production from feed production areas. Consequently, strategies for reducing

BOX 2.2

IMPACTS OF INCREASED CATTLE STOCKING RATE AND NITRATE LEACHING

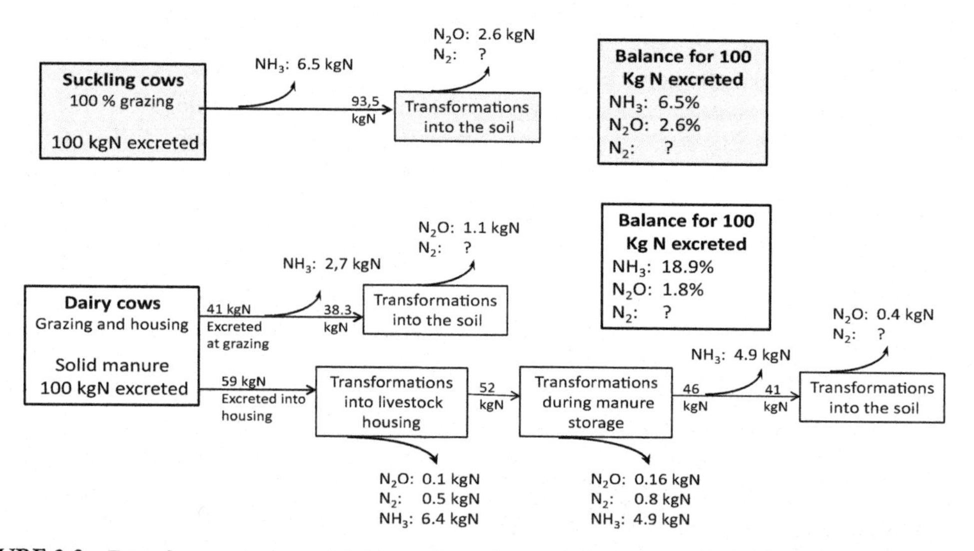

FIGURE 2.2 Fate of excreta and associated losses in cattle production systems. *In Delaby et al., 2014, From Gac, A., Béline, F., Bioteau, T., Maguet, K., 2007. A French inventory of gaseous emissions (CH_4, N_2O, NH_3) from livestock manure management using a mass-flow approach. Livestock Science, 112(3), 252–260, based on a large international review.* For details on other systems and manure types see Delaby et al. (2014).

High stocking rate is typically associated with increased nitrate leaching losses, and it can be used as an indicator of leaching risk (Fig. 2.2). Evidence from cattle-grazed pastures in Europe suggests that N leaching increases dramatically above a stocking rate threshold of 450–550 LSU·-day/ha per year; below this stocking rate, N leaching is usually low (less than 40 kgN/ha per year). Variation in leaching losses for a given stocking rate may reflect high denitrification in those conditions (difficult to quantify, in particular the N_2 part) or uncertainties on nitrate fluxes difficult to measure when ground water table is near the soil surface.

nutrient losses at the landscape scale rest on both agricultural and landscape levers (Table 2.6). Implementation of mitigation strategies requires a number of key steps: (1) Identification of stakeholders concerned by grasslands dynamics at different organizational scales within the catchment area (watershed); (2) knowledge of spatial indicators for grassland dynamics in cropping systems and land use (e.g., with RPG, remote sensing data); (3) understanding the drivers of

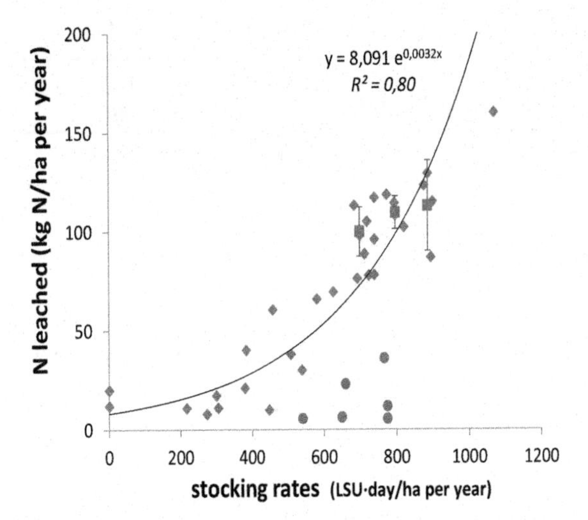

FIGURE 2.3 Relationship between stocking rates and nitrate leaching losses in grazed or mixed grassland. *Red symbols* (dark gray in print version) correspond to Irish results where grass-dominated fertilized grasslands are intensively grazed with high stocking rates. In free-draining soil conditions (*red squares* (dark gray in print version), McCarthy et al., 2015), N leaching increases with SR as predicted from all sites synthesis, while on poorly draining soils (*red circles* (dark gray in print version), Humphreys et al., 2008), very low values were observed for N leached, even with high SR, corresponding to another fate (denitrification and/or storage in soils) and difficulties to measure nitrate leaching in wet soils. *Completed from Simon, J.C., Vertès, F., Decau, M.L., Le Corre, L. 1997. Les flux d'azote au paturage. I- Bilans à l'exploitation et lessivage d'azote sous prairies. Fourrages 151, 249–262.*

present production systems and possible evolutions; (4) understanding technico-economic rationales and constraints (Osty et al., 1998); (5) interpretation of knowledge at different organization levels for a global diagnosis and co-construction of possible changes with stakeholders (e.g., Gascuel-Odoux et al., 2015).

Although distances between livestock and crop farms limit opportunities for recycling and recoupling cycles in space, spatial organization of farms and their fields can provide new opportunities for recoupling nutrient cycles (*see Chapter V-3*).

CONCLUSIONS

In livestock production systems, we have seen that animal feeding and digestion drives the biochemical uncoupling of C, N, P, and K. Divergence between the spatial distribution of feeding activities and the spatial distribution of nutrients that return to the field as excreta leads to spatial and temporal uncoupling of nutrient cycles at the field scale. This imbalance between nutrients in soil, plant, and animal compartments is compounded at the landscape scale where production systems rely heavily on nutrients produced elsewhere, i.e., grazing fields that require significant inputs of mineral N fertilizer or where the animal diet is supplemented by protein-rich concentrates. Consequently, grazing-based livestock production systems present a variety of economic and environmental benefits: (1) grazing is a cheap source of feed and can be applied to areas not suitable for crop cultivation, (2) grasslands can be used as part of schemes to limit nutrient losses via emissions and leaching (e.g., NL where slurry is largely spread on mown grasslands, IRL where wet soils limit nitrate losses and permanent grassland soils store N and C), and it can compensate farm emissions via C offsets, (3) the use of legumes provides a link between intensive dairy production and the soil (e.g., in Danish dairy systems, Rasmussen et al., 2012), and (4) grazing and locally produced animal feed improves nutrient recycling and minimizes losses by improving the spatial and temporal recoupling of nutrient cycles. Effective recoupling of C, N, P, and K cycles requires appropriate stocking rates and grassland management practices adapted to local bioclimatic conditions and soil properties; these management "best practices" form the basis of sustainable production systems and mitigation options that minimize the environmental implications of grassland intensification.

TABLE 2.5 Mitigation Options for Sustainable Grassland Management, and Associated Impacts on C and N Fluxes

Measure	Measure Examples	C (CO_2, CH_4)	N (NO_2, NH_3)	Effects	References
Optimization of grazing management	Increase in grazing intensity for low productive grasslands	↗↗	↗↙	C: increase in soil C sequestration for low productive grasslands	Skinner et al., 2013 Lemaire et al. (2011) Pellerin et al. (2013) Chapter 6
	Reduction in grazing intensity for intensively/overgrazed grasslands	↗	↙↙	N_2O: reduction in emission for intensively used grasslands C: increase in soil C sequestration by increasing C inputs through plants	Reeder et al. (2004) Luo et al. (2010)
	Prolongation of grazing season	↗↙	↙↙	GHG: reduction of emissions related to manure management and spreading	Lovett et al. (2008), Pellerin et al. (2013), Peyraud et al. (2010)
Pasture improvement	Use biologic fixation to reduce N fertilization and enteric CH_4	↗	↙↙	Reduction of mineral fertilizer application (up to 200 kgN/ha per year) CH_4: reduction through improved forage quality	N_2O: Ledgard et al. (2001) C: Lüscher et al. (2014), Derner and Schuman (2007) CH_4: Lee et al. (2004), FAO (2013)
	Restoration	↗↗	↗↗	Increase in soil C sequestration through improved biomass production Reduction of nutrient losses	Conant et al. (2001)
	Optimize N fertilization to adapt to vegetation requirements (i.e., timing and quantity)		↗↗	Reduction of N fertilizer application and N_2O emissions	Rees et al. (2013), Luo et al. (2010)

"↗↗" denotes a positive mitigative effect (i.e., emissions reduction, enhanced GHG removal or compensation by pasture C sequestration). "↙↙" denotes negative mitigative effect (i.e., an increase in emissions or suppression of GHG removal). "↗↙" denotes uncertain or variable response.

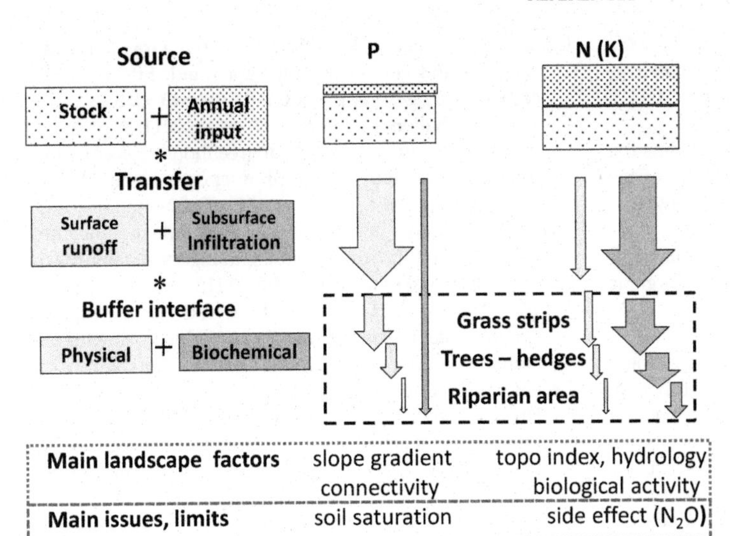

FIGURE 2.4 Drivers of N and P nutrient dynamics and the nutrient balance at the landscape scale. N and P sources are represented as *squares* (soil stock and inputs) and fluxes as *arrows* (size of arrows indicates the amounts of P and N transferred and the effects of buffer areas). *Based on Gascuel et al. (2009).*

TABLE 2.6 Possible Agricultural and/or Landscape Mitigation Strategies to Reduce N and P Pollution: Levers and Processes

Strategies and Possible Changes	Levers	Processes
Land use: localization buffer effect in watersheds • Buffer areas localization ("source" and "sink") • Connectivity between fluxes pathways	Grasslands and crops mosaic organization in landscape Locate high N uptake crop in lowlands	Intercept P, SEM, DOM Nutrients uptake by crops in bottom areas
Extensification: decrease N and P balances • In agricultural area • In whole watershed	• Longer crop rotations, higher part of grasslands • Convert arable fields into extensive grasslands, in wooden areas	Decrease N(P) inputs and/or increase outputs and storage in soils Dilution of nutrients from AA

From Gascuel-Odoux, C., Christen, B., Dorioz, J-M., Moreau, P., Ruiz, L., Trévisan, D., Vertès, F., 2011. Change in land use and patterns. In: Schoumans, O.F., Chardon, W.J., (Eds.), Mitigation Options for Reducing Nutrient Emissions from Agriculture, Alterra report 2141, Wageningen, NL, 93–106.

References

Aarons, S.R., O'Connor, C.R., Gourley, C.J.P., 2004. Dung decomposition in temperate dairy pastures I. Changes in soil chemical properties. Australian Journal of Soil Research 42, 107–114.

Adler, P.B., Raff, D., Lauenroth, W.K., 2001. The effect of grazing on the spatial heterogeneity of vegetation. Oecologia 128, 465–479.

Anglade, J., Billen, G., Garnier, J., 2015. Relationships for estimating N-2 fixation in legumes: incidence for N balance of legume-based cropping systems in Europe. Ecosphere 6 (3), 1–24.

Bahn, M., Rodeghiero, M., Anderson-Dunn, M., et al., 2008. Soil respiration in European grasslands in relation to climate and assimilate supply. Ecosystems 11 (8), 1352–1367.

Bakker, E.S., Ritchie, M.E., Olff, H., Milchunas, D.G., Knops, J.M.H., 2006. Herbivore impact on grassland plant diversity depends on habitat productivity and herbivore size. Ecology Letters 9, 780–788.

Ball, P.R., Ryden, J.C., 1984. Nitrogen relationship in intensively managed grasslands. Plant and Soil 76 (1–3), 23–33.

Bardgett, R.D., Streeter, T.C., Bol, R., 2003. Soil microbes compete effectively with plants for organic-nitrogen inputs to temperate grasslands. Ecology 84 (5), 1277–1287.

Bardgett, R.D., Wardle, D.A., 2003. Herbivore-mediated linkages between aboveground and belowground communities. Ecology 84, 2258–2268.

Bardgett, R.D., Anderson, J.M., Behan-Pelletier, V., et al., 2001. The influence of soil biodiversity on hydrological pathways and the transfer of materials between terrestrial and aquatic ecosystems. Ecosystems 4 (5), 421–429.

Bardgett, R.D., Bowman, W.D., Kaufmann, R., et al., 2005. A temporal approach to linking aboveground and belowground ecology. Trends in Ecology and Evolution 20 (11), 634−641.

Billen, G., Thieu, V., Garnier, J., Silvestre, 2009. Modelling the N cascade in regional watersheds: the case study of the Seine, Somme and Scheldt rivers. Agriculture. Ecosystems & Environment 133 (3/4), 234−246.

Bloor, J.M.G., 2015. Additive effects of dung amendment and plant species identity on soil processes and soil inorganic nitrogen in grass monocultures. Plant and Soil 396 (1-2), 189−200.

Bloor, J.M.G., Pottier, J., 2014. Grazing and spatial heterogeneity: implications for grassland structure and function. In: Mariotte, P., Kardol, P. (Eds.), Grassland Biodiversity and Conservation in a Changing World. Nova Science Publishers, Inc., Hauppauge, New York, USA, pp. 135−162.

Bloor, J.M.G., Jay-Robert, P., Le Morvan, A., Fleurance, G., 2012. Dung of domestic grazing animals: characteristics and role for grassland function. INRA Productions Animals 25, 45−56.

Blouin, M., Hodson, M.E., Delgado, E.A., et al., 2013. A review of earthworm impact on soil function and ecosystem services. European Journal of Soils Science 64 (2), 161−182.

Bol, R., Amelung, W., Friedrich, C., Ostle, N., 2000. Tracing dung derived carbon in temperate grassland using ^{13}C natural abundance measurements. Soil Biology and Biochemistry 32, 1337−1343.

Bouraoui, F., Grizzetti, B., Adelskold, G., et al., 2009. Basin characteristics and nutrient losses: the EUROHARP catchment network perspective. Journal of Environmental Monitoring 11 (3), 515−525.

Butterbach-Bahl, K., Dannenmann, M., 2011. Denitrification and associated soil N_2O emissions due to agricultural activities in a changing climate. Current opinion in environmental sustainability 3 (5), 389−395.

Butterbach-Bahl, K., Baggs, E., Dannenmann, M., et al., 2013. Nitrous oxide emissions from soils: how well do we understand the processes and their controls? Philosophical Transactions of the Royal Society of Biological Sciences 368, 1621.

Chadwick, D., Sommer, S., Thorman, R., Fangueiro, D., Cardenas, L., Amon, B., Misselbrook, T., 2011. Manure management: implications for greenhouse gas emissions. Animal Feed Science and Technology 167, 514−531.

Cluzeau, D., Binet, F., Vertès, F., Simon, J.C., Rivière, J.M., Tréhen, P., 1992. Effects of intensive cattle trampling on soil-plant-earthworms system in two grassland types. Soil Biology and Biochemistry 24 (12), 1661−1665.

Conant, R.T., Paustian, K., Elliott, E.T., 2001. Grassland management and conversion into grassland: Effects on soil carbon. Ecological applications 11 (2), 343−355.

Crosson, P., Shalloo, L., O'Brien, D., et al., 2011. A review of whole farm systems models of greenhouse gas emissions from beef and dairy cattle production systems. Animal Feed Science and Technology 166−67, 29−45.

Decau, M.L., Simon, J.C., Jacquet, A., 2003. Fate of urine nitrogen in three soils throughout a grazing season. Journal of Environmental Quality 32, 1405−1413.

Di, H.J., Cameron, K.C., 2002. Nitrate leaching in temperate agroecosystems: sources, factors and mitigating strategies. Nutrient Cycling in Agroecosystems 46, 237−256.

Delaby, L., Dourmad, J.-Y., Béline, F., Lescoat, P., Faverdin, P., Fiorelli, J.-L., Vertès, F., Veysset, P., Morvan, T., Parnaudeau, V., Durand, P., Rochette, P., Peyraud, J.-L., 2014. Origin, quantities and fate of nitrogen flows associated with animal production. In: Nitrogen flows in livestock farming systems: Reduce losses, restore balance, 5. Advances in Animal Biosciences, pp. 28−48. Special issue 1.

Derner, J.D., Schuman, G.E., 2007. Carbon sequestration and rangelands: A synthesis of land management and precipitation effects. Journal of soil and water conservation 62 (2), 77−85.

Díaz, S., Lavorel, S., McIntyre, S., Falczuk, V., Casanoves, F., Milchunas, D.G., Skarpe, C., Rusch, G.M., Sternberg, M., Noy-Meir, I., Landsberg, J., Zhang, W., Clark, H., Campbell, B.D., 2007. Plant trait responses to grazing−a global synthesis. Global Change Biology 12, 1−29.

Doreau, M., van der Werf, H.M.G., Micol, D., et al., 2011. Enteric methane production and greenhouse gases balance of diets differing in concentrate in the fattening phase of a beef production system. Journal of Animal Science 89 (8), 2518−2528.

Dorioz, J.-M., Gascuel-Odoux, C., Stutter, M., Durand, P., Merot, P., 2011. Landscape management. In: Schoumans, O.F., Chardon, W.J. (Eds.), Mitigation Options for Reducing Nutrient Emissions from Agriculture, pp. 107−126. Alterra report 2141, Wageningen, NL.

Dumont, B., Carrère, P., Ginane, C., Farruggia, A., Lanore, L., Tardif, A., Decuq, F., Darsonville, O., Louault, F., 2011. Plant−herbivore interactions affect the initial direction of community changes in an ecosystem manipulation experiment. Basic and Applied Ecology 12, 187−194.

FAO (Food and Agriculture Organization), 2011. Successes and failures with animal nutritionpractices and technologies in developing countries. In: Makkar, H.P.S. (Ed.), Proceedings of the FAO Electronic Conference. FAO Animal Production and Health Proceedings, No. 11. Rome, Italy. www.fao.org/faostat/.

Faverdin, P., Peyraud, J.L., 2010. Nouvelles conduites d'élevage et conséquences sur le territoire : cas des bovins laitiers. Les colloques de l'Académie d'Agriculture de France (2009-04-28-2009-04-29) Paris (FRA). In: Elevages intensifs et environnement. Les effluents : menace ou richesse ? CR Acad.Agr, France, pp. 89–99.

Faverdin, P., Maxin, G., Chardon, X., Brunschwig, P., Vermorel, M., 2007. Modèle de prévision du bilan carbone d'une vache laitière. Rencontres autour des Recherches sur les Ruminants 14, 66.

Flechard, C.R., Ambus, P., Skiba, U., Rees, R.M., Hensen, A., van Amstel, A., van den Pol-van Dasselaar, A., Soussana, J.F., Jones, M., Clifton-Brown, J., Raschi, A., Horvath, L., Neftel, A., Jocher, M., Ammann, C., Leifeld, J., Fuhrer, J., Calanca, P., Thalman, E., Pilegaard, K., DiMarco, C., Campbell, C., Nemitz, E., Hargreaves, K.J., Levy, P.E., Ball, B.C., Jones, S.K., van de Bulk, W.C.M., Groot, T., Blom, M., Domingues, R., Kasper, G., Allard, V., Ceschia, E., Cellier, P., Laville, P., Henault, C., Bizouard, F., Abdalla, M., Williams, M., Baronti, S., Berretti, F., Grosz, B., 2007. Effects of climate and management intensity on nitrous oxide emissions in grassland systems across Europe. Agriculture, Ecosystems & Environment 121, 135–152.

Franck, D.A., Groffman, P.M., Evans, R.D., Tracy, B.F., 2000. Ungulate stimulation of nitrogen cycling and retention in Yellowstone Park grasslands. Oecologia 123, 116–121.

Gac, A., Béline, F., Bioteau, T., Maguet, K., 2007. A French inventory of gaseous emissions (CH_4, N_2O, NH_3) from livestock manure management using a mass-flow approach. Livestock Science 112 (3), 252–260.

Galloway, J.N., Townsend, A.R., Erisman, J.W., Bekunda, M., Cai, Z.C., Freney, J.R., Martinelli, L.A., Seitzinger, S.P., Sutton, M.A., 2008. Transformation of the nitrogen cycle: recent trends, questions, and potential solutions. Science 320 (5878), 889–892.

Gascuel-Odoux, C., Christen, B., Dorioz, J.-M., Moreau, P., Ruiz, L., Trévisan, D., Vertès, F., 2011. Change in land use and patterns. In: Schoumans, O.F., Chardon, W.J. (Eds.), Mitigation Options for Reducing Nutrient Emissions from Agriculture, pp. 93–106. Alterra report 2141, Wageningen, NL.

Gascuel-Odoux, C., Ruiz L., Vertès F. (coord.), 2015. Comment réconcilier agriculture et littoral ? vers une agroécologie des territoires. éditions Quae, coll. Matière à débattre & décider 151 pp.

Gillet, F., Kohler, F., Vandenberghe, C., Buttler, A., 2010. Effect of dung deposition on small-scale patch structure and seasonal vegetation dynamics in mountain pastures. Agriculture. Ecosystems & Environment 135, 34–41.

Godinot, O., Leterme, P., Vertès, F., Carof, M., 2016. Indicators to evaluate agricultural nitrogen efficiency of the 27 member states of the European Union. Ecological Indicators 66, 612–622.

Hamilton, E.W., Frank, A.D., 2001. Can plants stimulate soil microbes and their own nutrient supply? Evidence from a Grazing Tolerant Grass Ecology 82 (9), 2397–2402.

Haynes, R.J., Williams, P.H., 1993. Nutrient cycling and soil fertility in the grazed pasture ecosystem. Advances in Agronomy 49, 119–200.

Hinsinger, P., 2001. Bioavailability of soil inorganic phosphorus in the rhizosphere as affected bu root-indices chemical changes. A Review: Plant and Soil 237 (2), 173–195.

Hogh-Jensen, H., Schjoerring, J.K., 2010. Interactions between nitrogen, phosphorus and potassium determine growth and N2-fixation in white clover and ryegrass leys. Nutrient Cycling in Agroecosystems 87, 327–338.

Hoosbeek, M.R., Van Breemen, N., Vasander, H., et al., 2002. Potassium limits potential growth of bog vegetation under elevated atmospheric CO2 and N deposition. Global Change Biology 8 (11), 1130–1138.

Hristov, A.N., Oh, J., Lee, C., Meinen, R., Montes, F., Ott, T., Firkins, J., Rotz, A., Dell, C., Adesogan, A., Yang, W., Tricarico, J., Kebreab, E., Waghorn, G., Dijkstra, J., Oosting, S., 2013. Mitigation of greenhouse gas emissions inlivestock production – A review of technical options for non-CO_2 emissions. In: Gerber, P.J., Henderson, B., Makkar, H.P.S. (Eds.), FAO Animal Production and Health Paper No. 177. FAO, Rome, Italy.

Humphreys, J., O'Connell, K., Casey, I.A., 2008. Nitrogen flows and balances in four grassland-based systems of dairy production on a clay-loam soil in a moist temperate climate. Grass and Forage Science 63 (4), 467–480.

Hutchings, N.J., Olesen, J.E., Petersen, B.M., Berntsen, J., 2007. Modelling spatial heterogeneity in grazed grassland and its effects on nitrogen cycling and greenhouse gas emissions. Agriculture. Ecosystems & Environment 121, 153–163.

Huyghe, C., De Vliegher, A., van Gils, B., Peeters, A. (coord.), 2014. Grasslands and Herbivore Production in Europe and Effects of Common Policies. Quae 320 pp http://www.quae.com/fr/r3371-grasslands-and-herbivore-production-in-europe-and-effects-of-common-policies.html.

INRA, 2010. Alimentation des Bovins, Ovins, Caprins: Besoins des animaux, valeurs des aliments. Editions Quae. ISBN : 978-2-7592-0874-6, 315 p.

Jarvis, S., Hutchings, N., Brentrup, F., Olesen, J.E., van de Hoek, K.W., 2011. Nitrogen flows in farming systems across Europe. In: Sutton, M.A., Howard, C.M., Erisman, J.W., Billen, G., Bleeker, A., Grennfelt, P., van Grinsven, H., Grizzetti, B. (Eds.), The European Nitrogen Assessment. Sources, Effects and Policy Perspectives. Cambridge University Press, Cambridge, pp. 211–228.

Jones, M.B., Donnelly, A., 2004. Carbon sequestration in temperate grassland ecosystems and the influence of management, climate and elevated CO_2. New Phytologist 164 (3), 423–439.

Jouany, C., Cruz, P., Petibon, P., et al., 2004. Diagnosing phosphorus status of natural grassland in the presence of white clover. European Journal of Agronomy 21 (3), 273–285.

Kayser, M., Isselstein, J., 2005. Potassium cycling and losses in grassland systems: a review. Grass and Forage Science 60, 213–224.

Kindler, R., Siemens, J., Kaiser, K., et al., 2011. Dissolved carbon leaching from soil is a crucial component of the net ecosystem carbon balance. Global Change Biology 17 (2), 1167–1185.

Klumpp, K., Soussana, J.-F., Falcimagne, R., 2007. Effects of past and current disturbance on carbon cycling in grassland mesocosms. Agriculture, Ecosystems and Environment 121 (1–2), 59–73.

Klumpp, K., Fontaine, S., Attard, E., et al., 2009. Grazing triggers soil carbon loss by altering plant roots and their control on soil microbial community. Journal of Ecology 97 (5), 876–885.

Ledgard, S.F., Penno, J.W., Sprosen, M.S., 1999. Nitrogen inputs and losses from clover/grass pastures grazed by dairy cows, as affected by nitrogen fertilizer application. Journal of Agricultural Science 132, 215–225.

Ledgard, S.F., Sprosen, M.S., Penno, J.W., et al., 2001. Nitrogen fixation by white clover in pastures grazed by dairy cows: Temporal variation and effects of nitrogen fertilization. Plant and Soil 229 (2), 177–187.

Lee, J.M., Woodward, S.L., Waghorn, G.C., Clark, D.A., 2004. Methane emissions by dairy cows fed increasing proportions of white clover (Trifolium repens) in pasture. Proceedings of the New Zealand Grassland Association 66, 151–155.

Leip, A., Billen, G., Garnier, J., Grizzetti, B., Lassaletta, L., Reis, S., Simpson, D., Sutton, M.A., De Vries, W., Weiss, F., Westhoek, H., 2015. Impacts of European livestock production: nitrogen, sulphur, phosphorus and GHG s emissions, land-use, water eutrophication and biodiversity. Environmental Research Letters 10 (11), 14 pp.

Lemaire, G., Chapman, D., 1996. Tisue flows in grazed communities. In: Hodgson, J., Illius, A.W. (Eds.), The Ecology and Management of Grazing Systems. CAB International Wallingford, UK, pp. 3–35.

Lemaire, G., Salette, J., 1984. Relationship between growth and nitrogen uptake in a pure grass stand I Environmental effects. Agronomie 4 (5), 423–430.

Lemaire, G., Gastal, F., Plenet, D., 1997. Dynamics of N uptake and N distribution in plant canopies. In: Lemaire, G., Burns, I.G. (Eds.), Diagnosis Procedure for Crop N Management. Colloques INRA 19-20 no 1996, 82, 15–29.

Lemaire, G., Hodgson, J., Moraes, A.D., Carvalho, P.D.F., Nabinger, C., 2000. Grassland Ecophysiology and Grazing Ecology. CABI.

Lemaire, G., Hodgson, J., Chabbi, A. (Eds.), 2011. Grassland Productivity and Ecosystems Services. CAB International, 287 pp.

Lemaire, G., Gastal, F., Franzluebbers, A., Chabbi, A., 2015. Grassland-cropping rotations: an avenue for agricultural diversification to reconcile high production with environmental quality. Environmental Management 56 (5), 1065–1077.

Leterme, P., Barré, C., Vertès, F., 2003. The fate of ^{15}N from dairy cow urine under pasture receiving different rates of N fertiliser. Agronomie 23 (7), 609–616.

Liebisch, F., Buenemann, E.K., Huguenin-Elie, O., et al., 2013. Plant phosphorus nutrition indicators evaluated in agricultural grasslands managed at different intensities. European Journal of Agronomy 44, 67–77.

Lovett, D.K., Shalloo, L., Dillon, P., et al., 2008. Greenhouse gas emissions from pastoral based dairying systems: The effect of uncertainty and management change under two contrasting production systems. Livestock Science 116 (1-3), 260–274.

Luo, J., de Klein, C.A.M., Ledgard, S.F., Saggar, S., 2010. Management options to reduce nitrous oxide emissions from intensively grazed pastures: a review. Agriculture, Ecosystems & Environment 136, 282–291.

Lüscher, A., Mueller-Harvey, I., Soussana, J.F., Rees, R.M., Peyraud, J.L., 2014. Potential of legume-based grassland – livestock systems in Europe: a review. Grass and Forage Science 69, 206–228.

Maxin, G., Faverdin, P., 2006. Modélisation des bilans entrée/sortie des éléments carbone, azote, eau et minéraux chez la vache laitière. Mémoire de fin d'études ESITPA/INRA de Rennes, 52 pages.

McCarthy, J., Delaby, L., Hennessy, D., et al., 2015. The effect of stocking rate on soil solution nitrate concentrations beneath a free-draining dairy production system in Ireland. Journal of Dairy Science 98 (6), 4211–4224.

Milchunas, D.G., Sala, O.E., Lauenroth, W.K., 1988. A generalized model of the effects of grazing by large herbivores on grassland community structure. The American Naturalist 132, 87–106.

Moraine, M., Duru, M., Nicholas, P., et al., 2014. Farming system design for innovative crop-livestock integration in Europe. Animal 8 (8), 1204–1217.

Moran, D., Macleod, M., Wall, E., Eory, V., McVittie, A., Barnes, A., Rees, R., 2011. Marginal abatement cost curves for UK agricultural greenhouse gas emissions. Journal of Agricultural Economics 62 (1), 93–118.

Morel, C., Tunney, H., Plenet, D., Pellerin, S., 2000. Transfer of phosphate ions between soil and solution. Journal of Environmental Quality 29 (1), 50–59.

Murphy, D.V., Recous, S., Stockdale, E., Fillery, I.R.P., Jensen, L.S., Hatch, D.J., Goulding, K.W.T., 2003a. Gross Nitrogen Fluxes in Soil : Theory, Measurement and Application of 15N Pool Dilution Techniques. In: Advances in Agronomy, vol. 79. Academic Press, pp. 69–118.

Murphy, D.V., Stockdale, E.A., Brookes, P.C., Goulding, K.W.T., 2003b. Impact of microorganisms on chemical transformations in soils. In: Abott, L.K., Murphy, D.V. (Eds.), Soil Biological Fertility. Kluwer Academic Publishers, pp. 37–59.

Orwin, K.H., Bertram, J.E., Clough, T.J., Condron, L.M., Sherlock, R.R., O'Callagha, M., 2009. Short-term consequences of spatial heterogeneity in soil nitrogen concentrations caused by urine patches of different sizes. Applied Soil Ecology 42, 271–278.

Parsons, A.J., Dumont, B., 2003. Spatial heterogeneity and grazing processes. Animal Research 52, 161–179.

Parsons, A.J., Carrère, P., Schwinning, S., 2000. Dynamics of heterogeneity in a grazed sward. In: Lemaire, G. (Ed.), Grassland Ecophysiology and Grazing Ecology. CABI, pp. 289–316.

Parsons, A., Rowarth, J., Thornley, J., Newton, P., 2011. Primary productin of grassalnds, herbage accumumation and use, and impacts of climate change. In: Lemaire, G., Hodgson, J., Chabbi, A. (Eds.), Grassland Productivity and Ecosystems Services. CABI, pp. 3–18.

Peeters, A., Beaufoy, G., Canals, R.M., De Vliegher, A., Huyghe, C., Isselstein, J., Jones, G., Kessler, W., Kirilov, A., Mosquera-Losada, M.R., Nilsdotter-Linde, N., Parente, G., Peyraud, J.-L., Pickert, J., Plantureux, S., Porqueddu, C., Rataj, D., Stypinski, P., Tonn, B., van den Pol–van Dasselaar, A., Vintu, V., Wilkins, R.J., 2014. Grassland term definitions and classifications adapted to the diversity of European grassland-based systems. Grassland Science in Europe 19, 743–750.

Pellerin, S., Bamière, L., Angers, D., Béline, F., Benoît, M., Butault, J.P., Chenu, C., Colnenne-David, C., De Cara, S., Delame, N., Doreau, M., Dupraz, P., Faverdin, P., Garcia-Launay, F., Hassouna, M., Hénault, C., Jeuffroy, M.H., Klumpp, K., Metay, A., Moran, D., Recous, S., Samson, E., Savini, I., Pardon, L., 2013. Quelle contribution de l'agriculture française à la réduction des émissions de gaz à effet de serre ? Potentiel-d'atténuation et coût de dix actions techniques. Synthèse du rapport d'étude, INRA (France). https://www6.paris.inra.fr/depe/Projets/Agriculture-et-GES.

Pellerin, Sylvain, Bamiere, Laure, Angers, Denis, et al., 2017. Identifying cost-competitive greenhouse gas mitigation potential of French agriculture. Environmental Science and Policy 77, 130–139.

Peyraud, J.L., Van den Pol-van Dasselaar, A., Dillon, P., Delaby, L., 2010. Producing milk from grazing to reconcile economic and environmental performances. In "Grassland in a changing world" Proc. 23th General Meeting of EGF, Kiel, Germany, pp. 865–879.

Peyraud, J.L., Peeters, A. 2016. The role of grassland based production system in the protein Security. EGFGrassland Science in Europe, Vol. 21 – The multiple roles of grassland in the European bioeconomy, 29–43.

Peyraud, J.-L., Taboada, M., Delaby, L., 2014. Integrated crop and livestock systems in Western Europe and South America: a review. European Journal of Agronomy 57, 31–42.

Quideau, P., 2010. Les effluents d'élevage, les co-produits de traitement et leur incidences environnementales. In: Espagnol, S., Leterme, P. (Eds.), Elevage et Environnement. Educagri-Quae Editions, pp. 119–186.

Rees, R.M., Baddeley, J.A., Bhogal, A., et al., 2013. Nitrous oxide mitigation in UK agriculture. Soil Science and Plant nutrition 59 (1), 3–15. SI, 2013.

Reeder, J.D., Schuman, G.E., Morgan, J.A., et al., 2004. Response of x`organic and inorganic carbon and nitrogen to long-term grazing of the shortgrass steppe. Environmental Management 33 (4), 485–495.

Rumpel, C., Crème, A., Ngo, N.T., Velásquez, G., Mora, M.L., Chabbi, A., 2015. The impact of grassland management on biogeochemical cycles involving carbon, nitrogen and phosphorus. Journal of Soil Science and Plant Nutrition 15, 353–371.

Ryden, J.C., Ball, P.R., Garwood, E.A., 1984. Nitrate leaching from grassland. Nature (London) 311, 50–53.

Schils, R.L.M., Eriksen, J., Ledgard, S.F., Vellinga, T.V., Kuikman, P.J., Luo, J., Petersen, S.O., Velthof, G.L., 2013. Strategies to mitigate nitrous oxide emissions from herbivore production systems. Animal 7, 29–40.

Schoumans, O.F., Chardon, W.J., Bechmann, M.E., Gascuel-Odoux, C., Hofman, G., Kronvang, B., Rubaek, G.H., Ulen, B., Dorioz, J.-M., 2014. Mitigation options to reduce phosphorus losses from the agricultural sector and improve surface water quality: a review. Science of the Total Environment 468, 1255–1266.

Schwinning, S., Parsons, A.J., 1996. Analysis of the coexistence mechanisms for grasses and legumes in grazing systems. Journal of Ecology (Oxford) 84, 799–813.

Shahzad, T., Chenu, C., Genet, P., Genet, P., Barot, S., Perveen, N., Mougin, C., Fontaine, S., 2015. Contribution of exudates, arbuscular mycorrhizal fungi and litter depositions to the rhizosphere priming effect induced by grassland species. Soil Biology and Biochemistry 80, 146–155.

Simon, J.C., Vertès, F., Decau, M.L., Le Corre, L., 1997. Les flux d'azote au paturage. I- Bilans à l'exploitation et lessivage d'azote sous prairies. Fourrages 151, 249–262.
Skinner, R.H., 2013. Nitrogen fertilization effects on pasture photosynthesis, respiration, and ecosystem carbon content. Agric. Ecosyst. Environ 172, 35–41.
Soussana, J.-F., Lemaire, G., 2014. Coupling carbon and nitrogen cycles for environmentally sustainable intensification of grasslands and crop-livestock systems. Agriculture, Ecosystems and Environment 190, 9–17.
Soussana, J.-F., Loiseau, P., Vuichard, N., Ceschia, E., Balesdent, J., Chevallier, T., Arrouays, D., 2004. Carbon cycling and sequestration opportunities in temperate grasslands. Soil Use and Management 20, 219–230.
Soussana, J.F., Tallec, T., Blanfort, V., 2010. Mitigating the greenhouse gas balance of ruminant production systems through carbon sequestration in grasslands. Animal 4 (3), 334–350.
Stehfest, E., Bouwman, L., 2006. N_2O and NO emission from agricultural fields and soils under natural vegetation: summarizing available measurement data and modeling of global annual emissions. Nutrient Cycling in Agroecosystems 74 (3), 207–228.
Thieu, V., Billen, G., Garnier, J., et al., 2011. Nitrogen cycling in a hypothetical scenario of generalised organic agriculture in the Seine, Somme and Scheldt watersheds. Regional Environmental Change 11 (2), 359–370.
Tilman, D., Wedin, D., Knops, J., 1996. Productivity and sustainability influenced by biodiversity in grassland ecosystems. Nature 379, 718–720.
Van Groenigen, J.W., Velthof, G.L., van der Bolt, F.J.E., Vos, A., Kuikman, P.J., 2005. Seasonal variation in N_2O emissions from urine patches: effects of urine concentration, soil compaction and dung. Plant and Soil 273, 15–27.
Velthof, G.L., van Beusichem, M.L., Raijmakers, W.M.F., et al., 1998. Relationship between availability indices and plant uptake of nitrogen and phosphorus from organic products. Plant and Soil 200 (2), 215–226.
Vérité, R., Delaby, L., 2000. Relation between nutrition, performances and nitrogen excretion in dairy cows. Annales de Zootechnie 49, 217–230.
Vertès, F., Mary, B., 2014. Part of grassland in ley-arable rotations is a proxy for predicting long term soil organic matter dynamics. In: Proceedings of the 18th Nitrogen Workshop, 30 June – 3 July 2014, Lisboa, pp. 347–348.
Vertès, F., Simon, J.C., Le Corre, L., Decau, M.L., 1997. Les flux d'azote au pâturage. II. Etude des flux et de leurs effets sur le lessivage. Fourrages 151, 263–280.
Vertès, F., Hatch, D., Velthof, G., Taube, F., Laurent, F., Loiseau, P., Recous, S., 2007. Short-term and cumulative effects of grassland cultivation on nitrogen and carbon cycling in ley-arable rotations. Grassland Science in Europe 12, 227–246.
Vertès, F., Simon, J.C., Giovanni, C., Grignani, M., Corson, M.S., Durand, P., Peyraud, J.L., 2008. Flux de nitrate dans les élevages bovins et qualité de l'eau : variabilité des phénomènes et diversité des conditions. In: Académie D'Agriculture, Mai 2008, pp. 6–26. IDELE.
Vertès, F., Trévisan, D., Gascuel-Odoux, C., Dorioz, J.M., 2009. Capacities and limits of two French grassland systems (intensive and extensive) to comply with WFD—developing tools to improve grassland management. Tearman 7, 161–173.
Voisin, A.S., Gastal, F., 2015; Nutrition azotée et fonctionnement agro-physiologique spécifique des légumineuses. In Les légumineuses pour des systèmes agricoles et alimentaires durables. Schneider, A. Huyghe, C. (coord.), Editions Quae, 79–138.
Wardle, D.A., Bardgett, R.D., Klironomos, J.N., Setälä, H., van der Putten, W.H., Wall, D.H., 2004. Ecological linkages between aboveground and belowground biota. Science 304, 1629–1633.
Whitehead, D.C., 2000. Nutrient Elements in Grassland. Soil-plant-animal Relationship. Cabi Publishing, 369 pp.
Whitehead, D.C., Bristow, A.W., Lockyer, D.R., 1990. Organic matter and nitrogen in the unharvested fractions of grass swards in relation to the potential for nitrate leaching after ploughing. Plant and Soil 123, 39–49.

Further Reading

Bol, R., Dunn, R.B., Pilgrim, E.S., 2011. Managing C and N in grassland systems: the adaptive cycle theory perspective. In: Lemaire, G., Hodgson, J., Chabbi, A. (Eds.), Grassland Productivity and Ecosystems Services. CABI, pp. 73–82.
Jarvie, H.P., Withers, P.J.A., Bowe, M.J., Palmer-Felgate, E.J., Harper, D.M., Wasiak, K., 2010. Streamwater phosphorus and nitrogen across a gradient in rural-agricultural land use intensity. Agriculture, Ecosystems & Environment 135, 238–252.
Jarvis, S.C., 2000. Progress in studies of nitrate leaching from grassland soils. Soil Use and Management 16, 152–156.

CHAPTER

3

C–N–P Decoupling Processes Linked to Arable Cropping Management Systems in Relation With Intensification of Production

Sylvie Recous[1], *Gwenaëlle Lashermes*[1], *Isabelle Bertrand*[2], *Michel Duru*[3], *Sylvain Pellerin*[4]

[1]FARE laboratory, INRA, Université Reims Champagne-Ardenne, Reims, France; [2]Eco&Sols, INRA, Univ Montpellier, CIRAD, IRD, Montpellier SupAgro, Montpellier, France; [3]AGIR, INRA, Université de Toulouse, Auzeville, France; [4]INRA, Bordeaux Sciences Agro, Univ. Bordeaux, ISPA, Villenave-d'Ornon, France

INTRODUCTION

Increasing world population, rising environmental concerns, and scarcity of resources have placed agriculture in a dual challenge: the need to maintain, and even increase, the primary productivity of cropped ecosystems to meet the increasing demand for food and nonfood biomass, while limiting environmental impacts and sustainably managing the world's critical resources for agricultural production. Over the past 50 years, the 2.5-fold increase in world food production has been accompanied by an eightfold increase in the amount of nitrogen (N) fertilizer input and an approximately fourfold increase in the amount of phosphate fertilizer input, using mining of phosphate sedimentary rocks. However, the P inputs have remained more or less constant since 1990, contributing to disproportionally higher N:P ratio inputs in most ecosystems (Peñuelas et al., 2012). In addition, atmospheric N deposition in terrestrial ecosystems is about 20 times more important than P deposition, thus contributing to the N:P imbalance, even in less anthropic ecosystems (Peñuelas et al., 2012).

The atmosphere is the main source of N, in the form of N_2, since it contains 78% by volume. But atmospheric N, however abundant, is not directly assimilated by plants because of its chemical stability that makes it a nonreactive species. The only biologic transformation of molecular N

Agroecosystem Diversity
https://doi.org/10.1016/B978-0-12-811050-8.00003-0

into organic N is the fixation of atmospheric N, produced by very specific microorganisms, free or in symbiosis with certain plants such as legumes. At the beginning of the 20th century (1910), the development of the Haber-Bosch process, consisting of an industrial conversion of atmospheric N into ammoniacal N and then nitric acid, led to considerable development of the N fertilizer industry. In contrast to N, weathering of primary rock minerals provides the dominant source of total P in natural terrestrial ecosystems (Chadwick et al., 1999). P fertilizer comes from nonrenewable phosphate sources, provided by only five countries, which hold 90% of the world P reserves, and P resources may end up becoming limited for agricultural lands because of the increasing demand (FAO, 2015). It is estimated that worldwide, a total of 5.7 billion hectares of land contain low levels of plant-available P (Cakmak, 2002).

Widespread use of synthetic N and P fertilizers led to an unprecedented increase in agricultural production, and the use of legume crops has decreased since fertilizer was available and cheap. As a consequence, agricultural systems and regions became more specialized, and the N and P cycles became locally "decoupled" (Billen et al., 2011), leading to losses to the atmosphere or hydrosphere with dramatic impact on the environment.

The end of the 1970s was marked by a rising awareness of the environmental consequences of intensive agricultural production with growing concerns about eutrophication, which was initially only related to P inputs, while controlling both N and P inputs to freshwaters was later deemed necessary (Conley et al., 2009). Because N is more mobile in soils than P, the diffuse pollution of aquifers by agricultural nitrate (e.g., exceeding the threshold of 50 mg/L water of nitrates (NO_3), for drinking water set by the European directive in 1980) was a strong environmental problem. Following the Rio summit in 1992 and the Kyoto summit in 1997, the role of global carbon (C) and N cycles in forecasting future climate scenarios and carbon storage potential in the continental biosphere became evident. The concept of "nitrogen cascade" (Galloway et al., 2003), which accounts for pollution transfers, emerged in the early 2000s. It is inseparable from the concept of "reactive N" which expresses that, as opposed to atmospheric diazote (dinitrogen, N_2), which is chemically inert, all the N compounds involved in biologic, photochemical, and radiative processes present in the atmosphere and the biosphere constitute the so-called reactive N, i.e., the reduced mineral forms of N (such as NH_3 or NH_4^+), oxidized mineral forms (NOx, N_2O, NO_3^-), and organic compounds. Since the discovery of the Haber-Bosch process, the use of synthetic N fertilizers has doubled the circulation of reactive N on a planetary scale (Erisman et al., 2008; Sutton et al., 2011), resulting in a range of effects associated with increasing concentrations or excess in agrosystems and natural, terrestrial, and aquatic ecosystems (leading to eutrophication of watersheds and sea coasts) and in the atmosphere (leading to air pollution and climate change). For example, on a European scale, agricultural soils account for 54% of N_2O emissions (Hertel et al., 2012), and for France, 81% of total N_2O emissions come from agriculture, of which 82% are from fertilized croplands and 4.5% from livestock (CITEPA, 2016).

In contrast to the tendency for an increase of the N:P ratio at the global scale (Peñuelas et al., 2012), P is polluting some agricultural areas with intense pasture grazing and application of animal excreta slurry for fertilization, with high inputs of P from off-farm concentrates, mostly imported (e.g., soybeans), leading to a decrease of the N:P ratio of lakes and streams. In areas with crops receiving high rates of P fertilizers, P tends to remain and accumulate in soils because P is strongly bound in insoluble forms to soil particles, which strongly decrease fertilizer P efficiency. The potential is much greater for soil P retention than for N retention, and there are fewer loss pathways for P than for N. These are the reasons for significant accumulation of P in many areas under intensive crop production and pasture grazing in Europe and the United States (McLauchlan, 2006). Ringeval et al. (2014) using computations at country level over the period 1945–2009, with France chosen as a typical western European country with intensive agriculture, found that, on average, 82% of soil P (sum of labile and stable pools) was anthropogenic, i.e., derived from P fertilizers. They showed that the contribution of anthropogenic P to food production was similar, at 84%, indicating the high human perturbations of the P cycle in agrosystems. Moreover, P accumulation in soils may lead to significant P transfers to aquatic ecosystems by soil erosion, with subsequent negative effect due to eutrophication.

Within these chains—or cascades—the mineral nutrition of the crops and the fertilization play a major role. However, the influence of agriculture on biogeochemical cycles and their environmental impacts is not limited to the direct effects of fertilization on crop production and nutrients losses, and on exports or recycling of elements contained in crops or crop residues. All the practices that affect the type of plant covers, the dates and conditions of the establishment of crops, and soil management, while not deliberately targeting the control of mineral nutrition, have important repercussions on the dynamics of absorption of mineral elements by plants and on the possible loss of these nutrients. Jarvis et al. (2011) clearly showed the impacts of farming systems, as a whole, on farm N budgets and the consequences on N leakages. Moreover, the decoupling of C and nutrient cycles is likely to boost these leakages and losses from the system.

In this context, this chapter develops the basis of coupling C, N, and P cycles in cropped soils and explains how management practices in intensive arable systems led to decoupling of these cycles. Most of the examples will concern N, considering the quantitative importance of this element, but many findings apply similarly to P and to a lesser extent to sulfur (S).

PROCESSES INVOLVED IN THE COUPLING OF C, N, AND P CYCLES IN SOILS

The cycles of carbon, nitrogen, and phosphorus are "naturally" coupled in terrestrial ecosystems by elemental stoichiometry of plant autotrophy and soil microbial heterotrophy (Soussana and Lemaire, 2014). In natural ecosystems, the N requirements for plants and soil organisms, coupled with relatively low levels of available N in many temperate and boreal forests ecosystems, causes atmospheric C uptake by plants and storage on land (soil + plants) to be tightly regulated by the N cycle (Vitousek and Howarth, 1991). Increases in N in ecosystems (e.g., via atmospheric deposition or symbiotic N fixers) is usually matched by significant increases in C storage, until N limitation of plants growth is alleviated (Asner et al., 1997) and if no other limiting factor prevails.

Role of Soil Microorganisms and Drivers

The heterotrophic soil microorganisms, bacteria, and fungi, also called "microbial biomass," mainly derive their energy and the elements necessary for their growth from the decomposition of organic matter. This results in mineralization of part of the carbon in the ultimate form of CO_2 emitted to the atmosphere, the other part being assimilated in microbial bodies. The microorganisms are subject to death and microbial recycling or are, in turn, consumed by predator organisms. Therefore the partition between mineralization and assimilation occurs later during microbial recycling, ultimately leading to most of the C being mineralized and the other part, the biomass-derived C, being stabilized through mineral association.

Therefore, the activity of heterotrophic organisms directly affects nutrient cycles, through the proportions of these nutrients in different compartments, i.e., the added (sometimes called "fresh" or "exogenous") and humified (sometimes called "native" or "endogenous" or "stabilized") organic matters, both substrates mineralized by the heterotrophic organisms. These proportions of elements, called stoichiometric ratios (C:N:P:S) and their variations across the different compartments, their stability or possible flexibility, govern the intensity of fluxes between compartments but also reflect the strong coupling between the nutrient cycles throughout the different levels of the food webs in soils (Manzoni et al., 2012; Soussana and Lemaire, 2014). While the stoichiometry ratio can vary greatly in plants according to species, maturity, conditions of growth and nutrition (Sardans et al., 2012), substrate origin, and processing types for organic products (Kallenbach and Grandy, 2011), the stoichiometry of soil organisms is considered highly constrained (Cleveland and Liptzin, 2007). Microbial processing narrows the elemental ratio during litter decomposition and the build-up of SOM (Mooshammer et al., 2014), while the structure and function of microbial communities and predominating life strategy adapt to the altered resources (Fanin and Bertrand, 2016). In a meta-analysis using 186 published observations on C, N, and P in the soil biomass, Cleveland and Liptzin (2007) showed that, even though a large range of nutrient concentration was observed in the microbial biomass, element ratios scaled isometrically and were constrained: C:N ratios averaged 8.6 ± 0.3, and N:P ratios averaged 6.9 ± 0.4. Based on this analysis, the authors estimated that the average soil microbial biomass C:N:P ratio is 60:7:1. More recently, Xu et al. (2013) estimated over a wider range of data and biomes that the average C:N:P stoichiometry for soil microbial biomass is 42:6:1.

Among several factors influencing C degradation and N or P mineralization or immobilization, one of the most important ones is the quality of the substrate to decompose, i.e., its biochemical and physical characteristics, which determine the intrinsic biodegradability and dynamics of decomposition. This factor received considerable attention in the past decades due to its strong effect on nutrient dynamics in soils in the short term, and on soil organic matter dynamics in the medium to long term. There is a large diversity of plant litter according to plant species, plant maturity, and plant parts and diversity in the composition of organic wastes and manures, according to the substrate's origin and process. This makes it difficult to predict the C mineralization and nutrient release from these organic substrates recycled to soils, despite a considerable amount of research on this topic (Trinsoutrot et al., 2000; Jensen et al., 2005; Justes et al., 2009; Bertrand et al., 2006; Lashermes et al., 2010).

The nature of the resource and the microbial activities affects the intensity of processes that consume or feed mineral and organic forms, i.e., mineralization (also called gross mineralization to distinguish it from net mineralization) and microbial assimilation (also called gross immobilization), the balance of which determines the net mineralization, i.e., the availability of mineral forms in the soils. C:N, C:P, and C:S ratios of plant litters are considered indicators of whether N, P, or S is mineralized or immobilized during decomposition. Net immobilization of N, P, and S is likely to occur if residues added to soil have a C:N ratio greater than 40 (Vigil and Kissel, 1991), C:P ratio greater than 300 (Curtin et al., 2003), and C:S ratio greater than 400, respectively (Niknahad-Gharmakher et al., 2012). Güsewell and Gessner (2009) observed that critical N:P ratios, which can limit the decomposition of litter, vary according to nutrient richness, and no critical single N:P ratio can be determined. Their results suggest that the litter N:P ratio contributes to determining the relative importance of bacteria and fungi in the decomposition process, with low N:P ratios promoting bacteria and high N:P ratios promoting fungi.

In soils, the gross mineralization flux is primarily driven by the amount, the chemical recalcitrance, and the physical accessibility of the organic matter sources (particularly with humus) to enzymes degradation. The gross immobilization flux is directly linked to the microbial assimilation requirements during C decomposition. Therefore the two opposite fluxes, which form the mineralization–immobilization turnover, are influenced differently by agricultural practices, while both of them are driven by environmental conditions (temperature, moisture) that affect microorganism growth and activities. Using isotopic N techniques, it has been observed that gross N immobilization rates were very different in soils under annual crops and grassland systems (Booth et al., 2005; Attard et al., 2016), grassland systems having much higher N fluxes thanks to high litter and root inputs and fast turnover of fresh organic matter and microbial biomasses, while the net N mineralization rates were in the same range (Hart et al., 1994). But the "classical view" of the nutrient availability to plant uptake, being the balance between gross mineralization and gross immobilization, was questioned in early 2000. Schimel and Bennet (2004) and others, showed that plants can compete successfully for N even in an N-limited environment. Indeed, plants may be poor competitors with soil microorganisms at the root segment, local scale, but due to microsite heterogeneity, they can be successful at the whole root system/plant scale. There are microsites where net N mineralization dominates and others where net N immobilization dominates. Therefore, the existence of microsite heterogeneity allows plant roots and mycorrhizae access to N that would otherwise be taken up by microbes. Therefore, the continuous process of mineralization of organic C as CO_2 in soils by heterotrophic microbial activity leads inevitably to extra mineral N production, released in the soil. When an organic substrate enters a soil, the N demand of the decomposers temporarily increases with these needs being met by two sources: the direct or indirect (after mineralization) assimilation of N derived from the added organic substrate and the assimilation of the available soil mineral N. In situations of high concentration of labile C from substrates in the soil, e.g., due to exudation in the close vicinity of roots, or at the interface between soil and particles of decomposing crop residues, in the detritusphere or in a mulch, the nutrient availability might be scarce compared to the microbial demand of decomposers. There are several consequences of this limitation for the soil and the plant: the substrate decomposition may be slowed down, leading temporarily to accumulation of organic C in the soil, the microbial heterotrophic community may change to adapt to nutrient limitation, i.e., degrading C with less N, or the competition for N between the soil microbes and plant roots may impact the N nutrition and growth of the crop. Surprisingly, these situations are often encountered in cropped soils, even in highly fertilized growing situations where systems are considered saturated with N, because either carbon or nutrients are not available together either spatially or temporally. These principles are fundamentally similar for the P (and S) cycling with mineralization and immobilization fluxes associated with C dynamics and heterotrophic microbial growth and recycling, even much less studied (Ha et al., 2008; Niknahad-Gharmarkher et al., 2012).

In conclusion, while gross mineralization is a step that potentially decouples the C and nutrient cycles by releasing mineral forms whose fate may be distinct, the intensity of this decoupling will depend on several physiologic parameters of the microorganisms (carbon and nutrient use efficiencies, turnover), on the stoichiometric relationships of the substrates, and of the organisms along the trophic chain, and on the respective intensity of the different processes; microorganisms can even assimilate organic N, avoiding the mineral N release in the soil: mineralization–immobilization. Fluxes can be very high (such as under grassland) without any accumulation in the mineral N pool; plants may have access to mineral N by preempting directly N released by ammonification in the close vicinity of their roots.

Role of Plant Traits

There is an increasing recognition of direct and indirect effects of crops on the cycling of C and nutrients in soils through their impact on soil microbial communities (Jangid et al., 2008). The direct effects are linked to rhizosphere interactions with the physical and chemical characteristics of the surrounding soil, the release of exudates, and the depletion of nutrients and water and the presence of symbiotic microorganisms for N-fixing species where appropriate (see Chapter 1). Indirect effects are linked to crop development and the dynamics of nutrient accumulation during crop life, above- and belowground litter traits and their afterlife effects on C and nutrient dynamics in soil after senescence or harvest (De Deyn et al., 2008; Freschet et al., 2012; Soussana and Lemaire, 2014, Prieto et al., 2015, 2016). For example, crop type directly impacts the availability of soil mineral N: duration of the crop growth cycle, amount and dynamics of nutrient requirements by the vegetation cover, and amount and nature of the above- and belowground plant litters recycled. While the question has been raised over the last 20 years on the relationship between cultural practices, the structure of soil communities, and their impact on functions (Zak et al., 2003), research on links between soil communities and soil functions have clarified this. Comparing situations with low and high N richness of the plant–soil system, Fontaine et al. (2011) showed that low N availability fosters soil communities with enzymes adapted to degradation of humified organic matter such as fungi, which exploit soil organic matter rich in N (known as "mining" or destocking). In a situation of greater N availability and C source to decompose (plant litters), microorganisms assimilate available N during the decomposition, which subsequently contributes to organic matter storage through humification. Crops that have a high growth rate and fast N uptake guide the soil–plant system toward saprophyte functioning, involving higher rates of mineralization of nutrients (De Deyn et al., 2008). Klumpp et al. (2009) showed in grasslands that slow-growing plants adapted to low disturbance have coarse roots with high tissue density, which reduces Gram(+) bacteria abundance, litter decomposition, and N availability, while increasing soil C sequestration.

DECOUPLING FACTORS IN INTENSIVE ARABLE PRODUCTION

Shifting from a natural ecosystem to cultivation causes an immediate and rapid loss of soil organic carbon, reducing the soil organic carbon (SOC) pools in agricultural soils to 70% of their original level on average, until reaching a new equilibrium (McLauchlan, 2006). The main mechanisms involved are the alteration of C inputs to soils (by removal of biomass by harvest and the recycling of more labile-type vegetation) and the increase in C outputs (mineralization following physical disturbance of soil). The conversion of native vegetation to cropping generally also results in decreases in soil N, P, and S stocks (Kopittke et al., 2017). Murty et al. (2002) showed that the conversion of forest soils to agricultural use resulted in an average loss of 15% of total N, but less than loss of soil organic C, indicating that the C:N ratio narrows when agriculture is introduced (McLauchlan, 2006; Butterbach-Bahl et al., 2011). The linkages between carbon and N cycles, determined by stoichiometry of autotrophic and heterotrophic organisms, are altered because cropped systems typically receive excess N from fertilizer, and an overall decrease in OM inputs compared to native systems prevents the N inputs from resulting in long-term carbon storage (Burke et al., 1989). Several management practices linked to intensive arable production had led to a partial or strong decoupling of C–N–P cycling.

Natural ecosystems (including many grasslands) exhibit a large degree of synchrony and synlocation between release and uptake potentials, biotic regulations play a greater role in maximizing nutrient recycling, losses are generally small (Soussana and Lemaire, 2014), and such systems are considered "closed." In contrast, most arable cropping systems are generally "open," primarily because annual crops with a large N demand during vegetative growth are used (Jarvis et al., 2011) and such open systems are characterized by high external N inputs and outputs (harvest), but also periods with large C supply and periods of C starvation. Management also introduces physical disturbance of the soil structure through tillage, and it affects hydrology through drainage and irrigation. These systems are considered leaky and open farming systems (Jarvis et al., 2011; Pearson, 2007). The processes involved are developed thereafter.

Farm and Agricultural Area Specialization

Analysis of agricultural evolution during the 20th century showed both at global and at local (country) levels that the result of agriculture intensification was high specialization in many regions. This was particularly emphasized by analysis of trade fluxes of proteins at the global scale (Billen et al., 2014), showing international specialization of agriculture, where a small number of countries supply the feed requirements of many deficit countries. It has been shown that the international disconnection of food production and consumption leads to a loss of efficiency in the use of N (Billen et al., 2014). Le Noe et al. (2016) showed for French agriculture that N surplus, defined as the balance between total inputs and total outputs at the time scale of a crop cycle, which is a robust indicator of N losses, ranged from 16 to 176 kg N/ha per year across the 33 French agricultural regions. The highest N surpluses were found in regions with high livestock density, which received both high inputs of animal manures and synthetic fertilizers, while regions with high N synthetic inputs only and high crop production (mostly annual crops) had low N surplus, indicative of good N use efficiency. P budget ranged from −6.4 to 41 kg P/ha per year, showing also large disparities between regions. The negative balance resulted from very intensive crop production, with inputs of P fertilizers lower than the requirements of crop growth (this was made possible without severe crop growth limitation owing to the legacy of accumulated stocks of P from past excessive P fertilization). Senthilkumar et al. (2012) also showed the impact of production systems on P flows (through feed, fodder, and animal excretion), and particularly influenced by the livestock density. As for N, P surplus was due to partial substitution between animal manure and mineral fertilizer. Specialized agricultural systems (either cropping or livestock farming systems) are associated with the highest environmental losses (surplus) and resource consumption per unit of agricultural surface (Le Noë et al., 2017). Therefore the decoupling arises mostly from disconnection of crop and livestock farming, with nutrients extracted from croplands and exported far away for livestock feeding not returning to the croplands as manure and being replaced by synthetic fertilizers, while the crop residues cannot be used by livestock. For example, in Europe, most feed comes from oilcake—rich in N—imported from Argentina or Brazil (soybeans) and more recently from rapeseed (byproduct of oil used for biofuel) and produced in arable farming systems. With organic farming, Nowak et al. (2015) showed that (1) *the local supply* (defined as the ratio of the amount of nutrients from exchanges among farms plus the amount of N from atmospheric sources, to the sum of inputs to organic farms) and (2) *the cycling index* (defined as the fraction of nutrients flowing at least twice through the same farms) were greater in the mixed districts (association of crops and livestock) than in specialized districts. As concluded by Le Noë et al. (2017), the opening of the N or P cycle associated with agricultural activity is not only a matter of agricultural practices, it is also the consequence of a highly specialized agrifood system very focused on long-distance trade, without seeking functional complementarities between close regions.

Simplified Crop Rotations and Reduction of Crop Diversity

The first immediate effect of arable cropping is the change in land use with a switch from natural vegetation to managed crops and a gradual change associated with "agricultural intensification." The consequence of intensification is the specialization in the production process, which leads to the reduction of the number of different plant species on a given piece of land. Historically the initial objective of such specialization was the improved efficiency due to the reduction of plant competition. Therefore, monoculture has been perceived as economically more efficient and easiest to manage and harvest, with the assumption that a greater efficiency can be achieved by growing a single species with crop-specific methods of management. System biodiversity of such agricultural systems was split into two categories according to Vandermeer et al. (1998): the planned biodiversity driven deliberately by farmers and the associated (unplanned) diversity, which is the consequence of the former, such as diversity of aboveground and belowground recycled plant litter, soil communities, weeds, etc. This associated diversity represents an actual component of the biodiversity but is more difficult to identify and quantify. In agricultural systems, the nonharvested plant parts serve important ecologic roles, as discussed earlier. For example, their characteristics directly impact the amount and quality of organic inputs to the soil, the microbial diversity and activity, and the subsequent dynamics of nutrient mineral and organic forms (Jangid et al., 2008; Trivedi et al., 2016). Nonharvested components are also key elements of habitats and resources for soil biota and determine biophysical parameters such as soil cover and soil physical properties (aggregation, compaction, water infiltration, and transport), which in turn affect erosion, water balance, and nutrient retention. For example, Ferchaud et al. (2016) compared changes in soil C stocks after 5–6 years under perennial, semiperennial, and annual bioenergy crops in the same pedoclimatic conditions, and they observed a much higher amount of crop residues left in soil and at the surface of the soil, at time of SOC measurements, with perennial crops (miscanthus and switchgrass) compared to other crop species, and different vertical stratification in the soil. However, the only significant change in soil organic C stocks was obtained with semiperennial crops treatment (fescue and alfalfa), with a significant effect of the type of rotation, but not of N fertilization. This example shows the important impacts of crops diversity and agricultural practices on nonharvested components of the system, and the potential consequences on soil biology and nutrient cycling. At the spatial scales on which the plant community functions, differences in the phenology of plant species, ground cover by vegetation, root channels, litter layers, soil organic pool size, and distribution, all have contributory effects to carbon, nutrient, and water fluxes in soils (Swift and Anderson, 1994; Østergård et al., 2009). However, it always remains difficult to disentangle the specific effects of plant diversity from those of agricultural management applied to various crops that can mask the influence of reduction of plant diversity.

Mineral and Organic Fertilization

Consequences of Fertilization at the Annual and Crop Rotation Scales

Until the middle of the 20th century, the main N input to European farms was via fixation by legumes and subsequent recycling by animal manure spread on arable fields. Then the importance of manure-N in farms continuously decreased during a 30–40 year period, because of the availability of cheap mineral fertilizer-N and the increasing demand for N created by increasing potential yields (Billen et al., 2011). For decades, nutrient management emphasized improving the delivery of fertilizer to the plant roots through fertilization practices and the increase in fertilizer use efficiency as reflected by metrics that describe the fate of fertilizer-N and P in the current growing season (Machet et al., 1987; Cassman et al., 2002; Drinkwater and Snapp, 2007). The low average recovery efficiency of the added mineral nutrients by crops, about 30%–50% of the added fertilizer for N (Tilman et al., 2002; Jarvis et al., 2011) and even less for P (Vance, 2001), but strongly dependent on the timing of application, is a consequence of (1) temporal asynchrony and (2) spatial separation between applied nutrients and crop demand (Drinkwater and Snapp, 2007) and (3) mostly for P, fixation on the soil matrix.

The consequences of such increase in the use for mineral fertilizers was that agricultural systems were placed in a state of "nutrient saturation" and inherently leaky because surplus additions of N and P are created at certain periods of the growth cycle to cope with climatic hazards while meeting yield objectives. For example, a decision-support tool for fertilization based on the balance-sheet method and largely used in France aims at calculating a priori optimal conditions for crop nutrition while minimizing residual N at harvest (Meynard et al., 1997; Machet et al., 2017). Even in areas having high availability of animal manures, farmers still apply mineral fertilizers to overcome uncertainty in the use efficiency of organic-added N, which creates high N surplus (Ministère de la transition écologique et solidaire, 2015).

The reliance on mineral fertilizer (combined with chemical control of weeds) has led to a series of consequences on agrosystems management: (1) simplified rotations became possible because mineral fertilization made it unnecessary to grow cover crops and forages (including legumes) in a sequence that alternated with cash crops; (2) removal of winter annuals has increased the presence of bare fallows, which has led to a reduction of the timeframe of living plant cover and C-fixation and decrease in OM inputs and SOM stocks (Drinkwater and Snapp, 2007). Crop rotations affect C dynamics and coupling/decoupling in several ways. First, the rotation defines the sequence of crops with various N demands and with various amounts of N in residues returned to the soil. Second, the crop sequence defines the time of break between the different crops, time of tillage (or no-till) operations, and possibilities (and needs) for growing cover crops (Justes et al., 1999). Third, the crop sequence affects crop growth by modifying soil properties and preventing weeds and diseases. For example, in such simplified rotations of western Europe, where bare fallows can be maintained from a few weeks up to 9 months depending on the rotation (the duration of fallow is higher with spring crops), the risk of nitrate losses is high, particularly during winter. The shape of relationship between N fertilizer application rate and nitrate leaching is often described as a "broken stick": relatively small changes in N leaching for rate of N applications up to the economic optimum rate, followed by large N losses thereafter. This is due to the fact that residual mineral N at harvest time changes only slightly with fertilizer applications below the crop's optimum application, but it increases dramatically when a crop is overfertilized (Shepherd and Chambers, 2007; Justes et al., 1999). Similar trends were observed for N_2O emissions (van Groenigen et al., 2010; Shcherbak et al., 2014). The risk of nitrate loss during fallow periods is also due to active soil organic matter, manure, or cover crop residue mineralization during wet and warm periods (i.e., autumn in much of northern hemisphere) combined with the lack of significant crop N uptake. For example, agricultural statistics in France over the agricultural season 2010–11 (Agreste, 2014) indicated that the areas remaining totally bare during the winter accounted for 20% of the French area for field crops, while 63% of the soils were planted in winter crops and 17% were covered by cover crops to trap nitrate pending spring crop sowing. The percentage of bare soils in winter was particularly important for corn grain and sunflower crops, where it accounts for two-thirds of the national area.

During periods of bare soils or low plant demand, microbial assimilation is the only major route for biologically mediated retention of residual added fertilizer-N and mineralized soil N. The reduction of plant-driven sinks in space and time, in such a rotation-simplified system, combined with the great irregularity in C supply to soil, which does not allow for sufficient immobilization of N by soil microorganisms, creates agricultural systems that maximize nutrient losses. This is proposed as one of the major reasons why conventional agricultural systems are leakier in nutrients than more diverse ecosystems or alternative agrosystems where the permanent cover of soils by living plants is fostered (e.g., conservation agriculture). The concept of ecosystem-scale N saturation, expanded to agricultural systems by Drinkwater and Snapp (2007), expresses the fact that the availability of N (and P) exceeds the capacity of the ecosystem to cycle or store the nutrients in internal reservoirs that can be accessed by plants and microorganisms.

Directs Effects of Fertilization on Soil N Biologic Processes

The frequent inputs of N, either as mineral N fertilizer or animal manure, have direct impacts on soil functioning, particularly on the mineralization–immobilization turnover described earlier (which includes nitrification) and which determines the rates of transformation between the different reactive forms of N. One important issue is the nitrification process, which determines in soil the rate at which any ammonium ion can be oxidized into nitrate, which has many consequences in terms of fate of added N. Nitrification rates have been shown to be higher in soils with a history of high levels of N fertilizer application due to the frequent stimulation of ammonium-oxidizing bacteria by their substrate, than in soils receiving low levels of N fertilizer application (Watson and Mills, 1998). Booth et al. (2005) in their meta-analysis of agricultural factors affecting gross N processes showed that soil C losses (due to agricultural practices, changes in vegetation, or climatic factors) may increase the ability of nitrifiers to compete with heterotrophic microorganisms for ammonium, which, in consequence, leads to higher N losses due to leaching. Lu et al. (2011) showed also through a meta-analysis the effect of anthropogenic N addition on the terrestrial N cycle. Their study showed that N addition to soils increased the fluxes of N mineralization by 25%, nitrification by 154%, and denitrification by 84% compared to nonenriched N situations, NH_4^+ concentration being one of the most important factors in determining nitrification rate (Robertson, 1989), while immobilization remained unchanged. Thus a key role in potential N retention and loss from a terrestrial ecosystem has been attributed to the balance of inorganic N immobilization by microbes and autotrophic nitrification (Tietema and Wessel, 1992; Stockdale et al., 2002).

These changes, i.e., decreases in plant C inputs and SOC concentrations combined with frequent N saturation, have a direct impact on the structure and functioning of the soil community and typically result in a substantial reduction of soil microbial biomass. Kallenbach and Grandy (2011) did a meta-analysis comparing microbial biomass data, where systems with organic wastes (animal manures, composts, etc.) were compared to the same system with inorganic fertilizer, in a large range of agrosystems, soil types, and climate (in total 297 comparisons). On average, C in the microbial biomass increased by +36% and microbial N by 27% compared to systems without organic amendment, but the ratio of microbial C to microbial N was fairly constant in all circumstances (between 6 and 11), despite a wide variation in the manure and plant amendment C:N ratio across studies. The authors confirmed the findings that in agricultural soils the microbial biomass is C-limited in most circumstances (and not N limited), and the addition of organic amendments may alleviate such limitation. They also confirmed a strong homeostatic relationship between microbial C and N, which persist even under conditions of intensive resource additions and irrespective of the elemental stoichiometry of those resources (Cleveland and Liptzin, 2007).

Effects of Irrigation

Irrigated agricultural land accounts for 16% of the world's total agricultural land and 40% of agricultural production (FAOSTAT). Irrigation is developed in arid and semiarid regions. In these regions, irrigation is a major contributor to the diffuse contamination of waters by nitrate due to excessive application of water and lack of matching between irrigation supply and crop needs, inducing water loss and nutrient leaching below the rooting zone. Mechanisms are well known, and the excess water supply reduces N recovery by crops, which in turn leads to higher fertilizer doses, defined as a vicious circle (Quemada et al., 2013). Generally, on irrigated crops, the fertilizer recovery efficiency is weaker than on the same nonirrigated land. Therefore, specific practices are required in irrigated areas to increase the efficiency of water and nitrogen supplied. This is a major challenge for ensuring food production (Tilman et al., 2002). The meta-analysis performed by Quemada et al. (2013) showed that for irrigated land, improved water management was the most effective strategy to reduce nitrate leaching, and it did not equate to a decrease in yield and nitrogen efficiency, while a deficit in water (by decreasing irrigation) decreased both yield and N recovery. With a deficit of irrigation, residual N in soil may increase, enhancing the risk of nitrate leaching during the nongrowing season. Therefore, irrigation contributes to leaky nutrient cycles when the availability of water and nutrients do not match simultaneously the crop requirements.

Effects of Soil Tillage

Tillage is one of the practices that distinguish highly intensified agricultural systems from less intensive and natural systems. There are many combinations of soil tillage techniques (for seedbed preparation, weed control, residue management, erosion management), and usually, they are selected according to the local pedo-climatic, technical, and socioeconomic constraints (Guérif et al., 2001). Regarding the coupling-decoupling of biogeochemical cycles, does the tillage that modifies C storage in the soils and its stratification impact in turn the fate of N and P in cropping systems? Do the changes in the soil communities and their activities modify significantly the nutrients fluxes and fate in agrosystems? Indeed, most of our knowledge to answer these questions comes from reverse situations, i.e., situations with reduction or abandonment of soil plowing, or through the comparison of agricultural (tilled) and grassland (no-till) systems. Agriculture with reduced or no-till systems developed greatly in the past 2 decades at places where the common model was full inversion tillage, first to reduce soil erosion and loss of soil fertility, then to increase soil biologic quality.

Tillage has a direct effect on soil structure by fragmentation, the initial placement of crop residues in the soil, and the amount and spatial distribution of soil organic matter and nutrients, seeds of weeds, and certain pathogens. Tillage also has many indirect effects via environmental characteristics (temperature, humidity, and aeration), plant rooting, physicochemical properties, and water availability. These factors influence in turn the composition and activity of living communities (Holland and Coleman, 1987). Among the different tillage techniques, deep plowing is the technique whose impact is probably the most important, due to the volume of soil concerned and the inversion of the horizons that it entails (Guérif et al., 2001). Following the cessation of plowing, the formation of a vertical gradient of organic matter content is observed (C and nutrient, especially for nonmobile nutrients like P), with a greater concentration of these nutrients in the surface horizon (usually the few top centimeters), whereas the total nutrient content of the soil is not always modified depending of the type of soil and the cropping systems (Constantin et al., 2010; Dimassi et al., 2013). The distribution of organic matter in the soil profile has an impact on a range of properties and processes in soil, particularly on the mineralization of humus. Reduced tillage increased formation of macroaggregates, which in turn influences the concentration and turnover of new soil organic matter, by limiting the accessibility of C substrates to microbes and fauna (Angers et al., 1997; Trivedi et al., 2015). Aggregation is providing physical protection to organic matter (Balesdent et al., 2000), therefore decreasing the rate of N mineralization. However, lower mineralization of organic matter could be due also to colder and drier conditions in the soil surface layer of no-tilled situations (Balesdent et al., 2000). While the effects of soil tillage on composition and abundance of decomposers has been extensively documented (e.g., Mbuthia et al., 2015), the consequences on nutrient dynamics is not very clear, sometimes nil or negligible, probably because a change in tillage practice in cropping system usually does not occur alone, and it is accompanied by several other major changes, including the characteristics of rotation and the use of pesticides (Mary et al., 2015). In several long-term experiments dedicated to the comparison of direct sowing and conventional deep plowing, Oorts et al. (2007) observed no significant difference on the net N mineralization of soil, with two sites differentiated for 33 and 12 years, respectively. To distinguish the direct effect of soil tillage on accumulation of soil organic matter and subsequent N mineralization, from the indirect effect of tillage on soil temperature and moisture, the authors expressed N mineralization as the product of (1) the potential rate of net N mineralization (Vp), at a given reference temperature and moisture, which is considered to be specific to microbial populations, to the accessibility of organic matter and the stock of organic matter; and (2) the time expressed as the number of days at reference temperature and moisture (JN), which relates to the effect of temperature and moisture regimes on microbial activities. In the situations explored, there was no significant difference in rate of mineralization between no-till and plowed plots. However, there were differences in the number of normalized days, indicating an effect of tillage on physical conditions of the mineralizing soil layer. The number of normalized days was higher in tilled plots in early spring, indicating more favorable conditions of mineralization, while the reverse situation was observed in early summer. This means that there is "seasonality" for the influence of tillage on mineralization conditions, but that, overall, the differences observed are small. Giacomini et al. (2010) looking at the fate of fertilizer-N in rotations of pea-wheat and maize-wheat sequences, observed little or no influence of tillage practice (reduced tillage vs. deep plowing) on the recovery of ^{15}N-labelled fertilizer-N in the wheat crops nor in the fraction of fertilizer-N immobilized into the soil.

Another direct effect of introducing tillage, is the management of crop residues and organic products that lead to the incorporation of these organic sources into soil by tillage (either partially or fully). The initial location of crop residues, e.g., the presence or absence of mulch at the soil surface, the clustering of crop residues, and the spreading of fragments in the soil modify many soil physical parameters, which in turn have a range of effects on decomposition and microbial properties. In no-till or reduced-till systems, plant litters/crop residues are left at the soil surface. In such situations, soil residue contact is limited, which makes spatially distant nutrients (mainly in the soil) and added carbon (at the soil surface): crop residues that have low nutrient concentration (high C:N, C:P) tend to decompose more slowly than when incorporated into the soil, which is the case, for example, for straw of cereals harvested at maturity. The presence of a decomposing organic layer at the soil surface also translates into a higher net N mineralization of the underneath soil layer due to the absence of the residue C incorporation and therefore lower microbial N immobilization due to decomposition (Coppens et al., 2006). More important is the modification of soil water and thermal regimes in the presence of residue mulch at the soil surface, which improves water penetration into the soil and decreases water evaporation (Quemada et al., 2013). However, these effects are different according to local climatic conditions (Guérif et al., 2001).

Thus the effect of tillage as a practice considered individually is mainly connected to the physical functioning of the soil that it includes: soil structure, moistening–desiccation regime, water retention and transport, and spatial distribution of plant derived-C substrate. In climatic situations favorable to decomposition, incorporation of crop residues and disturbance of the root system by plowing accelerates decomposition and the subsequent net mineralization of nutrients. It also accelerates the mineralization of native organic matter by the annual mechanical disturbance of soil. Attard et al. (2016) studied the effects of reciprocal shifts from rotation of annual crops to temporary grassland and vice-versa, which is a common feature of ley-arable cropping systems. They observed that plowing of the grasslands plots immediately caused an increase in the CO_2–C emitted, and in mineralization of N, followed by a general decrease in C and N fluxes due to destruction of the grass, mechanical disturbance of the no-till soil, and in the longer term, to the cessation of high organic matter inputs from grassland. Conversely the cessation of tillage following the establishment of grass did not modify significantly soil C and N pools and mineralization fluxes during the first 3 years after the shifts, time needed for the grass to establish its perennial root system, and accumulate extra C into the soil. This work and others clearly showed the asymmetric responses of cropping systems to reciprocal shifts in land use and the strong disturbance that represents soil tillage. The reduction or suppression of tillage has been advocated to be responsible for increased N_2O emissions (due to the combination of high concentrations in C, N, and moisture, in the soil top layer) and in some circumstances in higher risk of nutrient leaching due to high water retention in the soil profile. However, the reduction or suppression of tillage proved to foster many other important physical and biologic characteristics, which, combined with higher and more diverse plant litter inputs, may improve soil fertility and, in the longer term, decrease the needs for fertilizers. The reduction of the biologic activity in soil by tillage has been acknowledged for decades (e.g., Dick, 1992), due to reduction of macro aggregation, but there is no clear evidence of significant impact on mineralization and nutrient dynamics. Therefore the effects of soil tillage or absence of tillage are complex because of the many physical and biologic interactions in soil. Tillage leads to incorporation of plant residues, which foster microbial activity; the presence/absence of a residue layer at the soil surface strongly modifies temperature and moisture regimes, which may be favorable or unfavorable to microbial activities. Lastly, some nitrate losses might be enhanced or reduced via nitrate leaching or N_2 and N_2O emitted.

CONCLUSIONS

The intensification of agriculture in cropped systems has led to an opening of the nutrients cycles and an increase in nutrient saturation of systems, which both contributed to a partial or significant decoupling with the carbon cycle. This was first promoted by the introduction of intensive N and P mineral fertilization, which led to the decoupling of nutrient supply and demand: (1) in time, which deeply altered the succession of crops, reduced the occupancy of soils by growing plants, and made marginal the contribution of N fixation and organic matter mineralization to nutrients supply; and (2) in space, where fertilization made possible specialized farming systems on a regional scale, and agriculture on a world scale, increasing global nutrient flows and associated losses. Mineral fertilization, reduced crop diversity, and soil disturbance by tillage have also changed the soil physical properties, the soil communities, and the functions they perform. Thus, the potential for coupling of biogeochemical cycles through the degradation of organic matter (mineralization and immobilization) have been altered in intensive cropped systems, but they also became saturated by excess nutrients.

The necessary evolution of cropping systems toward greater energy and nutrient autonomy, to reduce chemical inputs, to save resources, and to reduce environmental impacts, must therefore find solutions inspired from less anthropized ecosystems. Hufnagl-Eichiner et al. (2011) distinguished the measures improving the efficiency of mineral and organic inputs, from measures involving the redesign of cropping systems. Improved fertilization practices, which aim at increasing the recovery and efficiency of the applied N, contribute to the efficiency of the systems (eco-efficient options) but do not mobilize coupling of C and nutrient cycles. The optimization of agricultural practices is certainly essential to improve the closure of biogeochemical cycles and limit the nuisances toward the environment. Nevertheless, the coupling of C and nutrient cycles must lead to an intensification of work toward the redesign of efficient cropping systems based on knowledge and valorization of biologic interactions in soils and the diversification of plant species (Duru et al., 2015). The increase in legume crops in cropping systems (Siddique et al., 2012) allows the removal of synthetic mineral fertilizers in certain phases of rotation. Practices of soil cover during intercropping periods are also means of enhancing permanent interactions between crops and soils communities, and synchrony of N uptake and supply, thereby reducing N leaching losses reducing nutrients leakage. The introduction of perennials and semiperennials in agrosystems, and development of agroforestry systems mixing spatially trees and annual crops, also appear as promising agrosystems that enhance in time and space the biotic and abiotic interactions. Permanent cover of soils by living and dead plant biomasses combined with reduced soil tillage constitute the foundation of conservation agriculture (Scopel et al., 2013).

A rediversification of crops at the scale of fields and cropping systems at the scales of landscapes also appears as a major lever to increase the sustainability of agricultural production systems, by reducing inputs (water, pesticides, and nitrogen fertilizer), increasing the heterogeneity of habitat mosaics, or reducing yield losses due to too frequent returns of the same species.

Increasingly, research suggests that the level of internal regulation of function in agrosystems is largely dependent on the level of plant and animal biodiversity present (Kremen and Miles, 2012; Dwidedi et al., 2017).

It is important to stress that the future cannot be limited to redo what has been undone before. The new insights reported here on processes, particularly knowledge on the processes of coupling and the role of soil microorganisms, the importance of the interactions between soil physical, chemical, and biologic functioning allow us to go beyond reproducing the past. Also, new techniques such as precision farming technologies and models are able to help both limiting leakages and designing the "right" assemblages of plants over time and the necessary exchanges of nutrients and biomasses flows at different spatial scales. The options considered in different types of agriculture and under different soil and climate conditions are discussed in other chapters of this book.

References

Agreste, 2014. Enquête Pratiques culturales 2011. http://agreste.agriculture.gouv.fr/-IMG/pdf/dossier21_integral.pdf.

Angers, D., Recous, S., Aita, C., 1997. Fate of carbon and nitrogen in water-stable aggregates during decomposition of wheat straw. European Journal of Soil Science 48, 295–300.

Asner, G.P., Seastedt, T.R., Townsend, A.R., 1997. The decoupling of terrestrial carbon and nitrogen cycles. BioScience 47 (4), 226–234.

Attard, E., Le Roux, X., Charrier, X., Delfosse, O., Guillaumaud, N., Lemaire, G., Recous, S., 2016. Delayed and asymmetric responses of soil C pools and N fluxes to grassland/cropland conversions. Soil Biology and Biochemistry 97, 31–39.

Balesdent, J., Chenu, C., Balabane, M., 2000. Relationship of soil organic matter dynamics to physical protection and tillage. Soil and Tillage Research 53, 215–230.

Bertrand, I., Chabbert, B., Kurek, B., Recous, S., 2006. Can the biological features and histology of wheat residues explain their decomposition in soil? Plant and Soil 281, 291–307.

Billen, G., Lassaletta, L., Garnier, J., 2014. A biogeochemical view of the global agro-food system: nitrogen flows associated with protein production, consumption and trade. Global Food Security 3, 209–219.

Billen, G., Silvestre, M., Grizzetti, B., Leip, A., Garnier, J., Voss, M., Howarth, R., Bouraoui, F., Lepisto, A., Kortelainen, P., Johnes, P., Curtis, C., Humborg, C., Smedburg, E., Kaste, O., Ganeshram, R., Beusen, A., Lancelot, C., 2011. Nitrogen flows from European watersheds to coastal marine waters. In: Sutton, M.A. (Ed.), The European Nitrogen Assessment. Cambridge University Press, Cambridge, ISBN 9781107006126, pp. 271–297.

Booth, M.S., Stark, J.M., Rastetter, E., 2005. Controls on nitrogen cycling in terrestrial ecosystems: a synthetic analysis of literature data. Ecological Monographs 75, 139–157.

Burke, I.C., Yonker, C.M., Parton, W.J., Cole, C.V., Flach, K., Schimel, D.S., 1989. Texture, climate, and cultivation effects on soil organic matter content in U.S. Grassland Soils. Soil Science Society of America Journal 53, 800–805.

Butterbach-Bahl, K., Gundersen, P., Ambus, P., Augustin, J., Beier, C., Boeckx, P., Dannenmann, M., Sanchez Gimeno, B., Ibrom, A., Kiese, R., et al., 2011. Nitrogen processes in terrestrial ecosystems. In: Sutton, M.A., Howard, C.M., Erisman, J.W., Billen, G., Bleeker, A., Grennfelt, P., van Grisven, H., Grizzetti, B. (Eds.), The European Nitrogen Assessment: Sources, Effects and Policy Perspectives. Cambridge University Press, Cambridge, UK, ISBN 9781107006126, pp. 99–125.

Cakmak, I., 2002. Plant nutrition research: priorities to meet human needs for food in sustainable ways. Plant and Soil 247, 3–24.

Cassman, K.G., Dobermann, A., Walters, D.T., 2002. Agroecosystems, nitrogen-use efficiency, and nitrogen management. AMBIO: A Journal of the Human Environment 31, 132–140.

Chadwick, O.A., Derry, L.A., Vitousek, P.M., Huebert, B.J., Hedin, L.O., 1999. Changing sources of nutrients during four million years of ecosystem development. Nature 397, 491–497.

CITEPA, 2016 (last consultation 29/07/2017). https://www.citepa.org/fr/air-et-climat/polluants/effet-de-serre/protoxyde-d-azote-n2o.

Cleveland, C.C., Liptzin, D., 2007. C:N:P stoichiometry in soil: is there a Redfield ratio for the microbial biomass ? Biogeochemistry 85, 235–252.

Conley, D.J., Paerl, H.W., Howarth, R.W., Boesch, D.F., Seitzinger, S.P., Havens, K.E., Lancelot, C., Likens, G.E., 2009. Controlling Eutrophication: Nitrogen and Phosphorus. Science 323 (5917), 1014–1015. https://doi.org/10.1126/science.1167755.

Constantin, J., Mary, B., Laurent, F., Aubrion, G., Fontaine, A., Kerveillant, P., Beaudoin, N., 2010. Effects of catch crops, no till and reduced nitrogen fertilization on nitrogen leaching and balance in three long-term experiments. Agriculture, Ecosystems & Environment 135, 268–278.

Coppens, F., Garnier, P., De Gryze, S., Merckx, R., Recous, S., 2006. Soil moisture, carbon and nitrogen dynamics following incorporation and surface application of labelled crop residues in soil columns. European Journal of Soil Science 57, 894–905.

Curtin, D., McCallum, F.M., Williams, P.H., 2003. Phosphorus in light fraction organic matter separated from soils receiving long-term applications of superphosphate. Biology and Fertility of Soils 37, 280–287.

De Deyn, G.B., Cornelissen, J.H.C., Bardgett, R.D., 2008. Plant functional traits and soil carbon sequestration in contrasting biomes. Ecology Letters 11, 516–531.

Dick, R.P., 1992. A review: long-term effects of agricultural systems on soil biochemical and microbiological parameters. Agriculture, Ecosystems & Environment 40, 25–36.

Dwivedi, S.L., Lammerts van Bueren, E.T., Ceccarelli, S., Grando, S., Upadhyaya, H.D., Ortiz, R., 2017. Diversifying food systems in the Pursuit of sustainable food production and healthy diets. Trends in Plant Science 22, 842–856.

Dimassi, B., Cohan, J.P., Labreuche, J., Mary, B., 2013. Changes in soil carbon and nitrogen following tillage conversion in a long-term experiment in Northern France. Agriculture, Ecosystems & Environment 169, 12–20.

Drinkwater, L., Snapp, S., 2007. Nutrients in agroecosystems: rethinking the management paradigm. Advances in Agronomy 92, 163.

Duru, M., Therond, O., Fares, M., 2015. Designing agroecological transitions: a review. Agronomy for Sustainable Development 35 (4), 1237–1257.

Erisman, J.W., Sutton, M.A., Galloway, J., Klimont, Z., Winiwarter, W., 2008. How a century of ammonia synthesis changed the world. Nature Geoscience 1, 636–639.

FAO, 2015. World Fertilizer Trends and Outlook to 2018, ISBN 978-92-5-108692-6 (last read 29/09/2017). http://www.fao.org/3/a-i4324e.pdf.

Fanin, N., Bertrand, I., 2016. Aboveground litter quality is a better predictor than belowground microbial communities when estimating carbon mineralization along a land-use gradient. Soil Biology and Biochemistry 94, 48–60.

Ferchaud, F., Vitte, G., Mary, B., 2016. Changes in soil carbon stocks under perennial and annual bioenergy crops. Global Change Biology Bioenergy 8, 290–306.

Fontaine, S., Henault, C., Aamor, A., Bdioui, N., Bloor, J.M.G., Maire, V., Mary, B., Revaillot, S., Maron, P.A., 2011. Fungi mediate long term sequestration of carbon and nitrogen in soil through their priming effect. Soil Biology and Biochemistry 43 (1), 86–96.

Freschet, G.T., Aerts, R., Cornelissen, J.H.C., 2012. A plant economics spectrum of litter decomposability. Functional Ecology 26, 56–65.

Galloway, J.N., Aber, J.D., Erisman, J.W., Seitzinger, S.P., Howarth, R.W., Cowling, E.B., Cosby, B.J., 2003. The nitrogen cascade. BioScience 53, 341–356.

Giacomini, S.J., Machet, J.M., Boizard, H., Recous, S., 2010. Dynamics and recovery of fertilizer ^{15}N in soil and winter wheat crop under minimum versus conventional tillage. Soil and Tillage Research 108, 51–58.

Guérif, J., Richard, G., Dürr, C., Machet, J.M., Recous, S., Roger-Estrade, J., 2001. A review of tillage effects on crop residues management, seedbed conditions and seedling establishment. Soil and Tillage 61, 13–32.

Güsewell, S., Gessner, M.O., 2009. N: P ratios influence litter decomposition and colonization by fungi and bacteria in microcosms. Functional Ecology 23, 211–219.

Ha, K.V., Marschner, P., Bünemann, E.K., 2008. Dynamics of C, N, P and microbial community composition in particulate soil organic matter during residues decomposition. Plant and Soil 303, 253–264.

Hart, S.C., Nason, G.E., Myrold, D.D., Perry, D.A., 1994. Dynamics of gross nitrogen transformations in an old-growth forest: the carbon connection. Ecology 75, 880–891.

Hertel, O., Skjøth, C.A., Reis, S., Bleeker, A., Harrison, R.M., Cape, J.N., Fowler, D., Skiba, U., Simpson, D., Jickells, T., Kulmala, M., Gyldenkærne, S., Sørensen, L.L., Erisman, J.W., Sutton, M., 2012. Governing processes for reactive nitrogen compounds in the European atmosphere. Biogeosciences 9, 4921–4954.

Holland, G.A., Coleman, D.C., 1987. Litter placement effects on microbial and organic matter dynamics in agro ecosystems. Ecology 62, 425–433.

Hufnagl-Eichiner, S., Wolf, S.A., Drinkwater, L.E., 2011. Assessing social–ecological coupling: agriculture and hypoxia in the Gulf of Mexico. Global Environmental Change 21 (2), 530–539.

Jangid, K., Williams, M., Franzluebbers, A., Sanderlin, J., Reeves, J.H., Jenkins, M.B., Endale, D.M., Coleman, D.C., Whitman, W.B., 2008. Relative impacts of land-use, management intensity and fertilization upon soil microbial community structure in agricultural systems. Soil Biology and Biochemistry 40, 2843–2853.

Jarvis, S., Hutchings, N., Brentrup, F., Olesen, J.E., van der Hoek, K.W., 2011. Nitrogen flows in farming systems across Europe. In: Sutton, M.A., Howard, C.M.,

Erisman, J.W., Billen, G., Bleeker, A., Grennfelt, P., van Grinsven, H., Grizzetti, B. (Eds.), The European Nitrogen Assessment: Sources, Effects and Policy Perspectives. Cambridge University Press, ISBN 978-1-107-00612-6, pp. 211–228.

Jensen, L.S., Salo, T., Palmason, F., Breland, T.A., Henriksen, T.M., Stenberg, B., 2005. Influence of biochemical quality on C and N mineralisation from a broad variety of plant materials in soil. Plant and Soil 273, 307–326.

Justes, E., Mary, B., Nicolardot, B., 1999. Comparing the effectiveness of radish cover crop, oilseed rape volunteers and oilseed rape residues incorporation for reducing nitrate leaching. Nutrient Cycling in Agroecosystems 55, 207–220.

Justes, E., Mary, B., Nicolardot, B., 2009. Quantifying and modelling C and N mineralization kinetics of catch crop residues in soil: parameterization of the residue decomposition module of STICS model for mature and non-mature residues. Plant and Soil 325, 171–185.

Kallenbach, C., Grandy, A.S., 2011. Controls over soil microbial biomass responses to carbon amendments in agricultural systems: a meta-analysis. Agriculture, Ecosystems & Environment 144, 241–252.

Klumpp, K., Fontaine, S., Attard, E., Le Roux, X., Gleixner, G., Soussana, J.-F., 2009. Grazing triggers soil carbon loss by altering plant roots and their control on soil microbial community. Journal of Ecology 97, 876–885.

Kopittke, P.M., Dalal, R.C., Finn, D., Menzies, N.W., 2017. Global changes in soil stocks of carbon, nitrogen, phosphorus, and sulphur as influenced by long-term agricultural production. Global Change Biology 23, 2509–2519.

Kremen, C., Miles, A., 2012. Ecosystem Services in biologically diversified versus conventional farming systems: benefits, externalities, and trade-offs. Ecological and Society 17 (4).

Lashermes, G., Nicolardot, B., Parnaudeau, V., Thuries, L., Chaussod, R., Guillotin, M.L., Lineres, M., Mary, B., Metzger, L., Morvan, T., Tricaud, A., Villette, C., Houot, S., 2010. Typology of exogenous organic matters based on chemical and biochemical composition to predict potential nitrogen mineralization. Bioresource Technology 101, 157–164.

Le Noë, J., Billen, G., Garnier, P., 2017. How the structure of agro-food systems shapes nitrogen, phosphorus and carbon fluxes: the generalized representation of agro-food system applied at the regional scale in France. The Science of the Total Environment 586, 42–55.

Le Noë, J., Billen, G., Lassaletta, L., Silvestre, M., Garnier, P., 2016. La place du transport de denrées agricoles dans le cycle biogéochimique de l'azote en France: un aspect de la spécialisation des territoires. Cahiers Agricultures 25, 15004.

Lu, M., Yang, Y., Luo, Y., Fang, C., Zhou, X., Chen, J., Yang, X., Li, B., 2011. Responses of ecosystem nitrogen cycle to nitrogen addition: a meta-analysis. New Phytologist 189, 1040–1050.

Machet, J.-M., Pierre, D., Recous, S., Rémy, J.C., 1987. Signification du coefficient réel d'utilisation et conséquences pour la fertilisation azotée des cultures. Comptes-rendus de l'Académie d'Agriculture de France 73 (3), 39–55.

Machet, J.-M., Dubrulle, P., Damay, N., Duval, R., Julien, J.-L., Recous, S., 2017. A dynamic decision-making tool for calculating the optimal rates of N application of 40 annual crops while minimising the residual level of mineral N at harvest. Agronomy 7 (4), 73.

Manzoni, S., Taylor, P., Richter, A., Porporato, A., Ågren, G.I., 2012. Environmental and stoichiometric controls on microbial carbon-use efficiency in soils. New Phytologist 196, 79–91.

Mary, B., Cohan, J.-P., Dimassi, B., Recous, S., Laurent, F., 2015. Effets du travail du sol sur les cycles biogéochimiques de l'azote et du carbone : de la compréhension des mécanismes aux conséquences pour la gestion des pratiques agricoles. In: Labreuche, J., Laurent, F., Roger-Estrade, J. (Eds.), Faut-il travailler le sol ? Acquis et Innovations pour une agriculture durable. Savoir Faire (Quae), Versailles, FRA. Editions Quae, Arvalis-Institut du Végétal. 192 p.https://prodinra.inra.fr/record/397638.

Mbuthia, L.W., Acosta-Martínez, V., DeBruyn, J., Schaeffer, S., Tyler, D., Odoi, E., Mpheshea, M., Walker, F., Eash, N., 2015. Long term tillage, cover crop, and fertilization effects on microbial community structure, activity: Implications for soil quality. Soil Biology and Biochemistry 89, 24–34. https://doi.org/10.1016/j.soilbio.2015.06.016.

McLauchlan, K., 2006. The Nature and longevity of agricultural impacts on soil carbon and nutrients: a review. Ecosystems 9, 1364–1382.

Meynard, J.M., Justes, E., Machet, J.-M., Recous, S., 1997. Fertilisation azotée des cultures annuelles de plein champ. Les Colloques 83. In: Lemaire et, G., Nicolardot, B. (Eds.), Maîtrise de l'azote dans les agrosystèmes. INRA, Paris, pp. 183–200.

Ministère de la transition écologique et solidaire, France, 2015 (last read online 03/08/2017). http://www.statistiques.developpement-durable.gouv.fr/lessentiel/ar/2396/0/surplus-phosphore-france.html.

Mooshammer, M., Wanek, W., Zechmeister-Boltenstern, S., Richter, A., 2014. Stoichiometric imbalances between terrestrial decomposer communities and their resources: mechanisms and implications of microbial adaptations to their resources. Frontiers in Microbiology 5, 22.

Murty, D., Kirschbaum, M.U.F., McMurtrie, R.E., McGilvray, A., 2002. Does conversion of forest to agriculture land change soil carbon and nitrogen? A review of the literature. Global Change Biology 8, 105–123.

Niknahad-Gharmakher, H., Piutti, S., Machet, J.-M., Benizri, E., Recous, S., 2012. Mineralization-immobilization of sulphur in a soil during decomposition of plant residues of varied chemical composition and S content. Plant and Soil 360, 391–404.

Nowak, B., Nesme, T., David, C., Pellerin, C., 2015. Nutrient recycling in organic farming is related to diversity in farm types at the local level. Agriculture. Ecosystems & Environment 204, 17–26.

Oorts, K., Laurent, F., Mary, B., Thiébeau, P., Labreuche, J., Nicolardot, B., 2007. Experimental and simulated soil mineral N dynamics for long-term tillage systems in northern France. Soil and Tillage Research 94, 441–456.

Østergård, H., Finckh, M.R., Fontaine, L., Goldringer, I., Hoad, S.P., Kristensen, K., Lammerts van Bueren, E.T., Mascher, F., Munk, L., Wolfe, M.S., 2009. Time for a shift in crop production: embracing complexity through diversity at all levels. Journal of the Science of Food and Agriculture 89, 1439–1445.

Pearson, C.J., 2007. Regenerative, semiclosed systems: a priority for twenty-first-century agriculture. BioScience 57, 409417.

Peñuelas, J., Sardans, J., Rivas-Ubach, A., Janssens, I.A., 2012. The human-induced imbalance between C, N and P in Earth's life system. Global Change Biology 18, 3–6.

Prieto, I., Roumet, C., Cardinael, R., Dupraz, C., Jourdan, C., Kim, J.H., Maeght, J.L., Mao, Z., Pierret, A., Portillo, N., Roupsard, O., Thammahacksa, C., Stokes, A., Cahill, J., 2015. Root functional parameters along a land-use gradient: evidence of a community-level economics spectrum. J Ecol 103, 361–373. https://doi.org/10.1111/1365-2745.12351.

Prieto, I., Stokes, A., Roumet, C., Mommer, L., 2016. Root functional parameters predict fine root decomposability at the community level. J Ecol 104, 725–733. https://doi.org/10.1111/1365-2745.12537.

Quemada, M., Baranski, M., Nobel-de Lange, M.N.J., Vallejoa, A., Cooper, J.M., 2013. Meta-analysis of strategies to control nitrate leaching in irrigated agricultural systems and their effects on crop yield. Agriculture, Ecosystems & Environment 174, 1–10.

Ringeval, B., Nowak, B., Nesme, T., Delmas, M., Pellerin, S., 2014. Contribution of anthropogenic phosphorus to agricultural soil fertility and food production, Global Biogeochemical. Cycles 28, 743–756.

Robertson, G.P., 1989. Nitrification and denitrification in humid tropical ecosystems: potential controls on nitrogen retention. In: Proctor, J. (Ed.), Mineral nutrients in Tropical Forest and Savanna Ecosystems. Blackwell Scientific, Cambridge, Massachusetts, USA, pp. 55–69.

Sardans, J., Rivas-Ubach, A., Peñuelas, J., 2012. The C: N:P stoichiometry of organisms and ecosystems in a changing world: a review and perspectives. Perspectives in Plant Ecology, Evolution and Systematics 14 (1), 33–47.

Schimel, J.P., Bennett, J., 2004. Nitrogen mineralization challenges of a changing paradigm. Ecology 85, 591–602.

Scopel, E., Triomphe, B., Affholder, F., Da Silva, F.A.M., Goulet, F., Corbeels, E., Sabourin, E., Xavier, J.H.V., Lhamar, R., Recous, S., Bernoux, M., Blanchart, E., de Carvalho Mendes, I., De Tourdonnet, S., 2013. Conservation agriculture cropping systems in temperate and tropical conditions, performances and impacts. A review. Agronomy for Sustainable Development 33, 113–130.

Senthilkumar, K., Nesme, T., Mollier, A., Pellerin, S., 2012. Regional-scale phosphorus flows and budgets within France: the importance of agricultural production systems. Nutrient Cycling in Agroecosystems 92, 145–159.

Shcherbak, I., Millar, N., Robertson, G.P., 2014. Global meta-analysis of the nonlinear response of soil nitrous oxide (N_2O) emissions to fertilizer nitrogen. Proceedings of the National Academy of Sciences 111 (25), 9199–9204.

Shepherd, M., Chambers, B., 2007. Managing nitrogen on the farm: the devil is in the detail. Journal of the Science of Food and Agriculture 87, 558–568.

Siddique, K.H.M., Johansen, C., Turner, N.C., Jeuffroy, M.H., Hashem, A., Sakar, D., Alghamdi, S.S., 2012. Innovations in agronomy for food legumes. A review. Agronomy for Sustainable Development 32, 45–64.

Soussana, J.F., Lemaire, G., 2014. Coupling carbon and nitrogen cycles for environmentally sustainable intensification of grasslands and crop-livestock systems. Agriculture, Ecosystems & Environment 190, 9–17.

Stockdale, E.A., Hatch, D.J., Murphy, D.V., Ledgard, S.F., Watson, C.J., 2002. Verifying the nitrification to immobilisation ratio (N/I) as a key determinant of potential nitrate loss in grassland and arable soils. Agronomie 22, 831–838.

Sutton, M.A., Oenema, O., Erisman, J.W., Leip, A., van Grinsven, H., Winiwarter, W., 2011. Too much of a good thing. Nature 472, 159–161.

Swift, M.J., Anderson, J.M., 1994. Biodiversity and ecosystem function in agricultural soils. In: Schulze, E.D., Mooney, H.A. (Eds.), Biodiversity and Ecosystem Function, Ecological Studies, vol. 99. Springer-Verlag, pp. 15–41.

Tietema, A., Wessel, W.W., 1992. Gross nitrogen transformations in the organic layer of acid forest ecosystems subjected to increased atmospheric nitrogen input. Soil Biology and Biochemistry 24, 943–950.

Tilman, D., Cassman, K.G., Matson, P.A., Naylor, R., Polasky, S., 2002. Agricultural sustainability and intensive production practices. Nature 418, 671–677.

Trinsoutrot, I., Recous, S., Bentz, B., Linères, M., Cheneby, D., Nicolardot, B., 2000. Relationships between biochemical quality of crop residues and C and N mineralisation kinetics under non-limiting N conditions. Soil Science Society of America Journal 64, 918–926.

Trivedi, P., Delgado-Baquerizo, M., Anderson, I.C., Singh, B.K., 2016. Response of soil properties and microbial communities to agriculture: implication for primary productivity and soil health indicators. Frontiers of Plant Science 7, 990.

Trivedi, P., Rochester, I.J., Trivedi, C., Van Nostrand, J.D., Zhou, J., Karunaratne, S., Anderson, I.C., Sing, B.K., 2015. Soil aggregate size mediates the impacts of cropping regimes on soil carbon and microbial communities. Soil Biology and Biochemistry 91, 169–181.

Van Groenigen, J.W., Velthof, G.L., Oenema, O., Van Groenigen, K.J., Van Kessel, C., 2010. Towards an agronomic assessment of N_2O emissions: a case study for arable crops. European Journal of Soil Science 61, 903–913.

Vance, C.P., 2001. Symbiotic nitrogen fixation and phosphorus acquisition. Plant nutrition in a world of declining renewable resources. Plant Physiology 127, 390–397.

Vandermeer, J., van Noordwijk, M., Anderson, J., Ong, C., Perfecto, I., 1998. Global change and multi-species agroecosystems: concepts and issues. Agriculture, Ecosystems & Environment 67, 1–22.

Vigil, M.F., Kissel, D.E., 1991. Equations for estimating the amount of nitrogen mineralized from crop residues. Soil Science Society of America Journal 55, 757–761.

Vitousek, P.M., Howarth, R.W., 1991. Nitrogen limitation on land and in the sea: how can it occur? Biogeochemistry 13, 87–115.

Watson, C.J., Mills, C.L., 1998. Gross nitrogen transformations in grassland soils as affected by previous management intensity. Soil Biology and Biochemistry 30, 743–753.

Xu, X.F., Thornton, P.E., Post, W.M., 2013. A global analysis of soil microbial biomass carbon, nitrogen and phosphorus in terrestrial ecosystems. Global Ecology and Biogeography 22, 737–749.

Zak, D.R., Holmes, W.E., White, D.C., Peacock, A.D., Tilman, D., 2003. Plant diversity, soil microbial communities, and ecosystem function: are there any links? Ecology 84, 2042–2050.

CHAPTER

4

Potential of Increased Temporal Crop Diversity to Improve Resource Use Efficiencies: Exploiting Water and Nitrogen Linkages

Leah L.R. Renwick[1], *Timothy M. Bowles*[2], *William Deen*[3], *Amélie C.M. Gaudin*[1]

[1]University of California Davis, Department of Plant Sciences, Davis, California, USA; [2]University of California Berkeley, Department of Environmental Science, Policy, and Management, Berkeley, California, USA; [3]University of Guelph, Department of Plant Agriculture, Guelph, Ontario, Canada

RETHINKING THE ROLE OF ROTATION COMPLEXITY IN RESOURCE USE EFFICIENCY

Growing evidence indicates that diverse agroecosystems (here referring to complex, long-term rotations with a variety of cash crops and cover crops) can capitalize on water and nitrogen (N) linkages with positive effects on both productivity and water and N use efficiencies (WUE, NUE). Crops grown in rotation tend to have higher yields in conventional and organic systems (Pittelkow et al., 2015; Ponisio et al., 2015) under both optimal and stressful soil moisture conditions (Gaudin et al., 2015b) and at suboptimal N input levels (Constable et al., 1992; Gaudin et al., 2015a). Along with higher yields, diversification has the potential to lower economically optimal rates of N fertilizer compared to monoculture, resulting in greater NUE and WUE (Fig. 4.1). In this review, we synthesize evidence from various agronomic cash crops growing in temperate, rainfed agroecosystems to show that more diverse rotations improve both NUE and WUE (Constable et al., 1992; Gaudin et al., 2015a; Zhang et al., 2016) and examine the underlying mechanisms. Diversification strategies are seldom integrated with common single-crop (e.g., breeding for N fertilizer recovery efficiency or single root traits), single-resource (N or water), and single-season strategies to improve resource use efficiency. We argue that a departure is required to improve system-level, multiyear resource use efficiencies

Agroecosystem Diversity
https://doi.org/10.1016/B978-0-12-811050-8.00004-2

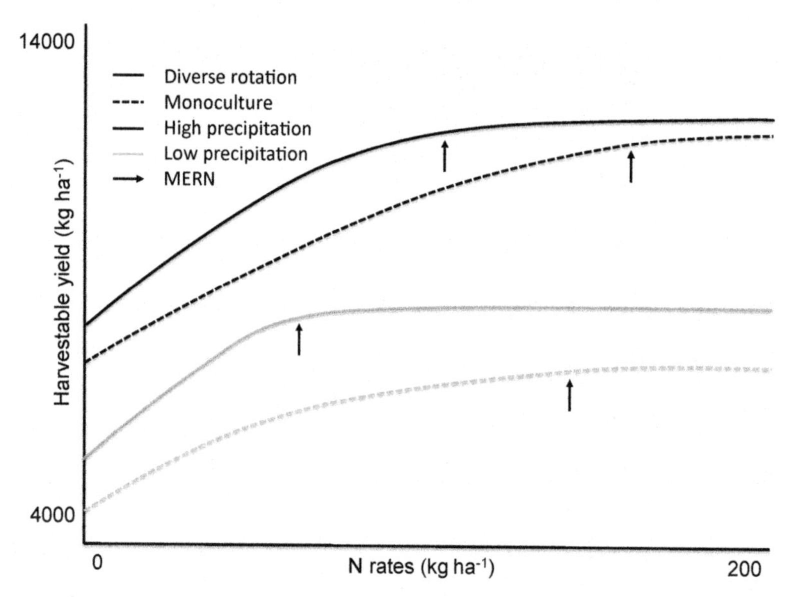

FIGURE 4.1 Hypothesized differences in nitrogen (N) response curves varying with rotation diversity and precipitation. The economically optimum N fertilizer rate is lower in diverse rotations compared to monoculture and in low precipitation scenarios compared to high precipitation scenarios. Effects of rotation and precipitation on yield response to N are adapted from Gaudin et al. (2015a), Quemada and Gabriel (2016), and Deen (unpublished data).

and that management approaches that consider coregulation of N and water dynamics offer the opportunity to significantly improve both NUE and WUE (Quemada and Gabriel, 2016).

In rainfed, arable cropping systems, yield is frequently limited by N and water (Sinclair and Rufty, 2012), which are often managed separately and are seen as having independent, additive effects (Carlson et al., 1996; Vanotti and Bundy, 1994). However, water and N availability interact and colimit productivity across agricultural and natural systems (Hooper and Johnson, 1999; Sadras, 2005). Evapotranspiration use efficiency (the ratio between crop yield or biomass and evapotranspiration) and N use efficiency have been shown to be positively related in rainfed cereal systems (Quemada and Gabriel, 2016). Cropping systems generally benefit during the growing season from positive interactions between water and N, particularly in its dominant plant-available, water-soluble nitrate (NO_3^-) form (Miller and Cramer, 2005). These interactions result in greater yield response to applied N at higher rainfall compared to lower rainfall (Fig. 4.1).

The need for improved NUE is underscored by current low use efficiency of N inputs in agricultural systems. A meta-analysis of experiments tracing stable isotope ^{15}N through crop and soil systems found that, on average, only two-thirds of applied N is retained in temperate grain cropping systems (e.g., harvested grain: 21.1%; crop residues: 12.3%; soil: 28.9%), while the remaining third is lost from the system (Gardner and Drinkwater, 2009). However, N losses are likely even higher, as such studies often do not account for short-term soil N retention followed by losses or for losses of nonlabeled N applied in previous seasons. Such significant N losses negatively impact the environment through NO_3^- leaching into groundwater (Burow et al., 2010; Zhang et al., 1996), eutrophication of fresh and marine water (Anderson et al., 2002), and emissions of the potent greenhouse gas nitrous oxide (N_2O). These N losses are tightly linked to water availability and movement through soil despite asynchronous soil N and water inputs in rainfed systems and water and N retention, which vary with soil properties and crop sequence. Yet, the overriding importance of water for N cycling and losses is often not considered when strategizing to improve N retention.

In this review, we synthesize knowledge on how shifts in water and N assimilation, retention, and loss pathways associated with rotation complexity depend on better exploitation of soil, plant, and microbial processes across temporal (weeks to years) and spatial (rhizosphere to landscape) scales to improve NUE and WUE. We use system-scale definitions of NUE and WUE encompassing whole crop rotation cycles, where NUE is defined as the proportion of applied N retained in soil and crop residue and removed in harvested biomass (Cassman et al., 2002), and WUE is defined as the proportion of precipitation retained in soil and transpired by the crop canopy. In particular, we explore improvements in both the soil supply and crop recovery/demand aspects of NUE and WUE: (1) how gains in soil organic carbon (SOC) affect soil water and N retention and availability, and (2) how root and rhizosphere processes shift, how NUE and WUE interact at the plant scale, and how pest pressure impacts crop demand for resources under both optimal and stressful conditions.

ROTATION DIVERSIFICATION OPTIMIZES WATER AND NITROGEN RETENTION AND CYCLING

Temporal crop diversity impacts WUE and NUE directly by altering fluxes of nutrients and water over the course of the rotation and indirectly through long-term changes in soil chemical and physical properties and microbial communities and processes. We propose that gains in SOC associated with rotation diversification help regulate and improve N retention and cycling and water retention and storage, thus reducing system losses and increasing WUE and NUE.

Water and Nitrogen Retention via Shifts in Soil Properties

A meta-analysis of 122 field studies showed that adding one or more cash crops to rotation increased SOC by 3.6% and total soil N by 5.3% on average compared to a monoculture, with even greater positive effects when a cover crop is included (7.8% and 12.8% increase to SOC and total soil N, respectively) relative to a monoculture (McDaniel et al., 2014b). Although a meta-analysis approach cannot differentiate between the effects of crop diversity versus compositional effects (e.g., adding a legume into a grain rotation), it is clear that increasing the number of crops in rotation generally increases SOC.

Several interacting mechanisms may be at work in this diversity effect. Firstly, adding cover crops into rotations increases SOC. Adding a cover crop into rotation can increase total plant residue inputs to soil, including root exudates that contribute disproportionately to the stable SOC pool (Austin et al., 2017). Even when total residue inputs are similar or lower, diversified rotations with cover crops maintain higher soil SOC due to changes in microbial C use efficiency (the proportion of total C uptake used for growth as opposed to respiration), changes that ultimately increase microbially derived stable SOC (Kallenbach et al., 2016, 2015). Cover crops are major drivers of increases in SOC, and rotation diversification only leads to significant increases in SOC when cover crops are included (McDaniel et al., 2014b). Secondly, cash crop type can also influence the effectiveness of rotation diversification, with higher SOC in diverse rotations compared to soybean (11%) and sorghum (8%) monocultures, minor increases compared to wheat monoculture (3%), and no gains for continuous corn transitioned to rotation (typically corn-soybean). The latter effect may be due to a reduction in crop residue inputs overall, especially corn inputs with high C:N and subsequent chemical recalcitrance (McDaniel et al., 2014a; West and Post, 2002).

Greater soil organic matter (SOM; consisting of approximately 50% SOC in most soils (Pribyl, 2010)) leads to greater soil water retention due to its effects on soil structure and bulk density near field capacity and on adsorption at low water potentials across most soils (Hudson, 1994; Rawls et al., 2003; Yang et al., 2014). Crop rotation can also improve soil aggregate stability (Congreves et al., 2015; Karlen et al., 2006; Wienhold et al., 2006) by up to twofold, as demonstrated by a three cash crop rotation (maize-soybean-wheat) with a rye/clover cover crop compared to a maize monoculture in Michigan (Tiemann et al., 2015). Improved aggregation, in turn, improves infiltration rates and decreases surface runoff and erosion, especially during extreme precipitation events (Barthes and Roose, 2002) that can otherwise result in significant lateral N losses in surface runoff (Quinton et al., 2010).

Labile C and Microbial Processes for N Retention and Mineralization

Diverse rotations also affect soil microbial biomass and activity via changes in readily available soil C that increases N retention and mineralization. Compared to monoculture, rotations of two or more crops have 20.7% and 26.1% higher microbial biomass C and N, respectively, an effect that does not vary with cover crop inclusion in the rotation or cash crop type (McDaniel et al., 2014b). Recent experimental evidence (Santonja et al., 2017; Tiemann et al., 2015) also links plant species diversity to higher microbial functional diversity and activity and faster residue decomposition rates under water-limited conditions, even when crop residue was of poor quality (McDaniel et al., 2014a), as well as higher activity of microbial enzymes involved in SOM breakdown and N mineralization (Acosta-Martinez et al., 2014; Carpenter-Boggs et al., 2000; Dodor and Tabatabai, 2003; Ekenler and Tabatabai, 2002; Klose and Tabatabai, 2000; McDaniel et al., 2014a). Such changes are driven mainly by higher "active" (labile) soil C pools in more diverse crop rotations (McDaniel and Grandy, 2016; Tiemann et al., 2015), which in turn support a larger and more active microbial biomass with a greater demand for N.

These findings are important because microbial biomass is the major biologic N sink when plant roots are not present or when plant N demand is low (Drinkwater and Snapp, 2007), for instance, after spring fertilizer applications when plants are young. If there is sufficient active C available, then microbes can temporarily immobilize inorganic N and incorporate it into organic N pools, which have a much longer mean residence time and can later be mineralized. Although this strategy may temporarily decrease the N available for crop uptake, over the longer term (weeks to months), plants benefit from microbial turnover and N mineralization as available C is depleted (Harrison et al., 2007; Hodge, 2004; Jackson et al., 2008; Schimel and Bennett, 2004). Because temperature and soil moisture are both fundamental controls on microbial processes and plant growth, N mineralized via microbial processes may be more synchronous with plant demand than N added as fertilizer. Impacts of diverse crop rotations on labile soil C thus drive the microbial processes that support greater N retention and mineralization, i.e., they support greater microbial N cycling capacity.

Improved water infiltration and retention in complex rotations may also indirectly affect microbial N cycling processes and synchronicity with root N uptake via positive effects of soil moisture availability on microbial processes. Although soil N cycling is understood to be moisture driven (Porporato et al., 2003), the combined effects of increased organic N, microbial biomass, and soil water retention on components of the soil N cycle have not been fully disentangled. At low soil moisture content, connectivity between water-filled soil pores decreases, which reduces not only mass flow of inorganic N to crop roots but also diffusion rates of enzymes and substrates regulating microbially mediated N cycling. This process may result in slower diffusion of microbial enzymes to SOM and, subsequently, dissolved organic matter products back to microbes, limiting microbial activity (Manzoni et al., 2016) and mineralization of organic N (Larsen et al., 2011). Slower diffusion of ammonium (NH_4^+) substrate to nitrifying bacteria might also decrease nitrification (oxidation of NH_4^+ to NO_3^-) (Stark and Firestone, 1995). These mechanisms imply that improvements in soil physical properties (e.g., aggregation, aeration, infiltration, water holding capacity) in diversified rotations may mitigate the negative effects of moisture limitation on microbial activity and thus decrease potential for N losses by tightening cycling between SOM, microbial, and mineral N pools. Reduced moisture limitation, together with higher "active" soil C, may help explain the links between observed increases in soil aggregate stability, SOC and total soil N, and microbial activity along a gradient of increasing rotation diversity and residue inputs (Tiemann et al., 2015).

Conversely, saturated soils (approximately 70%–90% water-filled pore space) (Dobbie et al., 1999) with anaerobic conditions (Zhu et al., 2013) lead to N_2O emissions primarily through denitrification (reduction of NO_3^- to dinitrogen (N_2)). N_2O:N_2 ratios peak at high percent water-filled pore space and low oxygen availability (i.e., field capacity) (Bateman and Baggs, 2005), implying that soil structure changes observed in complex rotations (Congreves et al., 2015; Karlen et al., 2006; Tiemann et al., 2015; Wienhold et al., 2006) could help decrease N_2O production through increased soil aeration (Zhu et al., 2013).

Cover Crop Effects on Resource Use Efficiency

Cover crop diversification strategies can directly alter soil water dynamics through changes in evaporation, transpiration, and loss pathways over the course of the rotation. Some evidence shows that cover crops can deplete soil water, but their net impact on soil water balance is variable (Carlson and Stockwell, 2013; Unger and Vigil, 1998; Zhang et al., 2016). Other evidence indicates that cover crops positively impact soil water content by (1) reducing runoff and improving infiltration and soil water storage capacity, especially deeper in the soil profile in both extremely dry and wet years, and (2) decreasing evaporative losses through a mulching effect, both following cover crop termination and during cover crop growth, as long as sufficient periodic precipitation occurs when the cover crop is actively transpiring (Basche et al., 2016b; Daigh et al., 2014; Qi et al., 2011; Reese et al., 2014). Cover crops can also help farmers reduce delays in crop planting during excessively wet springs while also reducing runoff, soil erosion, and leaching potential by increasing transpiration during periods of oversupply (Basche et al., 2016a, 2016b; Qi and Helmers, 2010).

One well-characterized benefit of crop rotations that include year-long soil cover is the ability to capture and retain residual N as well as newly mineralized N (Gardner and Drinkwater, 2009; Kaspar et al., 2012). Cereal cover crops are extremely effective at reducing NO_3^- leaching and water-dependent emissions of N_2O compared to legumes (Basche et al., 2014; Quemada et al., 2013). Findings on the net effect of crop rotation on N_2O emissions are mixed, with increases in N_2O emissions reported for addition of soybean to continuous corn but reductions in global warming potential and total greenhouse gas intensity (yield-scaled global warming potential) reported for diversification of small grain systems with legumes, differences that were attributed to soil-plant N cycling shifts (Sainju, 2016). Finally, fixed N from addition of leguminous cover crops can offset a portion of fertilizer N inputs for subsequent cash crops and therefore enhance N use efficiency of the whole crop rotation (Gentry et al., 2013), an effect that may be especially pronounced in diverse rotations with improved soil structure, water availability, and thus cover crop growth and N uptake.

In general, higher soil water and N availability and reduced probability of scarcity, when coupled with greater crop demand, creates potential for increased crop resource uptake and assimilation in diversified rotations.

ROTATION DIVERSIFICATION INCREASES CROP UPTAKE OF WATER AND N

Reported yield gains from rotation diversification (Pittelkow et al., 2015; Ponisio et al., 2015) can be attributed not only to supply-side shifts in soil properties and microbial ecology but also to demand-side root processes, crop water-N interactions, and pest pressure.

Root Foraging and Rhizosphere Processes

Crop rotation diversity regulates key aspects of root ability to forage and take up N and water from the soil matrix through shifts in (1) soil physical and chemical properties and (2) C, N, and water availability and distribution in time and space in the rhizosphere.

Improvements in soil properties associated with greater rotation diversity such as aggregation, reduced compaction, and creation of varied biopore structures have been shown to affect root developmental patterns and depth distribution in cereals (Bengough, 2003; Y. L. Chen et al., 2014; Han et al., 2015; Hatano et al., 1988; Perkons et al., 2014). More diverse rotation sequences typically include both deep- and shallow-rooted species and thus help form pathways for rapid root growth into deeper soil layers, assist in overcoming physical barriers in high density soil layers, and allow roots to reenter and forage subsoil to increase uptake and thus decrease leaching potential. By permitting less metabolically demanding deep root proliferation and thus improving synchrony of root foraging and N and water supply across the entire soil profile, diverse rotations might also promote expression of root phenes shown to improve water and N uptake and use efficiency across precipitation regimes and soil depths, such as steeper root angle, deep taproot/primary root development, dimorphic phenotypes with both shallow and deep roots, and effective lateral root branching at a low metabolic cost (Dathe et al., 2016; Lynch, 2015, 2013; Postma et al., 2014; Postma and Lynch, 2011; Saengwilai et al., 2014).

Cereals cultivated in rotation with taprooted crops benefit from enhanced uptake of water and nutrients during early growth stages (Han et al., 2015; Jakobsen and Dexter, 1988) or during mild dry spells (Gaiser et al., 2013; Jakobsen and Dexter, 1988). Additional belowground niche complementarities can improve nutrient and water acquisition and resource remobilization from the subsoil over the course of the rotation. This process includes nutrient and water mobilization by deeper-rooted crops, nutrient accumulation via root decay in the topsoil (Kautz, 2014; Kautz et al., 2013), and direct nutrient transfer during intercropping periods (Burity et al., 1989; Kunelius et al., 1992). Such mechanisms might help sustain increases in crop N and water demand while decreasing losses below the root zone in more diversified rotations.

Species arrangements in time and space also have the potential to regulate microbial rhizosphere functions essential for enhancing WUE and NUE. Although not directly experimentally tested, shifts in duration and severity of drying and rewetting cycles (Y.L. Chen et al., 2014; Gaudin et al., 2013; Lotter et al., 2003; Pimentel et al., 2005; Syswerda et al., 2012; Syswerda and Robertson, 2014), root traits (Bodner et al., 2014; Calonego and Rosolem, 2010; Chen and Weil, 2011), microbial communities (McDaniel et al., 2014a), and the amount and quality of mineralizable substrate (McDaniel et al., 2014a) likely shape the intensity and ecologic outcomes of rhizosphere microbial processes and losses. Greater colocation of organic N and water at the root surface may have a large impact on rhizosphere microbial activity and N mineralization and availability.

More heterogeneous crop residues and abundant dead roots with greater physiologic and morphologic trait diversity may also affect decomposition rates and rhizosphere priming (Berthrong et al., 2013; Chen and Brassard, 2013; Maul et al., 2014; McDaniel et al., 2014a, 2014b; Silver and Miya, 2001). Positive rhizosphere priming effects (increased SOM decomposition rates in the presence of live roots and their exudates) can stimulate microbial demand for organic N and thus accelerate N cycling, potentially increasing N supply to the crop over time (Zhu et al., 2014). As root exudates to rhizosphere microbial communities including mycorrhizal fungi represent a significant investment of belowground photosynthate (Kaiser et al., 2015), benefits to NUE are likely greatest in systems where synthetic N fertilizers inputs are partially offset by organic matter inputs and reliance on microbial N cycling is higher.

Finally, more diverse crop rotations tend to include cover crops and soil is thus less disturbed with longer periods with living plant roots. Mycorrhizal fungi and other soil microbes that are negatively impacted by disturbances, compaction, and fallow may become more abundant under diversified rotations (Bowles et al., 2016; Brito et al., 2012; Deguchi et al., 2007; Frey et al., 1999; Kabir, 2005; Lehman et al., 2012), thereby enhancing ecologic processes that may partially substitute for inputs (George et al., 1995), especially in water-stressed environments (Tobar et al., 1994).

Crop Water and Nitrogen Interactions

Diverse rotations have exhibited higher NUE and higher yield across a range of precipitation levels (Gaudin et al., 2015a, 2015b). Crops with fewer yield-limiting factors (e.g., with higher soil N and water availability and reduced pest pressure), such as those in complex rotations, can cumulatively take up more N and water (Sinclair, 2012; Lassaletta et al., 2014). Although the impact of rotation diversity on well-characterized water-N interactions at the crop scale is not fully understood, several interacting N-water mechanisms may contribute to WUE and NUE gains in diverse rotations.

In high input, rainfed agroecosystems, precipitation often limits yield, resulting in poor fertilizer recovery efficiency at lower rainfall and higher optimal N fertilizer rates at higher rainfall (Quemada and Gabriel, 2016). Poor yield response to N under low rainfall is due to reduced transpiration, which subsequently limits bulk flow of water and soluble nutrients to roots and reduces NO_3^- uptake once biomass accumulation decreases (Eck and Musick, 1979; Imsande and Touraine, 1994; Jenne et al., 1958). Through these processes, precipitation and thus soil moisture regulate fertilizer N recovery and thus system NUE in rainfed cropping systems.

Conversely, adequate N nutrition creates positive feedback for WUE by ensuring that photosynthesis, growth, and yield are not N-limited. Optimal N nutrition in conjunction with sufficient water supply results in greater canopy development and faster canopy closure. This process decreases the proportion of evapotranspiration (ET) lost as evaporation and increases the proportion of ET partitioned to transpiration.

Pest Pressure

Reductions in weed pressure in more complex rotations decrease competition for water and N as well as negative effects of weeds on plant growth and root properties. Weeds directly compete with the crop for resources, thereby reducing N and water availability and restricting uptake and yield (Vanheemst, 1985). Changes in crop leaf angle and reduced root-to-shoot ratio in the presence of weeds may also reduce the potential for water and N uptake at later growth stages. The advent of herbicides and herbicide-resistant crops have decreased historical reliance on crop rotation for weed management, and simple rotations are subject to increasingly prevalent herbicide resistance (Mortensen et al., 2012; Shaner, 2014). System diversification reduces the negative impacts of weed competition on crop N and water recovery via various mechanisms: (1) improved crop competitive ability due to fewer yield-limiting factors, (2) added weed suppression through increased cover crop options, (3) diversification of both the timing and nature of aboveground and belowground competition of crop with weeds, (4) reduction of weed seed banks, (5) increased tillage options, and (6) increased herbicide options, timing of application, and modes of actions over the length of the rotation (Liebman et al., 2008; Swanton et al., 2008). Diverse crop rotations can also increase crop yields by reducing the presence of weeds early in the season and altering crop morphology even before resource competition occurs (Rajcan et al., 2004).

Simple rotations or monoculture provide a consistent supply of host material for common pests and diseases, which increases the probability and severity of reinfestation. More diverse rotations can therefore lessen negative effects of pest damage on crop demand for resources through improvements in stand density and crop functions important for water and N uptake, N fixation, and yield (Oerke, 2006). Soil-borne pests such as take-all in wheat (Cook, 2003), legume foot and root rot in peas (Oyarzun et al., 1993), corn rootworm in corn, and soybean cyst nematode in soybean (Mock et al., 2012) compromise crop root structures, growth, and resource demand, and their reduction or elimination is beneficial to WUE and NUE.

SIGNIFICANCE OF CROP ROTATION DIVERSIFICATION FOR WUE AND NUE IN A CHANGING CLIMATE

Increased crop rotation diversity merits consideration as a strategy for mitigating the impacts of drought episodes on crop production, particularly given recent yield reductions due to drought (Lesk et al., 2016) and projected agronomic and socioeconomic consequences of climate change (Parry et al., 2004). As climate change progresses, predicted shifts in the magnitude and frequency of precipitation (Kirtman et al., 2013) may necessitate the use of practices such as cover cropping to retain water and N in the off-season (Congreves et al., 2016; Robertson et al., 2013). Improvements in NUE and WUE will become increasingly critical to decouple agricultural intensification from its current environmental footprint and build system resilience to limiting water and N resources associated with predicted changes in weather patterns.

Water and N Availability and Uptake

Diversified crop rotations with greater soil water holding capacity can help buffer against drought (Gaudin et al., 2015b; Williams et al., 2016), although rotation cover crop selection and management must be optimized to capitalize on increased infiltration and storage capacity while minimizing transpiration losses (Pala et al., 2007; Reese et al., 2014). Positive rotation and residue diversity effects on microbially mediated residue decomposition and N cycling persist under dry conditions and may compensate for lower decomposition rates under moisture stress (Acosta-Martinez et al., 2014; Santonja et al., 2017). Shifts in soil microbial communities typically found with rotation diversification, including higher diversity (Jiang et al., 2016; Tiemann et al., 2015) and higher relative abundance of fungi compared to bacteria (Bünemann et al., 2004; González-Chávez et al., 2010; Suzuki et al., 2012), have been linked to functional resilience or resistance to disturbance (de Vries et al., 2012; Griffiths et al., 2000). Rhizosphere microbes and the extracellular polysaccharide- and protein-based compounds they produce have also been shown to improve soil aggregation and porosity (Bronick and Lal, 2005) and regulate aquaporin expression and abscisic acid production (Augé, 2004, 2001; Feeney et al., 2006; Groppa et al., 2012), which may help maintain hydraulic conductivity in drying soil and maximize water uptake. Some 1-aminocyclopropane-1-carboxylic acid (ACC) deaminase-producing rhizobacteria can also prevent drought-induced yield reductions through local and systemic pathways (Belimov et al., 2009). Such mechanisms may help create positive feedback loops for WUE and NUE over the course of entire crop rotations under both optimal and stressful conditions.

Reduced Losses via Continuous Living Soil Cover

Periods of drought stress when low soil moisture inhibits plant growth, plant N demand, and microbial N cycling (Dijkstra et al., 2012) may decrease synchrony between fertilizer N availability and plant and microbial N demand and increase potential for time-lagged effects on N losses during subsequent rainfall events (Loecke et al., 2017). By providing a plant N sink to take up residual and newly mineralized soil NO_3^- following a cash crop, cover crops (Basche et al., 2014; Gardner and Drinkwater, 2009; Kaspar et al., 2012) or pasture phases (Kunrath et al., 2015) within the crop rotation cycle can reduce system vulnerability to N losses by increasing the proportion of the year with living soil cover. For instance, cover cropping with rye was found to effectively reduce erosion and sediment-bound and soluble nutrients under current and near-future (e.g., 2050) climatic conditions, which predict higher late fall and winter precipitation in the Upper Mississippi River Basin of North America (Panagopoulos et al., 2014).

Although climate change mitigation via increased soil C sequestration could be substantial, potential synergies and tradeoffs between rotation diversification and other management strategies for the net greenhouse gas budget and nonclimate-related resource conservation are also important to consider. For instance, net global warming potential depends on crop rotation composition as well as tillage (Sainju, 2016), and the combination of reduced or zero tillage with rotations optimized for emissions reduction may provide a pathway to reduce harmful above (greenhouse gases) and belowground (NO_3^-) N and C losses.

Sustained Yield Under Stressful Conditions

In some regions, sustaining and increasing yields may be part of the solution to feed a growing population and sustain farmer livelihoods in a changing climate (Foley et al., 2011; Parry et al., 2004). Maintaining yield gains under climate change scenarios will likely require larger amounts of water to meet greater evapotranspirative demand driven by higher temperatures, even when elevated atmospheric carbon dioxide is accounted for (Lobell et al., 2013). While genetic improvement has done much for both theoretical and actual yield response to inputs, improving availability and crop uptake capacity of water and N is equally important to close the yield gap between actual and potential yields (Sinclair and Rufty, 2012). Adoption of multifunctional management strategies such as rotation diversification to help conserve water and N while enhancing resource availability and foraging capacity under stress will be of prime importance under future climate scenarios.

KNOWLEDGE GAPS FOR INTEGRATING ROTATION DIVERSITY INTO RESOURCE USE EFFICIENCY IMPROVEMENT STRATEGIES

Increases in N and water availability, crop resource demand, and crop foraging and uptake capacity in more complex rotations provide an integrated approach to improve WUE and NUE. Although recent evidence collectively highlights the potential of rotation diversity to improve crop-soil-microbe system ability to capture soil resources and minimize losses (Fig. 4.2), the underlying mechanisms and quantitative, simultaneous benefits for WUE and NUE, as well as economic returns as a function of

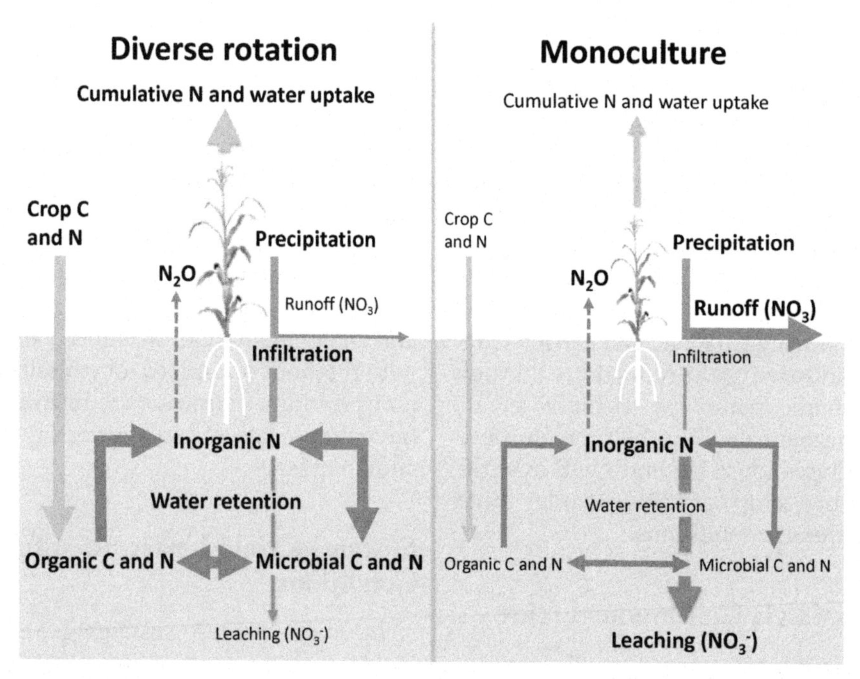

FIGURE 4.2 Generalized influence of crop rotation complexity (diversification of cash crops and addition of cover crops) on water, nitrogen (N), and carbon (C) fluxes that affect N and water use efficiency. Arrow width indicates magnitude of flux. In the diverse rotation scenario, increased organic C inputs, via higher cash and/or cover crop biomass, and greater soil water infiltration and retention reduce nitrate (NO_3^-) losses and drive soil N cycling. Higher crop N recovery, greater partitioning of evapotranspiration to transpiration, reduced pest pressure, and enhanced root foraging capacity leads to greater cumulative uptake of water and N. *Dotted line* indicates mixed effects of crop rotation (partially dependent on component crops) on nitrous oxide (N_2O) emissions.

adoption of other best management practices (e.g., conservation or zero tillage), soil, climate, and component crops still require thorough assessment at the system and watershed scale.

At the crop scale, characterization of the impact of system diversity on foraging ability, niche complementarities, and the underlying root and rhizosphere mechanisms over the length of rotations will open new frontiers for WUE and NUE improvement, particularly as knowledge of the rhizosphere and bulk soil microbiomes expands. Given differences in soil properties and water and nutrient distribution within the soil profile and over the duration of the distinct crop rotations, root ideotypes developed for conventional systems may be of less benefit in diversified systems (Lilley and Kirkegaard, 2016). Elucidating how roots regulate C, N, and water availability in the rhizosphere and interact with rhizosphere and bulk soil microbes will help define new crop ideotypes for breeding for more diversified systems, thereby bridging the gap between breeding targets and agronomic goals.

At the farm scale, the impacts of rotation diversity-driven shifts in overall agroecosystem N and water retention and recovery at the end of the growing season remain to be effectively integrated with fertilization strategies. Establishing coupled water and N budgets that account for higher NUE in diverse rotations would help redefine optimum N rates (Fig. 4.1) and optimize fertilizer rate calculators to further the benefits of improving NUE to farmers and the environment. Assessing NUE over the length of a calendar year or entire crop rotations rather than single cropping seasons may provide a useful framework for quantifying of the cumulative long-term significance of higher single-season NUE. Better understanding of rotation impacts on microbial processes governing N mineralization rates could help improve prediction of SOM N credits for N budgets in terms of quantity, consistency, and synchronicity relative to plant demand (Grandy et al., 2012). Using multiyear approaches that include market fluctuations in farm gate returns under both optimal and stressful conditions could also enhance cost-benefit analyses of rotation diversification and ultimately help identify and address barriers to adoption.

At the landscape scale, significant improvements in quantifying the ecologic, agronomic, and economic mitigation potential of diversification scenarios (e.g., cash crop diversification, cover cropping) relative to the magnitude and spatial distribution of water and N losses and their regional environmental and economic impact (e.g., Sutton et al., 2011) is required. For instance, advances in incorporating national-level soil data including SOC and soil water storage capacity into yield risk and crop insurance models may eventually permit the impact of management practices (e.g., rotation diversification, cover cropping) on soil properties to be integrated into crop insurance rating systems as a strategy to reduce errors in assessing yield risk and thus assigning insurance rates and incentivize adoption of such practices (Woodard and Verteramo-Chiu, 2017). At the regional scale, ongoing direct and remote sensing–based monitoring seeks to understand the impact of a cost-share program to incentivize expansion of cover cropped acreage to reduce soil erosion, nutrient loss and leaching, and water quality degradation in the Chesapeake Bay. Wintertime vegetation on corn fields in four Maryland counties increased from between 26% and 36% in 2010 to between 52% and 75% in 2013 (Hively et al., 2015), and modeling shows the potential for cover crop adoption to reduce NO_3^- leaching by 27%–76% at the watershed scale, or as high as 93% with early planted rye (Yeo et al., 2014).

Rotation diversification has the potential to improve WUE and NUE by exploiting the inextricable linkages between water and N use efficiencies (Quemada and Gabriel, 2016), although the issues mentioned earlier highlight some gaps remaining before this goal can be achieved. To optimize socioeconomic and environmental benefits of crop diversity, systems approaches must be used to assess how crop rotation diversification impacts water and N codynamics under current and expected future climate scenarios.

References

Acosta-Martinez, V., Moore-Kucera, J., Cotton, J., Gardner, T., Wester, D., 2014. Soil enzyme activities during the 2011 Texas record drought/heat wave and implications to biogeochemical cycling and organic matter dynamics. Applied Soil Ecology 75, 43–51. https://doi.org/10.1016/j.apsoil.2013.10.008.

Anderson, D.M., Glibert, P.M., Burkholder, J.M., 2002. Harmful algal blooms and eutrophication: nutrient sources, composition, and consequences. Estuaries 25, 704–726. https://doi.org/10.1007/BF02804901.

Augé, R.M., 2004. Arbuscular mycorrhizae and soil/plant water relations. Canadian Journal of Soil Science 84, 373–381. https://doi.org/10.4141/S04-002.

Augé, R.M., 2001. Water relations, drought and vesicular-arbuscular mycorrhizal symbiosis. Mycorrhiza 11, 3–42. https://doi.org/10.1007/s005720100097.

Austin, E.E., Wickings, K., McDaniel, M.D., Robertson, G.P., Grandy, A.S., 2017. Cover crop root contributions to soil carbon in a no-till corn bioenergy cropping system. Global Change Biology Bioenergy 9, 1252–1263. https://doi.org/10.1111/gcbb.12428.

Barthes, B., Roose, E., 2002. Aggregate stability as an indicator of soil susceptibility to runoff and erosion; validation at several levels. Catena 47, 133–149. https://doi.org/10.1016/S0341-8162(01)00180-1.

Basche, A.D., Archontoulis, S.V., Kaspar, T.C., Jaynes, D.B., Parkin, T.B., Miguez, F.E., 2016a. Simulating long-term impacts of cover crops and climate change on crop production and environmental outcomes in the Midwestern United States. Agriculture, Ecosystems & Environment 218, 95–106. https://doi.org/10.1016/j.agee.2015.11.011.

Basche, A.D., Kaspar, T.C., Archontoulis, S.V., Jaynes, D.B., Sauer, T.J., Parkin, T.B., Miguez, F.E., 2016b. Soil water improvements with the long-term use of a winter rye cover crop. Agricultural Water Management 172, 40–50. https://doi.org/10.1016/j.agwat.2016.04.006.

Basche, A.D., Miguez, F.E., Kaspar, T.C., Castellano, M.J., 2014. Do cover crops increase or decrease nitrous oxide emissions? A meta-analysis. Journal of Soil and Water Conservation 69, 471–482. https://doi.org/10.2489/jswc.69.6.471.

Bateman, E.J., Baggs, E.M., 2005. Contributions of nitrification and denitrification to N_2O emissions from soils at different water-filled pore space. Biology and Fertility of Soils 41, 379–388. https://doi.org/10.1007/s00374-005-0858-3.

Belimov, A.A., Dodd, I.C., Hontzeas, N., Theobald, J.C., Safronova, V.I., Davies, W.J., 2009. Rhizosphere bacteria containing 1-aminocyclopropane-1-carboxylate deaminase increase yield of plants grown in drying soil via both local and systemic hormone signalling. New Phytologist 181, 413–423. https://doi.org/10.1111/j.1469-8137.2008.02657.x.

Bengough, A.G., 2003. Root growth and function in relation to soil structure, composition, and strength. In: de Kroon, H., Visser, E.J.W. (Eds.), Root Ecology. Springer, Berlin Heidelberg, pp. 151–171. https://doi.org/10.1007/978-3-662-09784-7_6.

Berthrong, S.T., Buckley, D.H., Drinkwater, L.E., 2013. Agricultural management and labile carbon additions affect soil microbial community structure and interact with carbon and nitrogen cycling. Microbial Ecology 66, 158–170. https://doi.org/10.1007/s00248-013-0225-0.

Bodner, G., Leitner, D., Kaul, H.P., 2014. Coarse and fine root plants affect pore size distributions differently. Plant and Soil 380, 133–151. https://doi.org/10.1007/s11104-014-2079-8.

Bowles, T.M., Jackson, L.E., Loeher, M., Cavagnaro, T.R., 2016. Ecological intensification and arbuscular mycorrhizae: a meta-analysis of tillage and cover crop effects. Journal of Applied Ecology. https://doi.org/10.1111/1365–2664.12815.

Brito, I., Goss, M.J., De Carvalho, M., 2012. Effect of tillage and crop on arbuscular mycorrhiza colonization of winter wheat and triticale under mediterranean conditions. Soil Use & Management 28, 202–208. https://doi.org/10.1111/j.1475-2743.2012.00404.x.

Bronick, C.J., Lal, R., 2005. Soil structure and management: a review. Geoderma 124, 3–22. https://doi.org/10.1016/j.geoderma.2004.03.005.

Bünemann, E.K., Bossio, D.A., Smithson, P.C., Frossard, E., Oberson, A., 2004. Microbial community composition and substrate use in a highly weathered soil as affected by crop rotation and P fertilization. Soil Biology and Biochemistry 36, 889–901. https://doi.org/10.1016/j.soilbio.2004.02.002.

Burity, H.A., Ta, T.C., Faris, M.A., Coulman, B., 1989. Estimation of nitrogen-fixation and transfer from alfalfa to associated grasses in mixed swards under field conditions. Plant and Soil 114, 249–255. https://doi.org/10.1007/BF02220805.

Burow, K.R., Nolan, B.T., Rupert, M.G., Dubrovsky, N.M., 2010. Nitrate in groundwater of the United States, 1991–2003. Environmental Science and Technology 44, 4988–4997. https://doi.org/10.1021/es100546y.

Calonego, J.C., Rosolem, C.A., 2010. Soybean root growth and yield in rotation with cover crops under chiseling and no-till. European Journal of Agronomy 33, 242–249. https://doi.org/10.1016/j.eja.2010.06.002.

Carlson, R.E., Todey, D.P., Taylor, S.E., 1996. Midwestern corn yield and weather in relation to extremes of the southern oscillation. Journal of Production Agriculture 9, 347–352.

Carlson, S., Stockwell, R., 2013. Research priorities for advancing adoption of cover crops in agriculture-intensive regions. Journal of Agriculture, Food Systems, and Community Development 3, 125–129. https://doi.org/10.5304/jafscd.2013.034.017.

Carpenter-Boggs, L., Pikul, J.L., Vigil, M.F., Riedell, W.E., 2000. Soil nitrogen mineralization influenced by crop rotation and nitrogen fertilization. Soil Science Society of America Journal 64, 2038–2045. https://doi.org/10.2136/sssaj2000.6462038x.

Cassman, K.G., Dobermann, A., Walters, D.T., 2002. Agroecosystems, nitrogen-use efficiency, and nitrogen management. AMBIO A Jornal of Humman Environment 31, 132–140. https://doi.org/10.1579/0044-7447-31.2.132.

Chen, G., Weil, R.R., 2011. Root growth and yield of maize as affected by soil compaction and cover crops. Soil and Tillage Research 117, 17–27. https://doi.org/10.1016/j.still.2011.08.001.

Chen, G., Weil, R.R., Hill, R.L., 2014. Effects of compaction and cover crops on soil least limiting water range and air permeability. Soil and Tillage Research 136, 61–69. https://doi.org/10.1016/j.still.2013.09.004.

Chen, H.Y.H., Brassard, B.W., 2013. Intrinsic and extrinsic controls of fine root life span. CRC. Critical Reviews in Plant Sciences 32, 151–161. https://doi.org/10.1080/07352689.2012.734742.

Chen, Y.L., Palta, J., Clements, J., Buirchell, B., Siddique, K.H.M., Rengel, Z., 2014. Root architecture alteration of narrow-leafed lupin and wheat in response to soil compaction. Field Crops Research 165, 61–70. https://doi.org/10.1016/j.fcr.2014.04.007.

Congreves, K.A., Dutta, B., Grant, B.B., Smith, W.N., Desjardins, R.L., Wagner-Riddle, C., 2016. How does climate variability influence nitrogen loss in temperate agroecosystems under contrasting management systems? Agriculture, Ecosystems & Environment 227, 33–41. https://doi.org/10.1016/j.agee.2016.04.025.

Congreves, K.A., Hayes, A., Verhallen, E.A., Van Eerd, L.L., 2015. Long-term impact of tillage and crop rotation on soil health at four temperate agroecosystems. Soil and Tillage Research 152, 17–28. https://doi.org/10.1016/j.still.2015.03.012.

Constable, G.A., Rochester, I.J., Daniells, I.G., 1992. Cotton yield and nitrogen requirement is modified by crop rotation and tillage method. Soil and Tillage Research 23, 41–59. https://doi.org/10.1016/0167-1987(92)90004-U.

Cook, R.J., 2003. Take-all of wheat. Physiol. Molecular Plant Pathology 62, 73–86. https://doi.org/10.1016/S0885-5765(03)00042-0.

Daigh, A.L., Helmers, M.J., Kladivko, E., Zhou, X., Goeken, R., Cavdini, J., Barker, D., Sawyer, J., 2014. Soil water during the drought of 2012 as affected by rye cover crops in fields in Iowa and Indiana. Journal of Soil and Water Conservation 69, 564–573. https://doi.org/10.2489/jswc.69.6.564.

Dathe, A., Postma, J., Postma-Blaauw, M., Lynch, J., 2016. Impact of axial root growth angles on nitrogen acquisition in maize depends on environmental conditions. Annals of Botany 118, 401–414. https://doi.org/10.1093/aob/mcw112.

de Vries, F.T., Liiri, M.E., Bjørnlund, L., Bowker, M., a., Christensen, S., Setälä, H.M., Bardgett, R.D., 2012. Land use alters the resistance and resilience of soil food webs to drought. Nature Climate Change 2, 276–280. https://doi.org/10.1038/nclimate1368.

Deguchi, S., Shimazaki, Y., Uozumi, S., Tawaraya, K., Kawamoto, H., Tanaka, O., 2007. White clover living mulch increases the yield of silage corn via arbuscular mycorrhizal fungus colonization. Plant and Soil 291, 291–299. https://doi.org/10.1007/s11104-007-9194-8.

Dijkstra, F.A., Augustine, D.J., Brewer, P., von Fischer, J.C., 2012. Nitrogen cycling and water pulses in semiaridgrasslands: are microbial and plant processes temporally asynchronous? Oecologia 170, 799–808,. https://doi.org/10.1007/s00442-012-2336-6.

Dobbie, K.E., McTaggart, I.P., Smith, K.A., 1999. Nitrous oxide emissions from intensive agricultural systems: variations between crops and seasons, key driving variables, and mean emission factors. Journal of Geophysical Research 104, 26891–26899. https://doi.org/10.1029/1999JD900378.

Dodor, D.E., Tabatabai, M.A., 2003. Amidohydrolases in soils as affected by cropping systems. Applied Soil Ecology 24, 73–90. https://doi.org/10.1016/S0929-1393(03)00067-2.

Drinkwater, L.E., Snapp, S., 2007. Nutrients in agroecosystems: rethinking the management paradigm. Advances in Agronomy 92, 163–186. https://doi.org/10.1016/S0065-2113(04)92003-2.

Eck, H.V., Musick, J.T., 1979. Plant water stress effects on irrigated grain sorghum. II. Effects on nutrients in plant tissues. Crop Science 19, 592–598. https://doi.org/10.2135/cropsci1979.0011183X001900050010x.

Ekenler, M., Tabatabai, M.A., 2002. β-glucosaminidase activity of soils: effect of cropping systems and its relationship to nitrogen mineralization. Biology and Fertility of Soils 36, 367–376. https://doi.org/10.1007/s00374-002-0541-x.

Feeney, D.S., Crawford, J.W., Daniell, T., Hallett, P.D., Nunan, N., Ritz, K., Rivers, M., Young, I.M., 2006. Three-dimensional microorganization of the soil−root−microbe system. Microbial Ecology 52, 151−158. https://doi.org/10.1007/s00248-006-9062-8.

Foley, J.A., Ramankutty, N., Brauman, K.A., Cassidy, E.S., Gerber, J.S., Johnston, M., Mueller, N.D., O'Connell, C., Ray, D.K., West, P.C., Balzer, C., Bennett, E.M., Carpenter, S.R., Hill, J., Monfreda, C., Polasky, S., Rockstrom, J., Sheehan, J., Siebert, S., Tilman, D., Zaks, D.P.M., 2011. Solutions for a cultivated planet. Nature 478, 337−342. https://doi.org/10.1038/nature10452.

Frey, S.D., Elliott, E.T., Paustian, K., 1999. Bacterial and fungal abundance and biomass in conventional and no-tillage agroecosystems along two climatic gradients. Soil Biology and Biochemistry 31, 573−585. https://doi.org/10.1016/s0038-0717(98)00161-8.

Gaiser, T., Perkons, U., Küpper, P.M., Kautz, T., Uteau-Puschmann, D., Ewert, F., Enders, A., Krauss, G., 2013. Modeling biopore effects on root growth and biomass production on soils with pronounced sub-soil clay accumulation. Ecological Modelling 256, 6−15. https://doi.org/10.1016/j.ecolmodel.2013.02.016.

Gardner, J.B., Drinkwater, L.E., 2009. The fate of nitrogen in grain cropping systems: a meta-analysis of 15N field experiments. Ecological Applications 19, 2167−2184. https://doi.org/10.1890/08-1122.1.

Gaudin, A.C., Westra, S., Loucks, C.E., Janovicek, K., Martin, R.C., Deen, W., 2013. Improving resilience of northern field crop systems using inter-seeded red clover. Agronomy 3, 148−180. https://doi.org/10.3390/agronomy3010148.

Gaudin, A.C.M., Janovicek, K., Deen, B., Hooker, D.C., 2015a. Wheat improves nitrogen use efficiency of maize and soybean-based cropping systems. Agriculture, Ecosystems & Environment 210, 1−10. https://doi.org/10.1016/j.agee.2015.04.034.

Gaudin, A.C.M., Tolhurst, T.N., Ker, A.P., Janovicek, K., Tortora, C., Martin, R.C., Deen, W., 2015b. Increasing crop diversity mitigates weather variations and improves yield stability. PLoS One 10, e0113261. https://doi.org/10.1371/journal.pone.0113261.

Gentry, L.E., Snapp, S.S., Price, R.F., Gentry, L.F., 2013. Apparent red clover Nitrogen credit to corn: evaluating cover crop introduction. Agronomy Journal 105, 1658−1664. https://doi.org/10.2134/agronj2013.0089.

George, E., Marschner, H., Jakobsen, I., 1995. Role of arbuscular mycorrhizal fungi in uptake of phosphorus and nitrogen from soil. Critical Reviews in Biotechnology 15, 257−270. https://doi.org/10.3109/07388559509147412.

González-Chávez, Ma del Carmen A., et al., 2010. "Soil Microbial Community, C, N, and P Responses to Long-Term Tillage and Crop Rotation." Soil and Tillage Research 106.2, 285−293.

Grandy, A., Kallenbach, C., Loecke, T.D., Snapp, S.S., Smith, R.G., 2012. The biological basis for nitrogen management in agroecosystems. In: Cheeke, T.E., Coleman, D.C., Wall, D.H. (Eds.), Microbial Ecology in Sustainable Agroecosystems. CRC Press, pp. 113−132.

Griffiths, B.S., Ritz, K., Bardgett, R.D., Cook, R., Christensen, S., Ekelund, F., Sorensen, S.J., Baath, E., Bloem, J., de Ruiter, P.C., Dolfing, J., Nicolardot, B., 2000. Ecosystem response of pasture soil communities to fumigation-induced microbial diversity reductions: an examination of the biodiversity-ecosystem function relationship. Oikos 90, 279−294. https://doi.org/10.1034/j.1600-0706.2000.900208.x.

Groppa, M.D., Benavides, M.P., Zawoznik, M.S., 2012. Root hydraulic conductance, aquaporins and plant growth promoting microorganisms: a revision. Applied Soil Ecology 61, 247−254. https://doi.org/10.1016/j.apsoil.2011.11.013.

Han, E., Kautz, T., Perkons, U., Uteau, D., Peth, S., Huang, N., Horn, R., Köpke, U., 2015. Root growth dynamics inside and outside of soil biopores as affected by crop sequence determined with the profile wall method. Biology and Fertility of Soils 51, 847−856. https://doi.org/10.1007/s00374-015-1032-1.

Harrison, K.A., Bol, R., Bardgett, R.D., 2007. Preferences for different nitrogen forms by coexisting plant species and soil microbes. Ecology 88, 989−999. https://doi.org/10.1890/06-1018.

Hatano, R., Iwanaga, K., Okajima, H., Sakuma, T., 1988. Relationship between the distribution of soil macropores and root elongation. Soil Science & Plant Nutrition 34, 535−546. https://doi.org/10.1080/00380768.1988.10416469.

Hively, W.D., Duiker, S., McCarty, G., Prabhakara, K., 2015. Remote sensing to monitor cover crop adoption in southeastern Pennsylvania. Journal of Soil and Water Conservation 70, 340−352. https://doi.org/10.2489/jswc.70.6.340.

Hodge, A., 2004. The plastic plant: root responses to heterogeneous supplies of nutrients. New Phytologist 162, 9−24. https://doi.org/10.1111/j.1469-8137.2004.01015.x.

Hooper, D.U., Johnson, L., 1999. Nitrogen limitation in dryland ecosystems: responses to geographical and temporal variation in precipitation. Biogeochemistry 46, 247−293. https://doi.org/10.1023/A:1006145306009.

Hudson, B.D., 1994. Soil organic matter and available water capacity. Journal of Soil and Water Conservation 49, 189−194.

Imsande, J., Touraine, B., 1994. N demand and the regulation of nitrate uptake. Plant Physiology 105, 3–7. https://doi.org/10.1016/j.crvi.2008.11.005.

Jackson, L.E., Burger, M., Cavagnaro, T.R., 2008. Roots, nitrogen transformations, and ecosystem services. Annual Review of Plant Biology 59, 341–363. https://doi.org/10.1146/annurev.arplant.59.032607.092932.

Jakobsen, B.E., Dexter, A.R., 1988. Influence of biopores on root growth, water uptake and grain yield of wheat (*Triticum aestivum*) based on predictions from a computer model. Biology and Fertility of Soils 6, 315–321. https://doi.org/10.1007/BF00261020.

Jenne, E.A., Rhoades, H.F., Yien, C.H., Howe, O.W., 1958. Change in nutrient element accumulation by corn with depletion of soil moisture. Agronomy Journal 50, 71–74. https://doi.org/10.2134/agronj1958.00021962005000020004x.

Jiang, Y., Liang, Y., Li, C., Wang, F., Sui, Y., Suvannang, N., Zhou, J., Sun, B., 2016. Crop rotations alter bacterial and fungal diversity in paddy soils across East Asia. Soil Biology and Biochemistry 95, 250–261. https://doi.org/10.1016/j.soilbio.2016.01.007.

Kabir, Z., 2005. Tillage or no-tillage: impact on mycorrhizae. Canadian Journal of Plant Science 85, 23–29. https://doi.org/10.4141/P03-160.

Kaiser, C., Kilburn, M.R., Clode, P.L., Fuchslueger, L., Koranda, M., Cliff, J.B., Solaiman, Z.M., Murphy, D.V., 2015. Exploring the transfer of recent plant photosynthates to soil microbes: mycorrhizal pathway vs direct root exudation. New Phytologist 205, 1537–1551. https://doi.org/10.1111/nph.13138.

Kallenbach, C.M., Grandy, A., Frey, S.D., 2016. Direct evidence for microbial-derived soil organic matter formation and its ecophysiological controls. Nature Communications 7, 13630. https://doi.org/10.1038/ncomms13630.

Kallenbach, C.M., Grandy, A.S., Frey, S.D., Diefendorf, A.F., 2015. Microbial physiology and necromass regulate agricultural soil carbon accumulation. Soil Biology and Biochemistry 91, 279–290. https://doi.org/10.1016/j.soilbio.2015.09.005.

Karlen, D.L., Hurley, E.G., Andrews, S.S., Cambardella, C.A., Meek, D.W., Duffy, M.D., Mallarino, A.P., 2006. Crop rotation effects on soil quality at three northern corn/soybean belt locations. Agronomy Journal 98, 484–495. https://doi.org/10.2134/agronj2005.0098.

Kaspar, T.C., Jaynes, D.B., Parkin, T.B., Moorman, T.B., Singer, J.W., 2012. Effectiveness of oat and rye cover crops in reducing nitrate losses in drainage water. Agricultural Water Management 110, 25–33. https://doi.org/10.1016/j.agwat.2012.03.010.

Kautz, T., 2014. Research on subsoil biopores and their functions in organically managed soils: a review. Renewable Agriculture and Food Systems 30, 318–327. https://doi.org/10.1017/s1742170513000549.

Kautz, T., Amelung, W., Ewert, F., Gaiser, T., Horn, R., Jahn, R., Javaux, M., Kemna, A., Kuzyakov, Y., Munch, J.-C., Pätzold, S., Peth, S., Scherer, H.W., Schloter, M., Schneider, H., Vanderborght, J., Vetterlein, D., Walter, A., Wiesenberg, G.L.B., Köpke, U., 2013. Nutrient acquisition from arable subsoils in temperate climates: a review. Soil Biology and Biochemistry 57, 1003–1022. https://doi.org/10.1016/j.soilbio.2012.09.014.

Kirtman, B., Power, S., Adedoyin, J., Boer, G., Bojariu, R., Camilloni, I., Doblas-Reyes, F., Fiore, A., Kimoto, M., Meehl, G., Prather, M., Sarr, A., Schär, C., Sutton, R., van Oldenborgh, G., Vecchi, G., Wang, H., 2013. Near-term climate change: projections and predictability. In: Stocker, T., Qin, D., Plattner, G., Tignor, M., Allen, S., Boschung, J., Nauels, A., Xia, Y., Bex, V., Midgley, P. (Eds.), Climate Change 2013-The Physical Science Basis. Contribution of Working Group I to the Fifth Assessment Report of the Intergovernmental Panel on Climate Change. Cambridge University Press, Cambridge, UK, pp. 953–1028. https://doi.org/10.1017/CBO9781107415324.023.

Klose, S., Tabatabai, M.A., 2000. Urease activity of microbial biomass in soils as affected by cropping systems. Biology and Fertility of Soils 31, 191–199. https://doi.org/10.1007/s003740050645.

Kunelius, H.T., Johnston, H.W., Macleod, J.A., 1992. Effect of undersowing barley with Italian ryegrass or red clover on yield, crop composition and root biomass. Agriculture, Ecosystems & Environment 38, 127–137. https://doi.org/10.1016/0167-8809(92)90138-2.

Kunrath, T.R., de Berranger, C., Charrier, X., Gastal, F., de Faccio Carvalho, P.C., Lemaire, G., Emile, J.C., Durand, J.L., 2015. How much do sod-based rotations reduce nitrate leaching in a cereal cropping system? Agricultural Water Management 150, 46–56. https://doi.org/10.1016/j.agwat.2014.11.015.

Larsen, K.S., Andresen, L.C., Beier, C., Jonasson, S., Albert, K.R., Ambus, P., Arndal, M.F., Carter, M.S., Christensen, S., Holmstrup, M., Ibrom, A., Kongstad, J., Van Der Linden, L., Maraldo, K., Michelsen, A., Mikkelsen, T.N., Pilegaard, K., Priemé, A., Ro-Poulsen, H., Schmidt, I.K., Selsted, M.B., Stevnbak, K., 2011. Reduced N cycling in response to elevated CO_2, warming, and drought in a Danish heathland: synthesizing results of the CLIMAITE project after two years of treatments. Global Change Biology 17, 1884–1899. https://doi.org/10.1111/j.1365-2486.2010.02351.x.

Lassaletta, L., Billen, G., Grizzetti, B., Anglade, J., Garnier, J., 2014. 50 year trends in nitrogen use efficiency of world cropping systems: the relationship between yield and nitrogen input to cropland. Environmental Research Letters 9. https://doi.org/10.1088/1748-9326/9/10/105011.

Lehman, R.M., Taheri, W.I., Osborne, S.L., Buyer, J.S., Douds, D.D., 2012. Fall cover cropping can increase arbuscular mycorrhizae in soils supporting intensive agricultural production. Applied Soil Ecology 61, 300–304. https://doi.org/10.1016/j.apsoil.2011.11.008.

Lesk, C., Rowhani, P., Ramankutty, N., 2016. Influence of extreme weather disasters on global crop production. Nature 529, 84–87. https://doi.org/10.1038/nature16467.

Liebman, M., Gibson, L.R., Sundberg, D.N., Heggenstaller, A.H., Westerman, P.R., Chase, C.A., Hartzler, R.G., Menalled, F.D., Davis, A.S., Dixon, P.M., 2008. Agronomic and economic performance characteristics of conventional and low-external-input cropping systems in the central corn belt. Agronomy Journal 100, 600–610. https://doi.org/10.2134/agronj2007.0222.

Lilley, J.M., Kirkegaard, J.A., 2016. Farming system context drives the value of deep wheat roots in semi-arid environments. Journal of Experimental Botany 67, 3665–3681. https://doi.org/10.1093/jxb/erw093.

Lobell, D.B., Hammer, G.L., McLean, G., Messina, C., Roberts, M.J., Schlenker, W., 2013. The critical role of extreme heat for maize production in the United States. Nature Climate Change 3, 497–501. https://doi.org/10.1038/nclimate1832.

Loecke, T.D., Burgin, A.J., Riveros-Iregui, D.A., Ward, A.S., Thomas, S.A., Davis, C.A., Clair, M.A.S., 2017. Weather whiplash in agricultural regions drives deterioration of water quality. Biogeochemistry 133, 7–15. https://doi.org/10.1007/s10533-017-0315-z.

Lotter, D.W., Seidel, R., Liebhardt, W., 2003. "The Performance of Organic and Conventional Cropping Systems in an Extreme Climate Year." American Journal of Alternative Agriculture 18.3, 146–154.

Lynch, J.P., 2015. Root phenes that reduce the metabolic costs of soil exploration: opportunities for 21st century agriculture. Plant, Cell & Environment 38, 1775–1784. https://doi.org/10.1111/pce.12451.

Lynch, J.P., 2013. Steep, cheap and deep: an ideotype to optimize water and N acquisition by maize root systems. Annals of Botany 112, 347–357. https://doi.org/10.1093/aob/mcs293.

Manzoni, S., Moyano, F., Kätterer, T., Schimel, J., 2016. Modeling coupled enzymatic and solute transport controls on decomposition in drying soils. Soil Biology and Biochemistry 95, 275–287. https://doi.org/10.1016/j.soilbio.2016.01.006.

Maul, J.E., Buyer, J.S., Lehman, R.M., Culman, S., Blackwood, C.B., Roberts, D.P., Zasada, I.A., Teasdale, J.R., 2014. Microbial community structure and abundance in the rhizosphere and bulk soil of a tomato cropping system that includes cover crops. Applied Soil Ecology 77, 42–50. https://doi.org/10.1016/j.apsoil.2014.01.002.

McDaniel, M.D., Grandy, A.S., 2016. Soil microbial biomass and function are altered by 12 years of crop rotation. Soil 2, 583–599. https://doi.org/10.5194/soil-2016-39.

McDaniel, M.D., Grandy, A.S., Tiemann, L.K., Weintraub, M.N., 2014a. Crop rotation complexity regulates the decomposition of high and low quality residues. Soil Biology and Biochemistry 78, 243–254. https://doi.org/10.1016/j.soilbio.2014.07.027.

McDaniel, M.D., Tiemann, L.K., Grandy, A.S., 2014b. Does agricultural crop diversity enhance soil microbial biomass and organic matter dynamics? A meta-analysis. Ecological Applications 24, 560–570. https://doi.org/10.1890/13-0616.1.

Miller, A.J., Cramer, M.D., 2005. Root nitrogen acquisition and assimilation. Plant and Soil 274, 1–36. https://doi.org/10.1007/s11104-004-0965-1.

Mock, V.A., Creech, J.E., Ferris, V.R., Faghihi, J., Westphal, A., Santini, J.B., Johnson, W.G., 2012. Influence of winter annual weed management and crop rotation on soybean cyst nematode (*Heterodera glycines*) and winter annual weeds: years four and five. Weed Science 60, 634–640. https://doi.org/10.1614/WS-D-11-00192.1.

Mortensen, D.A., Egan, J.F., Maxwell, B.D., Ryan, M.R., Smith, R.G., 2012. Navigating a critical juncture for sustainable weed management. BioScience 62, 75–84. https://doi.org/10.1525/bio.2012.62.1.12.

Oerke, E.C., 2006. Crop losses to pests. The Journal of Agricultural Science 144, 31–43. https://doi.org/10.1017/S0021859605005708.

Oyarzun, P., Gerlagh, M., Hoogland, A.E., 1993. Relation between cropping frequency of peas and other legumes and food and root-rot in peas. Netherlands Journal of Plant Pathology 99, 35–44. https://doi.org/10.1007/BF01974783.

Pala, M., Ryan, J., Zhang, H., Singh, M., Harris, H.C., 2007. Water-use efficiency of wheat-based rotation systems in a Mediterranean environment. Agricultural Water Management 93, 136–144. https://doi.org/10.1016/j.agwat.2007.07.001.

Panagopoulos, Y., Gassman, P.W., Arritt, R.W., Herzmann, D.E., Campbell, T.D., Jha, M.K., Kling, C.L., Srinivasan, R., White, M., Arnold, J.G., 2014. Surface water quality and cropping systems sustainability under a changing climate in the Upper Mississippi River Basin. Journal of Soil and Water Conservation 69, 483–494. https://doi.org/10.2489/jswc.69.6.483.

Parry, M.L., Rosenzweig, C., Iglesias, A., Livermore, M., Fischer, G., 2004. Effects of climate change on global food production under SRES emissions and socio-economic scenarios. Globle Environment Change 14, 53–67. https://doi.org/10.1016/j.gloenvcha.2003.10.008.

Perkons, U., Kautz, T., Uteau, D., Peth, S., Geier, V., Thomas, K., Lütke Holz, K., Athmann, M., Pude, R., Köpke, U., 2014. Root-length densities of various annual crops following crops with contrasting root systems. Soil and Tillage Research 137, 50–57. https://doi.org/10.1016/j.still.2013.11.005.

Pimentel, D., Hepperly, P., Hanson, J., Douds, D., Seidel, R., 2005. Environmental, energetic, and economic comparisons of organic and conventional farming systems. BioScience 55, 573–582. https://doi.org/10.1641/0006-3568(2005)055[0573:EEAECO]2.0.CO;2.

Pittelkow, C.M., Liang, X., Linquist, B.A., van Groenigen, K.J., Lee, J., Lundy, M.E., van Gestel, N., Six, J., Venterea, R.T., van Kessel, C., 2015. Productivity limits and potentials of the principles of conservation agriculture. Nature 517, 365–368. https://doi.org/10.1038/nature13809.

Ponisio, L.C., M'Gonigle, L.K., Mace, K.C., Palomino, J., de Valpine, P., Kremen, C., 2015. Diversification practices reduce organic to conventional yield gap. Proceedings of the Royal Society B 282, 20141396. https://doi.org/10.1098/rspb.2014.1396.

Porporato, A., D'Odorico, P., Laio, F., Rodriguez-Iturbe, I., 2003. Hydrologic controls on soil carbon and nitrogen cycles. I. Modeling scheme. Advances in Water Resources 26, 45–58. https://doi.org/10.1016/S0309-1708(02)00094-5.

Postma, J.A., Dathe, A., Lynch, J.P., 2014. The optimal lateral root branching density for maize depends on nitrogen and phosphorus availability. Plant Physiology 166, 590–602. https://doi.org/10.1104/pp.113.233916.

Postma, J.A., Lynch, J.P., 2011. Root cortical aerenchyma enhances the growth of maize on soils with suboptimal availability of nitrogen, phosphorus, and potassium. Plant Physiology 156, 1190–1201. https://doi.org/10.1104/pp.111.175489.

Pribyl, D.W., 2010. A critical review of the conventional SOC to SOM conversion factor. Geoderma 156, 75–83. https://doi.org/10.1016/j.geoderma.2010.02.003.

Qi, Z., Helmers, M.J., 2010. Soil Water dynamics under winter rye cover crop in central Iowa. Vadose Zone Journal 9, 53–60. https://doi.org/10.2136/vzj2008.0163.

Qi, Z., Helmers, M.J., Kaleita, A.L., 2011. Soil water dynamics under various agricultural land covers on a subsurface drained field in north-central Iowa, USA. Agricultural Water Management 98, 665–674. https://doi.org/10.1016/j.agwat.2010.11.004.

Quemada, M., Baranski, M., Nobel-de Lange, M.N.J., Vallejo, A., Cooper, J.M., 2013. Meta-analysis of strategies to control nitrate leaching in irrigated agricultural systems and their effects on crop yield. Agriculture, Ecosystems & Environment 174, 1–10. https://doi.org/10.1016/j.agee.2013.04.018.

Quemada, M., Gabriel, J.L., 2016. Approaches for increasing nitrogen and water use efficiency simultaneously. Global Food Security 9, 29–35. https://doi.org/10.1016/j.gfs.2016.05.004.

Quinton, J.N., Govers, G., Van Oost, K., Bardgett, R.D., 2010. The impact of agricultural soil erosion on biogeochemical cycling. Nature Geoscience 3, 311–314. https://doi.org/10.1038/ngeo838.

Rajcan, I., Chandler, K.J., Swanton, C.J., 2004. Red-far-red ratio of reflected light: a hypothesis of why early-season weed control is important in corn. Weed Science 52, 774–778. https://doi.org/10.1614/WS-03-158R.

Rawls, W.J., Pachepsky, Y.A., Ritchie, J.C., Sobecki, T.M., Bloodworth, H., 2003. Effect of soil organic carbon on soil water retention. Geoderma 116, 61–76. https://doi.org/10.1016/S0016-7061(03)00094-6.

Reese, C.L., Clay, D.E., Clay, S.A., Bich, A.D., Kennedy, A.C., Hansen, S.A., Moriles, J., 2014. Winter cover crops impact on corn production in semiarid regions. Agronomy Journal 106, 1479–1488. https://doi.org/10.2134/agronj13.0540.

Robertson, G.P., Bruulsema, T.W., Gehl, R.J., Kanter, D., Mauzerall, D.L., Rotz, C.A., Williams, C.O., 2013. Nitrogen–climate interactions in US agriculture. Biogeochemistry 114, 41–70. https://doi.org/10.1007/s10533-012-9802-4.

Sadras, V.O., 2005. A quantitative top-down view of interactions between stresses: theory and analysis of nitrogen-water co-limitation in Mediterranean agro-ecosystems. Australian Journal of Agricultural Research 56, 1151–1157. https://doi.org/10.1071/AR05073.

Saengwilai, P., Tian, X., Lynch, J.P., 2014. Low crown root number enhances nitrogen acquisition from low-nitrogen soils in maize. Plant Physiology 166, 581–589. https://doi.org/10.1104/pp.113.232603.

Sainju, U.M., 2016. A global meta-analysis on the impact of management practices on net global warming potential and greenhouse gas intensity from cropland soils. PLoS One 11, e0148527. https://doi.org/10.1371/journal.pone.0148527.

Santonja, M., Fernandez, C., Proffit, M., Gers, C., Gauquelin, T., Reiter, I.M., Cramer, W., Baldy, V., 2017. Plant litter mixture partly mitigates the negative effects of extended drought on soil biota and litter decomposition in a Mediterranean oak forest. Journal of Ecology 105, 801–815. https://doi.org/10.1111/1365-2745.12711.

Schimel, J.P., Bennett, J., 2004. Nitrogen mineralization: challenges of a changing paradigm. Ecology 85, 591–602. https://doi.org/10.1890/03-8002.

Shaner, D.L., 2014. Lessons learned from the history of herbicide resistance. Weed Science 62, 427–431. https://doi.org/10.1614/WS-D-13-00109.1.

Silver, W.L., Miya, R.K., 2001. Global patterns in root decomposition: comparisons of climate and litter quality effects. Oecologia 129, 407–419. https://doi.org/10.1007/s004420100740.

Sinclair, T.R., 2012. Is transpiration efficiency a viable plant trait in breeding for crop improvement? Functional Plant Biology 39, 359–365. https://doi.org/10.1071/FP11198.

Sinclair, T.R., Rufty, T.W., 2012. Nitrogen and water resources commonly limit crop yield increases, not necessarily plant genetics. Global Food Security 1, 94–98. https://doi.org/10.1016/j.gfs.2012.07.001.

Stark, J.M., Firestone, M.K., 1995. Mechanisms for soil moisture effects on activity of nitrifying bacteria. Applied and Environmental Microbiology 61, 218–221.

Sutton, M., Howard, C., Erisman, J., 2011. The European Nitrogen Assessment: Sources, Effects and Policy Perspectives. Cambridge Univ. Press 612. https://doi.org/10.1017/CBO9780511976988.

Suzuki, C., Takenaka, M., Oka, N., Nagaoka, K., Karasawa, T., 2012. A DGGE analysis shows that crop rotation systems influence the bacterial and fungal communities in soils. Soil Science & Plant Nutrition 58, 288–296. https://doi.org/10.1080/00380768.2012.694119.

Swanton, C.J., Mahoney, K.J., Chandler, K., Gulden, R.H., 2008. Integrated weed management: knowledge-based weed management systems. Weed Science 56, 168–172. https://doi.org/10.1614/WS-07-126.1.

Syswerda, S.P., Basso, B., Hamilton, S.K., Tausig, J.B., Robertson, G.P., 2012. Long-term nitrate loss along an agricultural intensity gradient in the Upper Midwest USA. Agriculture, Ecosystems & Environment 149, 10–19. https://doi.org/10.1016/j.agee.2011.12.007.

Syswerda, S.P.P., Robertson, G.P.P., 2014. Ecosystem services along a management gradient in Michigan (USA) cropping systems. Agriculture, Ecosystems & Environment 189, 28–35. https://doi.org/10.1016/j.agee.2014.03.006.

Tiemann, L.K., Grandy, A.S., Atkinson, E.E., Marin-Spiotta, E., McDaniel, M.D., 2015. Crop rotational diversity enhances belowground communities and functions in an agroecosystem. Ecology Letters 18, 761–771. https://doi.org/10.1111/ele.12453.

Tobar, R.M., Azcon, R., Barea, J.M., 1994. The improvement of plant N Acquisition from an ammonium-treated, drought-stressed soil by the fungal symbiont in arbuscular mycorrhizae. Mycorrhiza 4, 105–108.

Unger, P.W., Vigil, M.F., 1998. Cover crop effects on soil water relationships. Journal of Soil and Water Conservation 53, 200–207.

Vanheemst, H.D.J., 1985. The influence of weed competition on crop yield. Agricultural Systems 18, 81–93. https://doi.org/10.1016/0308-521X(85)90047-2.

Vanotti, M.B., Bundy, L.G., 1994. Corn nitrogen recommendations based on yield response data. Journal of Production Agriculture 7, 249–256. https://doi.org/10.2134/jpa1994.0249.

West, T.O., Post, W.M., 2002. Soil organic carbon sequestration rates by tillage and crop rotation. Soil Science Society of America Journal 66, 1930–1946. https://doi.org/10.2136/sssaj2002.1930.

Wienhold, B.J., Pikul, J.L., Liebig, M.A., Mikha, M.M., Varvel, G.E., Doran, J.W., Andrews, S.S., 2006. Cropping system effects on soil quality in the great plains: synthesis from a regional project. Renewable Agriculture and Food Systems 21, 49–59. https://doi.org/10.1079/RAF2005125.

Williams, A., Hunter, M.C., Kammerer, M., Kane, D.A., Jordan, N.R., Mortensen, D.A., Smith, R.G., Snapp, S., Davis, A.S., 2016. Soil water holding capacity mitigates downside risk and volatility in US rainfed maize: time to invest in soil organic matter? PLoS One 11, e0160974. https://doi.org/10.1371/journal.pone.0160974.

Woodard, J.D., Verteramo-Chiu, L.J., 2017. Efficiency impacts of utilizing soil data in the pricing of the federal crop insurance program. American Journal of Agricultural Economics 757–772. https://doi.org/10.1093/ajae/aaw099.

Yang, F., Zhang, G.L., Yang, J.L., Li, D.C., Zhao, Y.G., Liu, F., Yang, R.M., Yang, F., 2014. Organic matter controls of soil water retention in an alpine grassland and its significance for hydrological processes. Journal of Hydrology 519, 3010–3027. https://doi.org/10.1016/j.jhydrol.2014.10.054.

Yeo, I.Y., Lee, S., Sadeghi, A.M., Beeson, P.C., Hively, W.D., McCarty, G.W., Lang, M.W., 2014. Assessing winter cover crop nutrient uptake efficiency using a water quality simulation model. Hydrology and Earth System Sciences 18, 5239–5253. https://doi.org/10.5194/hess-18-5239-2014.

Zhang, D., Yao, P., Na, Z., Cao, W., Zhang, S., Li, Y., Gao, Y., 2016. Soil water balance and water use efficiency of dryland wheat in different precipitation years in response to a green manure approach. Scientific Reports 6, 26856. https://doi.org/10.1038/srep26856.

Zhang, W.L., Tian, Z.X., Zhang, N., Li, X.Q., 1996. Nitrate pollution of groundwater in northern China. Agriculture, Ecosystems & Environment 59, 223–231. https://doi.org/10.1016/0167-8809(96)01052-3.

Zhu, B., Gutknecht, J.L.M., Herman, D.J., Keck, D.C., Firestone, M.K., Cheng, W., 2014. Rhizosphere priming effects on soil carbon and nitrogen mineralization. Soil Biology and Biochemistry 76, 183–192. https://doi.org/10.1016/j.soilbio.2014.04.033.

Zhu, X., Burger, M., Doane, T.A., Horwath, W.R., 2013. Ammonia oxidation pathways and nitrifier denitrification are significant sources of N_2O and NO under low oxygen availability. Proceedings of the National Academy of Sciences of the United States of America 110, 6328–6333. https://doi.org/10.1073/pnas.1219993110.

CHAPTER

5

Negative Impacts on the Environment and People From Simplification of Crop and Livestock Production*

Scott L. Kronberg[1], *Julie Ryschawy*[2]

[1]USDA – Agricultural Research Service, Northern Great Plains Research Laboratory, Mandan, ND, United States; [2]Université de Toulouse, AGIR UMR 1248, INRA, INPT-ENSAT, Auzeville, France

INTRODUCTION

Although it goes against selfish and short-sighted aspects of human nature, it would be better for communities of people and their environments if current socioeconomic systems had excellent human and environmental health as high priorities (see Ikerd, 2005). With this reprioritization, we could see consistent improvement in the quality and adequacy of food produced and consumed by people, reduced environmental impact of agriculture that leads eventually to more resilient and sustainable food production, and improved quality of life for all people including those working in agriculture. It is probably not realistic to expect all of these things to improve in a linear manner, but unfortunately, we appear to be in a multidecade period of generally going backward in respect to the environmental impacts of agriculture (Foley et al., 2011) and perhaps also in respect to dietary nutritional quality and human well-being (Tilman and Clark, 2014; Springmann et al., 2016), although this is more complicated in that people may consume enough calories and protein but not enough micronutrients such as iron, potassium, zinc, vitamins B-12 and D nor enough healthful phytochemicals and polyunsaturated fatty acids (Spencer et al., 2003; Key et al., 2006; Craig, 2009; Provenza et al., 2015). We are probably also going backward in respect to the percentage of people with access

* The United States Department of Agriculture (USDA) prohibits discrimination in all its programs and activities on the basis of race, color, national origin, age, disability, and where applicable, sex, marital status, family status, parental status, religion, sexual orientation, genetic information, political beliefs, reprisal, or because all or part of an individual's income is derived from any public assistance program. (Not all prohibited bases apply to all programs.) USDA is an equal opportunity provider and employer.

Agroecosystem Diversity
https://doi.org/10.1016/B978-0-12-811050-8.00005-4

to high-quality food (Andrieu et al., 2006; Drewnowski, 2009, 2010) and/or with land, materials, and knowledge to produce high-quality foodstuffs and fibers (Kraskinikov and Tabor, 2003; Vogl et al., 2005; Rigg, 2006; Burgess and Morris, 2009; Gómez-Baggethun et al., 2010; Robertson and Pinstrup-Andersen, 2010; Sklenicka, 2016; Lasanta et al., 2017). This is a problematic time to be going backward in respect to any of these factors given the challenge of sustainably nourishing many billions of people on the Earth now, with more to come.

Mixed crop-livestock farms were common agricultural systems in Western Europe and North America until the middle of the 20th century (Mazoyer and Roudart, 2006). The smaller scale and limited technology available to farmers to the middle of the 20th century may have reduced their negative impact on the environment (per units of foods produced) through stronger linkages between crop and livestock production. However, agricultural impacts on the environment can be local as well as global. Negative environmental impacts from agricultural production were significant in some areas in the past, while the global impact of agriculture was reduced with less food and livestock feed produced for a smaller global human population in preindustrial times. Examples of past localized impacts include massive soil degradation for crop and livestock production in the Middle East, Greece, Italy, and North Africa by ancient cultures (Montgomery, 2007). Elimination of millions of bison as well as wild sheep, wapiti, wolves, brown bears, and other wildlife on the Great Plains of North America in the 1800s followed by cattle, sheep, and soil-degrading annual crop production, and elimination of millions of wild salmon in rivers of northwestern North America in the 1900s due to construction of dams, in part, for crop irrigation are more recent examples of relatively localized environmental degradation associated with agriculture (Samson and Knopf, 1994; Waples et al., 2007). In more recent times, with agricultural activities increased to support billions of additional people, there are negative environmental impacts on a global scale associated with food production for all these people plus additional negative environmental impacts associated with more widespread specialization of contemporary agriculture, for example, pollution of fresh and salt waters in Europe, Japan, New Zealand, United States (USA), and Vietnam from intensive livestock grazing or confined feeding operations (or shrimp production; Novak et al., 2000; Mallin and Cahoon, 2003; Basset-Mens and van der Werf, 2005; Monaghan et al., 2008; Kato et al., 2009; Thi Anh et al., 2010; Zonderland-Thomassen et al., 2014). Consequently, Tilman et al. (2002) and Foley et al. (2005) argued that the current state of land use around the world is a serious threat to many ecosystem services that people depend on for good quality of life.

Beginning in the 1960s, farmers began to simplify, specialize, and enlarge their farms to capture economies of scale and increase production levels. This was driven by the Green Revolution of farming with increased use of synthetic fertilizers, herbicides, pesticides, and hybrid seeds for more uniform, higher yielding crops. General agricultural policies were interlinked with increasing market demand for food industry standards and low-cost inputs on larger farms through economies of scale (Steiner and Franzluebbers, 2009). In the European Union (EU), market globalization of food and agricultural inputs and output-based subsidies for cash crops with the European Common Agriculture Policy (CAP) favored segregation and specialization of crops or livestock at farm and regional levels. These combined with other driving forces have encouraged specialization, enlargement, and intensification of farms, which have been major causes of environmental problems for the last several decades (Stoate et al., 2009; Plieninger et al., 2016). More recently, regional rather than individual farm specialization has been identified as a major cause for environmental degradation, with simplification across entire agricultural landscapes negatively impacting biodiversity and water regulation (Gaigné et al., 2012; Peyraud et al., 2014). Crop specialization occurs in the most favorable soil-climatic conditions, and high-input livestock production is concentrated in the remaining regions in Europe that have greater access to shipping for trading inputs and products on the global market. Permanent grasslands are now found mostly in less favorable conditions such as mountainous or very wet or dry areas. The situation is similar in North America. However, overlapping different lands uses could provide a heterogeneous landscape, which would be favorable to multiple ecosystem services (Foley et al., 2005; Lemaire et al., 2014; Peyraud et al., 2014; Sabatier et al., 2015). Despite creation of a rural-environmental second pillar of the CAP, other drivers for simplification of farming systems and focused regions have remained in the EU, with associated environmental problems. In the USA, conservation programs have been part of national farm legislation for decades including the most recent farm bill, but recent conservation subsidy programs have problems with design and implementation that reduce their effectiveness (Lichtenberg and Smith-Ramirez, 2011; Lichtenberg, 2014). However, awareness is increasing about the many negative impacts of specialization of farms and farming regions on water and air quality, soil fertility, and invertebrate animals, as well as on wild birds and mammals, livestock, and people (Tilman et al., 2002; Stoate et al., 2009). After discussing the negative aspects of contemporary agriculture impacts on the environment, we will discuss its significant negative impacts on human well-being.

NEGATIVE IMPACTS LINKED TO THE CONCENTRATION OF PRODUCTION

Others have argued for intensified specialization and consequently land sparing (Green et al., 2005; Borlaug, 2007) and their ideas certainly require further consideration and evaluation because intensified specialization will likely expand in parts of the world and because intensification and land sharing approaches will probably occur simultaneously to increase food production and conserve wild species (Green et al., 2005). However, Tscharntke et al. (2012) argued that the land sparing argument does not always account for real-world complexity and this approach can use livestock feed and human food inefficiently. Desquilbet et al. (2017) argued that while intensive farming tends to lead to lower food prices, which is good in some respects but may encourage waste, intensive farmers have incentives to encroach on land zoned (actively spared) for conservation of biodiversity-rich nature.

Large-scale specialized production of only one or two crops or species of livestock is associated with poor recycling of organic matter and various nutrients from livestock back to the locations where livestock feedstuffs were produced (Tilman et al., 2002; Vitousek et al., 2009), as well as harmful effects of pesticides on biodiversity and biological control potential (Geiger et al., 2010; Meehan et al., 2016). Depletion of organic matter is widespread in cropland soils worldwide, especially for farms that do not add organic matter to their soils via animal manure, cover crops, or green "manure" crops (Liebig and Doran, 1999). Concentrating livestock in pastures, outdoor feedlots or indoor facilities can create significant concentrations of pathogens and airborne nutrients such as ammonia, and deposition of their excrement on nearby fields can result in waterborne nutrients such as nitrate and various phosphorus compounds, which if not managed carefully can escape to the general environment and cause significant problems (Monaghan et al., 2008; Peyraud et al., 2014; Thorne, 2007). This type of pollution can be difficult and costly to clean up and it can be difficult to identify the extent of pollution by individual producers. Farmers and specialized livestock producers have historically not had to pay the costs of their pollution, which economists consider to be externalized costs (Osterberg and Wallinga, 2004). There are currently some regulations and incentives for practices to reduce pollution, which helps to internalize these costs. It is an interesting question of whether specialized food production would be less profitable, more unprofitable, or perhaps eventually more profitable (for producers that survive), if farmers were required to prevent negative environmental impacts (e.g. air and water pollution and soil degradation), but they were also compensated by the market or by government payments for positive environmental impacts they provided such as maintaining or enhancing wildlife habitat or improving soils. But, in general these costs have been externalized to the environment and society. (Lubowski et al., 2006; Sumner et al., 2009; Dumont et al., 2012; Walters et al., 2012; Sumner, 2014; Summer and Zulauf, 2012).

NEGATIVE IMPACTS LINKED TO SIMPLIFICATION OF LAND USE

Creating new cropland for production of annual crops such as wheat, maize, and soybeans by destroying natural grassland or savanna (rangelands) with diverse mixtures of plant and animal species has been occurring for thousands of years and, unfortunately, continues in many parts of the world (Hoekstra et al., 2005). This conversion results in the loss of soil organic matter (stored carbon) if natural grasslands and rangelands are tilled to grow annual crops (Davidson and Ackerman, 1993) and loss of biodiversity associated with these land types (McCracken, 2005; Fahrig et al., 2011; Teillard et al., 2015a,b). Carefully managed grassland-based livestock production (hay and grazing) is the only long-term type of agriculture that has been shown to be possible without degrading soil fertility while providing habitat for many species of wild plants and animals (Brejda et al., 2000; Tilman et al., 2002; Wienhold et al., 2004; Farigone et al., 2009; Culman et al., 2010; Wright and Wimberly, 2013). Peters et al. (2016) concluded that under a range of ways to utilize all agricultural lands in the United States (grazing land, cultivated cropland, and perennial cropland), diets with low to modest amounts of meat provided estimates of greater human carrying capacity (food provisioning) than did a vegan diet, with an ovo-lacto vegetarian diet providing the second highest carrying capacity, and a vegetarian diet including dairy products providing the greatest human carrying capacity. However, intensification of grassland-based agriculture can also negatively impact wildlife (Lemaire, 2012; Teillard et al., 2015a,b). Food produced from grazing is currently mostly red meat or milk from cattle, sheep, or goats, and there has been pressure from human health specialists, environmentalists, and animal welfare groups to decrease or eliminate red meat and dairy foods from the human diet (Peters et al., 2016). Compounding this situation is the fact that the most efficient herbivorous livestock (rabbits and geese) are not popular foods in many regions of the world. Unfortunately, much less research has been conducted to decrease the cost of these meat animals and increase their integration with grazing and croplands.

Air quality is negatively impacted when various air pollutants, which cannot be effectively recycled and/or metabolized after returning to land surfaces by wet or dry deposition, are produced by agricultural operations. These include but are not limited to dust (particulate matter), hydrogen sulfide, organic acids, nitrogen oxides, ammonia, and greenhouse gases carbon dioxide, methane, and nitrous oxide that are often emitted in the production of livestock confined in confined animal feeding operations (CAFOs; methane and nitrous oxide emissions are also associated with grazing livestock). These emissions include those from farmland and machinery to provide feed, the animals, their excrement and the machinery used to feed, haul and process them and clean up their excrement (Aneja et al., 2009). Crop production for direct human consumption can also be an important source of ammonia and greenhouses gases such as nitrous oxide from various fertilized crops and methane from rice production (Aneja et al., 2009).

Water quality is negatively affected by wet and dry deposition of airborne compounds described earlier and by surface and groundwater contamination of water from both animal and crop production including pesticides (Gillom et al., 2007). These contaminants are discussed in more detail later in this chapter because they have negative implications for human and animal health. Much has been written about agricultural nonpoint pollution of water with compounds containing phosphorus and nitrogen (e.g., Sharpley et al., 1994; Di and Cameron, 2002). Many papers and reports have discussed in detail serious agriculture-related water quality problems such as the large dead hypoxia zone in the Gulf of Mexico near the mouth of the Mississippi River Basin in the USA (Alexander et al., 2008; Mississippi River Gulf of Mexico Watershed Nutrient Task Force, 2015), polluted waterways and estuaries from CAFO lagoon spills (Mallin and Cahoon, 2003; Burkholder et al., 2007), impacts of pesticide toxins and nitrates in the Salinas River (from production of lettuce, strawberries, artichokes, crucifers, and other crops), which drains into the Pacific Ocean in California (Anderson et al., 2003; Harter et al., 2012), and effects of nitrates in fertilizes and from livestock excreta on the green slimy beaches of France, Belgium, and the Netherlands (Peyraud et al., 2014; Gaigné et al., 2012).

NEED FOR IMPROVED DIVERSIFIED SYSTEMS

Regarding negative impacts of agriculture on soil fertility, it is important to distinguish between the negative effects on soil fertility that have resulted from simplification of crop and livestock production and those that occurred either when mixed crop-livestock production was common but less sophisticated (e.g., western Europe post World War II to boost food security; Vitousek et al., 2009) or currently where soil fertility is being degraded while practicing mixed crop-livestock production (e.g., cattle grazing of maize residue for multiple years has increased runoff losses of sediment, carbon, and nutrients; Blanco-Canqui et al., 2016). For example, a dairy farm could be raising all feed for their cattle and returning manure from their cattle back to their crop fields, but they could be using crop and/or livestock production practices that allow non-sustainable levels of soil erosion and concentration of macro and micro nutrients from livestock urine into pen soils where they sit non-recycled. The lack of judicious management of macro and micro nutrients is one of the two biggest problems of simplified contemporary agriculture (Jones et al., 2013), with the other being poor management of soil organic matter. These two problems are interrelated, especially in respect to integration or lack thereof of ruminant livestock with crop production, because when cattle, sheep, and goats are consuming moderate to high amounts of roughage, they excrete considerable amounts of organic matter as well as macro and micro nutrients (Bannink et al., 1999; Whalen et al., 2000; Aoyama and Kumakura, 2001). To have sustained (at least hundreds of years) production of food for people and livestock that contains adequate amounts of essential macro and micro nutrients, it may become essential that nutrients in livestock and human excreta are thoughtfully (timely) and carefully (not excessively) recycled back to cropland, and the excreta needs to be free of contaminants such as antibiotics, hormones, flame retardants, pharmaceuticals or their metabolites, and personal care and cleaning products, unless these compounds can be quickly metabolized by soil microbes.

Finally, there may be more sophisticated and ecologically sustainable ways to manage undesirable plants on crop and grazing lands that allow for reduced use of herbicides (Bonaudo et al., 2014; Liebman et al., 2016). Plus, herbicide-resistant weeds are a growing problem in crop fields (Powles and Yu, 2010), and herbicides are the most common way to control crop weeds, especially since the wide adoption of genetically engineered glyphosate-tolerant crops in many countries (Duke and Powles, 2008). However, herbicide-resistant weeds can be a problem in mixed crop-livestock farming operations, as well as crop-only farms, unless they are using some form of tillage for weed control, which can also cause ecological problems. Likewise, use of insecticide treatment of crop seeds has greatly increased the use of neonicotinoid insecticides in the USA (Douglas and Tooker, 2015), and this and/or other forms of application of neonicotinoids are likely harming desirable terrestrial and aquatic insects as well as birds and other wildlife that consume these insects (Douglas et al., 2014; Hallmann et al., 2014; Gibbons et al., 2015). These seed treatments can currently be used on mixed crop-livestock farms in the United States and other countries as long as the farms are not organic (biologique or BIO) producers. Mixed crop-livestock producers may provide more habitat for wildlife if they have pastures for livestock grazing, but if the pastures are composed of simple mixtures of introduced grasses, their value as wildlife habitat will be limited. The point is that just simple coexistence between crops and livestock is not a panacea for reducing environmental impacts and improving economic performance (Moraine et al., 2014). What is needed is not only diverse, but carefully managed and productive mixed crop-livestock farms with multiple species of annual and perennial crops (including tree crops) and multiple species of livestock interacting in synergistic ways similar to how natural grassland and savanna ecosystems functioned in the past before most were replaced with annual crops. This is asking a lot given the current state of agriculture in much of the world or at least the industrially developed regions. But, given the pervasive influence that people are now having on our planet, we need to be much more sustainable at integrating people and their needs into agroecosystems, and this includes recycling nutrients in our excreta safely and effectively back to land rather than contributing to significant water pollution problems. We also need to be mindful of the serious challenges associated with concentrating ourselves in high-density living arrangements just as we are doing with our livestock.

NEGATIVE IMPACTS ON PEOPLE FROM SIMPLIFICATION OF CROP AND LIVESTOCK PRODUCTION

We think of the environment as the air, land, water, sunlight, and other living and nonliving things around us, but we often exclude other people from our environment unless perhaps they have a contagious disease. Additionally, there is our internal environment, and as first proposed by the French physiologist Claude Bernard, maintenance of a constant and desirable internal environment is necessary for a person to survive and prosper in a varying external environment. Our external and internal environments interact a great deal, and it is difficult to separate disturbance and damage to one and not the other. There is a considerable amount written about the negative aspects of large-scale simplified, nonintegrated crop and livestock production on our external environment, but less about the negative aspects of this form of food production on our internal environment. Large-scale simplification of crop and livestock production can have negative effects on people, and this has significant ramifications for rural as well as urban communities and nation states.

Individuals that work in large CAFOs for multiple hours per day are frequently exposed to a variety of potentially unhealthful conditions including excessive noise (Sieben, 1997) from loud machinery and/or from many excited animals wanting feed. These workers can also be excessively exposed to air pollutants such as ammonia, hydrogen sulfide, methane, a variety of microbes or microbial components such as fungal spores and endotoxins, animal dander, and fecal and/or feed dust, and this exposure is associated with a higher prevalence of occupational respiratory problems (Whyte, 2002; Charavaryamath and Singh, 2006; Mitloehner and Calvo, 2008).

Globally, about 700,000 people per year die from infections by antibiotic-resistant "superbugs," and in the USA alone, more than 2 million people per year are sickened by drug-resistant bacteria (Review of Antimicrobial Resistance, 2016). Although certainly not the only contributor to the increasing incidence of antibiotic-resistant bacteria, there is evidence for a direct link between antibiotic use with livestock and the spread of antibiotic-resistant bacteria to people, especially in respect to enteric microorganisms (Levy and Marshall, 2004; Marshall and Levy, 2011; Spellberg et al., 2016). The EU banned the use of subtherapeutic amounts of antibiotics for growth promotion. Although this practice has not been banned in the USA, it is now under more scrutiny. A recent final recommendations report from the Review on Antimicrobial Resistance (2016), commissioned by the United Kingdom (UK) Government and the Wellcome Trust, estimates that by 2050, 10 million lives per year and a 100 trillion US dollars-worth of economic output are at risk due to the rise of drug-resistant infections if significant solutions cannot be found to greatly slow the increase in antibiotic resistance. Van Boeckel et al. (2015) estimated the global average (for high-income countries) annual consumption of antimicrobials per kg of animal produced was 45, 148, and 172 mg/kg for cattle, chicken, and pigs, respectively. Van Boeckel et al. (2015) also estimated that the global consumption of antimicrobials could increase 67% from about 63,000 to 106,000 tons between 2010 and 2030, respectively, with a third of the increase due to shifting production practices in middle-income countries. In these countries, extensive farming production will be replaced by large-scale intensive production, which routinely use antimicrobials for disease prevention and growth promotion. With as much as 75% of the antibiotics given to livestock excreted in their urine and feces (Chee-Sanford et al., 2001), the compounds can be transported out from the confined animal feeding facility when animal manure is applied to crop fields, via runoff in water, or via airborne particulate matter that blows away from feeding facilities in windy weather (Chapin et al., 2005; McEachran et al., 2015). Once the antibiotics are out in the environment, they can facilitate the development of additional resistant bacteria that can be picked up by livestock or people.

Emergence of pathogenic fungi that are resistant to anti-fungal drugs is a new and very serious concern for human health and food production (Fisher et al., 2018). This resistance is particularly worrisome because a limited number of anti-fungal compounds are used to treat people, animals and crops as well as for other purposes such as wood preservation.

Besides antibiotics and anti-fungal drugs, livestock can be treated with exogenous steroid hormones to promote growth. Endogenous hormones of animal origin have been in the environment for thousands of years but have increased as livestock populations have grown (Lange et al., 2002). Recently the hormonal disrupting activities of endogenous hormones, exogenous steroid growth promoters, and other endocrine-disrupting chemicals in the environment have been of great concern for people and wildlife. There is evidence that exogenous steroid growth promoters given to cattle as well as the endogenous hormones they excrete can be transported away from CAFOs via effluent and surface runoff (Gadd et al., 2010; Bartelt-Hunt et al., 2012), and exogenous steroid growth promoters have been found in particulate matter in cattle feed yards in arid and semiarid parts of the USA where this particulate matter is transported by wind (Blackwell et al., 2015). In the United Kingdom, where the combined farm animal population in much larger than the human population, livestock generate about four times more estrogens that do the people (Johnson et al., 2006). Excretion of steroid hormones by livestock into the environment needs to be managed carefully to keep them out of freshwater. Similar to nonagricultural influences on antibiotic resistance, there are many endocrine-disrupting compounds that do not originate from agriculture. However, specialists in human growth and reproduction have argued that estradiol residues in meat from livestock treated with exogenous steroid hormones may be less safe for prepubertal children, who are extremely sensitive to estradiol (Andersson and Skakkebæk, 1999; Aksglaede et al., 2006). These reproductive health specialists have argued that great caution should be taken to avoid exposure of fetuses and children to exogenous sex steroids and endocrine disruptors even at very low levels.

The transmission of the viral diseases avian and swine influenza is another potential negative impact on people, especially if one or both became pandemic influenza. These diseases can arise from small mixed crop and livestock farms, but CAFOs are considered of greater threat for transferring them to people because of large numbers of animals in close proximity and consequent rapid transmission and mixing of viruses, especially if poultry and swine facilities are close together (Gilchrist et al., 2007). People who work closely with poultry and pigs are considered to have especially high risk of contracting as well as increasing the dangerous evolution of the viruses (Meyers et al., 2006; Meyers et al., 2007a,b.

Intensive production of maize and milk have resulted in nitrate leaching into ground and surface waters in many farming areas of the world (Schepers et al., 1991; Clark et al., 2007; Cui et al., 2008). Intake of too much nitrate in water by children and adults is linked with more serious health problems than just the one most well-known: methemoglobinemia in infants (blue baby syndrome [Ward et al., 2005]; and children as well [Sadeq et al., 2008]). The additional health problems that are linked with high levels of nitrate in drinking water include various cancers (Ward et al., 2005; Yang et al., 2007) and thyroid dysfunction in children and pregnant women (van Maanen et al., 1994; Gatseva and Argirova, 2008).

People are also affected by simple, nondiverse food production systems through their daily food, nutrient, fiber, and phytochemical intake or lack thereof (Ludwig, 2011). Simple, nondiverse food production systems are more vulnerable to plant diseases, insect infestation, drought, and other undesirable weather, which can lead to inadequate availability and consumption of food by large numbers of people, perhaps especially in less developed areas of the world (Rosenzweig et al., 2001; Reidsma and Ewert, 2008). This can lead to civil unrest and emigration, and there is some evidence that this is one driver of conflicts in the Middle East and Africa (Couttenier and Soubeyran, 2013; Hammer, 2013).

In addition to these serious concerns, simplified food production systems provide commonly consumed foods that contain abundant amounts of calories and/or protein but inadequate amounts of micronutrients (Frazao, 1996; Welch and Graham, 1999; Welch, 2002) and phytochemicals (Mozaffarian and Ludwig, 2010; Ludwig, 2011; Provenza et al., 2015). Because contemporary agriculture has focused primarily on increasing the quantity, appearance, and transportability of their farm products rather than their quality, flavor, and overall healthfulness, micronutrient deficiencies are a major public health problem in many countries and considered a major contributor to the global disease burden (Knez and Graham, 2013). The valuable phytochemical richness and flavor of many fruits, vegetables, and grains has declined substantially over the past 40 years (Provenza et al., 2015). In developing countries, inadequate micronutrient intake by children and adults is associated with many basic health problems with devastating consequences including death of many children (Knez and Graham, 2013). Ames (2010) has argued that even modest micronutrient deficiencies, which occur in many if not most of the world's people in developed and developing countries, accelerate molecular aging and age-associated diseases such as cancer. Ames (2010) hypothesized that when ingestion of a micronutrient is inadequate to insure our short-term survival, our body adjusts its metabolism to deal with this inadequacy, but this adjustment is at the expense of more ideal metabolism. The ideal metabolism can have long-term benefits such as reducing our risk for cancer, immune dysfunction, cognitive decline, cardiovascular disease, and stroke. Regardless, there is evidence that maintaining a healthy body weight and consuming a diet that is primarily plant-based, low in red and especially processed meats, low in simple sugars and refined carbohydrates, with limited alcohol intake and food-focused nutrient intake (rather than supplement pills) can help prevent cancer (Bail et al., 2016; Mayne et al., 2016). Unfortunately, all fresh red meat is lumped together in World Cancer Research Fund (2007) recommendations without differentiation by how the livestock were raised. However, the method of livestock production may make a difference in red meat healthfulness (Provenza et al., 2015; Kronberg et al., 2017).

In the USA, childhood and adult obesity rates are about 17% and 38%, respectively, and diabetes rates have nearly doubled in the past 20 years from 5.5 in 1994 to 9.3% in 2012. Twenty-nine million Americans have diabetes and 86 million are prediabetic. At current rates, one-third of Americans are predicted to have diabetes by 2050, and this disease already accounts for $245 billion in medical costs and lost productivity each year in the USA. One-third of American adults have high blood pressure, which is a leading cause of stroke and also appears to be associated with cognitive decline and Alzheimer's disease (Goldstein et al., 2013) which, is also a huge and rapidly growing problem affecting millions of Americans. A simplified, industrialized food system and government policies are probably major drivers of these health problems (Ludwig, 2011; Mozaffarian and Ludwig, 2010).

Additionally, stress-related health problems are associated with simplified food production. As pointed out in the US's National Research Council's publication "Toward Sustainable Agricultural Systems in the 21st Century" (NRC, 2010), conventional economics suggests that specialized production systems have strong economic rewards. The economies of scale of the specialized systems are encouraged by new technologies, lower capital requirements, and greater labor efficiencies; however, expensive new technologies and frequent overproduction of specific crops or livestock can severely stress farmers and farm families when more money is spent producing only one, two, or a few products than received for their sale. This can cause a variety of stress-related health problems and even suicides by farmers. This situation can be especially acute if a period of unprofitable prices follows a period of very profitable prices with some farmers borrowing considerable amounts of money during the profitable times to buy expensive equipment and/or land and then having difficulty paying off their loans during the unprofitable times (Gregoire, 2002; Fraser et al., 2005). Interestingly, inadequate intake of some foods may make the brain less resistant to aging and stress (Denis et al., 2013; Parletta et al., 2013), and simplified food production and consumption may be linked in some people to mental health problems.

In summary, contemporary, simplified, intensive agriculture may have contributed to many negative impacts on our environment and us. Specialization of farms and farming regions have increased many environmental problems including greater reductions in air and water quality, less availability of water, more soil degradation, and more biodiversity losses. Negative effects on our environment have probably been exacerbated by complex interactions among specialization, intensification and enlargement of farms driven by industrialization and globalization of markets and public policies such as CAP in the EU and publicly-subsidized farm programs in the USA. This approach to farming has also created concerns about food quality and human health. Potential negative effects on people include unhealthful working conditions in concentrated animal feeding operations, excessive exposure to pesticides, supporting development of antibiotic and anti-fungal-resistant microbes, adding to the amount of steroid hormones that children and adults are exposed to by environmental contamination with exogenous steroid growth promoters given to livestock, increasing the risk of pandemic avian and swine influenza, increasing exposure to excessive levels of nitrates in drinking water, reducing micronutrient availability in foods and in some cases reducing availability of affordable healthful food, and increasing stress on farming families.

References

Aksglaede, L., Juul, A., Leffers, H., Skakkebæk, N.E., Andersson, A.-M., 2006. The sensitivity of the child to sex steroids: possible impact of exogenous estrogens. Human Reproduction Update 12, 341–349.

Alexander, R.B., Smith, R.A., Schwarz, G.E., Boyer, E.W., Nolan, J.V., Brakebill, J.W., 2008. Differences in phosphorus and nitrogen delivery to the Gulf of Mexico from the Mississippi River Basin. Environmental Science & Technology 42, 822–830.

Ames, B.N., 2010. Optimal micronutrients delay mitochondrial decay and age-associated diseases. Mechanisms of Aging and Development 131, 473–479.

Anderson, B.S., Hunt, J.W., Phillips, B.M., Nicely, P.A., Gilbert, K.D., de Vlaming, V., Connor, V., Richard, N., Tjeerdema, R.S., 2003. Ecotoxicologic impacts of agricultural drain water in the Salinas River, California, USA. Environmental Toxicology & Chemistry 22, 275–2384.

Andersson, A.-M., Skakkebæk, N.E., 1999. Exposure to exogenous estrogens in food: possible impact on human development and health. European Journal of Endocrinology 140, 477–485.

Andrieu, E., Darmon, N., Drewnowski, A., 2006. Low-cost diets: more energy, less nutrients. European Journal of Clinical Nutrition 60, 434–436.

Aneja, V.P., Schlesinger, W.H., Erisman, J.W., 2009. Effects of agriculture upon the air quality and climate: research, policy, and regulations. Environmental Science & Technology 43, 4234–4240.

Aoyama, M., Kumakura, N., 2001. Quantitative and qualitative changes of organic matter in an Ando soil induced by mineral fertilizer and cattle manure applications for 20 years. Soil Science & Plant Nutrition 47, 241–252.

Bail, J., Meneses, K., Demark-Wahnefried, W., 2016. Nutritional status and diet in cancer prevention. Seminars in Oncology Nursing 32, 206–214.

Bannink, A., Valk, H., Van Vuuren, A.M., 1999. Intake and excretion of sodium, potassium, and nitrogen and the effects on urine production by lactating dairy cows. Journal of Dairy Science 82, 1008–1018.

Bartelt-Hunt, S.L., Snow, D.D., Kranz, W.L., Mader, T.L., Shapiro, C.A., van Donk, S.J., Shelton, D.P., Tarkalson, D.D., Zhang, T.C., 2012. Effect of growth promotants on the occurrence of endogenous and synthetic steroid hormones on feedlot soils and in runoff from beef cattle feeding operations. Environmental Science & Technology 46, 1352–1360.

Basset-Mens, C., van der Werf, H.M.G., 2005. Scenario-based environmental assessment of farming systems: the case of pig production in France. Agriculture, Ecosystems & Environment 105, 127–144.

Blackwell, B.R., Wooten, K.J., Buser, M.D., Johnson, B.J., Cobb, G.P., Smith, P.N., 2015. Occurrence and characterization of steroid growth promoters associated with particulate matter originating from beef cattle feedyards. Environmental Science & Technology 49, 8796–8803.

Blanco-Canqui, H., Stalker, A.L., Rasby, R., Shaver, T.M., Drewnoski, M.E., van Donk, S., Kibet, L., 2016. Does cattle grazing and baling of corn residue increase water erosion. Soil Science Society of America Journal 80, 168–177.

Bonaudo, T., Bendahan, A.B., Sabatier, R., Ryschawy, J., Bellon, S., Leger, F., Magda, D., Tichit, M., 2014. Agroecological principles for the redesign of integrated crop-livestock systems. European Journal of Agronomy 57, 43–51.

Borlaug, N., 2007. Feeding a hungry world. Science 318, 359.

Brejda, J.J., Moorman, T.B., Karlen, D.L., Dao, T.H., 2000. Identification of regional soil quality factors and indicators: I. Central and Southern High Plains. Soil Science Society of America Journal 64, 2115–2124.

Burgess, P.J., Morris, J., 2009. Agricultural technology and land use futures: the UK case. Land Use Policy 26S, S222–S229.

Burkholder, J., Libra, B., Weyer, P., Heathcote, S., Kolpin, D., Thorne, P.S., Wichman, M., 2007. Impacts of waste from concentrated animal feeding operations on water quality. Environmental Health Perspectives 115, 308–312.

Charavaryamath, C., Singh, B., 2006. Pulmonary effects of exposure to pig barn air. Journal of Occupational Medicine and Toxicology 1, 10. Available from: http://www.occup-med.com/content/1/1/10 [25 September 2016].

Chapin, A., Rule, A., Gibson, K., Buckley, T., Schwab, K., 2005. Airborne multidrug-resistant bacteria isolated from a concentrated swine feeding operation. Environmental Health Perspectives 113, 137–142.

Chee-Sanford, J.C., Aminov, R.I., Krapac, I.J., Garrigues-Jeanjean, N., Mackie, R.I., 2001. Occurrence and diversity of tetracycline resistance genes in lagoons and groundwater underlying two swine production facilities. Applied and Environmental Microbiology 67, 1494–1502.

Clark, D.A., Caradus, J.R., Monaghan, R.M., Sharp, P., Thorrold, B.S., 2007. Issues and options for future dairy farming in New Zealand. New Zealand Journal of Agricultural Research 50, 203–221.

Couttenier, M., Soubeyran, R., 2013. Drought and civil war in sub-Saharan Africa. The Economic Journal 124, 201–244.

Craig, W.J., 2009. Health effects of vegan diets. American Journal of Clinical Nutrition 89, 1627S–1633S.

Cui, Z., Zhang, F., Miao, Y., Sun, Q., Li, F., Chen, X., Li, J., Ye, Y., Yang, Z., Zhang, Q., Liu, C., 2008. Soil nitrate-N levels required for high yield maize production in the North China Plain. Nutrient Cycling Agroecosystems 82, 187–196.

Culman, S.W., DuPont, S.T., Glover, J.D., Buckley, D.H., Fick, G.W., Ferris, H., Crews, T.E., 2010. Long-term impacts of high-input annual cropping and unfertilized perennial grass production on soil properties and belowground food webs in Kansas, USA. Agriculture, Ecosystems & Environment 137, 13–24.

Davidson, E.A., Ackerman, I.L., 1993. Changes in soil carbon inventories following cultivation of previously untilled soils. Biogeochemistry 20, 161–193.

Denis, I., Potier, B., Vancassel, S., Heberden, C., Lavialle, M., 2013. Omega-3 fatty acids and brain resistance to ageing and stress: body of evidence and possible mechanisms. Ageing Research Reviews 12, 579–594.

Desquilbet, M., Dorin, B., Couvet, D., 2017. Land sharing vs land sparing to conserve biodiversity: how agricultural markets make a difference. Environ Model Assess 22, 185–200.

Di, H.J., Cameron, K.C., 2002. Nitrate leaching in temperate agroecosystems: sources, factors and mitigating strategies. Nutrient Cycling Agroecosystems 46, 237–256.

Douglas, M.R., Rohr, J.R., Tooker, J.F., 2014. Neonicotinoid insecticide travels through a soil food chain, disrupting biological control of non-target pests and decreasing soya bean yield. Journal of Applied Ecology 52, 250–260.

Douglas, M.R., Tooker, J.F., 2015. Large-scale deployment of seed treatments have driven rapid increase in use of neonicotinoid insecticides and preemptive pest management in U.S. field crops. Environmental Science & Technology 49, 5088–5097.

Drewnowski, A., 2009. Obesity, diets, and social inequalities. Nutrition Reviews 67, S36–S39.

Drewnowski, A., 2010. The cost of US foods as related to their nutritional value. American Journal of Clinical Nutrition 92, 1181–1188.

Duke, S.O., Powles, S.B., 2008. Glyphosate: a once-in-a-century herbicide. Pest Management Science 64, 319–325.

Dumont, B., Fortun-Lamthe, L., Jouven, M., Thomas, M., Tichit, M., 2012. Prospects from agroecology and industrial ecology for animal production in the 21st century. Animal 7, 1028–1043.

Fahrig, L., Baudry, J., Brotons, L., Burel, F.G., Crist, T.O., et al., 2011. Functional landscape heterogeneity and animal biodiversity in agricultural landscapes. Ecology Letters 14, 101–112.

Fargione, J.E., Cooper, T.R., Flaspohler, D.J., Hill, J., Lehman, C., McCoy, T., McLeod, S., Nelson, E.J., Oberhauser, K.S., Tilman, D., 2009. Bioenergy and wildlife: threats and opportunities for grassland conservation. BioScience 59, 767–777.

Fisher, M.C., Hawkins, N.J., Sanglard, D., Gurr, S.J., 2018. Worldwide emergence of resistance to antifungal drugs challenges human health and food security. Science 360, 739–742.

Foley, J.A., DeFries, R., Asner, G.P., Barford, C., Bonan, G., Carpenter, S.R., Chapin, F.S., Coe, M.T., Daily, G.C., Gibbs, H.K., Helkowski, J.H., Holloway, T., Howard, E.A., Kucharik, J., Monfreda, C., Patz, J.A., Prentice, I.C., Ramankutty, N., Snyder, P.K., 2005. Global consequences of land use. Science 309. https://doi.org/10.1126/science.1111772.

Foley, F.A., Ramankutty, N., Brauman, K.A., Cassidy, E.S., Gerber, J.S., Johnston, M., Mueller, N.D., O'Connell, C., Ray, D.K., West, P.C., Balzer, C., Bennett, E.M., Carpenter, S.R., Hill, J., Monfreda, C., Polasky, S., Rockstrom, J., Sheehan, J., Siebert, S., Tilman, D., Zaks, D.P.M., 2011. Solutions for a cultivated planet. Nature 478, 337–342.

Fraser, C.E., Smith, K.B., Judd, F., Hmphreys, J.S., Fragar, L.J., Henderson, A., 2005. Farming and mental health problems and mental illness. International Journal of Society Psychiatry 51, 340–349.

Frazao, E., 1996, Jan/Apr. The American diet: a costly health problem. Food Rev 2–6.

Gadd, J.B., Tremblay, L.A., Northcott, G.L., 2010. Steroid estrogens, conjugated estrogens and estrogenic activity in farm dairy shed effluents. Environmental Pollution 158, 730–736.

Gaigné, C., Chatellier, V., Bossuat, H., 2012. Logique économique de la spécialisation productive des territoires agricoles. In: CIAG, Associer productions animales et végétales pour des territoires agricoles performants, 24 Octobre 2012, (Poitiers, France).

Gatseva, P.D., Argirova, M.D., 2008. High-nitrate levels in drinking water may be a risk factor for thyroid dysfunction in children and pregnant women living in rural Bulgarian areas. International Journal of Hygiene and Environmental Health 211, 555–559.

Geiger, F., Bengtsson, J., Berendse, F., Weisser, W.W., Emmerson, M., Morales, M.B., Ceryngier, P., Liira, J., Tscharntke, T., Winqvist, C., Eggers, S., Bommarco, R., Pärt, T., Bretagnolle, V., Plantegenest, M., Clement, L.W., Dennis, C., Palmer, C., Oñate, J.J., Guerrero, I., Hawro, V., Aavik, T., Thies, C., Flohre, A., Hänke, S., Fischer, C., Goedhart, P.W., Inchausti, P., 2010. Persistent negative effects of pesticides on biodiversity and biological control potential on European farmland. Basic and Applied Ecology 11, 97–105.

Gibbons, D., Morrissey, C., Mineau, P., 2015. A review of the direct and indirect effects of neonicotinoids and fipronil on veterbrate wildlife. Environmental Science & Pollution Research 22, 103–118.

Gilchrist, M.J., Greko, C., Wallinga, D.B., Beran, G.W., Riley, D.G., Thorne, P.S., 2007. The potential role of concentrated animal feeding operations in infectious disease epidemics and antibiotic resistance. Environmental Health Perspectives 115, 313–316.

Gillom, R.J., Barbash, J.E., Crawford, C.G., Hamilton, P.A., Martin, J.D., Nakagaki, N., Nowell, L.H., Scott, J.C., Stackelberg, P.E., Thelin, G.P., Wolock, D.M., 2007. The Quality of Our Nation's Waters: Pesticides in the Nation's Streams and Ground Water, 1992-2001. Circular 1291. US Department of Interior and US Geological Survey. Available from: http://pubs.usgs.gov/circ/2005/1291/ [7 October 2016].

Goldstein, F.C., Levey, A.I., Steenland, N.K., 2013. High blood pressure and cognitive decline in mild cognitive impairment. J Am Geriatr Soc 61, 67–73.

Gómez-Baggethun, E., Mingorría, S., Reyes-García, V., Calvet, L., Montes, C., 2010. Traditional ecological knowledge trends in the transition to a market economy: empirical study in the Doñana natural areas. Conservation Biology 24, 721–729.

Green, R.E., Cornell, S.J., Scharlemann, J.P.W., Balmford, A., 2007. Farming and the fate of wild nature. Science 307, 550–555.

Gregoire, A., 2002. The mental health of farmers. Occupational Medicine 52, 471–476.

Hallmann, C.A., Foppen, R.P.B., van Turnhout, C.A.M., de Kroon, H., Jongejans, E., 2014. Declines in insectivorous birds are associated with high neonicotinoid concentrations. Nature 511, 341–343.

Hammer, J., 2013. Is a Lack of Water to Blame for the Conflict in Syria? Smithsonian Magazine June. Available from: http://www.smithsonianmag.com/innovation/is-a-lack-of-water-to-blame-for-the-conflict-in-syria-72513729/?no-ist [25 September 2016].

Harter, T., Lund, J.R., Darby, J., Fogg, G.E., Howitt, R., Jessoe, K.K., Pettygrove, G.S., Quinn, J.F., Viers, J.H., Boyle, D.B., Canada, H.E., DeLaMora, N., Dzurella, K.N., Fryjoff-Hung, A., Hollander, A.D., Honeycut, K.L., Jenkins, M.W., Jensen, V.B., King, A.M., Kourakos, G., Liptzin, D., Lopez, E.M., Mayzelle, M.M., McNally, A., Medellin-Azuara, J., Rosenstock, T.S., 2012. Addressing Nitrate in California's Drinking Water with a Focus on Tulare Lake Basin and Salinas Valley Groundwater. Report for the State Water Resources Control Board Report to the Legislature. Available from: http://groundwaternitrate.ucdavis.edu/files/138956.pdf> [25 September 2016].

Hoekstra, J.M., Boucher, T.M., Ricketts, T.H., Roberts, C., 2005. Confronting a biome crisis: global disparities of habitat loss and protection. Ecology Letters 8, 23–29.

Ikerd, J., 2005. Sustainable Capitalism. Lynne Rienner Publishers, Inc, Boulder.

Johnson, A.C., Williams, R.J., Matthiessen, P., 2006. The potential steroid hormone contribution of farm animals to freshwaters, the United Kingdom as a case study. Science of the Total Environment 362, 166–178.

Jones, D.L., Cross, P., Withers, P.J.A., DeLuca, T.H., Robinson, D.A., Quilliam, R.S., Harris, I.M., Chadwick, D.R., Edward-Jones, G., 2013. Nutrient stripping: the global disparity between food security and soil nutrient stocks. Journal of Applied Ecology 50, 851–862.

Kato, T., Kuroda, H., Nakasone, H., 2009. Runoff characteristics of nutrients from an agricultural watershed with intensive livestock production. Journal of Hydrology 368, 79–87.

Key, T.J., Appleby, P.N., Rosell, M.S., 2006. Health effects of vegetarian and vegan diets. Proceedings of the Nutrition Society 65, 35–41.

Knez, M., Graham, R.D., 2013. The impact of micronutrient deficiencies in agricultural soils and crops on the nutritional health of humans. In: Selinus, O., Alloway, B., Centeno, J.A., Finkelman, R.B., Fuge, R., Lindh, U., Smedley, P. (Eds.), Essentials of Medical Geology: Revised Edition. Springer Science+Business Media, Dordrecht, pp. 517–533.

Krasilnikov, P.V., Tabor, J.A., 2003. Perspectives on utilitarian ethnopedology. Geoderma 111, 197–215.

Kronberg, S.L., Scholljegerdes, E.J., Maddock, R.J., Barcelo-Coblijn, G., Murphy, E.J., 2017. Rump and shoulder muscles from grass and linseed fed cattle as important sources of n-3 fatty acids for beef consumers. Eur J Lipid Sci Tech 119, 1600390.

Lange, I.G., Daxenberger, A., Schiffer, B., Witters, H., Ibarreta, D., Meyer, H.H.D., 2002. Sex hormones originating from different livestock production systems: fate and potential disrupting activity in the environment. Analytica Chimica Acta 473, 27–37.

Lasanta, T., Arnez, J., Pascual, N., Ruiz-Flaño, P., Errea, M.P., Lana-Renault, N., 2017. Space-time process and drivers of land abandonment in Europe. Catena 149, 810–823.

Lemaire, G., 2012. Intensification of animal production from grassland and ecosystem services: a trade-off. CAB International Reviews 7, 1–7.

Lemaire, G., Franzluebbers, A., de Faccio Carvalho, P.C., Dedieu, B., 2014. Integrated crop-livestock systems: strategies to achieve synergy between agricultural production and environmental quality. Agriculture, Ecosystems & Environment 190, 4–8.

Levy, S.B., Marshall, B., 2004. Antibacterial resistance worldwide: causes, challenges and responses. Nature Medicine 10, S122–S129.

Lichtenberg, E., Smith-Ramirez, R., 2011. Slippage in conservation cost sharing. Amer J Agric Econ 93, 113–129.

Lichtenberg, E., 2014. Conservation, the farm bill, and U.S. agri-environmental policy. Choices 23, 1–6.

Liebig, M.A., Doran, J.W., 1999. Impact of organic production practices on soil quality indicators. Journal of Environmental Quality 28, 1601–1609.

Liebman, M., Baraibar, B., Buckley, Y., Childs, D., Christensen, S., Cousens, R., Eizenberg, H., Heijting, S., Loddo, D., Merotto Jr., A., Renton, M., Riemens, M., 2016. Ecologically sustainable weed management: how do we get from proof-of-concept to adoption? Ecological Applications 26, 1352–1369.

Lubowski, R.N., Bucholtz, S., Claassen, R., Roberts, M.J., Cooper, J.C., Gueorguieva, A., Johansson, R., 2006. Environmental Effects of Agricultural Land-use Change: The Role of Economics and Policy. USDA Economic Research Service. Economic Research Report No. ERR-25. Available from: http://www.ers.usda.gov/publications/pub-details/?pubid=45621 [29 March 2018].

Ludwig, D.S., 2011. Technology, diet, and the burden of chronic disease. JAMA 305, 1352–1353.

Mallin, M.A., Cahoon, L.B., 2003. Industrialized animal production—a major source of nutrient and microbial pollution to aquatic ecosystems. Population and Environment 24, 369–385.

Marshall, B., Levy, S.B., 2011. Food animals and antimicrobials: impact on human health. Clinical Microbiology Reviews 24, 718–733.

Mayne, S.T., Playdon, M.C., Rock, C.L., 2016. Diet, nutrition, and cancer: past, present and future. Nature Rev Clin Oncol 13, 504–515.

Mazoyer, M., Roudart, L., 2006. A History of World Agriculture from the Neolitic, Age to the Current Crisis. Earthscan, New York.

McCracken, J.D., 2005. Where the bobolinks roam: the plight of the North America's grassland birds. Biodiversity 6, 20–29.

McEachran, A.D., Blackwell, B.R., Hanson, J.D., Wooten, K.J., Mayer, G.D., Cox, S.B., Smith, P.N., 2015. Antibiotics, bacteria, and antibiotic resistance genes: aerial transport from cattle feed yards via particulate matter. Environmental Health Perspectives 123, 337–343.

Meehan, T.D., Werling, B.P., Landis, D.A., Gratton, C., 2016. Agricultural landscape simplification and insecticide use in the Midwestern United States. Proceedings of the National Academy of Science. Available from: http://www.pnas.org/cgi/doi/10.1073/pnas.1100751108/ [1 October 2016].

Meyers, K.P., Olsen, C.W., Setterguist, S.F., Capuano, A.W., Donham, K.J., Thacker, E.L., Merchant, J.A., Gray, G.C., 2006. Are swine workers in the United States at increased risk of infection with zoonotic influenza virus? Clinical Infectious Diseases 42, 14–20.

Meyers, K.P., Olsen, C.W., Gray, G.C., 2007a. Cases of swine influenza in humans: a review of the literature. Clinical Infectious Diseases 44, 1084–1088.

Meyers, K.P., Setterquist, S.F., Capuano, A.W., Gray, G.C., 2007b. Infection due to 3 avian influenza subtypes in United States veterinarians. Clinical Infectious Diseases 45, 4–9.

Mississippi River - Gulf of Mexico Watershed Nutrient Task Force, 2015. 2015 Report to Congress. Available from: http://www.epa.gov/sites/production/files/2015-10/documents/htf_report_to_congress_final_-_10.1.15.pdf [26 September 2016].

Mitloehner, F.M., Calvo, M.S., 2008. Worker health and safety in concentrated animal feeding operations. Journal of Agricultural Safety and Health 14, 163–187.

Monaghan, R.M., de Klein, C.A.M., Muirhead, R.W., 2008. Prioritisation of farm scale remediation efforts for reducing losses of nutrients and faecal indicator organisms to waterways: a case study of New Zealand dairy farming. Journal of Environmental Management 87, 609–622.

Montgomery, D.R., 2007. Dirt: the erosion of civilizations. University of California Press, Berkeley.

Moraine, M., Duru, M., Nicholas, P., Leterme, P., Therond, O., 2014. Farming system design for innovative crop-livestock integration in Europe. Animal 8, 1204–1217.

Mozaffarian, D., Ludwig, D.S., 2010. Dietary guidelines in the 21st century—a time for food. JAMA 304, 681–682.

National Research Council, 2010. Toward Sustainable Agricultural Systems in the 21st Century. The National Academies Press, Washington, DC.

Novak, J.M., Watts, D.W., Hunt, P.G., Stone, K.C., 2000. Phosphorus movement through a coastal plain soil after a decade of intensive swine manure application. Journal of Environmental Quality 29, 1310–1315.

Osterberg, D., Wallinga, D., 2004. Addressing externalities from swine production to reduce public health and environmental impacts. American Journal of Public Health 94, 1703–1708.

Parletta, N., Milte, C.M., Meyer, B.J., 2013. Nutritional modulation of cognitive function and mental health. The Journal of Nutritional Biochemistry 24, 725–743.

Peters, C.J., Picardy, J., Darrouzet-Nardi, A.F., Wilkins, J.L., Griffin, T.S., Fick, G.W., 2016. Carrying capacity of U.S. agricultural land: ten diet scenarios. Elementa: Sci Anthropocene 4, 1–15.

Peyraud, J.-L., Taboada, M., Delaby, L., 2014. Integrated crop and livestock systems in Western Europe and South America: a review. European Journal of Agronomy 57, 31–42.

Plieninger, T., Draux, H., Fagerholm, N., Bieling, C., Burgi, M., Kizos, T., Kuemmerle, T., Primdahl, J., Verburg, P.H., 2016. The driving forces of landscape change in Europe: a systematic review of the evidence. Land Use Policy 57, 204–214.

Powles, S.B., Yu, Q., 2010. Evolution in action: plants resistant to herbicides. Annual Review of Plant Biology 61, 317–347.

Provenza, F.D., Meuret, M., Gregorini, P., 2015. Our landscapes, our livestock, ourselves: restoring broken linkages among plants, herbivores, and humans with diets that nourish and satiate. Appetite 95, 500–519.

Reidsma, P., Ewert, F., 2008. Regional farm diversity can reduce vulnerability of food production to climate change. Ecology and Society 13 (38). http://www.ecologyandsociety.org/vol13/iss1/art38/ [29 March 2018].

Review on Antimicrobial Resistance, 2016. Tackling Drug-resistant Infections Globally: Final Report and Recommendations. Wellcome Trust and HM Government. Available from: https://amr-review.org [24 September 2016].

Rigg, J., 2006. Land, farming, livelihoods, and poverty: rethinking the links in the rural south. World Development 34, 180–202.

Robertson, B., Pinstrup-Andersen, P., 2010. Global land acquisition: neo-colonialism or development opportunity? Food Security 2, 271–283.

Rosenzweig, C., Iglesias, A., Yang, X.B., Epstein, P.R., Chivian, E., 2001. Climate change and extreme weather events: implications for food production, plant diseases, and pests. Global Change & Human Health 2, 90–104.

Sabatier, R., Teillard, F., Rossing, W.A.H., Doyen, L., Tichit, M., 2015. Trade-offs between pasture production and farmland bird conservation: exploration of options using a dynamic farm model. Animal 9, 899–907.

Sadeq, M., Moe, C.L., Attarassi, B., Cherkaoui, I., ElAouad, R., Idrissi, L., 2008. Drinking water nitrate and prevelance of methemoglobinemia among infants and children 1-7 years in Moroccan areas. International Journal of Hygiene and Environmental Health 211, 546–554.

Samson, F., Knopf, F., 1994. Prairie conservation in North America. BioScience 44, 418–421.

Schepers, J.S., Moravek, M.G., Alberts, E.E., Frank, K.D., 1991. Maize production impacts on groundwater quality. Journal of Environmental Quality 20, 12–16.

Sharpley, A.N., Chapra, S.C., Wedepohl, R., Sims, J.T., Daniel, T.C., Reddy, K.R., 1994. Managing agricultural phosphorus for protection of surface waters: issues and options. Journal of Environmental Quality 23, 437–451.

Sieben D., 1997. Noise and hearing loss in agriculture, forestry, and fisheries. p 56-66, In: Safety and Health in Agriculture, Forestry, and Fisheries. eds Langley R.L., McLymore, R.L., Meggs, W.H., Roberson, G.R., Government Institutes, Rockville. p 56-66.

Sklenicka, P., 2016. Classification of farmland ownership fragmentation as a cause of land degradation: a review of typology, consequences, and remedies. Land Use Policy 57, 694–701.

Spellberg, B., Hansen, G.R., Kar, A., Cordova, C.D., Price, L.B., Johnson, J.R., 2016. Antibiotic resistance in humans and animals. National Academy of Medicine. Available from: https://nam.edu/wp-content/uploads/2016/Antibiotic-Resistance-in-Humans-and-Animals.pdf [23 September 2016].

Spencer, E.A., Appleby, P.N., Davey, G.K., Key, T.J., 2003. Diet and body mass index in 38 000 EPIC-Oxford meat-eaters, fish-eaters, vegetarians and vegans. International Journal of Obesity 27, 728–734.

Springmann, M., Mason-D'Croz, D., Robinson, S., Garnett, T., Godfray, H.C.J., Gollin, D., Rayner, M., Ballon, P., Scarborough, P., 2016. Global and regional health effects of future food production under climate change: a modelling study. Lancet 387, 1937–1946.

Steiner, J.L., Franzluebbers, A.J., 2009. Farming with grass—for people, for profit, for production, for protection. Journal of Soil and Water Conservation 64, 75A–80A.

Stoate, C., Báldi, A., Beja, P., Boatman, N.D., Herzon, I., van Doorn, A., de Snoo, G.R., Rakosy, L., Ramwell, C., 2009. Ecological impacts of early 21st century agricultural change in Europe—a review. Journal of Environmental Management 91, 22–46.

Sumner, D.A., Alston, J.M., Glauber, J.W., 2009. Evolution of the economics of agricultural policy. American Journal of Agricultural Economics 92, 403–423.

Sumner, D.A., Zulauf, C., 2012. Economic and environmental effects of agricultural insurance programs. The Council on Food, Agricultural and Resource Economics. Available from: http://ageconsearch.tind.io/record/156622/files/Sumner-Zulauf_Final.pdf [29 March 2018].

Sumner, D.A., 2014. American farms keep growing: size, productivity, and policy. The Journal of Economic Perspectives 28, 147–166.

Teillard, F., Jiguet, F., Tichi,t, M., 2015a. The response of farmland bird communities to agricultural intensity as influenced by its spatial aggregation. PLoS One 10, e0119674. https://doi.org/10.1371/journal.pone.0119674.

Teillard, F., Anton, A., Dumont, B., Finn, J.A., Henry, B., De Souza, D.M., Manzano, P., Milà i Canals, L., Phelps, C., Said, M., Vijn, S., White, S., 2015b. A Review of Indicators and Methods to Assess Biodiversity: Application to Livestock Production at Global Scale. Livestock Assessment and Performance Partnership. FAO, Rome. Available from. <http://www.fao.org/3/a-av151e.pdf>.

Thi Anh, P., Kroeze, C., Bush, S.R., Mol, A.P.J., 2010. Water pollution by intensive brackish shrimp farming in south-east Vietnam. Agric Water Mgmt 97, 872–882.

Thorne, P.S., 2007. Environmental health impacts of concentrated animal feeding operations: anticipated hazards—searching for solutions. Environmental Health Perspectives 115, 296–297.

Tilman, D., Cassman, K.G., Matson, P.A., Naylor, R., Polasky, S., 2002. Agricultural sustainability and intensive production practices. Nature 418, 671–677.

Tilman, D., Clark, M., 2014. Global diets link environmental sustainability and human health. Nature 515, 518–522.

Tscharntke, T., Clough,, Y., Wanger, T.C., Jackson, L., Motzke, I., Perfecto, I., Vandermeer, J., Whitbread, A., 2012. Global food security, biodiversity conservation and the future of agricultural intensification. Biol Conserv 151, 53–59.

Van Boeckel, T.P., Brower, C., Gilbert, M., Grenfell, B.T., Levin, S.A., Robinson, T.P., Teillant, A., Laxminarayan, R., 2015. Global trends in antimicrobial use in food animals. Proceedings of the National Academy of Science 112, 5649–5654.

van Maanen, J.M.S., van Dijk, A., Mulder, K., de Baets, M.H., Menheere, P.C.A., van der Heide, D., Mertens, P.L.J.M., KleinJans, J.C.S., 1994. Consumption of drinking water with high nitrate levels causes hypertrophy of the thyroid. Toxicology Letters 72, 365–374.

Vitousek, P.M., Naylor, R., Crews, T., David, M.B., Drinkwater, L.E., Holland, E., Johnes, P.J., Katzenberger, J., Martinelli, L.A., Matson, P.A., Nziguheba, G., Ojima, D., Palm, C.A., Robertson, G.P., Sanchez, P.A., Townsend, A.R., Zhang, F.S., 2009. Nutrient imbalances in agricultural development. Nature 324, 1519–1520.

Vogl, C.R., Kilcher, L., Schmidt, H., 2005. Are standards and regulations of organic farming moving away from small farmers' knowledge? Journal of Sustainable Agriculture 26, 5–26.

Walters, C.G., Shumway, C.R., Chouinard, H.H., Wandschneider, P.R., 2012. Crop insurance, land allocation, and the environment. Journal of Agricultural and Resource Economics 37, 301–320.

Waples, R.S., Zabel, R.W., Scheuerell, M.D., Sanderson, B.L., 2007. Evolutionary responses by native species to major anthropogenic changes to their ecosystems: Pacific salmon in the Columbia River hydropower system. Molecular Ecol 17, 84–96.

Ward, M.H., deKok, T.M., Levallois, P., Brender, J., Gulis, G., Nolan, B.T., VanDerslice, J., 2005. Workgroup report: drinking-water nitrate and health—recent findings and research needs. Environmental Health Perspectives 113, 1607–1614.

Welch, R.M., Graham, R.D., 1999. A new paradigm for world agriculture: meeting human needs productive, sustainable, nutritious. Field Crops Res 60, 1–10.

Welch, R.M., 2002. The impact of mineral nutrients in food crops on global human health. Plant Soil 247, 83–90.

Whalen, J.K., Chang, C., Clayton, G.W., Carefoot, J.P., 2000. Cattle manure amendments can increase the pH of acid soils. Soil Science Society of America Journal 64, 962–966.

Whyte, R.T., 2002. Occupational exposure of poultry stockmen in current barn systems for egg production in the United Kingdom. British Poultry Science 43, 364–373.

Wienhold, B.J., Andrews, S.S., Karlen, D.L., 2004. Soil quality: a review of the science and experiences in the USA. Environmental Geochemistry and Health 26, 89–95.

World Cancer Research Fund/American Institute for Cancer Research Fund, 2007. Food, Nutrition, Physical Activity, and the Prevention of Cancer: A Global Perspective. Available from: http://www.aicr.org/assets/docs/pdf/reports/Second_Expert_Report.pdf/ [7 October 2016].

Wright, C.K., Wimberly, M.C., 2013. Recent land use change in the western corn belt threatens grasslands and wetlands. Proceedings of the National Academy of Sciences 110, 4134–4139.

Yang, C.-Y., Wu, D.-C., Chang, C.-C., 2007. Nitrate in drinking water and risk of death from colon cancer in Taiwan. Environment International 33, 649–653.

Zonderland-Thomassen, M.A., Lieffering, M., Ledgard, 2014. Water footprint of beef cattle and sheep produced in New Zealand: water scarcity and eutrophication impacts. Journal of Cleaner Production 73, 253–262.

Further Reading

Gaudou, B., Therond, O., Sibertin-Blanc, C., Amblard, F., Auda, Y., Arcangeli, J.P., Balestrat, M., Charron-Moirez, M.H., Gondet, E., Hong, Y., Lardy, R., Louail, T., Mayor, E., Panzoli, D., Sauvage, S., Sanchez-Perez, J.M., Taillandier, P., Van Bai, N., Vavasseur, M., Mazzega, P., 2014. The MAELIA multi-agent platform for integrated assessment of low-water management issues. In: Jamal Alam, S., Van Dyke Parunak, H. (Eds.), Proceedings of the 14th International Workshop on Multi-agent-based Simulation, Minnesota, USA 6–7 May 2013, Lecture Notes in Artificial Intelligence (8235), pp. 85–110.

Lüscher, G., Jeanneret, P., Schneider, M.K., Turnbull, L.A., Arndorfer, M., Balázs, K., Báldi, A., Bailey, D., Bernhardt, K.G., Choisis, J.-P., Elek, Z., Frank, T., Friedel, J.K., Kainz, M., Kovács-Hostyánszki, A., Oschatz, M.-L., Paoletti, M.G., Papaja-Hülsbergen, S., Sarthou, J.-P., Siebrecht, N., Wolfrum, S., Herzog, F., 2014. Responses of plants, earthworms, spiders and bees to geographic location, agricultural management and surrounding landscape in European arable fields. Agriculture, Ecosystems & Environment 186, 124–134.

Matson, P.E., Parton, W.J., Power, A.G., Swift, M.J., 1997. Agricultural intensification and ecosystem properties. Science 277, 504–509.

Murgue, C., Therond, O., Leenhardt, D., 2015. Towards sustainable water and agricultural land management: participatory design of spatial distributions of cropping systems in a water-deficit basin. Land Use Policy 45, 52–63.

Waples, R.S., Zabel, R.W., Scheuerell, M.D., Sanderson, B.L., 2008. Evolutionary responses by native species to major anthropogenic changes to their ecosystems: Pacific salmon in the Columbia River hydropower system. Molecular Ecology 17, 84–96.

SECTION II

INCREASING DIVERSITY WITHIN AGROECOSYSTEMS FOR REDUCING ENVIRONMENTAL EMISSIONS

CHAPTER

6

Using Crop Diversity and Conservation Cropping to Develop More Sustainable Arable Cropping Systems

John Hendrickson[1], *Juan Cruz Colazo*[2]

[1]Northern Great Plains Research Laboratory, USDA-Agricultural Research Service, Mandan, North Dakota, United States; [2]EEA San Luis, INTA & National University of San Luis, Villa Mercedes, Argentina

INTRODUCTION

Review of Successes of Modern Agriculture

The world population has increased from slightly over 2.5 billion in 1950 to over 7 billion by 2012 (US Census Bureau, 2016), and agriculture has done an exceptional job of meeting the basic food requirements of this increased population. Between 1950 and 2000, global average food availability per person rose from <2400 to >2700 calories (Ruttan, 1999). The driver behind the increase in calorie availability has been a 150% increase in cereal crop yields while estimates of land area used in agriculture has remained relatively constant (Trewavas, 2002). For example, between 1950 and 2015, US maize production (*Zea mays* L.) has increased 390%, while the harvested area of maize has only increased 15% (NASS, 2016a,b). However, Foley et al. (2011) pointed out that yield increases may be overestimated and land area in agriculture underestimated because of multiple cropping, fewer fallow acres, and lower crop failures.

Multiple factors have led to yield increases. Globally, agriculture has incorporated more off-farm inputs. Global fertilizer use ($N + P_2O_5 + K_2O$) is projected to grow 1.8% annually and expected to reach over 200 M tonnes by 2018 (FAO, 2015). Pesticide use also increased globally between 1990 and 2007 before declining through 2011 (Fig. 6.1) because pesticides were used on a majority of the cropland (Osteen and Fernandez-Cornejo, 2013) and changes in chemical formulations lowered application rates (Benbrook, 2012). Despite the decline, global pesticide use was 2.5 times greater in 2011 than in 1990 (Fig. 6.1).

Besides increased input use, genetics also contributed to enhance yields. In both maize

Agroecosystem Diversity
https://doi.org/10.1016/B978-0-12-811050-8.00006-6

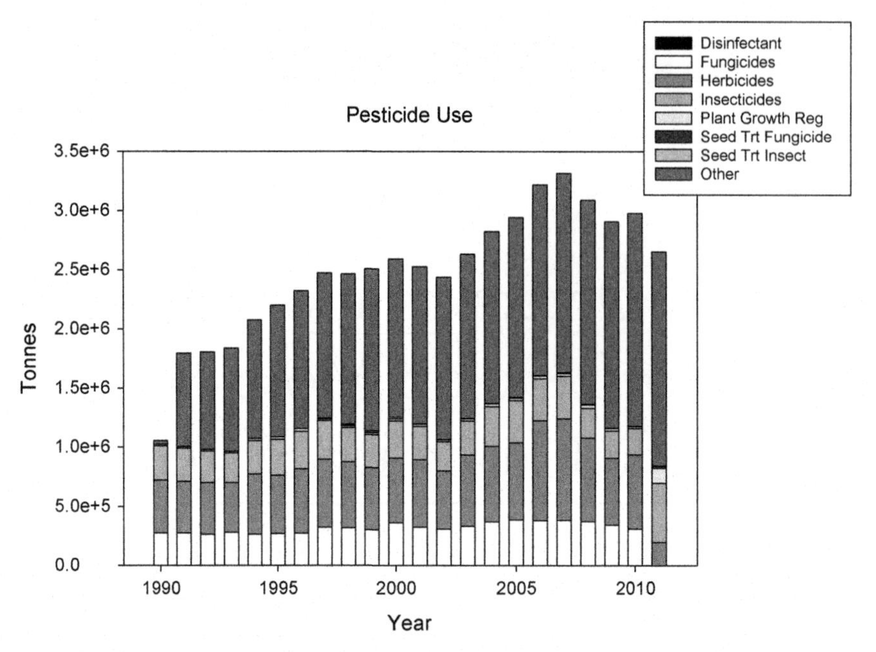

FIGURE 6.1 Data on tonnes of pesticides used globally between 1990 and 2011. *Data from FAOSTAT.*

(Egli, 2008) and cereals (Cassman, 1999), yield increases have been propelled by some extent by genetics. Often, genetics and management interact (Duvick and Cassman, 1999; Tollenaar and Lee, 2002). For example, in India, yield increases for genetically modified cotton (*Gossypium* spp.) were greater than for countries with an established history of using pest programs (Qaim and Zilberman, 2003).

Increased Use of Conservation Tillage Provides Benefits

Conservation tillage is the main principle of conservation agriculture. It is defined as any tillage sequence, the object of which is to minimize or reduce loss of soil and water: operationally, a tillage or tillage and planting combination that leaves a 30% or greater cover of crop residue on the surface (SSSA, 2017). Reducing soil disturbance by tillage in agricultural land began in the Great Plains in the United States in the 1930s in response to the devastation caused by prolonged drought. Initial research on "conservation" or reduced tillage involved early versions of a chisel plow, and later, stubble mulch farming became a forerunner of no-tillage farming. This collection of practices led to what became known as conservation tillage, although no-till (NT) systems by definition avoid soil disturbance by no-till seed drilling and maintain an organic mulch cover on the soil surface (Kassam et al., 2014).

The use of an NT cropping and management system has the potential in many cases to reduce water erosion rates to much more acceptable levels (Nearing et al., 2017). In tropical regions, conservation agriculture based on NT reduces soil erosion loss by a factor of 2–20, depending on the slope and soil texture. Under these conditions, mulching with crops residues is efficient in controlling erosion processes through the dissipation of the energy of rainfall impact and the diminution of runoff fluxes. Under temperate condition, the range of reduction is lower, from a factor of 2–10, and it is more related to an increase of the stability of topsoil aggregates and water infiltration rates (Scopel et al., 2013).

However, when NT is practiced in the absence of effective soil mulch cover, the effects can be disastrous, with rapid surface sealing leading to increased runoff and accelerated soil erosion (Giller et al., 2015).

NT had lower wind erosion rates than conventional tillage (CT) in cropping systems of Argentina, the United States, and China (Gao et al., 2016; Mendez and Buschiazzo, 2010; Sharrat et al., 2017). The main factors controlling wind erosion are the soil cover, which reduces surface wind speed (Fryrear, 1985; Mendez and Buschiazzo, 2010), and the reduction of the wind erodible fraction of topsoil (Hevia et al., 2007). Also, if NT increases the topsoil soil organic carbon (SOC) content, a higher dry aggregate stability of topsoil reduces wind erodibility (Colazo and Buschiazzo, 2010).

Soil C sequestration refers to the increase in C stored in the soil by capturing atmospheric CO_2 as a result of changes in land use or management (Palm et al., 2014). Increasing carbon stocks in agricultural soils can be achieved by reducing C mineralization, increasing C input, or both (Debaerke et al., 2017). No tillage can lead to a genuine carbon sequestration if it slows the rate of decomposition of existing SOC or promotes the stabilization of increasing organic carbon (Powlson et al., 2016).

Recent meta-analysis and literature reviews reveal an inconsistent effect of conservation agriculture in C sequestration in soil (Giller et al., 2015; Powlson et al., 2016). Reduced tillage or NT may increase soil C compared with CT, but these increases are often confined to near-surface layers (<10 cm). At deeper depths, soil C may be equal or even lower compared with CT (Luo et al., 2010; Palm et al., 2014). An answer to whether conversion from CT to NT merely redistributes SOC in the profile rather than enhances it lies in a critical assessment of the equilibrium SOC pool under new management and its depth-distribution in relation to several exogenous (climate, physiography, and biome) and endogenous (soil properties) factors (Lal, 2015b).

Water conservation generally is improved by using no tillage, according to literature review (Unger et al., 2010). Water conservation benefits of conservation tillage involving residues result from several factors, with the benefits generally improving with increasing amounts of surface residues. The benefits result from (1) protecting the surface against raindrop impact, thus reducing soil aggregate dispersion and surface sealing, which decrease water infiltration; (2) retarding the rate of water flow across the surface, thus providing more time for infiltration; and (3) reducing soil water evaporation by shading and cooling the soil and by reducing wind speed at the soil surface (Unger, 2002).

The rate of water infiltration in NT can be almost double of CT. No tillage, however, does not always result in the most infiltration from a given precipitation event. Infiltration may be greater into a tillage-loosened soil than a no-tillage soil, when precipitation amounts do not exceed the temporary storage capacity of the loosened soil layer. Also, soil infiltration is already high when precipitation occurs, thereby limiting opportunities for additional water storage (Unger et al., 2010).

The presence of cover residues decreases the first stage of evaporation by 10%–50%. This is particularly important under the hot and sunny conditions of the tropics. As a result, water is stored more quickly in the soil profile under CA systems during the beginning of the rainy season in the tropics and during winter and spring the temperate regions, which can act as a buffer against the effects of an eventual dry spell in the early stage of crop cycle (Scopel et al., 2013).

Negative Impacts of Modern Agriculture

Lower Crop Diversity

While the increases in yield have been impressive, the methods used to achieve these results have often resulted in system simplification (Hendrickson et al., 2008a). During the 20th century, commodity numbers per farm dropped from an estimated 5 to 1 (Dimitri et al., 2005). These simpler systems have minimal interactions between resources (Hendrickson et al., 2008a) and require less management input (Hendrickson et al., 2008b). One impact of adopting a simpler system in agriculture is a loss in diversity. In the United States, estimated crop diversity, aggregated at the county level, declined over the past 30 plus years (Aguilar et al., 2015). In particular, the historical corn production region (the Corn Belt) is growing as corn and soybean production expand into nontraditional production areas such as North Dakota (Aguilar et al., 2015).

Over 90% of the corn and soybean acreage is planted to genetically modified crops (ERS, 2018). Proponents of biotechnology have suggested that genetically modified crops may reduce pesticide use (Benbrook, 2001). However, in the United States, the use of genetically modified crops has resulted in the spread of herbicide-resistant weeds and a subsequent increase in herbicide applications that have overshadowed the benefits of genetic engineering on reducing insecticide applications (Benbrook, 2012).

Nitrogen Impacts

The decoupling of crops and livestock systems has presented additional challenges. One of the most pressing is the disruption in links in nitrogen cycling between crops and livestock (Brummer, 1998). Historically, manure from livestock could be applied to fields and planted forages, including legumes, needed for livestock, would provide nitrogen for crops (Honeyman, 1996). However, currently fertilizer inputs, including nitrogen, into agriculture exceed outputs in the form of production (Carpenter et al., 1998), and over half of the synthetic fertilizer produced occurred during the 15 years prior to 2008 (Howarth, 2008). This increase in synthetic fertilization has resulted in agriculture being a primary source of coastal nitrogen pollution (Howarth, 2008).

A flip side of the decoupling of agriculture has been the growth in confined animal feeding operations (CAFOs). The US production locus (the point that measures farm size from which half the production comes from smaller farms and half from larger farms) increased by 2000% in swine, 240% in dairy, 100% in fed-cattle, and 60% in broilers between 1987 and 2002 (MacDonald and McBride, 2009). While a majority of US farms have adequate land for spreading manure, those that cannot account for 60% of manure nitrogen and 70% of the manure phosphorus (Gollehon et al., 2001). Because of this, CAFOs have been linked to increases in nitrogen pollution in coastal areas (Mallin and Cahoon, 2003) and in well water (Zirkle et al., 2016).

Phosphorus Impacts

CAFOs have also had an impact on phosphorus pollution. Best manure management practices may not be enough to protect water resources from livestock waste (Burkholder et al., 2007). Weather events can exacerbate the issue. For example, approximately 10% of the CAFOs in eastern North Carolina, United States, may have had flooded waste pits after Hurricane Floyd in 1999 (Wing et al., 2002).

Phosphorus impacts have also been linked to other forms of agriculture. Between 1958 and 1998, global phosphorus accumulation in agricultural areas was about 8 Tg yr (Bennett et al., 2001). Simulation modeling indicated that phosphorus originating on cultivated cropland was the major source of phosphorus in the Mississippi River Basin (White et al., 2014). Increased phosphorus in the Mississippi River Basin has been linked to a hypoxic zone in the Gulf of Mexico.

Potassium Impacts

Potassium interacts with soil matrix, which limits its ability to leach into groundwater

(Balderacchi et al., 2013). However, potassium leaching has been reported on sandy soils, with increasing calcium concentration (Kolahchi and Jalali, 2007). Although relatively immobile, soil erosion from agriculture or overgrazing can also move potassium into the environment (Balderacchi et al., 2013).

Loss of Ecosystem Services

Generally, with a few exceptions, crop diversity in the United States has declined over time (Aguilar et al., 2015). The loss of crop diversity can result in more simplified landscapes, which may not promote ecosystem services as well as more complex landscapes (Tscharntke et al., 2005a). Simplification of the agricultural landscape impacts ecosystem services. For example, landscape simplification was correlated with increased pesticide use, suggesting increased pest pressure in the midwestern United States (Meehan et al., 2011).

DIVERSIFICATION OF CROP ROTATIONS

Benefits of Diversity

Less crop diversity often reduces the use of crop rotations. This can impact not only yields but also the nitrogen, phosphorus, and potassium cycling. Use of forage crops in extended rotations reduced the need for nitrogen inputs while increasing grain yields (Stanger and Lauer, 2008). Alfalfa, for example, increased the amount of available N_2 each year of stand life and also proved to be effective at accessing deep NO_3^- (Kelner et al., 1997). Incorporating annual legumes into nonlegume cropping systems may help to balance the nitrogen cycle also (Canfield et al., 2010).

Crop diversity can impact phosphorus cycles through either physical transport off site or through change in the soil microbial community. Soil erosion accounts for 75%–90% of the phosphorus movement in cultivated lands (Sharpley et al., 1993). Therefore, evaluating the impact of diverse crop rotation on phosphorus needs to focus on reduction of soil erosion. Incorporating oats and red clover into a corn rotation reduced soil erosion to about half that of continuous corn in Iowa (Johnston et al., 1943). However, soil loss under a corn-soybean rotation was greater than the loss under continuous corn (Van Doren et al., 1984). Soil aggregate stability, which does provide insight into soil erosion potential, was greater when at least 3 years of forage crops were included in the rotation (Karlen et al., 2006).

There have been suggestions that increased soil microbial activity may increase phosphorus availability (Bullock, 1992). Increased plant species richness has been shown to increase microbial biomass in natural systems (Zak et al., 2003), but this increase may be linked to greater plant productivity. The impact in cropping systems, however, is less clear. Crop residue and soil organic matter quality can impact diversity of the soil microbial community (Bending et al., 2002), but it appears that with a wheat-maize rotation, keeping residue on the field is the biggest determinant of soil microbial biomass (Govaerts et al., 2007). However, the management objective can impact the desired characteristics of the residue. For example, Cadisch and Giller (2001) suggested that the best residue quality for synchronizing N supply with plant demand would be mixed species residue with a high C:N ratio and high soluble C.

There is not much information on the impact of crop diversity on soil potassium levels. A study of cropping systems in Michigan indicated that under organic management, potassium increased when crop diversity increased beyond a monoculture (Snapp et al., 2010). However, the study authors indicated that an organic management system rather than increased diversity was the primary driver of changes in soil properties. The use of a winter legume as a cover crop redistributed potassium from lower soil depths to the surface (Hargrove, 1986).

Limitations Impacts of Diversity

Global food supplies are becoming more homogeneous (Khoury et al., 2014), while within the United States, crop diversity has declined over the past 30 plus years (Aguilar et al., 2015). This suggests that despite the many advantages of crop diversity, there are strong reasons for its lack of adaptation. There are many factors that influence producers' management decisions including a producer's goals, management concerns, and exogenous factors. Hendrickson et al. (2008b) indicated that as complexity of agricultural systems increased, management intensity also increased. Cropping systems that are relatively simple, such as the corn-soybean rotation in the United States, may require less management decision-making, which could appeal to producers. Also, the lack of complexity in these systems means that research and product development can focus on yield-limiting factors on relatively few crops, resulting in large impacts in productivity.

There have been conflicting reports about the impact of diversity on yield. In many perennial systems, increasing species diversity has been considered to increase biomass yield (Bullock et al., 2001; Cardinale et al., 2007; Tilman et al., 2001). In annual crops, the results are mixed. A review (Letourneau et al., 2011) suggested that use of polycultures to increase crop diversity and ecosystem services resulted in a decrease in primary crop yield, but Iverson et al. (2014) indicated that a win-win situation could be achieved by substituting large quantities of the primary crop for a compatibly harvested secondary crop. Crop yield is an important determinant in net farm income, and producers may not be willing to risk potential yield reduction.

Ways to Increase Crop Diversity

In the northern Great Plains of the United States, researchers have pioneered a dynamic cropping system (Tanaka et al., 2002; Hanson et al., 2007). This system emphasizes flexibility in crop production decisions rather than monoculture or fixed crop rotation (Hanson et al., 2007). Increased emphasis on the multifunctionality of agriculture (Boody et al., 2005) could emphasize the importance of agriculture in producing ecosystem services other than productivity. There is strong evidence that biodiversity can enhance these services (Schläpfer and Schmid, 1999; Isbell et al., 2011), which may result in economically or policy-driven increases in crop diversity.

CONSERVATION CROPPING SYSTEMS

Benefits of Conservation Agriculture

Tillage practices influence soil physical and biologic properties that are important for hydrologic processes and soil N transformations (Rasouli et al., 2014). Long-term studies show an increase in organic nitrogen under NT compared to CT (Varvel and Wilhelm, 2011; Weber et al., 2016). As for SOC, greater N concentrations can be expected in the surface layer under no tillage than under CT. Due to the mulch, NT generally provide temperature and moisture conditions that favor a more regular soil organic matter decomposition throughout the crop cycle and higher nitrogen (N) availability for crops (Scopel et al., 2013). The role of no tillage in denitrification is still controversial. However, a recent meta-analysis between CT and NT shows that NT did not alter N_2O emissions compared with CT. However, NT significantly reduced N_2O emissions in experiments >10 years, especially in dry climates (Van Kessel et al., 2013).

Immobile nutrients in the soil such as P and K tend to be affected substantially by the adoption of reduced-tillage management. Most notable is their tendency to become concentrated near the surface of the soil in response to the reduced soil disturbance (Sellles et al., 2002). The reduced tillage offers limited but positive impacts on soil P. Reduced tillage increased labile P and P sorption after 9 years in an Oxisol (Margenot et al., 2017). Generally, extractable P is higher in the topsoil in NT than in CT. This effect is related to SOC surface accumulation, which leads to differences in microbial activities, as well as to improvements in P cycling and soil quality (Zibilske et al., 2002; Zibilske and Bradford, 2003; Abdi et al., 2014). As for P, there is little evidence of changes on K. Soil profile under NT shows a higher vertical stratification of K (Wright et al., 2007).

An important benefit of conservation tillage is the diminution of nutrient loss by erosion. Associated with lower erosion rates, N, P, and K decreases more slowly under conservation compared to CT (Tan et al., 2015). The analysis of sediments from wind-eroded fields indicates a lesser loss of N and P in NT than in CT (Buschiazzo et al., 2007).

Limitations With Conservation Agriculture

A recent meta-analysis relates the reduction of yield following NT adoption to N fertilization practice (Lundy et al., 2015). No-till yield declines are more extreme in tropical than temperate regions. At low N rates, strongest NT yield declines in tropical/subtropical regions during first years of NT. These authors suggest among the hypothesis of these to soil N immobilization from residues. Also, under temperate conditions, nitrate losses may occur in nontillage soils when significant macropore flow relocates the nutrients into subsurface layers (Scopel et al., 2013). In paddy rice systems, a meta-analysis indicated that NT reduced N uptake and nitrogen use efficiency (NUE) of the rice by 5.4% and 16.9%, while it increased the N and P exports via runoff by 15.4% and 40.1% compared with CT, respectively (Liang et al., 2016).

No-till, and even reduced tillage, is associated with well-documented trade-offs when it comes to diffuse P pollution, particularly in cropping systems where fertilizers or manures are broadcast onto the soil surface, as opposed to banded at time of application, injected, or otherwise. While particulate P losses in runoff are largely curtailed with no-till, dissolved P losses can increase with no-till. No-till can exacerbate the direct transfer of broadcast fertilizer or manure P to runoff (Kleinman et al., 2015). Without irrigation, asynchronous availability of soil water and K in stratified soils is thought to restrict K uptake and reduce soybean yields, especially in dry years. However, under evenly distributed intermittent rainfall conditions in no-till soils, vertical stratification of soil K might not limit soybean yield (Fernández et al., 2008).

Ways to Increase Use

Increasing adoption of conservation cropping systems requires prudent strategies to address limitations and uncertainties of NT (Lal, 2015). It is often suggested that the risk for short-term decreases in crop productivity represents a major barrier for farmers considering conservation agriculture. A recent meta-analysis shows that negative impacts of no-till are minimized when permanent cover and crop rotation are present; even in this condition a yield benefit is expected under dry climate (Pittelkow et al., 2015). These authors suggest that instead of implementing no-till as the first step toward conservation agriculture in cropping systems, farmers should focus in residue retention and crop rotation previously. No-till after legumes, and especially perennial legumes, can be a good way to start no-till without expecting a reduction in yield (Derpsch et al., 2014).

Less than 1% of smallholder farmers worldwide have adopted conservation agriculture (Derpsch et al., 2010). The key to furthering conservation agriculture among them is development and deployment of affordable and effective minimum tillage implements (Johanssen et al., 2012). However, overcoming social and cultural factor may also be essential for an extensive adoption (Lal, 2015). Adequate intervention strategies and extension services are needed (Derpsch et al., 2016).

COMBINING CONSERVATION TILLAGE AND INCREASED DIVERSITY

A long-term study of crop diversity in the United States (Aguilar et al., 2015) indicated a decline in crop diversity but did identify central North Dakota, United States, as a region where crop diversity increased. The authors suggested there were multiple reasons for this increase including conservation tillage, disease pressure, and policy changes. They indicated that this region of North Dakota was an early adapter of conservation tillage technology. This increased soil water allowed for crops to be grown every year rather than every other year. With the increase in production, disease pressure began to grow, and this made producers look for alternatives to the dominant small grain production systems that were in place. These increases in diversity were helped by changes in US policy that no longer tied production to base acreages.

Constraints With Combining Conservation Agriculture With Increased Crop Diversity

For small holder farmers, NT may imply costly or improper changes due to the increase of labor burden to control weed in low-input systems. In addition to labor, a second problem is the competition for crop residues for soil mulching or livestock feed (Gillies et al., 2015). At higher scales, NT systems have been responsible for a significant increase in the use of chemicals, mainly herbicides (Scopel et al., 2013). The repeated reliance on specific herbicides such as glyphosate has led to rapid evolution of herbicide-resistant weeds (Giller et al., 2015).

Synergies in Combining Conservation Agriculture With Increased Crop Diversity

As noted earlier, conservation tillage can provide opportunities for crop diversification by enhancing available soil water, especially in dry areas. Increased soil water can potentially enhance crop diversity in two ways. First, because of the greater precipitation use efficiency, crops can be grown every year rather than every other year, as in a crop-fallow system (Peterson et al., 1996). This helps to promote diversity in semiarid regions as producers look for alternative crops to replace the small grains-fallow rotation. Secondly, better precipitation use efficiency under conservation tillage may provide opportunities to raise crops that typically would not be part of the producer's crop portfolio because of water stress. However, with better genetics and technologies, producers may select crop rotations that limit diversity. This happened in the southeast corner of North Dakota, United States, where producers began a corn-soybean rotation and crop diversity decreased (Aguilar et al., 2015).

Methods for Combining Conservation Agriculture With Increased Crop Diversity

In the present, the literature is replete with merits, limitations, and uncertainties of conservation tillage that the focus now is on conservation agriculture as a system. The way to increase biodiversity in a system-based approach to transform no-till into conservation agriculture is by cover cropping, complex crop rotations, and integration of systems (Lal, 2015; Sanderson et al., 2013). Economic and environmental performances will only improve concurrently if the reduction in tilling does not become equivalent to a "simplified cropping system." In other words, farmers will need technical reference materials, assessment tools, and agronomic rationale to adapt conservation agriculture principles to their particular situation (Giller et al., 2015; Schaller, 2013).

A cover crop (CC) is close-growing crop that provides soil protection, seeding protection, and soil improvement between periods of normal crop production, or between trees in orchards and vines in vineyards. When plowed under and incorporated into the soil, CCs may be referred to as green manure crops (SSSA, 2017). When combined with improved management systems such as no-till, CCs can enhance and increase the magnitude of benefits of current no-till and reduced-tillage practices relative to these same tillage systems without CCs. A recent review indicates that CCs can provide numerous ecosystem services including control of water and wind erosion, improvement in soil physical, chemical, and biologic properties, sequestration of soil organic C, nutrient cycling, suppression of weeds, improvement in wildlife habitat and diversity, and potential provision of both forage for livestock and feedstock for cellulosic biofuel production (Blanco-Canqui et al., 2015).

Crop selection and sequencing can serve as a critical component in conservation of agricultural systems. Among the many options available to select and sequence crops, a fixed-sequence system, whereby crops are sequenced in a consistent, unchanging pattern, is the most simple (Sanderson et al., 2013). Intercropping is particularly important on small farms in the tropics (Giller et al., 2015). Intercropping is a farming practice involving two or more crop species, or genotypes, growing together and coexisting for a time. On the fringes of modern intensive agriculture, intercropping is important in many subsistence or low-input/resource-limited agricultural systems. By allowing genuine yield gains without increased inputs, or greater stability of yield with decreased inputs, intercropping could be one route to delivering "sustainable intensification." In the short term, perhaps the most straightforward approach is simply to trial new combinations of crops to exploit beneficial mechanisms that have already been identified, for example, new combinations of cereals and legumes (a widespread focus for current research). Rapid improvements are also possible through the development of new agronomic practices, including the mechanization of intercropping systems and improved nutrient management (Brooker et al., 2015).

Diverse multiple crop/pasture systems are required rather than crop rotation alone (Giller et al., 2015). Crop–livestock integration diversifies landscape mosaics, enhancing biodiversity (Lemaire et al., 2014). The fundamental role of grasslands on the reduction of environmental fluxes to the atmosphere and hydrosphere operates through the coupling of C and N cycles within vegetation, soil organic matter, and soil microbial biomass. Therefore, close association of grassland systems with cropping systems should help mitigate negative environmental impacts resulting from intensification of cropping systems and improve the quality of grasslands through periodic renovations (Lemaire et al., 2014).

INSIGHTS

Loss of Diversity

Most biodiversity research has focused on natural areas (Tscharntke et al., 2005b), despite agriculture's large spatial extent. In the continental United States, land used for crop production occupies 22% of the land area (165 million ha) (Nickerson et al., 2012). Biodiversity in agroecosystems provides a variety of ecosystem services such as food production, nutrient recycling, and microclimate regulation (Altieri, 1999), and adding diversity to agroecosystems may provide more resilience to climatic change (Lin, 2011). Loss of crop diversity can increase landscape simplification and potentially increase pest pressure and insecticide use (Meehan et al., 2011). Agricultural landscapes also impact desirable or beneficial species. For instance, landscape structure or the amount of heterogeneity in the landscape is an important consideration in butterfly diversity and species composition (Weibull et al., 2000). Pollinators are essential for many of fruit crops that provide essential nutrients for a healthy diet (Klein et al., 2007). It is unclear if the decline in pollinators is from a decline in plant diversity or if plant diversity is declining because of a lack of pollinators (Potts et al., 2010). While crop diversification appears to benefit ecosystem services, increasing agricultural biodiversity can increase productivity over a range of environments and enhance diet diversity while providing a more resilient agroecosystem (Frison et al., 2011)

Benefits of Conservation Cropping

The potential contribution of no-till to the sustainable of agriculture is limited to the combination of the other two principles of conservation agriculture, residue retention and crop rotation (Pittelkow et al., 2015). In this sense, this is the importance of diversification of crops with crop rotation and CCs (Derpsch et al., 2014).

An approach based only in no-till is not enough to prevent water and wind erosion (Reicosky, 2015; Nearing et al., 2017; Merril et al., 2006). Under NT, water loss through runoff is more associated with the management of surface cover, mainly the number of months of the year occupied by crops, than with an improvement in the physical properties related to porosity and internal soil water movement (Sasal et al., 2010).

CCs can be used to reduce the yield gap between conservation agriculture and intensive tillage. The main drivers of the yield decrease were reduced nitrogen availability and increased weed infestation. Moreover, the inclusion of nitrogen fixing CCs in the organic production systems led to increased yields and could substantially contribute to decrease the yield gaps compared to the conventional systems (Wittwer et al., 2017). A continuous no-till system with a CC is a promising way to increase the NUE of maize, and consequently to reduce both the use and the loss of N fertilizers without any yield penalty (Habbib et al., 2016).

Intensification of cropping systems with high above- and belowground biomass (i.e., deep-rooted plant species) input may enhance CA systems for storing soil C relative to CT (Luo et al., 2010; Palm et al., 2014). The presence of perennial legumes in integrated crop-livestock systems indicated that soil C may be sequestered at depth, suggesting that roots provide a substantial contribution to organic matter (de Dios Herrero et al., 2016; Franzluebber et al., 2014; Giller et al., 2015). Also, grasslands couple C–N cycles and reduce environmental fluxes. Efficient recycling of animal manures is a key aspect of sustainable nutrient management, both to avoid pollution and to maintain soil organic matter crop nutrient supply (Lemaire et al., 2014; Giller et al., 2015).

Pathways Forward

While agricultural systems have previously focused on production, interest is increasing in alternative foci such as ecosystem services (Termorshuizen and Opdam, 2009; Willemen et al., 2010) and climate-smart agriculture (Lipper et al., 2014). Conservation cropping systems offer a diversity of options that can be combined to build climate-smart agriculture for adaptation and mitigation to climate change. Changing crop species and cultivars to take advantage of opportunities offered by climate change or to minimize the vulnerability to extreme conditions is an additional strategy that should be combined consistently with the change of planting date (Debaeke et al., 2017). Conservation agriculture can also enhance water infiltration, which allows producers access to a wider crop portfolio and may increase crop diversity. Including other techniques such as CC and integration of crop-livestock systems can also enhance diversity, resilience, and ecosystem services.

The combination of conservation tillage and a dynamic approach to crop sequencing based in a long-term strategy of annual crop sequencing that optimizes crop and soil use options to attain production, economic, and resource conservation goals by using sound ecologic management principles has been purported to be more economically and environmentally sustainable (Tanaka et al., 2002). The improved combined management practice that included a no-till, diversified cropping system (crop rotation, increased cropping intensity, and perennial crop) and reduced N rate decreased global warming potential and net greenhouse gas intensity I by 70%–88% compared with the traditional combined practice that included a conventional till, less diversified cropping system (Sainju, 2016).

Increasing adoption of conservation cropping systems requires prudent strategies to address limitations and uncertainties of NT (Lal, 2015). More specific studies to overcome such limitations and how to enhance farmers' adoption and adaptation are thus necessary (Scopel et al., 2013). In this sense, institutional support, extension services, information from long-term experiments, adaptation of machinery, and access to market are critically essential, especially with small landholders (Derpsch et al., 2016; Lal, 2015).

Agriculture is crucial to the health and stability of individuals, communities, and nations. Enhancing the resiliency of agriculture by using conservation agriculture techniques and improving diversity will help to ensure the role of agriculture in providing these functions in the future.

References

Abdi, D., Cade-Menun, B.J., Ziadi, N., Parent, L.E., 2014. Long-term impact of tillage practices and phosphorus fertilization on soil phosphorus forms as determined by 31P nuclear magnetic resonance spectroscopy. Journal of Environmental Quality 43, 1431–1441.

Aguilar, J., Gramig, G.G., Hendrickson, J.R., Archer, D.W., Forcella, F., Liebig, M.A., 2015. Crop species diversity changes in the United States: 1978–2012. PLoS One 10, e0136580.

Altieri, M.A., 1999. The ecological role of biodiversity in agroecosystems. Agriculture, Ecosystems and Environment 74, 19–31.

Balderacchi, M., Benoit, P., Cambier, P., Eklo, O.M., Gargini, A., Gemitzi, A., Gurel, M., Kløve, B., Nakic, Z., Predaa, E., 2013. Groundwater pollution and quality monitoring approaches at the European level. Critical Reviews in Environmental Science and Technology 43, 323–408.

Benbrook, C., 2001. Do GM crops mean less pesticide use? Pesticide Outlook 12, 204–207.

Benbrook, C.M., 2012. Impacts of genetically engineered crops on pesticide use in the US—the first sixteen years. Environmental Sciences Europe 24, 1.

Bending, G.D., Turner, M.K., Jones, J.E., 2002. Interactions between crop residue and soil organic matter quality and the functional diversity of soil microbial communities. Soil Biology and Biochemistry 34, 1073–1082.

Bennett, E.M., Carpenter, S.R., Caraco, N.F., 2001. Human Impact on Erodable Phosphorus and Eutrophication: a Global Perspective: increasing accumulation of phosphorus in soil threatens rivers, lakes, and coastal oceans with eutrophication. BioScience 51, 227–234.

Blanco-Canqui, H., Shaver, T.M., Lindquist, J.L., Shapiro, C.A., Elmore, R.W., Francis, C.A., Hergert, G.W., 2015. Cover crops and ecosystem services: insights from studies in temperate soils. Agronomy Journal 107, 2449–2474.

Boody, G., Vondracek, B., Andow, D.A., Krinke, M., Westra, J., Zimmerman, J., Welle, P., 2005. Multifunctional agriculture in the United States. BioScience 55, 27–38.

Brooker, R.W., Bennet, A.E., Cong, W.F., Daniell, T.J., George, T., Hallett, P.D., Hawes, C., Iannetta, P.P., Jones, H.G., Karley, A.J., Li, L., Mckenzie, B.M., Pakeman, J., Paterson, E., Schöb, C., Shen, J., Squire, G., Watson, C.A., Zhang, C., Zhang, F., Zhang, J., White, P.J., 2015. Improving intercropping: a synthesis of research in agronomy, plant physiology and ecology. New Phytologist 206, 107–117.

Brummer, E.C., 1998. Diversity, stability, and sustainable American agriculture. Agronomy Journal 90, 1–2.

Bullock, J.M., Pywell, R.F., Burke, M.J., Walker, K.J., 2001. Restoration of biodiversity enhances agricultural production. Ecology Letters 4, 185–189.

Bullock, D.G., 1992. Crop rotation. Critical Reviews in Plant Sciences 11, 309–326.

Burkholder, J., Libra, B., Weyer, P., Heathcote, S., Kolpin, D., Thorne, P.S., Wichman, M., 2007. Impacts of waste from concentrated animal feeding operations on water quality. Environmental Health Perspectives 115, 308–312.

Buschiazzo, D.E., Zobeck, T.M., Abascal, S.A., 2007. Wind erosion quantity and quality of an Entic Haplustoll of the semi-arid pampas of Argentina. Journal of Arid Environment 69, 29–39.

Cadisch, G., Giller, K., 2001. Soil organic matter management: The Roles of Residue Quality in C Sequestration and N Supply. Sustainable Management of Soil Organic Matter p. 97–111.

Canfield, D.E., Glazer, A.N., Falkowski, P.G., 2010. The evolution and future of Earth's nitrogen cycle. Science 330, 192–196.

Cardinale, B.J., Wright, J.P., Cadotte, M.W., Carroll, I.T., Hector, A., Srivastava, D.S., Loreau, M., Weis, J.J., 2007. Impacts of plant diversity on biomass production increase through time because of species complementarity. Proceedings of the National Academy of Sciences 104, 18123–18128.

Carpenter, S.R., Caraco, N.F., Correll, D.L., Howarth, R.W., Sharpley, A.N., Smith, V.H., 1998. Nonpoint pollution of surface waters with phosphorus and nitrogen. Ecological Applications 8, 559–568.

Cassman, K.G., 1999. Ecological intensification of cereal production systems: yield potential, soil quality, and precision agriculture. Proceedings of the National Academy of Sciences 96, 5952–5959.

Colazo, J.C., Buschiazzo, D.E., 2010. Soil dry aggregate stability and wind erodible fraction in a semiarid environment of Argentina. Geoderma 159, 228–236.

De Dios Herrero, J.M., Colazo, J.C., Guzmán, M.L., SAENZ, C.A., Sager, R., Sakadevan, K., 2016. Soil organic carbon assessments in cropping systems using isotopic techniques. EGU General Assembly Conference Abstracts 18, 4009.

Debaeke, P., Pellerin, S., Scopel, E., 2017. Climate-smart cropping systems form temperate and tropical agriculture: mitigation, adaptation and trade-offs. Cahiers Agricultures 26, 34022.

Derpsch, R., Friedrich, T., Kassam, A., Hongwen, L., 2010. Current status of adoption of No-till farming in the world and some of its main benefits. International Journal of Agricultural and Biological Engineering 3, 1–25.

Derpsch, R., Franzluebbers, A.J., Duiker, S.W., Reicosky, D.C., Koeller, K., Friedrich, T., 2014. Why do we need to standardize no-tillage research? Soil and Tillage Research 137, 16–22.

Derpsch, R., Lange, D., Birbaumer, G., Moriya, K., 2016. Why do medium- and large-scale farmers succeed practicing CA and small-scale farmers often do not? Experiences from Paraguay. International Journal of Agricultural Sustainability 14, 269–281.

Dimitri, C., Effland, A.B., Conklin, N.C., 2005. The 20th Century Transformation of US Agriculture and Farm Policy, US Department of Agriculture. Economic Research Service, Washington, DC.

Duvick, D., Cassman, K.G., 1999. Post–green revolution trends in yield potential of temperate maize in the North-Central United States. Crop Science 39, 1622–1630.

Egli, D., 2008. Comparison of corn and soybean yields in the United States: historical trends and future prospects. Agronomy Journal 100. S-79–S-88.

ERS, 2018. Adoption of genetically engineered crops in the U.S. Economic Research Service-USDA. Available at. https://www.ers.usda.gov/data-products/adoption-of-genetically-engineered-crops-in-the-us/. (Accessed 12 September 2018).

Fao, 2015. World Fertilizer Trends and Outlook to 2018. Food and Agriculture Organization of the United Nations, Rome, Italy.

Fernández, F.G., Brouder, S.M., Beyrouty, C.A., Volenec, J.J., Hoyum, R., 2008. Assessment of plant-available potassium for No-Till, rainfed soybean. Soil Science Society of America Journal 72, 1085–1095.

Foley, J.A., Ramankutty, N., Brauman, K.A., Cassidy, E.S., Gerber, J.S., Johnston, M., Mueller, N.D., O'Connell, C., Ray, D.K., West, P.C., Balzer, C., 2011. Solutions for a cultivated planet. Nature 478, 337–342.

Franzluebbers, A., Sawchik, J., Taboada, M., 2014. Agronomic and environmental impacts of pasture–crop rotations in temperate North and South America. Agriculture, Ecosystems and Environment 190, 18–26.

Frison, E.A., Cherfas, J., Hodgkin, T., 2011. Agricultural biodiversity is essential for a sustainable improvement in food and nutrition security. Sustainability 3, 238–253.

Fryrear, D.W., 1985. Soil cover and wind erosion. Transactions of ASAE 28, 781.

Gao, Y., Dang, X., Yu, Y., Liu, Y., Wang, J., 2016. Effects of tillage methods on soil carbon and wind erosion. Land Degradation and Development 27, 583–591.

Giller, K., Andersson, J.A., Corbeels, M., Kirkergaard, J., Mortensen, D., Erestien, O., Vanlauwe, B., 2015. Beyond conservation agriculture. Frontier in Plant Science 6, 870.

Gollehon, N., Caswell, M., Ribaudo, M., Kellogg, R., Lander, C., Letson, D., 2001. Confined Animal Production and Manure Nutrients. Resource Economics 0.37 0.37Division.

Govaerts, B., Mezzalama, M., Unno, Y., Sayre, K.D., Luna-Guido, M., Vanherck, K., Dendooven, L., Deckers, J., 2007. Influence of tillage, residue management, and crop rotation on soil microbial biomass and catabolic diversity. Applied Soil Ecology 37, 18–30.

Habbib, H., Verzeaux, J., Roger, D., Lacoux, J., Catterou, M., Hirel, B., Dubois, F., Tétu, T., 2016. Conversion to No-till Improves Maize Nitrogen Use Efficiency in a Continuous Cover Cropping System.

Hargrove, W., 1986. Winter legumes as a nitrogen source for no-till grain sorghum. Agronomy Journal 78, 70–74.

Hanson, J.D., Liebig, M.A., Merrill, S.D., Tanaka, D.L., Krupinsky, J.M., Stott, D.E., 2007. Dynamic cropping systems. Agronomy Journal 99, 939–943.

Hendrickson, J., Sassenrath, G., Archer, D., Hanson, J., Halloran, J., 2008a. Interactions in integrated US agricultural systems: the past, present and future. Renewable Agriculture and Food Systems 23, 314–324.

Hendrickson, J.R., Hanson, J., Tanaka, D.L., Sassenrath, G., 2008b. Principles of integrated agricultural systems: introduction to processes and definition. Renewable Agriculture and Food Systems 23, 265–271.

Hevia, G.G., Méndez, M.J., Buschiazzo, D.E., 2007. Tillage affects soil aggregation parameters linked with wind erosion. Geoderma 140, 90–96.

Honeyman, M., 1996. Sustainability issues of US swine production. Journal of Animal Science 74, 1410–1417.

Howarth, R.W., 2008. Coastal nitrogen pollution: a review of sources and trends globally and regionally. Harmful Algae 8, 14–20.

Isbell, F., Calcagno, V., Hector, A., Connolly, J., Harpole, W.S., Reich, P.B., Scherer-Lorenzen, M., Schmid, B., Tilman, D., Van Ruijven, J., Weigelt, A., 2011. High plant diversity is needed to maintain ecosystem services. Nature 477, 199–202.

Iverson, A.L., Marín, L.E., Ennis, K.K., Gonthier, D.J., Connor-Barrie, B.T., Remfert, J.L., Cardinale, B.J., Perfecto, I., 2014. Review: do polycultures promote win-wins or trade-offs in agricultural ecosystem services? A meta-analysis. Journal of Applied Ecology 51, 1593–1602. https://doi.org/10.1111/1365-2664.12334.

Johansen, C., Haque, M.E., Bell, R.W., Tierfelder, C., Esdaile, R.J., 2012. Conservation agriculture for smallholder rainfed farming: opportunities and constraints of new mechanized seeding systems. Field Crops Research 132, 18–32.

Johnston, J.R., Browning, G.M., Russell, M.B., 1943. The effect of cropping practices on aggregation, organic matter content, and loss of soil and water in the Marshall Silt Loam1. Soil Science Society of America Journal 7, 105–107.

Karlen, D.L., Hurley, E.G., Andrews, S.S., Cambardella, C.A., Meek, D.W., Duffy, M.D., Mallarino, A.P., 2006. Crop rotation effects on soil quality at three northern corn/soybean belt locations. Agronomy Journal 98, 484–495.

Kassam, A., Derpsch, R., Friedrich, T., 2014. Global achievements in soil and water conservation: the case of conservation agriculture. International Soil and Water Conservation Research 2, 5–13.

Kelner, D.J., Vessey, J.K., Entz, M.H., 1997. The nitrogen dynamics of 1-, 2-and 3-year stands of alfalfa in a cropping system. Agriculture, Ecosystems and Environment 64, 1–10.

Khoury, C.K., Bjorkman, A.D., Dempewolf, H., Ramirez-Villegas, J., Guarino, L., Jarvis, A., Rieseberg, L.H., Struik, P.C., 2014. Increasing homogeneity in global food supplies and the implications for food security. Proceedings of the National Academy of Sciences of the United States of America 111, 4001–4006.

Klein, A.M., Vaissiere, B.E., Cane, J.H., Steffan-Dewenter, I., Cunningham, S.A., Kremen, C., Tscharntke, T., 2007. Importance of pollinators in changing landscapes for world crops. Proceedings of the Royal Society B: Biological Sciences 274, 303–313.

Kleinman, P.J.A., Sharpley, A.N., Withers, P.J.A., Bergström, L., Johnson, L.T., Doody, D.G., 2015. Implementing agricultural phosphorus science and management to combat eutrophication. Ambio 44, S297–S310.

Kolahchi, Z., Jalali, M., 2007. Effect of water quality on the leaching of potassium from sandy soil. Journal of Arid Environments 68, 624–639.

Lal, R., 2015. A system approach to conservation agriculture. Journal of Soil and Water Conservation 70, 82–84.

Lal, R., 2015b. Sequestering carbon and increasing productivity by conservation agriculture. Journal of Soil and Water Conservation 70, 55–62.

Lemaire, G., Franzluebbers, A., De Fassio Carvalho, P., Dedieu, B., 2014. Integrated crop–livestock systems: strategies to achieve synergy between agricultural production and environmental quality. Agriculture, Ecosystems and Environment 190, 4–8.

Letourneau, D.K., Armbrecht, I., Rivera, B.S., Lerma, J.M., Carmona, E.J., Daza, M.C., Escobar, S., Galindo, V., Gutiérrez, C., López, S.D., Mejía, J.L., Rangel, A.M.A., Rangel, J.H., Rivera, L., Saavedra, C.A., Torres, A.M., Trujillo, A.R., 2011. Does plant diversity benefit agroecosystems? A synthetic review. Ecological Applications 21, 9–21. https://doi.org/10.1890/09-2026.1.

Liang, X., Zhang, H., He, M., Yuan, J., Xu, L., Tian, G., 2016. No-tillage effects on grain yield, N use efficiency and nutrient runoff losses in paddy fields. Environmental Science and Pollution Research 23, 21451–21459.

Lin, B.B., 2011. Resilience in agriculture through crop diversification: adaptive management for environmental change. BioScience 61, 183–193.

Lipper, L., Thornton, P., Campbell, B.M., Baedeker, T., Braimoh, A., Bwalya, M., Caron, P., Cattaneo, A., Garrity, D., Henry, K., Hottle, R., Jackson, L., Jarvis, A., Kossam, F., Mann, W., Mccarthy, N., Meybeck, A., Neufeldt, H., Remington, T., Sen, P.T., Sessa, R., Shula, R., Tibu, A., Torquebiau, E.F., 2014. Climate-smart agriculture for food security. Nature Climate Change 4, 1068–1072.

Lundy, M.E., Pittelkow, C.M., Linquist, B.A., Liang, X., Van Groenigen, K.J., Lee, J., Six, J., Venterea, R.T., 2015. Nitrogen fertilization reduces yield declines following no-till adoption. Field Crops Research 183, 204–210.

Luo, Z., Wang, E., Sun, O.J., 2010. Can no-tillage stimulate carbon sequestration in agricultural soils? A meta-analysis of paired experiments. Agriculture, Ecosystems and Environment 139, 224–231.

Macdonald, J.M., Mcbride, W.D., 2009. The Transformation of US Livestock Agriculture Scale, Efficiency, and Risks. Economic Information Bulletin.

Mallin, M.A., Cahoon, L.B., 2003. Industrialized animal production—a major source of nutrient and microbial pollution to aquatic ecosystems. Population and Environment 24, 369–385.

Margenot, A.J., Paul, B.K., Sommer, R.R., Pulleman, M.M., Parikh, S.J., Jackson, L.E., Fonte, S.J., 2017. Can conservation agriculture improve phosphorus (P) availability in weathered soils? Effects of tillage and residue management on soil P status after 9 years in a Kenyan Oxisol. Soil and Tillage Research 166, 157–166.

Meehan, T.D., Werling, B.P., Landis, D.A., Gratton, C., 2011. Agricultural landscape simplification and insecticide use in the Midwestern United States. Proceedings of the National Academy of Sciences of the United States of America 108, 11500–11505.

Mendez, M.J., Buschiazzo, D.E., 2010. Wind erosion risk in agricultural soils under different tillage systems in the semiarid Pampas of Argentina. Soil and Tillage Research 106, 311–316.

Merrill, S.D., Krupinsky, J.M., Tanaka, L.D., Anderson, R.L., 2006. Soil coverage by residue as affected by ten crop species under no-till in the northern Great Plains. Journal of Soil and Water Conservation 61, 7–13.

NASS, 2016a. USDA– National Agricultural Statistics Service, Quick Stats. https://www.nass.usda.gov/Statistics_by_Subject/?sector=CROPS.

NASS, 2016b. USDA—National Agricultural Statistics Service, Quick Stats. https://quickstats.nass.usda.gov/results/ECE6ABAA-02AD-37F1-9DCC-D74DCA58A6EB?pivot=-short_desc. https://quickstats.nass.usda.gov/results/ED AD73E0-02AB-38BE-BD79-6FAE7A7F62A0?pivot=short_desc.

Nearing, M.A., Xie, Y., Liu, B., Ye, Y., 2017. Natural and anthropogenic rates of soil erosion. International Soil and Water Conservation Research 5, 77–84.

Nickerson, C., Morehart, M., Kuethe, T., Beckman, J., Ifft, J., Williams, R., 2012. Trends in US Farmland Values and Ownership. US Department of Agriculture, Economic Research Service.

Osteen, C.D., Fernandez-Cornejo, J., 2013. Economic and policy issues of US agricultural pesticide use trends. Pest management science 69, 1001–1025.

Palm, C., Blanco Canqui, H., Declerk, F., Gatere, L., Grace, P., 2014. Conservation agriculture and ecosystem services: an overview. Agriculture, Ecosystems and Environment 187, 87–105.

Peterson, G., Schlegel, A., Tanaka, D., Jones, O., 1996. Precipitation use efficiency as affected by cropping and tillage systems. Journal of Production Agriculture 9, 180–186.

Pittelkow, C.M., Liang, X., Linquist, B.A., Van Groening, K.J., Lee, J., Lundy, M.E., Van Gestel, N., Six, J., Venterea, R.T., Van Kessel, C., 2015. Productivity limits and potentials of the principles of conservation agriculture. Nature 517, 365–370.

Potts, S.G., Biesmeijer, J.C., Kremen, C., Neumann, P., Schweiger, O., Kunin, W.E., 2010. Global pollinator declines: trends, impacts and drivers. Trends in Ecology and Evolution 25, 345–353.

Powlson, D.S., Stirling, C.M., Thierfelder, C., White, R.P., Jat, M.L., 2016. Does conservation agriculture deliver climate change mitigation through soil carbon sequestration in tropical agro-ecosystems? Agriculture, Ecosystems and Environment 220, 164–174.

Qaim, M., Zilberman, D., 2003. Yield effects of genetically modified crops in developing countries. Science 299, 900–902.

Rasouli, S., Whalen, J.K., Madramootoo, A., 2014. Review: reducing residual soil nitrogen losses from agroecosystems for surface water protection in Quebec and Ontario, Canada: best management practices, policies and perspectives. Canadian Journal of Soil Science 94, 109–127.

Reicosky, D.C., 2015. Conservation tillage is not conservation agriculture. Journal of Soil and Water Conservation 70, 103A–108A.

Ruttan, V.W., 1999. The transition to agricultural sustainability. Proceedings of the National Academy of Sciences of the United States of America 96, 5960–5967.

Sainju, U.M., 2016. A global meta-analysis on the impact of management practices on net global warming potential and greenhouse gas intensity from cropland soils. PLoS One 11, e0148527.

Sanderson, M.A., Archer, D., Hendrickson, J., Kronberg, S., Liebig, K.N., Schmer, M., Tanaka, D., Aguilar, J., 2013. Diversification and ecosystem services for conservation agriculture: Outcomes from pastures and integrated crop–livestock systems. Renewable Agriculture and Food Systems 28, 129–144.

Sasal, M.C., Castiglioni, M.G., Wilson, M.G., 2010. Effect of crop sequences on soil properties and runoff on natural-rainfall erosion plots under no tillage. Soil and Tillage Research 108, 24–29.

Schaller, N., 2013. Centre for Studies and Strategic Foresight. Conservation Agriculture. Analysis, 61.

Schläpfer, F., Schmid, B., 1999. Ecosystem effects of biodiversity: a classification of hypotheses and exploration of empirical results. Ecological Applications 9, 893–912.

Scopel, E., Triomphe, B., Affholder, F., DA Silva, F.A.M., Corbeels, M., Xavier, J.H.V., Lahmar, R., Recous, S., Bernoux, M., Blanchat, E., De Carvahlo Mendes, I., De Tourdonnet, S., 2013. Conservation agriculture cropping systems in temperate and tropical conditions, performances and impacts. A review. Agronomy for Sustainable Development 33, 113–130.

Selles, F., Grant, C., Johnston, A., 2002. Conservation tillage effects on soil phosphorus distribution. Better Crops 86, 1–6.

Sharpley, A.N., Daniel, T., Edwards, D., 1993. Phosphorus movement in the landscape. Journal of Production Agriculture 6, 492–500.

Sharrat, B., Young, F., Feng, G., 2017. Wind erosion and PM10 emissions from No-Tillage cropping systems in the Pacific Northwest. Agronomy Journal 109, 1303–1311.

Snapp, S.S., Gentry, L.E., Harwood, R., 2010. Management intensity – not biodiversity – the driver of ecosystem services in a long-term row crop experiment. Agriculture, Ecosystems and Environment 138, 242–248.

SSSA (Soil Science Society of America), 2017. Glossary of Soil Science Terms. https://www.soils.org/publications/soils-glossary.

Stanger, T.F., Lauer, J.G., 2008. Corn grain yield response to crop rotation and nitrogen over 35 years. Agronomy Journal 100, 643.

Tan, C., Cao, X., Yuan, S., Wang, W., Feng, Y., Qiao, B., 2015. Effects of long-term conservation tillage on soil nutrients in sloping fields in regions characterized by water and wind erosion. Scientific Reports 5, 17592.

Tanaka, D.L., Krupinsky, J.M., Liebig, M.A., Merrill, S.D., Ries, R.E., Hendrickson, J.R., Johnson, H.A., Hanson, J.D., 2002. Dynamic cropping systems: an adaptable approach to crop production in the Great Plains. Agronomy Journal 94, 957–961.

Termorshuizen, J.W., Opdam, P., 2009. Landscape services as a bridge between landscape ecology and sustainable development. Landscape Ecology 24, 1037–1052.

Tilman, D., Reich, P.B., Knops, J., Wedin, D., Mielke, T., Lehman, C., 2001. Diversity and productivity in a long-term grassland experiment. Science 294, 843–845.

Tollenaar, M., Lee, E., 2002. Yield potential, yield stability and stress tolerance in maize. Field Crops Research 75, 161–169.

Trewavas, A., 2002. Malthus foiled again and again. Nature 418, 668–670.

Tscharntke, T., Klein, A.M., Kruess, A., Steffan-Dewenter, I., Thies, C., 2005a. Landscape perspectives on agricultural intensification and biodiversity â€ ecosystem service management. Ecology Letters 8, 857–874.

Tscharntke, T., Klein, A.M., Kruess, A., Steffan-Dewenter, I., Thies, C., 2005b. Landscape perspectives on agricultural intensification and biodiversity–ecosystem service management. Ecology Letters 8, 857–874.

U.S. Census bureau, 2016. U.S. Census Bureau, International Database. Updated August 2016. http://www.census.gov/population/international/data/worldpop/table_population.php.

Unger, P.W., Kirkham, M.B., Nielsen, D.C., 2010. Water conservation for agriculture. In: Zobeck, T.M., Schillinger, W.F. (Eds.), Soil and Water Conservation Advance in the United States, vol. 60. SSSA Special Publication, pp. 1–45.

Unger, P.W., 2002. Conservation tillage for improving dryland crop yields. Ciencia del Suelo (Argentina) 20, 1–8.

Van Doren, D., Moldenhauer, W., Triplett, G., 1984. Influence of long-term tillage and crop rotation on water erosion. Soil Science Society of America Journal 48, 636–640.

Van Kessel, C., Venterea, R., Six, J., Adviento-Borbe, M.A., Linquist, B., VAN Groenigen, K.J., 2013. Climate, duration, and N placement determine N_2O emissions in reduced tillage systems: a meta-analysis. Global Change Biology 19, 33–44.

Varvel, G.E., Wilhelm, W.W., 2011. No-tillage increases soil profile carbon and nitrogen under long-term rainfed cropping systems. Soil and Tillage Research 114, 28–36.

Weber, M.A., Mielniczuk, J., Tornquist, C.G., 2016. Changes in soil organic carbon and nitrogen stocks in long-term experiments in southern Brazil simulated with century 4.5. Revista Brasileira de Ciência do Solo 40, 1–17.

Weibull, A.C., Bengtsson, J., Nohlgren, E., 2000. Diversity of butterflies in the agricultural landscape: the role of farming system and landscape heterogeneity. Ecography 23, 743–750.

White, M.J., Santhi, C., Kannan, N., Arnold, J.G., Harmel, D., Norfleet, L., Allen, P., Diluzio, M., Wang, X., Atwood, J., 2014. Nutrient delivery from the Mississippi River to the Gulf of Mexico and effects of cropland conservation. Journal of Soil and Water Conservation 69, 26–40.

Willemen, L., Hein, L., Verburg, P.H., 2010. Evaluating the impact of regional development policies on future landscape services. Ecological Economics 69, 2244–2254.

Wing, S., Freedman, S., Band, L., 2002. The potential impact of flooding on confined animal feeding operations in eastern North Carolina. Environmental Health Perspectives 110, 387.

Wittwer, R.A., Dorn, B., Werner, J., Van Der Heijen, M.G.A., 2017. Cover crops support ecological intensification of arable cropping systems. Scientific Reports 7, 41911.

Wright, A.L., Hons, F.M., Lemon, R.G., Mcfarland, M.L., Nichols, R.L., 2007. Stratification of nutrients in soil for different tillage regimes and cotton rotations. Soil and Tillage Research 96, 19–27.

Zak, D.R., Holmes, W.E., White, D.C., Peacock, A.D., Tilman, D., 2003. Plant diversity, soil microbial communities, and ecosystem function: are there any links? Ecology 84, 2042–2050.

Zibilske, L.M., Bradford, J., 2003. Tillage effect on phosphorous mineralization and microbial activity. Soil Science 168, 677–685.

Zibilske, L.M., Bradford, J.M., Smart, J.R., 2002. Conservation tillage induced changes in organic carbon, total nitrogen and available phosphorus in a semi-arid alkaline subtropical soil. Soil and Tillage Research 66, 153–163.

Zirkle, K.W., Nolan, B.T., Jones, R.R., Weyer, P.J., Ward, M.H., Wheeler, D.C., 2016. Assessing the relationship between groundwater nitrate and animal feeding operations in Iowa (USA). The Science of the Total Environment 566, 1062–1068.

CHAPTER

7

Building Agricultural Resilience With Conservation Pasture-Crop Rotations

Alan J. Franzluebbers[1], *Francois Gastal*[2]

[1]USDA – Agricultural Research Service, Raleigh, NC, United States; [2]FERLUS, INRA, Lusignan, France

INTRODUCTION

Agriculture in industrialized countries has become increasingly specialized in response to political and economic pressures to meet market demands of an ever-larger food and fiber processing sector (Russelle et al., 2007; Hendrickson et al., 2008). Shipping and global marketing are common throughout the world, allowing grain, fiber, and animal products to be transported large distances. Mass production with low profit margin has been the result, forcing producers to either expand or face economic extinction. Small-scale farmers that do remain in business often have to find alternative niche markets to compete. Community-based agricultural production has largely been replaced with the global production and marketing model.

As evidence of this change, specialization in the United States has been accompanied by a dramatic decline in the number of farms from >6 million in 1920 to <2 million in 2009, however, with an amazing increase in productivity (Hanson and Hendrickson, 2009). Large family and nonfamily farms account for the majority of agricultural sales. The majority of farmers in the United States also worked off the farm, and nearly half of them worked more than half of the year at an off-farm job, primarily to obtain sufficient income and employer-subsidized health insurance (USDA-NASS, 2009). Livestock production has changed, too. Compared with many small, diversified family farms in 1950, today, there are larger specialized row crop and/or concentrated animal feeding operations (Singer et al., 2009).

Similarly, in many western European countries, agriculture was based on mixed crop/livestock production systems from the first agricultural revolution in the 15th century until the middle of the 19th century (Mazoyer and Roudart, 1997). Since then, crop production has become highly specialized in areas like northern, central, and southwestern France, eastern Germany, and eastern England, while livestock production has become concentrated in areas, such as western France, Netherlands, and Denmark (Peyraud et al., 2014). Today, only ~15% of European farms are operating with some sort of mixed crop/livestock system

Agroecosystem Diversity
https://doi.org/10.1016/B978-0-12-811050-8.00007-8

(Peyraud et al., 2014). For both farm operations and food product processing, economy of scale has played a major role in the move toward specialization (Peyraud et al., 2014; Lemaire et al., 2017). For the most part, European farms specialized in intensive livestock production systems have become entangled in major environmental problems resulting from animal concentration and subsequent manure management and water quality issues.

In all these situations, farms specialized in intensive crop production face threats from soil and surface or underground water quality, loss of biodiversity, and many other environmental issues (Tilman et al., 2002). Conservation agriculture systems that integrate crops and livestock could provide opportunities to naturally capture ecologic interactions to make agricultural ecosystems more efficient at cycling of nutrients, rely more on renewable natural resources, and improve the inherent functioning of soils, while achieving acceptable or improved economic returns for the farmer (Franzluebbers, 2007; Russelle et al., 2007; Franzluebbers et al., 2011). Diversifying agricultural production enterprises might also utilize labor more efficiently for many on-farm tasks (Hoagland et al., 2010).

Although it is more ecologically efficient to consume calories and proteins from crops than from meat, and although some livestock production systems have contributed to environmental degradation, livestock play an extremely important role in sustainable agricultural systems because they can utilize crops and residues not suitable as food and fiber for humans (Russelle et al., 2007), as well as transform plant-bound nutrients into readily mineralizable substrates through passage in the rumen to improve soil fertility (Fortuna et al., 2011). In addition, forages as part of a livestock enterprise diversify agroecosystems and thus provide multiple ecosystem services, such as (1) air purification and climate regulation through greenhouse gas exchange, (2) soil formation and retention, (3) water cycling, quality, and infiltration, (4) nutrient cycling, (5) habitat provision, (6) aesthetic experience, (7) recreation, and (8) reflection (Franzluebbers, 2013a).

With recent emphasis on biomass harvest for renewable energy technologies, agriculture can be expected to provide an even wider range of beneficial ecosystem services to society, other than the traditional role of food production (Fike et al., 2006). Perennial biomass feedstocks for biofuel production could help avoid land degradation and improve land utilization of marginal areas. Design of sustainable landscapes to accommodate food, feed, fiber, and fuel production has biophysical limitations, but it also requires the input of a diversity of stakeholders, including farmers, communities, industries, and government policymakers. Attention to soil ecology and management to optimize multiple functions must be a part of the discussion (Franzluebbers, 2015).

LAND DEGRADATION AND SOIL QUALITY

Agriculture depends upon soil to serve as a medium for plant growth, as a reservoir of nutrients and water, and as a filter to detoxify chemical inputs. Well-functioning soil contributes to the production of abundant, high-quality food and fiber (Lehman et al., 2015). Unfortunately, poor management can exhaust soil (Franzluebbers et al., 2006). Excessive soil disturbance and minimal inputs of carbon via plant cover/residues and animal manures are leading factors causing poor soil health. Soil loss by erosion (wind, water, and tillage) has been, and continues to be, a major threat to soil sustainability around the world. Intensive tillage has been a traditional agricultural practice, but it is clear that it is not appropriate in many areas of North America, because it contributes to a high risk of erosion and losses of soil organic C and other nutrients. Soil productivity is also threatened by salinization from irrigation, contamination of soil with heavy metals, pesticides, and industrial byproducts, and suburban encroachment. Although soil resists degradation from some of these pressures, there is a point at which soil simply becomes exhausted and cannot continue to function normally. Economically, even a small loss of soil productivity could be detrimental for farmers operating on a small profit margin.

Grasslands and cultivated perennial forages provide a wealth of conservation and environmental quality benefits for improving soil functions (Singer et al., 2009). The value of forages in stopping soil erosion is unquestionably great and far reaching. Best management practices to control soil erosion include forage plantings across entire landscapes, in buffer strips at the edge of cultivated fields, and particularly in waterways where overland flow of water is at its greatest. Perennial roots of grasses are essential binding agents that keep soil in place and help develop stable soil structure and strong porosity to allow water to infiltrate the profile, thereby improving the hydrologic flow of water through soil and into seepage and groundwater outlets. With deep root systems and associated biologic life (particularly earthworms), grasslands and perennial forage species improve soil structure and soil permeability, facilitate water infiltration, and significantly reduce the risk of anoxia with hydromorphy (Lamande et al., 2003).

One of the key soil characteristics of land that has been in perennial forages for decades is the high concentration of organic matter near the soil surface compared with cultivated cropland (Franzluebbers, 2010a), as well as potentially greater concentrations with depth (Liebig et al., 2005). In the eastern United States, surface soil organic matter under grass approaches that under forest conditions, but deeper in the profile, it can exceed that under forest because of a more extensive and fibrous rooting system.

Significant research over the past few decades has been conducted to illustrate the positive effects of perennial forages on soil properties and processes. For example, water runoff and soil loss were considerable following moldboard plowing of an alfalfa–smooth bromegrass (*Bromus inermis* Leyss.) field, but nearly nonexistent if the converted cropland were managed with no tillage or if the field remained in forage production (Lindstrom et al., 1998). Water runoff and soil loss were reduced at the edge of cropland fields with a grass buffer strip compared with unprotected cropland fields, but more so when cropland was managed with clean tillage than with conservation tillage having presence of surface residues (Gilley et al., 2000). These studies exemplify that whether forages are planted across an entire field or simply in strips within a field, they can significantly reduce water runoff and soil loss.

Soil organic C and N and its biologically active components are often much greater under long-term perennial forages than under cropland. As one of many examples, soil organic C, soil microbial biomass C, and potentially mineralizable C were all greater in the surface 20 cm of soil under grass pasture than under cultivated cropland (Franzluebbers et al., 1998). Additionally, total soil N and stability of soil aggregates were greater under grass than under conventional tillage cropping in New Zealand (Haynes, 1999). Greater concentration of organic matter, larger aggregates, and longer residence time of soil C fractions were also observed under perennial forages than under annual crop in France (Panettieri et al., 2017). These few examples illustrate the more general conclusions regarding the positive role of grasslands and perennial forages on soil C storage, as observed across a number of different studies (Guo and Gifford, 2002; Soussana and Lemaire, 2014). Frequently, research has shown the strong relationship of soil organic C and N fractions on potential N mineralization (Griffin, 2008; Schomberg et al., 2009). These studies demonstrate that soil under perennial forages is enriched in organic C and N fractions, stable in structure, and inherently able to supply a greater quantity of nutrients to plants from the stored soil organic matter.

Accumulation of soil organic matter with perennial grasses and legumes can also prevent leakage of nitrate from soils. Some perennial forage (particularly alfalfa) species have roots that can penetrate deep into the soil (e.g., depth of 150 cm) to create a larger suction zone than in croplands to access nitrate before it leaches below the root zone. The perennial root system of grasslands and forage crops allows nitrate uptake during most of the year, as contrasted with annual crops that have a lag phase in uptake. Furthermore, nitrate that is accumulated by plants is converted to protein, which is subsequently deposited onto the soil surface in the form of organic N as senescent plant parts in unharvested systems, or it is removed mechanically or by grazing animals and deposited as manure in harvested grass systems.

Nitrate lost from tiles draining alfalfa or conservation grassland fields is often only a fraction of the nitrate lost from fields with annual crops of corn and soybean (Randall and Mulla, 2001). Additionally, when land under conservation grassland is converted to cropland, crop yields can be enhanced during the first few years due to release of nutrients stored in soil organic matter, as well as because of improved soil biologic and physical properties. Unfortunately, release of nutrients following grassland termination may not be in synchrony with subsequent crop demand, thereby potentially contributing to nutrient loss if appropriate rotation or tillage systems are not used. Despite the potential risk of nitrate leaching following grassland termination, a long-term pasture-crop rotation experiment in France demonstrated that nitrate concentration in drainage water when integrated over the entire rotation was much lower than in rotations without a grassland phase. Overall, nitrate leaching was inversely related to the duration of the grassland phase (Kunrath et al., 2015). Diversifying crop rotations with species that have different rooting habits, using cover crops, and reducing the disturbance of surface soil with reduced or no-tillage practices can lower the intensity of nitrate production from decomposition of soil organic matter. Several factors other than just deep and long-lasting root structures of perennial forages may contribute to reduced nitrate leaching, including alteration in N mineralization dynamics, recalcitrance of organic matter, biochemical properties of different plant species, and inputs of N into a system.

Pasture-crop rotations may be one of the most viable strategies to enhance soil fertility and store soil organic C in traditional cropland regions. Total and particulate organic C fractions often increase during pasture periods, but decrease during the first 3–4 years when land is in the crop production phase (Díaz-Zorita et al., 2002; García-Prechác et al., 2004; Gentile et al., 2005; Franzluebbers and Stuedemann, 2008). In rotations with <7 years of conventional cropping alternated with >3 years of pasture, soil properties were maintained within acceptable limits to avoid soil degradation (Studdert et al., 1997). Soil organic C tended to decrease in crop phases and increase in pasture phases of the rotation. Soil organic C accumulation during 11 years of cropping following pasture termination was positively related to the quantity of crop residue C returned to the soil, as well as timing of tillage (Studdert and Echeverria, 2000). In the long-term pasture-crop rotation experiment in La Estanzuela, Uruguay, soil organic C declined nearly constantly with continuous cropping since establishment of the experiment in 1963, with greater decline without fertilizer than with fertilizer input (Franzluebbers et al., 2014). With 33% pasture and 67% cropping, soil organic C was relatively stable over the long term, and with 50% pasture and 50% cropping, soil organic C increased gradually with time. In both rotation systems, oscillations in soil organic C could be detected during rotation phases, as soil organic C accumulated during the pasture phase and declined during the tilled cropping phase. This experiment shows clearly the value of long-term pasture-crop rotations for maintaining soil organic C.

The long-lasting effect of perennial forages in rotation with crops has also been illustrated in various regions, if subsequent cropland were managed without soil disturbance. In the Great Plains of the United States, chemical termination of 12-year-old conservation grassland sown to smooth bromegrass and direct planting of corn for 6 years revealed no change over time in soil organic C content at 0–5, 5–10, and 10–30 cm depths (Follett et al., 2009). However, there was a progressive increase in C4–C content (reflective of corn plant residue contribution to soil organic matter) and a progressive decline in C3–C content (reflective of bromegrass contribution) in the 0–5 and 5–10 cm depths, but no change in the 10–30 cm depth. In the eastern United States, soil organic C content was maintained for at least 3 years under no-tillage cropping following termination of tall fescue pasture, but it declined with time under conventional tillage (Franzluebbers and Stuedemann, 2008).

What we can glean from the literature is that there may be an enormous opportunity to recapture much of the C lost from decades to centuries of agricultural cultivation by rotating crops with perennial grasses/legumes when crops are managed with continuous no-tillage production strategies. In combination with frequent return of animal manure amendments, our thesis in the remainder of this chapter is that pasture-crop rotations utilizing various conservation tillage management approaches will greatly improve the functioning of soil and mitigate many of the environmental concerns with agriculture, including contamination of water bodies with nutrients, agrochemicals, and sediment and of air quality from dust and greenhouse gas emissions.

ENERGIZING AGRICULTURAL LANDSCAPES WITH A BIOLOGIC APPROACH TO SOIL MANAGEMENT USING CONSERVATION PASTURE-CROP ROTATIONS

Our hypothesis is that pastures (i.e., any form of perennial grass/legume species mixture as a fodder for ruminant animals or as feedstock for biofuel production) in rotation with crops managed with some form of conservation tillage on the majority of agricultural lands around the world will lead to improvement in soil organic C and N contents, so long-term fertility can be restored and environmental quality can be significantly improved to meet the challenges for greater quantity and quality of food production, sustenance of human health, maintenance of wildlife diversity, and balancing our human footprint with nature's capacity to serve our needs.

We realize that this is an audacious outlook based on the many competing social and political drivers present in today's society, but without a technically sound ground-based approach, other noble internationally recognized supplications to improve the human condition for those still suffering from hunger and to ameliorate our rapidly declining global environmental condition will not be reasonably achievable. Conservation pasture-crop rotations are suggested to be a key step in transforming agriculture from a burden on the environment and struggling to meet global food demands to a system that produces a diversity and abundance of food crops while fortifying one of our most precious natural resources: soil. This approach is considered necessary to bridge the current dichotomy between food productivity and environmental quality (Lemaire et al., 2014, 2015).

Long-term data from the Morrow Plots in Illinois, United States, (Nafziger and Dunker, 2011) serve as a useful example of how agronomic productivity can be maintained or enhanced while addressing soil health. Relative variation in yield among the 24 data points in Fig. 7.1 was 39% due to amendments (with and without manure, lime, fertilizer), 38% due to technological developments with time (among different eras as a result of genetics, equipment, weather), and 15% due to crop sequence (as reflected in strong association with soil organic C). This analysis clearly shows that technologic developments over time and sufficient nutrient amendments are key factors in grain production, and these factors should never be ignored. In addition, the analysis hints to the important role of soil organic matter as a determinant of overall soil fertility that allowed grain production to be enhanced significantly further beyond technologic advancements and nutrient amendments alone. Although the difference in soil organic C between monoculture corn and long-rotation corn was only 7.5 + 0.9 g/kg, this difference resulted in 1.8 + 0.5 times greater grain production over the various eras and amendment strategies in this experiment. As a more outcome-based relationship, grain yield increased 0.37 + 0.18 Mg/ha for each unit change in soil organic C (g/kg). Despite improvements in technology, the effect of soil organic C on grain yield continued to increase with time, averaging 0.27 Mg/ha per unit SOC during 1905–54, 0.22 Mg/ha per unit SOC during 1955–67, 0.33 Mg/ha per unit SOC during 1968–97, and 0.66 Mg/ha per unit SOC during 1998–2009.

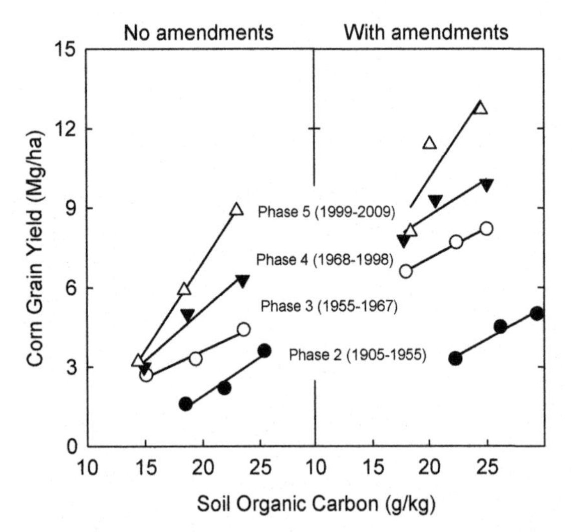

FIGURE 7.1 Corn grain yield as affected by soil organic C and amendment history during four phases of experimentation. Lowest value in each regression was from monoculture corn, intermediate value was from corn in rotation with oat (2-year rotation), and highest value was from corn in rotation with oat-clover hay (3-year rotation). Amendments were manure, lime, and rock or steamed bone P during Phase 2; manure, lime, and inorganic NPK during Phase 3; manure, lime, and low or high NPK during Phase 4; and manure, lime, and low NPK only (switched from moldboard plow to chisel plow in both unamended and amended treatments) during Phase 5. *Data from Nafziger, E.D., Dunker, R.E., 2011. Soil organic carbon trends over 100 years in the Morrow Plots. Agronomy Journal 103, 261–267.*

There are a few other examples of how sod-based rotation affects grain production, but most have been conducted for much shorter periods of time, so the full impact of multiple rotation cycles may not be mature. In the Hunter Rotation Experiment in Pennsylvania, United States, corn grain yield was analyzed during a 16-year period (Grover et al., 2009). Grain yield was 10%–12% greater under longer rotations (4-year rotation of corn-oat/wheat, 2-year red clover/timothy, and 8-year rotation of 4-year corn/4-year alfalfa) than under continuous corn. More diverse rotations also had lower variability in corn grain yield than continuous corn. In France, termination of grassland is considered to provide 20–140 kg N/ha through N mineralization to subsequent maize or wheat crops; the amount is dependent on grassland age and time of termination (spring vs. autumn) more than on forage composition (i.e., grass/legume proportion) (Vertès et al., 2007; Jeuffroy et al., 2015; Comifer, 2013). Alfalfa has high ability to fix and store N in soil, and to release it for subsequent crops over several years, although with more lasting effect than many other forage species. Net N mineralization from alfalfa residues can be 100–150 kg N/ha over 2 years (Justes et al., 2001). Mineralization of N from alfalfa residues declines progressively with years, but it still provides significant N to subsequent crops for 3–4 years (Angus et al., 2006).

We illustrate the dramatic effects of how soil organic C impacts yield potential in the following. Using the average increase in grain yield per unit of soil organic C of 0.37 Mg/ha as a low estimate and 0.66 Mg/ha as a high estimate (based on most current technologies and amendments in the Morrow Plots), farms with 6.7 Mg/ha corn grain yield on soils with 10 g/kg soil organic C could be expected to achieve 10.4–13.3 Mg/ha if soil organic C were to increase to 20 g/kg. If soil organic C increased further to 30 g/kg, then corn grain yield of 14.1–19.9 Mg/ha could be expected. In this example, yields could be expected to increase by 1.6–3.0 times the assumed current level (e.g., 6.7 Mg/ha was mean corn grain yield on 1.2 M ha in the southeastern US states of Alabama, Florida, Georgia, North Carolina, South Carolina, Tennessee, and Virginia in 2012 (USDA-NASS, 2014)). Since land would have to be devoted to pasture production for 3–6 years out of 10, total grain production could be expected to remain the same (e.g., 67 Mg/ha during 10 years under business as usual and 42–93 Mg/ha during a 10-year period with 4–7 years of corn cropping in a pasture-crop rotation system with 10 g/kg greater soil organic C), while the production of forage for grazing, feeding, or biofuel harvest would be additional, as well as development of a much healthier and vibrant soil resource. These additional benefits from adoption of a conservation pasture-crop rotation will be significant and potentially dramatic, as forages can be fed or sold to promote further economic gain, as well as to reduce external energy demands of the production system. The following sections outline several economic and environmental benefits attainable through adoption of a conservation pasture-crop rotation approach.

ECONOMIC LIMITATIONS AND OPPORTUNITIES OF CONSERVATION PASTURE-CROP ROTATIONS

Although free market forces are prevalent in many industrialized countries, the opportunities to enhance economic return with strategic investments leading to social goods of high-quality food production and environmental mitigation might draw sufficient government, farmer organizations, and/or individual producers to take the risk of adopting a conservation pasture-crop rotation approach. We are not advocating for a single type of conservation pasture-crop approach. The choice of food crops, forages, and animal species and classes should be derived from a menu of possibilities. Based on some initial research on pasture-crop rotations (e.g., Studdert et al. (1997), García-Préchac et al. (2004), and Ernst and Siri-Prieto (2009), who showed that 33%–50% of a decade should be occupied by forages to be most effective), we recommend that 3 years of forage could be followed by 3–7 years of cropping using some form of conservation tillage with minimal soil disturbance during the transition and during cropping. To maximize C inputs to the system, animal manures could be applied during cropping and forage could be grazed by ruminants whenever possible. During the cropping phase, cover cropping or double cropping should be a goal to keep a living crop growing for as long as possible, depending on the climatic conditions (such a strategy is mandatory in most agriculturally intensive western countries of Europe). In semiarid environments, residue management should be a high priority to keep the soil covered during the noncrop-growing season. All of these decisions have important economic implications, but there are already producers making such efforts in a variety of environments. Sharing these farmer experiences and economic implications of choices should become a goal of agricultural extension programs.

Efficient utilization of external inputs in conservation pasture-crop rotations needs to be a priority. Fertilizer recommendations will need to be revised to accommodate the potentially large reduction in N possible in pasture-crop rotations, especially when legume pastures and pulses can be incorporated into the system. The pasture phase offers an excellent opportunity to introduce more legumes that can symbiotically fix atmospheric N, particularly in crop rotations where annual grain legumes are difficult to grow. Additionally, it is envisioned that N losses via runoff, leaching, volatilization, and denitrification might be significantly reduced in the pasture phase, as well as in the cropping phase if N carryover via N mineralization of soil organic matter can be efficiently accounted.

Reports on economic benefits of grassland-crop rotations are scarce in the literature. A study in the Poitou-Charentes region of France aimed to quantify the economics of how different alfalfa introduction scenarios might fit into the main regional crop rotations with wheat, canola, and maize. Taking into account all related economic costs and benefits, including the indirect costs of alfalfa introduction (i.e., reduction in crop N fertilization, herbicides, and pesticides), the study showed that over the entire cropping-alfalfa rotation period, introduction of alfalfa was profitable in situations of intermediate agronomic potential, although not in situations of high agronomic performance (intensive irrigated maize) (S. Minette, personal communication). Including indirect costs/benefits from differences in water quality, soil C storage, greenhouse gas emissions, and soil biodiversity might further alter the balance. More investigations of ecologic economics are needed.

Perennial forages, and particularly legume or grass-legume mixtures as well as grain legumes, are not only favorable to the cropping phase, but they also can contribute to a substantial increase in protein self-sufficiency and associated economic benefits of livestock production (Peyraud et al., 2014).

Several recent case studies have been conducted in France to evaluate profit from mixed crop/livestock systems at the farm level. A case study conducted in Brittany showed that mixed crop/livestock farms enjoyed profitability, as long as they developed a strategy of self-sufficiency and reduced their dependence on external inputs (Bonaudo et al., 2014). Another case study from the French "Coteaux de Gascogne" analyzed the economics of crop, livestock, and mixed crop/livestock farms (Ryschawy et al., 2012). Although not necessarily having greater gross margin, mixed crop/livestock farms were less sensitive to fluctuation of inputs and product prices than specialized dairy or crop farms. In all of these case studies, profitability of mixed crop/livestock farms was systematically accompanied by substantial environmental benefits (although environmental values were not included in these economic calculations).

ENVIRONMENTAL BENEFITS OF CONSERVATION PASTURE-CROP ROTATIONS

Numerous environmental benefits are expected with development of conservation pasture-crop rotations. Previous reviews of pasture and surface residue management effects on soil organic matter, erosion, and nutrient losses have documented this important potential of conservation pasture-crop rotations (Franzluebbers et al., 2012, 2014). Mitigating soil and nutrient losses from agriculture must be a continual high priority topic despite decades of strong soil and water mitigation efforts in many countries, as these issues continue to occur (Cox et al., 2011; Garcia et al., 2016; Mulkey et al., 2016). Previous sections in this chapter have outlined the convincing effects of perennial forages on soil organic C and N stocks, which are keystone properties that affect a wide variety of other important soil properties and processes, including erosion control. More specifically, soil organic matter affects soil physical properties (color, water retention, structure, bulk density), chemical properties and processes (nutrient availability, cation exchange capacity, pH buffering capacity, chelation of metals, interactions with xenobiotics), and biologic properties and processes (reservoir of metabolic energy, mineralization of macronutrients, enzymatic activities, ecosystem resilience (Franzluebbers, 2010b).

Surface accumulation of soil organic matter is a common feature in many natural ecosystems, managed grasslands, and long-term conservation-tilled croplands (Franzluebbers, 2005, 2010a). Stratification of soil organic C and N fractions with depth generally has positive implications on soil C storage, but also on other important soil functions, such as water infiltration, mitigation of runoff, compaction alleviation from equipment and animal traffic, and development of habitat for a diversity of soil organisms. Depth stratification of soil organic C and N fractions has been proposed as an indicator of soil health and/or ecosystem functioning (Franzluebbers, 2002, 2013b). High surface soil organic matter accumulation resists erosion, fosters development of biopores to assist in water infiltration, and creates a more suitable habitat for a diversity of soil organisms and storing organically bound nutrients. We suggest that conservation pasture-crop rotation systems will also have high stratification of soil organic C and N fractions, but long-term data and data from a diversity of soil types are lacking. Although depth stratification of soil organic matter is viewed as a positive attribute, we do not know the implications of infrequent (e.g., once every decade), deep moldboard plowing of such systems on soil C storage and environmental outcomes. Occasional redistribution of soil organic matter may be beneficial in some soil types and environments to broaden in the soil profile the influence of the concentrated layer of organic matter or reduce surface accumulation of potentially erodible nutrients. The issue of stratification and how to enhance it by simultaneously improving subsoil C has great relevance for many different soil types, but particularly in warm regions, such as the southeastern United States, where depth stratification is dramatic and accumulation of soil organic C and N below the surface 10 cm is difficult to achieve.

Conservation pasture-crop rotations are expected to have greatly reduced requirements for external N inputs compared with traditional high-intensity cropping systems, because (1) conservation of N in soil organic matter should occur with high soil C accumulation, (2) leguminous species that have biologic N_2 fixation should be incorporated into pasture mixtures when possible, and (3) losses of N should be minimized through the biologically active surface soil organic matter that limits runoff, leaching, and gaseous emissions. Effects of enhanced soil organic matter when concentrated near the surface on greenhouse gas emissions require further investigation, especially with regard to a range of N inputs, as high C and N concentration combined with potential changes in soil pH might alter microbial community dynamics, leading to positive or negative impacts on nitrous oxide emissions and ammonia volatilization. Another research area that requires refinement is quantifying the change in mineralization of soil organic N as influenced by pasture and crop species in the rotation sequences. A rapid assay to determine soil biologic activity and relationship to N mineralization is currently being evaluated (Franzluebbers, 2016).

Although data are not available for total greenhouse gas emissions (particularly including N_2O) from conservation pasture-crop rotations, it is generally considered that N fertilization is a major determinant of N_2O emissions from croplands and grasslands (Davidson, 2009). Furthermore, it is also now considered that legumes lead to relatively lower N_2O emissions than moderately fertilized nonleguminous species (Rochette and Janzen, 2005). Therefore, it is likely that grasslands and perennial forages that contain a substantial proportion of legume species and are generally lightly fertilized result in lower rates of emission of N_2O, as indicated in several studies (Gregorich et al., 2005; Senapati et al., 2016). Variations in N_2O emissions may occur during pasture-crop transitions, particularly in association with the burying and senescence of grassland residues (Senapati et al., 2016). It is possible that use of conservation tillage approaches rather than inversion tillage will minimize greenhouse gas emissions, as soil disturbance and peaks in nitrate accumulation have been associated with greenhouse gas emissions in some studies (Cavigelli and Parkin, 2012).

Introduction of grassland and perennial forage phases into crop rotations may be an efficient strategy to control weed development, as well as to reduce the use of herbicides during the crop phase and to mitigate their potentially detrimental consequences on soil and water environments (Mediene et al., 2011). Annual crops and perennial forages allow different weed communities to develop, e.g., more annual and perennial weed species, respectively. Alternating between annual crops and perennial forages can break the conditions for development of specific weed spectra (Meiss et al., 2010; Mediene et al., 2012).

FURTHER NEEDS TO IMPROVE CONSERVATION PASTURE-CROP ROTATIONS

At what spatial scale should crop and livestock be integrated? Crop/livestock integration is typically considered at the farm level. As described by Bonaudo et al. (2014) in the context of existing crop/livestock integration, there are opportunities to improve integration at the farm level to achieve a higher level of positive interactions, such as reducing external inputs (i.e., increasing farm self-sufficiency) and improving environmental benefits. However, as pointed out by Lemaire et al. (2003) in the context of highly specialized production systems, it may be difficult to introduce animal production onto crop farms, considering the structural, technical, and social changes that would be necessary. Instead, integration of crop and livestock systems may need to be considered at a larger spatial scale, i.e., the territory or even higher level. All the beneficial economic and environmental synergies discussed earlier between cropping and livestock systems can be considered not only at the farm, but also at the territory level (Martin et al., 2016; Garnier et al., 2016). Such an approach was explored by Moraine et al. (2016) in a case study in the Aveyron Valley of France, where uplands have been more dedicated to animal production systems, while lowlands with deeper soils and more access to irrigation water are more dedicated to cropping systems. In their study, which directly involved farmers in the conceptualization of the approach, alfalfa was identified as a key cross-over crop between cropping and livestock farms of the territory. Alfalfa was introduced into crop rotations in the lowlands, thereby reducing N fertilizers and water use to result in positive economic and environmental benefits. Alfalfa hay and cereal straw were exported to livestock farms, thereby reducing feed purchase cost and allowing more grassland to be kept for grazing, as well as reducing work load, which is a difficulty often identified by farmers in mixed crop/livestock systems. Beyond these technical and economic aspects, integration of crops and livestock created social benefits at the territory level. Collective learning and social acceptance of different agricultural activities were additional outcomes (Martin et al., 2016).

References

Angus, J.F., Bolger, T.P., Kirkegaard, J.A., Peoples, M.B., 2006. Nitrogen mineralisation in relation to previous crops and pastures. Australian Journal of Soil Research 44, 355–365.

Bonaudo, T., Bendahan, A.B., Sabatier, R., Ryschawy, J., Bellon, S., Leger, F., Magda, D., Tichit, M., 2014. Agroecological principles for the redesign of integrated crop-livestock systems. European Journal of Agronomy 57, 43–51.

Cavigelli, M.A., Parkin, T.B., 2012. Cropland management contributions to greenhouse gas flux: central and eastern U.S. In: Liebig, M.A., Franzluebbers, A.J., Follett, R.F. (Eds.), Managing Agricultural Greenhouse Gases: Coordinated Agricultural Research through GRACEnet to Address Our Changing Climate. Academic Press, San Diego, CA, pp. 129–165.

Cox, C., Hug, A., Brzelius, N., 2011. Losing Ground. Environmental Working Group Report. Accessed from. http://www.ewg.org/losingground/, 36 pp.

Comifer, 2013. Calcul de la fertilisation azotée. In: Le Diamant, A. (Ed.), Guide méthodologique pour l'établissement des prescriptions locales. Paris.

Davidson, E.A., 2009. The contribution of manure and fertilizer nitrogen to atmospheric nitrous oxide since 1860. Nature Geoscience 2, 659–662.

Díaz-Zorita, M., Duarte, G.A., Grove, J.H., 2002. A review of no-till systems and and soil management for sustainable crop production in the subhumid and semiarid Pampas of Argentina. Soil and Tillage Research 65, 1–18.

Ernst, O., Siri-Prieto, G., 2009. Impact of perennial pasture and tillage systems on carbon input and soil quality indicators. Soil & Tillage Research 105, 260–268.

Fike, J.H., Parrish, D.J., Wolf, D.D., Balasko, J.A., Green Jr., J.T., Rasnake, M., Reynolds, J.H., 2006. Long-term yield potential of switchgrass-for-biofuel systems. Biomass and Bioenergy 30, 198–206.

Follett, R.F., Varvel, G.E., Kimble, J.M., Vogel, K.P., 2009. No-till corn after bromegrass: effect on soil carbon and soil aggregates. Agronomy Journal 101, 261–268.

Fortuna, A.M., Honeycutt, C.W., Vandemark, G., Griffin, T.S., Larkin, R.P., He, Z., Wienhold, B.J., Sistani, K.R., Albrecht, S.L., Woodbury, B.L., Torbert, H.A., Powell, J.M., Hubbard, R.K., Eigenberg, R.A., Wright, R.J., Alldredge, J.R., Harsh, J.B., 2011. Links among nitrification, nitrifier communities, and edaphic properties in contrasting soils receiving dairy slurry. Journal of Environmental Quality 41, 262–272.

Franzluebbers, A.J., 2002. Soil organic matter stratification ratio as an indicator of soil quality. Soil and Tillage Research 66, 95–106.

Franzluebbers, A.J., 2005. Soil organic carbon sequestration and agricultural greenhouse gas emissions in the southeastern USA. Soil and Tillage Research 83, 120–147.

Franzluebbers, A.J., Follett, R.F., Johnson, J.M.F., Liebig, M.A., Gregorich, E.G., Parkin, T.B., Smith, J.L., Del Grosso, S.J., Jawson, M.D., Martens, D.A., 2006. Agricultural exhaust: A reason to invest in soil. Journal of Soil & Water Conservation 61, 98A–101A.

Franzluebbers, A.J., 2007. Integrated crop-livestock systems in the southeastern USA. Agronomy Journal 99, 361–372.

Franzluebbers, A.J., 2010a. Achieving soil organic carbon sequestration with conservation agricultural systems in the southeastern United States. Soil Science Society of America Journal 74, 347–357.

Franzluebbers, A.J., 2010b. Will we allow soil carbon to feed our needs? Carbon Manage 1, 237–251.

Franzluebbers, A.J., 2013a. Ecosystem services from forages. In: Bittman, S., Hunt, D. (Eds.), Cool Forages: Advanced Management of Temperate Forages. Pacific Field Corn Association, Agassiz, BC, Canada, pp. 39–43.

Franzluebbers, A.J., 2013b. Pursuing robust agroecosystem functioning through effective soil organic carbon management. Carbon Management 4, 43–56.

Franzluebbers, A.J., 2015. Farming strategies to fuel bioenergy demands and facilitate essential soil services. Geoderma 259–260, 251–258.

Franzluebbers, A.J., 2016. Should soil testing services measure soil biological activity? Agricultural & Environmental Letters 1, 150009. https://doi.org/10.2134/ael2015.11.0009.

Franzluebbers, A.J., Hons, F.M., Zuberer, D.A., 1998. In situ and potential CO_2 evolution from a Fluventic Ustrochrept in south-central Texas as affected by tillage and cropping intensity. Soil and Tillage Research 47, 303–308.

Franzluebbers, A.J., Owens, L.B., Sigua, G.C., Cambardella, C.A., Haney, R.L., 2012. Soil organic carbon under pasture management. In: Liebig, M.A., Franzluebbers, A.J., Follett, R.F. (Eds.), Managing Agricultural Greenhouse Gases. Elsevier, San Diego, CA, pp. 93–110.

Franzluebbers, A.J., Sawchik, J., Taboada, M., 2014. Agronomic and environmental impacts of pasture–crop rotations in temperate North and South America. Agriculture, Ecosystems & Environment 190, 18–26.

Franzluebbers, A.J., Sulc, R.M., Russelle, M.P., 2011. Opportunities and challenges for integrating North-American crop and livestock systems. In: Lemaire, G., Hodgson, J., Chabbi, A. (Eds.), Grassland Productivity and Ecosystem Services. CAB International, Wallingford, UK, pp. 208–218.

Franzluebbers, A.J., Stuedemann, J.A., 2008. Early response of soil organic fractions to tillage and integrated crop-livestock production. Soil Science Society of America Journal 72, 613–625.

García, A.M., Alexander, R.B., Arnold, J.G., Norfleet, L., White, M.J., Robertson, D.M., Schwarz, G., 2016. Regional effects of agricultural conservation practices on nutrient transport in the Upper Mississippi River Basin. Environmental Science & Technology 50, 6991–7000.

García-Prechac, F., Ernst, O., Siri-Prieto, G., Terra, J.A., 2004. Integrating no-till into crop-pasture rotations in Uruguay. Soil and Tillage Research 77, 1–13.

Garnier, J., Anglade, J., Benoit, M., Billen, G., Puech, T., Ramarson, A., Passy, P., Silvestre, M., Lassaletta, L., Trommenschlager, J.M., Schott, C., Tallec, G., 2016. Reconnecting crop and cattle farming to reduce nitrogen losses to river water of an intensive agricultural catchment (Seine basin, France): past, present and future. Environmental Science & Policy 63, 76–90.

Gentile, R.M., Martino, D.L., Entz, M.H., 2005. Influence of perennial forages on subsoil organic carbon in a long term rotation study in Uruguay. Agriculture, Ecosystems & Environment 105, 419–423.

Gilley, J.E., Eghball, B., Kramer, L.A., Moorman, T.B., 2000. Narrow grass hedge effects on runoff and soil loss. Journal of Soil & Water Conservation 55, 190–196.

Gregorich, E.G., Rochette, P., VandenBygaart, A.J., Angers, D.A., 2005. Greenhouse gas contributions of agricultural soils and potential mitigation practices in eastern Canada. Soil and Tillage Research 83, 53–72.

Griffin, T.S., 2008. Nitrogen availability. In: Schepers, J.S., Raun, W.R. (Eds.), Nitrogen in Agricultural Soils, *Agron. Monogr. 49, Am. Soc. Agron., Crop Sci. Soc. Am., Soil Sci. Soc. Am., Madison, WI*, pp. 616–646.

Grover, K.K., Karsten, H.D., Roth, G.W., 2009. Corn grain yields and yield stability in four long-term cropping systems. Agronomy Journal 101, 940–946.

Guo, L.B., Gifford, R.M., 2002. Soil carbon stocks and land use change: a meta-analysis. Global Change Biology 8, 345–360.

Haynes, R.J., 1999. Labile organic matter fractions and aggregate stability under short-term, grass-based leys. Soil Biology and Biochemistry 31, 1821–1830.

Hanson, J.D., Hendrickson, J.R., 2009. Toward a sustainable agriculture. In: Franzluebbers, A.J. (Ed.), Farming with Grass: Achieving Sustainable Mixed Agricultural Landscapes. Soil Water Conserv. Soc, Ankeny, IA, pp. 26–36.

Hendrickson, J., Sassenrath, G.F., Archer, D., Hanson, J., Halloran, J., 2008. Interactions in integrated US agricultural systems: the past, present and future. Renewable Agriculture and Food Systems 23, 314–324.

Hoagland, L., Hodges, L., Helmers, G.A., Brandle, J.R., Francis, C.A., 2010. Labor availability in an integrated agricultural system. Journal of Sustainable Agriculture 34, 532–548.

Jeuffroy, M.H., Schneider, A., Biarnès, V., Cohan, J.P., Gastal, F., Corre-Hellou, G., Vertès, F., Landé, N., Louarn, G., Valantin-Morison, M., Plantureux, S., 2015. Performances agronomiques et gestion des légumineuses dans les systèmes de productions végétales. Chapitre III. In: Schneider, A., Huyghe, A. (Eds.), Les Légumineuses pour des Systèmes Agricoles et Alimentaires Surables. Éditions Quæ, Versailles, pp. 139–224.

Justes, E., Thiébeau, P., Cattin, G., Larbre, D., Nicolardot, B., 2001. Libération d'azote après retournement de luzerne. Un effet sur deux campagnes. Perspectives Agricoles 264, 22–28.

Kunrath, T.R., de Berranger, C., Charrier, X., Gastal, F., Carvalho, P.C.F., Lemaire, G., Emile, J.C., Durand, J.C., 2015. How much do sod-based rotations reduce nitrate leaching in a cereal cropping system? Agricultural Water Management 150, 46–56.

Lamande, M., Hallaire, V., Curmi, P., Peres, G., Cluzeau, D., 2003. Changes of pore morphology, infiltration and earthworm community in a loamy soil under different agricultural managements. Catena 54, 637–649.

Lehman, R.M., Acosta-Martinez, V., Buyer, J.S., Cambardella, C.A., Collins, H.P., Ducey, T.F., Halvorson, J.J., Jin, V.L., Johnson, J.M.F., Kremer, R.J., Lundgren, J.G., Manter, D.K., Maul, J.E., Smith, J.L., Stott, D.E., 2015. Soil biology for resilient, healthy soil. Journal of Soil & Water Conservation 70, 12A–18A.

Lemaire, G., Benoit, M., Vertès, F., 2003. Rechercher de nouvelles organisations à l'échelle d'un territoire pour concilier autonomie protéique et préservation de l'environnement. Fourrages 175, 303–318.

Lemaire, G., Franzluebbers, A.J., Carvalho, P.C.F., Dedieu, B., 2014. Integrated crop–livestock systems: strategies to achieve synergy between agricultural production and environmental quality. Agriculture, Ecosystems & Environment 190, 4–8.

Lemaire, G., Gastal, F., Franzluebbers, A.J., Chabbi, A., 2015. Grassland–cropping rotations: an avenue for agricultural diversification to reconcile high production with environmental quality. Environmental Management. https://doi.org/10.1007/s00267-015-0561-6.

Lemaire, G., Ryschawy, J., Carvalho, P.C.F., Gastal, F., 2017. Agricultural intensification and diversity for reconciling production and environment: the role of integrated crop–livestock systems. In: Gordon, J., Prins, H.T., Squire, G.R. (Eds.), Food Production and Nature Conservation: Conflicts and Solutions. Routledge Edition, New York, pp. 113–132.

Liebig, M.A., Johnson, H.A., Hanson, J.D., Frank, A.B., 2005. Soil carbon under switchgrass stands and cultivated cropland. Biomass and Bioenergy 28, 347–354.

Lindstrom, M.J., Schumacher, T.E., Cogo, N.P., Blecha, M.L., 1998. Tillage effects onwater runoff and soil erosion after sod. Journal of Soil & Water Conservation 53, 59–63.

Martin, G., Moraine, M., Ryschawy, J., Magne, M.A., Asai, M., Sarthou, J.P., Duru, M., Therond, O., 2016. Crop-livestock integration beyond the farm level: a review. Agronomy for Sustainable Development 36, 53.

Mazoyer, M., Roudart, L., 1997. Histoire des agricultures du monde: Du Néolithique à la crise contemporaine. Collection Pont Histoire, vol. 2002. Editions Seuil, Paris, 705 pp.

Mediene, S., Valantin-Morison, M., Sarthou, J.P., de Tourdonnet, S., Gosme, M., Bertrand, M., Roger-Estrade, J., Aubertot, J.N., Rusch, A., Motisi, N., 2011. Agroecosystem management and biotic interactions: a review. Agronomy for Sustainable Development 31, 491–514.

Médiene, S., Zhang, W., Doisy, D., Charrier, X., 2012. Temporary grasslands impact weed abundance and diversity. In: 12th Congress of European Society for Agronomy, Helsinki, Finland, pp. 70–71.

Meiss, H., Mediene, S., Waldhardt, R., Caneil, J., Bretagnolle, V., Reboud, X., Munier-Jolain, N., 2010. Perennial lucerne affects weed community trajectories in grain crop rotations. Weed Research 50, 331–340.

Moraine, M., Grimaldi, J., Murgue, C., Duru, M., Therond, O., 2016. Co-design and assessment of cropping systems for developing crop-livestock integration at the territory level. Agricultural Systems 147, 87–97.

Mulkey, A.S., Coale, F.J., Vadas, P.A., Shenk, G.W., Bhatt, G.X., 2016. Revised method and outcomes for estimating soil phosphorus losses from agricultural land in the Chesapeake Bay watershed model. Journal of Environmental Quality. https://doi.org/10.2134/jeq2016.05.0201.

Nafziger, E.D., Dunker, R.E., 2011. Soil organic carbon trends over 100 years in the Morrow Plots. Agronomy Journal 103, 261–267.

Panettieri, M., Rumpel, C., Dignac, M.-F., Chabbi, A., 2017. Does grassland introduction into cropping cycles affect carbon dynamics through changes of allocation of soil organic matter within aggregate fractions? Science of the Total Environment 576, 251–263.

Peyraud, J.L., Taboada, M., Delaby, J.L., 2014. Integrated crop and livestock systems in Western Europe and South America: a review. European Journal of Agronomy 57, 41–42.

Randall, G.W., Mulla, D.J., 2001. Nitrate nitrogen in surface waters as influenced by climatic conditions and agricultural practices. Journal of Environmental Quality 30, 337–344.

Rochette, P., Janzen, H.H., 2005. Towards a revised coefficient for estimating N2O emissions from legumes. Nutrient Cycling in Agroecosystems 73, 171–179.

Ryschawy, J., Choisis, N., Choisis, J.P., Joannon, A., Gibon, A., 2012. Mixed crop-livestock systems: an economic and environmental-friendly way of farming? Animal 6, 1722–1730.

Russelle, M.P., Entz, M.H., Franzluebbers, A.J., 2007. Reconsidering integrated crop-livestock systems in North America. Agronomy Journal 99, 325–334.

Schomberg, H.H., Wietholter, S., Griffin, T.S., Reeves, D.W., Cabrera, M.L., Fisher, D.S., Endale, D.M., Novak, J.M., Balkcom, K.S., Raper, R.L., Kitchen, N.R., Locke, M.A., Potter, K.N., Schwartz, R.C., Truman, C.C., Tyler, D.D., 2009. Assessing indices for predicting potential nitrogen mineralization in soils under different management systems. Soil Science Society of America Journal 73, 1575–1586.

Senapati, N., Chabbi, A., Giostri, A.F., Yeluripati, J.B., Smith, P., 2016. Modelling nitrous oxide emissions from mown-grass and grain-cropping systems: testing and sensitivity analysis of DailyDayCent using high frequency measurements. The Science of the Total Environment 572, 955–977.

Singer, J.W., Franzluebbers, A.J., Karlen, D.L., 2009. Grass-based farming systems: soil conservation and environmental quality. In: Wedin, W.F., Fales, S.L. (Eds.), Grassland: Quietness and Strength for a New American Agriculture. Am. Soc. Agron., Crop Sci. Soc. Am., Soil Sci. Soc. Am, Madison, WI, pp. 121–136.

Soussana, J.F., Lemaire, G., 2014. Coupling carbon and nitrogen cycles for environmentally sustainable intensification of grasslands and crop-livestock systems. Agriculture, Ecosystems & Environment 190, 9–17.

Studdert, G.A., Echeverria, H.E., 2000. Crop rotations and nitrogen fertilization to manage soil organic carbon dynamics. Soil Science Society of America Journal 64, 1496–1503.

Studdert, G.A., Echeverria, H.E., Casanovas, E.M., 1997. Crop-pasture rotation for sustaining the quality and productivity of a Typic Argiudoll. Soil Science Society of America Journal 61, 1466–1472.

Tilman, D., Cassman, K.G., Matson, P.A., Naylor, R., Polasky, S., 2002. Agricultural sustainability and intensive production practices. Nature 418, 671–677.

USDA-NASS (National Agricultural Statistics Service), 2009. 2007 Census of Agriculture Report. U.S. Dept. Agric, Washington, DC. Accessed at http://www.agcensus.usda.gov/.

USDA-NASS (National Agricultural Statistics Service), 2014. 2012 Census of Agriculture Report. U.S. Dept. Agric, Washington, DC. Available from. http://www.agcensus.usda.gov/.

Vertès, F., Hatch, D., Velthof, G., Taube, F., Laurent, F., Loiseau, P., Recous, S., 2007. Short-term and cumulative effects of grassland cultivation on nitrogen and carbon cycling in ley-arable rotations. Grassland Science European 12, 227–246.

Further Reading

NRC (National Research Council). 2010. Toward Sustainable Agricultural Systems in the 21st Century. Comm. Twenty-first Cent. Syst. Agric., Board Agric. Nat. Resources, National Academies Press, Washington, DC.

CHAPTER

8

The Contributions of Legumes to Reducing the Environmental Risk of Agricultural Production

Mark B. Peoples[1], *Henrik Hauggaard-Nielsen*[2], *Olivier Huguenin-Elie*[3], *Erik Steen Jensen*[4], *Eric Justes*[5,6], *Michael Williams*[7]

[1]CSIRO Agriculture and Food, Canberra, ACT, Australia; [2]Department of People and Technology, Roskilde University, Roskilde, Denmark; [3]Agroscope, Forage Production and Grassland Systems, Zurich, Switzerland; [4]Swedish University of Agricultural Sciences, Department of Biosystems and Technology, Alnarp, Sweden; [5]Formerly INRA, Joint Research Unit AGIR (Agroecology, Innovations, Territories), INRA-INPT-University of Toulouse, Castanet Tolosan, France; [6]Currently CIRAD, Joint Research Unit SYSTEM, CIRAD - INRA SupAgro, Montpellier, France; [7]Department of Botany, School of Natural Sciences, Trinity College Dublin, Dublin, Ireland

INTRODUCTION

Nitrogen (N) is a key component of plant amino and nucleic acids and chlorophyll, and provides the basis for the dietary N (protein) of all animals, including humans. The supply of N tends to be one of the most important factors regulating crop growth and yield and controlling the nutritional quality of plant products. The total amounts of N acquired by plants (either assimilated by roots or, in the case of legumes, fixed from the atmosphere) reflects ecosystem productivity and N availability. For example, the N uptake by irrigated cereals in highly productive systems may be as much as 450 kg N ha^{-1} per year, whereas in low rainfall (<300 mm per annum) environments, crop N uptake by rainfed (dryland) crops can be <100 kg N ha^{-1}. Every tonne (t, Mg; 10^6 g) of grain harvested by the major food crops grown for human consumption, wheat (*Triticum aestivum*), maize (corn; *Zea mays*), and rice (*Oriza sativa*), contain ~20, 15, and 12 kg N, respectively, and the current annual global removal of N in harvested grain of these three cereal crops represents ~40 million t (Tg; 10^{12} g) of N (Table 8.1; Peoples et al., 2009b).

Agroecosystem Diversity
https://doi.org/10.1016/B978-0-12-811050-8.00008-X

TABLE 8.1 Global Trends in Fertilizer N Consumption, Total Grain Production, and Average Yields of Major Cereal and Legume Crops Over the 50-Year Period Between 1965 and 2014 (FAOStat, 2016)

Fertilizer	**Consumption (Million t N)**					
	1965	1975	1985	1995	2005	2014
Nitrogen	19	44	70	78	85	109
Crop[a]	**Production (million t)**					
	1965	1975	1985	1995	2005	2014
Wheat	263.6	355.8	499.5	542.7	626.7	729.0
Maize	226.5	341.7	485.5	517.3	713.7	1021.6
Rice	254.1	357.0	468.2	547.4	634.3	740.9
Soybean	31.7	68.2	101.2	126.9	214.6	308.4
Groundnut	15.8	19.1	20.9	28.6	38.5	42.3
Pulses[b]	45.2	40.3	50.5	56.2	61.3	77.6
Crop	**Average grain yield (t/ha)**					
	1965	1975	1985	1995	2005	2014
Wheat	1.21	1.57	2.17	2.51	2.85	3.29
Maize	2.12	2.81	3.72	3.81	4.82	5.57
Rice	2.03	2.52	3.26	3.66	4.09	4.54
Soybean	1.22	1.66	1.91	2.03	2.32	2.62
Groundnut	0.80	0.96	1.13	1.30	1.60	1.65
Pulses[b]	0.66	0.64	0.76	0.79	0.85	0.91

[a] *Changes in global areas harvested (million ha) by the different crops between 1965 and 2014 represented: 217–220 for wheat, 106.7–184.8 for maize, 124.8–162.7 for rice, 25.8–117.5 for soybean, and 19.8–26.5 for groundnut. The 6 million ha sown to pulses in 2014 was the same as reported for 1965, although total area was only 4.1 million ha in 2005.*
[b] *The FAO crop category "Pulses - total" represents the combined production data for a range of crop legumes grown for grain.*

While modest increases in areas sown to cereals and improvements in genetics such as the introduction of semidwarf traits in wheat and hybrid technologies in maize both played key roles in increasing yield, the almost sixfold increase in the use of synthetic N fertilizer in agriculture since 1965, more than half of which tends to be supplied to cereals, was a major factor contributing to the more than threefold improvement in global cereal grain production over the past 50 years (Table 8.1). The advantage of N fertilizer is that it provides an immediately available source of N to support crop growth, and it enhances the profitability of individual farming operations. The desire by most farmers to ensure that productivity is not compromised by N supply has often resulted in N fertilizer been supplied in excess of crop requirements in many regions of the world since there is little economic incentive for farmers to avoid overapplication due to the relatively low cost of fertilizer N compared to the value of grain and other agricultural products (Cassman et al., 2002; Pannell, 2017). A notable exception to this generality is sub-Saharan Africa, where the challenge has been too little fertilizer N being applied for food production.

The timing of fertilizer application and the amounts of N supplied frequently do not match plant demand, and the capture and uptake of fertilizer N tends to be relatively poor, with often <50% of the fertilizer N applied to cereals being recovered in aboveground biomass (Crews and Peoples, 2005; Ladha et al., 2005). Given the labile nature of available (reactive) forms of N, the inorganic N not utilized by plants is susceptible to multiple loss processes (Peoples et al., 2004b). The implications for the environment arising from volatile losses of N fertilizer as ammonia (NH_3) and oxides of N such as nitrous oxide (N_2O) or the runoff and leaching of fertilizer-derived nitrate (NO_3^-) and organic N into groundwater, rivers, and coastal waters is well documented (Sutton et al., 2011). It has been estimated that synthetic N fertilizers are directly responsible for ~12% of the annual average 5180 million t of CO_2 equivalent greenhouse gas emissions calculated to be associated with agriculture over the 5-year period 2010–14 (FAOStat, 2016).

Manures, compost, and industrial wastes are also used to fertilize agricultural crops. However, they do not presently represent viable sources to replace N fertilizer, as there can be challenges in scaling up supply, and they tend to be more susceptible to N loss processes (Peoples et al., 2004b).

This chapter examines how the inclusion of legumes in redesigned cropping and forage systems might assist in reducing the environmental risk associated with the need to further increase the production of food, forage, and fiber to meet the ever increasing demand of a growing global population while addressing the current reliance of many of the world's agroecosystems on N fertilizer to maintain high crop yields.

THE IMPORTANCE OF LEGUMES TO GLOBAL AGRICULTURE

There are 20,000 species of annual and perennial legumes, several hundred of which have traditionally been used in subsistence and broadacre agriculture in different geographic regions of the world. These legumes are often grown as pure stands, but they are also included in mixtures of other species in pastures, mixed/intercropping, "cover cropping," and "alley" cropping systems that use eco-functional intensification for enhancing yields and land/resource use efficiency (Bedoussac et al., 2015; Hauggaard-Nielsen et al., 2016). Many legumes are harvested as fresh vegetables or grown for grain or forage. They can be used as green manures (fresh aboveground biomass mulched or incorporated into soil) or brown manures (legumes killed with herbicides mid-growing season prior to weed seed set) to enrich the nutrient status of soils and as a strategy to manage weeds for the benefit of future crops (Peoples et al., 2017). In the case of woody perennial species, in addition to their foliage providing a source of forage for livestock or green manure for crops, they may also be utilized for land restoration, fuel wood, or as "living" fences.

The ~230 million hectares (ha) of legume crops grown globally for commercial trade each year provide ~400 million t of grain (Tables 8.1 and 8.2). These crops are dominated by 15–20 legume species, of which just one, soybean (*Glycine max*), is grown over half the total area sown and provides 70% of the total legume grain produced (Table 8.2). Soybean is the fourth major grain crop after wheat, maize, and rice (Table 8.1), and it represents a major source of high-quality protein and, together with groundnut (peanut; *Arachis hypogea*), edible

TABLE 8.2 Average Contributions of Different Geographic Regions of the World to the Annual Global Production of Soybean, Groundnut, and Pulse Legume Grain Over the 5-Year Period 2010–14 (FAOStat, 2016)

Region	Soybean[a]	Groundnut[a]	Pulses[a]
North America	35% (USA 34%)	5% (USA 5%)	10% (Canada 7%)
Central America	1%	1%	3%
South America	50% (Brazil 28%; Argentina 18%)	3%	6%
Europe	2%	0	9%
Africa	1%	27% (Nigeria 8%; Sudan 4%)	23% (Nigeria 5%)
Asia	11% (China 5%; India 4%)	64% (China 39%; India 17%)	45% (India 24%; Myanmar 7%; China 6%)
Oceania	0	0.1%	4%
Average production (million t per year)	270.9	41.6	74.0

[a] *The five countries with the highest production and their relative contribution to the global total tonnage are indicated in parentheses.*

vegetable oil. Despite their relatively low grain yields (Table 8.1) a multitude of warm-season (e.g., common bean, *Phaseolus vulgaris*; cowpea, *Vigna unguicalata*; mung bean/green gram, *V. radiata*; pigeonpea, *Cajanus cajan*) and cool-season legume pulses (e.g., field pea, *Pisum sativum*; chickpea, *Cicer arientinum*; lentil, *Lens culinaris*; faba bean, *Vicia faba*; lupin, *Lupinus angustifolius*; Peoples et al., 2009a) collectively provide the fifth largest source of edible protein for humans and livestock and provide the basis of healthier humans diets that are less reliant upon meat as a source of protein. Different regions and countries tend to specialize in the production of various crop legumes as a result of climate, soils, farming systems, culinary traditions, or socioeconomic attributes that favor particular species (Table 8.2).

In the case of forage legumes, only 50 species are grown to any extent over large areas of land, with lucerne (alfalfa; *Medicago sativa*) being a component of about 30% of the total 110–120 million ha of legume-based pastures around the world. The cultivation of forage legume monocultures have markedly decreased in Europe during the last 50 years to now represent <5% of the total forage area (Voisin et al., 2014). By contrast, there has been a new impetus for grass-legume mixtures in recent years, and the contribution of white clover (*Trifolium repens*) and red clover (*T. pratense*) to forage production has been increasing (Lüscher et al., 2014).

The indeterminate growth patterns of many legumes and their multipurpose nature makes them a very versatile addition to any farming system. This could become particularly important in adapting to future climate change where agriculture will need to continue to produce nutritional human and animal food, generate income, and protect the soil resource in the face of increasing climatic variability (Peoples et al., 2014).

Effect of legumes on crop and forage production: Cereal growth and grain yields are commonly enhanced when legume "break crops," cover crops, or a legume-based pasture

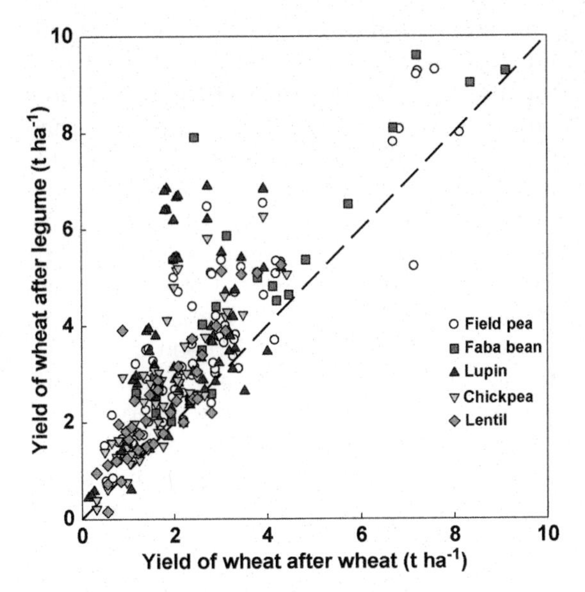

FIGURE 8.1 Grain yields of wheat grown following a crop legume compared with a wheat after wheat treatment grown in the same experiment. The *dashed line* represents equal yields. Any points above the dashed line indicate yield improvements when a legume is the preceding crop. Fitted regression: Grain yield (wheat after legumes) = 0.92 + 1.06 x (wheat after wheat) [$r^2 = 0.69$]. *Adapted from Angus, J.F., Kirkegaard, J.A., Hunt, J.R., Ryan, M.H., Ohlander, L., Peoples, M.B., 2015. Break crop and rotations for wheat. Crop and Pasture Science 66, 523–552.*

phase are included in cereal-dominated cropping sequences (Peoples et al., 2004a; Angus et al., 2015; Plaza-Bonilla et al., 2015). The data presented in Fig. 8.1, collated from 300 comparative field studies undertaken in Australia, Western Europe, North America, and West Asia, suggest that different pulse species can enhance wheat yields by 0.7–1.6 t ha^{-1} relative to wheat after wheat in the absence of N fertilizer. A recent example of the impact of legumes in arable rainfed cropping systems of southwest France was described by Plaza-Bonilla et al. (2017b), who observed 0.38 t ha^{-1} greater grain yield (8% increase) by durum wheat grown after winter pea over a 6-year experiment, compared to wheat following the traditional preceding crop sunflower (*Helianthus annuus*). Moreover, this improvement in gain yield was achieved with a mean reduction of 44 kg N ha^{-1} in fertilizer applications without any differences in grain protein concentration, which is crucial for producing high-quality semolina and pasta. Significant increases in wheat grain yield were reported in other environments by similar studies where data were collected from multiple years and/or sites range from an average of 0.27 t ha^{-1} (36% increase) in Pakistan after mung bean (Shah et al., 2003) to 1.48 t ha^{-1} (55% increase) in Chile after lupin and field pea (Espinoza et al., 2012).

European economic studies have demonstrated that such rotational benefits result in the farm profitability of legume-based rotations frequently being competitive with, or comparable to, alternative nonlegume systems (Preissel et al., 2015). Evidence from the northern Great Plains of North America also suggests that the diversified income derived from the inclusion of a pulse in a cropping sequence can also reduce the economic uncertainty of financial returns (Miller et al., 2015).

The rotational benefits of legumes are often attributed to an increased supply of plant-available forms of N in soil, raising the overall growth potential, and the effects of legumes on the N dynamics of agricultural systems will be a subject of discussion in subsequent sections. However, the economic value of the boost in yield of crops grown following legumes can be greater than the value of a reduced need for N fertilizer (Pannell, 2017), and it should be acknowledged that factors other than N will undoubtedly also be contributing to improvements in productivity. These could include an increased availability of nutrients other than N, changes in soil structure and chemistry, the carryover of unutilized soil water, a reduced incidence of pests and pathogens, a lowered weed burden, or increased abundance of beneficial soil biology (Peoples et al., 2009a; Angus et al., 2015).

Reviews of the international literature on the effects of legumes on grassland-based livestock systems indicate productivity benefits can be derived from improved forage quality and increased feed intake (Lüscher et al., 2014; Phelan et al., 2015). The beneficial impact of the inclusion of legumes on forage production was highlighted by the findings of a pan-European experiment consisting of 31 sites covering a wide pedo-climatic gradient, where grass-legume mixtures generated more biomass than the highest yielding forage species grown alone (transgressive overyielding) at 60% of the sites (Finn et al., 2013). Finally, there is a body of evidence that indicates that the presence of legumes positively affects ruminant productivity through improved forage quality, increased dietary intake, and higher stocking rates (Peoples and Baldock, 2001), and the quality of animal products also appear to be enhanced from the consumption of legumes rich in condensed tannins (Girard et al., 2015).

LEGUME EFFECTS ON N DYNAMICS OF AGROECOSYSTEMS

Inputs of fixed N by legumes: Legumes differ from other major crop and forage plant species in that they can provide oil, fiber, protein-rich food or feed, and/or high N organic residues without supplementary fertilizer N as a result of their ability to biologically fix atmospheric N_2 in a symbiotic relationship with the soil bacteria rhizobia. Thus legumes represent a source of renewable input of N that has the potential to improve the N supply to agroecosystems with less environmental and global warming impacts than fertilizer N derived from fossil fuels (Peoples et al., 2009b; Jensen et al., 2012).

The amount of N_2 fixed by legumes complements the uptake of soil mineral N derived from the mineralization of the soil organic matter and mineral N remaining in soil after the preceding crop. The inputs of fixed N_2 in any particular farming system will be subject to the presence of sufficient numbers of effective rhizobial species in the soil and influenced by a range of abiotic and biotic stresses or management practices that affect either the rhizobial microsymbiont and/or the leguminous host (Carlsson and Huss-Danell, 2003; Peoples et al., 2009a). In the absence of major constraints, and at low to moderate mineral N content in the upper layers of the soil, crop and forage legumes typically fix 15–25 kg of shoot N for every t of aboveground dry matter accumulated during growth across a range of environments (Carlsson and Huss-Danell, 2003; Peoples et al., 2009a, 2012; Espinoza et al., 2012). Commonly observed measures of N_2 fixation suggest that in the order of 100–200 kg shoot N per ha can be fixed each year or growing season by both annual and perennial legumes (Table 8.3). Legumes grown in association with a vigorous nonlegume species may be able to fix as much N per unit area as sole legume crops, because the stimulation of the legume's reliance upon N_2 fixation for growth by the presence of the strong competition for soil mineral N by the companion nonlegume can compensate for the lower legume abundance in the mixture (Carlsson and Huss-Danell, 2003; Bedoussac et al., 2015; Nyfeler et al., 2011). On the other hand, the stimulating effect of competition for N could be partly counterbalanced by competition for light, water, or other nutrients (Høgh-Jensen et al., 2002; Vitousek et al., 2013; Husse et al., 2016), and the relative abundance of the legume component of the mixture often declines significantly within a few years (Brophy et al., 2017). Although calculations of global inputs of fixed N are subject to gross approximations, it has been estimated that collectively symbiotic N_2 fixation by legumes may be contributing between 33 and 46 million t of above- and belowground N to agriculture each year (Herridge et al., 2008).

Legume effects on available soil N: Contributions of legume N to soil fertility may be achieved via above- and belowground organic residues remaining following grain harvest of crop legumes or from cover crops and green or

TABLE 8.3 Summary of the Amounts of Shoot N Derived From N_2 Fixation for Different Legume Systems[a]

Legume System	Total Range Measured	Values Commonly Observed[b]
	(kg N ha^{-1} per crop or per year)	(kg N ha^{-1} per crop or per year)
Crops	0–450	30–150
Annual and perennial forages	1–545	55–250
Woody perennial trees/shrubs	5–666	100–200
Green manures	7–324	50–150

[a] *Based on reviews of N_2 fixation by different legume species (Peoples et al. 1995, 2004a, 2009a, 2012).*
[b] *The range of commonly observed values reflect data collated for different legume species and/or farming systems in various geographic regions of the world.*

brown manures in annual arable cropping systems and perennial mixed farming systems. Since 75%–95% of the forage N ingested by animals is typically excreted, urine and feces can also be an important pathway for legume foliage N to impact on soil mineral N status in grazed livestock systems, in addition to the N accumulated in, and released from, the nodulated roots (Peoples and Baldock, 2001; Peoples et al., 2004a). In mixed communities of legumes and non-N_2-fixing species in annual intercropping systems and permanent pastures, some legume N can be transferred to the companion nonlegume plants within the growing season, although the amount is often limited (<10 kg N ha^{-1}; Rasmussen et al., 2007; Chalk et al., 2014). That the direct transfer of N between grain legume and companion cereal in annual intercrops is less important than generally expected is largely explained by niche complementarity and a better N nutrition of each cereal plant in intercrops than in sole cereal crop, since the intraspecific competition between two cereal plants for available water and N is greater than interspecific competition with a legume in an intercrop mix (e.g., Bedoussac et al., 2015).

The concentrations of plant-available forms of soil N (predominantly NH_4^+ and NO_3^-) depend upon the relative balance between factors that either favor the accumulation, or result in a reduction, of soil mineral N. An accumulation of available soil N can arise from the combined contribution of (1) the carryover of any mineral N not utilized by plant growth in the previous growing season (*spared N*) and (2) the total N mineralized from above- or belowground plant residues and the soil organic N pool by soil microbes (*total N released*). Factors that reduce soil mineral N on the other hand include (1) the extent to which plant residues or management influences the use of available soil N by soil microbes for growth (*N immobilized*), (2) assimilation of available N by weeds (*weed N uptake*), and (3) leaching, erosion, and gaseous losses (*N lost*). The concentrations of soil mineral N ultimately observed at the beginning of a growing season represent the net effect of all these variables according to the following conceptual equation (Peoples et al., 2017):

Soil mineral N = [(*spared N*) + (*total N released*)] − [(*N immobilized*) + (*weed N uptake*) + (*N lost*)]

Each of these processes can be influenced by the following:

1. the duration of the period of fallow between the end of one growing season and the

beginning of the next since this defines the time available for weed growth and mineralization or loss processes to occur,
2. rainfall amount and distribution during the fallow period between crops or growing seasons as soil moisture (and temperature) regulates soil microbial activity, determines the risk of N losses, and affects weed germination and growth,
3. the quantity of plant residues remaining at the end of the previous growing season and the concentration of N (or C:N ratio) and other biochemical components (polyphenols, etc.) of those residues. Residue N content determines the amount of N potentially available for mineralization, and C:N ratio influences whether a net release or immobilization of mineral N occurs.

The C:N ratio of green legume foliage is commonly <25, which is conducive to net mineralization in the short to medium term (3–12 months). By contrast, the mature, senesced residues of pulse legumes following grain harvest are commonly >30 (Peoples et al., 2009b, 2017), and the incorporation of such material can induce net N immobilization even for legume residues (Justes et al., 2009; Peoples et al., 2009b). However, since the senesced residues and stubbles of other plant species tend to have higher C:N ratios than legumes (typically 50–160; Peoples et al., 2017), incorporation of mature legume residues usually induces less net N immobilization and consequently results in higher net N mineralization than from the stubble of nonlegumes (Plaza-Bonilla et al., 2016). The net result is that the concentrations of plant-available forms of N in soil are frequently higher following sole-cropped legumes than after cereal or oilseed crops, grasses, or mixtures of grasses and legumes, and they are usually highest where legumes have been green or brown manured as a consequence of the larger inputs of organic N and low C:N (Peoples et al., 2017).

Average estimates of the apparent net mineralization of legume N derived from 25 years of experimentation in the rainfed cropping systems of southeastern Australia represented 0.15 kg N ha^{-1} per mm fallow rainfall, 9 kg N per t of stubble remaining after legume crops, and was equivalent to 28% of the total N estimated to be remaining in above- and belowground legume residues (Peoples et al., 2017). Such determinations of soil mineral N benefits following legume crops tend to be lower than calculated from studies with legume-based pastures in the same region (0.5 kg N ha^{-1} per mm rainfall following termination of a 3-year lucerne stand, 15 kg N per t of legume forage dry matter grown during a pasture phase, and 20 kg mineral N per t of legume forage dry matter grown in the year immediately prior to cropping: Peoples et al., 2004a; Peoples et al., 2017). This is perhaps not surprising given that pastures in southeastern Australia are generally grazed by livestock for at least 2 to 3 years before returning to a cropping phase, with the legume components of those pastures fixing atmospheric N_2, and foliage N being recycled back onto the pasture via livestock urine and feces, to contribute to the soil N fertility during the entire period (Peoples and Baldock, 2001), whereas legumes in a cropping sequence are present for only part of a single year. Furthermore, a much larger portion of accumulated plant N is generally removed at grain harvest than is removed in animal products or lost by NO_3^- leaching and N gaseous emissions from grazed pasture systems. Whether similar relationships can be developed to describe prospective soil mineral N benefits following legumes (or are relevant) in other environments remains to be tested.

For grass-legume mixtures, at least two mechanisms limit the increase of plant-available N in the soil in grasslands. First, N_2 fixation by the legume component of a mixed sward is largely regulated by the difference between the N demand of the plant community and the N availability in the soil (N sink strength;

Soussana and Tallec, 2010). In other words, the N input in the system through symbiotic fixation decreases when the level of plant-available N in the soil increases (Carlsson and Huss-Danell, 2003). Secondly, the grass component of the mixture is a strong sink for N and can capture as much soil mineral N as grass monocultures (Nyfeler et al., 2011). Because of the large N uptake by the grass component of the mixture, soil mineral N content is kept at low levels during the growing season. Legumes (especially perennial species) in permanent grasslands can have longer-term effects on soil N availability compared to the short-term effects that typically follow leys based on regenerating annual legumes that are grown for only a few years (Peoples and Baldock, 2001; Peoples et al., 2004a). However, the concentrations of soil mineral observed in both systems will be influenced by the legume content of the pasture sward, as the presence of legumes affects N cycling through the quantity and quality of plant residues derived from the mixture of species returned to the soil (Peoples et al., 2004a; Oelmann et al., 2011) and the intensity of grazing in livestock systems (Peoples and Baldock, 2001).

The elevated soil mineral N status following legume crops and legume-based forage systems can significantly reduce the need for fertilizer N by subsequent crops (Peoples et al., 1995; Angus et al., 2015; Plaza-Bonilla et al., 2017b; Peoples et al., 2017), but it can also constitute a risk of enhanced NO_3^- leaching or denitrification after legumes during the fallow period between crops in some environments, unless a cover-crop is included to scavenge the available soil NO_3^- (Jensen and Hauggaard-Nielsen, 2003; Plaza-Bonilla et al., 2015). The risks of N losses (NO_3^- leaching + NH_3, N_2O, and N_2 gaseous emissions) can be especially prevalent in livestock systems where urine represents 40% −80% of the excreted N (Peoples et al., 2004a). For example, Loiseau et al. (2001) measured similar levels of leaching under grass-legume mixtures and grass monocultures (mean over five drainage periods of 3 and 9 kg N ha^{-1}; respectively), but they observed higher NO_3^- leaching rates under legume monocultures (72 kg N ha^{-1}). Nonetheless, there are also examples where legumes can assist in the capture of leached soil NO_3^-. This includes deep-rooted perennial legumes such as lucerne in pastures and woody legume species in agroforestry or alley-cropping systems that can capture soil NO_3^- leached below the rooting zone of previous crops. Legume cover crops used alone or in mixtures with nonlegume crops in temperate systems or short duration legume "cover/catch crops" between two main cash crops (e.g., between two rice crops, or between an oilseed and cereal) can also help lower the risk of NO_3^- leaching and produce a source of N-rich green manure for the succeeding cash crop (Tribouillois et al., 2016). Legume cover crop recovery of soil NO_3^- may not be as effective as nonlegume species (gramineous or cruciferous species), but it is still better to sow a legume catch crop than having a bare soil during the fallow period to reduce NO_3^- leaching. However, when legumes are grown in association with nonlegumes in cover crop mixtures, they can be as effective as sole nonlegume cover crops at mitigating NO_3^- leaching and efficiently recycle N in the system (Tribouillois et al., 2016).

MITIGATION OF GREENHOUSE GAS EMISSIONS

CO_2 Emissions: It has been calculated that the varying efficiencies of different fertilizer manufacturing plants result in the release of between 0.7 and 1.0 kg of CO_2−C (equivalent to 2.6−3.7 kg CO_2 gas) per kg of NH_3−N produced (Jensen and Hauggaard-Nielsen, 2003). About half of the CO_2 generated during NH_3 production will be reused if the NH_3 is converted to urea, which is the most widely used form of N fertilizer applied to agroecosystems (around two-thirds of total fertilizer N consumed; Jensen et al., 2012). However, once the urea is applied to the soil, it is rapidly hydrolyzed by the enzyme urease to NH_3, and the CO_2 originally captured during urea production will be released. Consequently the current global production and consumption of around 110 million t of fertilizer N would be responsible for ~80–100 million t of CO_2–C released into the atmosphere each year. This compares to between 95 and 135 million t of CO_2–C likely to be respired annually from the nodulated roots of legumes (Jensen et al., 2012). But in contrast to fertilizer N where the CO_2 is derived wholly from fossil fuels, the CO_2 released from N_2-fixing legumes originates from photosynthesis, and any of the CO_2 that was not subsequently recaptured within the plant canopy would not represent a net contribution to atmospheric concentrations of CO_2.

Gaseous emissions of N: Nitrous oxide is a potent greenhouse gas with global warming potential equivalent to 250–296 times that of CO_2. Emissions of N_2O account for ~10% of total global greenhouse gas emissions, ~90% of which are derived from agriculture. Close to 60% of the greenhouse gas emissions released from all sources in agriculture are calculated to be derived from applications of fertilizer N to Asian farming systems (particularly in China [24%] and India [17%]), with a further 14%–15% from each North American and European systems (FAOStat, 2016). The release of N_2O is produced largely by the microbial process of denitrification of NO_3^- in soil under waterlogged or poorly oxygenated conditions, leading to incomplete reduction in N_2, and to a lesser extent as a by-product of nitrification of NH_4^+ to NO_3^-.

Experimental measurements of cumulative N_2O fluxes from legumes and N fertilized systems have been reported to vary enormously (0.03–7.1 and 0.09–18.2 kg N_2O-N ha^{-1}, respectively; Table 8.4). This reflects the data being collated using various methods (static and dynamic measurements) and from a diverse range of studies using different rates of fertilizer N inputs, as well as the large number of climatic, soil, and management variables known to influence net mineralization and denitrification, as well as the portion of the total N lost as N_2O rather than N_2 (Peoples et al., 2004b). Averaged across 71 site-years of data, soils under legumes emitted a total of 1.3 kg N_2O-N ha^{-1} during a growing season. This compared to a mean of 3.2 kg N_2O-N ha^{-1} from 67 site-years of data collected from N fertilized crops and pastures, and 1.2 kg N_2O-N ha^{-1} from 33 site-years of data from unplanted soils or unfertilized nonlegumes (Table 8.4). It was concluded from the data that there was no evidence that the process of symbiotic N_2 fixation by legumes substantially contributed to total N_2O emissions, and that

TABLE 8.4 Summary of Field Measurements of N_2O Emissions From Legume and N-Fertilized Systems[a]

Land Use	Number of Site Years of Data Collated	kg N_2O-N ha^{-1} per Growing Season or Year	
		Range	Mean
Legume-based pasture	25	0.10–4.57	1.38
N-fertilized grass pasture	19	0.30–18.16	4.49
Crop legume	46	0.03–7.09	1.02
N fertilized crop	48	0.09–12.67	2.71
No legume or added N	33	0.03–4.80	1.20

[a] *Derived from data presented by Jensen et al. (2012).*

losses of N_2O from soils under legume were generally lower than N fertilized systems, especially when high rates of N fertilizer typical of farmer practice were applied (Jensen et al., 2012). This conclusion has been confirmed by other comparisons of N_2O emissions from legume-based pastures or fertilized-based grass pastures in Ireland (Li et al., 2011), and comparisons of the N_2O emissions during both the growing season and the postharvest fallow period from N-fertilized canola (*Brassica napus*) with neighboring chickpea, faba bean, and field pea crops in Australia (Schwenke et al., 2015). Emissions of N_2O in the Australian cropping study were highest from canola during the growing season, whereas 75% of the emissions from the legume treatments occurred in the postharvest fallow period, presumably associated with residue decomposition, and the nitrification of NH_4^+ to NO_3^- and/or denitrification of NO_3^- derived from the residues. Interestingly, the total cumulative N_2O emitted over the entire 12 months of the investigation from canola (0.39 kg N_2O-N ha^{-1}) was still significantly greater than either the chickpea and faba bean (0.17 kg N_2O-N ha^{-1}) or field pea treatments (0.13 kg N_2O-N ha^{-1}; Schwenke et al., 2015). Jeuffroy et al. (2013) reported comparable findings from a similar 4-year study comparing canola and winter wheat cropping in France with field pea. However, there are exceptions to these findings, for example, in the study of Peyrard et al. (2016), who compared emissions of N_2O over a whole 3-year sequence of nonlegumes (sorghum-sunflower-durum wheat) with a rotation including faba bean instead of sorghum (sunflower-faba bean-durum wheat) in the arable cropping systems of southwestern France. Measured cumulative N_2O emissions were in the low range of values reported in the literature, but they were significantly higher on average with the inclusion of faba bean (1.12 kg N_2O-N ha^{-1} $year^{-1}$) than for the cereal-based cropping system (0.78 kg N_2O–N ha^{-1} $year^{-1}$). Patterns of emissions also differed between the two systems, with emissions taking place in the form of short-lived peaks only after N fertilization of the nonlegumes, while peaks occurred more frequently over the whole rotation with faba bean. The authors highlighted that both temperature and soil water content had also a strong influence on N_2O emissions. Plaza-Bonilla et al. (2017a) subsequently used the STICS soil-crop model, previously validated for simulating N_2O emission, in three Mediterranean European regions across a large rainfall gradient to investigate the magnitude of N_2O emissions in cereal-based cropping systems where a winter pea crop was inserted. The simulated N_2O emissions were found to be highly sensitive to the interannual variability in climatic conditions. Averaged across all cropping scenarios and the weather conditions tested, the incorporation of winter pea in the traditional cereal-based Mediterranean rotations was predicted to reduce N_2O emissions by about 22% the in the driest site (327 mm annual rainfall) without changing wheat yields. But in the wetter sites (450 and 685 mm), N_2O emissions were relatively unchanged since the emissions associated to the decomposition of low C:N ratio pea residues were similar to application of N fertilizer to cereal-based cropping systems (barley, wheat, and sunflower).

Consequently, the risk of elevated rates of N_2O losses from legume residues should not be ignored, especially under situations where there is a rapid build-up of high concentrations of NO_3^- in soil, such as after the termination of legume-based pastures or where legumes have been green-manured (Hauggaard-Nielsen et al., 2016).

Enteric methane production: Methane (CH_4) is another powerful greenhouse gas with a global warming potential equivalent to 28 times CO_2. Globally, livestock rumen enteric fermentation is responsible for generating ~80 million t of CH_4 annually, accounting for around one-third of all anthropogenic emissions of CH_4 and nearly 80% of agricultural emissions (Eckard et al., 2010; O'Mara, 2011). Around one-third of global enteric CH_4 emissions are generated by livestock in Asia (25% from China and India), 24% from Latin America, 15% from Africa, and 7%–8% from each Western Europe and North America (O'Mara, 2011). Enteric CH_4 is produced under anaerobic conditions in the rumen by methanogenic Archaea using CO_2 and hydrogen to form CH_4, the CH_4 being subsequently released to the atmosphere on the animal's breath. Cattle are the most numerous ruminant, and given their larger body size relative to the other most populous species, sheep and goats, they are the largest contributor to global enteric CH_4 emissions.

The inclusion of legumes in livestock diets can assist with CH_4 abatement by improving forage quality, which tends to increase the voluntary intake and reduces the retention time in the rumen, promoting energetically more efficient postruminal digestion and so reducing the proportion of dietary energy converted to CH_4. Methane emissions are also commonly lower with higher proportions of legumes in the diet, partly because of lower fiber content, the faster rate of passage, and in some cases the presence of condensed tannins and saponins that decrease cell wall digestion and alter rumen methanogenesis (Eckard et al., 2011). A meta-analysis comparing the effects of C4 and C3 grasses with warm and cool climate forage legumes on CH_4 production by ruminants indicated that livestock consuming C4 grasses generated 10%–17% more CH_4 than C3 grasses, while CH_4 production from warm climate legumes was 7%–22% lower than cool climate legumes and 20% less than where animals were fed C4 grasses (Archimède et al., 2011). However, differences between C3 grasses and cool climate legumes was less conclusive. There are reports where livestock fed lucerne produced less CH_4 than grasses, whereas other studies have found inconsistent responses in comparisons where diets have contained different mixtures of clover and grass.

EFFECT OF LEGUMES ON SOIL ORGANIC C

Storage of soil organic carbon (SOC) is an important ecosystem function directly linked to plant productivity, increased soil C inputs from root biomass, and enhanced mineralization of plant residues, all of which may be pronounced for soils associated with legume cropping or high-diversity plant systems (Cong et al., 2014).

For a change in land use to result in a net increase in soil C sequestration, inputs of C from plant vegetation must exceed outputs (i.e., the sum of soil respiration, dissolved C leached from the soil, and soil C lost through wind or water erosion). Legume residues can represent an important source of organic C for soils, particularly when legumes are introduced into grasslands (Guan et al., 2016; Li et al., 2016) and low-input (Shah et al., 2003) or no-tillage (NT) cropping systems (Frasier et al., 2016).

A meta-analysis of grassland management and soil carbon stocks undertaken by Conant et al. (2017) calculated that the largest increases in SOC were associated with land use change (conversion from cultivation; 0.87 t of C ha^{-1} $year^{-1}$), sowing legumes (0.66 t of C ha^{-1} $year^{-1}$), and fertilization (0.57 t of C ha^{-1} $year^{-1}$), all of which promote higher rates of plant productivity and root turnover. Soil C sequestration for legume systems is benefitted by increased biomass production offered by biologic N_2 fixation, enhanced plant diversity, and functional complementarity (Wang et al., 2017), higher root biomass (especially fine root biomass), and greater rates of root decomposition (Fornara et al., 2009). Increasing grassland plant diversity by including legumes may lead to increases in soil C storage of the order of 80% (Wu et al., 2017), and in temperate steppe grasslands the optimum grass-legume ratio for soil C sequestration was determined to be approximately 1:1 (Li et al., 2016).

Both the C:N ratio of legume root residues and their nutrient content may further promote soil C sequestration. Long-term C storage in soils is associated with humus, which typically represents 40%–60% of SOC. With the exception of charcoal, all other forms of SOC are more labile than humus and decompose within months or years. Nutrients such as N, phosphorus (P), and sulphur (S) are tied up along with C in humus, and SOC cannot be effectively sequestered unless adequate amounts of these nutrients are available (Kirkby et al., 2011). Legume root residues are richer in N compared to nonlegumes and tend to have higher P concentrations (Jensen et al., 2012), both of which may stimulate SOC formation at depth (Beniston et al., 2014; Guan et al., 2016).

The use of legumes in innovative cropping, cover cropping, forage, and agroforestry systems has been demonstrated to either slow the decline in SOC compared to that measured under the nonlegume control or encourage C sequestration (Jensen et al., 2012; Plaza-Bonilla et al., 2016). However, observations of increased SOC following the inclusion of legumes are not universal. Several studies have found either no change in SOC or a decline in SOC sequestration over time (Bell et al., 2012; Zotarelli et al., 2012; Robertson et al., 2015; Plaza-Bonilla et al., 2016). Accepting that the studies were of a sufficient duration for any measurable changes in SOC to be accurately quantified, these responses could have reflected lower biomass production and C inputs by legumes than fertilized crops (Plaza-Bonilla et al., 2016), but under some conditions, it may also be that SOC breakdown was enhanced by a positive priming effect (PE) due to the nature of the C input and characteristics of the soil itself.

PEs are short-term changes in SOC turnover due to additions of external organic C (Kuzyakov et al., 2000) and may be positive (accelerating the decomposition of SOC) or negative. A recent meta-analysis of PEs incorporating data from 250 studies on SOC decomposition offers some explanation as to the variable results of soil C sequestration for legume systems grown on nutrient-rich soils (Zhang et al., 2013). The authors found that the strength of the PE was dependent on the chemical properties of the soil, namely its SOC content, total N (TN) content, C:N ratio, and the C content of the residue added. Soils with high SOC (>20 g/kg) and high TN (>2 g/kg) had higher overall decomposition rates. Highest positive PEs on addition of organic C though were found for soils with high SOC but low TN (<2 g/kg) and high C:N ratios (>10). Here the soil microbial activity caused by organic C addition most likely involved scavenging of N from the soil organic matter, resulting in higher PEs (Zhang et al., 2013). In terms of the residue added, the higher concentrations of organic C would promote more microbial activity, which would increase the scale of the PE.

Interestingly, a recent short-term study has indicated an inhibitory effect of legume roots on nonlegume root degradation, which if it influenced the production of humus in the long term, would represent negative priming by legume residues (Saar et al., 2016). More research needs to be done to fully understand the complex interplay of soil and plant residues on SOC accumulation.

THE CONSUMPTION OF NONRENEWABLE ENERGY RESOURCES

A century after its invention, the Haber-Bosch process of NH_3 production for synthesis of fertilizer N essentially remains unchanged. Ammonia is generated from a 3:1 volume mixture of H_2 and N_2 at elevated temperature and pressure in the presence of an iron catalyst. All the N_2 used is obtained from the air, but the H_2 is generated from fossil fuels. It has been estimated that the fossil energy requirements associated with providing the high temperature and pressures and the generation of H_2 feedstock required for the synthesis of N fertilizer represents 1%–2% of the total world energy consumption (Jensen and Hauggaard-Nielsen, 2003).

Fossil fuels are also used in both legume and nonlegume cropping and forage systems in the production of seed for sowing, by on-farm machinery for tillage, sowing, and harvesting of agricultural produce, and in the manufacture, transport, and application of fertilizers and other agrichemical inputs used to either supplement crop nutritional requirements or for crop protection against pests, diseases, and weeds. Direct comparisons of energy consumption by legume and nonlegume systems indicate that legume crops and legume-based pastures use 35%–60% less fossil energy than N fertilized cereals or grasslands, and the inclusion of legumes in cropping sequences reduced the average annual energy usage over a rotation by 12%–34% (Jensen et al., 2012). Much of the reduced energy use is primarily due to the removal of the need to apply N fertilizer to legumes and the subsequently lower N fertilizer requirements of crops grown following legumes. However, the life cycle energy balances of legume-based rotations can also be assisted in some cases by a lower use of agrichemicals for crop protection, as diversification of cropping sequences frequently reduces the incidence of cereal pathogens and pests and can assist with weed control (Angus et al., 2015). It is important to note that differences in fossil energy use between legumes and N-fertilized systems will be somewhat less if energy use is expressed per unit of biomass or grain produced rather than on the basis of total energy usage. Moreover, for forage production, the energy demand for the storage and application of on-farm produced livestock manure in intensive systems where this is an option is quite low compared to the total energy demand of the forage production (Nemecek et al., 2011), so a lower fertilization of legume-rich grasslands compared to grass-dominated grasslands would only marginally reduce the energy consumption of forage production if the N was supplied in the form of livestock manure.

BIODIVERSITY AND ECOSYSTEM SERVICES

Legumes are known to be an important feed source to a range of pollen and nectar feeders, and they can represent the main pollen source to many bumblebee species (Williams and Osborne, 2009). Indeed, legumes could be considered a key element of pollinator conservation strategies in agroecosystems. For example, the use of legume species such as red clover, birdsfoot trefoil (*Lotus corniculatus*), and sainfoin (*Onobrychis viciifolia*) in seed mixtures for establishing flower-rich field margins can provide some mid-to late-season floral resources at a time when most of the mass flowering crops have lost their flowers. Consequently, this strategy can improve the temporal continuity of floral resources that is so crucial for populations of pollinators (Pywell et al., 2011; Hardman et al., 2016). White clover can also be a valuable species, as it produces flowers during most of the growing season, even in frequently defoliated grasslands, although in the case of other legume species, specific grazing or cutting strategies may be required to manage both forage production for livestock and flower resources for pollinators (Farruggia et al., 2012). Further measures may also be required to safeguard nesting habitats to maintain pollinator populations.

Although the understanding of the roles of legumes on soil fauna is still very patchy, there is evidence of positive effects of the presence of N-rich legume roots and residues on the abundance of earthworms, which is likely to be beneficial for soil fertility and structure (e.g., Milcu et al., 2008; van Eekeren et al., 2009). Crotty et al. (2015) also observed a significantly larger earthworm abundance under a white clover sward than under a red clover, indicating that the species of legume might play a further role in modulating earthworm abundance.

Beside their direct role as feed resources for soil micro- and macrofauna, legumes facilitate the maintenance of multiannual pastures and leys for the production of forage in mixed cropping-livestock agroecosystems (this book). Diversifying the agricultural landscape with different annual legume crops or multispecies grasslands containing legumes promotes biodiversity in the agroecosystems through two mechanisms. First, they increase habitat heterogeneity in the agricultural landscape (Bruel et al., 2013), and secondly, they maintain the continuity of multiannual habitats, which can be particularly important for nature conservation objectives in intensive cropping and cereal-dominated systems (Bretagnolle et al., 2011a). Examples of documented benefits include arthropods (Vasseur et al., 2013), bird populations (skylark, Miguet et al., 2013; little bustard, Bretagnolle et al., 2011b), and small mammals (common vole, Bonnet et al., 2013).

CONCLUSIONS

The preceding sections provide numerous examples of where legumes could assist in reducing the risks of environmental damage associated with applications of fertilizer N to support food, fiber, and fodder production, while also promoting a range of valuable ecosystems services. It is clear that to achieve any meaningful impact in terms of lowering agriculture's environmental footprint the areas grown to legumes would need to be greatly expanded to reduce or replace inputs of fertilizer N in countries whose agricultural systems now have a high reliance on N fertilizers, particularly in Asia, North America, and Europe. But to achieve this, it will be necessary to reverse a long-term global trend that has seen a decline in the use of legumes in rotations and reductions in areas sown to legume crops and forages in many regions of the world since the mid-1960s (Meynard et al., 2013; Hauggaard-Nielsen et al., 2016). The only exceptions to this general trend have been the expansion in areas of soybean in South America (from 24.1 million ha in 2000 to 55.7 million ha in 2014) and the progressive increase in areas sown to lentil and field pea in Canada since the 1990s (from 0.2 million ha in 1990 to 2.7 million ha in 2014; FAOStat, 2016).

The initial decline in legumes was driven largely by the widespread availability of cheap N fertilizers that removed the need to include a legume-based fertility restoration phase in farm rotations. Fluctuating grain prices and volatile markets, the absence (or removal) of governmental subsidies to encourage their production (unlike cereals) in some countries, and changes in consumer preferences toward animal rather than plant-based sources of dietary protein further contributed to their demise (Meynard et al., 2013). Other factors included the higher levels of management and agronomic skills required to grown successful legume crops compared to cereals, a general susceptibility of legumes to a range of pests and diseases, leading to poor yield potential, and a stagnation in yield improvement due to a lower level of plant breeding effort directed toward legumes in general, and pulses in particular, compared to cereals (Table 8.1).

Increased breeding effort will certainly be necessary to generate more disease- and pest-tolerant, higher yielding cultivars of established legume species. Yet, there may also be opportunities to introduce new species options in areas where legumes have either been disadopted or not previously been grown. It has long been recognized that beyond the 60–70 species that currently dominate agriculture, there are several hundred underexploited legumes that hold considerable promise of providing valuable sources of grain, vegetables, fruits, root crops, forage, or green manure with further domestication and selection (Peoples et al., 2014). Many of these underexploited species are already adapted across a wide range of environments, including dry climates and impoverished soils, and they represent an untapped pool of potentially valuable genetic material. In principle, there should be enough diversity within the large number of legume species to provide alternative cropping and forage options or with suitable tolerances to temperature extremes, diseases, or pests in regions where climate change threatens the future reliability of food supply (Peoples et al., 2014). A further strategy to decrease the reliance on synthetic N fertilizer would be to promote the utilization of grass-legume mixtures for forage production, at the expense of grass pure swards that still cover large areas worldwide. This strategy is compatible with economically competitive high levels of production (Lüscher et al., 2014; Phelan et al., 2015), and it has been an approach that has been applied in Switzerland for several decades (Kessler and Suter, 2005), although the stability and persistence of the legume in sufficient abundance in such mixtures remains problematic.

The generation of more robust legume germplasm and the development of new alternative legume options are just two hurdles to be overcome. Unfortunately, the prevailing management practices, rules, regulations, and commodity markets that pervade the existing cereal-based systems essentially reinforce the status quo and create barriers to the introduction of legumes despite the well-documented consequences and risks of N-fertilized cereal monocultures for the environment and ecosystems resilience (Jensen et al., 2012; Meynard et al., 2013). How the current situation can be counted so that legumes can be more widely adopted by farmers in global farming systems and what new or novel agricultural practices might still need to be developed to facilitate this remain major challenges.

To be successful from an agronomic point of view, the reintroduction of legumes in arable cropping systems must be carefully designed to address the weaknesses of pulses. For example, Plaza-Bonilla et al. (2016) demonstrated that in the temperate climatic conditions of France, the insertion of grain legumes in conventionally tilled rotations should be accompanied by cover crops between the cash crops to (1) mitigate potential losses of soil organic C and N, (2) increase the value of all the available N coming from the soil mineralization and N_2 fixation, and (3) increase the N use efficiency at a cropping system level without reducing productivity.

Many players need to be engaged if the desired changes are to be realized. This includes both upstream (farmers, advisors, and consultants, agricultural machinery and agribusiness input suppliers, researchers, plant breeders) and downstream participants (marketers, feed and food industries, and consumer organizations) because of their varying roles across value chains and their prevailing practices and biases. Each of these sectors may differ in their opinions concerning the major constraints to the wider adoption of legumes or how the necessary knowledge or technologies required to overcome them should be delivered. However, any transition path is likely to require a set of interconnected changes that reinforce each other and that recognize that initially most new innovations struggle to compete in the market against existing and long-established technologies (Geels and Schot, 2007).

Ultimately, farmers will be the key segment that needs to be consulted, informed, influenced, and supported, so they have the confidence to adopt technological change. Direct conversations will be required with farmers to identify the main drivers for those who regularly include legumes in their cropping or forage systems and to discover the dominant reasons why many other farmers do not. This should provide useful insights into farmer perspectives about the production and economic risks of growing legumes and indicate how this influences their decision-making. Marketing and economic issues are likely to be prominent factors as the value of the legumes is often lower and more volatile than other crops and creates strong disincentives for farmer adoption (Preissel et al., 2015; Zander et al., 2016), even though the evidence suggests legumes, and a greater diversification of cropping systems, contribute to the longer-term stability of profit from a multiple-year perspective (Miller et al., 2015).

Collectively the information gathered through farmer consultation could be used to design the "value-proposition" that might ultimately be needed to persuade farmers to increase the frequency of legumes in their otherwise cereal-dominated cropping systems. Bio-economic modeling could well assist in this process by providing predictions about the prospects for future adoption of new practices that in turn can lead to the development of a framework to identify what further research, development, or extension effort and/or market opportunities might be necessary (Pannell, 2017). Any new initiative will have to be negotiated and adjusted locally by relevant stakeholders to be effective. Participatory research and mutual learning approaches that take stakeholder-driven objectives as a starting point, but also challenge prevailing perceptions, would undoubtedly help overcome the disconnect between research and practice, and will allow progress to be made toward closing the gap between potential yield and what is typically achieved in farmers' fields.

The exchange of technical knowledge is not the only prerequisite. Global and national economic growth, the broader political situation, and deeply embedded cultural values need to be considered to target the necessary technological, social, economic, and/or institutional changes (Geels, 2002). Additional factors such as government intervention through policy, legislation, infrastructure, and market development may be required to facilitate change (Klerkx et al., 2012). These might be in the form of specific farm-level economic initiatives to directly encourage legume cultivation and the development of the related value chains (Zander et al., 2016), but to meet the desired goal of reducing the environmental impact of agriculture, focus must also be given to the consequences of current practices beyond farm boundaries, so the "external costs" of water pollution and greenhouse gas emissions associated with N fertilizer are considered (Geels, 2002). Since the farmers who generate these external costs do not bear the costs themselves, there are currently no appropriate incentives to prevent or repair the costs. Government legislation might well be required, so these external costs are either borne by farmers (e.g., through regulation to specifically reduce emissions) or by the broader community as part of an agri-environmental program (e.g., if farmers were paid to reduce emissions; Pannell, 2017). While it is acknowledged that policy mechanisms sometimes result in unexpected and perverse outcomes, it would be hoped that government interventions that aim to reduce N pollution and other external costs associated with the current dominant agriculture paradigm of a high reliance on N fertilizer to increase crop yields would in turn result in legumes and legume-based systems that have an inherently lower requirement for fertilizer N becoming an increasingly attractive option, especially if they are supported with proactive and well-targeted extension and education programs and viable marketing options.

References

Angus, J.F., Kirkegaard, J.A., Hunt, J.R., Ryan, M.H., Ohlander, L., Peoples, M.B., 2015. Break crop and rotations for wheat. Crop and Pasture Science 66, 523–552.

Archimède, H., Eugène, M., Magdeleine, C.M., Boval, M., Martin, C., Morgavi, D.P., Lecomte, P., Doreau, M., 2011. Comparison of methane production between C3 and C4 grasses and legumes. Animal Feed Science and Technology 166, 59–64.

Bedoussac, L., Journet, E.-P., Hauggaard-Nielsen, H., Naudin, C., Corre-Hellou, G., Jensen, E.S., Prieur, L., Justes, E., 2015. Ecological principles underlying the increase of productivity achieved by cereal-grain legume intercrops in organic farming. A review. Agronomy for Sustainable Development 35, 911–935.

Bell, L.W., Sparling, B., Tenuta, M., Entz, M.H., 2012. Soil profile carbon and nutrient stocks under long-term conventional and organic crop and alfalfa-crop rotations and re-established grassland. Agriculture, Ecosystems and Environment 158, 156–163.

Beniston, J.W., DuPont, S.T., Glover, J.D., Lal, R., Dungait, J.A.J., 2014. Soil organic carbon dynamics 75 years after land-use change in perennial grassland and annual wheat agricultural systems. Biogeochemistry 120, 37–49.

Bonnet, T., Crespin, L., Pinot, A., Bruneteau, L., Bretagnolle, V., Gauffre, B., 2013. How the common vole copes with modern farming: insights from a capture-mark-recapture experiment. Agriculture, Ecosystems and Environment 177, 21–27.

Bretagnolle, V., Gauffre, B., Meiss, H., Badenhausser, I., 2011a. The role of grassland areas within arable cropping systems for the conservation of biodiversity at the regional level. In: Lemaire, G., Hodgson, J., Chabbi, A. (Eds.), Grassland Productivity and Ecosystem Services. CAB International, pp. 251–260.

Bretagnolle, V., Villers, A., Denoufoux, L., Cornulier, T., Inchausti, P., Badenhausser, I., 2011b. Rapid recovery of a depleted population of Little Bustards Tetrax tetrax following provision of alfalfa through an agri-environment scheme. Ibis 153, 4–13.

Brophy, C., Finn, J.A., Lüscher, A., Suter, M., Kirwan, L., Sebastià, M.-T., Helgadóttir, A., Baadshaug, O.H., Bélanger, G., Black, A., Collins, R.P., Čop, J., Dalmannsdottir, S., Delgado, I., Elgersma, A., Fothergill, M., Frankow-Lindberg, B.E., Ghesquiere, A., Golinska, B., Golinski, P., Grieu, P., Gustavsson, A.-M., Höglind, M., Huguenin-Elie, O., Jørgensen, M., Kadziuliene, Z., Kurki, P., Llurba, R., Lunnan, T., Porqueddu, C., Thumm, U., Connolly, J., 2017. Major shifts in species' relative abundance in grassland mixtures alongside positive effects of species diversity in yield: a continental-scale experiment. Journal of Ecology. https://doi.org/10.1111/1365-2745.12754.

Burel, F., Aviron, S., Baudry, J., Le Féon, V., Vasseur, C., 2013. The structure and dynamics of agricultural landscapes as drivers of biodiversity. In: Fu, B., Jones, K.B. (Eds.), Landscape Ecology for Sustainable Environment and Culture. Springer Science+Business Media Dordrecht, pp. 285–308.

Carlsson, G., Huss-Danell, K., 2003. Nitrogen fixation in perennial forage legumes in the field. Plant and Soil 253, 57–67.

Cassman, K.G., Dobermann, A., Walters, D.T., 2002. Agroecosystems, nitrogen-use efficiency, and nitrogen management. Ambio 31, 132–140.

Chalk, P.M., Peoples, M.B., McNeill, A.M., Boddey, R.M., Unkovich, M.J., Gardener, M.J., Siva, C.F., Chen, D., 2014. Methodologies for estimating nitrogen transfer between legumes and companion species in agro-ecosystems: a review of ^{15}N-enriched techniques. Soil Biology and Biochemistry 73, 10–21.

Conant, R.T., Cerri, C.E.P., Osbourne, B.B., Paustian, K., 2017. Grassland management impacts and soil carbon stocks: a new synthesis. Ecological Applications 27, 662–668.

Cong, W., van Ruijven, J., Mommer, L., De Deyn, G.B., Berendse, F., Hoffland, E., 2014. Plant species richness promotes soil carbon and nitrogen stocks in grasslands without legumes. Journal of Ecology 102, 1163–1170.

Crews, T.E., Peoples, M.B., 2005. Synchrony of nitrogen supply and demand in legume versus fertilizer-based agroecosystems and potential improvements with perennials. Nutrient Cycling in Agroecosystems 72, 101–120.

Crotty, F.V., Fychan, R., Scullion, J., Sanderson, R., Marley, C.L., 2015. Assessing the impact of agricultural forage crops on soil biodiversity and abundance. Soil Biology and Biochemistry 91, 119–126.

Eckard, R.J., Grainger, C., de Klein, C.A.M., 2010. Options for the abatement of methane and nitrous oxide from ruminant production: a review. Livestock Science 130, 47–56.

Espinoza, S., Ovalle, C., Zagal, E., Matus, I., Tay, J., Peoples, M.B., del Pozo, A., 2012. Contribution of legumes to wheat productivity in Mediterranean environments of central Chil. Field Crops Research 133, 150–159.

FAOStat, 2016. FAO Crop Statistics On-line. http://faostat3.fao.org/browse/Q/QC/E.

Farruggia, A., Dumont, B., Scohier, A., Leroy, T., Pradel, P., Garel, J.-P., 2012. An alternative rotational stocking management designed to favour butterflies in permanent grasslands. Grass and Forage Science 67, 136–149.

Finn, J.A., Kirwan, L., Connolly, J., Sebastià, M.T., Helgadottir, A., Baadshaug, O.H., Bélanger, G., Black, A., Brophy, C., Collins, R.P., Čop, J., Dalmannsdóttir, S., Delgado, I., Elgersma, A., Fothergill, M., Frankow-Lindberg, B.E., Ghesquiere, A., Golinska, B., Golinski, P., Grieu, P., Gustavsson, A.-M., Höglind, M., Huguenin-Elie, O., Jørgensen, M., Kadziuliene, Z., Kurki, P., Llurba, R., Lunnan, T., Porqueddu, C., Suter, M., Thumm, U., Lüscher, A., 2013. Ecosystem function enhanced by combining four functional types of plant species in intensively managed grassland mixtures: a 3-year continental-scale field experiment. Journal of Applied Ecology 50, 365–375.

Fornara, D.A., Tilman, D., Hobbie, S.E., 2009. Linkages between plant functional composition, fine root processes and potential soil N mineralisation rates. Journal of Ecology 97, 48–56.

Frasier, I., Quiroga, A., Noellemeyer, E., 2016. Effect of different cover crops on C and N cycling in sorghum NT systems. The Science of the Total Environment 562, 628–639.

Geels, F.W., Schot, J., 2007. Typology of sociotechnical transition pathways. Research Policy 36, 399–417.

Geels, F.W., 2002. Technological transitions as evolutionary reconfiguration processes: a multi-level perspective and a case study. Research Policy 31, 1257–1274.

Girard, M., Dohme-Meier, F., Silacci, P., Ampuero Kragten, S., Kreuzer, M., Bee, G., 2015. Forage legumes rich in condensed tannins may increase n-3 fatty acid levels and sensory quality of lamb meat. Journal of the Science of Food and Agriculture 96, 1923–1933.

Guan, X.K., Turner, N.C., Song, L., Gu, Y.J., Wang, T.C., Li, F.M., 2016. Soil carbon sequestration by three perennial legume pastures is greater in deeper soil layers than in the surface soil. Biogeosciences 13, 527–534.

Hardman, C.J., Norris, K., Nevard, T.D., Hughes, B., Potts, S.G., 2016. Delivery of floral resources and pollination services on farmland under three different wildlife-friendly schemes. Agriculture, Ecosystems and Environment 220, 142–151.

Hauggaard-Nielsen, H., Lachouani, P., Knudsen, M.T., Ambus, P., Boelt, B., Gislum, R., 2016. Productivity and carbon foot-print of perennial grass-forage legume intercropping strategies with high or low fertilizer input. The Science of the Total Environment 541, 1339–1347.

Herridge, D.F., Peoples, M.B., Boddey, R.M., 2008. Marschner review: global inputs of biological nitrogen fixation in agricultural systems. Plant and Soil 311, 1–18.

Høgh-Jensen, H., Schjoerring, J.K., Soussana, J.-F., 2002. The influence of phosphorus deficiency on growth and nitrogen fixation of white clover plants. Annals of Botany 90, 745–753.

Husse, S., Huguenin-Elie, O., Buchmann, N., Lüscher, A., 2016. Larger yields of mixtures than monocultures of cultivated grassland species match with asynchrony in shoot growth among species but not with increased light interception. Field Crops Research 194, 1–11.

Jensen, E.S., Hauggaard-Nielsen, H., 2003. How can increased use of biological N_2 fixation in agriculture benefit the environment? Plant and Soil 252, 177–186.

Jensen, E.S., Peoples, M.B., Boddey, R.M., Gresshoff, P.M., Hauggaard-Nielsen, H., Alves, B.J.R., Morrison, M., 2012. Legumes for mitigation of climate change and the provision of feedstock for biofuels and biorefineries. A review. Agronomy for Sustainable Development 32, 329–364.

Jeuffroy, M.H., Baranger, E., Carrouée, B., De Chezelles, E., Gosme, M., Hénault, C., Schneider, A., Cellier, P., 2013. Nitrous oxide emissions from crop rotations including wheat, oilseed rape and dry peas. Biogeosciences 10, 1787–1797.

Justes, E., Mary, B., Nicolardot, B., 2009. Quantifying and modelling C and N mineralization kinetics of catch crop residues in soil: parameterization of the residue decomposition module of STICS model for mature and non mature residues. Plant and Soil 325, 171–185.

Kessler, W., Suter, D., 2005. The role of grass-clover mixtures in Swiss agriculture. In: Frankow-Lindberg, B.E., Collins, R.P., Lüscher, A., Sébastia, M.T., Helgadottir, A. (Eds.), Adaptation and Management of Forage Legumes – Strategies for Improved Reliability in Mixed Swards. SLU Service/Repro, Uppsala, Sweden, pp. 13–20.

Kirkby, C.A., Kirkegaard, J.A., Richardson, A.E., Wade, L.J., Blanchard, C., Batten, G., 2011. Stable soil organic matter: a comparison of CNPS ratios in Australian and international soils. Geoderma 163, 197–208.

Klerkx, L., Schut, M., Leeuwis, C., Kilelu, C., 2012. Advances in knowledge brokering in the agricultural sector: towards innovation system facilitation. IDS Bulletin 43, 53–60.

Kuzyakov, Y., Friedel, J.K., Stahr, K., 2000. Review of mechanisms and quantification of priming effects. Soil Biology and Biochemistry 32, 1485–1498.

Ladha, J.K., Pathak, H.P., Krupnik, T.J., Six, J., van Kessel, C., 2005. Efficiency of fertilizer nitrogen in cereal production: retrospects and prospects. Advances in Agronomy 87, 85–156.

Li, D., Lanigan, G., Humphreys, J., 2011. Measured and simulated nitrous oxide emissions from ryegrass-and ryegrass/white clover-based grasslands in a moist temperate climate. PLoS One 6 (10), e26176. https://doi.org/10.1371/journal.pone.0026176.

Li, Q., Yu, P., Li, G., Zhou, D., 2016. Grass-legume ratio can change soil carbon and nitrogen storage in a temperate steppe grassland. Soil and Tillage Research 157, 23–31.

Loiseau, P., Carrere, P., Lafarge, M., Delpy, R., Dublanchet, J., 2001. Effect of soil-N and urine-N on nitrate leaching under pure grass, pure clover and mixed grass/clover swards. European Journal of Agronomy 14, 113–121.

Lüscher, A., Mueller-Harvey, I., Soussana, J.F., Rees, R.M., Peyraud, J.L., 2014. Potential of legume-based grassland–livestock systems in Europe: a review. Grass and Forage Science 69, 206–228.

Meynard, J.M., Messéan, A., Charlier, A., Charrier, F., Farès, M., Le Bail, M., Magrini, M.B., Savini, I., Réchauchère, O., 2013. Crop Diversification: Obstacles and Levers. Study of Farms and Supply Chains. INRA.

Synopsis of the study report. https://www6.paris.inra.fr/depe/Projets/Diversification-des-cultures.

Miget, P., Gaucherel, C., Bretagnolle, V., 2013. Breeding habitat selection of Skylarks varies with crop heterogeneity, time and spatial scale, and reveals spatial and temporal crop complementation. Ecological Modelling 266, 10–18.

Milcu, A., Partsch, S., Scherber, C., Weisser, W.W., Scheu, S., 2008. Earthworms and legumes control litter decomposition in a plant diversity gradient. Ecology 89, 1872–1882.

Miller, P.R., Bekkerman, A., Jones, C.A., Burgess, M.A., Holmes, J.A., Engel, R.E., 2015. Pea in rotation with wheat reduced uncertainty of economic returns in southwest Montana. Agronomy Journal 107, 541–550.

Nemecek, T., Huguenin-Elie, O., Dubois, D., Gaillard, G., Schaller, B., Chervet, A., 2011. Life cycle assessment of Swiss farming systems: II. Extensive and intensive production. Agricultural Systems 104, 233–245.

Nyfeler, D., Huguenin-Elie, O., Suter, M., Frossard, E., Lüscher, A., 2011. Grass-legume mixtures can yield more nitrogen than legume pure stands due to mutual stimulation of nitrogen uptake from symbiotic and non-symbiotic sources. Agriculture, Ecosystems and Environment 140, 155–163.

O'Mara, F.P., 2011. The significance of livestock as a contributor to global greenhouse gas emissions today and in the near future. Animal Feed Science and Technology 166, 7–15.

Oelmann, Y., Buchmann, N., Gleixner, G., Habekost, M., Roscher, C., Rosenkranz, S., Schulze, E.D., Steinbeiss, S., Temperton, V.M., Weigelt, A., Weisser, W., Wilcke, W., 2011. Plant diversity effects on aboveground and belowground N pools in temperate grassland ecosystems: development in the first 5 years after establishment. Global Biogeochemistry Cycles 25, GB2014.

Pannell, D.J., 2017. Economic perspectives on nitrogen in farming systems: managing trade-offs between production, risk and the environment. Soil Research 55, 473–478.

Peoples, M.B., Baldock, J.A., 2001. Nitrogen dynamics of pastures: nitrogen fixation inputs, the impact of legumes on soil nitrogen fertility, and the contributions of fixed nitrogen to Australian farming systems. Australian Journal of Experimental Agriculture 41, 327–346.

Peoples, M.B., Herridge, D.F., Ladha, J.K., 1995. Biological nitrogen fixation: an efficient source of nitrogen for sustainable agricultural production? Plant and Soil 174, 3–28.

Peoples, M.B., Angus, J.F., Swan, A.D., Dear, B.S., Hauggaard-Nielsen, H., Jensen, E.S., Ryan, M.H., Virgona, J.M., 2004a. Nitrogen dynamics in legume-based pasture systems. In: Mosier, A.R., Syers, K., Freney, J.R. (Eds.), Agriculture and the Nitrogen Cycle. Island Press, Washington DC, pp. 245–260.

Peoples, M.B., Boyer, E.W., Goulding, K.W., Heffer, P., Ochwoh, V.A., Vanlauwe, B., Wood, S., Yagi, K., van Cleemput, O., 2004b. Pathways of nitrogen loss and their impacts on human health and the environment. In: Mosier, A.R., Syers, K., Freney, J.R. (Eds.), Agriculture and the Nitrogen Cycle. Island Press, Washington DC, pp. 53–69.

Peoples, M.B., Brockwell, J., Herridge, D.F., Rochester, I.J., Alves, B.J.R., Urquiaga, S., Boddey, R.M., Dakora, F.D., Bhattarai, S., Maskey, S.L., Sampet, C., Rerkasem, B., Khan, D.F., Hauggaard-Nielsen, H., Jensen, E.S., 2009a. The contributions of nitrogen-fixing crop legumes to the productivity of agricultural systems. Symbiosis 48, 1–17.

Peoples, M.B., Hauggaard-Nielsen, H., Jensen, E.S., 2009b. The potential environmental benefits and risks derived from legumes in rotations. In: Emerich, D.W., Krishnan, H.B. (Eds.), Nitrogen Fixation in Crop Production, vol. 52. ASA CSSA SSSA Agronomy Monograph, pp. 349–385.

Peoples, M.B., Brockwell, J., Hunt, J.R., Swan, A.D., Watson, L., Hayes, R.C., Li, G.D., Hackney, B., Nuttall, J.G., Davies, S.L., Fillery, I.R.P., 2012. Factors affecting the potential contributions of N_2 fixation by legumes in Australian pasture systems. Crop and Pasture Science 63, 759–786.

Peoples, M.B., Khan, D.F., Fillery, I.R.P., 2014. The use of isotopic techniques to quantify the potential contribution of legumes to the mitigation of climate change and future food security. In: Heng, L.K., Sakadevan, K., Dercon, G., Nguyen, M.L. (Eds.), Managing Soils for Food Security and Climate Change Adaptation and Mitigation. FAO, Rome, Italy, pp. 55–60.

Peoples, M.B., Swan, A.D., Goward, L., Kirkegaard, J.A., Hunt, J.R., Li, G.D., Schwenke, G.D., Herridge, D.F., Moodie, M., Wilhelm, N., Potter, T., Denton, M.D., Browne, C., Phillips, L.A., Khan, D.F., 2017. Soil mineral nitrogen benefits derived from legumes and comparisons of the apparent recovery of legume or fertiliser nitrogen by wheat. Soil Research 55, 600–615.

Peyrard, C., Mary, B., Perrin, P., Véricel, G., Gréhan, E., Justes, E., Léonard, J., 2016. N_2O emissions of low input cropping systems as affected by legume and cover crops use. Agriculture, Ecosystem and Environment 224, 145–156.

Phelan, P., Moloney, A.P., McGeough, E.J., Humphreys, J., Bertilsson, J., O'Riordan, E.G., O'Kiely, P., 2015. Forage legumes for grazing and conserving in ruminant production systems. Critical Reviews in Plant Sciences 34, 281–326.

Plaza-Bonilla, D., Nolot, J.-M., Raffaillac, D., Justes, E., 2015. Cover crops mitigate nitrate leaching in cropping systems including grain legumes: field evidence and model simulations. Agriculture, Ecosystem and Environment 212, 1–12.

Plaza-Bonilla, D., Nolot, J.-M., Passot, S., Raffaillac, D., Justes, E., 2016. Grain legume-based rotations managed under conventional tillage need cover crops to mitigate soil organic matter losses. Soil and Tillage Research 156, 33–43.

Plaza-Bonilla, D., Léonard, J., Peyrard, C., Mary, B., Justes, E., 2017a. Precipitation gradient and crop management affect N_2O emissions: simulation of mitigation strategies in rainfed Mediterranean conditions. Agriculture, Ecosystem and Environment 238, 89–103.

Plaza-Bonilla, D., Nolo,t, J.-M., Raffaillac, D., Justes, E., 2017b. Innovative cropping systems to reduce N inputs and maintain wheat yields by inserting grain legumes and cover crops in southwestern France. European Journal of Agronomy 82, 331–341.

Preissel, S., Reckling, M., Schläfke, N., Zander, P., 2015. Magnitude and farm-economic value of grain legume pre-crop benefits in Europe: a review. Field Crops Research 175, 64–79.

Pywell, R.F., Meek, W.R., Hulmes, L., Hulmes, S., James, K.L., Nowakowski, M., Carvell, C., 2011. Management to enhance pollen and nectar resources for bumblebees and butterflies within intensively farmed landscapes. Journal of Insect Conservation 15, 853–864.

Rasmussen, J., Eriksen, J., Jensen, E.S., Esbensen, K.H., Høgh-Jensen, H., 2007. In situ carbon and nitrogen dynamics in ryegrass–clover mixtures: transfers, deposition and leaching. Soil Biology and Biochemistry 39, 804–815.

Robertson, F., Armstrong, R., Partington, D., Perris, R., Oliver, I., Aumann, C., Crawford, D., Rees, D., 2015. Effect of cropping practices on soil organic carbon: evidence from long-term field experiments in Victoria, Australia. Soil Research 53, 636–646.

Saar, S., Semchenko, M., Barel, J.M., De Deyn, G.B., 2016. Legume presence reduces the decomposition rate of non-legume roots. Soil Biology and Biochemistry 94, 88–93.

Schwenke, G.D., Herridge, D.F., Scheer, C., Rowlings, D.W., Haigh, B.M., McMullen, K.G., 2015. Soil N_2O emissions under N_2-fixing legumes and N-fertilised canola: a reappraisal of emissions factor calculations. Agriculture, Ecosystems and Environment 202, 232–242.

Shah, Z., Shah, S.H., Peoples, M.B., Schwenke, G.D., Herridge, D.F., 2003. Crop residues and fertiliser N effects on nitrogen fixation and yields of legume-cereal rotations and soil organic fertility. Field Crops Research 83, 1–11.

Soussana, J.F., Tallec, T., 2010. Can we understand and predict the regulation of biological N_2 fixation in grassland ecosystems? Nutrient Cycling in Agroecosystems 88, 197–213.

Sutton, M.A., Oenema, O., Erisman, J.W., Leip, A., van Grinsven, H., Winiwarter, W., 2011. Too much of a good thing. Nature 472, 159–161.

Tribouillois, H., Cohan, J.-P., Justes, E., 2016. Cover crop mixtures including legume produce ecosystem services of nitrate capture and green manuring: assessment combining experimentation and modelling. Plant and Soil 401, 347–364.

van Eekeren, N., van Liere, D., de Vries, F., Rutgers, M., de Goede, R., Brussaard, L., 2009. A mixture of grass and clover combines the positive effects of both plant species on selected soil biota. Applied Soil Ecology 42, 254–263.

Vasseur, C., Joannon, A., Aviron, S., Burel, F., Meynard, J.-M., Baudry, J., 2013. The cropping systems mosaic: how does the hidden heterogeneity of agricultural landscapes drive arthropod populations? Agriculture, Ecosystems and Environment 166, 3–14.

Vitousek, P.M., Menge, D.N.L., Reed, S.C., Cleveland, C.C., 2013. Biological nitrogen fixation: rates, patterns and ecological controls in terrestrial ecosystems. Philosophical Transactions of the Royal Society B: Biological Sciences 368, 20130119.

Voisin, A.-S., Guéguen, J., Huyghe, C., Jeuffroy, M.-H., Magrini, M.-B., Meynard, J.-M., Mougel, C., Pellerin, S., Pelzer, E., 2014. Legumes for feed, food, biomaterials and bioenergy in Europe: a review. Agronomy for Sustainable Development 343, 361–380.

Wang, Y., Ji, H., Wang, R., Guo, S., Gao, C., 2017. Impact of root diversity upon coupling between soil C and N accumulation and bacterial community dynamics and activity: result of a 30 year rotation experiment. Geoderma 292, 87–95.

Williams, P.H., Osborne, J.L., 2009. Bumblebee vulnerability and conservation world-wide. Apidologie 40, 367–387.

Wu, G., Liu, Y., Tian, F., Shi, Z., 2017. Legumes functional group promotes soil organic carbon and nitrogen storage by increasing plant diversity. Land Degradation and Development 28, 1336–1344.

Zander, P., Amjath-Babu, T.S., Preissel, S., Reckling, M., Bues, A., Schläfke, N., Kuhlman, T., Bachinger, J., Uthes, S., Stoddard, F., Murphy-Bokern, D., Watson, C., 2016. Grain legume decline and potential recovery in European agriculture: a review. Agronomy for Sustainable Development 36. https://doi.org/10.1007/s13593-016-0365-y. Article 26.

Zhang, W., Wang, X., Wang, S., 2013. Addition of external organic carbon and native soil organic carbon decomposition: a meta-analysis. PLoS One 8 (2), e54779. https://doi.org/10.1371/journal.pone.0054779.

Zotarelli, L., Zatorre, N.P., Boddey, R.M., Urquiaga, S., Jantalia, C.P., Franchini, J.C., Alves, B.J., 2012. Influence of no-tillage and frequency of a green manure legume in crop rotations for balancing N outputs and preserving soil organic stocks. Fields Crop Research 132, 185–195.

CHAPTER

9

Can Silvoarable Systems Maintain Yield, Resilience, and Diversity in the Face of Changing Environments?

Gerry Lawson[1], *Christian Dupraz*[2], *Jeroen Watté*[3]

[1]Centre for Ecology and Hydrology, Edinburgh, Scotland; [2]INRA, UMR System, University of Montpellier, France; [3]Werkgroep voor Rechtvaardige en Verantwoorde Landbouw, Brussels, Belgium

INTRODUCTION

Agroforestry systems bring structural and ecological diversity to farmed landscapes. In some circumstances, they also "square the circle" by providing both high agricultural yields and enhanced ecosystem services. "Agricultural intensification" is gained by "ecologic intensification" or more complete use of resources of light, water, and nutrients at a plot level, with the trees acquiring resources that would not otherwise be available to the crop (Cannell et al., 1996). In this context, agroforestry can be defined at the level of a unit of land (Nair, 1993):

> *Agroforestry is a collective name for land-use systems and technologies where woody perennials (trees, shrubs, palms, bamboos, etc.) are deliberately used on the same land-management units as agricultural crops and/or animals, in some form of spatial arrangement or temporal sequence.*

Agroforestry can also be defined in terms of wider social, economic benefits at a landscape level (Leakey, 1996):

> *A dynamic, ecologically based, natural resource management system that, through the integration of trees in farm- and rangeland, diversifies and sustains smallholder production for increased social, economic and environmental benefits.*

Or it can be defined in a simple way that emphasizes that it forms part of a continuum with agriculture and forestry and that it is found both on *both* types of land (van Noordwijk et al., 2016):

> *Agroforestry, a contraction of the terms agriculture and forestry, is land use that combines aspects of both, including the agricultural use of trees*

This chapter focuses on the biophysical aspects of agroforestry, largely at a plot scale. We also discuss the role of agroforestry in mitigation of

Agroecosystem Diversity
https://doi.org/10.1016/B978-0-12-811050-8.00009-1

climate change and its possible use within the European Union Climate and Energy Policy Framework (European Council, 2014), which calls for both "sustainable intensification of agriculture" and "afforestation" as methods to reduce land-based greenhouse gas (GHG) emissions:

> *... the multiple objectives of the agriculture and land use sector, with their lower mitigation potential, should be acknowledged, as well as the need to ensure coherence between the EU's food security and climate change objectives. The European Council invites the Commission to examine the best means of encouraging the* ***sustainable intensification of food production, while optimising the sector's contribution to greenhouse gas mitigation and sequestration****, including through afforestation.*

The impacts of agroforestry can be analyzed using the Common International Classification of Ecosystem Services (Haines-Young and Potschin, 2013), which involves identification of the following:

- *provisioning services* related to human **nutrition** (crops, animals, wild plants, and water supply for drinking), **materials** (fiber, chemical feedstocks, water not for drinking), and **energy** (wood fuel, straw, etc.);
- *regulating services* like **waste mediation** (bioremediation, filtration, abatement), **mediation of flows** (solid flows: control of erosion and landslides; water flows: hydrological balances, flood protection; air flows: storm protection, control of ventilation); **maintenance of physical, chemical, and biologic conditions** (habitat and genepool protection; pest and disease control; soil formation and composition; water condition; atmospheric condition and climate regulation);
- *cultural services* including **physical and intellectual interactions** with biota and ecosystems (scientific, educational, heritage, aesthetic values); **spiritual and/or emblematic values**.

Fig. 9.1 uses this approach to list the **provisioning services** of agroforestry related to products (timber, crops, animals, wild foods, and services), management practices (thinning, pruning, pollarding, hedging, and plowing to favor either the timber or crop components) and economic value of the products, while the **regulating services** of agroforestry includes soil conservation (deeper rooting profiles, soil organic matter, improved soil structure, and nutrition), water regulation (amelioration of extremes and access to greater resources), climate modification (at micro, meso, and macro scales, giving adaptation options to climate change), carbon sequestration (in soils, organic matter, and harvested wood products) and diversity/resilience (biodiversity, landscape diversity, pollution, and disease resilience).

This chapter does not consider the economics of agroforestry (Atangana et al., 2014a; Mercer et al., 2014), nor its cultural services, and it focuses largely on silvoarable systems in temperate regions.

PRODUCTION FUNCTIONS

Agroforestry Products

Most conventional forestry and agricultural products can be grown in agroforestry systems. Less conventional products can also be exploited, including "wild" tropical tree species, which are selected and improved, or "domesticated," to give a range of outputs including fruit, fuelwood, timber, resins, and medicines (Simons and Leakey, 2004). Temperate agroforestry also provides a diverse range of products. The *dehesas/montados* of Spain and Portugal, for example, produce pig meat, beef, fighting bulls, sheep meat and milk, goat meat and milk, poultry, honey, game meat, fish products, cereal, agrotourism, active tourism, bird watching, hunting, firewood, timber, cork, acorns, wool, pasture,

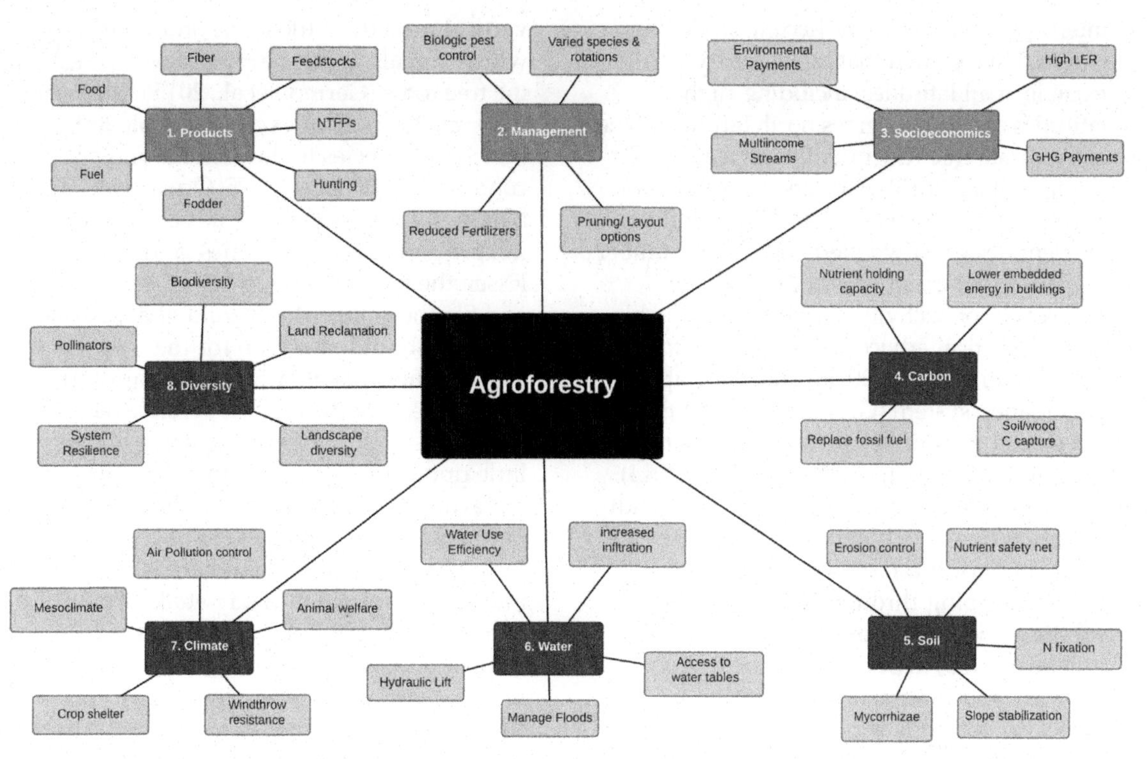

FIGURE 9.1 The ecosystem services of agroforestry systems in terms of provisioning functions (orange [dark gray in print version] denotes agricultural and forestry products, economics, bioenergy) and regulating functions (red [gray in print version] denotes carbon, soil, water, climate, diversity).

mushrooms, asparagus, acorn liqueur, medicinal plants, and aromatic plants (Gaspar et al., 2016).

Orchards and silvopastoral parklands are a traditional land use in temperate areas, and new studies are improving silvopastoral management methods for poultry (Lauri et al., 2016; Rosati et al., 2016; Timmermans and Bestman, 2016), sheep (Decousin, 2015), pigs (Buehrer and Grieshop, 2014), and cattle (Novak et al., 2016).

Management Options

While agroforestry *can* produce all the commercial products associated with conventional forestry or agriculture, farmers need to acquire new skills and modify their normal management practices. Silvoarable farmers need to plan the orientation, spacing, species, protection, and pruning of tree rows carefully.

- **Orientation** of rows in temperate latitudes is normally north–south since this provides more homogeneous light penetration to crop alleys (Dufour et al., 2013). However, in the tropics, the preference is often east–west (Nygren and Jiménez, 1993) since the sun is higher and shading less of a problem. Models can help with this decision and also take account of the effect of slope on light interception and erosion. Recent modelling studies have shown that agroforestry is feasible at all latitudes, including high latitudes above 50 degrees north latitude, if the appropriate tree row orientation is adopted (Dufour et al., 2013). Farmers using boundary agroforestry systems need to focus on the efficient design of wooded hedges and timber strips for shelter, structural diversity, and landscape connectivity. Guides are available giving practical advice on the layouts to use (Wight and Straight, 2015), and geographic information system (GIS) models are being developed to optimize the use of trees in agricultural landscapes (Welsch et al., 2014).

- **Spacing** of tree rows should match the width of agricultural machinery like harvesters and spray booms. If a farmer wants to maintain arable cropping through the whole tree rotation, then wider alleys like 30 or 40 m will be needed, depending on the tree species (Dupraz and Liagre, 2009). Planting at narrower spacings is possible if the arable crop is replaced by grass lays in the last third of the rotation. These narrower spacings assume mixed farming systems. These have become less common in many temperate regions, but scope exists for integrated crop-livestock systems to be introduced by collaboration between arable and livestock farmers (Moraine et al., 2014). Fodder from tree prunings can provide an important part of animal diets (Emile et al., 2016; Green, 2016), particularly when grass forage is restricted by water shortages.
- **Species** selection is both a challenge and opportunity for agroforestry. Tree species need to match the site characteristics, the commercial priorities of the owner, and the requirements of the intercrop. The species of crop used will also affect the early growth of trees. Some crops, like maize, favor the early vertical growth of the trees, and others, like winter cereals, induce deeper development of the tree roots (Germon et al., 2015a; Mulia and Dupraz, 2006). Guides are available to help match tree species to the edaphic and climatic conditions (Ellis et al., 2005; Orwa et al., 2015), although farmers are well advised to use mixtures of species, varieties, and clones to lessen the risk of damage from disease (Millar and Stephenson, 2015; Stenlid et al., 2011). Farmers should also plan for the consequences of climate change. Agroforestry projects require the highest possible quality of planting stock, since wide spacing allows little opportunity for selective thinning, and the available seedlings and rooted cuttings have seldom been selected to have good stem form characteristics when there is little intertree competition (Fady et al., 2003; Palmer et al., 1998).
- **Protection** is vital for young trees, especially in silvopastoral systems where they need shelter against grazing or herbicides (Tubex, 2015). Solid wall tubes impose extra costs but enhance the early growth of most tree species (Balandier and Dupraz, 1999a; Tuley, 1985), providing that ventilation is sufficient (Balandier and Dupraz, 1999b; Dupraz and Bergez, 1999; Tuley, 1985).
- **Pruning** of the lower branches of trees is needed to provide a clean bole of around 30%–50% of tree height, and it allows the timber to be exploited as knot-free veneer logs or high-quality carpentry logs. Pruning also limits stem taper and reaction wood and increases light penetration to crops (Morhart et al., 2015). With conifers, it is true that fast growth at wide spacing can reduce timber quality, but the quality of diffuse-porous temperate broadleaves like maple, cherry, and poplar is little changed by fast growth. The timber quality of ring-porous species like

ash, oak, black locust, catalpa, sweet chestnut, elm, hickory, and mulberry can actually be improved when they are grown rapidly at wide spacing (Coulson, 2016).

REGULATORY FUNCTIONS

Soil Fertility and Conservation

Silvoarable systems have profound effects on both soil nutrients and soil conservation. Agroforestry has long been used to improve soil fertility and conservation in tropical (Young, 1989) and temperate regions due to enhanced organic matter input, greater biosequestration of carbon in the soil, enhanced nutrient conservation, and the potential for symbiotic associations with microorganisms (Moreno and López-Díaz, 2009; Nerlich et al., 2012; Tsonkova et al., 2012).

Soil Nutrients

In conventional agriculture a combination of plowing and leaving the soil exposed for part of the year often leads to high oxidative losses (aerobic respiration) of soil organic matter (SOM) to the atmosphere. Many plowed arable soils have low carbon content since the bulk of crop root and leaf residues are lost in this way and not accumulated or sequestered into more stable organic compounds in the soil. The leaf and root litter of trees tends to take longer than crop litter to decompose, especially in deeper soil layers. Carbon from trees therefore has a longer residence time in the soil than that of crops. Exudates from tree roots also directly add carbon compounds to the soil and increase its aggregate stability (Paudel et al., 2012). Furthermore, the shade and shelter from tree canopies can reduce soil respiration by lowering daytime temperatures and keeping soils moister. Agroforestry is therefore widely recognized to be the land use system with greatest potential to sequester carbon worldwide (Lorenz and Lal, 2014; Watson et al., 2000).

However, trees provide only half the solution to enhancing SOM in silvoarable systems. The optimum crop rotation in such systems should include a catch or cover crop to maximize nitrogen and carbon inputs to the soil and to ensure that soil is not left bare for long (AFAF, 2015; Canet and Schreiber, 2015).

The increased SOM in agroforestry soils directly improves their water and nutrient storage capacity. Increased SOM improves the soil's physical properties and cation exchange capacity. Aggregate stability, soil hydraulic conductivity (macropores), and water-holding capacity are increased, and extremes of water content and temperature reduced. Increasing SOM, in turn, stimulates microbial activity, and it causes more rapid release or "mineralization" of nitrogen in organic matter to the growing roots of trees or crops (Cardinael et al., 2015a).

The effectiveness of silvoarable trees in reducing nitrogen leaching from intercrops can be strong but is difficult to quantify. Many factors are at work:

- Uptake of nitrate (NO_3^-) and ammonium (NH_4^+) is greatest when trees and crops are actively growing; for example, Andrianarisoa et al. (2015) found, during the growing season, that soil mineral nitrogen in the top meter of soil of an agroforestry plot was less than half the level observed in a crop control.
- Uptake is increased by the "safety net" of tree roots growing under the roots of crops grown in alleys (Bergeron et al., 2011), and in temperate areas, tree root growth can take place at depth (e.g., >2.5 m), even in winter when leaves are absent (Germon et al., 2015b).

- Uptake is increased if the root architecture of trees and crops is as complementary as possible, for example, if tree roots have been encouraged to grow more deeply using a succession of winter cereals to dry the surface layers (Cardinael et al., 2015b; Mulia and Dupraz, 2006) and/or by deep plowing before planting trees to break up any soil pans and/or by regular plowing of alleys to induce deeper rooting (Newaj et al., 2013).
- Turnover of fine roots is greater for agroforestry trees in forest monocultures (Lehmann and Zech, 1998) because of greater disturbance and aboveground pruning, which increases fine root turnover and inputs SOM and nutrients to the soil.
- Mineralization of nitrogen in organic matter to ammonium is stimulated by temperature, moisture, pH, and the "quality" of the SOM itself; most deciduous tree leaves and fine roots decompose rapidly (Prieto et al., 2016), but branches and coarse roots provide slower release of nutrients.
- Microclimate extremes are reduced in agroforestry systems; this can have positive or negative effects on mineralization of SOM, but it is common to observe higher carbon and nitrogen under trees in wood pastures (Williams et al., 1999), tropical grasslands (Bernardi et al., 2016), savannas (Wilson and Wild, 1991), and Mediterranean dehesas (Uribe et al., 2015).
- Tree roots with access to groundwater will take advantage of the often high nitrate concentrations (Grimaldi et al., 2012; Wang et al., 2012), and some of this nitrogen will be recycled in the crop rooting zone by a combination of tree root turnover and "hydraulic lift" (Prieto et al., 2012; Sun et al., 2013), the latter being important in dry zones.

The roots of agroforestry trees spread extremely widely, and their extent, when projected to the surface, can be at least four times greater than the projected crown area. Dupraz et al. (1995) found that a trial with 100 trees per hectare, of 6 m crown diameter and 10 m height, completely colonized the surface layers of soil. Dupraz et al. (2011) modeled nitrate leaching from a stand of 50 mature walnut trees per ha, and a crown cover of 30%, on deep and well-drained soils, cropped with a winter cereal rotation (durum wheat), and they showed that nitrate leaching was effectively stopped by the trees, providing that the rainfall events in autumn were not too heavy.

Trees at the edges of fields and in riparian buffers can also play an important role in absorbing nitrates from the horizontal flow surface waters and of saturated groundwater toward watercourses (Grandgirard et al., 2014; Mayer, 2005), particularly if sited some distance away from the stream itself (Komor and Magner, 1996). Trees can act as a "pump" for nutrients that may have moved horizontally into their rooting zone through groundwater flow (van Noordwijk et al., 2003) or that have been translocated from weathering of the bedrock. In an agroforestry system, this nutrient pump may bring long-term benefit to the intercrop. Long-lived trees are frequently observed to enhance soil fertility, particularly under their crowns (Obrador Olan et al., 2004).

Trees in silvoarable systems can also benefit the nutrition of crops through associations with symbiotic organisms in their root systems. These symbionts can be nitrogen-fixing bacteria or mycorrhizal fungi.

- Nitrogen-fixing bacteria come in two forms. Leguminous trees like *Robinia*, *Acacia*, *Calliandra*, *Erythrina*, *Gliricidia*, *Inga*, *Mimosa*, *Leucaena*, and *Sesbania* form root nodules that are symbiotic associations with rhizobia bacteria. Leguminous trees are widespread in the tropics but rather rare in temperate latitudes. Nonleguminous plants also form root nodules, but in association with N_2-fixing actinomycetes in the genus *Frankia*. Actinorhizal species are mostly trees and are found in 24 genera, including *Alnus*, *Casuarina*, *Paulownia*, and *Allocasuarina*. *Alnus* has been recorded to fix up to 300 kg N/ha/yr, which is similar to the highest rate recorded in legumes (Benson et al., 2000). Both types of N-fixing tree have been widely used in agroforestry systems to improve soil fertility and crop yields (Dommergues, 1987). There are concerns that nitrogen fixation will lead to increased emissions of the greenhouse gas nitrous oxide (N_2O), but this is unlikely since the additional ammonium and nitrate is produced in close proximity to the tree and crop roots (Rosenstock et al., 2014).

- Mycorrhizal symbionts are fungi, especially Glomeromycetes, which improve the nutrition and water supply of crops and trees and can also form a connection between the two root systems (Ingleby et al., 2007; van der Heijden et al., 2015). Mycorrhizae can provide protection to the plants against toxic compounds and pathogens (Cameron et al., 2013), and tree roots may serve as a reservoir for species of arbuscular mycorrhizal species, which are killed by exudates from roots of brassicas like rape, but which are beneficial to the growth of subsequent cereal crops (Dodd, 2000). Much work remains to be done researching the optimal endomycorrhizal associations that benefit both trees and crops in silvoarable systems and the extent to which their use can allow conventional inputs to crops be reduced (de Carvalho et al., 2010).

Soil Structure and Conservation

A recent study identified a mean water erosion rate in eight European countries of 2.46 t/ha/yr, with 4 million hectares of croplands having a rate of more than 5 t/ha/yr (Panagos et al., 2015b). Additionally, around 42 Mha of European soils are affected by significant wind erosion (EEA, 2014).

Exudates from tree roots improve the stability of soil aggregates (Monnier, 2015; Mosquera-Losada et al., 2015) and soil hydraulic conductivity (Akdemir et al., 2016; Leung et al., 2015). Trees on their own do not necessarily reduce erosion, however, and some of the highest erosion losses observed in Europe are from olive plantations where the soil is continually plowed and left bare (e.g., 61−184 t/ha/yr for olive plantations in southern Spain (Vanwalleghem et al., 2011)). Furthermore, leaves in the canopy can increase the kinetic energy of throughfall from rain and sometimes increase erosion (Goebes et al., 2015). However, the herbaceous or shrub vegetation in the tree strips do have a significant effect in reducing erosion, particularly when these follow the slope contours (Panagos et al., 2015a).

Thus, agroforestry and other agricultural practices like reduced-tillage, intercropping, contour farming, contour ditches, and grass-mulch cover management have major role in erosion reduction, even in temperate latitudes. Modern planning techniques involving GIS mapping and modeling can identify erosion "hot-spots" where tree-strips can have the greatest impact (Dosskey et al., 2015; Vallebona et al., 2016).

Water and Flooding

Trees compete with crops for water, but they also access deeper resources of water and can make water use more efficient. Their effects on the scale and management of floods are entirely positive.

Water Use

The water use of silvoarable trees is influenced by factors such as tree density, canopy opening, topography, soil water-holding capacity, and competition from crops in the alleys and herbaceous vegetation in the tree rows.

Presence or absence of a water table that is accessible by tree roots in summer is important. Species like poplar depend on access to the water table: others, like walnut, are more flexible in their water requirements. Trees planted at wide spacing tend to have rapid root growth, and their fine roots quickly fill the surface soil horizons, particularly when water and nutrients are not limiting. However, the very dense and competitive fine roots of crops like winter cereals can deplete resources in surface soil horizons and induce the trees to produce roots at greater depth (Fig. 9.2).

In parklands and wood pastures, such as in the dehesas of Spain, tree roots can extend horizontally to many times that of the crown width (seven times in the case of Holm oak), and the majority of tree roots are found below the roots of the herbaceous understory (Moreno Marcos et al., 2007, 2005).

Arbuscular mycorrhizal fungi are also important for water uptake in agroforestry (Schroth, 1998), particularly in drought conditions, but also when the soil water content seems not to be limiting plant growth (Augé et al., 2015; Meddich et al., 2000).

Hydraulic lift from deep roots is potentially important in dry conditions when the trees have access to deep water. It enables them to exude water during the night to drier surface areas of the soil (Kurz-Besson et al., 2006) to be taken up the following day by surface roots. Hydraulic lift is seldom large enough to benefit intercrops, although this benefit is sometimes recorded (Caldwell et al., 1998).

Since agroforestry tends to produce more biomass than sole crops, it will also extract more water and nitrogen from the soil, especially when the mixed species have very different physiology and phenology. In a Mediterranean climate, silvoarable systems are likely to demonstrate the following (Talbot, 2011):

- Water uptake in the summer is increased and leads to drier soils at depth, which in turn increases the storage capacity of the soil column prior to the rains of autumn and winter. Dupraz et al. (2011) estimated this additional capacity to be 100 mm for a stand of 100/ha 12-year-old walnut trees of 8 m height, or 160 mm for a similar stand of 12-year-old 125 poplars of 25 m height.

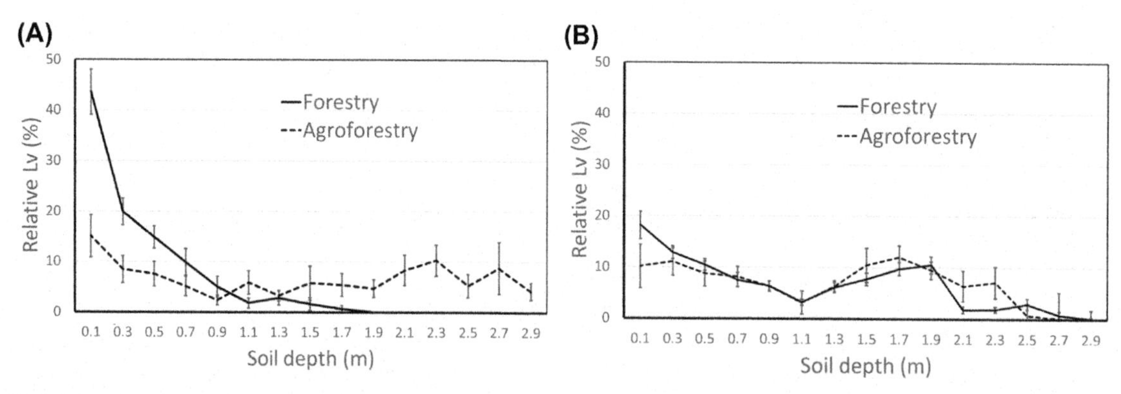

FIGURE 9.2 Root length density distribution profiles for 10-year-old walnut (A) and poplar (B) trees in agroforestry and forestry plots in southern France planted with or without winter intercrops, showing the influence of competition with crop roots on the depth profile of tree roots (Mulia and Dupraz, 2006). The *gray bars* in (B) indicate sandy layers unfavorable to rooting.

- Water uptake by the trees is largely from the deeper horizons, which are not accessible to the crop, and it does not penalize the establishment of winter crops since the soils are rewetted by autumn rains.
- Crop and soil evaporation are reduced in the shade of the trees, which conserves soil water and increases the water use efficiency of the system.

Thus, the main benefits of agroforestry in relation to hydrology are (1) accessing deeper reserves of water that are not available to crop monocultures, (2) improving the use of water through the year, and (3) increasing the proportion of rain that percolates in the ground rather than being lost as runoff. There are also circumstances, particularly in arid conditions, in which the water use efficiency can be improved. This depends on the tree species or genotype however: some species respond to dry air by closing their stomata, whereas others will transpire continuously as long as they have access to soil water (Gebrekirstos et al., 2010). It is likely that trees that are regularly pruned will have higher water use efficiencies since the higher and younger leaves have lower maintenance respiration requirements (Droppelmann et al., 2000).

Flooding

Trees in the upper catchments of rivers reduce the size and intensity of floods compared to catchments covered by grass. This is explained by a combination of (1) increased interception of rainfall on tree canopies, (2) greater roughness of the canopy leading to increased evapotranspiration, (3) greater transpiration in drier weather when trees can access deeper water resources, and (4) increased storage of rainfall caused by improved soil porosity and previous drying of the soil (Grace et al., 2003). Silvopastoral areas in the uplands share these advantages with forests (Bird et al., 2003; Jackson et al., 2008) while maintaining animal yields. Silvoarable plots in the lowlands also increase the evaporation and infiltration of rainfall compared to grassland or cropland, but they play an additional role by helping to direct floodwaters away from major rivers and providing temporary storage of peak flows.

Most climate change models predict increased rainfall and storm frequency in temperate latitudes, and the planting of trees in fields and field-boundaries will be increasingly important to mitigate and adapt to these changes (Servair, 2007; Watté, 2014) (Fig. 9.3).

During the growing season, mature trees will dry soils in areas liable to flooding and may reduce the need for field drains. However, where field drains are used, the tree roots cause a significant risk of blockage. Tree species with "water loving" superficial roots, like poplar and willow, should be avoided, and tree rows should be planted as far away as possible from drains, and should run in the same direction as them (Leuty, 2012).

Climate

Agroforestry trees cause important changes in microclimate (immediately under and adjacent to trees), mesoclimate (tens to hundreds of square meters away from the trees), and macroclimate (at a landscape scale of tens to hundreds of square meters). Planting trees at wide spacing also helps them to develop more stable root systems to resist damage from storms, and regular pruning of lower branches helps to avoid windthrow.

Microclimate

The crops beneath rows of trees will experience lower daytime temperatures, higher nighttime temperatures, lower wind speed, higher humidity, and lower light levels than crops outside the canopy (Corlett et al., 1989). However, the scale of these changes will depend on the density, height, and leaf area of the trees, and

FIGURE 9.3 (Left) Swale established on pasture in Heuvelland, Flanders. It is mulched and planted with fruit trees. The following step is fencing to avoid compaction by the cattle and planting perennial vegetables on the berm, as well as fruit shrubs. (Right) Swale on same farm, after being planted with strawberries. The white line represents the slope. A peak rainfall of 30 mm/h increased the water level in the ditch for a couple of hours, after which every drop was infiltrated. *(Left) Photo: Bert Reubens. (Right) Photo: Jeroen Watté (2014).*

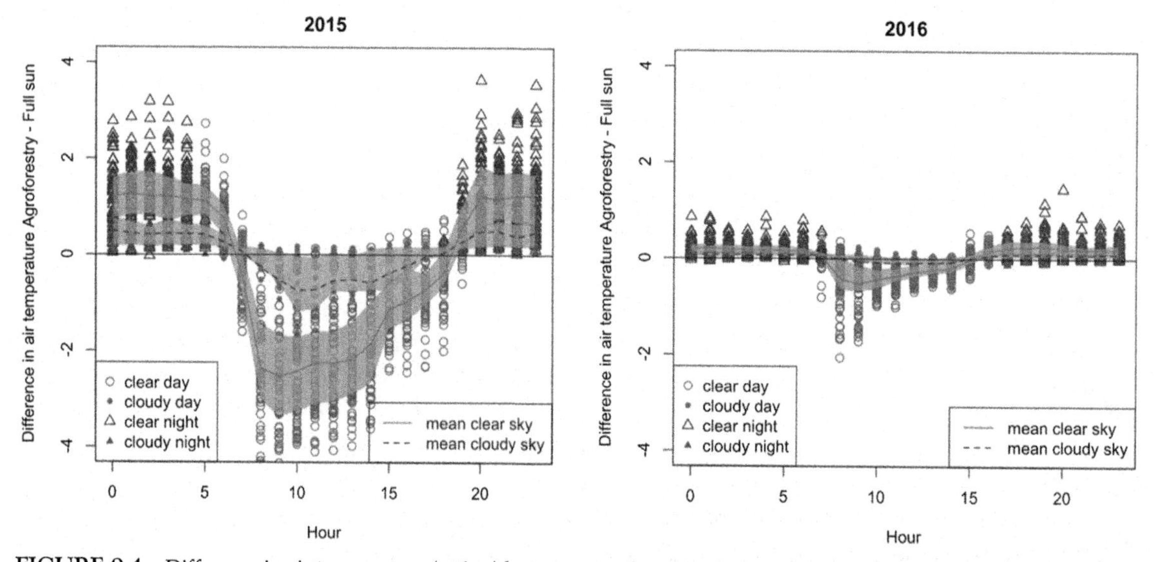

FIGURE 9.4 Difference in air temperature (at 1 m) between agroforestry plots in the south of France (30-m-high, 15-year-old poplar in 2015 and 15-m-high, 20-year-old ash in 2016) and full sun conditions, as a function of time of day in the growing seasons of 2015 (left) and 2016 (right). Observations made during the day are shown as *circles*, and at night as *triangles*. Days were classified as clear (*open symbols*) or cloudy (*solid symbols*) to compute the mean (line) and 25% and 75% percentiles (*shaded areas*) separately for the clear days (*solid line*) and the cloudy days (*dashed line*) (Gosme et al., 2016).

also on the weather conditions. Gosme et al. (2016) only found temperature effects in agroforestry on clear days, and these were dependent on the stand characteristics (Fig. 9.4). In 2015, daytime air temperatures at crop level were up to 4°C lower within 20-m-tall poplar plots than in full sun, whereas measurements in 2016, beneath less dense canopy of 15-m ash trees caused a reduction of only up to 2°C. Lower daytime temperature for most cereals are likely to be beneficial for yields, providing they happen before anthesis. However, higher nighttime temperatures will increase dark respiration and reduce yields. The net result will depend on the timing of these effects and the phenology of the trees and the crop (Grimaldi et al., 2016; Kanzler et al., 2016).

In the semi-open space of an agroforestry plot, trees are likely to transpire more than in a dense forest, because there is more light, greater air turbulence, and increased access to soil water at depth. However, because of reduced wind speed, more humid air, and lower temperatures, the transpiration of the crop or herbaceous layer is likely to be less than in the open. Surface soils may therefore dry out more slowly than outside the agroforest, and this is one explanation of the green areas under trees in Mediterranean silvopastures (Joffre et al., 1999).

Mesoclimate

Shelterbelts give the best example of the effects of trees on mesoclimate, with well-known windspeed reductions and increases in air temperature and atmospheric humidity (Caborn, 1957; van Eimern, 1964). Shelterbelts also influence the distribution of snow and rainfall. Evaporative demand is generally proportional to the square root of windspeed. Good manuals are available to assist in the design of shelterbelts (Brandle et al., 2012; Cornelis and Gabriels, 2005; Ticknor, 1988) and make predictions of the effects on yields (Vityi and Frank, 2016).

Reductions in wind speed can normally be expected at distances downwind from the shelterbelt of 5–20 times its height. Porous shelterbelts are more effective, but their aerodynamics are complicated. Models are available to predict the effect of shelterbelts, depending on their structure (length, height, width, and cross-sectional shape) and porosity (amount and arrangement of vegetative surface area and the shape and volume of individual vegetative elements) (Zhou et al., 2005). However, the effect within an agroforestry system with wide-spaced pruned trees is difficult to predict and winds can sometimes be accelerated by Venturi effects. Turbulence may be the cause of some localized damage to crops in silvoarable plots.

Macroclimate

At the macro scale the cumulative shelterbelt effects of agroforestry trees across a landscape may be important but are difficult to document. Shelter will be more effective in multiple small windbreaks than in a smaller number of larger blocks, but these landscape-scale effects have seldom been measured. One exception is Jensen (1954), who showed that the average velocity of wind blowing from the North Sea was reduced by almost 50% over a transect through the woods and shelterbelts of mid-Jutland, whereas the same length of transect in the open landscape of southern Denmark caused only a 20% reduction.

It used to be suggested by water engineers that the greater loss of water evaporated from rainfall intercepted on tree canopies or transpired through their deep root systems would lead to reduced water yield from upland catchments and threaten aquifer recharge and water supply in reservoirs (Rennie, 1957). This is partially counterbalanced by the increased catchment water-holding capacity mentioned in Section Mesoclimate, and it is partially answered by modern assessments of rainfall on a regional scale. Thus much of the rain evapotranspired from a block of forest or agroforest will simply fall again as rainfall a few hundred kilometers downwind (Ellison et al., 2017/3).

Carbon and Other Greenhouse Gases

Afforestation is often suggested as a climate mitigation option, but agroforestry has the potential to simultaneously establish trees, maintain agricultural production, *and* enhance carbon sequestration. Agroforestry also has implications for emissions of other greenhouse gases, like nitrous oxide and methane, and provides opportunities for the adaptation of crop production to climate change.

Mitigation: Carbon Sequestration

Most data on C sequestration in agroforestry comes from the tropics (Atangana et al., 2014b; Jose and Bardhan, 2012; Kumar and Nair, 2011), but comprehensive studies are now being undertaken in temperate agroforestry (Table 9.1).

Thus, the available estimates of total C sequestration potential in agroforestry in temperate regions vary between 1 and 12 Mg C/ha/yr, depending on species, climate, soil, management, and rotation. Even at the lower end of the range the potential to mitigate climate change is significant. Aertsens et al. (2013) assumed an average of 2.75 Mg C/ha/yr and multiplied it by 90 M ha for the area in Europe with potential for productive silvoarable systems (Reisner et al., 2007), and 50 M ha for potentially productive silvopasture. This gave an estimate of 486 Gg carbon $year^{-1}$ for potential C sequestration in new agroforestry, or 1.5 Tgs of CO_2 equivalent $year^{-1}$. This is the equivalent of 37% of EU CO_2 emissions.

International interest in the potential Land Use Land Use Change and Forestry (LULUCF) to sequester carbon is rising, and more than 100 countries included agricultural carbon sinks in their UNFCCC Intended Nationally Determined Contributions (INDCs) (Richards et al., 2015). Post 2021, most signatories will report carbon sequestration by trees outside forests as part their returns for "Cropland Management" and "Grassland Management." In the European Union, Member States are planning to calculate net emissions using high-resolution geospatial data from the annual Integrated Administration and Control System (IACS) returns made by farmers. Member States could potentially use these geospatial systems to inform individual farmers of the assumed GHG emissions from their farms (de Gruijter et al., 2016; Tuomisto et al., 2015) and offer them incentives to take mitigation actions.

Mitigation: Other Greenhouse Gases

The previous section emphasized the advantages of agroforestry in terms of nitrogen fertilization. Internal recycling of N within agroforestry systems reduces N losses in gaseous forms like nitrogen oxides (NOx, i.e., mononitrogen oxides like NO and NO_2), ammonia (NH_3), and nitrous oxide (N_2O). The latter is 300 times more potent as a greenhouse gas than CO_2. The best way of controlling N_2O is to limit the conditions needed for its emission from the soil. This means reducing the nitrate concentration in soil water and keeping soils less saturated (Skiba et al., 1993). There is a negative correlation between N_2O emissions and the presence of fine roots of fast-growing trees (Weintraub et al., 2014), and with the C:N ratio in the soil (Pimentel et al., 2015). Plentiful fine roots of trees should therefore contribute to lowering reactive nitrogen and moisture levels in the soil, limiting the conditions that favor denitrification and N_2O production. This positive effect of agroforestry on the N_2O balance is not clear in all circumstances however (Kim et al., 2016; Shvaleva et al., 2015).

Trees are effective scavengers of both gaseous and particulate pollutants from the atmosphere. Both rural and urban trees filter ammonia from the atmosphere (Nilsson et al., 2005). Bealey et al. (2014) used a coupled turbulence and deposition model to examine the effects of trees on three types of ammonia emitters: animal houses, slurry lagoons surrounded by tree belts, and intensive silvopasture. They predicted that a maximum of 27% of the emitted ammonia could be captured by trees in strips around animal houses, but that trees in silvopasture systems could capture up to 60% of the ammonia emitted by animals grazing beneath their canopies.

TABLE 9.1 Observed Carbon Sequestration Rates in Tree-Based Agricultural Systems

Location	Tree Species	System	Density Tree/ha	Age (years)	Total Mg C/ha/yr	References
	Poplar	Silvoarable	111		13.2	
Ontario, Canada	Spruce			13	1.1	Peichl et al. (2006)
	Sole crop control	Monoculture	0		−2.9	
	Populus sp.				2.1	
	Picea abies				1.6	
Ontario, Canada	*Quercus rubra*	Silvoarable	111	25	0.8	Wotherspoon et al. (2014)
	Juglans nigra				1.8	
	Thuja occidentalis				1.4	
	Sole soybean control	Monoculture	0	25	−1.2	
Gard, France	Poplar	Silvoarable	140	13	6.5	Hamon et al. (2009)
Montpellier, France	Hybrid walnut		80	14	3.1	
Charante-Maritime, France	Walnut	Silvoarable	70	30	1	Gavaland and Burnel (2005)
Chateaudun, France	*Juglans regia x nigra*		34	6	0.31	
Melle, France	*Juglans regia x nigra*		35	6	0.12	
Saint Jean D-Angély, France	*Juglans nigra*	Silvoarable	102	41	6.01	Cardinael et al. (2017)
Vézénobres, France	*Juglans regia x nigra*		100	18	7.17	
Restincliers, France	*Juglans regia x nigra*		110	18	3.34	
Theix, France	*Prunus avium*	Silvopastoral	200	26	9.62	
Extremadura, Spain	*Juglans major x nigra*	Silvopastoral	333	13	5.6−8.1	López-Díaz et al. (2016)
Extremadura, Spain	*Pinus radiata*	Silvopastoral	833−2500	11	9.4−12.5	
	Betula pubescens				3.9−4.1	Fernández-Núñez et al. (2010)
Nebraska, United States	*Populus sp*	Crops + windbreaks		10−30	2.44−4.69	Schoenberger et al. (2012)
Brandenburg, Germany	*Robinia pseudoacacia*	Short rotation coppice	9200	12	7	Quinkenstein et al. (2011)

Methane (CH_4) emissions from agricultural soils are low and are unlikely to be affected by the introduction of silvopastoral or silvoarable systems. Riparian buffer strips may be significant emitters, however, and as with N_2O, it is an advantage to have trees in these strips that dry the soil and oxidize methane. One study has predicted that the climate changes predicted for temperate forestry, with more nitrogen deposition, extreme rainfall events, and more intensive harvesting, are likely to move the balance of oxygenation/reduction toward greater soil saturation and production of methane and N_2O (Gundersen et al., 2012).

Another contribution that agroforestry can make to methane reduction is by changing the quality of forage ingested by ruminants in silvoarable systems or in cut fodder. Forage grown under partial shade tends to have a higher protein content than forage in the open (Lin et al., 2001; Norton et al., 1991), and many tree leaves and small twigs have a high tannin content that can balance the diet of ruminants and reduce their methane production (Blaxter and Clapperton, 1965; Martin et al., 2010).

Adaptation

The predicted rise in temperatures and reduction in rainfall will have dramatic effects on many temperate regions (UNECE, 2015). Some areas will experience greater drought and risk of fire; other areas will have more rainfall and storms. Climate change also affects the temporal and spatial dynamics of pest species, influencing the frequency, intensity, and consequences of outbreaks, as well as their spatial patterns, size, and geographic range. Coevolved relationships between hosts and their pests probably will be disturbed, and hosts will come in contact with novel pathogens and herbivores. Changes in the species composition of communities are also expected (Lindner et al., 2008).

Studies have suggested that agroforestry can both mitigate and adapt to climate change (Kandji et al., 2006; Schoenberger et al., 2012; Verchot et al., 2007). The most recent IPCC report on Agriculture, Forestry, and Other Land Uses lists the management options available in agroforestry as options to adapt to climate stress (Smith et al., 2014).

Shade from trees has several effects on crops. The main effect is negative due to reduced energy for photosynthesis, but there are positive effects since shade lengthens the life of leaves, reduces their surface temperature, and reduces evaporation. Thus the shade and shelter provided by a tree can compensate the crop for reduced photosynthesis, and yields can be higher and more stable than in monocultures (Talbot et al., 2014). Simulations have shown that agroforestry in French conditions can be more resilient to climate change than monocultures because the trees and crops often have offset phenology. Trees provide shade and shelter and limit the impact of extreme climatic events, like drought and storms, on crops.

Yield reductions in cereals are systematically correlated to climate stress in spring and early summer, which are key times for the growth of cereals. For winter cereals, a change in spring temperatures is thought to explain 80% of the stagnation in yields observed in France during the past 20 years (Brisson et al., 2010). High temperatures speed up the phenological development of crops, reducing the grain-filling period and yields. Current research is trying to assess whether cereal intercrops in agroforestry would be more resilient to climate hazards (Schuller et al., 2015).

Diversity and Risks

Trees do not only impact on crops through the sharing of light, water, and nutrients. They develop a range of accompanying organisms in their canopies and in the herbaceous vegetation of the tree rows that impact crop protection and the biodiversity of the plot.

Biodiversity

The tree rows in silvoarable systems create a refuge for species to survive that are not present in intensive arable systems and can have many benefits for biodiversity:

- Flowers of deciduous agroforestry tree species and herbaceous species in tree rows provide a source of nectar and pollen for insects (Kirk and Howes, 2012).
- The fruit and mast of tree species provide food for birds (Vickery and Arlettaz, 2012) and wild or domesticated animals.
- Insects are also a source of food for birds and bats, with the latter often needing connecting lines of trees to navigate across the landscape (BCT, 2009; Disca, 2003; Maas et al., 2015; Wolfrum, 2014).
- These landscape corridors or "green veins" offer benefits to many small mammals for foraging and migration (Grashof-Bokdam et al., 2008; Hilty et al., 2012).
- Vegetation in tree rows, and small mammal nests, nesting habitats for bumblebee queens (Kells et al., 2001; Lye et al., 2009).
- Wildflowers in the tree rows favor a range of pollinators like bumblebees and other bees, butterflies and moths, flies, and other insects such as beetles, wasps, and thrips (Hill and Webster, 1995; Nieto et al., 2014).

Biodiversity in agroforestry therefore happens both vertically and horizontally: from tree rows out into the crop and from the soil to the top of the tree crowns. At the landscape scale, agroforestry plots change the conditions for breeding and feeding of a wide range of species, from birds to large mammals, from bats to frogs.

Crop Protection

The tree rows can serve as a refuge for crop pests or weeds that are harmful to crops. Conversely, they can shelter predatory species, like ground beetles, which can prevent the development of epidemics or invasions (Holland and Luff, 2000; Lavorel et al., 2008).

The microclimate changes discussed earlier will impact on biological processes: for example, the development of many insect pests of crops is directly proportional to the temperature of the crops, and cooling by shade from the trees will slow down their development. Conversely, the increase of air humidity, or the duration of the dew, can promote the development of foliar diseases like mildews and rusts. Farmers know that these diseases are more frequent near the dense hedgerows.

To protect crops in silvoarable systems we can do the following:

- Select the best tree species to provide a habitat for predatory species of crop pests; for example, the rough and hairy leaves of *Sorbus domestica* and hackberry can host several predatory species of mite that feed on pests of vine and berry crops (Barbar et al., 2007).
- Manage the tree canopy and tree row habitat to favor the predatory species; for example, the wolf spider (*Pardosa hortensis*) responded positively to lower branch pruning in silvoarable plots in the south of France (Martin-Chave et al., 2016).

The canopies of agroforestry trees normally escape from the effects of crop pesticides and are home to many species, which are an asset to ecologic agriculture. Indeed, to combat the main pest of pear trees (a psyllid), farmers in Europe are advised to surround their orchards with hedges of hazel and lime trees, and ivy, which host auxiliary predators and protect against drift from pesticides (Herard, 1985). With the increasing occurrence of pesticide resistance in many pests, the International Organization for Biological Control recommends that countries should ensure that at least 5% of their arable land is covered with natural elements such as lines of trees and herbaceous species (Boller et al., 2004). This matches the proportion of land used by tree rows in many silvoarable systems.

Tree species differ in their ability to host these predators (Dix et al., 1995). Factors such as growing period, hairiness of the leaves, and provision of a food source are important. Some species provide a reserve food supply, others provide permanent or transitional shelter. In winter, many of these auxiliary insects abandon trees after leaf fall and seek refuge in species such as ivy and other evergreen hedgerow species. Thus, it is not only the tree that plays a major role in agroforestry; it is the whole ecosystem that should be managed, even if it means accepting the presence of species that are often wrongly considered undesirable or parasitic.

CONCLUSIONS

A recent study using Land Use and Coverage Area Frame Survey (LUCAS) data from 27 EU member states shows that agroforestry covers 15.4 million ha, which is equivalent to around 3.6% of the territorial area and 8.8% of the utilized agricultural area (den Herder et al., 2017). The World Agroforestry Centre uses Moderate Resolution Imaging Spectroradiometer (MODIS) data of percent tree crown cover at 250-m resolution to estimate that around 45% of the world's agricultural land has a tree crown cover greater than 10% (Zomer et al., 2016).

These are large areas. In the European Union, agroforestry has been implemented as a rural development measure within 35 of the 118 regions (Lawson et al., 2016). It is ranked as the highest impact "greening" measure within the direct support component of the CAP (Warner et al., 2016), and is increasingly viewed a frontrunner in terms of carbon sequestration within EU climate policies (Frelih-Larsen et al., 2014). Yet, it remains a term little known to farmers in Europe, particularly in cool to temperate areas. We hope that uptake can be increased by the research results discussed in this chapter, showing that agroforestry can bring both "sustainable intensification" and "ecosystem services" to agricultural landscapes.

The management challenges for farmers are considerable, however, and more "participative" research is needed. This paper has touched on the complexity of available management options and the need to focus on the highest possible quality of timber; the impacts of agroforestry on soil fertility and soil conservation; the influence of agroforestry trees on climate (at three scales), and on reducing the impact of floods and storms; and the role of agroforestry in climate mitigation and adaptation, and its role in enhancing biodiversity and landscape variety.

Hopefully, increasing research evidence on the positive features of agroforestry may allow policies to be implemented, especially in temperate regions, that overcome the current fiscal and financial disincentives.

References

Aertsens, J., De Nocker, L., Gobin, A., 2013. Valuing the carbon sequestration potential for European agriculture. Land Use Policy 31, 584–594.

AFAF, 2015. Des arbres et des sols éléments-clés de fertilité. Arbre & Paysage, p. 32.

Akdemir, E., Anderson, S.H., Udawatta, R.P., 2016. Influence of agroforestry buffers on soil hydraulic properties relative to row crop management. Soil Science 181, 368–376.

Andrianarisoa, K.S., Dufour, L., Bienaimé, S., Zeller, B., Dupraz, C., 2015. The introduction of hybrid walnut trees (Juglans nigra × regia cv. NG23) into cropland reduces soil mineral N content in autumn in southern France. Agroforestry Systems 90, 193–205.

Atangana, A., Khasa, D., Chang, S., Degrande, A., 2014a. Economics in agroforestry. In: Tropical Agroforestry. Springer, Netherlands, pp. 291–322.

Atangana, A., Khasa, D., Chang, S., Degrande, A., 2014b. Carbon sequestration in agroforestry systems. In: Tropical Agroforestry. Springer, Netherlands, pp. 217–225.

Augé, R.M., Toler, H.D., Saxton, A.M., 2015. Arbuscular mycorrhizal symbiosis alters stomatal conductance of host plants more under drought than under amply watered conditions: a meta-analysis. Mycorrhiza 25, 13–24.

Balandier, P., Dupraz, C., 1999a. Growth of widely spaced trees. A case study from young agroforestry plantations in France. Agroforestry Systems 43, 151–167.

Balandier, P., Dupraz, C., 1999b. Growth of widely spaced trees. A case study from young agroforestry plantations in France. In: Agroforestry for Sustainable Land-Use Fundamental Research and Modelling with Emphasis on Temperate and Mediterranean Applications, Forestry Sciences. Springer, Netherlands, pp. 151–167.

Barbar, Z., Tixier, M.S., Cheval, B., Kreiter, S., 2007. Effects of agroforestry on phytoseiid mite communities (Acari: Phytoseiidae) in vineyards in the South of France. Experimental and Applied Acarology 40, 175–188.

BCT, 2009. Encouraging Bats a Guide for Bat-friendly Gardening and Living. Bat Conservation Trust.

Bealey, W.J., Loubet, B., Braban, C.F., Famulari, D., Theobald, M.R., Reis, S., Reay, D.S., Sutton, M.A., 2014. Modelling agroforestry scenarios for ammonia abatement in the landscape. Environmental Research Letters 9, 125001.

Benson, D.R., Clawson, M.L., Triplett, E.W., Others, 2000. Evolution of the actinorhizal plant symbiosis. Prokaryotic Nitrogen Fixation: A Model System for the Analysis of a Biological Process 207–224.

Bergeron, M., Lacombe, S., Bradley, R.L., Whalen, J., Cogliastro, A., Jutras, M.F., Arp, P., 2011. Reduced soil nutrient leaching following the establishment of tree-based intercropping systems in eastern Canada. Agroforestry Systems 83, 321–330.

Bernardi, R.E., de Jonge, I.K., Holmgren, M., 2016. Trees improve forage quality and abundance in South American subtropical grasslands. Agriculture, Ecosystems and Environment 223, 227–231.

Bird, S.B., Emmett, B.A., Sinclair, F.L., Stevens, P.A., Reynolds, B., Nicholson, S., Jones, T., 2003. Pontbren: Effects of Tree Planting on Agricultural Soils and Their Functions. Countryside Commission for Wales.

Blaxter, K.L., Clapperton, J.L., 1965. Prediction of the amount of methane produced by ruminants. British Journal of Nutrition 19, 511–522.

Boller, E.F., Avilla, J., Joerg, E., Malavolta, C., Wijnanda, F.G., Esbjerg, P. (Eds.), 2004. Integrated Production: Objectives, Principles and Technical Guidelines, IOBC/WRPS Bulleting. International Organisation for Biological Control.

Brandle, J.R., Hintz, D.L., Sturrock, J.W., 2012. Windbreak Technology. Elsevier Science.

Brisson, N., Gate, P., Gouache, D., Charmet, G., Oury, F.-X., Huard, F., 2010. Why are wheat yields stagnating in Europe? A comprehensive data analysis for France. Field Crops Research 119, 201–212.

Buehrer, K.A., Grieshop, M.J., 2014. Postharvest grazing of hogs in organic fruit orchards for weed, fruit, and insect pest management. Organic Agriculture 4, 223–232.

Caborn, J.M., 1957. Shelterbelts and Microclimate. HMSO, Edinburgh.

Caldwell, M.M., Dawson, T.E., Richards, J.H., 1998. Hydraulic lift: consequences of water efflux from the roots of plants. Oecologia 113, 151–161.

Cameron, D.D., Neal, A.L., van Wees, S.C.M., Ton, J., 2013. Mycorrhiza-induced resistance: more than the sum of its parts? Trends in Plant Science 18, 539–545.

Canet, A., Schreiber, K., 2015. La couverture végétale des sols et les pratiques agroforestières, au service de territoires productifs et durables. AFAF (Association Française d'Agroforesterie).

Cannell, M.G.R., Van Noordwijk, M., Ong, C.K., 1996. The central agroforestry hypothesis: the trees must acquire resources that the crop would not otherwise acquire. Agroforestry Systems 34, 27–31.

Cardinael, R., Chevallier, T., Barthès, B.G., Saby, N.P., Parent, T., Dupraz, C., Bernoux, M., Chenu, C., 2015a. Impact of alley cropping agroforestry on stocks, forms and spatial distribution of soil organic carbon - a case study in a Mediterranean context. Geoderma 288–299. Accepted June 2015.

Cardinael, R., Chevallier, T., Cambou, A., Béral, C., Barthès, B.G., Dupraz, C., Durand, C., Kouakoua, E., Chenu, C., 2017. Increased soil organic carbon stocks under agroforestry: a survey of six different sites in France. Agriculture, Ecosystems and Environment 236, 243–255.

Cardinael, R., Mao, Z., Prieto, I., Stokes, A., Dupraz, C., Kim, J.H., Jourdan, C., 2015b. Competition with winter crops induces deeper rooting of walnut trees in a Mediterranean alley cropping agroforestry system. Plant and Soil 391, 219–235.

Corlett, J.E., Ong, C.K., Black, C.R., 1989. Microclimatic modification in intercropping and alley-cropping systems. In: International Workshop on the Applications of Meteorology to Agroforestry Systems Planning and Management, Nairobi (Kenya), 9–13 Feb 1987. Icraf.

Cornelis, W.M., Gabriels, D., 2005. Optimal windbreak design for wind-erosion control. Journal of Arid Environments 61, 315–332.

Coulson, J., 2016. What Difference Does Growth Rate Make to Timber Quality in Temperate Hardwoods? EURAF Newsletter. September, 7–9.

de Carvalho, A.M.X., de Castro Tavares, R., Cardoso, I.M., Kuyper, T.W., 2010. Mycorrhizal associations in agroforestry systems. In: Soil Biology and Agriculture in the Tropics, Soil Biology. Springer, Berlin, Heidelberg, pp. 185–208.

Decousin, J., 2015. Sheep and Orchards: A Promising Association for More Sustainable Cider Apple Production. Hereford Orchards Network of Excellence.

de Gruijter, J.J., McBratney, A.B., Minasny, B., Wheeler, I., Malone, B.P., Stockmann, U., 2016. Farm-scale soil carbon auditing. Geoderma 265, 120–130.

den Herder, M., Moreno, G., Mosquera-Losada, R.M., Palma, J.H.N., Sidiropoulou, A., Santiago Freijanes, J.J., Crous-Duran, J., Paulo, J.A., Tomé, M., Pantera, A., Papanastasis, V.P., Mantzanas, K., Pachana, P., Papadopoulos, A., Plieninger, T., Burgess, P.J., 2017. Current extent and stratification of agroforestry in the European Union. Agriculture, Ecosystems and Environment 241, 121–132.

Disca, T., 2003. Impact des pratiques agroforestières sur l'évolution de la biodiversité - Etude des chiroptères. In: Programme Intégré de Recherches En Agroforesterie à ResTinclières (PIRAT)- Rapport D'activité 2003. INRA-UMR System, Montpellier, pp. 92–104.

Dix, M.E., Johnson, R.J., Harrell, M.O., Case, R.M., Wright, R.J., Hodges, L., Brandle, J.R., Schoeneberger, M.M., Sunderman, N.J., Fitzmaurice, R.L., Young, L.J., Hubbard, K.G., 1995. Influences of trees on abundance of natural enemies of insect pests: a review. Agroforestry Systems 29, 303–311.

Dodd, J.C., 2000. The role of arbuscular mycorrhizal fungi in agro- and natural ecosystems. Outlook on Agriculture 29, 55.

Dommergues, Y.R., 1987. The role of biological nitrogen fixation in agroforestry. In: Steppler, H.A., Nair, P. (Eds.), Agroforestry: A Decade of Development. ICRAF, Nairobi, pp. 245–272.

Dosskey, M.G., Neelakantan, S., Mueller, T.G., Kellerman, T., Helmers, M.J., Rienzi, E., 2015. AgBufferBuilder: a geographic information system (GIS) tool for precision design and performance assessment of filter strips. Journal of Soil and Water Conservation 70, 209–217.

Droppelmann, K.J., Lehmann, J., Ephrath, J.E., Berliner, P.R., 2000. Water use efficiency and uptake patterns in a runoff agroforestry system in an arid environment. Agroforestry Systems 49, 223–243.

Dufour, L., Metay, A., Talbot, G., Dupraz, C., 2013. Assessing light competition for cereal production in temperate agroforestry systems using experimentation and crop modelling. Journal of Agronomy and Crop Science 199, 217–227.

Dupraz, C., Bergez, J.-E., 1999. Carbon dioxide limitation of the photosynthesis of *Prunus avium* L. seedlings inside an unventilated treeshelter. Forest Ecology and Management 119, 89–97.

Dupraz, C., Dauzat, M., Suard, B., Girardin, N., Olivier, A., 1995. Root extension of young wide-spaced *Prunus avium* trees in an agroforest as deduced from the water budget. In: Proceedings of the 4th North-American Agroforestry Conference, Growing a Sustainable Future. Association For Temperate Agroforestry, Boise, Idaho, pp. 46–50.

Dupraz, C., Liagre, F., 2009. Agroforesterie, des arbres et des cultures [Agroforestry, trees and crops]. Agricole, France, 413 p.

Dupraz, C., Liagre, F., Querne, A., Andrianarisoa, S., Talbot, G., 2011. L'agroforesterie peut-elle permettre de réduire les pollution diffuses azotées d'origine agricole ? (No. N° codique INRA: 24000236) INRA, Montpellier.

EEA, 2014. The European Environment: State and Outlook 2015: An Integrated Assessment of the European Environment. European Environment Agency.

Ellis, E.A., Nair, P.K.R., Jeswani, S.D., 2005. Development of a web-based application for agroforestry planning and tree selection. Computers and Electronics in Agriculture 49, 129–141.

Ellison, D., Morris, C.E., Locatelli, B., Sheil, D., Cohen, J., Murdiyarso, D., Gutierrez, V., Noordwijk, M., van, Creed, I.F., Pokorny, J., Gaveau, D., Spracklen, D.V., Tobella, A.B., Ilstedt, U., Teuling, A.J., Gebrehiwot, S.G., Sands, D.C., Muys, B., Verbist, B., Springgay, E., Sugandi, Y., Sullivan, C.A., 2017/3. Trees, forests and water: cool insights for a hot world. Global Environmental Change 43, 51–61.

Emile, J.C., Delagarde, R., Barre, P., Novak, S., 2016. Nutritive value and degradability of leaves from temperate woody resources for feeding ruminants in summer. In: Gosme, M., et al. (Eds.), 3rd European Agroforestry Conference – Montpellier, 23-25 May 2016. EURAF, pp. 410–413.

European Council, 2014. Conclusions on 2030 Climate and Energy Policy Framework (No. SN 79/14). European Union.

Fady, B., Ducci, F., Aleta, N., Becquey, J., Diaz Vazquez, R., Fernandez Lopez, F., Jay-Allemand, C., Lefèvre, F., Ninot, A., Panetsos, K., Paris, P., Pisanelli, A., Rumpf, H., 2003. Walnut demonstrates strong genetic variability for adaptive and wood quality traits in a network of juvenile field tests across Europe. New Forests 25, 211–225.

Fernández-Núñez, E., Rigueiro-Rodríguez, A., Mosquera-Losada, M.R., 2010. Carbon allocation dynamics one decade after afforestation with *Pinus radiata* D. Don and *Betula alba* L. under two stand densities in NW Spain. Ecological Engineering 36, 876–890.

Frelih-Larsen, A., MacLeod, M., Osterburg, B., Eory, A.V., Dooley, E., Kätsch, S., Naumann, S., Rees, B., Tarsitano, D., Topp, K.A.W., Metayer, N., Molnar, A.,

Povellato, A., Bochu, J.L., Lasorella, M.V., Longhitano, D., 2014. Mainstreaming Climate Change into Rural Development Policy Post 2013. Ecologic Institut, Berlin.

Gaspar, P., Escribano, M., Mesías, F.J., 2016. A participatory approach to develop new products that promote social valorization of agroforestry systems. In: Gosme, M., et al. (Eds.), 3rd European Agroforestry Conference – Montpellier, 23-25 May 2016. EURAF, pp. 156–158.

Gavaland, A., Burnel, L., 2005. Croissance et biomasse aérienne de noyers noirs. Chambres d'agriculture, pp. 20–21.

Gebrekirstos, A., van Noordwijk, M., Neufeldt, H., Mitlöhner, R., 2010. Relationships of stable carbon isotopes, plant water potential and growth: an approach to asses water use efficiency and growth strategies of dry land agroforestry species. Trees 25, 95–102.

Germon, A., Cardinael, R., Prieto, I., Mao, Z., Kim, J.H., Stokes, A., Dupraz, C., Laclau, J.-P., Jourdan, C., 2015a. Unexpected phenology and lifespan of shallow and deep fine roots of walnut trees grown in a Mediterranean agroforestry system. Plant and Soil 401, 409–426.

Germon, A., Cardinael, R., Prieto, I., Mao, Z., Kim, J.H., Stokes, A., Dupraz, C., Laclau, J.-P., Jourdan, C., 2015b. Unexpected phenology and lifespan of shallow and deep fine roots of walnut trees grown in a Mediterranean agroforestry system. Plant and Soil 401, 409–426.

Goebes, P., Bruelheide, H., Härdtle, W., Kröber, W., Kühn, P., Li, Y., Seitz, S., von Oheimb, G., Scholten, T., 2015. Species-specific effects on throughfall kinetic energy in subtropical forest plantations are related to leaf traits and tree architecture. PLoS One 10, e0128084.

Gosme, M., Dufour, L., Hd, I.A., Dupraz, C., 2016. Microclimatic effect of agroforestry on diurnal temperature cycle. In: Gosme, M., et al. (Eds.), 3rd European Agroforestry Conference – Montpellier, 23-25 May 2016. EURAF, pp. 183–186.

Grace, J., Moncrieff, J., McNaughton, K., 2003. Forests at the Land-atmosphere Interface. CABI, Wallingford.

Grandgirard, D., Combaud, A., Mercadal, L., Liagre, A.M., Bachevillier, Y., Marin, A., 2014. Agroforestry Systems Design at Parcel Scale within the Territory (FR): Towards a Spatial Decision Support System for Catched Water Quality Improvement. EURAF.

Grashof-Bokdam, C.J., Paul Chardon, J., Vos, C.C., Foppen, R.P.B., WallisDeVries, M., van der Veen, M., Meeuwsen, H.A.M., 2008. The synergistic effect of combining woodlands and green veining for biodiversity. Landscape Ecology 24, 1105–1121.

Green, T., 2016. Forgotten food – tree hay. In: Gosme, Al, et al. (Eds.), 3rd European Agroforestry Conference – Montpellier, 23-25 May 2016. EURAF, pp. 407–409.

Grimaldi, C., Fossey, M., Thomas, Z., Fauvel, Y., Merot, P., 2012. Nitrate attenuation in soil and shallow groundwater under a bottomland hedgerow in a European farming landscape. Hydrological Processes 26, 3570–3578.

Grimaldi, J., Fieuzal, R., Pelletier, C., Bustillo, V., Houet, T., Sheeren, D., 2016. Microclimate patterns in an agroforestry intercropped vineyard: first results. In: Gosme, M., et al. (Eds.), 3rd European Agroforestry Conference – Montpellier, 23-25 May 2016. EURAF, pp. 191–194.

Gundersen, P., Christiansen, J.R., Alberti, G., Brüggemann, N., Castaldi, S., Gasche, R., Kitzler, B., Klemedtsson, L., Lobo-do-Vale, R., Moldan, F., Others, 2012. The response of methane and nitrous oxide fluxes to forest change in Europe. Biogeosciences 9, 3999–4012.

Haines-Young, R., Potschin, M., 2013. Common International Classification of Ecosystem Services (CICES): Consultation on Version 4, August-December 2012.

Hamon, X., Dupraz, C., Liagre, G., 2009. L'Agroforesterie: Outil de Séquestration du Carbone en Agriculture (No. Report for Compte d'Affection Special pour le Développement Agricole (CASDRA) du Ministere de l'Agriculture, de l'Alimentation de de la Peche). CASDAR.

Herard, F., 1985. Analysis of parasite and predator populations observed in pear orchards infested by *Psylla pyri* (L.)(Hom.: Psyllidae) in France. Agronomie 5, 773–778.

Hill, D.B., Webster, T.C., 1995. Apiculture and forestry (bees and trees). Agroforestry Systems 29, 313–320.

Hilty, J.A., Lidicker, W.Z., Merenlender, A., Dobson, A.P., 2012. Corridor Ecology: The Science and Practice of Linking Landscapes for Biodiversity Conservation. Island Press.

Holland, J.M., Luff, M.L., 2000. The effects of agricultural practices on Carabidae in temperate agroecosystems. Integrated Pest Management Reviews 5, 109–129.

Ingleby, K., Wilson, J., Munro, R.C., Cavers, S., 2007. Mycorrhizas in agroforestry: spread and sharing of arbuscular mycorrhizal fungi between trees and crops: complementary use of molecular and microscopic approaches. Plant and Soil 294, 125–136.

Jackson, B.M., Wheater, H.S., Mcintyre, N.R., Chell, J., Francis, O.J., Frogbrook, Z., Marshall, M., Reynolds, B., Solloway, I., 2008. The impact of upland land management on flooding: insights from a multiscale experimental and modelling programme: impact of upland land management on flooding. Journal of Flood Risk Management 1, 71–80.

Jensen, M., 1954. Shelter Effect: Investigations into the Aerodynamics of Shelter and its Effects on Climate and Crops. Danish Technical Press, Copenhagen.

Joffre, R., Rambal, S., Ratte, J.P., 1999. The dehesa system of southern Spain and Portugal as a natural ecosystem mimic. Agroforestry Systems 45, 57–79.

Jose, S., Bardhan, S., 2012. Agroforestry for biomass production and carbon sequestration: an overview. Agroforestry Systems 86, 105–111.

Kandji, S.T., Verchot, L.V., Mackensen, J., Boye, A., Van Noordwijk, M., Tomich, T.P., Ong, C.K., Albrecht, A.,

Palm, C.A., Garrity, D.P., Others, 2006. Opportunities for Linking Climate Change Adaptation and Mitigation through Agroforestry Systems. World Agroforestry into the Future. World Agroforestry Centre-ICRAF, Nairobi, Kenya, pp. 113–121.

Kanzler, M., Böhm, C., Mirck, J., 2016. Microclimate effects of short rotation tree-strips in Germany. In: Gosme, M., et al. (Eds.), 3rd European Agroforestry Conference – Montpellier, 23-25 May 2016. EURAF, pp. 31–324.

Kells, A.R., Holland, J.M., Goulson, D., 2001. The value of uncropped field margins for foraging bumblebees. Journal of Insect Conservation 5, 283–291.

Kim, D.-G., Kirschbaum, M.U.F., Beedy, T.L., 2016. Carbon sequestration and net emissions of CH4 and N2O under agroforestry: synthesizing available data and suggestions for future studies. Agriculture, Ecosystems and Environment 226, 65–78.

Kirk, W.D.J., Howes, F.N., 2012. Plants for Bees: A Guide to the Plants that Benefit the Bees of the British Isles. IBRA.

Komor, S.C., Magner, J.A., 1996. Nitrate in groundwater and water sources used by riparian trees in an agricultural watershed: a chemical and isotopic investigation in southern Minnesota. Water Resources Research 32, 1039–1050.

Kumar, B.M., Nair, P.K.R., 2011. Carbon Sequestration Potential of Agroforestry Systems: Opportunities and Challenges. Springer Science & Business Media.

Kurz-Besson, C., Otieno, D., do Vale, R.L., Siegwolf, R., Schmidt, M., Herd, A., Nogueira, C., David, T.S., David, J.S., Tenhunen, J., Pereira, J.S., Chaves, M., 2006. Hydraulic lift in cork oak trees in a Savannah-type mediterranean ecosystem and its contribution to the local water balance. Plant and Soil 282, 361–378.

Lauri, P., Mézière, D., Dufour, L., Gosme, M., Simon, S., Gary, C., 2016. When chickens graze in olive orchards, the environmental impact of both chicken rearing and olive growing decreases. In: Gosme, M., et al. (Eds.), 3rd European Agroforestry Conference – Montpellier. EURAF, pp. 400–402.

Lavorel, S., Sarthou, J.-P., Carré, G., Chauvel, B., Cortet, J., Dajoz, I., Dupraz, C., Farruggia, A., Lavergne, S., Liagre, F., Lumaret, J.-P., Quétier, F., Roger-Estrade, J., Schmid, B., Simon, S., Steinberg, C., Tichit, M., Vaissière, B., Tuinen, D., van, Villenave, C., 2008. Intérêts de la biodiversité pour les services rendus par les écosystèmes. In: INRA (Ed.), Agriculture et Biodiversité: Des Synergies à Valoriser. Rapport de L'expertise Scientifique Collective Réalisée Par l'Inra à La Demande Du Ministère de l'Agriculture et de La Pêche (MAP) et Du Ministère de l'Écologie, de l'Énergie, Du Développement Durable et de l'Aménagement Du Territoire (MEEDDAT). INRA, Paris, p. 266.

Lawson, G.J., Balaguer, F., Palma, J., Papanastasis, V., 2016. Options for agroforestry in the CAP 2014-2020. In: Gosme, M., et al. (Eds.), 3rd European Agroforestry Conference – Montpellier, 23-25 May 2016. EURAF, pp. 425–428.

Leakey, R., 1996. Definition of Agroforestry Revisited. Agroforestry Today (ICRAF).

Lehmann, J., Zech, W., 1998. Fine root turnover of irrigated hedgerow intercropping in Northern Kenya. Plant and Soil 198, 19–31.

Leung, A.K., Garg, A., Coo, J.L., Ng, C.W.W., Hau, B.C.H., 2015. Effects of the roots of *Cynodon dactylon* and *Schefflera heptaphylla* on water infiltration rate and soil hydraulic conductivity. Hydrological Processes 29, 3342–3354.

Leuty, T., 2012. Farm Tile Drains and Tree Roots (No. AGDEX 555). Ontario Ministry of Agriculture Food and Rural Affairs.

Lin, C.H., McGraw, M.L., George, M.F., Garrett, H.E., 2001. Nutritive quality and morphological development under partial shade of some forage species with agroforestry potential. Agroforestry Systems 53 (3), 269–281.

Lindner, M., Garcia-Gonzalo, J., Kolström, M., Green, T., Reguera, R., Maroschek, M., Seidl, R., Lexer, M.J., Netherer, S., Schopf, A., Others, 2008. Impacts of Climate Change on European Forests and Options for Adaptation (No. AGRI-2007-G4-06 Report to the European Commission Directorate-General for Agriculture and Rural Development). EFI, BOKU, INRA, IAFS.

López-Díaz, M.L., Bertomeu, M., Arenas-Corralizas, B.R., Moreno, G., 2016. Carbon sequestration in intensive hardwood plantations: influence of management. In: Gosme, M., et al. (Eds.), 3rd European Agroforestry Conference – Montpellier, 23-25 May 2016. EURAF, pp. 179–182.

Lorenz, K., Lal, R., 2014. Soil organic carbon sequestration in agroforestry systems. A review. Agronomy for Sustainable Development 34, 443–454.

Lye, G., Park, K., Osborne, J., Holland, J., Goulson, D., 2009. Assessing the value of Rural Stewardship schemes for providing foraging resources and nesting habitat for bumblebee queens (Hymenoptera: Apidae). Biological Conservation 142, 2023–2032.

Maas, B., Karp, D.S., Bumrungsri, S., Darras, K., Gonthier, D., Huang, J.C.-C., Lindell, C.A., Maine, J.J., Mestre, L., Michel, N.L., Morrison, E.B., Perfecto, I., Philpott, S.M., Şekercioğlu, Ç.H., Silva, R.M., Taylor, P.J., Tscharntke, T., Van Bael, S.A., Whelan, C.J., Williams-Guillén, K., 2015. Bird and bat predation services in tropical forests and agroforestry landscapes. Biological reviews of the Cambridge Philosophical Society 91 (4), 1081–1101.

Martin-Chave, A., Mazzia, C., Beral, C., Capowiez, Y., 2016. How agroforestry microclimates could affect the daily-activity of major predatory arthropods in organic vegetable crops? In: Gosme, M., et al. (Eds.), 3rd European Agroforestry Conference – Montpellier, 23-25 May 2016. EURAF, pp. 62–65.

Martin, C., Morgavi, D.P., Doreau, M., 2010. Methane mitigation in ruminants: from microbe to the farm scale. Animal: An International Journal of Animal Bioscience 4 (3), 351–365.

Mayer, P.M., 2005. Riparian Buffer Width, Vegetative Cover, and Nitrogen Removal Effectiveness.

Meddich, A., Oihabi, A., Abbas, Y., Bizid, E., 2000. Rôle des champignons mycorhiziens à arbuscules de zones arides dans la résistance du trèfle (*Trifolium alexandrinum* L.) au déficit hydrique. Agronomie 20, 283–295.

Mercer, D.E., Cubbage, F.W., Frey, G.E., 2014. Economics of Agroforestry.

Millar, C.I., Stephenson, N.L., 2015. Temperate forest health in an era of emerging megadisturbance. Science 349, 823–826.

Monnier, Y., 2015. Could agroforestry be a way to limit soil erosion susceptibility under a temperate climate?. In: 3. Climate Smart Agriculture. HAL CCSD, p. 298.

Moraine, M., Duru, M., Nicholas, P., Leterme, P., Therond, O., 2014. Farming system design for innovative crop-livestock integration in Europe. Animal 8, 1204–1217.

Moreno, G., López-Díaz, M.L., 2009. Roles and functioning of agroforestry systems. In: AECID (Ed.), Agroforestry Systems as a Technique for Sustainable Land Management. AECID, ISBN 978-84-96351-59-2, pp. 197–208.

Moreno Marcos, G., Obrador, J.J., Cubera, E., Dupraz, C., 2005. Fine root distribution in dehesas of central-western Spain. Plant and Soil 277, 153–162.

Moreno Marcos, G., Obrador, J.J., García, E., Cubera, E., Montero, M.J., Pulido, F., Dupraz, C., 2007. Driving competitive and facilitative interactions in oak dehesas through management practices. Agroforestry Systems 70, 25–40.

Morhart, C., Sheppard, J., Douglas, G.C., Lunny Paris, P., Spiecker, H., Nahm, M., 2015. Management Guidelines for Valuable Wood Production in Agroforestry Systems. School of Forest Growth, University of Freiburg.

Mosquera-Losada, M.R., Rigueiro-Rodríguez, A., Ferreiro-Domínguez, N., 2015. Effect of liming and organic and inorganic fertilization on soil carbon sequestered in macro-and microaggregates in a 17-year old Pinus radiata silvopastoral system. Journal of Environmental Management 150, 28–38.

Mulia, R., Dupraz, C., 2006. Unusual fine root distributions of two deciduous tree species in southern France: what consequences for modelling of tree root dynamics? Plant and Soil 281, 71–85.

Nair, P.K.R., 1993. An Introduction to Agroforestry. Springer Science & Business Media.

Nerlich, K., Graeff-Hönninger, S., Claupein, W., 2012. Agroforestry in Europe: a review of the disappearance of traditional systems and development of modern agroforestry practices, with emphasis on experiences in Germany. Agroforestry Systems 87, 475–492.

Newaj, R., Dhyani, S.K., Alam, B., Prasad, R., Handa, A.K., Kumar, U., Neekher, S., 2013. Long term effect of root management practices on rooting pattern in *Dalbergia sissoo* and grain yield of mustard under agrisilviculture system. Range Management and Agroforestry 34, 47–50.

Nieto, A., Roberts, S., Kemp, J., Rasmont, P., Kuhlmann, M., García-Criado, M., Biesmeijer, J.C., Bogusch, P., Dathe, H.H., la Rúa, P.D., De Meulemeester, T., Dehon, M., Dewulf, A., Ortiz-Sánchez, F.J., Lhomme, P., Pauly, A., Potts, S.G., Praz, C., Quaranta, M., Radchenko, V.G., Scheuchl, E., Smit, J., Straka, J., Terzo, M., Tomozii, B., Window, J., Michez, D., 2014. European Red List of Bees. IUCN.

Nilsson, K., Konijnendijk, C., Randrup, T.B., 2005. Research on urban forests and trees in Europe. In: Konijnendijk, C., Nilsson, K., Randrup, T., Schipperijn, J. (Eds.), Urban Forests and Trees. Springer-Verlag, Berlin/Heidelberg, pp. 445–463.

Norton, B.W., Wilson, J.R., Shelton, H.M., Hill, K.D., 1991. The Effect of Shade on Forage Quality. Forages for Plantation Crops. ACIAR, Canberra, pp. 83–88.

Novak, S., Liagre, F., Emile, J.C., 2016. Integrating agroforestry into an innovative mixed crop-dairy system. In: Gosme, M. (Ed.), 3rd European Agroforestry Conference – Montpellier, 23-25 May 2016. EURAF, pp. 397–399.

Nygren, P., Jiménez, J.M., 1993. Radiation regime and nitrogen supply in modelled alley cropping systems of Erythrina poeppigiana with sequential maize-bean cultivation. Agroforestry Systems 21, 271–285.

Obrador Olan, J.J., Garcia Lopez, E., Moreno, G., 2004. Consequences of dehesa land use on the nutritional status of vegetation in Central Western Spain. Advances in Geoecology 327–340.

Orwa, C., Mutua, A., Kindt, R., Jamnadass, R., Simons, A., 2015. Agroforestree Database: A Tree Reference and Selection Guide Version 4.0. 2009. http://www.worldagroforestry.org/af/treedb/.

Palmer, H.E., Newton, A.C., Doyle, C.J., Thomson, S., Stewart, L.E.D., 1998. An economic evaluation of alternative genetic improvement strategies for farm woodland trees. Forestry 71, 333–347.

Panagos, P., Borrelli, P., Meusburger, K., Alewell, C., Lugato, E., Montanarella, L., 2015a. Estimating the soil erosion cover-management factor at the European scale. Land Use Policy 48, 38–50.

Panagos, P., Borrelli, P., Poesen, J., Ballabio, C., Lugato, E., Meusburger, K., Montanarella, L., Alewell, C., 2015b. The new assessment of soil loss by water erosion in Europe. Environmental Science and Policy 54, 438–447.

Paudel, B.R., Udawatta, R.P., Kremer, R.J., Anderson, S.H., 2012. Soil quality indicator responses to row crop, grazed pasture, and agroforestry buffer management. Agroforestry Systems 84, 311–323.

Peichl, M., Thevathasan, N.V., Gordon, A.M., Huss, J., Abohassan, R.A., 2006. Carbon sequestration potentials in temperate tree-based intercropping systems, southern Ontario, Canada. Agroforestry Systems 66, 243–257.

Pimentel, L.G., Weiler, D.A., Pedroso, G.M., Bayer, C., 2015. Soil N_2O emissions following cover-crop residues application under two soil moisture conditions. Journal of Plant Nutrition and Soil Science 178, 631–640.

Prieto, I., Armas, C., Pugnaire, F.I., 2012. Water release through plant roots: new insights into its consequences at the plant and ecosystem level. New Phytologist 193, 830–841.

Prieto, I., Stokes, A., Roumet, C., 2016. Root functional parameters predict fine root decomposability at the community level. Journal of Ecology 104, 725–733.

Quinkenstein, A., Böhm, C., da Silva Matos, E., Freese, D., Hüttl, R.F., 2011. Assessing the carbon sequestration in short rotation Coppices of *Robinia pseudoacacia* L. on marginal sites in Northeast Germany. In: Carbon Sequestration Potential of Agroforestry Systems, Advances in Agroforestry. Springer, Netherlands, pp. 201–216.

Reisner, Y., De Filippi, R., Herzog, F., Palma, J., 2007. Target regions for silvoarable agroforestry in Europe. Ecological Engineering 29, 401–418.

Rennie, P.J., 1957. Effect of the afforestation of catchment areas upon water yield. Nature 180, 663–664.

Richards, M., Gregersen, L., Kuntze, V., Madsen, S., Oldvig, M., Campbell, B., Vasileiou, I., 2015. Agriculture's Prominence in the INDCs.

Rosati, A., Boggia, A., Castellini, C., Paolotti, L., Rocchi, L., 2016. When chickens graze in olive orchards, the environmental impact of both chicken rearing and olive growing decreases. In: Gosme, M., et al. (Eds.), Proceedings 3rd European Agroforestry Conference. EURAF, pp. 400–402.

Rosenstock, T.S., Tully, K.L., Arias-Navarro, C., Neufeldt, H., Butterbach-Bahl, K., Verchot, L.V., 2014. Agroforestry with N2-fixing trees: sustainable development's friend or foe? Current Opinion in Environmental Sustainability 6, 15–21.

Schoenberger, M., Bentrup, G., Gooijer, H., Soolanayakanahally, R., Sauer, T., Brandle, J., Zhou, X., Current, D., 2012. Branching Out: Agroforestry as a Climate Change Mitigation and Adaptation Tool for Agriculture.

Schroth, G., 1998. A review of belowground interactions in agroforestry, focussing on mechanisms and management options. Agroforestry Systems 43, 5–34.

Schuller, A., Gosme, M., Talbot, G., Dupraz, C., 2015. A model-based assessment of the adaptation of Mediterranean agroforestry systems to climate change (Poster). In: Cirad-Inra (Ed.), Climate-Smart Agriculture. Montpellier.

Servair, M., 2007. Etude de la faisabilité de la mise en place d'agroforesterie sur la plaine du Vistre. ENESAD.

Shvaleva, A., Siljanen, H.M.P., Correia, A., Costa E Silva, F., Lamprecht, R.E., Lobo-do-Vale, R., Bicho, C., Fangueiro, D., Anderson, M., Pereira, J.S., Chaves, M.M., Cruz, C., Martikainen, P.J., 2015. Environmental and microbial factors influencing methane and nitrous oxide fluxes in Mediterranean cork oak woodlands: trees make a difference. Frontiers in Microbiology 6, 1104.

Simons, A.J., Leakey, R., 2004. Tree domestication in tropical agroforestry. Agroforestry Systems 61–2, 167–181.

Skiba, U., Smith, K.A., Fowler, D., 1993. Nitrification and denitrification as sources of nitric oxide and nitrous oxide in a sandy loam soil. Soil Biology and Biochemistry 25, 1527–1536.

Smith, P., Bustamante, M., Ahammad, H., Clark, H., Dong, H., Elsiddig, E.A., Haberl, H., Harper, R., House, J., Jafari, O., 2014. Agriculture, forestry and other land use (AFOLU). In: IPCC (Ed.), Climate Change 2014 Mitigation of Climate Change. Cambridge University Press, Cambridge, pp. 811–922.

Stenlid, J., Oliva, J., Boberg, J.B., Hopkins, A.J.M., 2011. Emerging diseases in European forest ecosystems and responses in society. Forests, Trees and Livelihoods 2, 486–504.

Sun, S.-J., Meng, P., Zhang, J.-S., Wan, X., 2013. Hydraulic lift by Juglans regia relates to nutrient status in the intercropped shallow-root crop plant. Plant and Soil 374, 629–641.

Talbot, G., 2011. L'intégration spatiale et temporelle des compétitions pour l'eau et la lumière dans un système agroforestiers noyers-céréales permet-elle d'en comprendre la productivité? Université de Montpellier 2.

Talbot, G., Roux, S., Graves, A., Dupraz, C., Marrou, H., Wery, J., 2014. Relative yield decomposition: a method for understanding the behaviour of complex crop models. Environmental Modelling and Software 51, 136–148.

Ticknor, K.A., 1988. Design and use of field windbreaks in wind erosion control systems. In: Windbreak Technology, pp. 123–132.

Timmermans, B., Bestman, M., 2016. Quality of apple trees and apples in poultry free range areas. In: Gosme, M., et al. (Eds.), 3rd European Agroforestry Conference – Montpellier, 23-25 May 2016. EURAF, pp. 420–423.

Tsonkova, P., Böhm, C., Quinkenstein, A., Freese, D., 2012. Ecological benefits provided by alley cropping systems for production of woody biomass in the temperate region: a review. Agroforestry Systems 85, 133–152.

Tubex and Euraf, 2015. A Guide to Agroforestry. European Agroforestry Federation, Montpellier.

Tuley, G., 1985. The growth of young oak trees in shelters. Forestry 58, 181–195.

Tuomisto, H.L., De Camillis, C., Leip, A., Nisini, L., Pelletier, N., Haastrup, P., 2015. Development and testing of a European Union-wide farm-level carbon calculator. Integrated Environmental Assessment and Management 11, 404–416.

UNECE, 2015. Forests in the ECE Region - Trends and Challenges in Achieving the Global Objectives on Forests (No. ECE/TIM/SP/37). UN Economic Commission for Europe.

Uribe, C., Inclan, R., Hernando, L., Roman, M., Clavero, M.A., Roig, S., Van Miegroet, H., 2015. Grazing, tilling and canopy effects on carbon dioxide fluxes in a Spanish dehesa. Agroforestry Systems 89, 305–318.

Vallebona, C., Mantino, A., Bonari, E., 2016. Exploring the potential of perennial crops in reducing soil erosion: a GIS-based scenario analysis in southern Tuscany, Italy. Applied Geography 66, 119–131.

van der Heijden, M.G.A., Martin, F.M., Selosse, M.-A., Sanders, I.R., 2015. Mycorrhizal ecology and evolution: the past, the present, and the future. New Phytologist 205, 1406–1423.

van Eimern, J., 1964. Windbreaks and Shelterbelts. Secretariat of the World Meteorological Organization.

van Noordwijk, M., Coe, R., Sinclair, F., 2016. Central hypotheses for the Third Agroforestry Paradigm within a Common Definition (No. 233). World Agroforestry Centre Working Paper.

van Noordwijk, M., Farida, A., Verbist, B., Tomich, T.P., 2003. Agroforestry and watershed functions of tropical land use mosaics. In: 2nd Asia Pacific Training Workshop on Ecohydrology: Integrating Ecohydrology and Phytotechnology into Workplans of Government. Private and Multinational Companies, pp. 1–10.

Vanwalleghem, T., Amate, J.I., de Molina, M.G., Fernández, D.S., Gómez, J.A., 2011. Quantifying the effect of historical soil management on soil erosion rates in Mediterranean olive orchards. Agriculture, Ecosystems and Environment 142, 341–351.

Verchot, L.V., Van Noordwijk, M., Kandji, S., Tomich, T., Ong, C., Albrecht, A., Mackensen, J., Bantilan, C., Anupama, K.V., Palm, C., 2007. Climate change: linking adaptation and mitigation through agroforestry. Mitigation and Adaptation Strategies for Global Change 12, 901–918.

Vickery, J., Arlettaz, R., 2012. The importance of habitat heterogeneity at multiple scales for birds in European agricultural landscapes. In: Fuller, R.J., Fuller, R.J. (Eds.), Birds and Habitat. Cambridge University Press, Cambridge, pp. 177–204.

Vityi, A., Frank, N., 2016. Shelterbelt as a best practice of improving agricultural production. In: Gosme, M., et al. (Eds.), 3rd European Agroforestry Conference – Montpellier, 23-25 May 2016. EURAF, pp. 212–213.

Wang, L., Duggin, J.A., Nie, 2012. Nitrate-nitrogen reduction by established tree and pasture buffer strips associated with a cattle feedlot effluent disposal area near Armidale, NSW Australia. Journal of Environmental Management 99, 1–9.

Warner, D.J., Tzilivakis, J., Green, A., Lewis, K.A., 2016. A guidance tool to support farmers with ecological focus areas—the benefits of agroforestry for ecosystem services and biodiversity. In: Gosme, M., et al. (Eds.), 3rd European Agroforestry Conference – Montpellier, 23-25 May 2016. EURAF, pp. 373–375.

Watson, R.T., Noble, I.R., Bolin, B., Ravindranath, N.H., Verardo, D.J., Dokken, D.J., 2000. Land Use, Land Use Change, and Forestry: Special Report of the Intergovernmental Panel on Climate Change, Special Report to IPCC. Cambridge University Press, Cambridge.

Watté, J., 2014. Featured farm: agroforestry on contour swales in Flanders: venturing into aquaculture. EURAF Newsletter 14, 7–9.

Weintraub, S.R., Russell, A.E., Townsend, A.R., 2014. Native tree species regulate nitrous oxide fluxes in tropical plantations. Ecological Applications 24, 750–758.

Welsch, J., Case, B.S., Bigsby, H., 2014. Trees on farms: investigating and mapping woody re-vegetation potential in an intensely-farmed agricultural landscape. Agriculture, Ecosystems and Environment 183, 93–102.

Wight, B., Straight, R., 2015. Chapter 6: windbreaks. In: Gold, M., Cernusca, M., Hall, M. (Eds.), Training Manual for Applied Agroforestry Practices. University of Missouri Center for Agroforestry, pp. 92–114.

Williams, D.G., Wallace, P., McKeon, G.M., Hall, W., Katjiua, M., Abel, N., 1999. Effects of Trees on Native Pasture Production on the Southern Tablelands (No. Publication 99/165). Joint Venture Agroforestry Program - RIRDC.

Wilson, J.R., Wild, D., 1991. Improvement of nitrogen nutrition and grass growth under shading. In: Shelton, H.M. (Ed.), Forages for Plantation Crops Part 3, Proceedings, vol. 32. ACIAR.

Wolfrum, A., 2014. Do agroforestry systems promote a thriving nightlife? Assessing bat activity with an easy to use standardized protocol. In: Palma, J.H.N., Chalmin, A. (Eds.), Proceedings of Second European Agroforestry Conference. European Agroforestry Federation, Montpellier, pp. 39–41.

Wotherspoon, A., Thevathasan, N.V., Gordon, A.M., Paul Voroney, R., 2014. Carbon sequestration potential of five tree species in a 25-year-old temperate tree-based intercropping system in southern Ontario, Canada. Agroforestry Systems 88, 631–643.

Young, A., 1989. Agroforestry for Soil Conservation.

Zhou, X.H., Brandle, J.R., Mize, C.W., Takle, E.S., 2005. Three-dimensional aerodynamic structure of a tree shelterbelt: definition, characterization and working models. Agroforestry Systems 63, 133–147.

Zomer, R.J., Neufeldt, H., Xu, J., Ahrends, A., Bossio, D., Trabucco, A., van Noordwijk, M., Wang, M., 2016. Global tree cover and biomass carbon on agricultural land: the contribution of agroforestry to global and national carbon budgets. Scientific Reports 6, 29987.

CHAPTER

10

Linking Arable Cropping and Livestock Production for Efficient Recycling of N and P

Christine A. Watson[1,4], *Cairistiona F.E. Topp*[2], *Julie Ryschawy*[3]

[1]Crop and Soil Systems, SRUC, Aberdeen, United Kingdom; [2]Crop and Soil Systems, SRUC, Edinburgh, United Kingdom; [3]Université de Toulouse, AGIR UMR 1248, INRA, INPT-ENSAT, Auzeville, France; [4]SLU, Box 7043, 75007, Uppsala, Sweden

INTRODUCTION

Over the last 100 years, crop and livestock production has become increasingly decoupled both geographically and managerially, resulting in many livestock units becoming heavily reliant on bought-in feed and straw and specialized arable units on purchased fertilizer. In some areas of Europe, there is evidence of declining soil fertility in arable agriculture, which may in part be due to declining reliance on animal manures (Lemaire et al., 2014; Peyraud et al., 2014). Straw has continued to be transported from arable areas to intensive livestock production systems, but manure has not been exported due to issues such as cost and transport. Thus manure is used within the regions where it is produced often exceeding plant-soil system demand and resulting in a localised build-up of nutrients in the soil and potential for losses. Furthermore, nutrients in manures/slurries are often not returned evenly across available land areas. Easy access means that the fields close to buildings are often managed intensively in terms of grazing and/or receive disproportionately large amounts of manure/slurry. These areas may also receive nutrients from leakages from yard washings and silos (McCormick et al., 2009). Livestock within the farming system creates a reason to diversify rotations to a wider variety of crops, including the production of protein from sources other than cereals, e.g., grain and forage legumes, oil seed rape, or grass. Without ruminants in the farming system, grass leys become uneconomic, so rotations tend to change to all arable systems without the soil fertility–building properties of leys. The use of inputs, such as fertilizers and pesticides, has helped to overcome the need for rotations to build soil fertility.

Integrated crop livestock systems (ICLS) potentially provide better resource utilization (e.g., energy, nutrients, land) than specialized

Agroecosystem Diversity
https://doi.org/10.1016/B978-0-12-811050-8.00010-8

systems, with the potential to reduce reliance on external inputs thanks to synergies between components (Schiere et al., 2002). Crops and crops residues can be fed to animals, while manure can fertilize crops and grasslands (Hendrickson et al., 2008; Moraine et al., 2014; Ryschawy et al., 2012; Wilkins, 2008). Coordination between crops and livestock is proven to allow a recoupling of nitrogen and carbon cycles (Soussana and Lemaire, 2014) and in particular better efficiency in the use of nutrients such as nitrogen and phosphorus. While the purpose of this chapter is to address nitrogen (N) and phosphorus (P) flows, the efficiency of N and P use is closely linked to C use. Management practices that improve soil structure through the addition of carbon will also influence cycling of both N and P through several mechanisms including the improvement of soil structure. In ICLS, compared to specialized systems, improvements in efficiency are linked to the degree of synergy between components. The extent of synergies between enterprises depends on the ability to integrate the operations of the farm enterprises (Moraine et al., 2014).

In this chapter, we explore how ICLS as an alternative to specialization could influence nitrogen and phosphorus cycling in European agriculture. We think about this in the context of whether ICLS can help to balance input and output budgets of nutrients while maintaining reservoirs of N and P that can be accessed via plant root systems and the associated soil biologic communities. We refer to global systems where crop products are transported across the globe and traded through markets for completeness but do not elaborate since there is currently little opportunity to link N and P cycling in these systems. Here, we examine three models of ICLS with different degrees of enterprise integration or synergy, resulting in different opportunities to optimize nutrient use within and between farms at a range of scales.

We discuss how increases in technical efficiency and improved synergies between enterprises could improve N and P cycling and the potential to reduce reliance on external inputs. We focus on the integration rather than attempting to discuss all the possible advantages and disadvantages of ICLS for nutrient management.

CURRENT STATUS OF N AND P IN SOILS AND WATER COURSES IN EUROPE WITH REFERENCE TO THE SPECIALIZATION OF AGRICULTURE

In many areas of Europe the annual diffuse agricultural emissions of nitrogen to freshwater is high (Fig. 10.1), often where farming is very specialized and/or the density of livestock is very high (Fig. 10.2). Consumption of plant material by ruminants effectively disrupts the stable balance of C:N in plant material and allows reactive N to enter the environment. Both ruminant and monogastric species kept indoors and outdoors contribute to this. The impact of housed livestock on soils and water will, of course, be dependent on the way in which manure produced is stored as well as disposed of to land. Groundwater in arable catchments often also contains unacceptably high levels of nutrients (e.g., Benoit et al., 2014).

The amount of P in agricultural topsoils in Europe has increased over time due to the use of fertilizers and manures with the highest P, resulting from manure spreading in intensive livestock systems (Lemercier et al., 2008). It has been suggested that there is a critical level of P saturation in topsoil beyond which the risk of P leaching increases (Vadas and Sims, 2014). Intensive grazing and intensive arable production can also lead to soil loss and hence P loss through erosion. Surface losses of P can also take place following manure. In ICLS where there is sufficient arable land available to recycle manure, P use efficiency is much higher than in very intensive livestock systems disconnected from land (Russelle et al., 2007; Sharpley et al., 2007). Detailed maps of agricultural regions in France

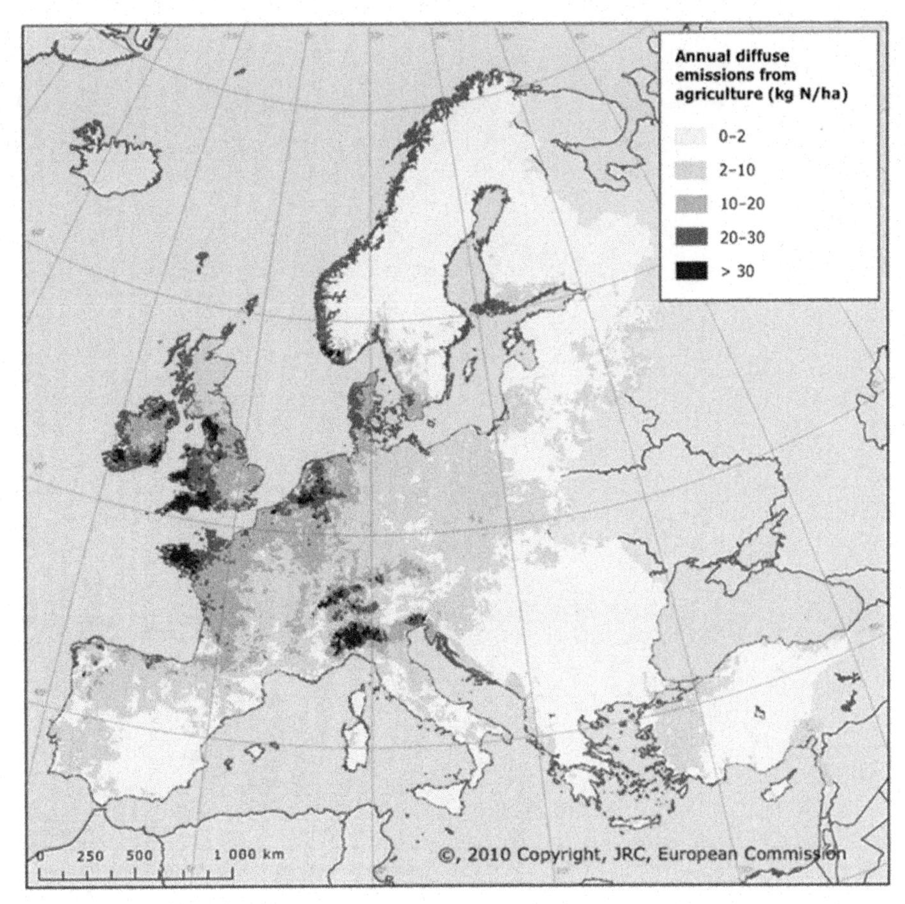

FIGURE 10.1 Annual diffuse agricultural emissions of nitrogen to freshwater (kg N per ha of total land area) (European Environment Agency, 2010).

show high levels of available P in the topsoil in regions (Fig. 10.3A) with high cattle densities (dairy and/or beef) (Fig. 10.3B).

Farm gate nutrient budgets, defined as the balance between inputs and outputs of nutrients, are a useful indicator of the nutrient use efficiency within a given management unit, for example, field, farm, or region (Simon et al., 2000). Nutrient budgets in agricultural systems are generally driven by fertilizer inputs, livestock density, and feed imports and in some systems by biologic fixation for N (Watson et al., 2002). At the farm level a number of studies of N and P balances have shown that nutrient use efficiency measured as output/input changes with farm type with arable systems > mixed > dairy (e.g., Domburg et al., 2000; Ryschawy et al., 2012; Watson et al., 2002). Using output/input as an indicator nutrient use efficiency increases on farms that focus more on crop export than on autonomy (Swensson, 2002; Schröder et al., 2003). Even within the same geographic area, nutrient budgets can characterize differences in farm management, as shown by Giustini et al. (2007), who found lower N surpluses in dairy farms in mountainous areas of Tuscany, Northern Italy, compared to lowland systems. Higher stocking rates and lower proportions of

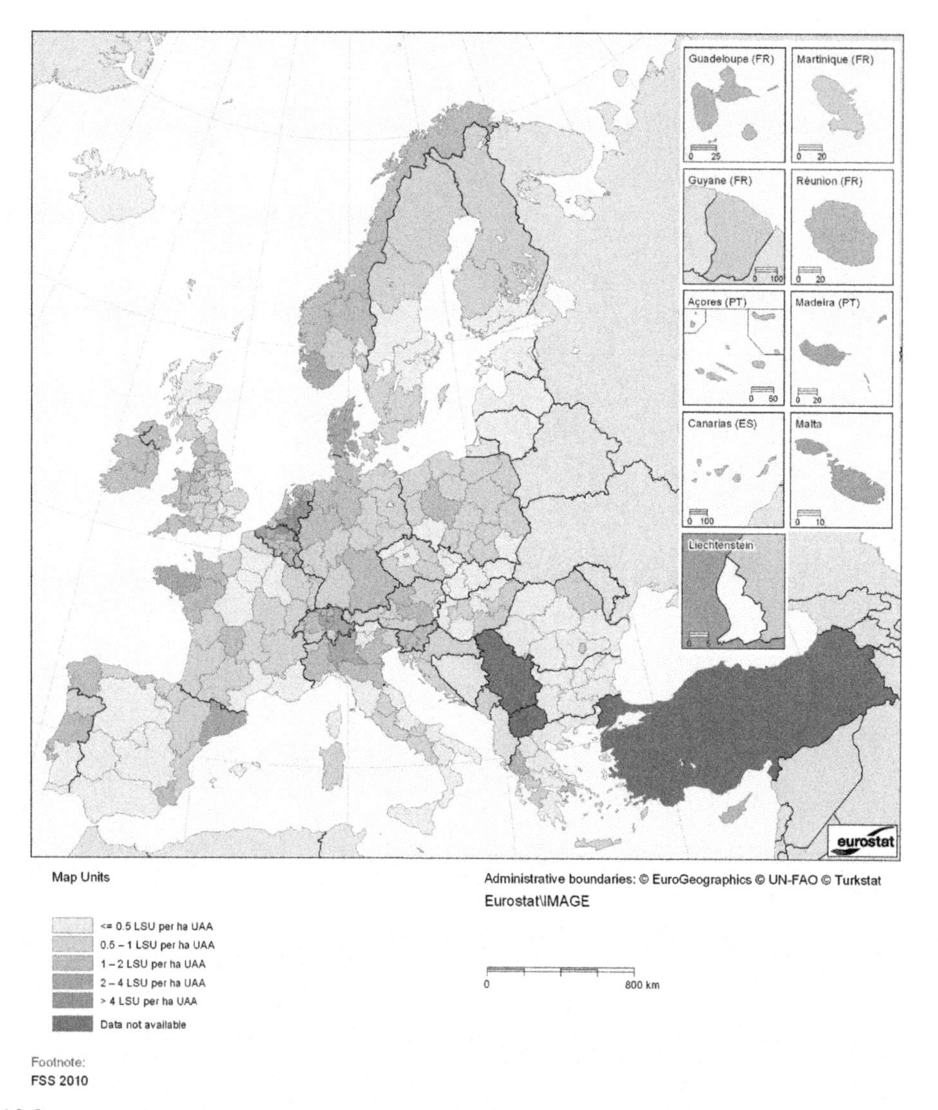

FIGURE 10.2 Livestock density in Europe expressed as total LU per ha (2010). *Source Eurostat website (http://ec.europa.eu/eurostat/statistics-explained/index.php/File:Livestock_pattern_-_Total_livestock_density,_EU-27,_IS,_NO,_CH,_ME_and_HR,_NUTS2,_2010_English.png).*

permanent fodder crops were linked to higher losses. Fangueiro et al. (2008) also found a strong positive relationship between stocking rate and N surplus on farms in NW Portugal. Farm nitrogen budgets have frequently shown large surpluses of N in regions dominated by intensive livestock production; e.g., the farm nitrogen surplus in Brittany was shown to be more than double the French average (Le Gall et al., 2005). However, overall balances often hide the detail of both the system and individual management practices within the system, which are important in fully understanding the issues of crop livestock integration.

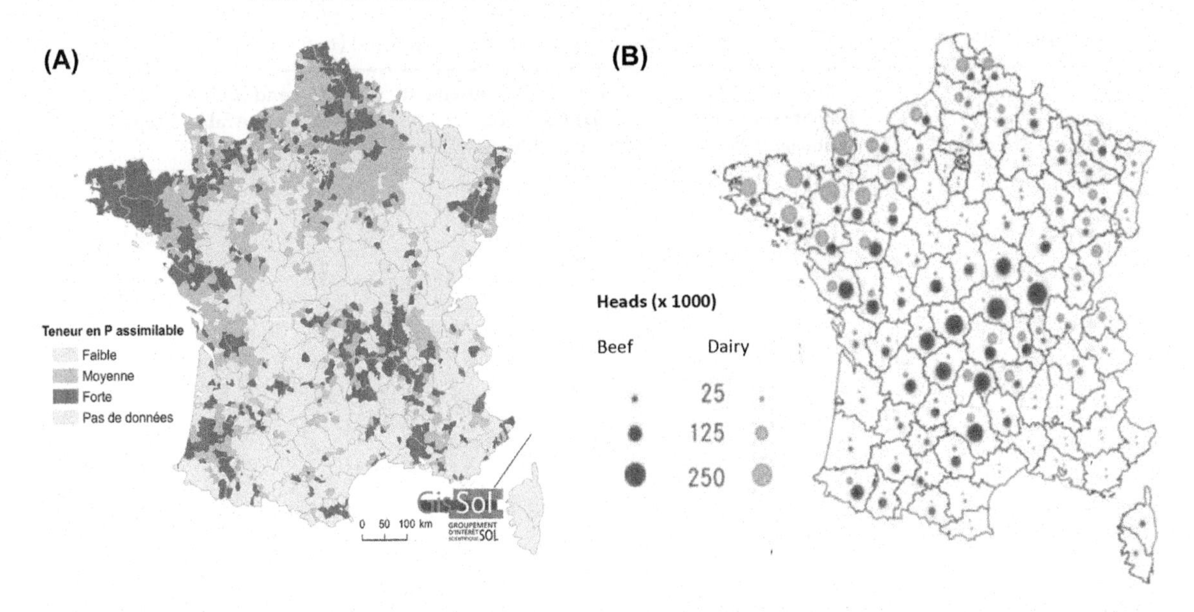

FIGURE 10.3 (A) Available P in topsoils by French districts in 2011 and (B) population of beef (green [light gray in print version]) and milk (pink [gray in print version]) cattle by 1000 head in French districts in 2010. *Source: Gis Sol, BDAT, 2011; IGN, Geofla®, 2006.*

A recent study of soil P budgets (calculated as P inflow − P outflow) in 21 different regions of France showed that predominantly arable regions could have a P budget very close to equilibrium, but these systems relied very heavily on fertilizer inputs (Senthilkumar et al., 2012). On the other hand, livestock regions often had a positive P budget, suggesting that continued use of P fertilizer may be unnecessary if more effective use was made of excreta P. In the same study, regions with ICLS showed P budgets that were both positive and close to neutral. In ICLS, there is an opportunity to manage budgets by manipulating the use of home-produced versus imported feeds and the relative reliance on manure versus purchased fertilizer both within single farms and in partnered farms. The problem of regional surpluses of P associated with farm specialization was also highlighted by Bateman et al. (2011) in the United Kingdom, who showed positive budgets for livestock-dominated regions in the west of the country and deficits in the crop-rich east (Table 10.1). On a country scale, they calculated that manure could supply 71% of the P requirement of crop production and that there was a need for an annual export of 4.7 1000 tons of P (or 2.8 million tons of manures) to balance P supply and demand across the country. The problem is further complicated by the temporal aspect of housing livestock of winter, meaning that 23.0 1000 tons P (15 million tons of manure) accumulated in manure stores over winter needs to be moved in spring for crop production. At a much bigger scale, Schipanski and Bennett (2012) showed some countries becoming net exporters of P nutrients, while others have become net importers, with important implications for the loss of soil fertility in exporting regions.

WHAT DO WE MEAN BY IMPROVED SYNERGY BETWEEN CROPS AND LIVESTOCK?

Synergy has been defined as "combined" or "cooperative" effects, literally, the effects

TABLE 10.1 Phosphorus Surpluses and Deficits on a Regional Scale in England (Bateman et al., 2011)

Region of England	Surplus (+) of Deficit (−) of P (Thousand Tons)	% Crop P Requirement That Could be Supplied by Manure	% Land under Annual Arable Crops in the Region[a]
Northwest	+2.9	143	14
Southwest	+1.7	110	29
West Midlands	−0.9	92	40
Northeast	−2.3	56	28
Yorkshire	−3.7	75	49
Southeast	−7.2	47	47
East Midlands	−9.2	52	61
Eastern	−14.4	44	74

[a] *Defra (2010).*

produced by things that "operate together" (parts, elements, or individuals) (Corning, 1995). Here, we concentrate on what von Eye et al. (1998) refer to as the classical conceptual type of synergy or the common ideotype, where the whole is greater than the sum of the parts (Schiere et al., 2002). In relation to nutrient cycling the combined effect desired relates to more efficient use of a given nutrient, leading to greater productivity (yield and product quality) and/or reduced loss to water courses (relating N and P) or the atmosphere (in the case of N only). This may also lead to economic benefits, such as economies of scope achieved thanks to diversification, by the combined elaboration of multiple products (Coquil et al., 2013; Ryschawy et al., 2012). This could also be beneficial in terms of compliance with environmental regulation.

Many options for improving the efficiency of nutrient cycling will come from technologies developed for specialist systems, e.g., developments in fertilizer technology (Baudon et al., 2005) or by altering the phytate content of monogastric diets (McGrath et al., 2010). These kinds of solutions have been reviewed for P from a plant perspective by Withers et al. (2014) and for livestock by Kebreab et al. (2012) and similar reviews are available for N, e.g., Chalova et al. (2016). Here, we aim to focus on those improvements that relate directly to synergies between crop and livestock. This requires slightly different thinking from a more specialist crop or livestock approach to improving efficiency. A good example of a development relevant to mixed systems comes from plant breeding in Australia and involves the development of dual-purpose crops that provide some grazing and still give an economic crop yield (e.g., Dove and Kirkegaard, 2014).

We focus here on available technologies and the ideas of system redesign within the farm or landscape context. However, in the future, it will be necessary to ask some more challenging and fundamental questions about our production systems. For example, breeding crops with lower P requirements than current varieties would reduce the requirement for P fertiliser which would in turn reduce manure P and the quantities of P flowing within our agricultural systems. This could potentially be achieved through reducing the uptake of P into storage compounds in plants (White and Veneklass, 2012). However, it is important that these developments are viewed from an interdisciplinary perspective to ensure that feed is matched with livestock P requirements.

INFLUENCE OF CROP LIVESTOCK INTEGRATION ON N AND P MANAGEMENT

To illustrate how closer integration of crops and livestock can influence N and P cycling by identifying and managing synergies between components, we use three models for nutrient flows moving from total disconnection toward total integration (Fig. 10.4).

In Fig. 10.5, we show how the three models relate to each other Fig. 10.5 and the classification of ICLS takes inspiration from the work of Moraine et al. (2014) but with a focus specifically on nutrient transfers and exchanges.

Coexistence Model

Within the *"coexistence model"* (Fig. 10.4A), crop and livestock systems are disconnected at a local level, e.g., a watershed or administrative region (described by Russelle et al. (2007) as "among farm"), but there are management options at a higher spatial scale through the market, such as movement of manure, feed or livestock

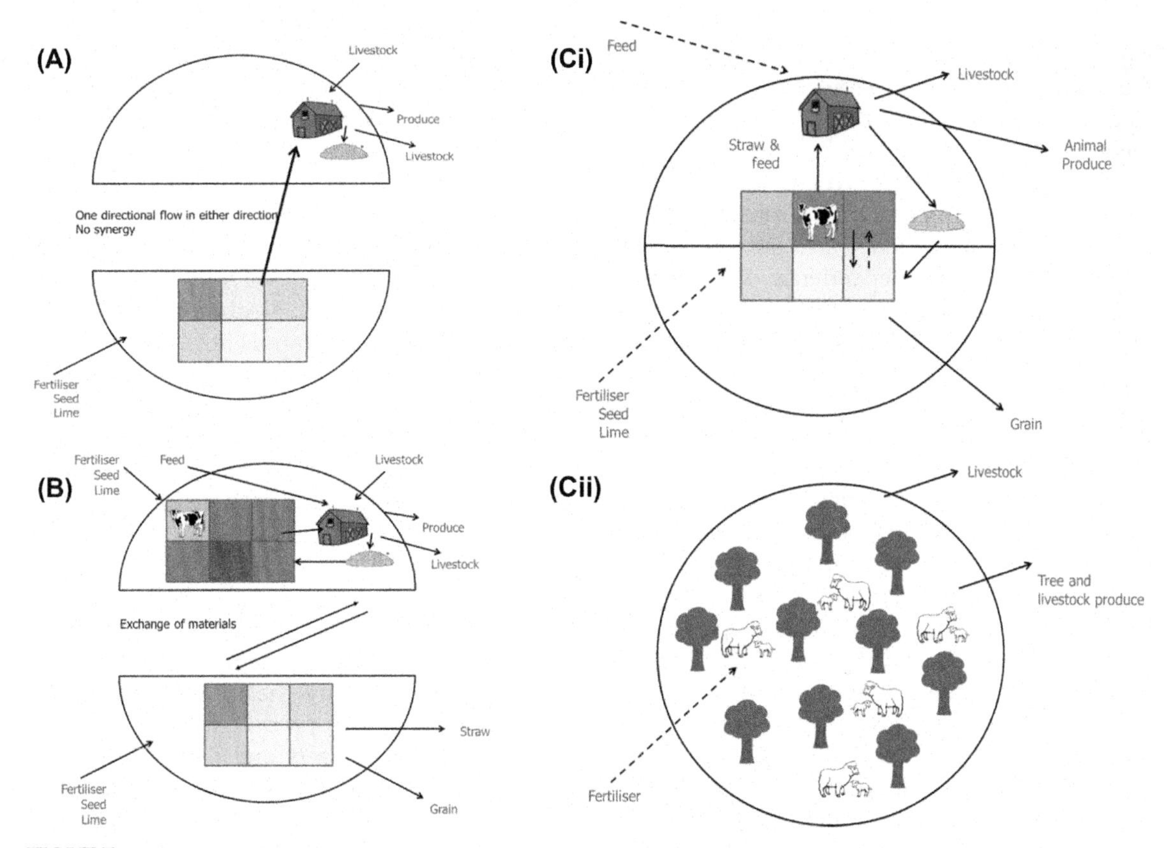

FIGURE 10.4 Nutrient flows in three farming system models according to a gradient of integration between crop and livestock: (A) coexistence, (B) complementarity, (Ci) synergy (partial), and (Cii) synergy (increased).

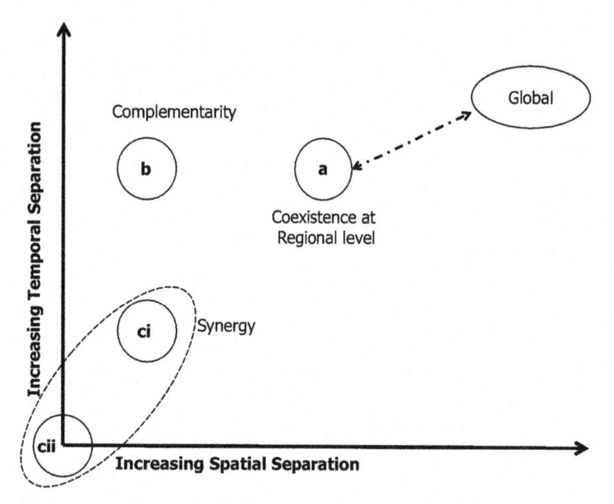

FIGURE 10.5 Temporal and spatial separation of models (A), (B), (Ci), and (Cii) shown in Fig. 10.4.

between farms. The need for feed to be imported is usually related to stocking rate with greater need for imported feed in more intensive systems. In Figure 5 we also show the position of global farming systems where commodities such as feed e.g. soybean are transported across vast distances where there is no clear possibility for reconnection. The dependence of European agriculture on imported soybean and soymeal has recently been discussed by Watson et al. (2017). There have been relatively few attempts to quantify the movement of N and P between farms at a regional scale, but Nowak et al. (2015) studying nutrient transfers between organic farms in three districts of France found mass flows of nutrients between farms were two to four times greater in areas that contained a mixture of crop and livestock farms compared with farms that mainly contained specialized farms. However, even within the mixed areas, these flows between farms were mainly associated with a one-way flow of nutrients and not with exchanges of nutrients. Another alternative mechanism for managing nutrient flows at the landscape level could be a fundamental redesign of the physical or socio-economic landscape, for example, planning control of where intensive livestock units are situated.

Complementarity Model

In the *"complementarity model"* (Fig. 10.4B) crop and livestock systems are spatially close together, and they develop some exchanges but are not fully integrated (defined as regional integration by Russelle et al., 2007). There are options to diversify the systems by introducing annual crops into the livestock systems or introducing alternative annual or perennial crops into grassland systems within the limits of the land capability. This then allows an increased flexibility in grazing options, manure management, and feeding strategies. Increased integration of livestock and/or livestock manures into arable farming systems in this model increases soil carbon contents, which in turn improves soil structure and water-holding capacity. This has the potential to reduce waterlogging and anaerobic zones in soils, which will also lead to a reduction in nitrous oxide production. Careful management of manure to limit inputs and match crop demand while improving soil fertility is another important option; e.g., legumes have, for instance, lower N requirement relative to P. Introducing a variety of crop types provides a wider series of options for spreading manures/slurries than in specialist farms as a result of variable tillage options (timing and technique) and ground covers (row crops vs. perennial sward).

Synergy Model

Within the *"synergy"* (Fig. 10.4C,i) or traditional "mixed farming" concept, N and P cycling can be altered either by changing management of the existing system or redesigning the system. Options for improving N and P recycling here include recycling of animal manure to fertilize crops, diversifying crop rotations to feed animals, and limiting inputs by altering the rotation or sequence design or the spatial arrangement of crops in the field. Described by Russelle et al. (2007) as farm-scale integration, this includes both integration through a rotation (Fig. 10.4Ci) and also potentially livestock grazing permanent grassland, which is not part of the rotation. A search for self-sufficiency in animal feeding allows diversification of the cropping system, through broadening the range of crops that could be fed to the animals (grasslands, cereals, protein, and oil crops) (Ryschawy et al., 2013). The choice of crops to export versus crops to feed on farm can also affect nutrient budgets as the N:P ratios of exported products as well as quantities affect requirements for imported nutrients as either feed or fertilizer and/or the efficiency of manure use on farm.

The "*synergy*" model currently occurs at the farm level, but there are also an increasing number of examples of this taking place between farms, where local synergies between specialized crop and livestock farmers could be seen as the highest level of coordination and described as "collective designed land use." The level of synergy differs, and models such as agroforestry where there is a crop product from the tree is an example where crops and livestock exist in exactly the same space and at the same time (Fig. 10.4Cii).

EXAMPLES OF PRACTICES THAT COULD IMPROVE N AND P MANAGEMENT THROUGH SYNERGIES BETWEEN CROPS AND LIVESTOCK

Considering these different models, there is a wide diversity of practices that could improve the integration of crops and livestock. Each model could evolve to the next one along the gradient through improved synergies between crops and livestock. For instance, a "*complementarity*" type could reinforce animal feed autonomy through diversified rotations and thus become more of a "*synergy*" model. We identified two main themes to favor crop-livestock synergies for improved N and P management: (1) manure/fertilizer management and (2) rotation/spatial design. They can take place within the different models and allow movement from less to more integration. Some examples of synergies are provided in Table 10.2 and organized according to the two main themes.

Manure and Fertilizer Management

Manure Treatment

The logistics and economics of manure movement are currently likely to be the limiting factors for the development of the "*complementarity*" model at a local scale (regional, watershed), but the development of separation and treatment procedures for manures to make them less bulky and less volatile in terms of likelihood of nutrient losses provides options. In existing "*complementarity*" models, the stimulus is often a need to meet environmental regulation rather than agronomic or economic drivers (Asai et al., 2014a). If the manure is to be transported for use on farms as a replacement to synthetic N fertilizer the emissions resulting from manure processing need to be reduced as far as possible, particularly in relation to NH_3 emission from manure processing and storage together with denitrification-related N_2O and NOx emissions from storage (De Vries et al., 2012). See Table 10.2 for more detailed information. Effective management of N and P in this context therefore requires consideration of nutrient management at a greater scale than the individual farms, i.e., a redrawing of the boundaries within which nutrient management plans or nutrient budgets are determined.

Nutrient Exchanges Between Farms

Movement of feed, straw and manure between farms all contribute to the net balance of

TABLE 10.2 Examples of How Different Management Practices Can Influence N and P Use in ICLS

Management Practice	Synergy	Model	Potential to Reduce External Inputs of N and P	Potential Influence Environmental Impacts (N and P Loss and/or Soil Fertility)	Example Reference
A) MANURE AND FERTILIZER MANAGEMENT					
Temporal and spatial dynamics of use of manure in rotations	Improved crop quality and yield; protein and nutrient content; Distribution of manure over wider land area as opposed to area close to farm buildings	Complementarity Synergy	Yes	Reduced N in runoff, N leaching, and nitrous oxide Reduced loss of P	Kronberg & Ryschawy (2017)
Pretreatment of manures, e.g., separation of solid and liquid	Separated manure fractions allows selection for materials with N:P ratios in line with the N:P ratio required by crop. Solid fractions are rich in organic N, so mineralization of this into plant-available forms is a prerequisite for adequate utilization of solid fraction N	Complementarity Synergy	Yes	Reduced loss of P fractionation may alter nitrous oxide emissions after spreading	Fangueiro et al. (2012); ten Hoeve et al. (2014); Hjorth et al. (2010); Rigolot (2010).
Accounting for legacy P fertilizer in soil	Matching P supply from manure and fertilizer with soil P resources	All	Yes	Reduced loss of P from soils	Withers et al. (2014)
B) ROTATION DESIGN					
Home-produced feed replaces imported feed	Reduced reliance on imported protein can reduce GHG emissions	Complementarity Synergy	Yes	Reduced GHGs	Battini et al. (2016)
Increased rotational diversity	Produces feeds with different nutritional profiles; more efficient use of soil resources; increased productivity; N and P available from incorporation of residues (precrop)	Coexistence Complementarity Synergy	Reduced fertilizer input	Reduce loss of P through erosion, leaching and GHGs, build soil fertility	Martin et al. (2016) and Moraine et al. (2017); Preissel et al. (2015)

TABLE 10.2 Examples of How Different Management Practices Can Influence N and P Use in ICLS—cont'd

Management Practice	Synergy	Model	Potential to Reduce External Inputs of N and P	Potential Influence Environmental Impacts (N and P Loss and/or Soil Fertility)	Example Reference
Include cover crops and green manures in arable rotations that may be grazed	Inserting legumes in between cash crops to increase feed self-sufficiency; more efficient use of soil resources; potential fodder and manure returns increased productivity	Coexistence Complementarity Synergy	Reduced fertilizer input	Reduce loss of P through erosion, leaching, and GHGs; build soil fertility	Franzluebbers and Stuedemann (2007) and Ryschawy et al. (2014)
Introduce mixed species leys to arable rotations that may be grazed	Increased N from biologic fixation; more efficient use of soil resources; potential fodder and manure returns increased productivity	Complementarity	Reduced fertilizer input	Reduce loss of P through erosion, leaching and GHGs, build soil fertility	Alard et al. (2002), Franzluebbers (2007) and Ryschawy et al. (2014)
Introduce alternative forage crops into grassland-only systems	Potential for high yield and N content of subsequent crop following plowing of grassland; cereal/grain legume intercrops provide high quality livestock feed	Complementarity Synergy	Reduces reliance on imported feed	Risk of increased N loss by leaching and GHG from plowing long-term grass	Rotz et al. (2005)
Making use of stubbles, residues	Stubble, e.g., maize grazed by cows	Complementarity Synergy	Reduces reliance on imported feed	Risk of increased N loss by leaching and GHG from urine and feces on stubble	Dumont et al. (2013); Gliessman (2006); Liebig et al. (2012)
Substitution of crops with different energy and nutrient profiles suited to livestock feed	Changing the balance between degradable protein and energy in the ration can reduce N loss from livestock, e.g., substituting maize silage for grass silage	Complementarity Synergy	Reduces reliance on imported feed	Reduced N loss	Rotz et al. (2005)
Increase the diversity within the agroecosystem	Introduce trees, to diversify the land parcels and feed animals	Complementarity Synergy	Reduces reliance on fertilizer	Reduced N and P loss through improved capture of nutrients	Bealey et al. (2014) and Patterson et al. (2008)

Actual losses will depend on the management practices and combinations of practices and their interaction with pedoclimatic conditions.

nutrients across farms in the "Complementarity" model. More efficient use of manures on farm applies to all of the ICLS models discussed, but we focus primarily on this topic within this section because of the opportunities for improvement in complementarity opened up here. Field level losses of nitrous oxide are very variable from both manures and synthetic fertilizers because they are highly related to soil and environmental conditions at the time of spreading. Introducing a greater diversity of crops is a likely outcome of a shift to ICLS, and if this includes species that take up N more efficiently, then this could reduce N_2O losses within livestock-based farming systems as a result of reducing the area needed to produce forage for silage and grazed grass (del Prado et al., 2010), although this will also depend on the fertilizer application rates and nitrogen demand and the yield of the different crop species.

A major risk in developing crop-livestock interactions at a collective level is the risk of losing self-sufficiency at the farm level to favor the collective one. If self-sufficiency of the systems is not considered, there is risk of intensification of the existing systems rather than improving agroecologic aspects. In three case studies of crop and livestock integration at the district level in Spain, Switzerland, and Netherlands, Regan et al. (2017) observed that dairy farmers, when accessing extra land from crop farmers to spread manure, intensified their production instead of developing environmental benefits. Farmers sought to develop their milk production while conforming to the Nitrates Directive by spreading manure on a broader area on crop farms areas. As a consequence, overall environmental pollution did not decrease as much as expected, but there could still be an advantage in reducing the intensity of emissions. Fundamental redesign of the landscape could potentially involve agronomists in land use planning, e.g., only allowing pig and poultry units to be sited within or close to arable units. Within the current EU Regulation on Organic Farming, if a poultry unit does not have an adequate area of land for safely disposing of and utilizing manure nutrients, then there must be a written agreement in place with another land owner. This approach could be applied to other production systems. There are also examples of collective "manure banks" being operated to broker nutrients (Lopez-Ridaura et al., 2009), where sharing of machinery and knowledge is an added incentive for the farmers involved.

Rotation or Spatial Design

Rotation Diversification

The introduction of ICLS with a greater reliance on legumes could reduce nitrous oxide emissions from ICLS through reduced use and manufacture of synthetic fertilizers. However, some losses will still occur as a result of nitrous oxide production during the decomposition of leguminous residues (Pappa et al., 2011). The return of N-rich legume residues provides a readily available substrate for conversion to nitrous oxide as well as increasing potential loss of nitrate by leaching. Nitrous oxide during the production of legumes is generally very small, and recent work from Jeuffroy et al. (2013) in France showed five to seven times less GHG per unit area from legumes than from other crops. It is widely accepted that there is in an inverse relationship between fixation rate and available mineral nitrogen, so the amount of on-farm nitrogen fixation will be influenced by fertilizer and manure management practices. Intercropping of grain legumes with cereals is an option for providing arable silage in ICLS, although the total amount of N fixed is generally lower in grain legumes grown in intercrops than in sole crops. The proportion of N derived from fixation (%Ndfa) is, however, generally higher in intercrops. At a European scale, both grain and forage legumes contribute significantly to the input of N into agriculture: grain legumes have been calculated to fix an average of 133 kg/ha, totalling 225 1000 tons of N in 2009 (Baddeley et al., 2014), of which the three major contributors were faba bean, pea, and soybean. Forage legumes were estimated to fix 586 1000 tons in Europe in the same study, although it is difficult to calculate this accurately as many European countries do not record areas of forage legumes produced.

Legumes have a lower N requirement relative to P than many other crops owing to their ability to fix N. This makes them a suitable crop for receiving a form of organic manure low in N relative to P (see Table 10.2a). By contrast, potatoes require a relatively large amount of P, as they are inefficient in their use of soil P (Sandaña, 2016). There may also be landscape design features that can improve N and P cycling. When redesigning systems for ICLS, the introduction of landscape-level features such as hedgerows and buffer strips strategically into the landscape can help to reduce sediment and therefore P losses (Dabney et al., 2012). The additional benefit could be nutritional properties of the hedgerow species for livestock nutrition.

Concerning the management of crop rotations, cover crops tend to restore the C–N coupling through the optimization of soil-vegetation interactions (Sousanna and Lemaire, 2014). Beneficial effects of cover crops can be further enhanced with the use of reduced or no tillage in a conservation agricultural approach (Franzluebbers, 2007; Liebig et al. 2012). Moreover, a large part of cover crop biomass production in a conservation system can be used for feeding ruminant livestock (Franzluebbers and Stuedemann, 2007), in particular including grazed cover crops in between two cash crops (Ryschawy et al., 2014). The inclusion of several species within a grass sward or as winter cover can reduce leaching losses compared to single species swards, as well as potentially providing nutritional benefits. The ability of monogastric animals to recycle a large diversity of waste or crop by-products contributes to feed self-sufficiency (Ermgassen et al., 2016; van Zanten et al., 2016), so the presence of more than one type of livestock in ICLS can contribute to a greater proportion of nutrients being recycled within the system. Crop breeding programs do not generally focus on the quality of crop residues, but this could be a future target in terms of providing residues with specific properties for grazing. There are also examples of crops being able to provide grazing prior to seed production, e.g., canola (Kirkegaard et al., 2008). Alternative feed resources such as millet, wheat, oat, and barley straws are other cheap resources that serve as supplemental feed for ruminants, horses, and donkeys in many agroecosystems around the world.

Spatial Design

There is also the potential for reduced nitrate leaching though the introduction of trees in the system, as a result of luxury uptake of N by trees (Bergeron et al., 2011), and by increasing the volume and depth of soil explored by roots. This "safety net hypothesis" (Jose et al., 2004) applies also in intercropping and multispecies pasture where deep-rooting crops are used. This could also reduce the soil N readily available for the production of N_2O. The use of either leguminous tree species or leguminous understory species can also reduce the need for fertilizer N. The potential for trees to reduce the transport of ammonia downwind of livestock production systems by capturing N is well established (e.g., Patterson et al., 2008; Bealey et al., 2014). Losses of N from permanent grassland are in any case generally very low unless the system is disturbed.

Livestock can also potentially increase tree growth and productivity through enhanced recycling of nutrients, as exemplified by geese grazing unfertilized grass growing between tree rows in a walnut plantation where tree growth increased by 6% and walnut production by 26% (Dubois et al., 2008). Leaves can also contribute to livestock nutrition by providing protein through browsing in situ or harvesting and feeding remotely, as described in tropical systems by Mafongoya et al. (2000).

Rotation or Sequence Design

Sod-based rotations or ley/arable systems are able to capture temporally the benefits of leys, providing nitrogen and cover to the soil, and thus minimizing environmental impacts in combination with periods of intensive cropping for integration of livestock within a cropping system (Allen et al., 2007; Franzluebbers, 2007). Watson et al. (2005) characterize three rotational phases in ley/arable systems in terms of nitrogen management objectives: pasture, plowing out, and subsequent arable cropping. The plowing out phase usually carries the highest risk of nitrogen loss, but it is important to note that the impacts of plowing of leys can last for several years, and the associated losses happen over more than one season. Models that capture the impact of rotations over the period of the entire rotation rather than a single year are now being developed that are able to capture the results of decision-making such as legume inclusion in the rotation (Reckling et al., 2016). It has been suggested that the greater the duration of the ley within the rotation, or the lower the frequency of plowing of leys, the greater the potential for soil C sequestration and mitigation of N losses to the environment (Franzluebbers, 2007). Vertès et al. (2010) suggest that losses per annum over the rotation decline when the ley duration exceeds 5 years. The crop follow plowing out of the ley should have a high N requirement, and in an ICLS, there will be a need to decide whether this should be a cash crop or one with a high value for animal feed, e.g., fodder beet. Undersowing of cereals to avoid bare ground in autumn after harvest works particularly well in mixed systems where grazing of small ruminants in autumn can help to establish a dense sward.

LOOKING TO THE FUTURE

Farmers and advisors require practical tools to be able to better manage nutrient cycling in ICLS. Tools have been developed and used with the farming community include Calcola N in Italy[1] and ERA tool in the Baltic Sea area[2]; nevertheless, there is a need for these types of tools to be more widely used by the industry. An extension of simple nutrient budgets has been developed by Godinot et al. (2015), where the efficiency of nutrient use is calculated relative to the potential maximum efficiency of the system. At the policy level, nutrient budgets at the regional level can be carried out using nationally collected information. Although this level of analysis is less detailed than a farm-level nutrient budget, it is still useful in identifying regions where there are likely to be surpluses or deficits of N and P. In addition, the budgeting approach can be used to explore how government policy decisions will influence these surpluses/deficits.

Farmers also need tools to optimize the management of manures, including reliable information on the N and P content of the manure. In a nitrate vulnerable zone, there is regulatory pressure on farmers to demonstrate compliance. In the United Kingdom, this is supported by the decision support tool PLANET.[3]

The diversity of farm types within a given area clearly increases opportunities for nutrient recycling between farms and thus reducing the reliance of a given area on fertilizer inputs. Moraine et al. (2017) studied scenarios to enhance exchanges between 24 organic specialized crop farms and livestock farms to integrate crops and livestock. All the farmers were interested in crop-manure exchanges, including in-

[1] http://aqua.crpa.it/nqcontent.cfm?a_id=12681&tt=t_law_market_www&aa=tool.

[2] http://beras.eu/what-we-do/era-guidelines/.

[3] http://www.planet4farmers.co.uk/.

vestment in collective equipment. While the scenarios increased impact and had positive environmental benefits, logistical aspects limited uptake in practice.

Ryschawy et al. (2017) went on to develop scenarios with a subgroup of seven neighboring organic farmers, where four had a dairy production system oriented around either cows, goats, or ewes. Collective organization to manage exchanges of crops and livestock between farms revealed specific logistical and social barriers that need to be explored further. Questions concerning the distance and scale for the exchanges have been studied by Asai et al. (2014b) and de Wit and Verhoog (2007) in the context of Northern Europe. Here, major logistical aspects were solved because of the trust established between farmers. The participatory process developed by Moraine et al. (2017) and Ryschawy et al. (2017) enabled farmers to highlight the different individual objectives, develop collective ones, and exchange knowledge. The major constraint posed by the need to establish trust should be considered when seeking to out-scale these results to other contexts. In this sense, smaller collective size seems more favorable for developing strong cooperation, based on knowledge exchange.

At the research level, dynamic and deterministic models of the farm-level processes have been developed. This allows the interactions between the crop, livestock, and manure management to be explored, and therefore the impact of policies and changes in farm management practices can be assessed. FASSET (Bernsten et al., 2003) and Melodie (Chardon et al., 2012) have been developed primarily to explore these interactions for N. Other models, for example, FARMAC,[4] which can be used to assess N flows through the system, have been primarily developed to quantify the effects of management of crop and livestock systems of greenhouse gas emissions, based on the IPCC tier II and III methodologies.

More recently, recognizing the importance of the skills and decision-making of farmers, specific decision support tools have been developed to be used in collaboration with farmers during participative meetings. These models are mainly based on offer-need balances in inputs such as feed (forage and concentrates) and crops, including the need for nitrogen and phosphorus. CLIFS is a model used at the farm scale to test prospective scenarios of crop-livestock integration at the farm scale (Ryschawy et al., 2014; Le Gal et al., 2013). Other tools are under development and used to conceive local crop-livestock integration between specialized farmers (Moraine et al., 2017; Ryschawy et al., 2017).

Apart from the development of ICLS decision support tools, research also needs to address technical issues of particular relevance to mixed systems, respecting that they could also benefit from more specialized systems, e.g., development of fertilizer recommendations that take account of precrop and undersowing techniques; livestock feeding recommendations for livestock that can account for on-farm by-products, e.g., stubbles; or development of dual-purpose crop varieties (either for cutting and grazing or to provide residues with particular properties). We also recognize that efficiencies in N and P use can come from improvement in individual components, such as plant breeding, improvements in technology, such as manure handling facilities, and through improved management decision-making such as rotation design.

Research at the farm scale is often limited by cost, but both spatial and temporal scales are important in determining the technical efficiency of a farming system. In a mixed system based on a crop rotation, some effects will only become apparent over the timescale of a rotation. The benefits of, for example, shelter belts or

[4] http://www.farmac.dk/.

agroforestry that result from a changed physical environment need to be measured but may be gradual. Planning also needs to take in the scale of the farm and the spatial arrangements of productive land, housing, hedgerow, forest, wetland, and wild areas. This allows for synergies between enterprises to be optimized.

Optimizing the synergies between the different crop and livestock components of ICLS will improve the efficiency of use of N, P, and other nutrients, and this will contribute to both the economic and environmental success of ICLS. However, many other challenges also exist for both the maintenance of existing ICLS and the introduction of ICLS over systems including weeds, pests, diseases, and climate change. Many factors other than nutrient management will determine the uptake of ICLS including socioeconomic factors, land capability for agriculture, and pedoclimatic conditions. The different models described here will be appropriate for different land capabilities and socioeconomic situations that are beyond the scope of this chapter. The success of these systems to meet environmental and production goals will depend on the motivation of the farmers to successfully integrate crops and livestock, which is in turn dependent on incentives, either policy or economic, as well as knowledge and thus support through education and advice.

In Europe, most of the agricultural land with favorable pedoclimatic conditions has allowed specialization and intensification either in cash crops or livestock production (Wilkins, 2008). ICLS are mostly found in unfavored areas, where they cannot produce the economic results that could be realized in better conditions. Thus it is difficult to quantify the overall potential advantages of ICLS, and it is important to recognize that comparisons of nutrient management and other parameters of ICLS have not always been made under equal pedoclimatic conditions, which may bias the results. On poorer land the management options are more limited, particularly by issues such as slope and soil depth as well as climate. In such circumstances, mixed farming may be able to provide self-sufficiency but at a low conversion of inputs to outputs. ICLS may also be able to provide specialist products or commodities, e.g., the use of very extensive mixed systems to maintain clean water for selling as bottled water. Extensive livestock systems can have extremely low surpluses of nutrients, as demonstrated for an upland farm with sheep and suckled cows at Redesdale in Northumberland, United Kingdom (Goulding et al., 2000). In areas with limited land available for production and where the land is of good quality, then very intensive specialist systems may be more efficient than ICLS, at least in the short term, and as long as natural resource quality can be maintained. The appropriateness of introducing different technologies to improve N and P cycling may also depend on the environmental context; for example, while the use of legumes can be very beneficial in low fertility situations, they may only serve to increase possible nutrient losses in environments that are already nutrient enriched. Thus in the future development of ICLS, it is important to recognize that "one size does not fit all" and that land capability plays an important role in the relative efficiency of both ICLS and specialized farming systems.

Acknowledgments

This book chapter was inspired by the EIP Focus Group on Mixed Farming. The SRUC authors were supported through the Scottish Government RESAS Strategic Research Programme.

References

Alard, V., Béranger, C., Journet, M., 2002. A la recherche d'une agriculture durable. Etude de systèmes économes en Bretagne. INRA Editions, Paris.

Allen, V.G., Brown, C.P., Segarra, E., Green, C.J., Wheeler, T.A., Acosta-Martinez, V., Zobeck, T.M., 2007. In search of sustainable agricultural systems for the LlanoEstacado of the U.S Southern high plains. Agriculture, Ecosystems and Environment 124 (1), 3–12.

Asai, M., Langer, V., Frederiksen, P., 2014a. Responding to environmental regulations through collaborative arrangements: social aspects of manure partnerships in Denmark. Livestock Science 167, 370–380.

Asai, M., Langer, V., Frederiksen, P., Jacobsen, B.H., 2014b. Livestock farmer perceptions of successful collaborative arrangements for manure exchange: a study in Denmark. Agricultural Systems 128, 55–65.

Baddeley, J.A., Jones, S., Topp, C.F.E., Watson, C.A., Helming, J., Stoddard, F.L., 2014. Biological Nitrogen Fixation (BNF) by Legume Crops in Europe. Legume Futures Report 1.5'. Available from: www.legumefutures.de [23 January 2017].

Bateman, A., van der Horst, D., Boardman, D., Kansal, A., Carliell-Marquet, C., 2011. Closing the phosphorus loop in England: the spatio-temporal balance of phosphorus capture from manure versus crop demand for fertiliser. Resources, Conservation and Recycling 55 (12), 1146–1153.

Battini, F., Agostini, A., Tabaglio, V., Amaducci, S., 2016. Environmental impacts of different dairy farming systems in the Po Valley. Journal of Cleaner Production 112, 91–102.

Baudon, E., Cottais, L., Leterme, P., Espagnol, S., Dourmad, J.Y., 2005. Optimisation environnementale des systèmes de production porcine. Journées de la recherche porcine en France 37, 325–332.

Bealey, W.J., Loubet, B., Braban, C.F., Famulari, D., Theobald, M.R., Reis, S., Reay, D.S., Sutton, M.A., 2014. Modelling agro-forestry scenarios for ammonia abatement in the landscape. Environmental Research Letters 9 (12), 125001. Available from: http://iopscience.iop.org/article/10.1088/1748-9326/9/12/125001/pdf.

Benoit, M., Garnier, J., Anglade, J., Billen, G., 2014. Nitrate leaching from organic and conventional arable crop farms in the Seine Basin (France). Nutrient Cycling in Agroecosystems 100 (3), 285–299.

Bergeron, M., Lacombe, S., Bradley, R., Whalen, J., Cogliastro, A., Jutras, M.F., Arp, P., 2011. Reduced soil nutrient leaching following the establishment of tree-based intercropping systems in eastern Canada. Agroforestry Systems 83 (3), 321–330.

Bernsten, J., Petersen, B.M., Jacobsen, J.H., Olesen, J.E., Hutchings, N.J., 2003. Evaluating nitrogen taxation scenarios using the dynamic whole farm simulation model FASSET. Agricultural Systems 76 (3), 817–839.

Chalova, V.I., Kim, J.H., Patterson, P.H., Ricke, S.C., Kim, W.K., 2016. Reduction of nitrogen excretion and emissions from poultry: a review for conventional poultry. World's Poultry Science Journal 7 (3), 509–520.

Chardon, X., Rigolot, C., Baratte, C., Espagnol, S., Raison, C., Martin-Clouaire, R., Rellier, J.-P., Le Gall, A., Dourmad, J.Y., Piquemal, B., Leterme, P., Paillat, J.M., Delaby, L., Garcia, F., Peyraud, J.L., Poupa, J.C., Morvan, T., Faverdin, P., 2012. MELODIE: a whole-farm model to study the dynamics of nutrients in dairy and pig farms with crops. Animal 6 (10), 1711–1721.

Coquil, X., Beguin, P., Dedieu, B., 2013. Transition to self-sufficient mixed crop–dairy farming systems. Renewable Agriculture and Food Systems 29 (3), 195–205.

Corning, P.A., 1995. Synergy and self-organization in the evolution of complex systems. Systems Research 12 (2), 89–121.

Dabney, S.M., Wilson, G.V., Mcgregor, K.C., Vieira, D.A., 2012. Runoff through and upslope of contour switchgrass hedges. Soil Science Society of America Journal 76, 210–219.

De Vries, J.W., Groenestein, M., De Boer, I.J.M., 2012. Environmental consequences of processing manure to produce mineral fertilizer and bio-energy. Journal of Environmental Management 102, 173–183.

De Wit, J., Verhoog, H., 2007. Organic values and the conventionalization of organic agriculture. NJAS - Wageningen Journal of Life Sciences 54 (4), 449–462.

Defrahttp://www.statistics.gov.uk/hub/regional-statistics/england/index.html.2010

del Prado, A., Chadwick, D., Cardenas, L., Misselbrook, T., Scholefield, D., Merino, P., 2010. Exploring systems responses to mitigation of GHG in UK dairy farms. Agriculture, Ecosystems and Environment 136, 318–332.

Domburg, P., Edwards, A.C., Sinclair, A.H., Chalmers, N.A., 2000. Assessing nitrogen and phosphorous efficiency at farm and catchment scale using nutrient budgets. Journal of the Science of Food and Agriculture 80, 1946–1952.

Dove, H., Kirkegaard, J., 2014. Using dual-purpose crops in sheep-grazing systems. Journal of the Science of Food and Agriculture 94 (7), 1276–1283.

Dubois, J.P., Bijja, M., Auvergne, A., Lavigne, F., Fernandez, X., Babilé, R., 2008. Qualité des parcours de palmipè des: choix des espèces végétales, rendement et résistance au piétinement. Actes des 8èmes Journées de la Recherche sur les Palmipèdes à Foie Gras 30–31. Octobre 2008, Arcachon, France.

Dumont, B., Fortun-Lamothe, L., Jouven, M., Thomas, M., Tichit, M., 2013. Prospects from agroecology and industrial ecology for animal production in the 21st century. Animal 7 (6), 1028–1043.

Ermgassen, E., Phalan, B., Green, R.E., Balmford, A., 2016. Reducing the land use of EU pork production: where there's swill, there's a way. Food Policy 58, 35–48.

European Environment Agency 2010 Available from: http://www.eea.europa.eu/data-and-maps/figures/annual-diffuse-agricultural-emissions-of.

Fangueiro, D., Coutinho, J., Cabral, F., Fidalgo, P., Bol, R., Trindade, H., 2012. Nitric oxide and greenhouse gases emissions following the application of different cattle slurry particle size fractions to soil. Atmospheric Environment 47, 373–380.

Fangueiro, D., Pereira, J., Coutinho, J., Moreira, N., Trindade, H., 2008. NPK farm-gate nutrient balances in dairy farms from Northwest Portugal. European Journal of Agronomy 28 (4), 625–634.

Franzluebbers, A.J., 2007. Integrated crop–livestock systems in the southeastern USA. Agronomy Journal 99 (2), 361–372.

Franzluebbers, A.J., Stuedemann, J.A., 2007. Crop and cattle responses to tillage systems for integrated crop–livestock production in the Southern Piedmont, USA. Renewable Agriculture and Food Systems 22 (3), 168–180.

Giustini, L., Acciaioli, A., Argenti, G., 2007. Apparent balance of nitrogen and phosphorus in dairy farms in Mugello (Italy). Italian Journal of Animal Science 6 (2), 175–185.

Gliessman, S.R., 2006. Agroecology: The Ecology of Sustainable Food Systems, second ed. CRC Press, Boca Raton, FL, USA.

Godinot, O., Leterme, P., Vertès, F., Faverdin, P., Carof, M., 2015. Relative nitrogen efficiency, a new indicator to assess crop livestock farming systems. Agronomy for Sustainable Development 35, 857–868.

Goulding, K.T., Stockdale, E.A., Fortune, S., Watson, C.A., 2000. Nutrient cycling on organic farms. Journal of the Royal Agricultural Society of England 161, 65–75.

Hendrickson, J.R., Hanson, J.D., Tanaka, D.L., Sassenrath, G., 2008. Principles of integrated agricultural systems: introduction to processes and definition. Renewable Agriculture and Food Systems 23 (4), 265–271.

Hjorth, M., Christensen, K.V., Christensen, M.L., Sommer, S.G., 2010. Solideliquid separation of animal slurry in theory and practice. A review. Agronomy for Sustainable Development 30 (1), 153–180.

Jeuffroy, M.H., Baranger, E., Carrouée, B., de Chezelles, E., Gosme, M., Hénault, C., Schneider, A., Cellier, P., 2013. Nitrous oxide emissions from crop rotations including wheat, oilseed rape and dry peas. Biogeosciences 10, 1787–1797. Available from: http://www.biogeosciences.net/10/1787/2013/bg-10-1787-2013.pdf.

Jose, S., Gillespie, A.R., Pallardy, S.G., 2004. Interspecific interactions in temperate agroforestry. Agroforestry Systems 61, 237–255.

Kebreab, E., Hansen, A.V., Strathe, A.B., 2012. Animal production for efficient phosphate utilization: from optimized feed to high efficiency livestock. Current Opinion in Biotechnology 23 (6), 872–877.

Kirkegaard, J.A., Sprague, S.J., Dove, H., Kelman, W.M., Marcroft, S.J., Lieschke, A., Howe, G.N., Graham, J.M., 2008. Dual-purpose canola—a new opportunity in mixed farming systems. Crop and Pasture Science 59 (4), 291–302.

Kronberg, S.L., Ryschawy, J., 2017. Integration of crop and livestock production in temperate regions to improve agroecosystem functioning, improve ecosystem services and reduce negative consequences of food production on animals, people and the environment. This volume (Submitted).

Le Gal, P.Y., Bernard, J., Moulin, C.H., 2013. Supporting strategic thinking of smallholder dairy farmers using a whole farm simulation tool. Tropical Animal Health and Production 45, 1119–1129.

Le Gall, A., Vertès, F., Pflimlin, A., Chambaut, H., Delaby, L., Durand, P., van der Werf, H., Turpin, N., Bras, A., 2005. Flux d'azote et de phosphore dans les fermes franç aiseslaitières et mise en oeuvre des règlementations environnementales. Rapport no.190533017. Collection "Résultats" Paris, France INRA. Institut de l'Elevage, p. 64.

Lemaire, G., Franzluebbers, A., de Faccio Carvalho, P.C., Dedieu, B., 2014. Integrated crop–livestock systems: strategies to achieve synergy between agricultural production and environmental quality. Agriculture, Ecosystems and Environment 190, 4–8.

Lemercier, B., Gaudin, L., Walter, C., Aurousseau, P., Arrouays, D., Schvartz, C., Saby, N.P.A., Follain, S., Abrassart, J., 2008. Soil phosphorus monitoring at the regional level by means of a soil test database. Soil Use and Management 24 (2), 131–138.

Liebig, M.A., Tanaka, D.L., Kronberg, S.L., Scholljegerdes, E.J., Karn, J.F., 2012. Integrated crops and livestock in central North Dakota, USA: agroecosystem management to buffer soil change. Renewable Agriculture and Food Systems 27 (2), 115–124.

Lopez-Ridaura, S., van der Werf, H., Paillat, J.M., Le Bris, B., 2009. Environmental evaluation of transfer and treatment of excess pig slurry by life cycle assessment. Journal of Environmental Management 90 (2), 1296–1304.

Mafongoya, P.L., Barak, P., Reed, J.D., 2000. Carbon, nitrogen and phosphorus mineralization of tree leaves and manure. Biology and Fertility of Soils 30 (4), 298–305.

Martin, G., Moraine, M., Ryschawy, J., Magne, M.A., Asai, M., Sarthou, J.P., Duru, M., Therond, O., 2016. Crop–livestock integration beyond the farm level: a review of prospects and issues. Agronomy for Sustainable Development 36 (53). Available from: http://link.springer.com/article/10.1007/s13593-016-0390-x.

McCormick, S., Jordan, C., Bailey, J.S., 2009. Within and between-field spatial variation in soil phosphorus in permanent grassland. Precision Agriculture 10, 262–276.

McGrath, J.M., Sims, J.T., Maguire, R.O., Saylor, W.W., Angel, C.R., 2010. Modifying broiler diets with phytase and vitamin D metabolite (25-OH D3): impact on phosphorus in litter, amended soils and runoff. Journal of Environmental Quality 39 (1), 324–332.

Moraine, M., Duru, M., Nicholas, P., Leterme, P., Therond, O., 2014. Farming system design for innovative crop-livestock integration in Europe. Animal 8 (8), 1204–1217.

Moraine, M., Melac, P., Ryschawy, J., Duru, M., Therond, O., 2017. A participatory method for the design and integrated assessment of crop-livestock systems in farmers' groups. Ecological Indicators 72, 340–351.

Nowak, B., Nesme, T., David, C., Pellerin, S., 2015. Nutrient recycling in organic farming is related to diversity in farm types at the local level. Agriculture, Ecosystems and Environment 204, 17–26.

Pappa, V.A., Rees, R.M., Walker, R.L., Baddeley, J.A., Watson, C.A., 2011. Nitrous oxide emissions and nitrate leaching in an arable rotation resulting from the presence of an intercrop. Agriculture, Ecosystems and Environment 141 (1), 153–161.

Patterson, P.H., Adrizal, A., Hulet, R.M., Bates, R.M., Despot, D.A., Wheeler, E.F., Topper, P.A., 2008. The potential for plants to trap emissions from farms with Laying Hens: 1 Ammonia. The Journal of Applied Poultry Research 17 (1), 54–63.

Peyraud, J.-L., Taboada, M., Delaby, L., 2014. Integrated crop and livestock systems in Western Europe and South America: a review. European Journal of Agronomy 57, 31–42.

Preissel, S., Reckling, M., Schl€afke, N., Zander, P., 2015. Magnitude and farm-economic value of grain legume pre-crop benefits in Europe: A review. Field Crops Research 175, 64–79.

Reckling, M., Hecker, J.-M., Bergkvist, G., Watson, C.A., Zander, P., Schläfke, N., Stoddard, F.L., Eory, V., Topp, C.F.E., Maired, J., Bachinger, J., 2016. A cropping system assessment framework and its application to legumes. European Journal of Agronomy 76, 186–197.

Regan, J.T., Marton, S., Barrantes, O., Ruane, E., Hanegraaf, M., Berland, J., Korevaar, H., Pellerin, S., Nesme, T., 2017. Does the recoupling of dairy and crop production via cooperation between farms generate environmental benefits? A case-study approach in Europe. European Journal of Agronomy 82, 342–356.

Rigolot, C., Espagnol, S., Robin, P., Hassouna, M., Béline, F., Paillat, J.M., Dourmad, J.Y., 2010. Modelling of manure production by pigs and NH_3,N_2O and CH_4 emissions. Part II: effect of animal housing, manure storage and treatment practices. Animal 4 (8), 1413–1424.

Rotz, C.A., Taube, F., Russelle, M.P., Oenema, J., Sanderson, M.A., Wachendorf, M., 2005. Whole-farm perspectives of nutrient flows in grassland agriculture. Crop Science 45 (6), 2139–2159.

Russelle, M.P., Entz, M.H., Franzluebbers, A.J., 2007. Reconsidering integrated crop-livestock systems in north America. Agronomy Journal 99 (2), 325–334.

Ryschawy, J., Choisis, N., Choisis, J.P., Joannon, A., Gibon, A., 2012. Mixed croplivestock systems: an economic and environmental friendly way of farming? Animal 6 (10), 1722–1730.

Ryschawy, J., Choisis, N., Choisis, J.P., Gibon, A., 2013. Paths to last in mixed crop-livestock farming: lessons from an assessment of farm trajectories of change. Animal 7 (4), 673–681.

Ryschawy, J., Choisis, J.-P., Joannon, A., Gibon, A., Le Gal, P.-Y., 2014. Participative assessment of innovative technical scenarios for enhancing sustainability of French mixed crop-livestock farms. Agricultural Systems 129, 1–8.

Ryschawy, J., Martin, G., Moraine, M., Duru, M., Therond, O., 2017. Designing crop–livestock integration at different levels: toward new agroecological models? Nutrient Cycling in Agroecosystems 1–16.

Sandaña, P., 2016. Phosphorus uptake and utilization efficiency in response to potato genotype and phosphorus availability. European Journal of Agronomy 76, 95–106.

Schiere, J.B., Ibrahim, M.N.M., Van Keulen, H., 2002. The role of livestock for sustainability in mixed farming: criteria and scenario studies under varying resource allocation. Agriculture, Ecosystems and Environment 90 (2), 139–153.

Schipanski, M.E., Bennett, E.M., 2012. The influence of agricultural trade and livestock production on the global phosphorus cycle. Ecosystems 15 (2), 256–268.

Schröder, J.J., Aarts, H.F.M., ten Berge, H.F.M., van Keulen, H., Neeteson, J.J., 2003. An evaluation of whole-farm nitrogen balances and related indices for efficient nitrogen use. European Journal of Agronomy 20, 33–44.

Senthilkumar, K., Nesme, T., Mollier, A., Pellerin, S., 2012. Regional-scale phosphorus flows and budgets within France: the importance of agricultural production systems. Nutrient Cycling in Agroecosystems 92 (2), 145–159.

Sharpley, A.N., Herron, S., Daniel, T., 2007. Overcoming the challenges of phosphorus-based management in poultry farming. Journal of Soil and Water Conservation 62 (6), 375–389.

Simon, J.C., Grignani, C., Jacquet, A., Corre, L.L., 2000. Typology of nitrogen balances on a farm scale: research of operating indicators. Agronomie 20 (2), 175–195.

Soussana, J.F., Lemaire, G., 2014. Coupling carbon and nitrogen cycles for environmentally sustainable intensification of grasslands and crop-livestock systems. Agriculture, Ecosystems and Environment 190, 9–17.

Swensson, C., 2002. Effect of manure handling system, N fertiliser use and area of sugar beet on N surpluses from dairy farms in southern Sweden. Journal of Agricultural Science, Cambridge 138 (4), 403–413.

ten Hoeve, M., Hutchings, N.J., Peters, G.M., Svanstr€om, M., Jensen, L.S., Bruun, S., 2014. Life cycle assessment of pig slurry treatment technologies for nutrient redistribution in Denmark. Journal of Environmental Management 132, 60–70.

Vadas, P., Sims, J.T., 2014. Soil Fertility: Phosphorus in Soils', Reference Module in Earth Systems and Environmental Sciences. Available from: http://www.sciencedirect.com/science/article/pii/B9780124095489091168.

Vertès, F., Benoit, M., Dorioz, J.M., 2010. Couverts herbacés perennes et enjeux envi-ronnementaux en particulier eutrophisation: atouts et limites. Fourrages 202, 83–94.

von Eye, A., Schuster, C., Rogers, W.M., 1998. Modelling synergy using manifest categorical variables. International Journal of Behavioral Development 22 (3), 537–557.

Watson, C., Bengtsson, H., Løes, A.-K., Myrbeck, A., Salomon, E., Schroder, J., Stockdale, E.A., 2002. A review of farm-scale nutrient budgets for organic farms as a tool for management of soil fertility. Soil Use and Management 18 (s1), 264–273.

Watson, C.A., Reckling, M., Preissel, S., Bachinger, J., Bergkvist, G., Kuhlman, T., Lindström, K., Nemecek, T., Topp, C.F.E., Vanhatalo, A., Zander, P., Murphy-Bokern, D., Stoddard, F.L., 2017. Grain legume production and use in European agricultural systems. Advances in Agronomy 144, 235–303.

Watson, C.A., Öborn, I., Eriksen, J., Edwards, A.C., 2005. Perspectives on nutrient management on mixed farms. Soil Use and Management 21 (s1), 132–140.

White, P.J., Veneklaas, E.J., 2012. Nature and nurture: the importance of seed phosphorus content. Plant and Soil 357 (1–2), 1–8.

Wilkins, R.J., 2008. Eco-efficient approaches to land management: a case for increased integration of crop and animal production systems. Philosophical Transactions of the Royal Society B: Biological Sciences 363 (1491), 517–525.

Withers, P.J., Sylvester-Bradley, R., Jones, D.L., Healey, J.R., Talboys, P.J., 2014. Feed the crop not the soil: rethinking phosphorus management in the food chain. Environmental Science and Technology 48 (12), 6523–6530.

van Zanten, H.H.E., Mollenhorst, H., Klootwijk, C.W., van Middelaar, C.E., De Boer, I.J.M., 2016. Global food supply: land use efficiency of livestock systems. International Journal of Life Cycle Assessment 21 (5), 747–758.

SECTION III

HETEROGENEITY WITHIN AND AMONG AGROECOSYSTEMS AND DYNAMICS OF BIODIVERSITY

CHAPTER

11

Can Increased Within-Field Diversity Boost Ecosystem Services and Crop Adaptability to Climatic Uncertainty?

Isabelle Litrico[1], *Christian Huyghe*[2]

[1]P3F UR 004 - INRA - Le Chêne RD150, F-86600, Lusignan, France; [2]Paris Siège - INRA - 147 rue de l'Université, F-75338, Paris, France

Faced with changing climate and an increasing incidence of extreme weather events (Stott et al., 2004) and the negative environmental impacts of a number of conventional agriculture inputs, a major challenge for agriculture is to develop more sustainable cropping systems. Over the last few decades, many studies have emphasized the benefits of species diversity on ecosystem function (Grace, 1991; Lortie et al., 2004; Craine, 2005; Brooker et al., 2008). Species diversity benefits have generally been explored within natural communities to better understand both the occurrence and the abundance of species. However, benefits of species diversity have also been demonstrated for stability of agricultural production systems (Allard, 1961; Rasmusson et al., 1967; Tilman et al., 1996; Hockett et al., 1983; Hector et al., 1999; Finckh et al., 2000; Grime, 2006; Nyfeler et al., 2009). These benefits apply particularly under conditions of climate variability (Tilahun, 1995; Lesica and Allendorf, 1999; Tilman et al., 2001). Increased plant diversity is necessary to increase the ability of a plant population or community to adapt to new environmental conditions (Burger and Lynch, 1995; Aitken et al., 2008; Hoffmann and Sgrò, 2011) and also to better meet environmental challenges (Frey and Maldonado, 1967; Hajjar et al., 2008; Vlachostergios et al., 2011). Hence, the introduction of greater genotypic and species diversity into cropping systems should allow increased sustainability of production under conditions of an increasingly variable climate (Bignal and McCracken, 2000) while also providing some valuable environmental benefits (Malézieux et al., 2009; Letourneau et al., 2011; Dore et al., 2011).

The positive effects of species diversity on the stability and functioning of a plant community can be explained on the basis of nonexclusive theories of community ecology. Species diversity may offer a buffer, with some promise of improved productivity through an increase in the probability that the community will contain genotypes and species better fitted to the new conditions and hence more productive (Huston

Agroecosystem Diversity
https://doi.org/10.1016/B978-0-12-811050-8.00011-X

and McBride, 2002). This effect is called the "sampling effect" (Huston, 1997; Tilman et al., 1997). The sampling effect may lead to greater community stability, i.e., the continuation of that community following an extreme climatic event or any other major disturbance linked to change in environment or in management. To explain the species diversity effect on community function, the theory of "niche complementarity" (Macarthur and Levins, 1967; Abrams, 1983; Silvertown, 2004) has been proposed. This theory is based on the determinist processes of species interaction within the community, particularly in regard to competition for resources. Niche complementarity explores the spatial and temporal complementarity of species regarding access to resources, trade-offs in their acquisition of different resources, and trade-offs in their abilities to colonize and compete. Complementarity among species in their patterns of resource acquisition and/or competition strategies leads to decreased competition and increased production and nutrient efficiency. The review of Hooper et al. (2005) concluded that some combinations of species are complementary in their uses and use-efficiencies of resources and that these interactions can improve overall community productivity (see Chapters 8 and 13 this volume). The notion of complementary leads to a functional perspective that emphasizes the concept of functional diversity to describe species diversity. Different species that share the same strategies, properties, and capacities can be combined into one functional group. Complementary effects suggest relationships between functional diversity (the number of functional groups) and the function of a community (Hooper, 1998; Hooper et al., 2005; Kremen, 2005). In the recent literature, a number of authors have stressed the potential benefits of functional diversity for community function (Balvanera et al., 2006; Díaz et al., 2006) and biomass production (Hector et al., 1999; Franco et al., 2015). The importance of the functional diversity effect supports the analysis of functional traits of species within a community to better understand the dynamics and productivity of the community (see Chapter 13 this volume). The functional traits of a species may be defined as a feature of this species that links its response to environmental pressures and their effects on the community. Based on "limiting similarity" (Macarthur and Levis, 1967), the assumption is that species filling different niches have different trait combinations and thus offer a greater probability of coexistence through the complementarity effect.

The sampling effect and the complementarity effect should be nonexclusive mechanisms (Loreau and Hector, 2001) explaining the positive diversity effects on a community. These two mechanisms have been used to explain the species and functional diversity effects. However, intraspecific diversity (a different diversity level) has been less addressed in community ecology studies. In spite of the significant benefits that intraspecific diversity likely offers to the stability and functioning of a community (Booth and Grime, 2003; Vellend, 2006; Fridley et al., 2007; Vellend and Litrico, 2008; Violle et al., 2012), few studies have addressed the effects of this diversity level. Recent studies (Jung et al., 2010; Albert et al., 2011) point out the importance of intraspecific diversity on species structure and functioning in plant communities. More recently, Prieto et al. (2015) have shown clearly how the benefits of genetic diversity of species belonging to a community accrue to the overall biomass stability of the community. The same assumptions that relate to the benefits from species diversity and functional diversity may explain the positive effects arising from genetic diversity (Prieto et al., 2015). The different diversity levels (genetic, species, and functional) may also play an additional role (with domino effect) in the effect of diversity in promoting the stability and functioning of a whole community. Thus, functional diversity may improve the functioning of a community, species diversity could help preserve functional groups within the community, while genetic diversity could help preserve species in community, particularly following a serious negative climatic event. Community functioning and stability are key to the provision of ecosystem services (Malézieux et al., 2009; Cadotte et al., 2011). Ecosystem services result from ecosystem functions based on a number of levels of diversity (Kremen, 2005; Díaz et al., 2006). Each level is needed if ecosystem services are to be improved, including yield, in a cropping system.

Various examples can be used to illustrate how diversity within and among species (in respect to functional traits) may provide distinct ecosystem services relating either to production, to crop protection (see Chapter 12 this volume), or to environmental factors. A first example illustrates a service resulting from genetic diversity for functional diversity. In oilseed rape, the rapeseed pollen beetle (*Meligethes aeneus*) causes severe damage as it grazes the flowering buds. Although several attempts have been made to use parasitoids to control the rapeseed pollen beetle (especially *Tersilochus heterocerus*), the presence of genetic diversity within the crop offers a very efficient alternative solution. Adding a small proportion (10% of the seed) of a very early-flowering variety to a sward of a productive variety proved to be a very efficient solution. When the early-flowering variety blooms, the beetles (a monovoltin species) fly in from the surrounding area, graze on the buds, and thus complete their biologic cycle and exit the crop before the productive variety come into flower. Thus, no damage occurs in the productive variety. This may be considered an "escape" strategy.

A second example is the significant benefits from increased genetic diversity in wheat when varieties having contrasting alleles for leaf-rust resistance are mixed (Lannou et al., 2013). The reduction in disease damage recorded in such mixtures is linked to a reduction in the rate of disease expansion. Here, the presence of multiple resistance alleles reduces the risk of resistance breakdown, even in the presence of a diverse population of pathogens. Moreover, in these studies, the authors demonstrate that the variety mixtures also improved both grain yield and quality. Wheat variety mixtures can also offer environmental services in respect to the diversity of collembola and other arthropods at field scale (Chateil et al., 2013).

A third example is provided by mixtures of crop species belonging to different functional groups, such as grasses and legumes (see Chapter 8 this volume). In grasslands, this is very well-known and is widely used. Mixtures of perennial forage grasses and legumes have been developed that lead to increased yield stability (Prieto et al., 2015) and increased biomass production that, on average, is similar to or higher than a pure grass sward maintained under a high-nitrogen fertilization regime (Hector et al., 1999; Kirwan et al., 2007). Thus, a complementarity effect is noticed for nitrogen nutrition: from mineral nitrogen for grasses and from symbiotic fixation for legumes. Similarly, benefits have also been observed when a cereal and a legume crop are mixed, such as wheat and pea, under a low mineral nitrogen fertilization regime (see Chapter 8 this volume). Obtaining the optimal species balance (i.e., the right proportion of each partner in the stand) is essential to obtain maximum yield and quality. As documented by Justes et al. (2014), a high land equivalent ratio (LER) may be achieved, with the highest LERs occurring under conditions of lowest soil nitrogen availability (Corre-Hellou et al., 2006). The LER indicator is widely used for evaluating the relative performance of associated crops compared to the pure crops. It may be used for biomass or protein production.

A final example of the services provided by increased diversity within a sward is cocropping winter rapeseed and annual legumes. In such a system, the annual legume is sown at the same time as the rapeseed. The legume species is selected to be killed by the winter frosts to eliminate later growth competition with rapeseed. Based on this physiologic trait, common vetch (*Vicia sativa*), purple vetch (*Vicia benghalensis*), *Lathyrus*, pea, lentil, and serradella (*Ornithopus sativus*) are potentially good legume candidates. In this co-cropping arrangement, the presence of an annual legume provides three ecosystem services. (1) The legume and rapeseed exhibit complementarity for nitrogen nutrition, with the legume preferentially using symbiotic fixation. When the legume is killed by winter frosts, the fixed nitrogen becomes available for the young rapeseed plants. It is estimated 40–60 kg N/ha can be provided to the rapeseed crop by the legumes (Lorin et al., 2016). (2) During their growth, the legumes compete very efficiently with weed species. As a result, an autumn chemical herbicide treatment is unnecessary. Herbicide sprays may, however, be needed in areas lacking winter frosts to kill the legumes. (3) it was found that co-cropping rapeseed and legumes during autumn was not favorable to the presence of the flea beetle (*Psylliodes chrysocephalus*). In a pure rapeseed crop, control of this beetle requires one insecticide spray, but with co-cropping, no insecticide spray is required. This complex interaction between rapeseed and an annual legume requires a higher level of understanding by the farmer. The optimal choice of legume species depends on local conditions. The increased functional diversity achieved by addition of nonproductive species in a sward provides outstanding ecosystem services. It also suggests new breeding opportunities for species not presently in the scope of any breeding program.

Introduction of diversity into agroecosystems appears to offer promising paths ahead for increasing adaptation to climate change and the provision of ecosystem services. However, for this, greater knowledge is required in a number of diverse areas of competence. Indeed, even in studies showing the benefits of diversity, it is still not always clear if the benefits accrue for all types of disturbance. Hence, further work is necessary to fully understand these relationships. Further, positive interactions between species and genotypes within a crop, must be studied from the perspective of complementarity for resources usage. Moreover, more precise identification is required of diversity for which traits and over what ranges. As previously described, the limiting similarity of traits approach should be used. This, based on the control of trait variability between and within plant species, with characterization of resources, should help us understand the mechanisms leading to the optimal benefits of diversity. In addition, this approach should emphasize the major traits involved in species and genotype interactions that should be adopted as essential selection criteria for plant breeding and selection schemes for species mixtures (Litrico and Violle, 2015). Indeed, the use of swards containing mixtures of cultivated species (plurispecies) has significant implications for plant breeding that to this point has focused mainly on pure (monospecific) swards. While promoting diversity, plant breeding should also revisit their selection schemes and should consider a number of new selection criteria (Litrico and Violle, 2015). Furthermore, it will be necessary to determine the available variability of the key traits in existing crop varieties and in ecotypes of cultivated species to fully exploit this aspect and thus obtain the greatest benefit. This offers two options in plant breeding: (1) reducing genotypic variability for matching to a narrow window of favorable environments obtained through a well-defined crop management; (2) increasing genotypic diversity for better facing wider and variable environmental conditions and ecosystem services provision. These two strategies may correspond to two types of agriculture systems and must also be assessed on the basis of the market sizes, the regulatory dimension, and environmental impact for agriculture sustainability. The variabilities of related species should also be explored to obtain a wider range of variability. Furthermore, new species, exhibiting new traits and properties, may also be worth introducing to breeding programs, specifically to target ecosystem services. This emphasizes the need to study the intrinsic properties of species in line with targeted ecosystem services. Numerous ecosystem services are based on regulation functions driven by the population dynamics of associated organisms (see Chapter 12 this volume). For instance, mycorrhizal diversity contributes to nutrient cycles (Brussaard et al., 2007), and soil stability is closely linked to the functional group diversity of soil organisms (Swift et al., 2004). This type of relationship is not restricted to the soil ecosystem; it applies also to the services provided by pollinators (Zhang et al., 2007). Improvement for multiservices creates the need to study a plant and also its closely associated phytobiome.

The introduction of field diversity currently aggregates the best species and varieties, based on their performances in pure culture. This knowledge will allow better designed species blends and the development of breeding strategies for new varieties specifically developed for use in mixtures.

References

Abrams, P., 1983. The theory of limiting similarity. Annual Review of Ecology, Evolution and Systematics 14, 359–376.

Aitken, S.N., Yeaman, S., Holliday, J.A., Wang, T., Curtis-McLane, S., 2008. Adaptation, migration or extirpation: climate change outcomes for tree populations. Evolutionary Applications 1, 95–111.

Albert, C.H., Grassein, F., Schurrd, F.M., Vieilledent, G., Violle, C., 2011. When & how should intraspecific variability be considered in trait-based plant ecology? Perspectives in Plant Ecology, Evolution & Systematics 13, 217–225.

Allard, R.W., 1961. Relationship between genetic diversity and consistency of performance in different environments. Crop Science 1, 127–133.

Balvanera, P., Pfisterer, A.B., Buchmann, N., He, J.S., Nakashizuka, T., Raffaelli, D., Schmid, B., 2006. Quantifying the evidence for biodiversity effects on ecosystem functioning and services. Ecology Letters 9, 1146–1156.

Bignal, E.M., McCracken, D.I., 2000. The nature conservation value of European traditional farming systems. Environmental Reviews 8, 149–171.

Booth, R.E., Grime, J.P., 2003. Effects of genetic impoverishment on plant community diversity. Journal of Ecology 91, 721–730.

Brooker, R.W., Maestre, F.T., Callaway, R.M., Lortie, C.L., Cavieres, L.A., Kunstler, G., Liancourt, P., Tielbörger, K., Travis, J.M.J., Anthelme, F., Armas, C., Coll, L., Corcket, E., Delzon, S., Forey, E., Kikvidze, Z., Olofsson, J., Pugnaire, F., Quiroz, C.L., Saccone, P., Schiffers, K., Seifan, M., Touzard, B., Michalet, R., 2008. Facilitation in plant communities: the past, the present, and the future. Journal of Ecology 96, 18–34.

Brussaard, L., De Ruiter, P.C., Brown, G.G., 2007. Soil Biodiversity for agricultural sustainability. Agriculture, Ecosystems and Environment 121, 233–244.

Burger, R., Lynch, M., 1995. Evolution & extinction in a changing environment: a quantitative-genetic analysis. Evolution 49, 151–163.

Cadotte, M.W., Carscadden, K., Mirotchnick, N., 2011. Beyond species: functional diversity and the maintenance of ecological processes and services. Journal of Applied Ecology 48, 1079–1087.

Chateil, C., Goldringer, I., Taralloa, L., Kerbirioua, C., Le Viola, I., Pongeb, J.F., Salmon, S., Gachet, S., Porcher, E., 2013. Crop genetic diversity benefits farmland biodiversity in cultivated fields. Agriculture, Ecosystems & Environment 171, 25–32.

Corre-Hellou, G., Fustec, J., Crozat, Y., 2006. Interspecific competition for soil N and its interaction with N2 fixation, leaf expansion and crop growth in pea-barley intercrops. Plant and Soil 282, 195–208.

Craine, J.M., 2005. Reconciling plant strategy theories of Grime and Tilman. Journal of Ecology 93, 1041–1052.

Díaz, S., Fargione, J., Chapin III, F.S., Tilman, D., 2006. Biodiversity loss threatens human well-being. PLoS Biology 4 (8), 1300–1305.

Dore, T., Makowski, D., Malezieux, E., Munier-Jolain, N., Tchamitchian, M., Tittonell, P., 2011. Facing up to the paradigm of ecological intensification in agronomy: revisiting methods, concepts & knowledge. European Journal of Agronomy 34, 197–210.

Finckh, M.R., Gacek, E.S., Goyeau, H., Lannou, C., Merz, U., Mundt, C., Munk, L., Nadziak, J., Newton, A., De Vallavieille-Pope, C., Wolfe, M., 2000. Cereal variety and species mixtures in practice, with emphasis on disease resistance. Agronomie 20, 813–837.

Franco, J.G., King, S.R., Masabni, J.G., Volder, A., 2015. Plant functional diversity improves short-term yields in a low-input intercropping system. Agriculture, Ecosystems & Environment 203, 1–10.

Frey, K.J., Maldonado, U., 1967. Relative productivity of homogeneous and heterogeneous oat cultivars in optimum and suboptimum environments. Crop Science 7, 532–535.

Fridley, J., Stachowicz, J.J., Naeem, S., Sax, D.F., Seabloom, E.W., Smith, M.D., Stohlgren, T.J., Tilman, D., Von Holle, B., 2007. The invasion paradox: reconciling pattern & process in species invasions. Ecology 88, 3–17.

Grace, J.B., 1991. A clarification of the debate between Grime and Tilman. Functional Ecology 5, 583–587.

Grime, J.P., 2006. Trait convergence and trait divergence in herbaceous plant communities: mechanisms and consequences. Journal of Vegetation Science 17, 255–260.

Hajjar, R., Jarvis, D.I., Gemmill-Herren, B., 2008. The utility of crop genetic diversity in maintaining ecosystem services. Agriculture, Ecosystems & Environment 123, 261–270.

Hector, A., Schmid, B., Beierkuhnlein, C., Caldeira, M.C., Diemer, M., Dimitrakopoulos, P.G., Finn, J.A., Freitas, H., Giller, P.S., Good, J., Harris, R., Högberg, P., Huss-Danell, K., Joshi, J., Jumpponen, A., Körner, C., Leadley, P.W., Loreau, M., Minns, A., Mulder, C.P.H., O'Donovan, G., Otway, S.J., Pereira, J.S., Prinz, A., Read, D.J., Scherer-Lorenzen, M., Schulze, E.D., Siamantziouras, A.S.D., Spehn, E.M., Terry, A.C., Troumbis, A.Y., Woodward, F.I., Yachi, S., Lawton, J.H., 1999. Plant diversity and productivity experiments in European grasslands. Science 286, 1123–1127.

Hockett, E.A., Eslick, R.F., Qualset, C.O., Dubbs, A.L., Stewart, V.R., 1983. Effects of natural selection in advanced generations of Barley composite cross II. Crop Science 23, 752–756.

Hoffmann, A.A., Sgrò, C.M., 2011. Climate change & evolutionary adaptation. Nature 470, 479–485.

Hooper, D.U., 1998. The role of complementarity and competition in ecosystem responses to variation in plant diversity. Ecology 79, 704–719.

Hooper, D.U., Chapin III, F.S., Ewel, J.J., Hector, A., Inchausti, P., Lavorel, S., Lawton, J.H., Lodge, D.M., Loreau, M., Naeem, S., Schmid, B., Setälä, H., Symstad, A.J., Vandermeer, J., Wardle, D.A., 2005. Effects of biodiversity on ecosystem functioning: a consensus of current knowledge. Ecological Monographs 75 (1), 3–35.

Huston, M.A., 1997. Hidden treatments in ecological experiments: re-evaluating the ecosystem function of biodiversity. Oecologia 110, 449–460.

Huston, M.A., McBride, A.C., 2002. Evaluating the relative strengths of biotic versus abiotic controls on ecosystem processes. In: Loreau, M., Naeem, S., Inchausti, P. (Eds.), Biodiversity of Ecosystem Functioning: Synthesis and Perspectives. Oxford University Press, Oxford, pp. 47–60.

Jung, V., Violle, C., Mondy, C., Hoffmann, L., Muller, S., 2010. Intraspecific variability and trait-based community assembly. Journal of Ecology 98, 1134–1140.

Justes, E., Bedoussac, L., Corre-Hellou, G., Fustec, J., Hinsinger, P., Journet, E.P., Louarn, G., Naudin, C., Pelzer, E., 2014. Les processus de complémentarité de niche et de facilitation déterminent le fonctionnement des associations végétales et leur efficacité pour l'acquisition des ressources abiotiques. Innovations Agronomiques 40, 1–24.

Kirwan, L., Luscher, A., Sebastia, M.T., Finn, J.A., Collins, R.P., Porqueddu, C., Helgadottir, A., Baadshaug, O.H., Brophy, C., Coran, C., Dalmannsdottir, S., Delgado, I., Elgersma, A., Fothergill, M., Frankow-Lindberg, B.E., Golinski, P., Grieu, P., Gustavsson, A.M., Hoglind, M., Huguenin-Elie, O., Iliadis, C., Jorgensen, M., Kadziuliene, Z., Karyotis, T., Lunnan, T., Malengier, M., Maltoni, S., Meyer, V., Nyfeler, D., Nykanen-Kurki, P., Parente, J., Smit, H.J., Thumm, U., Connolly, J., 2007. Evenness drives consistent diversity effects in intensive grassland systems across 28 European sites. Journal of Ecology 95, 530–539.

Kremen, C., 2005. Managing ecosystem services: what do we need to know about their ecology? Ecological Letters 8, 468–479.

Lannou, C., Papaïx, J., Monod, H., Raboin, L.M., Goyeau, H., 2013. Gestion de la résistance aux maladies à l'échelle des territoires cultivés. Innovations Agronomiques 29, 33–44.

Lesica, P., Allendorf, F.W., 1999. Ecological genetics and the restoration of plant communities: mix or match? Restoration Ecology 7 (1), 42–50.

Letourneau, D.K., Armbrecht, I., Rivera, B.S., Lerma, J.M., Carmona, E.J., Daza, M.C., Escobar, S., Galindo, V., Gutiérrez, C., Duque Lopez, S., Lopez Mejia, J., Acosta Rangel, A.M., Herrera Rangel, J., Rivera, L., Saavedra, C.A., Torres, A.M., Reyes Trujillo, A., 2011. Does plant diversity benefit agroecosystems? A synthetic review. Ecological Applications 21, 9–21.

Litrico, I., Violle, C., 2015. Diversity in plant breeding: a new conceptual framework. Trends in Plant Science 20 (10), 604–613.

Loreau, M., Hector, A., 2001. Partitioning selection and complementarity in biodiversity experiments. Nature 412, 72–76.

Lorin, M., Jeuffroy, M.H., Butier, A., Valantin-Morison, M., 2016. Undersowing winter oilseed rape with frost-sensitive legume living mulch: consequence on the cash crop nitrogen nutrition. Field Crop Research 193, 24–33.

Lortie, C.J., Brooker, R.W., Choler, P., Kikvidze, Z., Michalet, R., Pugnaire, F.I., Callaway, R.M., 2004. Rethinking plant community theory. Oikos 107, 433–438.

Macarthur, R.H., Levins, R., 1967. The limiting similarity, convergence and divergence of coexisting species. The American Naturalist 101, 377–385.

Malézieux, E., Crozat, Y., Dupraz, C., Laurans, M., Makowski, D., Ozier-Lafontaine, H., Rapidel, B., De Tourdonnet, S., Valantin-Morison, M., 2009. Mixing plant species in cropping systems: concepts, tools and models. A review. Agronomy for Sustainable Development 29, 43–62.

Nyfeler, D., Huguenin, O.E., Suter, M., 2009. Strong mixture effects among four species in fertilized agricultural grassland led to persistent & consistent transgressive overyielding. Journal of Applied Ecology 46, 683–691.

Prieto, I., Violle, C., Barre, P., Durand, J.L., Ghesquiere, M., Litrico, I., 2015. Complementary effects of species and genetic diversity on productivity and stability of sown. Nature Plants 1 (4), 15033.

Rasmusson, D.C., Beard, B.H., Johnson, F.K., 1967. Effect of natural selection on performance of a barley population. Crop Science 7, 543.

Silvertown, J., 2004. Plant coexistence and the niche. Trends in Ecology & Evolution 19, 605–611.

Stott, P.A., Stone, D.A., Allen, M.R., 2004. Human contribution to the European heatwave of 2003. Nature 432, 610–614.

Swift, M.J., Izac, A.M.N., Van Noordwijk, M., 2004. Biodiversity and ecosystem services in agricultural landscapes—are we asking the right questions? Agriculture, Ecosystems & Environment 104, 113–134.

Tilahun, A., 1995. Yield gain and risk minimization in maize (Zea mays) through cultivar mixtures in semi-arid zones of the rift valley in Ethiopia. Experimental Agriculture 31, 161–168.

Tilman, D., Lehman, C., Thompson, K., 1997. Plant diversity & ecosystem productivity: theoretical considerations. Proceedings of the National Academy of Sciences of the United States of America 94, 1857–1861.

Tilman, D., Reich, P., Knops, J., Wedin, D.A., Mielke, T., Lehman, C., 2001. Diversity and productivity in a long-term grassland experiment. Science 294, 843–845.

Tilman, D., Wedin, D., Knops, J., 1996. Productivity and sustainability influenced by biodiversity in grassland ecosystems. Nature 379, 718–720.

Vellend, M., 2006. The consequences of genetic diversity in competitive communities. Ecology 87, 304–311.

Vellend, M., Litrico, I., 2008. Sex & space destabilize intransitive competition within & between species. Proceedings of the Royal Society B: Biological Sciences 275, 1857–1864.

Violle, C., Enquist, B.J., McGill, B.J., Jiang, L., Albert, C.H., Hulshof, C., Jung, V., Messier, J., 2012. The return of the variance: intraspecific variability in community ecology. Trends in Ecology & Evolution 27, 244–252.

Vlachostergios, A., Lithourgidis, A., Korkovelos, A., Baxevanos, D., Lazaridou, T., Khah, A., Mavromatis, A., 2011. Mixing ability of conventionally bred common vetch (Vicia sativa L.) cultivars for grain yield under low-input cultivation. AJCS 5 (12), 1588–1594.

Zhang, W., Ricketts, T.H., Kremen, C., Carney, K., Swinton, S.M., 2007. Ecosystem services and dis-services to agriculture. Ecological Economics 64 (2), 253–260.

CHAPTER

12

The Future of Sustainable Crop Protection Relies on Increased Diversity of Cropping Systems and Landscapes

Jonathan Storkey[1], *Toby J.A. Bruce*[2], *Vanessa E. McMillan*[1], *Paul Neve*[1]

[1]**Rothamsted Research, Harpenden, United Kingdom;** [2]**School of Life Sciences, Keele University, Staffordshire, United Kingdom**

INTRODUCTION

Global market forces and government policies that incentivize the production of a few, profitable crops and encourage specialization and consolidation of land and management through economies of scale have led to a progressive loss of agroecosystem diversity since the end of World War II (Stoate et al., 2009). The dominant paradigm of large-scale production of a few major crops in large, contiguous areas of genetically similar, high-yielding cultivars supported by high agrochemical inputs has underpinned increases in productivity over recent decades, keeping pace with a rising population. However, increasingly, it is being recognized that these spatially and temporally homogeneous systems may not be sustainable in the long term, and an increasing yield gap between attainable and actual yields is now being observed across all the world's major crops, accompanied by a plateauing of yields (van Ittersum et al., 2013). In the short to medium term, the challenge of protecting crops against pests, weeds, and pathogens is emerging as the main vulnerability of modern cropping systems, threatening their sustainability. Diversity loss is manifested at multiple scales and has had consequences both for the spectrum of species that reduce crop yield and the predators that regulate their populations (Table 12.1).

At the level of the crop genotype, the high yielding crop varieties developed in the Green Revolution are widely perceived as having solved the world's agricultural production problems (Evenson and Gollin, 2003). While they have made a tremendous difference, saving at least 1 billion people from food insecurity and reducing the need for expansion of farmland area required to feed a growing world population (Borlaug, 2007), their development was in a pesticide-treated background.

Agroecosystem Diversity
https://doi.org/10.1016/B978-0-12-811050-8.00012-1

TABLE 12.1 Loss of Diversity in Agroecosystems at the Field, Farm and Landscape Scale and Implications for Management of Pests, Weeds and Pathogens

Loss of Agroecosystem Diversity	Implications for Crop Protection
FIELD SCALE	
Loss of within crop genetic diversity (limited number of cultivars bred for narrow range of traits driven by yield potential)	Loss of resilience at the population level against pest, weed, and disease pressure. In the case of GMHT crops, reliance on a single active ingredient for weed control
Reduced range of crop protection active ingredients through legislation and lack of new products on the market	Increased selection pressure to individual active ingredients and evolution of pesticide resistance
Increased fertilizer use	Homogenization of soil properties and selection for a smaller number of nitrophilous, competitive weed species
FARM SCALE	
Simplified crop rotations	Reduced spectrum of pests, weeds and diseases, increasing potential yield loss and pressure on fewer crop protection products
Block cropping	Increased distances between diverse crops and habitats impacting predation rates
Removal of field boundaries	Increasing distance from field edge, reducing pest predation by natural enemies using boundaries as a habitat
LANDSCAPE SCALE	
Loss of seminatural habitat	Loss of source habitats for natural enemy populations, less diverse weed communities
Consolidation of farm units	Simplification of cropping patterns and management at the landscape scale leading to fewer, easily dispersed, generalist pest, weed, and pathogen species
Specialization (separation of arable and livestock production)	Loss of pasture in arable landscapes reducing diversity of natural enemy communities

Without support from herbicides, fungicides, and insecticides the high yield of these crops is not realized (Oerke, 2006). The focus on crop yield and, to a lesser extent, nutritional quality, in high-input systems has meant breeders have selected for a restricted set of plant traits that are expressed in the absence of diseases and crop enemy attack. However, as pesticide resistance in weeds, pathogens, and insects gathers pace and with fewer crop protection products on the market, it will become increasingly important to breed into future cultivars a diversity of traits that have been lost from modern genetic material and confer resistance to pests and diseases and tolerance of weeds.

At the level of crop species, each individual crop represents a recruitment niche for communities of pest, weed, and pathogen species that are adapted to attack the crop directly or compete with it for resources. Contrasting crop types (broad-leaf vs. grass, autumn vs. spring sown, legumes vs. nonlegumes, annual vs. perennial) represent differences in this "ecologic template" that select for functionally discrete pest, weed, and pathogen communities (in the case of pests and pathogens, variability in this selection pressure can also operate at the level of the crop species or cultivar). The simplification of crop rotations has, therefore, led to a homogenization of this habitat niche and a trend toward a smaller number of more problematic species. For example, in northwestern Europe, with the generalized use of large pesticide applications, rotations have been highly simplified and are now dominated by annual cereal crops and a limited number of break crops, mainly oilseed rape (OSR). Examples of widespread, economically damaging weeds, pathogens, and pests in these crops include (respectively) black-grass (*Alopecurus myosuroides*) and Septoria leaf blotch in wheat and cabbage stem beetle (*Psylliodes chrysocephala*) in OSR, all of which have also now evolved resistance to pesticides (as discussed later) (Fig. 12.1).

One additional, indirect, consequence of the large-scale cultivation of a small number of major crops has been the channeling of research investment by agrochemical companies toward the largest markets, leading to a lack of crop protection products for minor crops that are necessary for rotation diversification, further constraining crop choice for growers. This has happened in tandem with two additional drivers of a reduction in the diversity of the crop protection armory available to farmers. Firstly, there has been a slowdown in the discovery pipeline for new crop protection active ingredients and authorization. According to a study by the European Crop Protection Association, there were 70 new active ingredients in the development pipeline in 2000, while there were only 28 in 2012 (McDougall, 2013). Secondly, there has been a recent change in EU legislation regulating the use of pesticides from a risk-based assessment to "hazard-based cut-off criteria" (Hillocks, 2012). This is encapsulated in EU Plant Protection Regulation 1107/2009 and has led to active ingredients being lost from the market because of their intrinsic properties (for example, endocrine disruption), irrespective of levels of exposure or risk. Not only are farmers growing a narrower range of crops with a narrower genetic diversity, therefore, but also the range of crop protection products used within them has also reduced further, intensifying selection pressure, leading to the dominance of a few well-adapted pest, weed, and pathogen biotypes.

As well as driving the spectrum of pest and weed communities directly through the simplification of crop sequences, crop genetic material, and management, the loss of diversity in agroecosystems has also impacted the populations of natural enemies and predators that regulate their populations. These beneficial organisms rely on a diversity of seminatural habitats in addition to cropped fields and so respond to processes operating at the landscape scale. A recent meta-analysis of control of pests by natural predators concluded that pest control was 46% lower in homogeneous landscapes dominated by cultivated land (Rusch et al., 2016), and landscape diversity has also been shown to increase weed seed predation by carabids (Trichard et al., 2013). The post World War II loss of habitat diversity in agricultural landscapes through the removal of seminatural features has been identified as a fundamental underlying cause of recent declines of biodiversity associated with agroecosystems, a component of which delivers the important ecosystem service of regulating pest and weed populations (Benton et al., 2003; Tscharntke et al., 2005). Boundaries between fields, and their associated field margins, are a key landscape feature in

FIGURE 12.1 Examples of a pest, weed, and pathogen that have become a major problem because of simplified rotations, the development of pesticide resistance, and removal of crop protection products through legislation: (A) cabbage stem beetle (*Psylliodes chrysocephalus*) is a major pest of oilseed rape; it is increasingly a problem because of the banning of noenicitinoid seed treatments; (B) black-grass (*Alopecurus myosuroides*) is a competitive weed of winter cereals; some populations have now become almost impossible to control chemically because of evolved cross-resistance to multiple herbicides; (C) Septoria leaf blotch is the most important foliar disease of wheat in the United Kingdom. It is caused by *Zymoseptoria tritici* (previously known as *Mycosphaerella graminicola* and by the previous asexual stage name, *Septoria tritici*). Yield losses of 30%–50% have been reported in susceptible crop varieties, and it is able to rapidly evolve resistance to new fungicides.

this regard, as they are integrated within the crops and can be managed to support ecosystem service providers (Holland et al., 2014). Restoring these habitats in the farmed landscape, based on a knowledge of the ecologic requirements and behavior of pests and natural enemies, will complement increased diversity of management at the crop scale and be an important component of future sustainable crop protection strategies. In the following sections, the impact of the loss of agroecosystem diversity on pests, weeds, and pathogens is reviewed along with a discussion of potential solutions implemented at these multiple scales.

PESTS

Insect pests undermine agricultural productivity by feeding damage that causes crop yield loss, by spreading diseases, and by reducing the quality of crops, so their market value is reduced (Bruce et al., 2011; Oerke, 2006; Savary et al., 2000). They are currently a major concern because insecticide availability is declining due to increasingly stringent legislation coupled with evolution of insecticide resistance (Bruce, 2012; Hillocks, 2012). To give one specific example, there are now serious problems with cabbage stem flea beetle, *Psylliodes chrysocephalus*, in OSR (Fig. 12.1) associated with large-scale block cropping and restrictions on the neonicotinoid pesticides in Europe that were previously used to control it (Scott and Bilsborrow, 2015). There is a need, therefore, to reintroduce some of the mechanisms that reduce populations of herbivores in wild ecosystems into agricultural ecosystems and a scientific challenge to rediversify agricultural systems to make them less vulnerable to pests (Bruce, 2015). This not only involves recruiting the defense mechanisms that exist in nature into crops but also understanding how spatial and temporal heterogeneity may influence pest performance through the contrasting nutritional profiles of different plants. Wetzel et al. (2016) recently found that plant heterogeneity in nutrient levels contributed to the suppression of herbivore populations. They suggest that pest outbreaks are more likely in simplified agricultural environments with low plant species diversity and crops bred to minimize variation and that increasing heterogeneity in plant nutrients may help reduce risk of outbreaks.

Agricultural habitats are simplified with large areas of one plant species that provides ideal conditions for adapted pests to flourish. There are various opportunities to use agricultural biodiversity to make cropped environments more resilient to insect pest attack, and these fall into two broad categories. To reduce pest pressure, habitat diversification can make crops less apparent to colonizing insects, and to increase mortality of pests that have colonized the crop, additional biodiversity can make arable environments more conducive for the natural enemies of pests. The use of natural populations of insect predators and parasitoids as an "ecosystem service" is referred to as conservation biologic control. Losey and Vaughan (2006) estimated, for the United States, that the value of ecologic services provided by natural enemies of agricultural pests in controlling their populations is $57 billion per annum. For aphid management, conservation of natural enemies such as parasitoid wasps and carabid beetles can reduce the need for insecticide use. Flowering field margins and other areas of uncropped land containing wildflower mixtures attract parasitoids and provide cover for other predators (Holland et al., 2008). Avoidance of pesticide use when aphid populations are low can help to conserve natural enemies. The use of insect-resistant crops can dramatically reduce the need for insecticide application; for example, with the orange wheat blossom midge, *Sitodiplosis mosellana*, resistant wheat varieties do not need to be treated with insecticide (Bruce et al., 2011) and have formed part of an Integrated Pest Management program in which pheromone traps are also used to monitor pest populations in susceptible wheat varieties and minimize insecticide use (Bruce and Smart, 2009).

Many studies of insect-plant interactions tend to look at only one insect and only one plant, and these are often in a laboratory environment. There is a need for further studies that look at combinations of plants and insects. An example that shows what could be done is a system of companion planting known as "push-pull" that uses different species of plants grown with the main crop to deliver semiochemicals (compounds that manipulate insect behavior) in the field (Cook et al., 2007). This approach was developed for smallholder agriculture and has been used with much success in maize and sorghum in eastern Africa (Hassanali et al., 2008; Khan et al., 2010) against two main pest problems, stem or stalk borers and the African witchweed, Striga. Use of constitutive emission of relevant semiochemicals from plant sources provides an appropriate solution for smallholder African agriculture, where synthetic chemicals would be too costly and logistically difficult to deliver.

The system uses a combination of an intercrop that releases semiochemicals that are repellent to pests (push) and a trap crop grown around the edges of the cropped area that releases semiochemicals that are attractive to pests of the main crop (pull). Semiochemicals released by the intercrop also attract natural enemies of pests (Khan et al., 1997). Intercrops constitutively release volatiles such as (*E*)-ocimene and (E)-4,8-dimethyl-1,3,7-nonatriene that are typically released from maize when attacked by chewing herbivores. These act to repel pests but attract their natural enemies. The main semiochemicals released from trap crops are produced in much larger amounts than the main crops of maize or sorghum, at nightfall when stemborer moths are active (Chamberlain et al., 2006). For stem borer control alone, the intercrop is molasses grass, but for controlling the Striga weed, the intercrop is silverleaf or greenleaf, two cattle forage legumes of Desmodium species, which also repel stem borers. The mechanism of Striga control consists of release of root exudate allelochemicals from Desmodium that induce suicidal germination of Striga seeds. The trap crop is preferably Napier grass, but Sudan grass can also be used. Besides controlling stem borers and the Striga weed, all of these companion crops are valuable as cattle forage, thereby improving livestock holdings in addition to producing a sustainable cereal harvest protected against pests and weeds.

WEEDS

The drive toward simplified crop rotations, less spatiotemporal diversity in weed management, and increasingly homogenous agricultural landscapes has demonstrable consequences for weed communities and weed evolution (Neve et al., 2009). At the level of weed communities, several large-scale surveys of the impact of the simplification of cropping systems and landscapes on weed communities in three European countries (France, Germany, and the United Kingdom) have drawn similar conclusions (Fried et al., 2009; Meyer et al., 2013; Potts et al., 2009). Firstly, weed species at the intrafield scale (α diversity) have decreased because of use of herbicides and fertilizers and simpler crop rotations. But, secondly, there has been a shift from species specialized to the arable fields toward more generalist species that are able to colonize a more diverse range of habitats. Consequently, recruitment into fields from neighboring habitats has become more important in maintaining field-level weed diversity, and it may be argued that these generalist species have a greater intrinsic capacity to adapt to weed control strategies, particularly herbicide use. Sixty years ago, around the time when the first synthetic herbicides were being introduced for weed control, Harper (1956) commented on the potential for herbicide-based management to select for resistance in weed populations. In broader context, Harper also noted that weed species are "selected by the very practices that were originally designed to suppress them." The capacity for rapid evolution in weed (and other pest) populations is clearly recognized (Barrett, 1988; Clements et al., 2004; Jordan and Jannink, 1997; Vigueira et al., 2013), and there is a direct relationship between the diversity of cropping systems and the potential for rapid evolutionary responses in weed populations.

Herbicide resistance is one of the most palpable examples of the impacts of a loss of cropping system and crop protection diversity on crop protection. Harper's foresight is now emphatically demonstrated, and globally, cases of evolved resistance to herbicides have been reported in 250 plant species, across 87 countries (Heap, 2016). In many cropping systems, for example, corn-cotton-soybean systems in the southern United States (Ward et al., 2013) and wheat-based production systems in the United Kingdom (Cummins et al., 2013), the failure of weed control and the resulting impacts on productivity and sustainability are driven by widespread herbicide resistance in a single weed species (*Amaranthus palmeri* and *Alopecurus myosuroides* (Fig. 12.1), respectively). The causes of these epidemics of herbicide resistance are clear and simple: overreliance on herbicides, narrow crop rotations, a lack of diversity in herbicide use, and in the United States particularly, the widespread adoption of genetically modified glyphosate-tolerant crops. The solutions to resistance problems in agriculture are equally simple and rely on increasing cropping and management diversity, be that through more diverse weed management that integrates cultural, physical, genetic, and chemical control, or via more diverse crop rotations that implicitly impact weed population dynamics and the diversity of management options. However, while being simple in principle, global market forces, agricultural policies, and the desire for simplified and time-efficient weed management strategies continue to limit the uptake of integrated weed management.

The potential for diverse cropping patterns to deliver crop protection that is robust to pest adaptation is rooted in ecologic and evolutionary principles. Spatial and temporal heterogeneity in the cropped landscape provides diverse ecologic niches for weed recruitment and establishment (Navas, 2012). In the case of agricultural weeds, many of which have a persistent soil seed bank, rotation of crops with different phenological and morphologic characteristics will favor different species in different phases of the rotation, meaning that no single species is able to dominate the weed flora or the seed bank. Diversity in crop rotation will, in turn, facilitate more diverse weed management systems that reduce selection pressure for weed adaptation. Additionally, more heterogeneous agricultural landscapes will limit the potential for gene flow between adjacent weed populations, slowing the spread of novel adaptive traits, such as herbicide resistance. The prerequisites for rapid weed evolution are large populations of a single dominant species, fast generation times, concerted selection pressure, and the interconnectedness of populations that facilitates rapid gene flow (Neve et al., 2014). Diverse cropping systems are able to moderate all of these drivers by ensuring a more diverse weed flora, limiting population growth of dominant species, and diversifying selection pressures. In this way, the multifaceted goals of weed management, i.e., to limit weed competition, slow weed adaptation, and to maintain the ecosystem services provided by diverse weed communities, are synergistic and can be realized by the adoption of more diverse cropping and management approaches.

PLANT PATHOGENS

Plant diseases caused by phytopathogenic viruses, bacteria, fungi, protists, and nematode species are considered a major threat to current and future food security with annual estimated crop harvest losses worldwide of 10%–16% (Strange and Scott, 2005; Oerke, 2006). Crop diseases can substantially reduce yield, quality, and resource-use efficiency, as well as pose a significant risk to human and animal health via mycotoxin contamination of food and feed. As agriculture has intensified, the increases in field size and host plant densities combined with the high genetic uniformity within crop species have significantly increased the risk of devastating plant pathogens emerging. New approaches are required to increase environmental and genetic diversity in agricultural systems at field, farm, and landscape scales to make crops more resilient against biotic threats (Stukenbrock and McDonald, 2008).

Agroecosystem diversity can be improved by growing more diverse and complex crop rotations including double cropping, intercropping, and use of cover crops. Within crop species, cultivar mixes have shown considerable potential for fungal disease control (See Chapter 11). In rice mixtures the severity of blast disease was significant reduced, and yields increased in farmer's fields compared to cultivar monocultures (Zhu et al., 2000). The cultivar mixtures also supported a more diverse population of rice blast isolates that could restrict or slow down the breakdown of resistance genes within the cultivar mix. Rotations with nonhost crops can reduce pathogen inoculum pressure carried over into subsequent crops, particularly for soil-borne pathogens. However, success relies upon a detailed understanding of the host-pathogen interaction and lifecycle at multiple scales. For example, the management of soil-borne pathogens by crop rotation can be more difficult when the pathogen is able to produce resting structures that allow them to survive in the soil in the absence of a host crop. In the case of *Sclerotinia sclerotiorum*, which causes stem rot in OSR (*Brassica napus*), sclerotia can survive in the soil and remain infective for over 5 years (Derbyshire and Denton-Giles, 2016). Consequently, in Australia, it is recommended that OSR is not grown more frequently than 1 year in 4, but care also needs to be given to the other species grown in the rotation. In the case of *Sclerotinia* a wide variety of other vegetable crops are very susceptible and can increase the risk of disease when grown in rotation with OSR.

A complementary strategy for plant disease management is to increase the genetic diversity within crop species. Genetic diversity is inadequate in modern cultivated crops due to the genetic bottlenecks that developed during domestication as well as the abandonment of diverse landraces in favor of more genetically uniform, high-yielding cultivars suitable for high-input mechanized agricultural systems. In an attempt to rectify genetic deficiencies, in many cases, single locus plant disease resistance (*R*) gene(s) have been introgressed into cultivars and deployed against plant pathogens across large land areas (Wulff and Moscou, 2014). This exerts a very strong selection pressure on the target pathogen and often leads to major break downs in resistance, and severe pathogen outbreaks ensue. For example, in Uganda in 1998 the emergence of a new race (Ug99) of *Puccinia graminis* f. sp. *tritici*, causal agent of stem rust in wheat (*Triticum* spp.), overcame the last functional race-specific resistance gene *Sr31* present in the majority of widely grown wheat cultivars and breeding programs (Singh et al., 2015). The Ug99 race group has continued to evolve and spread geographically, making it a major threat to world wheat production. Major efforts are underway to identify new sources of resistance against stem rust including deploying multiple race-specific *R* genes using marker-assisted breeding in combination with the identification of slow-rusting phenotypes conferred by major quantitative trait loci to reduce the risk of sudden resistance breakdowns (Getie et al., 2016; Prins et al., 2016). Using GM technologies, scientists are also developing strategies to deliver tailored *R*-gene cassettes containing different combinations of *R* genes specific to the particular agricultural ecosystem and local pathogen races present (Wulff and Moscou, 2014; Ellis et al., 2014).

In the case of soil-borne diseases, variety rotations could be a valuable future management strategy to reduce pathogen pressure. Low inoculum building varieties could be utilized to restrict the build-up of pathogen inoculum in the soil and so lower the risk of severe disease to the following crop (McMillan et al., 2011; Vadakattu et al., 2015). However, under favorable environmental conditions, considerable inoculum could still build-up over one or more seasons. In the case of take-all root disease of wheat, caused by *Gaeumannomyces graminis* var. *tritici*, a 1-year break crop away from wheat or other susceptible cereal crops could then be used to decrease inoculum of the soil-borne fungus to negligible levels. Alternatively, in cereal-based rotations, other host species that are more resistant and/or tolerant to take-all disease, e.g., rye (*Secale cereale*), triticale (x *Triticosecale*), barley (*Hordeum vulgare*), or oats (*Avena sativa*) could be substituted for wheat with a high risk of disease (Hornby et al., 1998).

In addition to increasing genetic diversity, there is also a need to increase the diversity of chemistries and their management used to control pathogens to maximize their durability. In the case of Septoria leaf blotch, the most important foliar disease of winter wheat in the United Kingdom and Western Europe (Fig. 12.1), there are now cases of multidrug-resistant *Zymospetoria tritici* pathogen strains (Omrane et al., 2015). Further work is now required to understand the impact of this on fungicide management strategies. Earlier experimental and modelling evidence suggests that mixtures of fungicides with different modes of action can help to reduce the selection for fungicide resistance in pathogen populations, including *Z. tritici* (van den Bosch et al., 2014). Increased understanding of host-pathogen interactions and the gene pathways regulating pathogenicity should also provide opportunities to develop completely new chemistry classes with novel modes of action including the application of small interfering RNAs (Lucas, 2011; Koch et al., 2016).

CONCLUSION

Although pests, weeds, and pathogens have contrasting phylogeny and biology, the loss of diversity in agroecosystems has made the control of all three groups more difficult because of the generic principles outlined before. In response, we need to rediversify agricultural systems to weaken the selection pressure on specific traits or genotypes and so avoid the emergence of intractable pest, weed, or pathogen populations that have evolved to survive a specific management practice. Such an approach, termed "integrated pest (or weed) management" (IPM), while less convenient and potentially costing more in the short term than strategies that rely on pesticides alone, is the only route to sustainable crop protection (Barzman et al., 2015).

References

Barrett, S.C.H., 1988. Genetics and evolution of agricultural weeds. In: Altieri, M.A., Liebman, M. (Eds.), Weed management in Agroecosystems: Ecological Approaches. CRC Press, Boca Raton, FL, USA, pp. 57–76.

Barzman, M., Barberi, P., Birch, A.N.E., et al., 2015. Eight principles of integrated pest management. Agronomy for Sustainable Development 35, 1199–1215.

Benton, T.G., Vickery, J.A., Wilson, J.D., 2003. Farmland biodiversity: is habitat heterogeneity the key? Trends in Ecology and Evolution 18, 182–188.

Borlaug, N., 2007. Feeding a hungry world. Science 318, 359.

Bruce, T.J.A., 2012. GM as a route for delivery of sustainable crop protection. Journal of Experimental Botany 63, 537–541.

Bruce, T.J.A., 2015. Interplay between insects and plants: dynamic and complex interactions that have coevolved over millions of years but act in milliseconds. Journal of Experimental Botany 66, 455–465.

Bruce, T.J.A., Smart, L.E., 2009. Orange wheat blossom midge, Sitodiplosis mosellana, management. Outlooks in Pest Management 20, 89–92.

Bruce, T.J.A., Smart, L.E., Pickett, J.A., 2011. Wheat insect pests - a worldwide perspective. In: William, A., Alain, B., Maarten, V.G. (Eds.), The World Wheat Book: A History of Wheat Breeding. Lavoisier, Paris, pp. 1109–1130.

Chamberlain, K., Khan, Z.R., Pickett, J.A., Toshova, T., Wadhams, L.J., 2006. Diel periodicity in the production of green leaf volatiles by wild and cultivated host plants of stemborer moths, Chilo partellus and Busseola fusca. Journal of Chemical Ecology 32, 565–577.

Clements, D.R., Ditommaso, A., Jordan, N., et al., 2004. Adaptability of plants invading north American cropland. Agriculture, Ecosystems and Environment 104, 379–398.

Cook, S.M., Khan, Z.R., Pickett, J.A., 2007. The use of push-pull strategies in integrated pest management. Annual Review of Entomology. In: Annual Review of Entomology, vol. 52, pp. 375–400.

Cummins, I., Wortley, D.J., Sabbadin, F., et al., 2013. Key role for a glutathione transferase in multiple-herbicide resistance in grass weeds. Proceedings of the National Academy of Sciences of the United States of America 110, 5812–5817.

Derbyshire, M.C., Denton-Giles, M., 2016. The control of sclerotinia stem rot on oilseed rape (*Brassica napus*): current practices and future opportunities. Plant Pathology 65, 859–877.

Ellis, J.G., Lagudah, E.S., Spielmeyer, W., Dodds, P.N., 2014. The past, present and future of breeding rust resistant wheat. Frontiers in Plant Science 5.

Evenson, R.E., Gollin, D., 2003. Assessing the impact of the green revolution, 1960 to 2000. Science 300, 758–762.

Fried, G., Petit, S., Dessaint, F., Reboud, X., 2009. Arable weed decline in Northern France: crop edges as refugia for weed conservation? Biological Conservation 142, 238–243.

Getie, B., Singh, D., Bansal, U., Simmonds, J., Uauy, C., Park, R.F., 2016. Identification and mapping of resistance to stem rust in the European winter wheat cultivars Spark and Rialto. Molecular Breeding 36, 11.

Harper, J.L., 1956. The evolution of weeds in relation to resistance to herbicides. In: Proceedings of the 1956 British Weed Control Conference. British Crop Protection Council, pp. 179–188.

Hassanali, A., Herren, H., Khan, Z.R., Pickett, J.A., Woodcock, C.M., 2008. Integrated pest management: the push-pull approach for controlling insect pests and weeds of cereals, and its potential for other agricultural systems including animal husbandry. Philosophical Transactions of the Royal Society B-Biological Sciences 363, 611–621.

Heap, I.M., 2016. The International Survey of Herbicide Resistant Weeds.

Hillocks, R.J., 2012. Farming with fewer pesticides: EU pesticide review and resulting challenges for UK agriculture. Crop Protection 31, 85–93.

Holland, J.M., Oaten, H., Southway, S., Moreby, S., 2008. The effectiveness of field margin enhancement for cereal aphid control by different natural enemy guilds. Biological Control 47, 71–76.

Holland, J.M., Storkey, J., Lutman, P.J.W., Birkett, T., Simper, J., Aebischer, N.J., 2014. Utilisation of agri-environment scheme habitats to enhance invertebrate ecosystem service providers. Agriculture, Ecosystems and Environment 183, 103–109.

Hornby, D., Bateman, G.L., Gutteridge, R.J., et al., 1998. Take-All Disease of Cereals: A Regional Perspective. CAB International.

Jordan, N.R., Jannink, J.L., 1997. Assessing the practical importance of weed evolution: a research agenda. Weed Research 37, 237–246.

Khan, Z.R., Ampongnyarko, K., Chiliswa, P., et al., 1997. Intercropping increases parasitism of pests. Nature 388, 631–632.

Khan, Z.R., Midega, C.A., Bruce, T.J.A., Hooper, A.M., Pickett, J.A., 2010. Exploiting phytochemicals for developing a 'push-pull' crop protection strategy for cereal farmers in Africa. Journal of Experimental Botany 61, 4185–4196.

Koch, A., Biedenkopf, D., Furch, A., et al., 2016. An RNAi-based control of *Fusarium graminearum* infections through spraying of long dsRNAs involves a plant passage and is controlled by the fungal silencing machinery. PLoS Pathogens 12, e1005901.

Losey, J.E., Vaughan, M., 2006. The economic value of ecological services provided by insects. Bioscience 56, 311–323.

Lucas, J.A., 2011. Advances in plant disease and pest management. Journal of Agricultural Science 149, 91–114.

Mcdougall, P., 2013. R&D trends for chemical crop protection products and the position of the European market. A consultancy study undertaken for ECPA. In: European Crop Protection Association.

Mcmillan, V.E., Hammond-Kosack, K.E., Gutteridge, R.J., 2011. Evidence that wheat cultivars differ in their ability to build up inoculum of the take-all fungus, *Gaeumannomyces graminis* var. *tritici*, under a first wheat crop. Plant Pathology 60, 200–206.

Meyer, S., Wesche, K., Krause, B., Leuschner, C., 2013. Dramatic losses of specialist arable plants in Central Germany since the 1950s/60s - a cross-regional analysis. Diversity and Distributions 19, 1175–1187.

Navas, M.-L., 2012. Trait-based approaches to unravelling the assembly of weed communities and their impact on agro-ecosystem functioning. Weed Research 52, 479–488.

Neve, P., Busi, R., Renton, M., Vila-Aiub, M.M., 2014. Expanding the eco-evolutionary context of herbicide resistance research. Pest Management Science 70, 1385–1393.

Neve, P., Vila-Aiub, M., Roux, F., 2009. Evolutionary-thinking in agricultural weed management. New Phytologist 184, 783–793.

Oerke, E.C., 2006. Crop losses to pests. Journal of Agricultural Science 144, 31–43.

Omrane, S., Sghyer, H., Audeon, C., et al., 2015. Fungicide efflux and the MgMFS1 transporter contribute to the multidrug resistance phenotype in Zymoseptoria tritici field isolates. Environmental Microbiology 17, 2805–2823.

Potts, G.R., Ewald, J.A., Aebischer, N.J., 2009. Long-term changes in the flora of the cereal ecosystem on the Sussex Downs, England, focusing on the years 1968-2005. Journal of Applied Ecology 47, 215–226.

Prins, R., Dreisigacker, S., Pretorius, Z., et al., 2016. Stem rust resistance in a geographically diverse collection of spring wheat lines collected from across Africa. Frontiers in Plant Science 7, 15.

Rusch, A., Chaplin-Kramer, R., Gardiner, M.M., et al., 2016. Agricultural landscape simplification reduces natural pest control: a quantitative synthesis. Agriculture, Ecosystems and Environment 221, 198–204.

Savary, S., Willocquet, L., Elazegui, F.A., Castilla, N.P., Teng, P.S., 2000. Rice pest constraints in tropical Asia: quantification of yield losses due to rice pests in a range of production situations. Plant Disease 84, 357–369.

Scott, C., Bilsborrow, P., 2015. An Interim Impact Assessment of the Neonicitinoid Seed Treatment Ban on Oilseed Rape Production in England. Newcastle University, UK. Rural Business REsearch.

Singh, R.P., Hodson, D.P., Jin, Y., et al., 2015. Emergence and spread of new races of wheat stem rust fungus: continued threat to food security and prospects of genetic control. Phytopathology 105, 872–884.

Stoate, C., Baldi, A., Beja, P., et al., 2009. Ecological impacts of early 21st century agricultural change in Europe - a review. Journal of Environmental Management 91, 22–46.

Strange, R.N., Scott, P.R., 2005. Plant disease: a threat to global food security. Annual Review of Phytopathology 43, 83–116.

Stukenbrock, E.H., Mcdonald, B.A., 2008. The origins of plant pathogens in agro-ecosystems. Annual Review of Phytopathology 46, 75–100.

Trichard, A., Alignier, A., Biju-Duval, L., Petit, S., 2013. The relative effects of local management and landscape context on weed seed predation and carabid functional groups. Basic and Applied Ecology 14, 235–245.

Tscharntke, T., Klein, A.M., Kruess, A., Steffan-Dewenter, I., Thies, C., 2005. Landscape perspectives on agricultural intensification and biodiversity - ecosystem service management. Ecology Letters 8, 857–874.

Vadakattu, G., Mckay, A., Ophel-Keller, K., Kirkegaard, J., Wilhelm, N., Roget, D., 2015. Root growth or resistance: Rhizoctonia solani AG8 Inoculum Buildup Varies with Crop Type. In: ISPR 6: Roots Down under. Australia International Society of Root Research 13, Canberra. ISRR9 Oral Abstracts.

Van Den Bosch, F., Paveley, N., Van Den Berg, F., Hobbelen, P., Oliver, R., 2014. Mixtures as a fungicide resistance management tactic. Phytopathology 104, 1264–1273.

Van Ittersum, M.K., Cassman, K.G., Grassini, P., Wolf, J., Tittonell, P., Hochman, Z., 2013. Yield gap analysis with local to global relevance-A review. Field Crops Research 143, 4–17.

Vigueira, C.C., Olsen, K.M., Caicedo, A.L., 2013. The red queen in the corn: agricultural weeds as models of rapid adaptive evolution. Heredity 110, 303–311.

Ward, S.M., Webster, T.M., Steckel, L.E., 2013. Palmer amaranth (Amaranthus palmeri): a review. Weed Technology 27, 12–27.

Wetzel, W.C., Kharouba, H.M., Robinson, M., Holyoak, M., Karban, R., 2016. Variability in plant nutrients reduces insect herbivore performance. Nature Advance. Online publication.

Wulff, B.B.H., Moscou, M.J., 2014. Strategies for transferring resistance into wheat: from wide crosses to GM cassettes. Frontiers in Plant Science 5.

Zhu, Y.Y., Chen, H.R., Fan, J.H., et al., 2000. Genetic diversity and disease control in rice. Nature 406, 718–722.

CHAPTER

13

Grassland Functional Diversity and Management for Enhancing Ecosystem Services and Reducing Environmental Impacts: A Cross-Scale Analysis

Michel Duru[1], *L.D.A.S. Pontes*[2], *J. Schellberg*[3], *J.P. Theau*[1], *O. Therond*[4]

[1]AGIR, INRA, Université de Toulouse, Auzeville, France; [2]IAPAR - Agronomic Institute of Paraná, Ponta Grossa-PR, Brazil; [3]Institute of Crop Science and Resource Conservation, University of Bonn, Bonn, Germany; [4]UMR LAE, INRA, Université de Lorraine, Colmar, France

INTRODUCTION

Focusing on relations between ecosystem services (ES) and agriculture, Zhang et al. (2007) highlighted that agroecosystems provide ES both for the whole of society and more particularly for farmers themselves. Agricultural management practices can promote many ES, mainly through increasing biodiversity from field to landscape levels (Duru et al., 2015a), but they also influence the potential for negative environmental impacts, sometimes considered "disservices" from agriculture, such as a decrease in habitat quality for biodiversity conservation as well as nitrogen (N) emissions, nutrient runoff, sedimentation of waterways, and pesticide poisoning of humans and nontarget species (Power, 2010).

Grassland ecosystems range from rangelands, seminatural and permanent grasslands to temporary sown grasslands (grass-legume mixtures, pure stands). They cover most agricultural land in less agriculturally favorable regions, but they form a relatively small component of the landscape mosaic in more agriculturally favorable regions. In the European Union, permanent and temporary grasslands covered around 57 and 10 million ha, respectively (https://ec.europa.eu/eurostat/fr/web/conferences/conf-2007). Grasslands are agricultural ecosystems that provide a low-cost source of forage for ruminants, especially when grazed.

Agroecosystem Diversity
https://doi.org/10.1016/B978-0-12-811050-8.00013-3

They are increasingly recognized for providing multiple ES, including pollination, pest control, soil conservation, resistance to weed invasion, regulation of soil fertility, and nutrient cycling, often at levels greater than those of arable crops (Cameron et al., 2013, and see chapters in Sections I and II in this volume). Although effects of grassland management practices on ES are increasingly documented, several limitations remain in implementing suitable management practices in a given context and in developing relevant agricultural and environmental policies.

- First, grassland ES and impacts are most often considered in a global manner, without indicating the scale(s) at which they are detectable, their beneficiaries, or the people harmed by impacts (Fischer et al., 2015).
- Second, the relative importance of ES varies greatly according to grassland composition, management practices, place within crop rotations, and locations within the landscape mosaic.
- Third, the farm level is rarely considered, even though it is the relevant level for management decisions.

To overcome these limitations, we developed an analytical framework linking socioecologic systems with functional ecology. Socioecologic systems are complex adaptive systems characterized by feedback across multiple interlinked scales that amplify or dampen changes. This framework provides a powerful analytical tool for understanding interlinked dynamics of environmental and societal changes. It also characterizes interactions between people and ecosystems that increase in scale, scope, and intensity (Fischer et al., 2015). Functional ecology provides conceptual and methodological frameworks to analyze relations between biodiversity, ecologic processes and function, and ES. The ability of an ecosystem to provide multiple ES depends on the state of key abiotic and biotic ecosystem properties, including organisms or groups of organisms (de Bello et al., 2010). To increase ES, it is necessary to identify the functional influence of plant strategies on resource acquisition and use (Lavorel and Grigulis, 2012) and of particular plant groups as N fixers (Spehn et al., 2000).

The next section describes the analytical frameworks: (1) a structured multiscale and multidomain socioecologic framework to analyze effects of management practices on the ES that grasslands provide and (2) a method based on plant growth strategies to characterize ES provided by permanent and temporary grasslands. Then, we examine relations between ES provided by grassland communities according to their management and briefly illustrate application to a case study. Finally, we discuss the relevance of functional diversity, the need to perform cross-scale analysis, and opportunities for improving ES.

HOW TO ANALYZE GRASSLAND DIVERSITY, MANAGEMENT, SERVICES, AND IMPACTS?

A Structured Multiscale and Multidomain Framework to Analyze ES Provided by Grasslands

ES Provided by Grassland Composition and Spatiotemporal Distribution

Some grassland ES depend mainly on grassland plant composition, whereas others depend strongly on the spatiotemporal distribution of grasslands. Among the nine ES and goods identified that grasslands provide, seven depend mostly on composition, and four depend mostly on grassland location within the landscape or crop rotation (Table 13.1).

Grassland plant composition determines ecosystem properties, i.e., the biophysical state, structure, and functioning of ecosystems. For example, grassland composition of multiyear grasslands strongly determines the levels of pollination (Delaney et al., 2015),

TABLE 13.1 Overview of Ecosystem Services (ES) Provided by Grasslands According Their Place in the Landscape and Crop Rotation, and Their Composition

		Grasslands Within Landscape (Composition and Configuration)	Grasslands Within Insertion in Crop Rotation	Functional Composition of Grasslands
Production	Forage production, flexibility in management	NE	+	+++
Input ES	Biologic regulations	+++	++	+
	Soil fertility (nutrients and structure regulations)	NE	+++	++
	Soil stability (erosion control)	+++	+++	(+)
	Pollination	+++	NE	+++
Other ES	Material and liquid flow regulations	+++	++	(+)
	Water quality regulations	+++	++	+
	Carbon sequestration	NE	+	+ –> +++
	Cultural value (aesthetic)	+++	+	++

NE, no effect; from high (+++) to light effect (x).

carbon (C) sequestration, and regulation of the N cycle (Soussana and Lemaire, 2014). An increase in species functional diversity benefits ES, as well as herbage production, seasonality, and interannual forage stability (Duru et al., 2013). Some grassland characteristics play an important role within food systems, such as species-rich grasslands that contain polyphenols found in dairy products (Duru et al., 2017). Furthermore, certain species-rich grasslands contain secondary metabolites with anthelmintic properties (e.g., tannins) that help regulate animal parasites, thus decreasing the use of commercial anthelmintics for animal health (Hoste et al., 2006). Finally, species-rich grasslands are a source of forage that may potentially decrease methane emissions in comparison with monospecific grasslands (Hammond et al., 2014).

Both grassland composition and spatial distribution contribute to the aesthetic value of landscapes, which generally increases as species richness and landscape heterogeneity increase (Lindemann-Matthies et al., 2010). As a component of the landscape mosaic, grasslands influence natural biologic control (pest and disease regulation services) and flows of matter and nutrients in watersheds (water-related services). The temporal distribution of grasslands in a crop rotation determines C sequestration, soil stabilization, and soil fertility. Both grassland temporal and spatial distribution determine the level of ES they provide to crops in a landscape; these ES to agriculture are called "input services" (Duru et al., 2015a).

Finally, grasslands, in themselves, enhance the quality of animal products (goods) because they provide feed richer in omega-3 fatty acids than cereal-based rations do (Pighin et al., 2016). This is important because people in Western countries receive only half of the recommended amount of omega-3 fatty acids (Molendi-Coste et al., 2011).

Ecosystem Services in Socioecologic Systems: From Crop and Livestock Systems to Landscapes, Food Systems, and the Planet

ES interact and generate trade-offs and synergies across levels according to the biophysical context (Anderson, 2015). Most services that grasslands provide as well as their impacts are controlled by farm management practices but are detectable at different levels (Duru et al., 2015b), such as the animal (e.g., health regulation), field (e.g., soil fertility), farm (e.g., biologic regulation), landscape (e.g., biologic, matter flow, and air quality regulations; N emissions), or planet (e.g., C sequestration). Since they determine the biophysical state of the human environment and the quality of agricultural products (e.g., composition of animal products), they also impact human health. Their impacts are detectable for different social groups at different places. Further, through "cascade" phenomena, effects at one level may have consequences at lower or higher levels (Fig. 13.1).

Most research attempting to analyze relations among management practices, plant community functional composition, ecologic processes, and impacts (e.g., N emissions) or services (C sequestration) is performed at the field and landscape levels (Benton et al., 2003). However, the level at which one should analyze trade-offs and synergies among ES (and impacts) depends on beneficiaries of the ES (e.g., farmers, society) (Fisher et al., 2009). Studying ES supply levels, trade-offs, and synergies in farming systems raises several issues. If considered at all, only the entire farm is studied (Smukler et al., 2010), although sublevels (land management units) are key for understanding ES (Rawnsley et al., 2013). Farmers combine fields of different forage crops into several groups that are used to feed

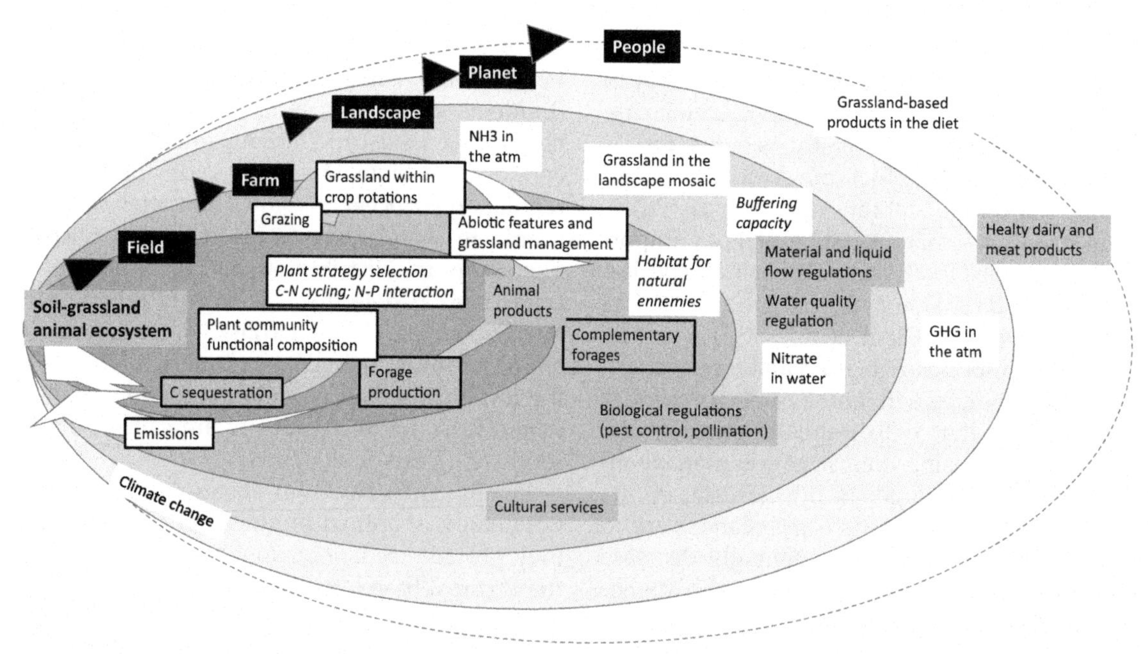

FIGURE 13.1 A multilevel framework (from field to people) for analyzing goods and ecosystem services (ES) provided by grasslands (and animal products) and impacts of management that are interconnected by gas fluxes, particles, organisms, and nutrients (*oval lines*), flowing within and between levels via upscaling and downscaling (only components with in a *black frame* were studied); ecologic conditions (*blue*); ES of goods (*green*); environmental impacts (*red*); *GHG*, greenhouse gas.

particular herds. The number and nature of these groups are designed to meet objectives, such as to increase self-sufficiency of the system (i.e., the ratio of forage production to consumption), reduce operational costs, or increase flexibility in work organization. Hence, it is necessary to investigate the farm at the level of such field groups. Most research conducted at the farm level is limited to the quantity and quality of forage that grasslands produce at an annual scale. Key challenges for farmers are seasonality of herbage production and insufficient complementarities among forages, especially in grazing-based feeding systems (Duru et al., 2015b). For example, Lugnot and Martin (2013) highlight that functional plant diversity in grasslands provides flexibility in the timing of grassland use and provides a "management service," since it helps farmers manage herbage growth. This is an example of an ES that is seldom identified, even though farmers promote it. Similarly, Nozière et al. (2011) suggest that within-farm plant diversity decreases production costs by better matching grassland types to animal nutritional requirements, which improves productivity of the livestock system.

Ecosystem Services and Plant Growth Strategies: Characterizing Grassland Functional Composition and Diversity

A Response and Effect Trait Framework for Linking Resources and Management Practices to Ecosystem Properties and Services

Ecosystem properties that provide ES depend largely on plant functional diversity (presence, abundance, and divergence of specific functional groups or traits) rather than on species diversity (Hooper et al., 2005). Plant functional diversity depends on plant functional traits, which are biologic characteristics with demonstrable links to a plant's functioning. Functional traits determine a plant's response to environmental and management factors (response trait) or its effects on ecosystem processes or services (effect trait) (Lavorel and Garnier, 2002). Functional groups are defined as groups of species that share common functional traits; they either respond in a similar manner to biotic and abiotic selective filters or exert similar effects on ecosystem functioning. Definition of groups depends greatly on the response or effect traits considered. The function of symbiotic N fixation often separates grasses and legumes, but boundaries are less clear-cut within each of these groups, mainly because species lie along a continuous gradient for most functional traits (Pontes et al., 2007). There is a growing consensus that the functional approach can help humans manage several of the ES that agricultural ecosystems provide (Duru et al., 2015a; Wood et al., 2015).

In response to management and abiotic and biotic environmental factors, species with different combinations of functional traits are most likely to coexist in the long term. It is often considered that ecosystem properties can be analyzed through community-weighted mean plant traits and the divergence and complementarity of functional groups. At least two main forms of complementarity exist between species (Loreau and de Mazencourt, 2013). Functional complementarity is a nontemporal form that captures short-term effects of species interactions on resource partitioning (e.g., species differ in resource uptake in space or chemical form) and facilitation, which allows for community overyielding. Temporal complementarity is use of the same resource by different species at different times, especially when species have asynchronous growth patterns (Husse et al., 2016).

In grasslands, soil resource availability, climate, and disturbance (grazing, cutting) can control ecosystem functioning either directly or indirectly via shifts in functional group composition (Lavorel et al., 2010). In permanent grasslands with constant management, plant and community traits represent responses to biotic and abiotic factors and determine the ecosystem functioning underpinning ES. In temporary grasslands, especially those of short duration, response and effect traits are strongly determined by the species and cultivars sown and later by the plants' biologic plasticity in response to biotic and abiotic environmental factors and management (Fig. 13.2). Sown and natural species can have an impact on soil biodiversity, with possible feedback on plant community functioning (Crotty et al., 2015).

Plant Growth Strategies and Plant Functional Types

Since soil resource availability or frequency and intensity of management are the two main factors that directly or indirectly control plant functional composition, two main functional plant strategies are traditionally considered. First, a plant strategy regarding resource acquisition and use distinguishes exploitative versus conservative plant types (or fast vs. slow growth strategy). They differ in leaf traits: thin, short-lived leaves with high photosynthetic rates for

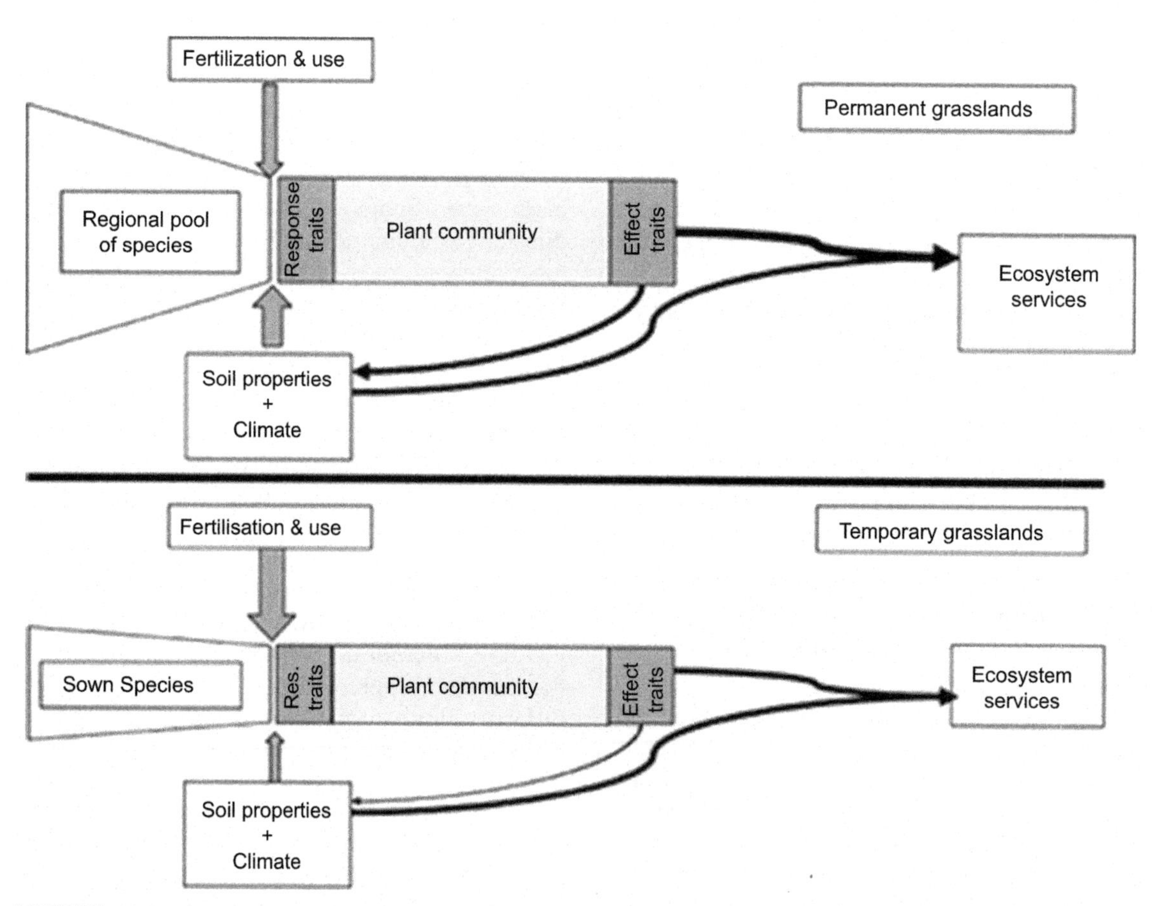

FIGURE 13.2 Conceptual framework for analyzing direct and indirect (via plant community functional composition) effects of management and soil properties on ecosystem properties underlying ecosystem services for permanent grasslands and temporary grasslands. Thickness of the *arrows* represents the intensity of the effect.

the former versus thicker, more fibrous, longer-lived leaves with lower photosynthetic rates for the latter (Wilson et al., 1999). The second strategy is related to plant growth-height trajectories, which captures plant competitive ability in response to disturbance. This has a direct relation to flowering phenology and early versus late grassland species (Sun and Frelich, 2011). According to these two strategies, grassland communities can be described by their community-weighted mean plant traits and their divergence regarding (1) resource capture and use capacity (fast vs. slow growth strategy) and (2) phenology of plant growth pattern (early vs. late). These two dimensions are rarely related to each other in permanent grasslands (Duru et al., 2013). Temporary grasslands based on species mixtures should aim to combine species and genotypes with contrasting plant heights, phenologies, and even resource capture, to increase functional and temporal complementarities (Huyghe et al., 2012). This clearly holds true for grass and legume mixtures, and even mixtures with dicots such as chicory. In improved grasslands, management should aim to develop or maintain multiple functional groups of species to increase temporal complementarity.

Characterization of Plant Growth Strategies

Several researchers highlight the need for a simple methodological framework for ES indicators that is useful for stakeholders (van Oudenhoven et al., 2012). Thus, we first developed an operational approach that groups grassland species into key plant functional types (PFTs). These groups distinguish species according to the first dimension of the grass resource capture and use strategy (Duru et al., 2013) extended to include legumes (Table 13.2).

Species are also distinguished according their plant growth pattern strategy. We

TABLE 13.2 Classification of Grass and Legume Species in Plant Functional Types (PFTs), Based on Leaf Dry Matter Content (LDMC in mg/g), for Acquisition and Use of Resources. Grass and Legume Species on the Same Line are Frequently Associated Within (Semi)Natural Grasslands

Grass Functional Type (GFT)		Legume Functional Type		PFT Scoring[a]
Examples	**LDMC**	**Examples (b)**	**LDMC (b)**	
E (*Lolium multiflorum*)	210			1
A (*Lolium perenne*)	230	White clover	205	
B (*Dactylis glomerata*)	260	Red clover, alfalfa	240	
b (*Agrostis capillaris*)	270			2
C (*Festuca rubra*)	290	Birdsfoot	260	
D (*Molinia caerulea*)	370			

[a] *For characterizing plant community-weighted PFT.*

Grass functional types are labeled from A to E: Adapted from Duru, M., Jouany, C., Le Roux, X., Navas, M.L., Cruz, P. 2013. From a conceptual framework to an operational approach for managing grassland functional diversity to obtain targeted ecosystem services: case studies from French mountains. Renewable Agriculture and Food Systems, 1–16; b, Ansquer, Unpublished; After Duru, M., Jouany, C., Le Roux, X., Navas, M.L., Cruz, P. 2013. From a conceptual framework to an operational approach for managing grassland functional diversity to obtain targeted ecosystem services: case studies from French mountains. Renewable Agriculture and Food Systems, 1–16.

extended the criteria beyond plant phenology (i.e., flowering time). We first considered the characteristic of the legume functional group: higher temperature thresholds for growth (5–25°C for the main legumes instead of 0–20°C for the main grasses). In addition, several legumes (and certain forbs) develop a deep root system that renders them less sensitive to water deficit than grasses. Consequently, the second strategy is a composite dimension based on plant phenology (early vs. late), plant group (grass vs. legume), and root system architecture (shallow vs. deep) (Table 13.3). Finally, we grouped species into functional types to simplify analysis of plant strategies at the field level, with functional composition of grasslands being described using the abundances of their functional types.

Plant functional types can be mapped according to these data (Tables 13.2 and 13.3) to place them within the two dimensions of plant growth strategy (Fig. 13.3).

Seminatural grasslands are composed of one or more of the six defined grass functional types (GFTs) (Tables 13.2 and 13.3). In Europe, typical multispecific temporary grasslands are a mixture of *Lolium perenne* (GFT-A) with *Trifolium repens, Dactylis glomerata,* or *Festuca arundinacea* (GFT-B) with *Medicago sativa* or *Trifolium pratense,* and more complex mixtures such as the typical Swiss mixture of *Festuca rubra* (GFT-C), *M. sativa,* and *Lotus corniculatus.*

TABLE 13.3 Classification of Plant Types by Their Growth Pattern According to Three criteria: (1) Duru et al. (2013); (2) Cruz, pers. comm.; (3) The First Author

	Flowering Time (degree-days)	Rating for Flowering Time (c)	Temperature Threshold (c)	Root System (c)	Sum[a]: PFT Scoring
Very late grasses (GFT-D)	2000a	3	1	1	5
Late grasses (GFT-B)	1700a	2.5	1	1	4.5
Early grasses (GFT-C)	1400a	2	1	1	4
Early grasses (GFT-B)	1250a	1.5	1	1	3.5
Early grasses (GFT-A)	850a	1	1	1	3
Very early grasses (GFT-E)	600a	0.5	1	1	2.5
Early flowering legumes with shallow root system	600b	0.5	2	1	3.5
Early flowering legumes with intermediate root system	600b	0.5	2	1.5	4
Late flowering legumes with deep root system	1100b	1.5	2	2	5.5
Early forbs	600b	0.5	1	1	2.5

GFT, grass functional type. Sum is the index of plant growth pattern.
[a] *Rating for characterizing plant community-weighted plant functional type (PFT).*

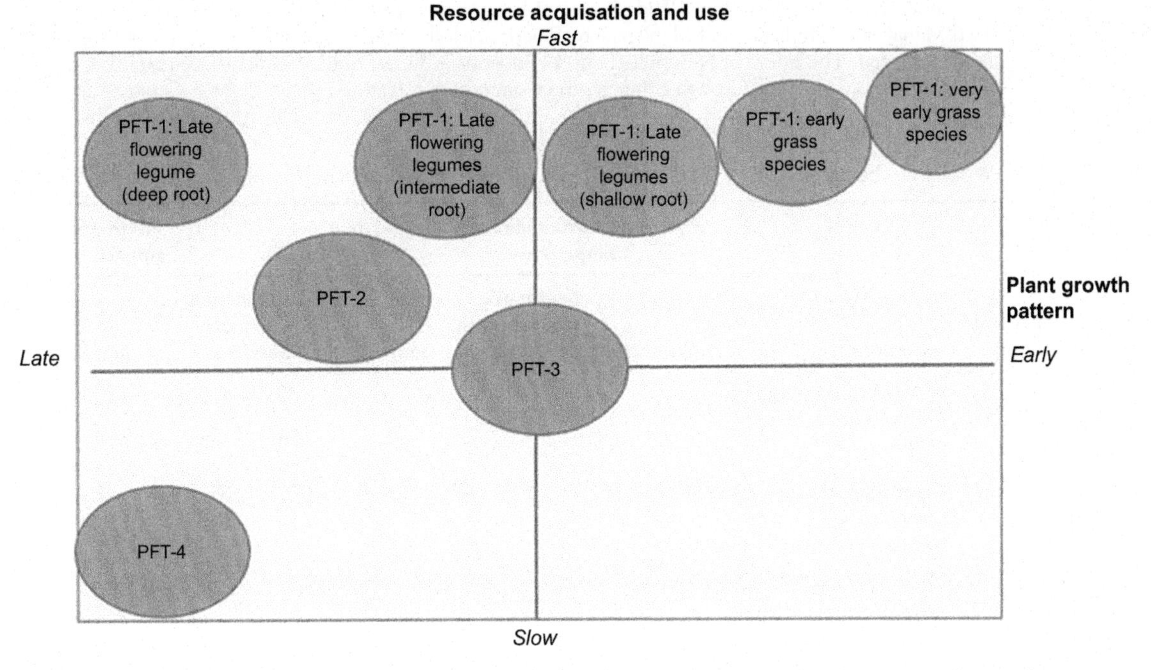

FIGURE 13.3 Ranking plant functional groups and types (PFT) for the two dimensions of plant growth strategies: resource acquisition and use (from slow to fast, see Table 13.2) and plant growth pattern in relation to phenology (early vs. late, see Table 13.3).

CHARACTERISTICS OF GOODS, ECOSYSTEM SERVICES, AND ENVIRONMENTAL IMPACTS ACCORDING TO GRASSLAND COMMUNITY TYPE AND ITS MANAGEMENT

Main Lessons From Functional Ecology

Abiotic factors (e.g., climate, fertilization, cutting, grazing) can have direct impacts on soil properties (e.g., water content, soil cover) and, consequently, on ES (e.g., soil fertility). As explained earlier, they also determine plant community functional composition, its functioning, and thus indirectly ES. An analysis of 14 ES from 150 seminatural grasslands in Germany demonstrated that indirect land use effects, i.e., those mediated by changes in functional composition, were generally as strong as direct effects (Allan et al., 2015).

Subsequently, we analyze relations between grassland functional composition, characteristics of goods (forage), and ES levels (Table 13.4). Consistent with ecologic research (e.g., Lavorel et al., 2010), grassland functional composition is described by the community-weighted mean PFT and an index of the diversity of PFT for the two dimensions of plant growth strategy. Community-weighted mean PFT and the diversity index are calculated as $\left(1 \Big/ n \sum_{1}^{n} Pi\right)$ and $\left(1 - \sum_{1}^{n} Pi2\right)$, respectively, where n is the number of PFT (Tables 13.2 and 13.3, last column) and Pi is the proportion of PFTi.

TABLE 13.4 Indicators (Rows) for Assessing a Wide Range of Ecosystem Services (ES) and Impacts (Columns). Two Management Indicators of Management Practices Modulating the Level of ES or Impacts Are Also Indicated. The Effect of N Supplied on N Emissions Is Threshold Dependent (see text). For the Two Dimensions of Plant Growth Strategy, the Community-Weighted Mean Plant Functional Type (PFT) and The Diversity Index for PFT Are Calculated as $1 / n \sum_{1}^{n} Pi = 1 - \sum_{1}^{n} Pi2$, Respectively, Where n Is the Number of PFT (Tables 13.2 and 13.3, Last Column) and Pi is the Proportion of PFTi

		Forage Production Components		Other ES		Environmental Impact
Type of Grassland Characterization	**Indicators**	**Production**	**Seasonality (Distribution of Biomass)**	**C Sequestration**	**Pollination Cultural Services**	**N Emissions**
Functional composition	Weighted plant strategy for resources capture (PFT-1)			x		
	Diversity index for functional complementarity (%PFT-1)	x				x
	Weighted plant strategy for asynchrony in growth		x			
	Diversity index for plant temporal complementarities	x	x	x		x
	Legumes content	x		x	(x)	x
	Forbs content	(x)			x	
Management	N supplied			x		
	Defoliation	x		(x)	(x)	x

Plant Functional Types and Characteristics of Forage Production: Productivity, Stability, Resilience, Flexibility, Seasonality, and Quality

Herbage yield depends mainly on N availability and on the degree of asynchrony between functional types, which determines distribution of herbage yield throughout the year (Fig. 13.3). Forbs (e.g., chicory, plantain) can increase herbage yield by evening out herbage growth during the growing season due to differences in their phenologies (or rooting depths, for summer growth) compared to those of grasses (Dhamala et al., 2014).

For seminatural grasslands, studies have shown that the percentage of grass functional types following a fast growth strategy helps to predict herbage yield at the field level or the potential stocking rate at the farm level in a given area (Duru et al., 2013). It was also shown that grasslands with high functional diversity that receive moderate amounts of N fertilizers yield more dry matter than those with less functional diversity (Duru et al., 2015b). Here, functional complementarity among species results in an "overyielding" effect. In addition, herbage digestibility and N content of seminatural grasslands can be predicted by their plant functional composition: those composed of species with a fast growth strategy have higher quality at the leafy stage (Duru et al., 2008). Grasslands receiving large amounts of N are dominated by grass functional types with a fast growth strategy (Duru et al., 2010), which weakens the stability of their production around the peak of herbage by decreasing species asynchrony (Zhang et al., 2016). Overall, dry matter yield and quality of functionally diversified grasslands vary little as a function of harvest date around the peak of herbage. In this way, they allow greater flexibility in the use of their herbage. Herbage growth pattern across seasons depends greatly on the influence of grass functional types with a late growth strategy (Michaud et al., 2012), which is generally negatively correlated with those with a fast growth strategy (Fig. 13.3) (Duru et al., 2013). Forbs often strengthen this effect due to their temporal complementarity (van Ruijven and Berendse, 2003). Many researchers found that regardless of grassland type, production of diversified grasslands remains more stable over several years due to their greater functional diversity (e.g., Sanderson, 2010). In extensive seminatural grasslands, communities tended to be dominated by slow-growing and slow-reproducing plants that are adapted to low nutrient conditions; they are identified as "stress tolerators" in the Grime competitive-stress-ruderal classification (Morecroft et al., 2016).

Carbon Sequestration

According to the most recent results, in general, grasslands in temperate zones are a sink for carbon, storing an average of 0.7 ± 0.1 t C/ha per year (Soussana and Lemaire, 2014). For permanent grasslands, the amount of C sequestered depends on several factors (climate, soil properties, grassland age, etc.). For any given location, we found that C sequestration is highest when functional diversity is high (Duru et al., 2013, 2015b). For temporary grasslands, legumes sequester more organic C in the soil than pure grass covers. This emphasizes the importance of legumes and their symbiotic N fixation for coupling C and N cycles and delivering the N needed to sequester C in soil organic matter (Lüscher et al., 2014). Temporary grasslands are considered to have a sequestration potential that is intermediate between those of permanent grassland and crops. An increase in the duration of temporary grasslands increases C sequestration. Restoration of grassland on soils previously cropped with annual species did not significantly change soil N or C characteristics within 24 months. Changes began to appear after 36 months (Attard et al., 2016). This time lag corresponded to the time that grasses needed to complete establishment of root systems.

Nitrogen Losses

Once production is maximized in a system, all further N inputs are lost to the environment through volatilization, denitrification, runoff, and leaching (Vertès et al., 2009). However, great differences occur among grasslands according to their management and composition. A synthesis of many experiments shows that N losses (nitrate) increase with the amount of N supplied (regardless of its origin: inorganic, symbiotic, urine, or manure), first slowly and then strongly for N supplied in amounts greater than 200 kg/ha for grazed grasslands and 400 kg/ha for cut grasslands (Vertès et al., 2009). Legume-based grasslands have the potential to reduce negative effects of nitrate and nitrous oxide (N_2O) emissions from livestock systems on the environment (Lüscher et al., 2014).

Pollination, Animal and Human Health, and Cultural Services

Flower-rich grasslands, like many seminatural grasslands, most often composed of b-GFT (Table 13.2) (Duru et al., 2013), are essential to sustain the abundance and diversity of insect pollinators in an intensively farmed agricultural landscape (Öckinger and Smith, 2007). These grasslands provide food and habitat for pollinators. According to Orford et al. (2016), spillover of pollinators from grasslands to surrounding habitats could improve pollination at the landscape scale. These authors found that modest increases in plant species richness in grasslands, mainly of forbs and legumes, are associated with significantly improved pollination services and potentially increased yields of pollination-dependent crops and wildflower reproduction in adjacent habitats.

Some specific effects of species are observed on animal health and animal products. For instance, legume species rich in tannins, especially *Onobrychis* spp. in sown grasslands, allow farmers to reduce or even avoid use of synthetic anthelmintics to fight ruminant gastrointestinal nematodes (Hoste et al., 2006). This helps circumvent development of resistance, but mainly prevents drug residues from appearing in manure (Peysson and Vulliet, 2013) and milk, in which it can impact human health (Tsiboukis et al., 2013). Species-rich grasslands, especially those containing forbs, have secondary plant metabolites that can influence rumen lipid metabolism in cows and thus affect the nutritional quality of dairy products and the flavor of cheese (Farrugia et al., 2014). These positive effects contribute to the heritage value of grasslands. As in other features with cultural value, the aesthetic side of grasslands is associated with biodiversity (Allan et al., 2015), especially grasslands containing many flowering species (composed of b-GFT-b).

Characterizing Trade-Offs and Synergies at the Field Level

Trade-offs among multiple ES (see Bennett et al., 2009 for a review) in grasslands are based on management intensity and grassland functional composition, and occur at different levels, from field to landscape (Fig. 13.4). The

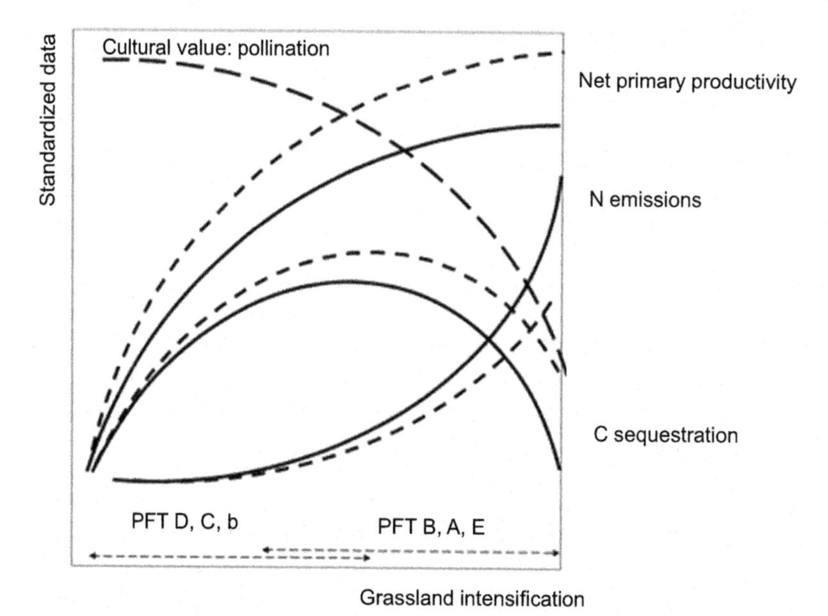

FIGURE 13.4 Effects of grassland intensification by N fertilizer application on net primary productivity, soil C sequestration, and N emissions. Responses are standardized to one for an un-intensified control grassland system prior to intensification. Response curves with low (*solid line*) and high (*dotted line*) legume content and functional diversity (inspired by Soussana and Lemaire, 2014); relations between grassland intensification and grass functional type (GFT in Table 13.2) are drawn at the bottom.

cross-scale analysis of studies presented in the previous section shows several trade-offs between ES, but some possible synergies exist and can be promoted.

Most often, land use intensification in seminatural grasslands increases herbage production and may cause considerable losses in biodiversity, which reduces the part of production obtained without exogenous inputs (Pan et al., 2014) and reduces cultural services (Allan et al., 2015). High intensification is usually but not always associated with low functional diversity, less flexibility, and less stability in plant community functional composition. As discussed before, N emissions increase with the amount of N supplied, especially above a given N threshold that is generally context dependent (Duru and Vertès, 2016), and depending mainly on management practice (cutting vs. grazing). A strong trade-off exists between herbage production and C sequestration, except under moderate grazing (Assmann et al., 2014) and a low level of N input. However, this trade-off depends on grassland functional diversity and management. It can be mitigated by management promoting legume content and increasing the level of functional diversity. Grass-legume mixtures with 30%–50% legumes, which rarely occurs in unsown grasslands, seems an optimal system with the following advantages: yielding high amounts of N from symbiosis, generating high net primary production, producing forages of high nutritive value, generating high voluntary intake and livestock performance, decreasing the risk of N losses to the environment, and storing more C than fertilized grass monocultures (Lüscher et al., 2014; Soussana et al., 2004).

In summary, a high level of ES (or a low level of impacts such as N emissions) occurs at an intermediate level of management intensity. Increased legume content decreases N emissions and increases C sequestration. Cutting is better for reducing N emissions (if N fertilization is monitored) and grazing for increasing C sequestration.

Characterizing Ecosystem Services and Negative Environmental Impacts: A Case Study

To demonstrate the utility of our analytical framework, we assessed ES provided by grasslands first at field and farm levels, then at the landscape level, in a case study in the southern Massif Central, France. It focuses on the "Segala" region (Moraine et al., 2016a), a lightly populated livestock-oriented region at an elevation of 600–900 m a.s.l.

Based on observations of botanical composition using a simplified method (Theau et al., 2010) in all grasslands ($n = 186$) of six farms that have seminatural and temporary grasslands, we calculated the plant functional composition of each grassland based on several indicators: plant community functional composition for the two dimensions of plant growth strategy (community-weighted mean PFT and the diversity), legume content, and two management variables: N input (inorganic, organic, N return from grazing) and defoliation regime (cutting vs. grazing). We distinguished short-term (≤5-year) versus long-term (>5-year) grasslands because it is well documented that certain species, such as *M. sativa*, tend to decline in long-term grasslands, and this threshold conforms to statistics usually used to describe land use.

Functional Characterization of Grasslands

Compared to long-term temporary and seminatural grasslands, short-term temporary grasslands generally contained higher percentages of plants exhibiting a resource capture strategy grass functional type (GFT-E, GFT-A, and GFT-B, Table 13.2). A positive correlation was found between plant strategy for resource capture and plant growth pattern, but only for seminatural grasslands. The highest indices of plant growth patterns (largest PFT, Table 13.3) were observed in short-term temporary grasslands composed of *M. sativa*. These grasslands provide steadier herbage growth throughout the growing season than those composed of *Lolium multiflorum,* which exhibit a spring-centered herbage production.

Short-term temporary grasslands displayed higher plant temporal complementarity than others, especially for the lowest and highest indices of plant growth pattern. This includes grasslands composed of *L. multiflorum* with either *L. hybridum* or *M. sativa,* respectively. For seminatural and long-term temporary grasslands, weighted mean indices of plant growth pattern and within-community plant diversity indices were positively correlated, indicating that functional diversity for these criteria is highest in plant communities with early spring-centered growth patterns.

Considering all grasslands, mean legume content was lowest in seminatural and long-term temporary grasslands and highest in short-term temporary grasslands. It decreased significantly with the age of temporary grassland ($r^2 = 0.43$; $P < .01$) and did not vary significantly with N supply. When larger amounts of inorganic N are supplied, N_2 fixation decreases (Unkovich, 2012); thus, the risk of N losses exists even with high legume content in the grassland. While short-term grasslands had significantly fewer species than long-term and permanent grasslands (8.3 vs. 13.4 and 13.0, respectively; $P < .05$), they had a higher index of functional complementarity (0.96 vs. 0.44 and 0.47, respectively; $P < .05$), probably due to higher legume content.

Evaluating Synergies and Trade-Offs Between Ecosystem Services at the Field Level

The summary representation (Figs. 13.3 and 13.4) can be used to determine the most and least suitable grassland compositions and management practices for providing ES and for limiting N emissions. We found that characteristics that provide ES (C sequestration) and limit N emissions (high ES) occurred mainly in temporary grasslands, especially when considering thresholds for both legume content and functional diversity. The configuration with a low level of ES (C sequestration) or a high risk of N emissions was found in permanently grazed grasslands when considering thresholds for both legume content and functional diversity.

Analyzing Grassland Diversity at the Farm Level

Based on the farm sample, we found that even a single farm can contain a wide range of within- and between-field functional plant diversity and that contrasting land use, including a combination of temporary and seminatural grasslands within a farm, can create a diversity of plant functional composition, as observed elsewhere (Rudmann-Maure et al., 2008). This provides forage complementarities that allow animals with different nutritional requirements (productive and unproductive cows, heifers) to be fed at low cost (Rawnsley et al., 2013; Duru et al., 2015b).

Spatiotemporal Grassland Distribution at the Landscape Level

Most grasslands in the study region were temporary (Fig. 13.5). Most temporary grasslands exist for 3–4 years and thus provide a high level of potential C sequestration and soil fertility. Because they are evenly spread over the entire region and are the dominant land use, the levels of ES that depend on landscape composition and configuration (e.g., regulation of matter and water flows) are expected to be high. Furthermore, their potentially low plant species richness (see earlier) does not necessarily impede ES such as pollination (Orford et al., 2016). They even

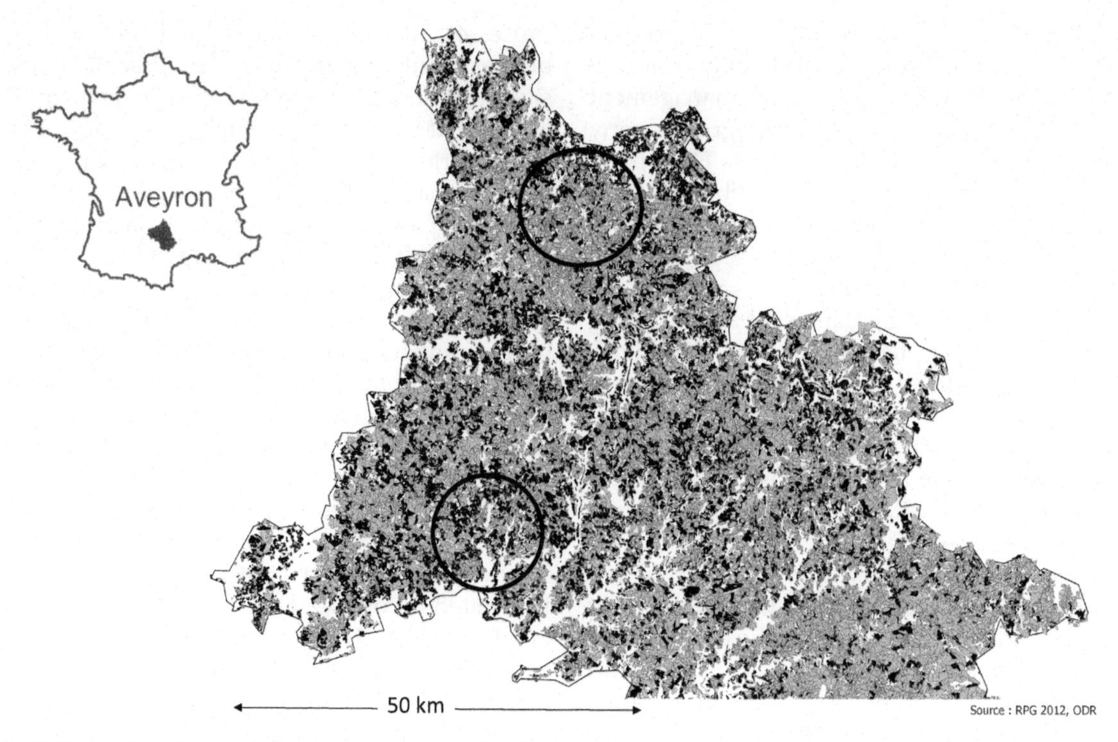

FIGURE 13.5 Mapping land use at the Eurostat NUTS 3 level (Aveyron department): permanent grasslands (*black*), temporary grasslands (*gray*), other land use (*white*); *black circles* indicate locations of the two groups of three farms studied. The spatial distribution of different types of grasslands in the "Segala" agricultural region (the area mapped) determines the level of generic and of some specific ecosystem services (ES, Table 13.1). The map shows dominant land use per European Union Common Agricultural Policy islet (one or several contiguous field plots). In Segala, about 75% of islets have only one field, i.e., only one crop field or one grassland (with possibly several paddocks). The six farms studied are located in the center of the region.

provide management flexibility. In this region historically focusing on livestock production, the dominance of grasslands ensures that the aesthetic dimension of the landscape will continue to meet expectations of local populations and tourists by providing a high level of associated cultural services. Improving assessment of (sub-)levels of ES at the landscape level requires obtaining data on spatiotemporal distribution, composition, and management of grasslands (Table 13.1).

SYNTHESIS

Grassland Functional Composition as Indicators of Ecosystem Goods, Services, and Environmental Impacts

Where grasslands are the main land use in a landscape, they are usually seminatural in mountains and permanent in lowlands. In other regions, the landscape mosaic is composed of permanent grasslands and temporary grasslands in rotation with crops. Besides forage production, grasslands provide considerably more ES than crops. However, grassland management has the potential to increase (or degrade) many of these ES (Finn et al., 2013). These ES depend on grassland spatiotemporal distribution (e.g., biologic, matter and water and nutrient flow regulations) or composition (e.g., C sequestration, pollination, or cultural value).

The theoretical (e.g., Maire et al., 2012; Zhang et al., 2016) and applied (Husse et al., 2016) literature shows many advantages of favoring diverse plant functional strategies in grassland communities, which we summarize as the ability to capture and use resources (fast vs. slow) and plant growth pattern (spring-centered vs. growing season-centered).

Ecologists studied herbage productivity and stability, while agronomists studied amount of forage production, focusing on the contribution of soil nutrients, seasonal distribution, and management flexibility. The case study illustrates that the framework helps to characterize all of these features. For example, we found that grasslands with high functional and/or temporal plant complementarity have greater management flexibility (e.g., changing the date of cutting or grazing without changing herbage yield or quality too much) and capture more resources. We show that functional diversity increases C sequestration and decreases N losses. Management practices (burying or not burying manure, plowing date of temporary grasslands, etc.) also can affect N losses greatly.

Cross-Scale Analysis of Ecosystem Services and Management Impacts

Cross-scale analyses are essential to identify synergies and trade-offs between ES because they are detectable only at a specific level. Trade-offs among ES should be considered in terms of spatial scale, temporal scale, and reversibility (Power, 2010). To analyze the underlying processes that determine matter flows (e.g., biomass, nutrient and C cycles; gaseous emissions) influenced by grassland management, it is essential to focus on the field level (Fig. 13.1). This clearly helps to identify ecosystem properties that shape delivery of ES as well as negative impacts of management practices at this level and, through cascade effects, at higher levels (Haines-Young and Potschin, 2010).

In contrast, management practices should be examined at the farm scale to understand the strategy for and consistency in animal feeding (forage self-sufficiency and complementarities). Livestock farmers can manage grasslands to provide forage, management flexibility, reduced production costs and increased ES such as soil fertility, soil retention, pollination, and pest control, as expected, in integrated crop-livestock systems (Moraine et al., 2016b). This is because these ES are provided to the farmer at the field and farm levels. Changing management practices usually influences forage production rapidly. However, studying ES such as water and matter flow regulations and C sequestration make sense at least at the landscape and global levels (Fig. 13.1). At these higher levels, however, positive or negative impacts concern other beneficiaries besides the farmer. Those who restore on-farm habitat complexity to increase pollination and pest control services may also affect a neighbor's farm (Power, 2010) or people who live elsewhere and consume agricultural products from that particular farm.

Another characteristic of ES provided by grasslands is the potentially long lag time until effects are observed. Effects may become "diluted" if the management practices causing the effects originate from a few farmers or few areas at watershed, regional, or higher levels. Thus, future transition toward agricultural practices providing ES from field to landscape levels, and even to food systems and the planet, has to be planned collectively by multiple stakeholders, such as farmers and those working in agri-food systems and natural resource management (Duru et al., 2015b).

Our research and other studies show that opportunities exist for win-win situations. Some of these situations occur at the farm level when within-farm grassland diversity is seen as an asset for efficient animal feeding (Rawnsley et al., 2013), reducing production costs (Duru et al., 2015b), or increasing farm sustainability (Sanderson et al., 2013). In this way, certain grasslands can be intensively managed, while those that are less intensively managed provide more ES to society (Benton et al., 2003). At the regional level, win-win situations can be developed by promoting exchanges between specialized crop farms and specialized livestock farms and encouraging crop farmers to grow legume-based grasslands in rotation with crops (Moraine et al., 2016a,b). Finally, trade-offs become more uncertain and difficult to manage when the temporal or spatial scale increases (Power, 2010). Overall, grassland is an invaluable land use for many purposes, mainly environmental, but also for indirect purposes related to human health (less air and water pollution, healthier agricultural products).

Management and Governance to Improve Ecosystem Services at the Local Level

Major issues in managing grasslands are uncertainty due to incomplete knowledge (e.g., for C sequestration), incomplete control (e.g., by grazing), and complex internal feedback (e.g., due to soil microbial activity, especially when temporary grasslands are grown in rotation with crops) (Duru et al., 2015a,b). In such situations, adaptive management is required as a structured approach that farmers and their advisors can use to design practices and assess trade-offs among competing ES (Birge et al., 2016) that occur at different spatial (field to farm) and temporal scales (e.g., productivity during a growing season vs. stability over several years). Useful knowledge needs to be developed to help select grassland species to be grown in temporary grasslands considering characteristics of the production situation (livestock system, climate, and soil) and expected ES, as is already somewhat available for cover-crop species (Damour et al., 2013). Development of "interactive" knowledge databases that provide generic knowledge from science and can incorporate farmer-experience feedback from a wide range of farming conditions seems a promising knowledge production and management strategy (Duru et al., 2015a; Duru and Vertès, 2016).

At the landscape scale, since experimentation is not possible, scenario planning is necessary to explore relations between spatiotemporal distribution and composition of grassland and expected ES at multiple levels (Birge et al., 2016). Simulation models are needed to assess land use scenarios ex ante, for example, to compare different landscape mosaics including grasslands to improve pest control (Parisey et al., 2016) or to create exchanges between specialized crop and livestock farms (Moraine et al., 2016a). These challenges often require a participatory design approach. Scenario planning can address the scientific uncertainty associated with land use change and also consider social relationships (Moraine et al., 2016b). Ultimately, because scenario planning encourages exploratory dialogue about critical uncertainties and risks that stakeholders may encounter, it serves as more than a decision-making model (Cobb and Thompson, 2012). Changing the agri-food chain to promote grassland-based animal products requires coordination among all stakeholders, especially when defining specifications for grassland management, product composition, or authorized processes, which guarantees to the consumer that products, as well as their related production processes, meet their expectations. Governance issues must be addressed when exploring such changes (Duru et al., 2015b).

References

Allan, E., Manning, P., Alt, F., et al., 2015. Land use intensification alters ecosystem multifunctionality via loss of biodiversity and changes to functional composition. Ecology Letters 18, 834–843.

Anderson, R.L., 2015. Integrating a complex rotation with no-till improves weed management in organic farming. A review. Agronomy for Sustainable Development 35, 967–974.

Assmann, T.S., de Bortolli, M.A., Assmann, et al., 2014. Does cattle grazing of dual-purpose wheat accelerate the rate of stubble decomposition and nutrients released? Agriculture, Ecosystems & Environment 190, 37–42.

Attard, E., Le Roux, X., Charrier, X., Delfosse, O., Guillaumaud, N., Lemaire, G., Recous, S., 2016. Delayed and asymmetric responses of soil C pools and N fluxes to grassland/cropland conversions. Soil Biology and Biochemistry 97, 31–39. https://doi.org/10.1016/j.soilbio.2016.02.016.

Bennett, E.M., Peterson, G.D., Gordon, L.J., 2009. Understanding relationships among multiple ecosystem services. Ecology letters 12 (12), 1394–1404.

Benton, T.G., Vickery, J.A., Wilson, J.D., 2003. Farmland biodiversity: is habitat heterogeneity the key? Trends in Ecology & Evolution 18, 182–188.

Birge, H.E., Allen, C.R., Garmestani, A.S., Pope, K.L., 2016. Adaptive management for ecosystem services. Journal of Environmental Management 1–10.

Cameron, K.C., Di, H.J., Moir, J.L., 2013. Nitrogen losses from the soil/plant system: a review. Annals of Applied Biology 162 (2), 145–173. https://doi.org/10.1111/aab.12014.

Cobb, A.N., Thompson, J.L., 2012. Climate change scenario planning: a model for the integration of science and management in environmental decision-making. Environmental Modelling & Software 38, 296–305.

Crotty, F.V., Fychan, R., Scullion, J., Sanderson, R., Marley, C.L., 2015. Assessing the impact of agricultural forage crops on soil biodiversity and abundance. Soil Biology and Biochemistry 91, 119–126.

Damour, G., Dorel, M., Quoc, H.T., Meynard, C., Risède, J.M., 2013. A trait-based characterization of cover plants to assess their potential to provide a set of ecological services in banana cropping systems. European Journal of Agronomy 52, 218–228.

de Bello, F., Lavorel, S., Díaz, S., et al., 2010. Towards an assessment of multiple ecosystem processes and services via functional traits. Biodiversity & Conservation 2873–2893.

Delaney, J.T., Jokela, K.J., Debinski, D.M., 2015. Seasonal succession of pollinator floral resources in four types of grasslands. Ecosphere 6 (11), 243.

Dhamala, N.R., Søegaard, K., Eriksen, J., 2014. Competitive forms in high-producing temporary grasslands with perennial ryegrass and red clover can increase plant diversity and herbage yield. Grassland Science in Europe, Vol. 20 – Grassland and forages in high output dairy farming 209–211.

Duru, M., Cruz, P., Al Haj Kaled, R., Ducourtieux, C., Theau, J.P., 2008. Relevance of plant functional types based on leaf dry matter content for assessing digestibility of native grass species and species-rich grassland communities in spring. Agronomy Journal 100, 1622–1630.

Duru, M., Cruz, P., Theau, J., 2010. Designing a simplified method for characterizing the agricultural value of species-rich grasslands through the functional composition of the vegetation. Crop & Pasture Science 61, 420–433.

Duru, M., Jouany, C., Le Roux, X., Navas, M.L., Cruz, P., 2013. From a conceptual framework to an operational approach for managing grassland functional diversity to obtain targeted ecosystem services: case studies from French mountains. Renewable Agriculture and Food Systems 1–16.

Duru, M., Therond, O., Martin, G., et al., 2015a. How to implement biodiversity-based agriculture to enhance ecosystem services : a review. Agronomy for Sustainable Development 35 (4), 1259–1281.

Duru, M., Theau, J.P., Martin, G., 2015b. A methodological framework to facilitate analysis of ecosystem servicesES provided by grassland-based livestock systems. International Journal of Biodiversity Science, Ecosystem Services & Management 1–17.

Duru, M., Vertès, F., 2016, Guide de la fertilization raisonée. Principes et outils pour organiser et piloter la fertilisation des prairies Chapter 7.5.1 Comifer 377–389, Ed France Agricole.

Duru, M., Bastien, D., Froidmont, E., Graulet, B., Gruffat, D., 2017. Importance qualitative et quantitative des produits issus de bovins au pâturage sur les apports nutritionnels et la santé du consommateur. Fourrages 230, 131–140.

Farruggia, A., Pomiès, D., Coppa, M., Ferlay, A., Verdier-Metz, I., Le Morvan, A., Martin, B., 2014. Animal performances, pasture biodiversity and dairy product quality: how it works in contrasted mountain grazing systems. Agriculture, Ecosystems & Environment 185, 231–244.

Finn, J.A., Kirwan, L., Connolly, J., et al., 2013. Ecosystem function enhanced by combining four functional types of plant species in intensively managed grassland mixtures: a 3-year continental-scale field experiment. Journal of Applied Ecology 50, 365–375.

Fischer, J., Gardner, T.A., Bennett, E.M., Balvanera, P., Biggs, R., Carpenter, S., Tenhunen, J., 2015. Advancing sustainability through mainstreaming a social–ecological systems perspective. Current Opinion in Environmental Sustainability 14, 144–149.

Fisher, B., Turner, R.K., Morling, P., 2009. Defining and classifying ecosystem services for decision making. Ecological Economics 68 (3), 643–653. https://doi.org/10.1016/j.ecolecon.2008.09.014.

Hammond, K.J., Humphries, D.J., Westbury, D.B., Thompson, A., Crompton, L.A., Kirton, P., Reynolds, C.K., 2014. The inclusion of forage mixtures in the diet of growing dairy heifers: impacts on digestion, energy utilisation, and methane emissions. Agriculture, Ecosystems & Environment 197, 88–95.

Haines-Young, R., Potschin, M., 2010. The links between biodiversity, ecosystem services and human well-being. Ecosystem Ecology: A New Synthesis 110–139.

Hooper, D.U., Chapin, F.S., Ewel, J.J., Hector, A., Inchausti, P., Lavorel, S., Schmid, B., 2005. Effects of biodiversity on ecosystem functioning: a consensus of current knowledge. Ecological Monographs 75 (1), 3–35.

Hoste, H., Jackson, F., Athanasiadou, S., Thamsborg, S.M., Hoskin, S.O., 2006. The effects of tannin-rich plants on parasitic nematodes in ruminants. Trends in Parasitology 22, 253–261.

Husse, S., Huguenin-Elie, O., Buchmann, N., Lüscher, A., 2016. Larger yields of mixtures than monocultures of cultivated grassland species match with asynchrony in shoot growth among species but not with increased light interception. Field Crops Research 194, 1–11.

Huyghe, C., Litrico, I., Surault, F., 2012. Agronomic value and provisioning services of multi-species swards. Grassland Science in Europe 17, 35–46.

Lavorel, S., Garnier, E., 2002. Predicting changes in community composition and ecosystem functioning from plant traits: revisiting the Holy Grail. Functional Ecology 16 (5), 545–556.

Lavorel, S., Grigulis, K., Lamarque, P., Colace, M.P., Garden, D., Girel, J., Douzet, R., 2010. Using plant functional traits to understand the landscape distribution of multiple ES. Journal of Ecology 99 (1), 135–147. https://doi.org/10.1111/j.1365-2745.2010.01753.x.

Lavorel, S., Grigulis, K., 2012. How fundamental plant functional trait relationships scale-up to trade-offs and synergies in ecosystem services. Journal of Ecology 100 (1), 128–140. https://doi.org/10.1111/j.1365-2745.2011.01914.x.

Lindemann-Matthies, P., Junge, X., Matthies, D., 2010. The influence of plant diversity on people's perception and aesthetic appreciation of grassland vegetation. Biological Conservation 143 (1), 195–202.

Loreau, M., de Mazancourt, C., 2013. Biodiversity and ecosystem stability: a synthesis of underlying mechanisms. Ecology Letters 16 (Suppl. 1), 106–115.

Lugnot, M., Martin, G., 2013. Biodiversity provides ecosystem services: scientific results versus stakeholders' knowledge. Regional Environmental Change 13 (6), 1145–1155.

Lüscher, A., Mueller-Harvey, I., Soussana, J.F., Rees, R.M., Peyraud, J.L., 2014. Potential of legume-based grassland-livestock systems in Europe: a review. Grass and Forage Science 69 (2), 206–228.

Maire, V., Gross, N., Börger, L., Proulx, R., Wirth, C., Pontes, L.D.A.S., Soussana, J.-F., Louault, F., 2012. Habitat filtering and niche differentiation jointly explain species relative abundance within grassland communities along fertility and disturbance gradients. New Phytologist 196 (2), 497–509.

Michaud, A., Andueza, D., Picard, F., Plantureux, S., Baumont, R., 2012. Seasonal dynamics of biomass production and herbage quality of three grasslands with contrasting functional compositions. Grass and Forage Science 67 (1), 64–76.

Molendi-Coste, O., Legry, V., Leclercq, I.A., 2011. Why and how meet n-3 PUFA dietary recommendations? Gastroenterology Research and Practice 2011, 364040. https://doi.org/10.1155/2011/364040.

Moraine, M., Grimaldi, J., Murgue, C., Duru, M., Therond, O., 2016a. Co-design and assessment of cropping systems for developing crop-livestock integration at the territory level. Agricultural Systems 147, 87–97.

Moraine, M., Melac, P., Ryschawy, J., Duru, M., Therond, O., 2016b. A participatory method for the design and integrated assessment of crop-livestock systems in farmers' groups. Ecological Indicators 72, 340–351.

Morecroft, M.D., Bealey, C.E., Scott, W.A., Taylor, M.E., 2016. Interannual variability, stability and resilience in UK plant communities. Ecological Indicators 68, 63–72.

Nozières, M.O., Moulin, C.H., Dedieu, B., 2011. The herd, a source of flexibility for livestock farming systems faced with uncertainties? Animal, (2010), pp. 1–16. http://doi.org/10.1017/S1751731111000486.

Öckinger, E., Smith, H.G., 2007. Semi-natural grasslands as population sources for pollinating insects in agricultural landscapes. Journal of Applied Ecology 44 (1), 50–59.

Orford, K.A., Murray, P.J., Vaughan, I.P., Memmott, J., 2016. Modest enhancements to conventional grassland diversity improve the provision of pollination services. Journal of Applied Ecology 53 (3), 906–915.

Pan, Y., Wu, J., Xu, Z., 2014. Analysis of the tradeoffs between provisioning and regulating services from the perspective of varied share of net primary production in an alpine grassland ecosystem. Ecological Complexity 17 (1), 79–86.

Parisey, N., Bourhis, Y., Roques, L., Soubeyrand, S., Ricci, B., Poggi, S., 2016. Rearranging agricultural landscapes towards habitat quality optimisation: in silico application to pest regulation. Ecological Complexity 28, 113–122.

Peysson, W., Vulliet, E., 2013. Determination of 136 pharmaceuticals and hormones in sewage sludge using quick, easy, cheap, effective, rugged and safe extraction followed by analysis with liquid chromatography-time-of-flight-mass spectrometry. Journal of Chromatography A 1290, 46–61.

Pighin, D., Pazos, A., Chamorro, V., Paschetta, F., Cunzolo, S., Godoy, F., Grigioni, G., 2016. A contribution of beef to human health : a review of the role of the animal production systems. ScientificWorldJournal. https://doi.org/10.1155/2016/8681491.

Pontes, L.D.A.S., Soussana, J.-F., Louault, F., Andueza, D., Carrère, P., 2007. Leaf traits affect the above-ground productivity and quality of pasture grasses. Functional Ecology 21, 844–853.

Power, A.G., 2010. Ecosystem services and agriculture: tradeoffs and synergies. Philosophical Transactions of the Royal Society B: Biological Sciences 365 (1554), 2959–2971.

Rawnsley, R.P., Chapman, D.F., Jacobs, J.L., Garcia, S.C., Callow, M.N., Edwards, G.R., Pembleton, K.P., 2013. Complementary forages—integration at a whole-farm level. Animal Production Science 53, 976–987.

Rudmann-Maurer, K., Weyand, A., Fischer, M., Stöcklin, J., 2008. The role of landuse and natural determinants for grassland vegetation composition in the Swiss Alps. Basic and Applied Ecology 9 (5), 494–503.

Sanderson, M.A., 2010. Stability of production and plant species diversity in managed grasslands: a retrospective study. Basic and Applied Ecology 11 (3), 216–224.

Sanderson, M.A., Archer, D., Hendrickson, J., Kronberg, S., Liebig, M., Nichols, K., Aguilar, J., 2013. Diversification and ES for conservation agriculture: outcomes from pastures and integrated crop–livestock systems. Renewable Agriculture and Food Systems 1–16.

Smukler, S.M.M., Sánchez-Moreno, S., Fonte, S.J.J., Ferris, H., Klonsky, K., O'Geen, A.T.T., Scow, K.M., Steenwerth, K.L., Jackson, L.E.E., 2010. Biodiversity and multiple ecosystem functions in an organic farmscape. Agriculture, Ecosystems & Environment 139 (1–2), 80–97.

Soussana, J.F., Lemaire, G., 2014. Coupling carbon and nitrogen cycles for environmentally sustainable intensification of grasslands and crop-livestock systems. Agriculture, Ecosystems & Environment 190, 9–17.

Soussana, J.F., Loiseau, P., Vuichard, N., Ceschia, E., Balesdent, J., Chevallier, T., Arrouays, D., 2004. Carbon cycling and sequestration opportunities in temperate grasslands. Soil use and management 20 (2), 219–230.

Spehn, E.M., Joshi, J., Schmid, B., Alphei, J., Körner, C., Basel, C., Umweltwis, I., 2000. Plant diversity effects on soil heterotrophic activity in experimental grassland ecosystems. Plant and Soil 224, 217–230.

Sun, S., Frelich, L.E., 2011. Flowering phenology and height growth pattern are associated with maximum plant height, relative growth rate and stem tissue mass density in herbaceous grassland species. Journal of Ecology 99, 991–1000.

Theau, J.P., Cruz, P., Fallour, D., Jouany, C., Lecloux, E., Duru, M., 2010. Une méthode simplifiée de relevé botanique pour une caractérisation agronomique des prairies permanentes. Fourrages 401, 19–25.

Tsiboukis, D., Sazakli, E., Jelastopulu, E., Leotsinidis, M., 2013. Anthelmintics residues in raw milk. Assessing intake by a children population. Polish Journal of Veterinary Sciences 16 (1), 85–91.

Unkovich, M., 2012. Nitrogen fixation in Australian dairy systems: review and prospect. Crop & Pasture Science 63 (9), 787–804.

van Oudenhoven, A.P., Petz, K., Alkemade, R., Hein, L., de Groot, R.S., 2012. Framework for systematic indicator selection to assess effects of land management on ecosystem services. Ecological Indicators 21, 110–122.

van Ruijven, J., Berendse, F., 2003. Positive effects of plant species diversity on productivity in the absence of legumes. Ecology Letters 6 (3), 170–175.

Vertès, F., Simon, J.C., Giovanni, C., Grignani, M., Corson, M.S., Durand, P., Peyraud, J.L., 2009. Flux de nitrate dans les élevages bovins et qualité de l'eau : variabilité des phénomènes et diversité des conditions. May 2008, Ed. Ac. Agr. - IDELE Presentation to the Académie d'Agriculture de France 6–26.

Wilson, P.J., Thompson, K., Hodgson, J.G., 1999. Specific leaf area and leaf dry matter content as alternative predictors of plant strategies. New Phytologist 143, 155–162.

Wood, S.A., Karp, D.S., DeClerck, F., et al., 2015. Functional traits in agriculture: agrobiodiversity and ecosystem services. Trends in Ecology & Evolution 1–9.

Zhang, W., Ricketts, T.H., Kremen, C., Carney, K., Swinton, S.M., 2007. Ecosystem services and dis-services to agriculture. Ecological Economics 64 (2), 253–260.

Zhang, Y., Loreau, M., Lü, X., He, N., Zhang, G., Han, X., 2016. Nitrogen enrichment weakens ecosystem stability through decreased species asynchrony and population stability in a temperate grassland. Global Change Biology 22 (4), 1445–1455.

CHAPTER

14

Local and Landscape Scale Effects of Heterogeneity in Shaping Bird Communities and Population Dynamics: Crop-Grassland Interactions

Vincent Bretagnolle[1], *Gavin Siriwardena*[2], *Paul Miguet*[1], *Laura Henckel*[1], *David Kleijn*[3]

[1]CEBC-CNRS, Beauvoir-sur-Niort, France; [2]Terrestrial Ecology, British Trust for Ornithology, The Nunnery, Norfolk, United Kingdom; [3]Plant Ecology and Nature Conservation, Wageningen University, Wageningen, The Netherlands

INTRODUCTION

Farmland landscapes support very high biodiversity (Pimentel et al., 1992), including functional species that provide ecosystem services (Tscharntke et al., 2005) and flagship species for wider ecosystems. However, over the past 50 years, biodiversity has strongly declined in agricultural areas, with major losses in plants, amphibians, reptiles, arthropods, mammals, and birds, and these losses have been attributed to agricultural intensification (Robinson and Sutherland, 2002; Inger et al., 2015). Agriculture intensification refers to the combination of rapid land use changes with, e.g., the replacement of natural habitats by crops, and more intensive use of existing farmland (Krebs et al., 1999; Robinson and Sutherland, 2002; Stoate et al., 2001). About 50% of all European bird species live in rural landscapes (Tucker, 1997), but farmland birds have declined in Europe much faster (−57% on the European Union farmland bird indicator between 1980 and 2013, EBCC, 2017) than in other ecosystems, i.e., birds were more or less stable in forests during the same period (EBCC, 2017). Farmland bird specialist species, even extremely common species such as the skylark *Alauda arvensis*, have declined by more than 50% in the past 30 years (Gregory et al., 2005, Voříšek et al., 2010). Although the main service expected in farmland is obviously food production, species and habitat conservation in agroecosystems are important issues, since farmland species provide other ecosystem

Agroecosystem Diversity
https://doi.org/10.1016/B978-0-12-811050-8.00014-5

services and are often the elements of biodiversity that are most accessible to humans close to urban areas (Power, 2010). Biodiversity loss has therefore additional consequences for ecosystem function and, ultimately, societal repercussions.

AGRICULTURAL INTENSIFICATION, HABITAT HETEROGENEITY, AND BIODIVERSITY

Recent empirical evidence strongly supports the hypothesis that plant and animal diversity decreases with increasing crop yield, a good proxy of agricultural intensification (see Geiger et al., 2010 and references therein), although acting through a range of specific ecologic processes that may not all be known. Processes proposed to date include loss of seminatural habitats such as field margins and hedgerows, intensification of in-field management (increased use and efficacy of pesticides and fertilizers), loss of fallow habitats (often crop stubbles) in winter due to increased winter sowing, and loss of crop diversity itself: only 30 species now provide more than 95% of the total human food consumption, and 75% of the genetic diversity of cultivated plants is already extinct (Rahmann, 2011). Those three processes all contributed to habitat homogenization (Tscharntke et al., 2005) through loss of habitat heterogeneity, at different spatial scales (i.e., countries, regions, farms, fields, and within fields; Benton et al., 2003). Mixed (arable and pastoral) farmland landscapes were replaced by homogeneous areas of either crop production or grassland, together with an increase in mean field size and the temporal simplification of arable crop rotations. Though, it is currently uncertain whether biodiversity loss at the regional scale results more from cropping intensification at the field scale, from loss of diversity within cropping system, or from the loss of natural elements in the landscape.

In their review of biodiversity in farmland, Benton et al. (2003) argued that habitat heterogeneity in farmlands is associated with higher biodiversity, and that the recent losses of farmland biodiversity are therefore due to homogenization of farmland at multiple spatial and temporal scales. Habitat heterogeneity is supposed to affect biodiversity at the level of communities (species richness and composition), but also at the level of populations (growth rate, dispersal, and connectivity). Landscape heterogeneity has been shown to increase species richness in butterflies (Weibull et al., 2000), birds (Berg et al., 2015), and plants (Belfrage et al., 2015). Landscape heterogeneity could influence species richness through either compositional or configurational heterogeneity (Fahrig et al., 2011). The former refers to the proportion of each cover type in a landscape, while the latter considers the spatial arrangement of these cover types and their neighborhood relationships. Heterogeneity can have a negative effect on individual species due to the decrease of the area of preferred habitat type (Benton et al., 2003; Hiron et al., 2015). Pickett and Siriwardena (2011) also showed that ground-nesting birds prefer homogeneous landscapes. But our ability to detect such effect is limited, first because there will always remain a hidden heterogeneity that cannot be directly detected (Vasseur et al., 2012), secondly because results will strongly depend on landscape characterization and spatial grain (see Fahrig, 2011). In general, heterogeneity effects depend on the species considered (Teillard et al., 2014), with species-level responses not necessarily reflecting those of communities.

More complex mechanisms may be further involved, such as resource complementation (Dunning et al., 1992), niche diversity, or species interactions (Danielson, 1991). More diverse communities are supposed to be more stable since ecologic functions are supported by more potential species and may also support more complex relationships involving edge effects and patch size. Relationships between biodiversity and heterogeneity may also be nonlinear: Concepcion et al. (2008) proposed, for instance, a sigmoidal response of biodiversity to landscape heterogeneity, with a threshold below which biodiversity does not increase and saturation at high level of landscape heterogeneity. Allouche et al. (2012) extended this view with a quadratic response, positive until an intermediate level of heterogeneity and then negative effect above this level, due to stochastic extinction of individual species when patch size becomes too small. Landscape heterogeneity also has important consequences for individual behavior and population dynamics, in particular through habitat selection, a process that links individual behavior and its demographic consequences. Habitat selection is a hierarchic behavioral process that involves multiple spatial but also temporal scales (Jones, 2001). In particular, when species require more than one resource to complete their life cycle sequentially or simultaneously (e.g., for nesting and foraging), their abundance is predicted to be higher when those resources are both present within the home range, a process called landscape complementation (Dunning et al., 1992). For example, winter cereals are optimal for the nesting of skylarks only early in the breeding season (Donald, 2004; Eggers et al., 2011), but they then shift to other crops once cereals are too tall and dense to allow nesting (Miguet et al., 2013) for other breeding attempts, a necessary condition for maintaining a stable population in intensive farmland habitat (Wilson et al., 1997; Siriwardena et al., 2001).

Quantifying landscape heterogeneity effects is, however, a difficult task in farmland landscapes, since it results from the interaction of green infrastructure (i.e., grasslands, seminatural elements) abundance and distribution, with crop configuration and composition (Fahrig et al., 2011). The resulting effect ultimately depends on the scale at which species respond to the landscape. Relationships between habitat heterogeneity and biodiversity may further depend on the level of agricultural intensification, yet very few empirical tests of this have been performed so far. Batary et al. (2011) found that in simple landscapes, less intensive agricultural practices improved biodiversity, though other studies found a higher effect of such practices on biodiversity in more complex landscapes (Duelli and Obrist, 2003), or no interaction at all (Winqvist et al., 2011). A major limitation of almost all studies to date is that they have considered landscape heterogeneity as the percentage of seminatural habitats (or seminatural elements) present (e.g., Chiron et al., 2010 on birds), or with arable fields being considered a single land use (Pickett and Siriwardena, 2011). Sometimes, the quantity of arable land or grasslands in the landscape has been used as a proxy of habitat-type diversity (Roschewitz et al., 2005; Rundlöf and Smith, 2006). Therefore, most studies so far have ignored an important source of landscape complexity, namely crop heterogeneity. Indeed, annual crop rotations, as well as the temporal variability in crop growth rates, result in heterogeneity in time and space that contributes to farmland landscape heterogeneity in addition to seminatural habitats. A few studies have shown an effect of crop diversity having a landscape complementation effect on biodiversity (Miguet et al., 2013; Kragten, 2011). An effect of configurational heterogeneity has also been shown (Fahrig et al., 2015), but effects of crop heterogeneity on wildlife populations have received relatively little attention so far, and the few studies that investigated crop heterogeneity did not attempt to identify the processes responsible for its effects (e.g., Siriwardena et al., 2012).

THE COMPONENTS OF FARMLAND HABITAT HETEROGENEITY AND THEIR EFFECT ON BIRDS

A commonly tested prediction is that complex landscapes can support more species due to a spatial (or temporal) landscape complementation effect (Fahrig et al., 2011). Landscape complementation may occur in particular between annual arable crops and permanent or temporary grasslands. At a landscape level, cereal-based agroecosystems are characterized by high spatial discontinuity and high temporal turnover (due to crop rotations, harvesting, mowing, and plowing). Plowing, introducing temporally asynchronous alterations of habitat quality, is a main human disturbance in these systems (see Gaba et al., 2014), barely considered so far, conversely to other components of heterogeneity. Though, in landscapes dominated by intensive cereal cropping, multiannual forage crops such as meadows, mown grasslands, and alfalfa can be considered "perennial habitats" because they are usually retained for 3–4 years (or even longer). They differ radically from all annual crops in terms of mechanical disturbance (soil tillage, sowing, cutting for harvest), and they also generally receive fewer pesticide inputs. Highly fragmented in space and time, intensive farmland landscapes interspaced with grasslands provide animal and plant populations with a spatial patchwork of habitats. Many studies have found, unsurprisingly, that grasslands were indeed important reservoirs of biodiversity within agricultural landscapes (Henckel et al., 2015; Bretagnolle and Gaba, 2015). Such undisturbed or at least less disturbed habitats may act as refuges or breeding habitats for many taxa, in particular prey of birds (insects, small mammals, or plants). From the refuge habitats, the prey population can then invade annual crops. Such source sink or metapopulation dynamics have been repeatedly suggested to maintain biodiversity in fragmented and unpredictable agroecosystems (Mouquet et al., 2006).

Indeed, grassland presence within intensive cereal landscapes is of utmost importance, not only for biodiversity perspective but also for many other agronomic and environmental services (see Lemaire et al., 2014). Due to the modern separation of crop and livestock production within agricultural businesses, current crop rotations are often only composed of annual crops. The inclusion of temporary grasslands (perennial forage crops) into crop rotations may have particularly strong impacts, especially in weed communities (Meiss et al., 2010; Gaba et al., 2010), just as areas of arable land use within pastoral systems can have positive effects on many species (Robinson et al., 2001). Weed persistence and abundance in farmland landscapes represent a critical keystone element for higher trophic levels (Bretagnolle and Gaba, 2015), including birds that feed mainly on weed seeds in winter (Moorcroft et al., 2002). This is also the case for grasshoppers (Orthoptera: Acrididae), which play a major trophic role, being the primary invertebrate herbivores in grassland habitats, and also because they are prey for other invertebrates, e.g., spiders, and vertebrates, notably, farmland birds that use them as food for chicks (Barker, 2004; Bretagnolle et al., 2011). Like many invertebrates, grasshoppers need perennial habitats due to the fact that their survival from one year to the other is achieved by eggs that are deposited in the soil. Maintaining grasshopper populations in the landscape can be achieved only through perennial habitats such as grasslands or field boundaries incorporating suitable grassy habitats. In addition, patches of perennial crops, as well as field margins and other noncrop habitats, may act as shelters for wildlife within the "matrix" of less favorable annual crops. With respect to biodiversity and trophic networks, the presence, abundance, and distribution of these perennial habitats may have strong impacts on metapopulation and metacommunity dynamics of various organisms (Hanski, 1999).

To summarize, testing the effect of cropland heterogeneity first needs the amount of seminatural habitat to be accounted for (Concepción et al., 2012), given its overwhelming effect. Fahrig et al. (2015) found a general positive effect of complex landscape configuration, independent of the effect of seminatural components. However, the effect of crop configuration for birds is particularly difficult to separate from the effect of seminatural habitat composition, especially hedges, because variations in the arrangement of field types are often confounded with lengths of such permanent field boundary vegetation, and the strongly structuring effect of hedges and small forest fragments for farmland passerines has been largely demonstrated (Hinsley and Bellamy, 2000). The latter can be positive (Macleod et al., 2004; Whittingham et al., 2001) or sometimes negative, e.g., for species that prefer more open habitats (Mason and MacDonald, 2000; Miguet et al., 2013). After controlling for this influence, a positive effect of crop compositional heterogeneity on birds has consistently been found (Firbank et al., 2008; Lindsay et al., 2013; Miguet et al., 2013), as in plants (Marshall, 2009) and insects. In addition, landscapes with a more complex configuration (smaller field size, with a higher linear length of field border) were found to stabilize community temporal variation (Henckel et al. submitted), and overall in birds, there may be a stronger effect of landscape configuration than composition (Fahrig et al., 2015).

SELECTED EXAMPLES FROM FLAGSHIP SPECIES

Black-Tailed Godwit Conservation in the Netherlands

Before humans started large-scale modifications of the landscape, the black-tailed godwit, *Limosa limosa*, is thought to have nested in bogs and littoral grasslands that naturally occurred in the riverine delta of what now is the Netherlands (Blankers and Kleijn, 2011). As natural habitats were increasingly cultivated, black-tailed godwits switched to extensively managed anthropogenic grasslands. Over the last century, the largest numbers and highest densities were found on agricultural grasslands in polder areas with high water tables and clay or peat soils in the western and northern parts of the country. Black-tailed godwits are typical of many ground-nesting farmland birds, in that they require large, open areas free of buildings or vertical landscape elements that can be used as a perch by birds of prey. At a large scale, they therefore require structurally simple landscapes that, to the human eye, are monotonous and of low diversity. Heterogeneity is nevertheless critically important (Verhulst et al., 2011) but is expressed in more subtle ways than gross landscape character. In the black-tailed godwit, habitat heterogeneity is mostly driven by water level that, in combination with natural relief, can create complex spatiotemporal gradients in soil moisture that, in turn, result in heterogeneity in vegetation height and composition. This creates a robust system in which godwits, and other so-called meadow birds, are able to find good nesting and chick-raising habitat under a variety of environmental conditions. With the acceleration of agricultural intensification since the 1960s, and particularly the large-scale drainage of agricultural grasslands in the 1970s and 1980s, the grasslands became increasingly monotonous, both in space and time, and the black-tailed godwit population started to decline rapidly. In the period 1990–2008, the national godwit population declined with approximately 3% per year, but in more recent years the decline seemed to increase to almost 7% per year in the period 2004–08 (Van Paasen and Teunissen, 2010). Conservation of black-tailed godwits is being done by means of national agri-environmental schemes on farmland, as well as in protected areas, with €21 million being spent on so-called meadow bird agreements with farmers and €4 million on meadow bird conservation in spatially much more restricted protected areas (2008 data; van Paassen and Teunissen, 2010). In the succession of Dutch agri-environmental programs up to 2016, meadow bird agreements at best delayed the first seasonal activities of farmers. This potentially protects clutches and chicks of nesting meadow birds (Beintema and Muskens, 1987; Schekkerman et al., 2008). However, because these schemes did not address key factors affecting habitat quality for meadow birds, such as fertilization, ground water level, openness of the landscape, and disturbance during the breeding season, they were largely ineffective (Kleijn et al., 2001; Verhulst et al., 2007; Schekkerman et al., 2008). Reproductive success, in particular chick survival, the key factor determining population dynamics, is higher in protected areas than on farmland with or without meadow bird agreements (Berendse et al., 2004; Kentie et al., 2013), but population trends are negative in many reserves as well. There are examples that show that, even under the difficult environmental conditions of modern agricultural landscapes, godwits can successfully and sustainably be conserved. These case studies suggest that, for this to be possible, areas have to be large (>200 ha), have to be free of vertical structures (e.g., trees, woodlots, buildings), need to have limited predator abundance, and need to have a significant proportion of the land area with raised water levels to create optimal chick-rearing habitat (Kleijn and Lammertsma, 2012). In the Netherlands, this is generally only possible by implementing agri-environment schemes on farmland in a buffer surrounding protected areas and by managing both the protected area and adjacent farmland as if they formed a single unit. Such efforts are generally only successful when an ornithologically, as well as agriculturally, skilled coordinator is present who oversees and coordinates all activities in the area and who has the policy tools and authority with farmers to optimize black-tailed godwit management by means of implementing last-minute management adjustments (Kleijn and Lammertsma, 2012).

Little Bustard and Other Threatened Top Predators

Insects and small mammals make up the bulk of biomass in food supply for higher trophic levels in agroecosystems. For example, the common vole, *Microtus arvalis*, represents a key resource in the trophic chain in agroecosystems (Lambin et al., 2006). In western France, the Montagu's harrier, *Circus pygargus*, appears to be highly dependent on common vole abundance, since its population density displays a numerical response to the cyclic dynamics of this prey species (Salamolard et al., 2000; Millon and Bretagnolle, 2008). Thus, the persistence of this flagship raptor species is conditional upon the maintenance of common vole populations. And the abundance, in particular the amplitude of the peak vole years, appears to be related to the presence of the grasslands at the landscape scale (Bonnet et al., 2013), although there is a complex interplay between dispersal and density dependence processes, likely linked to grassland habitat quality (Pinot et al., 2016).

Similarly, differences in grasshopper availability appear to be critical to little bustard, *Tetrax tetrax*, productivity. The little bustard population of western France, which winters in Spain (Villers et al., 2010), has undergone one of the steepest declines documented to date for a bird species in Europe: 7800 males in 1978 to 390 in 1996 (a decrease of 95% in 18 years: Inchausti and Bretagnolle, 2005, Bretagnolle et al., 2011b), and 250 in 2016. There is a strong positive relationship between annual average grasshopper abundance (calculated as the mean abundance over the surveyed grasslands for a given year) and total annual productivity of little bustard, as estimated by the number of fledglings counted in postnuptial groups (Bretagnolle et al., 2011). The case is not unique, as many bird chicks in cereal systems feed primarily on insects (e.g., Rands, 1986; Baines et al., 1996; Panek, 1997) and especially on Orthoptera. In western France, different agri-environment schemes (AESs) have been implemented to protect the little bustard. Most of these schemes were based on grassland restoration and modification of grassland management (Berthet et al., 2012, 2014). The strong decline in little bustard has been linked to a reduction in the areas of perennial habitats that are suitable for breeding, but also to strong decreases in insects (particularly grasshoppers) for feeding (Inchausti and Bretagnolle, 2005; Bretagnolle et al., 2011b). The latter resulted simultaneously from a decrease in the grasslands in which the grasshoppers breed, as well as the intensive use of insecticides and herbicides (reducing food availability for these insects). The conservation strategy for little bustard was therefore to counteract the loss of habitat and the low availability of food resources. Protection measures providing food resources and favorable nesting plots (reducing agricultural activities to minimize the risk of destruction of nests and incubating females), and more generally to encourage farmers to restore perennial vegetation covers, were developed. The little bustard population, which had shown an initial decrease by a factor of five in just 8 years (about 13% per year since 1996), has then recovered in no more than 5 years (Bretagnolle et al., 2011b). More recently, however, the wheat price increased (especially in 2008 and 2009), in addition to changes in AESs and rules following the last Common Agricultural Policy (CAP) reform (2014), and these two factors resulted in strong decrease of grasslands (by up to 45%) and AES contracts. Little bustard populations in western France started to decrease again.

Skylark and Heterogeneity at Multiple Scales

The skylark *Alauda arvensis* is one of the most iconic farmland birds across Europe, and its decline has been symbolic of wider declines in farmland biodiversity since the 1990s. Skylarks both breed and overwinter in grassland and arable habitats, feeding and nesting on the ground. Given that sufficient food is actually present, the species critically needs access to bare ground for foraging on seeds in the winter (adults) and on insects in the breeding season (for feeding to chicks): variation in accessibility can override food abundance (e.g., Atkinson et al., 2005). Agricultural intensification has potentially affected skylarks through (1) a loss of seed-rich winter habitat with autumn sowing, and increasingly effective herbicides and efficient harvesting, (2) reduced food availability and accessibility in pastures with high nitrogen inputs that have denser swards and much reduced weed floras, (3) loss of insect food for chicks from broad spectrum pesticides and loss of weed flora supporting them, and (4) loss of breeding opportunities, especially late in the breeding season, due to the high vegetation density in autumn-sown cereal crops (Donald, 2004; Wilson et al., 1997; Siriwardena et al., 1998a). However, of these potential effects, only one can be limiting for populations in a given location and season, and conservation solutions need to identify the limiting factor and put in management to reverse it. Demographic analysis has shown that skylark breeding success per nesting attempt is good, in common with most other farmland passerines, suggesting that effects on this demographic rate, such as effects on chick food, have not been important (Siriwardena et al., 2000). The paucity of ring-recovery data means, however, that knowledge of variation in skylark survival is poor, but the survival of many farmland granivores has been affected negatively by reduced seed availability (Siriwardena et al., 1998b), and while skylarks can eat green vegetation in winter cereal crops, this probably does not provide sufficient resources for them (Green, 1978). Analyses of population trends reveal a strong correlation with areas of winter cereal stubbles (Gillings et al., 2005): this could show the effect of seed availability in stubble on survival, but also could reflect the benefits of spring cropping allowing higher numbers of breeding attempts per pair per year. Evidence from studies at the field and farm scales supports an important role for numbers of attempts in skylark demography (Wilson et al., 1997; Donald, 2004). To inform potential solutions, both the potential winter and spring drivers of skylark population decline can be regarded as reflecting effects of a loss of heterogeneity in space or time. Multiple breeding attempts per year require both nest cover and access to food to be maintained over several months. In arable fields, this means fine-scale heterogeneity, so these resources can be found within the areas of individual territories, as well as this structure then being replicated in different fields through the growing season. Autumn-sown crops are poor for late season breeding, but they provide better cover than spring crops early in the breeding season, so a heterogeneous cover of each is beneficial (Wilson et al., 1997; Donald, 2004). In grassland, such structural heterogeneity needs to persist through the season, which means extensive management because improved pasture and silage fields will feature swards that are too dense and grow too fast.

Conservation management cannot just replace historical land use, because it has to work in the context of modern agriculture. Management manipulating heterogeneity in farmed habitats may be the key to population recovery. Limited crop areas with high seed density, such as individual fields within standard crop or grassland mosaics, can potentially support flocks made up of an entire local breeding population, plus winter migrants. Hence, heterogeneity at the landscape scale, interspersing winter-sown crops with seed-rich winter fallows, should address issues of winter limitation. For breeding birds, heterogeneity benefits need to be more dispersed, to allow territoriality. Hence, a key AES option to address the breeding season problem in arable farmland is "skylark plots" (uncropped areas of a few square meters in winter cereal crops), which effectively replace interfield heterogeneity with that within fields, providing access to nesting sites (Morris et al., 2004). This conservation option probably requires further development because its efficacy has been found to vary with habitat context, suggesting interactions with the facilitation of predation and specificity to farming systems (Morris et al., 2004; Berg and Kvarnbaeck, 2011); the evidence shows also that the option can succeed when placed in the right context (Morris et al., 2004; Schmidt et al., 2017; Dillon et al., 2009). It is therefore worthy of establishment at the landscape scale, as in English AESs, but needs to be monitored and revised if necessary. However, it can be unpopular with farmers who dislike actions that interfere with crop management, so there is also a role for enhanced engagement and communication.

CONCLUDING REMARKS AND FUTURE PROSPECTS

Many species of birds have undergone sharp declines in Europe (Gibbons et al., 1993; Potts, 1997). In Western Europe, about 1% of the avifauna of lowland landscapes disappears annually (Donald et al., 2001; Julliard et al., 2004). In France, between 1989 and 2003, bird populations (all species combined) have declined by 3%, while birds using agricultural habitats declined by 25% (Julliard et al., 2004). It has been repeatedly suggested that the environmental impact of intensive agriculture may rely more on oversimplification of landscapes at all spatial scales (fields, farm, region) than intensification of production from a given cropped area. We provided here some evidence that indeed landscape simplification has a strong impact on biodiversity, though the balance between local factors (e.g., intensive use of agrochemicals) and landscape scale factors is still debated.

Understanding the causes of historical population changes to aid prediction of the impact of future land use change and to suggest management actions relies on knowledge of the relationships between the abundance of species and habitat characteristics. In terms of conservation, AES and NATURA 2000 are the two main leverage tools available (if not the only ones) to try to mitigate the devastating effects of intensive agriculture on biodiversity. Grasslands have a critical role in shaping the distribution and abundance of organisms of different trophic levels including plants, grasshoppers, small mammals, and birds. Despite the evidence that increasing landscape diversity and/or restoring grassland habitats in intensive cereal systems can have major benefits, AESs have not commonly targeted the conservation and management of permanent and temporary grasslands, due to the difficulties in restoring a market for livestock where it has disappeared (Berthet et al., 2012). Organic farming is an interesting alternative (which is often included in AESs, for example, in France), although neither AESs nor organic farming always provide the positive effects on biodiversity that they are intended to deliver (Kleijn and Sutherland, 2003). The management of grassland habitat in such ecosystems is therefore critical for both the maintenance of ecosystem services such as those depending on functional biodiversity, as well as for the conservation of threatened species. For the latter, grassland must be managed at the regional rather than the local scale because many bird species forage on vast areas and occur at rather low densities, so that their population dynamics can operate at the regional scale. Moreover, most studies have dealt with only a few taxa, and there is a lack of studies investigating the effects of biodiversity loss on ecosystem services, e.g., soil conservation, nutriment cycling, groundwater purification, pollination, or biologic control (but see Geiger et al., 2010). Therefore, restoring crop diversity is itself not sufficient to increase biodiversity, but managing both Semi-Natural Elements (SNE) and crop diversity should be targeted. Grassland-arable cropping integration could be an important mechanism for diversifying agricultural systems.

Acknowledgments

Part of the research by VB was supported by ANR AGROBIOSE, ERANET ECODEAL, and Fondation LISEA. The PhD grant of L. Henckel was provided by ANR FARMLAND, while P. Miguet was supported by INRA.

References

Allouche, O., Kalyuzhny, M., Moreno-Rueda, G., Pizarro, M., Kadmon, R., 2012. Area-heterogeneity tradeoff and the diversity of ecological communities. Proceedings of the National Academy of Sciences 109, 17495–17500.

Atkinson, P.W., Fuller, R.J., Vickery, J.A., Conway, G.J., Tallowin, J.R.B., Smith, R.E.N., Haysom, K.A., Ings, T.C., Brown, V.K., 2005. Influence of agricultural management, sward structure and food resources on grassland field use by birds in lowland England. Journal of Applied Ecology 42 (5), 932–942.

Baines, D., Wilson, I.A., Beeley, G., 1996. Timing of breeding in black grouse Tetrao tetrix and capercaillie Tetrao urogallus and distribution of insect food for the chicks. IBIS 138, 181–187.

Barker, A.M., 2004. Insects as food for farmland birds: is there a problem? In: Van Emden, H.F., Rothschild, M. (Eds.), Insects and Bird Interactions. Intercept, Andover, pp. 37–50.

Batary, P., Baldi, A., Kleijn, D., Tscharntke, T., 2011. Landscape-moderated biodiversity effects of agri-environmental management: a meta-analysis. Proceedings of the Royal Society B: Biological Sciences 278, 1894–1902.

Beintema, A.J., Müskens, G.J.D.M., 1987. Nesting success of birds breeding in Dutch agricultural grasslands. Journal of Applied Ecology 24, 743–758.

Belfrage, K., Björklund, J., Salomonsson, L., 2015. Effects of farm size and on-farm landscape heterogeneity on biodiversity—case study of twelve farms in a Swedish landscape. Agroecology and Sustainable Food Systems 39 (2), 170–188.

Benton, T.G., Vickery, J.A., Wilson, J.D., 2003. Farmland biodiversity: is habitat heterogeneity the key? Trends in Ecology & Evolution 18, 182–188.

Berendse, F., Chamberlain, D., Kleijn, D., Schekkerman, H., 2004. Declining biodiversity in agricultural landscapes and the effectiveness of agri-environment schemes. AMBIO: A Journal of the Human Environment 33 (8), 499–502.

Berg, Å., Kvarnbaeck, O.L.L.E., 2011. Density and reproductive success of Skylarks Alauda arvensis on organic farms- an experiment with unsown Skylark plots on autumn sown cereals. Ornis Svecica 21 (1), 3–10.

Berg, Å., Wretenberg, J., Żmihorski, M., Hiron, M., Pärt, T., 2015. Linking occurrence and changes in local abundance of farmland bird species to landscape composition and land-use changes. Agriculture, Ecosystems & Environment 204, 1–7.

Berthet, E., Bretagnolle, V., Segrestin, B., 2012. Analyzing the design process of farming practices ensuring little bustard conservation: lessons for collective landscape management. Journal of Sustainable Agriculture 36, 319–336.

Berthet, E.T.A., Bretagnolle, V., Segrestin, B., 2014. Overcoming innovation bottlenecks through a collective approach. Reintroduction of alfalfa in a cereal farming area. Fourrages 217, 13–21.

Blankers, P., Kleijn, D., 2011. Weidevogels natuurlijk: nieuwe inzichten in het Nederlandse weidevogelbeheer. In: Schaminee, J., Jansen, J., Weeda, E. (Eds.), Gewapende vrede - Beschouwingen over plant-dierrelaties in de natuur. KNNV Uitgeverij, Zeist, pp. 130–149. In Dutch).

Bonnet, T., Crespin, L., Pinot, A., Bruneteau, L., Bretagnolle, V., Gauffre, B., 2013. How the common vole copes with modern farming: insights from a capture-mark-recapture experiment. Agriculture, Ecosystems & Environment 177, 21–27. https://doi.org/10.1016/j.agee.2013.05.005.

Bretagnolle, V., Gaba, S., 2015. Weeds for bees? A review. Agronomy for Sustainable Development 35 (3), 891–909.

Bretagnolle, V., Gauffre, B., Meiss, H., Badenhausser, I., 2011a. The role of grassland areas within arable cropping systems for the conservation of biodiversity at the regional level. In: Lemaire, G., Hodgson, J.A., Chabbi, A. (Eds.), Grassland Productivity and Ecosystem Services. CABI, pp. 251–260.

Bretagnolle, V., Villers, A., Denonfoux, L., Cornulier, T., Inchausti, P., Badenhausser, I., 2011b. Rapid recovery of a depleted population of Little Bustards following provision of Alfalfa through an agri-environment scheme. IBIS 153, 4–13.

Chiron, F., Filippi-Codaccioni, O., Jiguet, F., Devictor, V., 2010. Effects of non-cropped landscape diversity on spatial dynamics of farmland birds in intensive farming systems. Biological Conservation 143 (11), 2609–2616.

Concepción, E.D., Díaz, M., Baquero, R.A., 2008. Effects of landscape complexity on the ecological effectiveness of agri-environment schemes. Landscape Ecology 23 (2), 135–148.

Concepcion, E.D., Díaz, M., Kleijn, D., Baldi, A., Batary, P., Clough, Y., et al., 2012. Interactive effects of landscape context constrain the effectiveness of local agri-environmental management. Journal of Applied Ecology 49 (3), 695–705.

Danielson, B.J., 1991. Communities in a landscape: the influence of habitat heterogeneity on the interactions between species. The American Naturalist 138 (5), 1105–1120.

Dillon, I.A., Morris, A.J., Bailey, C.M., Uney, G., 2009. Assessing the vegetation response to differing establishment methods of 'Skylark Plots' in winter wheat at Grange Farm, Cambridgeshire, England. Conservation Evidence 6, 89–97.

Donald, P., Gree, R.E., Heath, M.F., 2001. Agricultural intensification and the collapse of Europe's farmland bird populations. Proceedings of the Royal Society (London) B 268, 25–29.

Donald, P., 2004. The Skylark. Bloomsbury Publishing.

Duelli, P., Obrist, M.K., 2003. Regional biodiversity in an agricultural landscape: the contribution of seminatural habitat islands. Basic and Applied Ecology 4 (2), 129–138.

Dunning, J.B., Danielson, B.J., Pulliam, H.R., 1992. Ecological processes that affect populations in complex landscapes. Oikos 65 (1), 169.

EBCC Indicators, 2017. http://www.ebcc.info/indicators2015.html.

Eggers, S., Unell, M., Pärt, T., 2011. Autumn-sowing of cereals reduces breeding bird numbers in a heterogeneous agricultural landscape, 144, 1137–1144.

Fahrig, L., Baudry, J., Brotons, L., Burel, F.G., Crist, T.O., Fuller, R.J., et al., 2011. Functional landscape heterogeneity and animal biodiversity in agricultural landscapes. Ecology Letters 14 (2), 101–112.

Fahrig, L., Girard, J., Duro, D., Pasher, J., Smith, A., Javorek, S., et al., 2015. Farmlands with smaller crop fields have higher within-field biodiversity. Agriculture, Ecosystems & Environment 200, 219–234.

Firbank, L., Petit, S., Smart, S., Blain, A., Fuller, R.J., 2008. Assessing the impacts of agricultural intensification on biodiversity: a British perspective. Philosophical Transactions of the Royal Society B 363, 777–787.

Gaba, S., Chauvel, B., Dessaint, F., Bretagnolle, V., Petit, S., 2010. Weed species richness in winter wheat increases with landscape heterogeneity. Agriculture, Ecosystems & Environment 138 (3), 318–323.

Gaba, S., Fried, G., Kazakou, E., et al., 2014. Agroecological weed control using a functional approach: a review of cropping systems diversity. Agronomy for Sustainable Development 34, 103–119.

Geiger, F., Bengtsson, J., Berendse, F., Weisser, W.W., Emmerson, M., Morales, M.B., et al., 2010. Persistent negative effects of pesticides on biodiversity and biological control potential on European farmland. Basic and Applied Ecology 11 (2), 97–105.

Gibbons, D.W., Reid, J.B., Chapman, R.A., 1993. The New Atlas of Breeding Birds in Britain and Ireland: 1988–1991.

Gillings, S., Newson, S.E., Noble, D.G., Vickery, J.A., 2005. Winter availability of cereal stubbles attracts declining farmland birds and positively influences breeding population trends. Proceedings of the Royal Society of London B Biological Sciences 272 (1564), 733–739.

Green, R., 1978. Factors affecting the diet of farmland skylarks, Alauda arvensis. Journal of Animal Ecology 913–928.

Gregory, R.D., van Strien, A., Voříšek, P., et al., 2005. Developing indicators for European birds. Philosophical Transactions of the Royal Society B 360, 269–288.

Hanski, I., 1999. Metapopulation Dynamics. Oxford University Press, Oxford.

Henckel, L., Börger, L., Meiss, H., Gaba, S., Bretagnolle, V., 2015. Organic fields sustain weed metacommunity dynamics in farmland landscapes. Proceedings of the Royal Society B: Biological Sciences 282, 1808.

Henckel, L., Mouquet, N., Devictor, V., Bretagnolle, V., submitted. Landscape configurational heterogeneity and wooded area increase temporal stability of bird community in intensive agricultural system.

Hinsley, S., Bellamy, P., 2000. The influence of hedge structure, management and landscape context on the value of hedgerows to birds: a review. Journal of Environmental Management 60 (1), 33–49.

Hiron, M., Berg, Å., Eggers, S., Berggren, Å., Josefsson, J., Pärt, T., 2015. The relationship of bird diversity to crop and non-crop heterogeneity in agricultural landscapes. Landscape Ecology 30, 2001–2013.

Inchausti, P., Bretagnolle, V., 2005. Predicting short-term extinction risk for the declining Little Bustard (Tetrax tetrax) in intensive agricultural habitats. Biological Conservation 122 (3), 375–384.

Inger, R., Gregory, R., Duffy, J.P., Stott, I., 2015. Common European birds are declining rapidly while less abundant species' numbers are rising. Ecology Letters 18 (1), 28–36.

Jones, J., 2001. Habitat selection studies in avian ecology: a critical review. The Auk: Ornithological Advances 118 (2), 557–562.

Julliard, R., Jiguet, F., Couvet, D., 2004. Common birds facing global changes: what makes a species at risk? Global Change Biology 10, 148–154.

Kentie, R., Hooijmeijer, J.C.E.W., Trimbos, K.B., Groen, N.M., Piersma, T., 2013. Intensified agricultural use of grasslands reduces growth and survival of precocial shorebird chicks. Journal of Applied Ecology 50, 243–25.

Kleijn, D., Lammertsma, D., 2012. Conserving the Black-tailed Godwit - Legislation, Implementation and Effectiveness of Conservation in Belgium, Germany and the Netherlands. Alterra Report 2366. Alterra, Wageningen.

Kleijn, D., Sutherland, W.J., 2003. How effective are European agri-environment schemes in conserving and promoting biodiversity? Journal of Applied Ecology 40, 947–969.

Kleijn, D., Berendse, F., Smit, S., Gilissen, N., 2001. Agri-environment schemes do not effectively protect biodiversity in Dutch agricultural landscapes. Nature 413, 723–725.

Kragten, S., 2011. Shift in crop preference during the breeding season by Yellow Wagtails Motacilla flava flava on arable farms in The Netherlands. Journal of Ornithology 152 (3), 751–757.

Krebs, J.R., Wilson, J.D., Bradbury, R.B., Siriwardena, G.M., 1999. The second silent spring? Nature 400, 611–612.

Lambin, X., Bretagnolle, V., Yoccoz, N.G., 2006. Vole population cycles in northern and southern Europe: is there a need for different explanations for single pattern? Journal of Animal Ecology 75, 340–349.

Lemaire, G., Gastal, F., Franzluebbers, A., Chabbi, A., 2014. Grassland-cropping rotation, an avenue for agriculture intensification to reconcile high production with environment quality. Environmental Management. https://doi.org/10.1007/s00267-015-0561-6.

Lindsay, K.E., Kirk, D.A., Bergin, T.M., Best, L.B., Sifneos, J.C., Smith, J., 2013. Farmland heterogeneity benefits birds in American Mid-west Watersheds. The American Midland Naturalist 170 (1), 121–143.

Macleod, C.J., Parish, D.M.B., Hubbard, S.F., 2004. Habitat associations and breeding success of the Chaffinch *Fringilla coelebs*: capsule population trends for Chaffinch on farmland are unlikely to be explained by their preference for non-crop habitats alone. Bird Study 51 (3), 239–247.

Marshall, E.J.P., 2009. The impact of landscape structure and sown grass margin strips on weed assemblages in arable crops and their boundaries. Weed Research 49 (1), 107–115.

Mason, C.F., Macdonald, S.M., 2000. Corn Bunting Miliaria calandra populations, landscape and land-use in an arable district of eastern England. Bird Conservation International 10 (2), 169–186.

Meiss, H., Lagadec, L.L., Munier-Jolain, N., Waldhardt, R., Petit, S., 2010. Weed seed predation increases with vegetation cover in perennial forage crops. Agriculture, Ecosystems & Environment 138 (1–2), 10–16.

Miguet, P., Gaucherel, C., Bretagnolle, V., 2013. Breeding habitat selection of Skylarks varies with crop heterogeneity, time and spatial scale, and reveals spatial and temporal crop complementation. Ecological Modelling 266, 10–18.

Millon, A., Bretagnolle, V., 2008. Predator population dynamics under a cyclic prey regime: numerical responses, demographic parameters and growth rates. OIKOS 117, 1500–1510.

Moorcroft, D., Whittingham, M.J., Bradbury, R.B., Wilson, J.D., 2002. The selection of stubble fields by wintering granivorous birds reflects vegetation cover and food abundance. Journal of Applied Ecology 39 (3), 535–547.

Morris, A.J., Holland, J.M., Smith, B., Jones, N.E., 2004. Sustainable arable farming for an improved environment (SAFFIE): managing winter wheat sward structure for skylarks Alauda arvensis. IBIS 146 (s2), 155–162.

Mouquet, N., Miller, T.E., Daufresne, T., Kneitel, J.M., 2006. Consequences of varying regional heterogeneity in source sink metacommunities. OIKOS 113, 481–488.

Panek, M., 1997. The effect of agricultural landscape structure on food resources and survival of grey partridge Perdrix perdrix chicks in Poland. Journal of Applied Ecology 34, 787–792.

Pickett, S.R.A., Siriwardena, G.M., 2011. The relationship between multi-scale habitat heterogeneity and farmland bird abundance. Ecography 34 (6), 955–969.

Pimentel, D., Stachow, U., Takacs, D.A., Brubaker, H.W., Dumas, A.R., Meaney, J.J., Corzilius, D.B., 1992. Conserving biological diversity in agricultural/forestry systems. BioScience 42 (5), 354–362.

Pinot, A., Barraquand, F., Tedesco, E., Lecoustre, V., Bretagnolle, V., Gauffre, B., 2016. Density-dependent reproduction causes winter crashes in a common vole population. Population Ecology 58, 395–405.

Potts, D., 1997. "Cereal farming, pesticides and grey partridges", farming and birds in Europe. In: Pain, D.J., Pienkowski, M.W. (Eds.), The Common Agricultural Policy and its Implications for Bird Conservation. Academic Press, London, pp. 150–177.

Power, A.G., 2010. Ecosystem services and agriculture: tradeoffs and synergies. Philosophical Transactions of the Royal Society B: Biological Sciences 365 (1554), 2959–2971.

Rahmann, G., 2011. Biodiversity and organic farming: what do we know? Landbauforschung 61 (3), 189–208.

Rands, M.R.W., 1986. The survival of gamebird (Galliformes) chicks in relation to pesticide use on cereals. IBIS 128, 57–64.

Robinson, R.A., Wilson, J.D., Crick, H.Q., 2001. The importance of arable habitat for farmland birds in grassland landscapes. Journal of Applied Ecology 38 (5), 1059–1069.

Robinson, R., Sutherland, W., 2002. Post-war changes in arable farming and biodiversity in great Britain. Journal of Applied Ecology 39, 157–176.

Roschewitz, I., Gabriel, D., Tscharntke, T., Thies, C., 2005. The effects of landscape complexity on arable weed species diversity in organic and conventional farming. Journal of Applied Ecology 42 (5), 873–882.

Rundlöf, M., Smith, H.G., 2006. The effect of organic farming on butterfly diversity depends on landscape context. Journal of Applied Ecology 43 (6), 1121–1127.

Salamolard, M., Butet, A., Leroux, A., Bretagnolle, V., 2000. Responses of an avian predator to variations in prey density at a temperate latitude. Ecology 81 (9), 2428–2441.

Schekkerman, H., Teunissen, W., Oosterveld, E., 2008. The effect of 'Mosaic management' on the demography of black-tailed godwit *Limosa limosa* on farmland. Journal of Applied Ecology 45, 1067–1075.

Schmidt, J.U., Eilers, A., Schimkat, M., Krause-Heiber, J., Timm, A., Nachtigall, W., Kleber, A., 2017. Effect of Sky Lark plots and additional tramlines on territory densities of the Sky Lark Alauda arvensis in an intensively managed agricultural landscape. Bird Study 64, 1–11.

Siriwardena, G.M., Baillie, S.R., Wilson, J.D., 1998a. Variation in the survival rates of British farmland passerines with respect to their population trends. Bird Study 45, 276–292.

Siriwardena, G.M., Baillie, S.R., Buckland, S.T., Fewster, R.M., Marchant, J.H., Wilson, J.D., 1998b. Trends in the abundance of farmland birds: a quantitative comparison of smoothed common birds census indices. Journal of Applied Ecology 35, 24–43.

Siriwardena, G.M., Baillie, S.R., Crick, H.Q.P., Wilson, J.D., 2000. The importance of variation in the breeding

performance of seed-eating birds for their population trends on farmland. Journal of Applied Ecology 37, 1–22.
Siriwardena, G.M., Wilson, J.D., Baillie, S.R., Crick, H.Q.P., 2001. Can the decline of the skylark be "recovered" using present-day habitat preferences and changes in agricultural land-use?. In: Proceedings of the RSPB/BTO Skylark Workshop, Southampton 1999.
Siriwardena, G.M., Cooke, I.R., Sutherland, W.J., 2012. Landscape, cropping and field boundary influences on bird abundance. Ecography 35 (2), 162–173.
Stoate, C., Boatman, N.D., Borralho, R.J., Carvalho, C.R., de Snoo, G.R., Eden, P., 2001. Ecological impacts of arable intensification in Europe. Journal of Environmental Management 63 (4), 337–365.
Teillard, F., Antoniucci, D., Jiguet, F., Tichit, M., 2014. Contrasting distributions of grassland and arable birds in heterogenous farmlands: implications for conservation. Biological Conservation 176, 243–251.
Tscharntke, T., Klein, A.M., Kruess, A., Steffan-Dewenter, I., Thies, C., 2005. Landscape perspectives on agricultural intensification and biodiversity—ecosystem service management. Ecology Letters 8, 857–874.
Tucker, G., 1997. "Priorities for bird conservation in Europe: the importance of the farmed landscape", farming and birds in Europe. In: Pain, D., Pienkowski, M.W. (Eds.), The Common Agricultural Policy and its Implications for Bird Conservation. Academic Press, San Diego, pp. 79–116.
Van Paassen, A., Teunissen, W., 2010. Weidevogelbalans 2010. Landschapsbeheer Nederland, Utrecht/SOVON Vogelonderzoek Nederland, Beek-Ubbergen.
Vasseur, C., Joannon, A., Aviron, S., Burel, F., Meynard, J.-M., Baudry, J., 2012. The cropping systems mosaic: how does the hidden heterogeneity of agricultural landscapes drive arthropod populations? Agriculture. Ecosystems & Environment 166, 3–14.
Verhulst, J., Kleijn, D., Berendse, F., 2007. Direct and indirect effects of the most widely implemented Dutch agri-environment schemes on breeding waders. Journal of Applied Ecology 44, 70–80.
Verhulst, J., Smit, C., Loonen, W., Kleijn, D., Berendse, F., 2011. Seasonal distribution of meadow birds in relation to in-field heterogeneity and management. Agriculture, Ecosystems and Environment 142, 161–166.
Villers, A. Millon, A., Jiguet, F., Lett, J.M., Attie, C., Morales, M.B., Bretagnolle, V., 2010. Migration of wild and captive-bred Little Bustards: releasing birds from Spain threatens attempts to conserve declining French populations.
Voříšek, P., Jiguet, F., van Strien, A., Škorpilová, J., Klvaňová, A., Gregory, R.D., 2010. Trends in abundance and biomass of widespread European farmland birds: how much have we lost?. In: BOU Proceedings—Lowland Farmland Birds III.
Weibull, A.-C., Bengtsson, J., Nohlgren, E., 2000. Diversity of butterflies in the agricultural landscape: the role of farming system and landscape heterogeneity. Ecography 23 (6), 743–750.
Whittingham, M.J., Bradbury, R.B., Wilson, J.D., Morris, A.J., Perkins, A.J., Siriwardena, G.M., 2001. Chaffinch *Fringilla coelebs* foraging patterns, nestling survival and territory distribution on lowland farmland. Bird Study 48 (3), 257–270.
Wilson, J.D., Evans, J., Browne, S.J., King, J.R., 1997. Territory distribution and breeding success of skylarks Alauda arvensis on organic and intensive farmland in southern England. Journal of Applied Ecology 1462–1478.
Winqvist, C., Bengtsson, J., Aavik, T., Berendse, F., Clement, L.W., Eggers, S., et al., 2011. Mixed effects of organic farming and landscape complexity on farmland biodiversity and biological control potential across Europe. Journal of Applied Ecology 48 (3), 570–579.

SECTION IV

DIVERSIFIED AGROECOSYSTEMS AT FARM LEVELS FOR MORE SUSTAINABLE AGRICULTURE PRODUCTION?

CHAPTER

15

Integration of Crop and Livestock Production in Temperate Regions to Improve Agroecosystem Functioning, Ecosystem Services, and Human Nutrition and Health[1]

Scott L. Kronberg[1], *Julie Ryschawy*[2]

[1]USDA - Agricultural Research Service, Northern Great Plains Research Laboratory, Mandan, ND, United States; [2]Université de Toulouse, AGIR UMR 1248, INRA, INPT-ENSAT, Auzeville, France

INTRODUCTION

After the many years of human existence, hopefully, most people including farmers would agree that the fundamental goals of food-producing agriculture are (1) producing a variety of foods that provide the nourishment that children and adults need to mature properly, reproduce healthy offspring, and have maximum potential to live long, disease-free lives and (2) do this without damaging our environment too much. However, our food production systems have evolved to specialize in growing foods, fibers, and other things that people need or want, and it is easy to forget and even ignore the two fundamental goals of agriculture (at great cost to our wellbeing) while producing just one or a few products with as little financial expense as possible. For example, prevalence of obesity continues to increase for

[1] Mandatory insert for USDA-ARS employees: The United States Department of Agriculture (USDA) prohibits discrimination in all its programs and activities on the basis of race, color, national origin, age, disability, and where applicable, sex, marital status, family status, parental status, religion, sexual orientation, genetic information, political beliefs, reprisal, or because all or part of an individual's income is derived from any public assistance program. (Not all prohibited bases apply to all programs.) USDA is an equal opportunity provider and employer.

Agroecosystem Diversity
https://doi.org/10.1016/B978-0-12-811050-8.00015-7

American adults, with 40% obese in 2015–16, and progress in reducing stroke and heart disease in Americans has slowed. The great challenge of reaching the two fundamental goals is compounded as human populations and communities become larger, with most people specializing on other endeavors besides food production while consuming foods that only a few people and companies produce. Also, incomplete knowledge on how all aspects of human nutrition (and intake of nonnutrient compounds such as phthalates, steroid hormones, and pesticides in air or water) interact to affect our growth, reproduction, and health compounds the great challenge. So, it is not surprising that food provisioning systems in industrialized parts of the world have evolved into highly specialized systems that emphasize economies of scale and deemphasize interdependent relationships (Kirschenmann, 2008), while other highly specialized (and expensive) medical systems try to keep us healthy or at least alive. Consequently, the specialized food provisioning services are not necessarily concerned with providing all the macro- and micronutrients that people require (as well as healthful but nonnutrient phytochemicals in fruits and vegetables) for long and healthy lives nor seriously addressing all the negative environmental impacts of contemporary agriculture.

Industrialized livestock production in temperate regions of the world and the complex trading and movement of animal products and nutrients among other regions have been criticized as major causes of environmental problems (Galloway et al., 2007; Steinfeld and Wassenaar, 2007). In respect to these and related consumer concerns, considerable numbers of people are now consuming meat-free lacto- or lacto-ovo vegetarian diets or even vegan diets, believing that these diets are more environmentally sustainable, kinder to animals, and more healthful for people. However, while diets with high amounts of animal products are likely not the ideal for sustainability, especially with the great and growing number of people on Earth (Foley et al., 2011; Dumont et al., 2013; Garnett et al., 2013), arguments for animal product–free or reduced diets seldom if ever include plans for ecologically essential and proper recycling of macro- and micronutrients in plant-based foods that need to be carefully recycled from the soil to plants to people and possibly other animals and then back to the soil. This makes diverse diets including some animal-based foods to have more potential, if properly organized, to support more sustainable and ecologically viable food production because, unfortunately, it is easier now to recycle manure and urine from livestock than from people.

Although it is known that some versions of agriculture have included some forms of relatively unsophisticated integration of crops and livestock for thousands of years, farming has had very significant negative environmental impacts (e.g., degradation of soil fertility from lack of proper nutrient recycling, soil erosion from tillage, and overgrazing by livestock) over the many centuries it has been practiced in various regions of the world (Hillel, 1991; Montgomery, 2007). These failures should make it very clear that mere coexistence of crops and livestock on a farm or interrelated group of farms is not sufficient. Still, improved forms of integration of crop and livestock production are more likely to be adaptable to sustainable agroecosystem functioning while providing more ecosystem services, compared to specialized, simpler and less diverse production systems, which have also been practiced by some farmers for at least the last 200 years (e.g., farming only tobacco or cotton in the early years of the United States) and probably for thousands elsewhere. So the focus of this chapter is on integrated crop-livestock production in temperate regions.

The traditional form of crop-livestock production of the past centuries and more recent past was either what we now consider organic agriculture or nearly organic agriculture in that no or only a small amount of artificial pesticides, fertilizers, and genetically modified organisms were used in the farming operation. However, we need to keep in mind that currently nonorganic mixed crop-livestock production (at least in the United States and several other countries) can include the use of artificial fertilizer and pesticides such as the herbicide glyphosate for no-till crop production, nonorganic insecticides for fly control on livestock, and genetically engineered glyphosate-tolerant corn and soybeans to aid in reducing weed problems with the herbicide glyphosate. We also need to keep in mind that traditional organic annual crop production has relied on soil tillage for weed control, and this makes the soil more vulnerable to wind and water erosion, hence the recent interest in organic crop production using no-tillage techniques. In this chapter, we will also argue that more ideal agriculture in respect to producing highly healthful food, which is more likely to support healthy children and adults and long disease-free lives, is probably a diverse and complex mixture of annual crops including grains, pulses, oilseeds, vegetables, fruits, perennial tree crops (nuts and fruits), herbaceous forage crops, and a variety of small and large livestock species including fish. Obviously, this type of farming requires team efforts of highly knowledgeable and skilled farmers, and there are plenty of intelligent and energetic unemployed people in world now that could be trained to become highly skilled farming specialists and, contrary to popular thinking, not lead lives of drudgery but rather have interesting and satisfying lives (Kirschenmann, 2008).

INTEGRATING CROP AND LIVESTOCK PRODUCTION TO IMPROVE AGROECOSYSTEM FUNCTIONING

There have been several good reviews covering various advantages associated with integrated crop-livestock farming systems in the temperate region (Russelle et al., 2007; Sulc and Tracy, 2007; Wilkins, 2008; Bell et al., 2014; Lemaire et al., 2014; Soussana and Lemaire, 2014; Sulc and Franzluebbers, 2014). Essentially, the objective of these farming systems is for self-sufficiency of feedstuffs for the animals and maximum possible nutrient recycling between soil, plants, and animals within the farming unit, so nutrients are not lost from the farm's soils, and nutrients are not imported into the farm unless they are deficient and needed in the farm's soils. Crop-livestock integration is gaining renewed interest by scientists and policymakers because it has the potential to provide more ecosystem services such as provisioning of food, fiber, construction materials, and wildlife habitat, and regulating populations of pests, water quality, decomposition, and detoxification while also improving economic performance of farming.

The specific advantages of integrated crops-livestock production systems include (but are not limited to) the following: (1) the option of feeding the crops produced on the farm to livestock produced on the farm without the added cost of transport and/or profit to a supplier of the feed, (2) less or no importing of expensive inputs such as pesticides, synthetic fertilizer, and livestock feed, which can also bring excess nutrients within the feed to the farm (e.g., nitrogenous and phosphorus compounds) that can become a costly environmental and economic problem (and poor nutrient management by not properly recycling a valuable nutrient back to its source), (3) use of excreta (feces and urine) from the livestock as a valuable source of nutrients, organic matter, microbes, and perhaps other constituents that improve soil fertility (Stukenholtz et al., 2002) rather than as a waste problem, as in many confined animal feeding operations, (4) use and conversion of crop residues/by-products by livestock, (5) encouragement for producing perennial forages for livestock in rotation with annual crops with associated benefits for insect pollinators, birds, and other wildlife,

(6) dual-purpose use (grazing) of cereals and brassicas for forage while vegetative then later harvesting seeds from these plants, (7) use of livestock for weed control in annual crop fields to reduce or eliminate herbicide spraying, (8) profitability in one or more aspects of a diverse crop-livestock farm when other aspects are less or not profitable (financial risk management), (9) readily available use of annual crops by livestock if yields or prices are too low for conventional harvesting, and (10) potentially less dependence on government payments to survive financially. There are also disadvantages including the potential for livestock traffic to compact soil for annual crop production as well as the extra knowledge and managerial requirements needed for success (Sulc and Franzluebbers, 2014), but we have tap-rooted cover crops such as forage radish and canola that can be grown to reduce compaction (bio-tillage), and there are groups of progressive farmers, at least in the United States and probably elsewhere, who are willing to teach and mentor farmers who are interested in become successful and sophisticated integrated crop-livestock producers.

INTEGRATING CROP AND LIVESTOCK PRODUCTION TO IMPROVE PROVISIONING OF HIGHER QUALITY FOOD

Unfortunately, at this point in time in human evolution, producing a variety of foods that provide the nourishment that children and adults need to grow and reproduce successfully and have maximum potential to live long disease-free lives is not one of the fundamental goals of industrialized agriculture because doing this is complicated and industrialized agriculture strives for simplicity (Mozaffarian and Ludwig, 2010; Ludwig, 2011). Provenza et al. (2015) have made a strong argument that people in industrialized counties, which are largely located in temperate regions of the world, would be healthier consuming much fewer processed, fortified, and enriched foods from specialized and industrialized food production systems and eating much more diverse diets of whole plant and animal-based foods from animals that are also consuming a diverse diet of whole foods. Assuming their proposition is correct, which is similar to what others are suggesting (Bail et al., 2016), and many people in the industrialized world need to eat a greater diversity of non- or less-processed foods, then there is an obvious need to produce these foods, assuming more would people buy them, as appears to be the case in the United States and Europe. This presents an opportunity for integrated crop-livestock system to become more diverse by adding and mixing more species of livestock and more types of crops together, including trees (Bell et al., 2014; Sulc and Franzluebbers, 2014), which could produce nuts, fruit, wood, forage, improve soil fertility, modify the weather for other organisms (e.g., shade for livestock on hot days), and possibly increase carbon sequestration (Udawatta and Jose, 2012). When integrated with pasture and annual crop production, this provides the ecosystem services of food provisioning and pollination because there is habitat for a variety of insects including various pollinator species (Shepard, 2013). Grazing pigs, turkeys, and chickens can also provide important ecosystem services including soil fertilization, insect and weed control, and food (Shepard, 2013). Although chicken production with pasture-based systems is becoming more common (Sossidou et al., 2011), integration of small herbivorous livestock, such as geese and rabbits, needs more consideration for integrated crop-livestock systems because of their ability to compete with granivorous/omnivorous chickens in respect to reproductive efficiency of females and growth rate of young animals (Large, 1973). Plus, free-ranging geese can be useful for control of some weeds and insect pests (Clark and Gage, 1996), and rabbits can be raised effectively on grassland like ruminant livestock (Martin et al., 2016). Finally, some farms with adequate water supply have the ability to integrate crops with fish and ducks or fish, pigs, and cattle production into efficient and highly productive integrated crop-livestock production systems (Furuno, cited in Kirschenmann, 2007; Bonaudo et al., 2014).

INTEGRATION RATHER THAN JUST COEXISTENCE OF CROPS AND LIVESTOCK

The ability of integrated crop-livestock production systems to provide multiple provisioning and regulating ecosystem services is a strong argument for using them. But in reality, it is not so simple. A macro-scale analysis led in the EUFP7 Cantogether project (Chambaut et al., 2015) highlighted large variability in mixed crop-livestock farms in terms of environmental performance. Moraine et al. (2014) defined a gradient of mixed farms according to the level of integration between crops and livestock in time and space (Fig. 15.1). In some mixed farms, a simple coexistence between crops and livestock was observed, with juxtaposed units interacting only through the market. At the opposite of the gradient, an agronomic integration allowed self-sufficiency of animal feeding through the produced crops and grasslands and fertilization of parcels with animal manure.

Evaluation of mixed farms by their level of integration between crops and livestock indicated that the more integrated farms were, the more environmentally friendly they were. This framework could be applied for regional integration between crop and livestock farmers. The limited use of external inputs has benefits for the environment and would have economic benefits balancing macro-scale analysis (Chambaut et al., 2015; European Commission, 2015), according to which integrated crop-livestock system did not have as good of economic results as specialized farms. Macro-scale studies on European Commission data highlighted that all mixed farms did not achieve the potential expected for sustainability in respect to specialized farms. According to Chambaut et al. (2015) the more integrated farms had fewer negative impacts on the environment. Only specific integrated crop-livestock systems should therefore be considered more ecologically effective systems.

TOWARD MORE AGROECOLOGICAL INTEGRATED CROP-LIVESTOCK SYSTEMS

Integration between crops and livestock is an opportunity to design more effective agroecological systems, in terms of enhancing synergies between components to favor higher environmental and economic performances. Considering that coexistence between crops and livestock was not allowing effective agroecological systems, Bonaudo et al. (2014) analyzed how agroecological principles could be adapted to integrated crop-livestock systems to redesign and improve the resilience, self-sufficiency, productivity, and efficiency of integration between crops and livestock. Considering the classification of Schiere et al. (2002), they considered that integration between crops and livestock was a possible key to moving from high external input agriculture (HEIA) to a new agroecological agriculture (so-called new conservation agriculture). They pointed out that new agroecologically integrated crop-livestock systems should benefit from diversified production and increased interactions between subsystems to offset trade-offs between agricultural production and environmental impacts observed in many integrated crop-livestock systems around the world (Bonaudo et al., 2014). Considering the trajectories of farms moving from old forms of integrated agriculture without inputs to

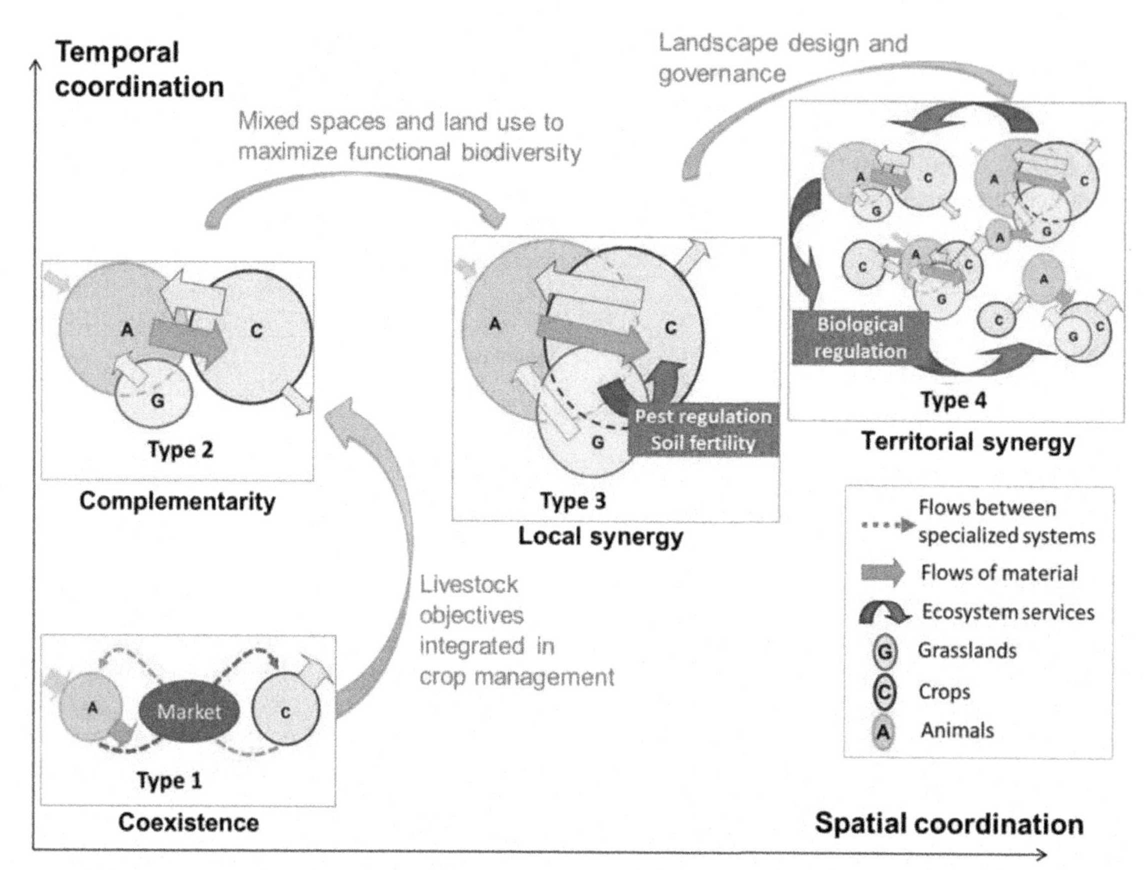

FIGURE 15.1 Generic representation of four crop-livestock system archetypes, distinguished according to the degree of spatial and temporal coordination between three spheres: crops (blue [gray in print version]), grasslands (green [light gray in print version]), and livestock (red [dark gray in print version]) either at farm or territorial level (Moraine, 2015). Each type has an illustrative name, and key drivers of integration necessary to pass from one type to another are noted. The "global coexistence type" corresponds to exchanges of grain, forage, straw, and manure between specialized farms, regulated by the markets. This type of spatially segregated coordination strongly limits expression of ecological benefits of crop-livestock integration. The "complementarity" type involves crop systems designed to produce the quantity and quality of crop products required for livestock production (in concentrates, forage, straw, etc.) and to use livestock manure as fertilizer. There is only a little spatial interaction among the three spheres to enhance the ecological processes and services. In the "farm-level (local) synergy," stronger temporal and spatial interaction among the three spheres allows stubble grazing, temporary grasslands in rotations, and intercropped forages. This farming system is designed to reduce input use by enhancing a wide range of ecosystem services at the local scale (e.g., soil quality enhancement, water and erosion regulation, maintaining biodiversity with landscape heterogeneity). Finally, in the "territory-level synergy," strong stakeholder coordination optimizes resource allocation and creates local diversified marketing chains that are adapted to specific characteristics of the territory. Exchanges within and between farms are organized to decrease input use and benefit farm-to-landscape level ecosystem services.

high-input agriculture, Teillard et al. (2015) argued that agroecologically integrated crop-livestock systems could be a possible solution to go beyond current systems in increasing agricultural production in respect to the output per hectare without increasing the intensity, which was shown to have negative effects on biodiversity (Fig. 15.2).

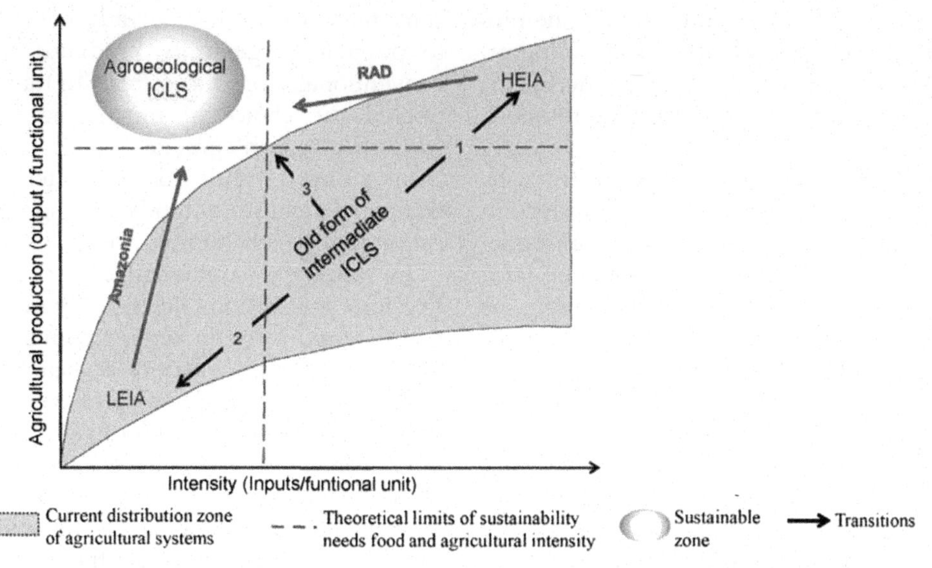

FIGURE 15.2 Trajectories of different integrated crop-livestock systems (ICLS) in the set of currently feasible systems. HEIA and LEIA stand for High and Low External Input Agriculture, the two extremes on a gradient of old forms of ICLS. *Arrows 1 and 2* illustrate the dynamics on the continuum of systems between the two extremes: arrow 1 corresponds to conventional intensification, and arrow 2 corresponds to ecologization. *Red arrows (dark gray in print version)* illustrate Amazonian and the RAD (Réseau Agriculture Durable; a network of sustainably managed French farms) case studies. Arrow 3 corresponds to orthogonal dynamics: agroecological transition that does not necessarily optimize production but decreases the input/production ratio. Agroecological ICLS could be a way to go beyond the current feasible systems in increasing the agricultural production without increasing the intensity, which has negative effects on biodiversity. New agroecological ICLS could thus be an opportunity to manage trade-offs between production and the environment while enhancing ecosystem services without declining production levels.

UNDERSTANDING THE DRIVERS AND ENCOURAGING THE DEVELOPMENT OF SOPHISTICATED INTEGRATED CROP-LIVESTOCK SYSTEMS

Even with potential advantages of agroecological integrated crop-livestock systems the current drivers in the political and market context are not favoring transitions toward such systems. For example, in Europe, despite a renewed interest, mixed farming systems are still declining and are only about 14% of agricultural systems across the European Union (European Commission, 2015). As mentioned before, most of the current mixed farms in Europe are just coexistence between crops and livestock with few additional advantages with regard to ecosystem services. Specialization and intensification of farms and regions in Europe has been strongly driven by economics and politics (Veysset et al., 2005). Global markets have favored cash crops, with high prices for cereals that encouraged abandonment of livestock when the soil and climatic conditions were favorable to cash crops. The first pillar of the EU's Common Agricultural Policy (CAP) helped with investments to modernize agriculture, such as for irrigation and land management improvements, which favored the intensification and specialization of cash crop production. This has had a drastic impact on mixed farming even with agro-environmental subsidies of the second pillar of CAP encouraging farmers to maintain grasslands and therefore livestock on their farm

(Ryschawy et al., 2013). As a result of economies of scale, farm size increased all over the European Union, and the agricultural workforce declined.

The current status of integrated crop-livestock systems in developed countries is particularly worrying with regard to labor opportunity costs. When farmers quit livestock or crop production due to lack of suitably skilled labor, reintroducing crop-livestock integration later at the farm scale is no longer a reasonable possibility, as the skills needed are no longer available (Peyraud et al., 2014; Ryschawy et al., 2013). Therefore, regional integration between crops and livestock could potentially be developed through exchanges between specialized crop and livestock farms (Moraine et al., 2014). These exchanges could have similar potential to provide more ecosystem services. Solutions for problems at the farm scale could be found through this regional approach, but other important problems arise (Moraine, 2015; Ryschawy et al., 2017). Coordination between farmers highlights new constraints that have to be addressed and would likely be more complex to deal with than farm-level approaches (e.g., logistics to transport and store feed and manure, trade-offs between individual and collective performances). Considering the past and current drivers favoring specialization, new policies could be developed to encourage the readoption or maintenance of mixed systems both at the farm and local scale. These policies could be considering, for instance, autonomy of farms for inputs, favoring recycling of nitrogen, phosphorous, and other nutrients within the farm. Agro-environmental measures could favor maintaining seminatural elements in a more incentivized way, as is the case in the current second pillar of the CAP. Finally, some payment for ecosystem services could be developed considering improvements in soil quality through diversified crop rotations, proper use of manure, or local self-sufficiency in inputs. These environmentally beneficial practices could be encouraged by public policies and/or paid for by consumers. Consumers already have the choice to purchase organically produced instead of conventionally produced food, so they could be offered the choice of purchasing food from state-of-the-art integrated crop-livestock farms, then indirectly pay for ecosystem services, through purchases of foods with specific certifications and labels (Simons, 2015).

In summary, sophisticated integrated crop-livestock production systems offer many advantages for reducing the environmental impact of producing a high quantity of healthful, nutritious food, and the diversity of crops and livestock produced is important in several respects, but not the only aspects to consider. The variety of ecosystems services provided by these systems is important too, as is the biodiversity they support. To adjust to the high environmental performance expected, these systems need to be truly integrated to maximize recycling of nutrients. There are some sophisticated integrations of crop and livestock production systems in temperate regions of the world that are currently unique. The unique systems likely have improved agroecosystem functioning, provide more ecosystem services, and in many cases more profitability for farmers, especially if they market their products directly to consumers who are willing to pay to support farming practices that externalize fewer of their costs to the environment.

References

Bail, J., Meneses, K., Demark-Wahnefried, W., 2016. Nutritional status and diet in cancer prevention. Seminars in Oncology Nursing 32, 206–214.

Bell, L.W., Moore, A.D., Kirkegaard, J.A., 2014. Evolution in crop-livestock integration systems that improve farm productivity and environmental performance in Australia. European Journal of Agronomy 57, 10–20.

Bonaudo, T., Bendahan, A.B., Sabatier, R., Ryschawy, J., Bellon, S., Leger, F., Magda, D., Tichit, M., 2014. Agroecological principles for the redesign of integrated crop-livestock systems. European Journal of Agronomy 57, 43–51.

Chambaut, H., Fiorelli, J.L., Espagnol, S., Foray, S., Maignan, S., Leterme, P., 2015. Enhancing the complimentarity between crops and livestock production on farms to improve the environmental sustainability of food production. Rencontres autour des Recherches sur les Ruminants. 22, 61–64.

Clark, M.S., Gage, S.H., 1996. Effects of free-range chickens and geese on insect pests and weeds in an agroecosystem. American Journal of Alternative Agriculture 11, 39–47.

Dumont, B., Fortun-Lamothe, L., Jouven, M., Thomas, M., Tichit, M., 2013. Prospects from agroecology and industrial ecology for animal production in the 21st century. Animal 7, 1028–1043.

European Commission, 2015. EU Farm Economics Overview Based on 2012 FADN Data. Report from the Directorate-General for Agriculture and Rural Development. Brussels, Belgium, 72 pp.

Foley, J.A., Ramankutty, N., Brauman, K.A., Cassidy, E.S., Gerber, J.S., Johnson, M., Mueller, N.D., O'Connell, C., Ray, D.K., West, P.C., Balzer, C., Bennett, E.M., Carpenter, S.R., Hill, J., Monfreda, C., Polasky, S., Rockström, J., Sheehan, J., Siebert, S., Tilman, D., Zaks, D.P.M., 2011. Solutions for a cultivated planet. Nature 478, 337–342.

Galloway, J.N., Burke, M., Bradford, G.E., Naylor, R., Falcon, W., Chapagain, A.K., Gaskell, J.C., McCullough, E., Mooney, H.A., Oleson, K.L.L., Steinfeld, H., Wassenaar, T., Smil, V., 2007. International trade in meat: the tip of the pork chop. Ambio 36, 622–628.

Garnett, T., Appleby, M.C., Balmford, A., Bateman, I.J., Benton, T.G., Bloomer, P., Burlingame, B., Dawkins, M., Dolan, L., Fraser, D., Herrero, M., Hoffmann, I., Smith, P., Thornton, P.K., Toulmin, C., Vermeulen, S.J., Godfray, H.C.J., 2013. Sustainable intensification in agriculture: premises and policies. Science 341, 33–34.

Hillel, D.J., 1991. Out of the Earth: Civilization and the Life of the Soil. The Free Press, New York.

Kirschenmann, F.L., 2007. Potential for a new generation of biodiversity in agroecosystems of the future. Agronomy Journal 99, 373–376.

Kirschenmann, F.L., 2008. Food as relationship. Journal of Hunger and Environmental Nutrition 3, 106–121.

Large, R.V., 1973. Factors affecting the efficiency of protein production by populations of animals. In: Jones, J.G.W. (Ed.), The Biological Efficiency of Protein Production. Cambridge University Press, London, pp. 183–199.

Lemaire, G., Franzluebbers, A., Carvalho, P.C.F., Dedieu, B., 2014. Integrated crop-livestock systems: strategies to achieve synergy between agricultural production and environmental quality. Agriculture, Ecosystems and Environment 190, 4–8.

Ludwig, D.S., 2011. Technology, diet, and the burden of chronic disease. Journal of the American Medical Association 305, 1352–1353.

Martin, G., Duprat, A., Goby, J.-P., Theau, J.-P., Roinsard, A., Descombes, M., Legendre, H., Gidenne, T., 2016. Herbage intake regulation and growth of rabbits raised on grasslands: back to basics and looking forward. Animal 10, 1609–1618.

Montgomery, D.R., 2007. Dirt: The Erosion of Civilizations. University of California Press, Berkeley.

Moraine, M., Duru, M., Nicholas, P., Leterme, P., Therond, O., 2014. Farming system design for innovative crop-livestock integration in Europe. Animal 8, 1204–1217.

Moraine, M., 2015. Conception et evaluation de systemes de production intégrant cultures et élevage à l'échelle du territoire (Ph.D. thesis, Toulouse, France), p. 200.

Mozaffarian, D., Ludwig, D.S., 2010. Dietary guidelines in the 21st century – a time for food. Journal of the American Medical Association 304, 681–682.

Peyraud, J.-L., Taboada, M., Delaby, L., 2014. Integrated crop and livestock systems in Western Europe and South America: a review. European Journal of Agronomy 57, 31–42.

Provenza, F.D., Meuret, M., Gregorini, P., 2015. Our landscapes, our livestock, ourselves: restoring broken linkages among plants, herbivores, and humans with diets that nourish and satiate. Appetite 95, 500–519.

Russelle, M.P., Entz, M.H., Franzluebbers, A.J., 2007. Reconsidering integrated crop-livestock systems in North America. Agronomy Journal 99, 325–334.

Ryschawy, J., Choisis, N., Choisis, J.P., Gibon, A., 2013. Paths to last in mixed crop-livestock farming: lessons from an assessment of farm trajectories of change. Animal 7, 673–681.

Ryschawy, J., Martin, G., Moraine, M., Duru, M., Therond, O., 2017. Designing crop-livestock integration at different levels: toward new agroecological models? Nutrient Cycling in Agroecosystems 108, 5–20.

Schiere, J.B., Ibrahim, M.N.M., van Keulen, H., 2002. The role of livestock for sustainability in mixed farming: criteria and scenario studies under varying resource allocation. Agriculture, Ecosystems and Environment 90, 139–153.

Shepard, M., 2013. Restoration Agriculture. Acres U.S.A., Austin.

Simons, L., 2015. Changing the Food Game: Market Transformation Strategies for Sustainable Agriculture. Greenleaf Publishing, Sheffield.

Sossidou, E.N., Dal Bosco, A., Elson, H.A., Fontes, C.M.G.A., 2011. Pasture-based systems for poultry production: implications and perspectives. World's Poultry Science Association 67, 47–58.

Soussana, J.-F., Lemaire, G., 2014. Coupling carbon and nitrogen cycles for environmentally sustainable intensification of grasslands and crop-livestock systems. Agriculture, Ecosystems and Environment 190, 9–17.

Steinfeld, H., Wassenaar, T., 2007. The role of livestock production in carbon and nitrogen cycles. Annual Review of Environment and Resources 32, 271–294.

Stukenholtz, P.D., Koenig, R.T., Hole, D.J., Miller, B.E., 2002. Partitioning the nutrient and nonnutrient contributions of compost to dryland-organic wheat. Compost Science and Utilization 10, 238–243.

Sulc, R.M., Franzluebbers, A.J., 2014. Exploring integrated crop-livestock systems in different ecoregions of the United States. European Journal of Agronomy 57, 21–30.

Sulc, R.M., Tracy, B.F., 2007. Integrated crop-livestock systems in the U.S. corn belt. Agronomy Journal 99, 335–345.

Teillard, F., Jiguet, F., Tichit, M., 2015. The response of farmland bird communities to agricultural intensity as influenced by its spatial aggregation. PLoS One. https://doi.org/10.1371/journal.pone.0119674.

Udawatta, R.P., Jose, S., 2012. Agroforestry strategies to sequester carbon in temperate North America. Agroforestry Systems 86, 225–242.

Veysset, P., Bebin, D., Lherm, M., 2005. Adaptation to Agenda 2000 (CAP reform) and optimisation of the farming system of French suckler cattle farms in the Charolais area: a model-based study. Agricultural Systems 83, 179–202.

Wilkins, R.J., 2008. Eco-efficient approaches to land management: a case for increased integration of crop and animal production systems. Philosophical Transactions of the Royal Society B 363, 517–525.

Further Reading

Jones, D.L., Cross, P., Withers, P.J.A., DeLuca, T.H., Robinson, D.A., Quilliam, R.S., Harris, I.M., Chadwick, D.R., Edward-Jones, G., 2013. Nutrient stripping: the global disparity between food security and soil nutrient stocks. Journal of Applied Ecology 50, 851–862.

Knez, M., Graham, R.D., 2013. The impact of micronutrient deficiencies in agricultural soils and crops on the nutritional health of humans. In: Selinus, O., Alloway, B., Centeno, J.A., Finkelman, R.B., Fuge, R., Lindh, U., Smedley, P. (Eds.), Essentials of Medical Geology, Revised Edition. Springer Science+Business Media, Dordrecht, pp. 517–533.

National Research Council, 2010. Toward Sustainable Agricultural Systems in the 21st Century. The National Academies Press, Washington, D.C.

CHAPTER

16

Integrated Crop-Livestock Systems as a Solution Facing the Destruction of Pampa and Cerrado Biomes in South America by Intensive Monoculture Systems

Anibal de Moraes[1], *Paulo César de F. Carvalho*[2], *Carlos Alexandre Costa Crusciol*[3], *Claudete Reisdorfer Lang*[1], *Cristiano Magalhães Pariz*[4], *Leonardo Deiss*[5], *R. Mark Sulc*[6]

[1]University Federal of Paraná (UFPR), Curitiba, Brazil; [2]University Federal of Rio Grande do Sul (UFRGS), Porto Alegre, Brazil; [3]São Paulo State University (UNESP), College of Agricultural Science, Botucatu, Brazil; [4]UNESP, School of Veterinary Medicine and Animal Science, Botucatu, Brazil; [5]Doutorando, University Federal of Paraná (UFPR), Curitiba, Brazil; [6]The Ohio State University, Columbus, OH, United States

INTRODUCTION

While global demand for food increases, agricultural expansion faces more stringent environmental preservation demands and sustainability laws aimed to prevent deforestation (Nascente and Crusciol, 2012). Integrated crop-livestock systems (ICLS) are diversified agroecosystems that can contribute to sustainable intensification, increasing food production while maintaining or improving environmental quality and preserving natural biodiversity. A conceptual representation of ICLS can be drawn from the description of how crop and livestock activities interact. From a field- to territorial-level scale (e.g., landscape, watershed, or region), crop and livestock production can be structurally independent or interact in time and space dimensions (Moraine et al., 2016). ICLS are designed to achieve synergy and emergent properties resulting from the spatial and temporal interactions among the components soil, plant, animal, and atmosphere (Moraes et al., 2014a,b).

When creating a spatial-temporal design for ICLS, one must consider the climate conditions (i.e., temperature and precipitation) and soil characteristics, which will govern the animal and plant species to be raised, as well as other

Agroecosystem Diversity
https://doi.org/10.1016/B978-0-12-811050-8.00016-9

agronomic practices that will allow their productive growth. At the same time, the corresponding land use and its capacity must be considered while mitigating risk and improving soil and water preservation (Herrero et al., 2010; Moraine et al., 2016). Moreover, conservation practices (Palm et al., 2014) may be part of the ICLS design. Finally, ICLS must be designed to increase climate-change resiliency through soil organic matter management, improved water harvest and conservation, and increased agrobiodiversity (Altieri et al., 2015). Therefore, when planning an ICLS, one cannot simply prioritize a given crop in years of favorable prices. A main principle is to work with diversity, maintaining rotations in a way that favors the whole system over the long term.

The best-characterized ICLS in South America are the ones that integrate activities within the farm rather than between farms, being described by alternating crop and pasture over time in the same area. ICLS in this region are predominantly characterized by the rotation, succession, or mixtures of annual pastures and crops, in no-tillage systems, where the pasture component is used to produce meat, milk, and wool. Nonetheless, a better characterization of ICLS at the territorial level in South America is still emerging.

Here, we will explore how agricultural production can be intensified with ICLS within the same area or at the territorial level, in two distinct biomes in South America. We will evaluate how this intensification can result in the same or greater agricultural production, increasing both biodiversity and the environmental quality of agroecosystems, concomitantly with avoiding destruction of intact biomes.

DEGRADATION OF PAMPAS AND CERRADOS BIOMES BY INTENSIVE MONOCULTURE SYSTEMS

Crop and livestock activities can compromise the sustainability of agriculture, especially when they lead to declining grain yields, soil fertility, and carrying capacity of pastures, while increasing soil erosion and pasture degradation. In Brazilian Cerrado, the intensive monoculture over the years in pasture areas caused the degradation of approximately 80 million hectares (Mha), and nearly 50 Mha have severe degradation with low forage yield, low animal stocking rate, and soil erosion (Pariz et al., 2011a). This unfortunate reality is also true for the Brazilian Pampas, from which ~11.5 Mha have been misplaced by soil degradation and/or land use conversion to sole-cropping systems or exotic pastures (CSR/IBAMA, 2010) (Fig. 16.1). In agricultural areas, the intensive monoculture over the years, without crop rotation using cover crops that lay straw on the soil surface to maintain or increase soil organic matter content that improves soil physical, chemical, and biologic characteristics, also has caused land degradation (Crusciol et al., 2016). In pasture and agricultural areas with sandy soils, the main problem is soil erosion (Fig. 16.2), and in areas with clay soil, the main problem is soil compaction, which compromises root growth and water infiltration, with subsequent reflection on soil erosion as well. Thus, potential exists for ICLS with no-tillage to recover these degraded areas while reducing production costs and enhancing utilization throughout the year, all of which could generate positive socioeconomic and environmental outcomes (Pariz et al., 2017a).

TYPES OF INTEGRATED CROP-LIVESTOCK SYSTEMS IN SOUTH AMERICA

Two main types of ICLS in South America exist: (1) within a farm and (2) among farms at the territorial level. Each type has many variants, and they will be described in the following paragraphs. Although these two types of integration can coexist and even interact, they do not have the same benefits regarding matter and energy

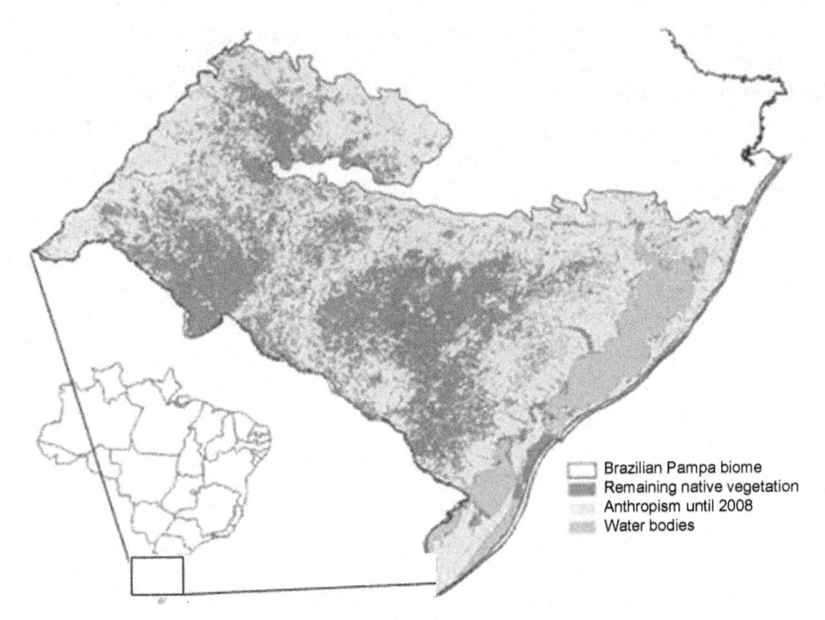

FIGURE 16.1 Degradation of Brazilian Pampas biome mainly promoted by intensive monoculture systems. *Adapted from Centro de Sensoriamento Remoto do Instituto Brasileiro do Meio Ambiente e dos Recursos Naturais Renováveis – CSR/IBAMA, 2010. Monitoramento do desmatamento nos biomas brasileiros por satélite. Ministério do Meio Ambiente, Brasília, 37 p.*

FIGURE 16.2 Pasture areas in Brazilian Cerrado with sandy soils erosion. *Credit: Cristiano M. Pariz.*

recycling, as well as efficiency of using biotic and abiotic resources (Moraine et al., 2014).

A farm under ICLS management must be holistically understood, considering the property as a whole, including all activities involved, in a manner that all productive components optimize the use of abiotic and biotic resources, and where land is used according to its capacity. Using a diversity of spaces and resources can increase the resilience of production systems to biotic and abiotic adversities and allows development of new rural activities (Darnhofer et al., 2010). Some crop and livestock activities can compromise the sustainability of agriculture, especially when they lead to declining grain yields, soil fertility, and carrying capacity of pastures, and increasing soil erosion and pasture degradation (Crusciol et al., 2010). Characterizing key interactions within these systems relies primarily on the knowledge about animals' direct and indirect effect (Bell and Moore, 2012). Biomass consumption and deposition in situ is considered a direct effect, for example, when animals are being raised in pastures. As an indirect effect, one can consider the exogenous manure application. However, in the latter case, synergic effects resulting from the soil-plant-animal interaction, such as greater soil organic carbon accumulation resulting from grazing stimuli on root and shoot growth, are not expected to be preeminent (Moraes et al., 2014a). Other indirect effects may arise, for example, soil fertility improvements from manure application, which can also result in greater soil organic carbon accumulation.

Within a farm, crop and livestock activities can occur at same or disjointed times. To attain simultaneous activities, a farm must have a combination of production systems functioning interactively, and it may allow ruminant animals to stay in the farm during all seasons across the years. In contrast, disjointed crop and livestock activities can occur through temporal variation between the activities. Therefore, in this case, animals may leave the farm before the crop activities begin. This latter class is out of scope for this investigation, but it has been adopted in many regions of South America, for example in Uruguay, where ICLS have been managed continuously for 2 or 3 years with the crop phase followed by 2 or 3 years of the pasture phase. They have observed that introducing the pasture phase in sole-cropping systems have increased crop grain yields, net income, and soil organic carbon stocks and improved other soil characteristics (Fernandez, 1992; García-Prechac et al., 2004; Ernst and Siri-Prieto, 2009).

The great majority of studies about ICLS have considered the environmental, economic, and social performance at the farm scale (Ryschawy et al., 2012; Botreau et al., 2014). Considering that introducing ruminant animals on specialized cropped farms is often not a simple task, some authors have proposed investigating the potential of integrating crop and livestock activities among farms, i.e., at the territorial level (Lemaire et al., 2014; Peyraud et al., 2014; Soussana and Lemaire, 2014). Our understanding of territorial level follows the definition proposed by Moraine et al. (2016): the geographic dimension where the ICLS and the natural resources management occur. In this scale, there remains to be solved many methodological issues and conceptual gaps about how to analyze and dimension interactions between crop and livestock activities (Tanaka et al., 2008; Randrianasolo et al., 2010; Moraine et al., 2016).

Farmer's resistance to the introduction of ruminant animals into cropped areas is a common phenomenon in South America. However, the many reasons cited by farmers and uninformed technical advisors are frequently based on empirical factors such as soil compaction, residue reduction for no-tillage system, and soil nutrient depletion. These factors can indeed happen as a result of livestock grazing in cropland, but they are due to the mismanagement of ruminant animals rather than the animal presence itself. A good example of animal mismanagement is overgrazing, which may reduce soil surface cover and increase soil compaction. In like fashion, the lack of replenishment of exported nutrients is one of the main reasons of soil degradation in South America, which may reduce both crop and pasture production, and consequently animal production, resulting in declining system performance as a whole (Anghinoni et al., 2013).

At the territorial level, the ICLS development will depend upon the direct relationships between farmers or indirect relationships mediated by intermediaries. Key ecologic and governmental attributes must be related to scale out ICLS among farms (Biggs et al., 2012). From the simpler to the more complex interactions,

ICLS at territorial level can occur as coexistence, complementarity, local synergy, and territorial synergy (Moraine et al., 2016). Typically, bigger and highly specialized farms, in the corporate model, can organize cooperation between crop and livestock activities following the concepts of industrial ecology and maintaining economies of scale, while medium to small farms can benefit from the diversity of land use and activities (Moraine et al., 2014).

INTEGRATED CROP-LIVESTOCK SYSTEMS IN THE PAMPA AND CERRADO BIOMES

In the following sections, we will describe the main agronomic characteristics of the Pampa and Cerrado biomes and discuss how ICLS can be used to preserve intact areas, while increasing both biodiversity and the environmental quality of agroecosystems.

Integrated Crop-Livestock Systems in the Pampa Biome

The "Pampa" biome is located in the subtropical region of South America, covering an area of ~18 million ha, including all of Uruguay and parts of Argentina and southern Brazil. The climate is characterized by hot summers and cold winters, without a definite dry season. Of the total biome area, 36% remains as the original vegetation. Despite the many possible crop rotations in a subtropical environment, the main farming systems found in this region are specialized sole-crop systems or degraded natural pastures. We will explore scenarios aiming to point out alternatives to mitigate the destruction of the original Pampa biome, while maintaining environmental quality when producing crops, livestock, or both.

Despite the Pampa biome having suitable climatic conditions for forest development, it is a grassland-dominated environment (Roesch et al., 2009) comprising at least seven different physiographic formations: savanna, steppe, steppe-savanna, coast, transition areas, and patches of seasonal deciduous and semideciduous forests. In spite of high diversity in the Pampa grasslands, studies on plant diversity and distribution within the Brazilian Pampa indicate a typical pattern of high species richness but with low relative abundance (Boldrini et al., 1998; Overbeck et al., 2006; Freitas et al., 2009). The main botanical families found in the grasslands of the Brazilian Pampa are Poaceae, Asteraceae, Cyperacea, Fabaceae, Apiacea, Oxalidaceae, Verbanaceae, and Iridaceae (Overbeck et al., 2006, 2007). According to the Brazilian Institute of Geography and Statistics (IBGE, 2004), from 1970 to 2002, the Brazilian Pampa area occupied by grassland decreased from 14 to 8.9 million hectares because native areas were converted into other uses. Such reductions of native grassland areas may have led to a sensitive loss of biodiversity in the Pampa biome (Roesch et al., 2009).

This land use change has been occurring due to low pasture productivity of improperly managed native grassland areas (e.g., high grazing intensity, in which pasture production cannot sustain ruminant demand for feed, along with no replenishment of exported nutrients). Moreover, the high prices that are sometimes paid for other crops, such as soybean or corn, makes poorly managed livestock less compelling. Driven by high prices of soybean, the soybean area increased by 210% between 2000 and 2010, at the expense of 2 million ha (5%) in Pampas and Campos grasslands of southern South America (Modernel et al., 2016). The Pampa grassland potential to produce biomass to feed ruminants occurs mainly in the summer. Therefore, if farmers want to rely on pasture production during winter, they need to introduce winter species. This kind of grassland is called improved native pastures. Fertility improvement is mandatory for both summer and winter phases, since the Pampa biome generally has low natural soil fertility (Roesch et al., 2009). However, fertilizer inputs must be added cautiously, in a not too intensive way, so the native species are not irreversibly suppressed by the exotic introduced species (e.g., *Avena strigosa*, *Lolium multiflorum*, and *Trifolium* sp.).

What we understand as the worst possible scenario for the Pampa biome is the conversion of grassland areas that can be cropped to being used exclusively for cropping activities (i.e., crop only). This could lead to a profound loss of original/native biodiversity, since choice of crops is typically governed by favorable prices; therefore, fewer crops are rotated, which reduces diversity of the agroecosystem. Another aspect that can also be negatively impacted is the soil organic carbon stock. After 45 years of continuous cropping in the Pampa region of Uruguay, without and with N and P fertilization, the soil organic carbon was reduced 45% and 29%, respectively (Morón and Sawchik, 2002; Sawchik et al., 2011).

To understand how an ICLS can be designed within a farm, in a hypothetical situation where animals will be fed on pasture only, it is necessary to measure or estimate both the herbage allowance and potential animal intake, i.e., the forage balance considering demand and supply. The forage balance can be designed for the short, medium, and long terms, and this allows farmers to make decisions regarding land use, animal number and category, pasture species, and inputs (Poli and Carvalho, 2001). Different pasture species have different genetic production potential and depend on the input levels (Table 16.1). In addition, one can predict the potential animal intake based on the animal number and its category. For example, cattle for meat production have a potential intake equivalent to 2.5% of animal live weight (NRC, 1985). Calculating the pasture area necessary to feed a given number of animals during a certain time, it is possible to designate other areas to another land use without compromising the animal live weight gain, therefore contributing to the sustainable intensification of the production system as a whole.

In the Pampa region, the climate favors growth of C3 grasses in the winter and C4 grasses in the summer, in mixture or not, and the overall biomass production will reflect the added inputs, especially nitrogen fertilizers (Table 16.1). The dynamic response of plant biomass production across seasons, resulting from the physiologic response of the plant species and the weather conditions during the different seasons, allows for maintenance of the same ruminant live weight carried in the winter on a smaller area during the summer. For this reason, where integration within a farm where simultaneous crop and livestock activities occur, ICLS have been designed to enable, at a minimum, pasture availability for the same ruminant live weight during all seasons of the year. This can be done by maintaining part of a farm's summer area (~25%–50%) with annual or perennial C4 pasture species (instead of sole-cropping all the area during the summer) to maintain the same ruminant live weight supported on all the farm's area during the winter pasture phase. Many arrangements can be established to take advantage of the environmental conditions (soil and climate) where the farm is located, along with the management conditions (investment degree on abiotic inputs). Finally, it is important to consider that summer pasture areas, including the ones occupied with perennial species, can vary across space within the farm over time (i.e., rotations of pastures).

Following the previous description, it is possible to design an ICLS within a farm that has simultaneous activities of crop and livestock, having animals present all year in the Pampa region, and at the same time, preserving part of the farm in native grassland. We understand that the partial conversion of native pasture areas to summer cropping is a better option than the complete conversion to cropland. Therefore, considering ~50%–75% of the summer area to

TABLE 16.1 Estimated Total Dry Matter Production and Temporal Distribution of the Main Forage Species for the Subtropical and Temperate Environments in the Pampa Region, Under Three Levels of Soil Fertility and/or Fertilizer Input

Climate and Pasture Species	Pasture Condition			Temporal Distribution of Pasture Production											
	Degraded	Intermediate	Optimal	JAN	FEB	MAR	APR	MAY	JUN	JUL	AUG	SEP	OCT	NOV	DEC
	Total Dry Matter Production (Mg ha^{-1})			Percent of Total Dry Matter Production											
SUBTROPICAL (CFA)															
Native Pampa grasslands	3–5	6–11	11–20	20	15	7	3	2	2	1	2	3	10	15	20
Summer perennial grasses	3–5	6–11	11–20	20	15	7	3	2	2	1	2	3	10	15	20
Summer annual grasses	3–5	6–11	11–20	25	15	10	0	0	0	0	0	0	0	20	30
Winter Annual Grasses															
Avena strigosa	2–3	4–6	7–10	0	0	0	0	20	30	30	20	0	0	0	0
Lolium multiflorum	2–3	4–6	7–10	0	0	0	0	5	25	30	20	15	5	0	0
A. Strigosa + *L. multiflorum*	2–3	4–6	7–10	0	0	0	0	10	20	25	25	15	5	0	0
Secale cereale	2–3	4–6	7–10	0	0	0	10	30	20	20	20	0	0	0	0
TEMPERATE (CFB)															
Native Pampa grasslands	3–5	6–11	11–20	22	18	8	2	0	0	0	0	2	5	21	22
Summer perennial grasses	3–5	6–11	11–20	22	18	8	2	0	0	0	0	2	5	21	22
Summer annual grasses	3–5	6–11	11–20	30	25	15	0	0	0	0	0	0	0	5	25
Winter Annual Grasses															
Avena strigosa	2–3	4–6	7–10	0	0	0	0	20	30	20	20	10	0	0	0
Lolium multiflorum	2–3	4–6	7–10	0	0	0	0	0	10	20	25	30	15	0	0
A. Strigosa + *L. multiflorum*	2–3	4–6	7–10	0	0	0	0	10	20	25	25	15	5	0	0
Secale cereale	2–3	4–6	7–10	0	0	0	10	30	20	20	20	0	0	0	0

Summer perennial grasses: *Brachiaria* spp., *Panicum* spp., and *Cynodon* spp. Summer annual grasses: *Sorghum* spp. and 273 millets.

be cropped or to have higher proportion of native pasture areas, many arrangements can be done to fit the farmer's wishes. In this way, it is possible to diversify the farm production through ICLS, while maintaining high levels of productivity of both animal and plant components and preserving at least ~25%–50% of the original biome. Moraes et al. (2014a) summarized many experimental results showing that summer cropping following winter pastures did not result in crop yield reduction, and in some cases, yield improvements were observed when compared with sole-cropping systems. It is important to state that most of those results reflect areas under moderate grazing intensity and adequate soil fertility. Those crop yield improvements have been attributed to emergent properties such as lower aluminum toxicity, due to a higher rate of the downward movement of lime into the soil, and higher microbiological functional diversity, which can improve nutrient cycling dynamics through improved mineralization processes (Carvalho et al., 2010; Deiss et al., 2016; Martins et al., 2014a,b, 2016).

If a farm is exclusively used for livestock production, we believe other alternatives may be used to sustain the animal demand for feed, without compromising native Pampa grasslands. Knowing that the same livestock live weight carried in summer cannot be maintained on the same area during the winter, due to lower pasture availability, we believe that the alternative is to do ICLS at the territorial level. This can be done by moving animals to finish their cycle in nearby areas where there is greater pasture availability. This is a practice already being done by farmers in the Pampa region of Brazil, in which ruminants are born and stay on the original farm for the beginning phase of their development during the summer phase because there is native pasture availability, then the animals are moved to other regions of Rio Grande do Sul state, namely, e.g., *Misões* or *Planalto Médio*, to continue the livestock cycle without compromising animal live weight gain during the winter phase.

This has allowed farmers to maintain livestock activities and preserve the Pampa biome. Therefore, if complementary cropping areas could be used to finish livestock production, then the excessive grazing pressure imposed on natural Pampa grasslands during winter that is responsible for vegetation degradation would be reduced. In the whole Rio Grande do Sul state in 2015–16, 14.7% of the summer cropped area was cultivated with winter crops (Table 16.2). So, introducing winter pastures through ICLS on the remaining 85.3% would provide an opportunity to move animals from the Pampa region to where there is greater feed availability and promote a better natural vegetation grassland utilization. Finally, this would lead to a "virtuous cycle" for vegetation restoration, and then ecosystem preservation. Along with preserving the Pampa natural vegetation, many ecosystem services can be provisioned, such as regulation services of herbivory, pollination, and seed or propagule dispersion, support services of habitat availability, nutrient cycling, and carbon sequestration, and cultural services of culture preservation. However, still needed are documented results that enable us to measure and explore the environmental, economic, and social impact of such practices in the Pampa region.

In summary, concrete alternatives to enhance agrobiodiversity (including other nearby regions to enable ICLS at territorial level) and to preserve native vegetation in the Pampa biome are the rotation or succession of summer crops (*Glycine max, Zea mays, Phaseolus vulgaris*, or *Oryza sativa*) with winter annual grazed species (mixed or solely *Avena strigosa, Lolium multiflorum*, and *Trifolium* sp.), which results in preservation of native grasslands in the summer phase, or to move animals from farms with exclusive livestock activities to nearby areas during the winter phase. These practices enable a

TABLE 16.2 Cultivated Area for Grain Production on Summer and Winter Seasons in Rio Grande do Sul State, Brazil, Harvest 2015–16 (CONAB, 2017)

Season	Crops	Planted Area (1000 ha)	Relative Contribution (%)
Summer	Common bean (*Phaseolus vulgaris*)	67.9	0.9
	Maize (*Zea mays*)	823.0	11.1
	Peanut (*Arachis hypogaea*)	3.4	0.0
	Rice (*Oryza sativa*)	1076.0	14.5
	Sorghum (*Sorghum bicolor*)	9.0	0.1
	Soybean (*Glycine max*)	5455.0	73.4
	Sugarcane (*Saccharum officinarum*)	1.2	0.0
	Total summer area	7435.5	100
Winter	Barley (*Hordeum vulgare*)	51.80	4.7
	Canola (*Brassica napus*)	41.2	3.8
	Oat (*Avena sativa*)	218.3	19.9
	Rye (*Secale cereale*)	1.5	0.1
	Triticale (× *Triticosecale*)	5.7	0.5
	Wheat (*Triticum aestivum*)	776.9	70.9
	Total winter area	1095.4	100
	Winter to summer area ratio (%)	14.73	

moderate grazing intensity in native Pampa grasslands and do not compromise environmental quality.

Integrated Crop-Livestock Systems in the Cerrado Biome

As a contrast, we now discuss possible scenarios in regions with dry winters and high temperatures throughout year, such as the Brazilian "Cerrado," where the use of ICLS is a viable option to increase food production during times of irregular rain and reduced pasture availability. The "Cerrado" biome covers an area of ~204 million ha, predominantly in central Brazil, where only 22% of the area remains under the original vegetation. The Cerrado Region of Brazil contains approximately 80 million hectares of cultivated pasture, and nearly 50 million hectares of the total area is severely degraded with low forage yield, low animal stocking rate (beef, dairy cattle, and sheep), low live weight gain, and low carcass yield per area (Pariz et al., 2011a). In this context, the use of ICLS combined with no-tillage has been used to remedy these negative effects, being a mixed system of utilization, characterized by diversification, crop rotation, intercropping systems, and activities of agriculture and livestock in the same area (Macedo, 2009). These ICLS are characterized by maintenance of permanent soil cover with cover crops (grazed or not) and main crops enabling cycling of nutrients, a process that is extremely important for the sustainability of tropical agriculture in this region where most soils are acidic, have low fertility, and low cation exchange capacities.

One basic principle of no-tillage systems is the use of methods that protect the environment, such as maintaining residue on the soil surface, using crop rotations, and limiting soil disturbance (Borghi et al., 2013a; Crusciol et al., 2015). Although no-tillage systems require residue on the soil surface, cash crops alone often do not produce enough straw to adequately cover the soil throughout the year (Mateus et al., 2016). In such warm conditions, crop residues decompose rapidly (Pariz et al., 2011b), negatively affecting the success of no-tillage systems. Thus, the introduction of intercropped tropical perennial grasses with grain crops could be a key strategy to enhance early establishment and successful production of winter-season forage for grazing in the Brazilian Cerrado, using ICLS combined with no-tillage to provide constant soil cover (Kluthcouski and Aidar, 2003).

There is currently a large array of studies with corn, sorghum, and soybean production intercropped with tropical perennial grasses (mainly *Urochloa* and *Panicum* genus) in the Brazilian Cerrado aimed at producing grain or silage, with subsequent production of forage for animal feed during autumn to spring (Crusciol et al., 2011, 2012, 2013, 2014; Mateus et al., 2011, 2016; Pariz et al., 2016, 2017a,b). A compilation of those results obtained in higher altitude Cerrado conditions (Botucatu, São Paulo, Brazil) demonstrated the feasibility of corn, sorghum, and soybean intercropped with tropical perennial grasses, aimed at producing grain in the summer/autumn and forage in the winter/spring (Table 16.3). The forage produced can be used for grazing, green chop, silage, silage followed by grazing, and hay (Kluthcouski and Aidar, 2003), which enables animal feeding on pasture, semifeedlot, or feedlot, and also provides sufficient residue on the soil surface for continuity of the no-tillage system (Costa et al., 2014; Pariz et al., 2011b; Mendonça et al., 2015; Pereira et al., 2016).

Sorghum is an excellent alternative to other summer grain crops for soils with poor fertility where corn does not thrive, particularly in regions with warm conditions and rain oscillations (e.g., the Brazilian Cerrado and African Savanna) (Mateus et al., 2011, 2016). Results of Borghi et al. (2013a), Costa et al. (2016), and Mateus et al. (2016) demonstrate that sorghum intercropped with tropical grasses resulted in a higher land equivalent ratio and relative nitrogen yield, and greater land use efficiency.

In recent literature about ICLS, it is noteworthy that only about 5% of the studies have directly measured animal responses (Moraes et al., 2014b). The animal component remains poorly studied given the complexity of the pasture cycle and grazing ecology. A diversity of field studies is needed with ruminant livestock, including beef, dairy cattle, and sheep, to adequately characterize the impacts of ICLS on animal performance and crop yield, because stocking rate and management practices can alter crop residue and quantity and quality of forage, as well as affect soil quality.

Results of Pariz et al. (2009, 2010, 2011c,d,e) and Costa et al. (2012, 2016) obtained in an irrigated area under center pivot in low altitude Cerrado conditions (Selvíria, Mato Grosso do Sul, Brazil) demonstrate the feasibility of corn and sorghum intercropped with tropical perennial grasses, aimed at producing grain or silage in the summer/autumn and forage in the winter/spring. Harvesting forage can be used for fodder production for animals (roughage component in the formulation of diets), as well as for silage production. Therefore, this ICLS has been important to enabling the forage production in quantity and quality, mainly for large feedlots, in which the daily requirement of roughage is high. Results of Costa et al. (2015) under the same edaphoclimatic conditions also demonstrated that corn and sorghum silages

TABLE 16.3 Grain Yield of Corn, Sorghum, and Soybean Intercropped With Tropical Grasses in the Summer/Autumn and Forage Dry Matter Production (FDMP) in the Winter/Spring in ICLS Conducted in High Altitude Cerrado Conditions (Botucatu, São Paulo, Brazil)

Cropping Systems	Grain Yield	FDMP
	(Mg ha^{-1})	
Corn + palisade grass (*U. brizantha* cv. Marandu)[a]		
Row spacing		
0.90 m	8.9	20.7
0.45 m	9.9	18.4
Crop management		
Sole corn	10.3	–
Corn intercropping	9.7	19.6
Corn + palisade grass[b]		
Palisade grass sowed simultaneously	10.9	18.2
Palisade grass sowed at topdressing time	10.2	12.4
Corn + guinea grass[b]		
Palisade grass sowed simultaneously	9.8	20.8
Palisade grass sowed at topdressing time	12.1	13.8
Sole corn[b]	10.0	–
Corn + palisade grass[c]		
Sole corn super-early cycle	6.4	–
Sole corn early cycle	7.0	–
Sole corn normal cycle	9.6	–
Sole corn late cycle	9.6	–
Intercropping corn super-early cycle	6.8	8.9
Intercropping corn early cycle	7.3	9.4
Intercropping corn normal cycle	8.9	8.3
Intercropping corn late cycle	6.9	6.6
Sorghum + palisade grass[d]		
Palisade grass sowed simultaneously	5.4	25.4
Palisade grass sowed at topdressing time	6.2	21.8
Sorghum + guinea grass[d]		
Palisade grass sowed simultaneously	5.6	32.3
Palisade grass sowed at topdressing time	6.1	25.5
Sole sorghum[d]	5.4	–
Sorghum + palisade grass[e]		
Sidedress nitrogen rate (0 kg ha^{-1})	3.1	2.5
Sidedress nitrogen rate (200 kg ha^{-1})	3.8	5.3
Sorghum + guinea grass[e]		
Sidedress nitrogen rate (0 kg ha^{-1})	1.7	2.5
Sidedress nitrogen rate (200 kg ha^{-1})	3.1	5.3
Sole sorghum[e]		
Sidedress nitrogen rate (0 kg ha^{-1})	3.1	–
Sidedress nitrogen rate (200 kg ha^{-1})	3.8	–
Soybean + palisade grass[f]		
Sole soybean super-early cycle	2.7	–
Sole soybean early cycle	2.9	–
Sole soybean normal cycle	3.4	–
Sole soybean late cycle	4.0	–
Intercropping soybean super-early cycle	2.9	18.0
Intercropping soybean early cycle	3.0	19.2
Intercropping soybean normal cycle	2.8	15.2
Intercropping soybean late cycle	2.2	12.8
Soybean + palisade grass[g]		
Sole soybean	3.0	–
Palisade grass sowed simultaneously	2.9	18.9
Palisade grass sowed 30 d after the emergence	3.0	18.4
Palisade grass oversowed at soybean stage R6	2.9	14.2

[a] *Borghi et al. (2012).*
[b] *Borghi et al. (2013b).*
[c] *Crusciol et al. (2013).*
[d] *Borghi et al. (2013a).*
[e] *Mateus et al. (2016).*
[f] *Crusciol et al. (2012).*
[g] *Crusciol et al. (2014).*

intercropped with palisade grass (*U. brizantha* cv. Marandu) and guinea grass (*P. maximum* cv. Tanzânia), even with high nutrient export and machine traffic, were efficient in maintaining and even improving soil fertility and soil C stocks. Over a 3-year period, the ICLS led to a reduction in soil compaction through the positive effect of increased macroporosity and total porosity and decreased resistance to penetration and bulk density, in the 0–0.10 and 0.10–0.20 m soil layers.

In a silage crop system, if the crop is cut low to the ground, survival and production of tropical perennial grasses intercropped with cash crops can be negatively affected. Low cutting height could remove valuable tillers, compromising pasture development. Silage harvest also imposes extensive traffic and heavy pressure from loaded equipment (Jaremtchuk et al., 2006). Silage harvest traffic could affect regrowth of grass and pasture formation, in addition to compacting soil. To reduce soil surface compaction, many farmers opt to till after silage harvest, a practice that jeopardizes soil conservation. Therefore, developing sustainable no-tillage silage production systems is needed, especially if combined with promising ICLS production strategies to improve overall agricultural functionality. In this context, results of Pariz et al. (2016) demonstrated the feasibility of corn silage intercropped with palisade grass cut at 0.45 m height rather than at 0.20 m height in the summer/autumn and forage production for grazing lambs in the winter/spring, as well as on production of subsequent silage soybean and soil responses in high-altitude Cerrado conditions (Botucatu, São Paulo, Brazil).

As an innovation, it is also possible to carry out the corn intercropping with palisade grass and pigeon pea (*Cajanus cajan*) or just palisade grass intercropped with pigeon pea, aiming at the production of grains or silage, and subsequent pasture (Garcia et al., 2016). As pigeon pea has a deep tap root system, it contributes to reducing soil compaction and increases the nitrogen supply in the soil via biologic fixation of atmospheric nitrogen; however, ICLS are still limited by the lack of nitrogen, having a high dependence on nitrogen fertilizer use for successful production (Mateus et al., 2016). Protein concentrates are usually added to the diets for animals supplemented with corn, sorghum, or tropical perennial grasses silages, raising the production cost. Thus, the rationale for the use of legumes associated with a grass is primarily to increase the crude protein content of the silage. Regions with dry winters (low and irregular rainfall), such as the Brazilian Cerrado, have large risks in growing a successful dry season crop, resulting in a long fallow period without productivity (Borghi et al., 2013a). Therefore, in these agricultural areas, only one crop per year is the norm. An option to the grain crops intercropped with tropical perennial grasses described before is the sowing of those grasses in February/March or September/October in succession or before the summer crop, respectively, as an alternative in ICLS (Pariz et al., 2011f). Depending on the producer planning, this sowing in spring or autumn can provide quality forage for animals and/or straw for the no-tillage system (Macedo, 2009). This system can also be deployed in agricultural areas with rotation of grain crops (bean, corn, sorghum, rice, cotton, and soybean). In addition, soil cover throughout the year and cover crops to cycle nutrients are also extremely important for sustainable tropical agriculture in the Brazilian Cerrado, where most soils are acidic, have low fertility, low soil organic matter (SOM), and low cation exchange capacities (CEC).

Soil acidity and low natural fertility are the main limiting factors for grain production in tropical regions such as the Brazilian Cerrado. Results of Crusciol et al. (2016) demonstrated that despite the tropical perennial grasses being more tolerant of soil acidity, the surface application of limestone in an ICLS increases the forage dry matter yield, crude protein concentration, and estimated meat production during winter/

spring in a corn-palisade grass intercropping, and it provided the highest total and mean net profit during the four growing seasons. In addition, increases were observed for the yield components of peanut, white oat, and corn, improving the long-term sustainability of tropical agriculture in the Brazilian Cerrado.

Few studies have evaluated soil fertility changes resulting from the cultivation of tropical perennial grasses in areas traditionally used to grow annual crops. Similarly, few studies have examined the effects of grass cover crops on subsequent cash crops. However, the introduction of palisade grass temporary pasture in an ICLS established into a traditional area of annual crop production has been reported to provide higher amounts of plant biomass (Pacheco et al., 2011; Borghi et al., 2008, 2013a,b; Garcia et al., 2008; Crusciol et al., 2012, 2013, 2014; Nascente et al., 2013) and is likely to improve soil fertility, thus improving conditions for the development of subsequent crops. Kluthcouski and Aidar (2003) and Crusciol et al. (2015) reported that soybean, corn, white oat, and common bean produced 15%–25% higher grain yields after palisade grass pasture cultivation than after fallow. Crusciol et al. (2015) attributed this result to the increased soil fertility, based on significantly higher values of pH, SOM, P, K_{ex}, Ca_{ex}, Mg_{ex}, SO4–S, and CEC when compared with areas cultivated without palisade grass pasture. These results demonstrated the great potential that exists for the inclusion of palisade grass pasture through ICLS in cash crop areas, which will allow increased food production from the land area, thereby benefiting the environment and increasing profits so farmers can diversify their activities in a sustainable agriculture system.

As described earlier, for the Cerrado biome, ICLS has been mostly explored regarding ecologic aspects within the farm level. However, there are also some emerging indicators about use of ICLS at the territorial level. Gil et al. (2015) emphasized that in Brazilian conditions, where there are large farms and highly specialized crop or livestock activities, cooperation between farms can help farmers adopt ICLS. According to them, the low adoption of ICLS within the farm has been negatively influenced by low labor availability and poor knowledge of farmers to introduce a new activity on the farm. Thus, coexistence or complementarity between farms where farmers combine their skills in different types of enterprises can improve agroecosystem sustainability, and later facilitate adoption of higher complexity levels of ICLS, at both territorial and field levels.

The potential exists for ICLS combined with no-tillage systems to recover degraded areas, reduce production costs, and enhance land utilization throughout the year, all of which could generate positive socioeconomic and environmental outcomes. Therefore, ICLS could be a key form of ecologic intensification needed for achieving future food security and environmental sustainability. Due to the productive, economic, and environmental benefits, ICLS are considered a "new green revolution in the tropics" and can be a solution facing the destruction of Cerrado biome in South America by intensive monoculture systems.

FINAL CONSIDERATIONS

Many ICLS benefits have been shown with respect to aspects of the soil–plant–animal system. There is evidence that such a system is not only a livestock–agriculture combination, but also a unique system reaching a new complexity threshold, resulting in emergent properties with novel functionalities, some of which have yet to be investigated.

We conclude that in subtropical and tropical regions of South America, ICLS provide an opportunity to intensify with sustainability, restoring the environmental quality and enhancing biodiversity of simplified and specialized agroecosystems, while preserving the original vegetation of these biomes.

References

Altieri, M.A., Nicholls, C.I., Henao, A., Lana, M.A., 2015. Agroecology and the design of climate change-resilient farming systems. Agronomy for Sustainable Development 35, 869–890.

Anghinoni, I., Carvalho, P.C.F., Costa, S.E.V.G.A., 2013. Abordagem sistêmica do solo em sistemas integrados de produção agrícola e pecuária no subtrópico brasileiro. In: Araújo, A.P., Avelar, B.J.R. (Eds.), Tópicos Em Ciência Do Solo, eighth ed. UFV, Viçosa, pp. 221–278.

Bell, L.W., Moore, A.D., 2012. Integrated crop–livestock systems in Australian agriculture: trends, drivers and implications. Agricultural Systems 111, 1–12.

Biggs, R., Schlüter, M., Biggs, D., Bohensky, E.L., Burnsilver, S., Cundill, G., Dakos, V., Daw, T.M., Evans, L.S., Kotschy, K., Leitch, A.M., Meek, C., Quinlan, A., Raudsepp- Hearne, C., Robards, M.D., Schoon, M.L., Schultz, L., West, P.C., 2012. Towards principles for enhancing the resilience of ecosystem services. Annual Review of Environment and Resources 37, 421–448.

Boldrini, I.I., Miotto, S.T.S., Longhi-Wagner, H.M., Pillar, V.D., Marzall, K., 1998. Aspectos florísticos e ecológicos da vegetação campestre do Morro da Polícia, Porto Alegre, RS, Brasil. Acta Botanica Brasilica 12, 89–100.

Borghi, E., Costa, N.V., Crusciol, C.A.C., Mateus, G.P., 2008. Influence of the spatial distribution of maize and *Brachiaria brizantha* intercropping on the weed population under no tillage (in Portuguese, with English abstract). Planta Daninha 26, 559–568.

Borghi, E., Crusciol, C.A.C., Nascente, A.S., Mateus, G.P., Martins, P.O., Costa, C., 2012. Effects of row spacing and intercrop on maize grain yield and forage production of palisade grass. Crop & Pasture Science 63, 1106–1113.

Borghi, E., Crusciol, C.A.C., Nascente, A.S., Souza, V.V., Martins, P.O., Mateus, G.P., Costa, C., 2013a. Sorghum grain yield, forage biomass production and revenue as affected by intercropping time. European Journal of Agronomy 51, 130–139.

Borghi, E., Crusciol, C.A.C., Mateus, G.P., Nascente, A.S., Martins, P.O., 2013b. Intercropping time of corn and palisade grass or guinea grass affecting grain yield and forage production. Crop Science 53, 629–636.

Botreau, R., Farruggia, A., Martin, B., Pomiès, D., Dumont, B., 2014. Towards an agroecological assessment of dairy systems: proposal for a set of criteria suited to mountain farming. Animal 8, 1349–1360.

Carvalho, P.C.F., Anghinoni, I., Moraes, A., Souza, E.D., Sulc, R.M., Lang, C.R., Flores, J.P.C., Lopes, M.L.T., Silva, J.L.S., Conte, O., Wesp, C.L., Levien, R., Fontaneli, R.S., Bayer, C., 2010. Managing grazing animals to achieve nutrient cycling and soil improvement in no-till integrated systems. Nutrient Cycling in Agroecosystems 88, 259–273.

Companhia Nacional de Abastecimento - CONAB, 2017. Safras - Séries Históricas. Available at: http://www.conab.gov.br/.

Costa, N.R., Andreotti, M., Buzetti, S., Lopes, K.S.M., Santos, F.G., Pariz, C.M., 2014. Macronutrient accumulation and decomposition of *Brachiaria* species as a function of nitrogen fertilization during and after intercropping with corn. Revista Brasileira de Ciência do Solo 38, 1223–1233.

Costa, N.R., Andreotti, M., Crusciol, C.A.C., Pariz, C.M., Lopes, K.S.M., Yokobatake, K.L.A., Ferreira, J.P., Lima, C.G.R., Souza, D.M., 2016. Effect of intercropped tropical perennial grasses on the production of sorghum-based silage. Agronomy Journal 108, 2379–2390.

Costa, N.R., Andreotti, M., Gameiro, R.A., Pariz, C.M., Buzetti, S., Lopes, K.S.M., 2012. Nitrogen fertilization in the intercropping of corn with two *Brachiaria* species in a no-tillage system (in Portuguese, with English abstract). Pesquisa Agropecuaria Brasileira 47, 1038–1047.

Costa, N.R., Andreotti, M., Lopes, K.S.M., Yokobatake, K.L., Ferreira, J.P., Pariz, C.M., Bonini, C.S.B., Longhini, V.Z., 2015. Soil properties and carbon accumulation in an integrated crop-livestock system under no-tillage (in Portuguese, with English abstract). Revista Brasileira de Ciência do Solo 39, 852–863.

Crusciol, C.A.C., Soratto, R.P., Borghi, E., Matheus, G.P., 2010. Benefits of integrating crops and tropical pastures as systems of production. Better Crops 94, 14–16.

Crusciol, C.A.C., Mateus, G.P., Pariz, C.M., Borghi, E., Costa, C., Silveira, J.P.F., 2011. Nutrition and sorghum hybrids yield with contrasting cycles intercropped with Marandu grass (in Portuguese, with English abstract). Pesquisa Agropecuaria Brasileira 46, 1234–1240.

Crusciol, C.A.C., Mateus, G.P., Nascente, A.S., Martins, P.O., Borghi, E., Pariz, C.M., 2012. An innovative crop-forage intercrop system: early cycle soybean cultivars and palisade grass. Agronomy Journal 104, 1085–1095.

Crusciol, C.A.C., Nascente, A.S., Mateus, G.P., Borghi, E., Leles, E.P., Santos, N.C.B., 2013. Effect of intercropping on yields of corn with different relative maturities and palisade grass. Agronomy Journal 105, 599–606.

Crusciol, C.A.C., Nascente, A.S., Mateus, G.P., Pariz, C.M., Martins, P.O., Borghi, E., 2014. Intercropping soybean and palisade grass for enhanced land use efficiency and revenue in a no till system. European Journal of Agronomy 58, 53–62.

Crusciol, C.A.C., Nascente, A.S., Borghi, E., Soratto, R.P., Martins, P.O., 2015. Improving soil fertility and crop yield in a tropical region with palisade grass cover crops. Agronomy Journal 107, 2271–2280.

Crusciol, C.A.C., Marques, R.R., Carmeis Filho, A.C.A., Soratto, R.P., Costa, C.H.M., Ferrari Neto, J., Castro, G.S.A., Pariz, C.M., Castilhos, A.M., 2016. Annual crop rotation of tropical pastures with no-till soil as affected by lime surface application. European Journal of Agronomy 80, 88–104.

Centro de Sensoriamento Remoto do Instituto Brasileiro do Meio Ambiente e dos Recursos Naturais Renováveis – CSR/IBAMA, 2010. Monitoramento do desmatamento nos biomas brasileiros por satélite. Ministério do Meio Ambiente, Brasília, 37 p.

Darnhofer, I., Bellon, S., Dedieu, B., Milestad, R., 2010. Adaptiveness to enhance the sustainability of farming systems. A review. Agronomy for Sustainable Development 30, 545–555.

Deiss, L., Moraes, A., Dieckow, J., Franzluebbers, A.J., Gatiboni, L.C., Sassaki, G.I., Carvalho, P.C.F., 2016. Soil phosphorus compounds in integrated crop- livestock systems of subtropical Brazil. Geoderma 274, 88–96.

Ernst, O., Siri-Prieto, G., 2009. Impact of perennial pasture and tillage systems on carbon input and soil quality indicators. Soil and Tillage Research 105, 260–268.

Fernandez, E., 1992. Análisis físico y económico de siete rotaciones de cultivos y pasturas en el suroeste de Uruguay (Economical and Physical Analysis of Seven Crop–Pasture Rotations in Southwestern Uruguay). No. 1 Revista INIA- Uruguay Inv. Agr., Tomo II, pp. 251–271.

Freitas, E.M., Boldrini, I.I., Müller, S.C., Verdum, R., 2009. Florística e fitossociologia da vegetação de um campo sujeito à arenizaçãono sudoeste do Estado do Rio Grande do Sul, Brasil. Acta Botanica Brasilica 23, 414–426.

Garcia, R.A., Crusciol, C.A.C., Calonego, J.C., Rosolem, C.A., 2008. Potassium cycling in a corn-brachiaria cropping system. European Journal of Agronomy 28, 579–585.

Garcia, C.M.P., Costa, C., Andreotti, M., Meirelles, P.R.L., Pariz, C.M., Freitas, L.A., Teixeira Filho, M.C.M., 2016. Yield of wet and dry corn grains from an intercrop with marandu grass and/or dwarf pigeon pea and nutritional value of the marandu grass in succession. Australian Journal of Crop Science 10, 1564–1571.

García-Prechac, F., Ernst, O., Siri-Prieto, G., Terra, J.A., 2004. Integrating no-till into crop–pasture rotations in Uruguay. Soil and Tillage Research 77, 1–13.

Gil, J., Siebold, M., Berger, T., 2015. Adoption and development of integrated crop– livestock–forestry systems in Mato Grosso, Brazil. Agriculture, Ecosystems & Environment 199, 394–406.

Herrero, M., Thornton, P.K., Notenbaert, A.M., Wood, S., MSangi, S., Freeman, H.A., Bossio, D., Dixon, J., Peters, M., Van de Steeg, J., Lynam, J., Parthasarathy, Rao, P., Macmillan, S., Gerard, B., Mcdermott, J., Seré, C., Rosegrant, M., 2010. Smart investments in sustainable food production: revisiting mixed crop–livestock systems. Science 327, 822–825.

Instituto Brasileiro de Geografia e Estatística (IBGE), 2004. Mapa de Biomas do Brasil. Available online: http://www.ibge.gov.br.

Jaremtchuk, A.R., Costa, C., Meirelles, P.R.L., Gonçalves, H.C., Ostrensky, A., Koslowski, L.A., Madeira, H.M.F., 2006. Yield, chemical composition and potassium soil removal by corn crops grown for silage production and harvested at two cut heights (in Portuguese, with English abstract). Acta Scientiarum. Agronomy 28, 351–357.

Kluthcouski, J., Aidar, H., 2003. Uso da integração lavoura-pecuária na recuperação de pastagens degradadas (in Portuguese). In: Kluthcouski, J., Stone, L.F., Aidar, H. (Eds.), Integração Lavoura-pecuária, first ed. Embrapa, Santo Antônio de Goiás, Goiás, Brazil, pp. 185–223.

Lemaire, G., Franzluebbers, A., Carvalho, P.C.F., Dedieu, B., 2014. Integrated crop- livestock systems: strategies to achieve synergy between agricultural production and environmental quality. Agriculture, Ecosystem & Environment 190, 4–8.

Macedo, M.C.M., 2009. Crop and livestock integration: the state of the art and the near future (in Portuguese, with English abstract). Revista Brasileira de Zootecnia 38, 133–146.

Martins, A.P., Anghinoni, I., Costa, S.E.V.G.A., Carlos, F.S., Nichel, G.H., Silva, R.A.P., Carvalho, P.C.F., 2014a. Amelioration of soil acidity and soybean yield after surface lime reapplication to a long-term no-till integrated crop-livestock system under varying grazing intensities. Soil and Tillage Research 144, 141–149.

Martins, A.P., Costa, S.E.V.G.A., Anghinoni, I., Kunrath, T.R., Balerini, F., Cecagno, D., Carvalho, P.C.F., 2014b. Soil acidification and basic cation use efficiency in an integrated no-till crop-livestock system under different grazing intensities. Agriculture, Ecosystems & Environment 195, 18–28.

Martins, A.P., Cecagno, D., Borin, J.B.M., Arnuti, F., Lochmann, S.H., Anghinoni, I., Bissani, C.A., Bayer, C., Carvalho, P.C.F., 2016. Long-, medium- and short-term dynamics of soil acidity in an integrated crop-livestock system under different grazing intensities. Nutrient Cycling in Agroecosystems 104, 67–77.

Mateus, G.P., Crusciol, C.A.C., Borghi, E., Pariz, C.M., Costa, C., Silveira, J.P.F., 2011. Nitrogen fertilization on sorghum intercropped with grass in a no-tillage system (in Portuguese, with English abstract). Pesquisa Agropecuaria Brasileira 46, 1161–1169.

Mateus, G.P., Crusciol, C.A.C., Pariz, C.M., Borghi, E., Costa, C., Martello, J.M., Franzluebbers, A.J., Castilhos, A.M., 2016. Sidedress nitrogen application rates to sorghum intercropped with tropical perennial grasses. Agronomy Journal 108, 433–447.

Mendonça, V.Z., Mello, L.M.M., Andreotti, M., Pariz, C.M., Yano, E.H., Pereira, F.C.B.L., 2015. Liberação de nutrientes da palhada de forrageiras consorciadas com milho e sucessão com soja (in Portuguese, with English abstract). Revista Brasileira de Ciência do Solo 39, 183–193.

Modernel, P., Rossing, W.A.H., Corbeels, M., Dogliotti, S., Picasso, V., Tittonell, P., 2016. Land use change and ecosystem service provision in Pampas and Campos grasslands of southern South America. Environmental Research Letters 11.

Moraes, A., Carvalho, P.C.F., Anghinoni, I., Lustosa, S.B.C., Costa, S.E.V.G.A., Kunrath, T.R., 2014a. Integrated crop-livestock systems in the Brazilian subtropics. European Journal of Agronomy 57, 4–9.

Moraes, A., Carvalho, P.C.F., Lustosa, S.B.C., Lang, C.R., Deiss, L., 2014b. Research on integrated crop-livestock systems in Brazil. Revista de Ciencias Agronomicas 45, 1024–1031.

Moraine, M., Duru, M., Nicholas, P., Leterme, P., Therond, O., 2014. Farming system design for innovative crop-livestock integration in Europe. Animal 1–14.

Moraine, M., Duru, M., Therond, O., 2016. A social-ecological framework for analyzing and designing integrated crop–livestock systems from farm to territory levels. Renewable Agriculture and Food Systems 1–14. https://doi.org/10.1017/S1742170515000526.

Morón, A., Sawchik, J., 2002. Soil quality indicators in a long-term crop–pasture rotation experiment in Uruguay. In: 17th World Congr. Soil Sci., Symp. No. 32, Paper 1327, Thailand.

Nascente, A.S., Crusciol, C.A.C., 2012. Cover crops and herbicide timing management on soybean yield under no-tillage system. Pesquisa Agropecuaria Brasileira 47, 187–192.

Nascente, A.S., Crusciol, C.A.C., Cobucci, T., 2013. The no tillage system and cover crops alternatives to increase upland rice yields. European Journal of Agronomy 45, 124–131.

National Research Concil - NRC, 1985. Nutrient Requirement of Sheep, sixth ed. National Academy of Science, Washington.

Overbeck, G.E., Müller, S.C., Pillar, V.D., Pfadenhauer, J., 2006. Floristic composition, environmental variation and species distribution patterns in burned grassland in southern Brazil. Brazilian Journal of Biology 66, 1073–1090.

Overbeck, G.E., Muller, S.C., Fidelis, A., Pfadenhauer, J., Pillar, V.D., Blanco, C.C., Boldrini, I.I., Both, R., Forneck, E.D., 2007. Brazil's neglected biome: the South Brazilian Campos. Perspectives in Plant Ecology, Evolution and Systematics 9, 101–116.

Pacheco, L.P., Leandro, W.M., Machado, P.L.O.A., Assis, R.L., Cobucci, T., Madari, B.E., Petter, F.A., 2011. Biomass production and nutrient accumulation and release by cover crops in the off-season (in Portuguese, with English abstract). Pesquisa Agropecuária Brasileira 46, 17–25.

Palm, C., Blanco-Canqui, H., Declerck, F., Gatere, L., Grace, P., 2014. Conservation agriculture and ecosystem services: an overview. Agriculture, Ecosystems & Environment 187, 87–105.

Pariz, C.M., Andreotti, M., Tarsitano, M.A.A., Bergamaschine, A.F., Buzetti, S., Chioderolli, C.A., 2009. Technical and economic performance of corn intercropped with *Panicum* and *Brachiaria* forage in crop-livestock integration system (in Portuguese, with English abstract). Pesquisa Agropecuária Tropical 39, 360–370.

Pariz, C.M., Andreotti, M., Azenha, M.V., Bergamaschine, A.F., Mello, L.M.M., Lima, R.C., 2010. Dry mass and chemical composition of four *Brachiaria* species sown in rows or spread, in intercrop with corn crop in no-tillage system (in Portuguese, with English abstract). Acta Scientiarum. Animal Sciences 32, 147–154.

Pariz, C.M., Carvalho, M.P., Chioderoli, C.A., Nakayama, F.T., Andreotti, M., Montanari, R., 2011a. Spatial variability of forage yield and soil physical attributes of a *Brachiaria decumbens* pasture in the Brazilian Cerrado. Revista Brasileira de Zootecnia 40, 2111–2120.

Pariz, C.M., Andreotti, M., Buzetti, S., Bergamaschine, A.F., Ulian, N.A., Furlan, L.C., Meirelles, P.R.L., Cavasano, F.A., 2011b. Straw decomposition of nitrogen-fertilized grasses after intercropping with corn crop in irrigated integrated crop-livestock system. Revista Brasileira de Ciência do Solo 35, 2029–2037.

Pariz, C.M., Andreotti, M., Bergamaschine, A.F., Buzetti, S., Costa, N.R., Cavallini, M.C., Ulian, N.A., Luiggi, F.G., 2011c. Yield, chemical composition and chlorophyll relative content of Tanzania and Mombaça grasses irrigated and fertilized with nitrogen after corn intercropping. Revista Brasileira de Zootecnia 40, 728–738.

Pariz, C.M., Andreotti, M., Bergamaschine, A.F., Buzetti, S., Costa, N.R., Cavallini, M.C., 2011d. Production, chemical composition and chlorophyll index of *Brachiaria* spp. after the intercrop with corn (in Portuguese, with English abstract). Archivos de Zootecnia 60, 1041–1052.

Pariz, C.M., Andreotti, M., Azenha, M.V., Bergamaschine, A.F., Mello, L.M.M., Lima, R.C., 2011e. Corn grain yield and dry mass of *Brachiaria* intercrops in the crop- livestock integration system (in Portuguese, with English abstract). Ciencia Rural 41, 875–882.

Pariz, C.M., Azenha, M.V., Andreotti, M., Araújo, F.C.M., Ulian, N.A., Bergamaschine, A.F., 2011f. Yield and chemical composition of forage in crop-livestock integration in different sowing time (in Portuguese, with English abstract). Pesquisa Agropecuaria Brasileira 46, 1392–1400.

Pariz, C.M., Costa, C., Crusciol, C.A.C., Meirelles, P.R.L., Castilhos, A.M., Andreotti, M., Costa, N.R., Martello, J.M., Souza, D.M., Sarto, J.R.W., Franzluebbers, A.J., 2016. Production and soil responses to intercropping of forage grasses with corn and soybean silage. Agronomy Journal 108, 2541–2553.

Pariz, C.M., Costa, C., Crusciol, C.A.C., Castilhos, A.M., Meirelles, P.R.L., Roça, R.O., Pinheiro, R.S.B., Kuwahara, F.A., Martello, J.M., Cavasano, F.A., Yasuoka, J.I., Sarto, J.R.W., Melo, V.F.P., Franzluebbers, A.J., 2017a. Lamb production responses to grass grazing in a companion crop system with corn silage and oversowing of yellow oat in a tropical region. Agricultural Systems 151, 1–11.

Pariz, C.M., Costa, C., Crusciol, C.A.C., Meirelles, P.R.L., Castilhos, A.M., Andreotti, M., Costa, N.R., Martello, J.M., Souza, D.M., Protes, V.M., Longhini, V.Z.J.R.W., Franzluebbers, A.J., 2017b. Production, nutrient cycling and soil compaction to grazing of grass companion cropping with corn and soybean. Nutrient Cycling in Agroecosystems 108, 35–54.

Pereira, F.C.B.L., Mello, L.M.M., Pariz, C.M., Mendonça, V.Z., Yano, E.H., Miranda, E.E.V., Crusciol, C.A.C., 2016. Autumn maize intercropped with tropical forages: crop residues, nutrient cycling, subsequent soybean and soil quality. Revista Brasileira de Ciência do Solo 40, e0150003.

Peyraud, J.L., Taboada, M., Delaby, L., 2014. Integrated crop and livestock systems in Western Europe and South America: a review. European Journal of Agronomy 57, 31–42.

Poli, C.H.E.C., Carvalho, P.C.F., 2001. Planejamento alimentar de animais: proposta de gerenciamento para o sistema de produção à base de pasto. Pesquisa Agropecuária Gaúcha 7, 145–156.

Randrianasolo, J., Lecomte, P., Salgado, P., Lepelley, D., 2010. Modeling crop- livestock integration systems on a regional scale in Reunion Island: sugar cane and dairy cow activities. Advances in Animal Biosciences 1, 498.

Roesch, L.F., Vieira, F.C., Pereira, V.A., Schünemann, A.L., Teixeira, I.F., Senna, A.J., Stefenon, V.M., 2009. The Brazilian Pampa: a fragile biome. Diversity 1, 182–198.

Ryschawy, J., Choisis, N., Choisis, J.P., Joannon, A., Gibon, A., 2012. Mixed crop- livestock systems: an economic and environmental-friendly way of farming? Animal 6, 1722–1730.

Sawchik, J., Quincke, A., Terra, J.A., 2011. Soil carbon and nitrogen contents under long-term cropping systems. In: Ann. Mtg. ASA-CSSA-SSSA, 16–19 October 2011, San Antonio, TX.

Soussana, J.-F., Lemaire, G., 2014. Coupling carbon and nitrogen cycles for environmentally sustainable intensification of grasslands and crop-livestock systems. Agriculture, Ecosystem & Environment 190, 9–17.

Tanaka, D.L., Karn, J.R., Scholljegerdes, E., 2008. Integrated crop/livestock systems research: practical research considerations. Renewable Agricultural and Food Systems 23, 80–86.

CHAPTER

17

Toward Integrated Crop-Livestock Systems in West Africa: A Project for Dairy Production Along Senegal River

Gilles Lemaire[1], *Bernard Giroud*[2], *Bagoré Bathily*[3], *Philippe Lecomte*[4], *Christian Corniaux*[4]

[1]INRA, Centre de Recherche Poitou-Charentes, 86600, Lusignan, France; [2]SAFE Nutrition, 8998 Sacré Coeur 3, Dakar, Sénégal; [3]Laiterie du Berger, Richard Toll, Sénégal; [4]CIRAD, Unité SELMET, Dakar, Sénégal

INTRODUCTION

Smallholder farmers in mixed crop-livestock systems produce about half of the world's foods (Herrero et al., 2012). Moreover, as stated by Thornton et al. (2002), climate change, mainly in Asia and in Africa, will have major impacts for numerous low-income farmers who depend on livestock as well as crops for their livelihoods. Regarding cattle and small ruminants, livestock farming systems in sub-Saharan Africa (SSA) are mainly of pastoral or agropastoral (crop-livestock) nature. Pastoralists are mainly transhumant, adapting the feeding of local breeds to resource availability with more and more often partial sedentarization of the people's and main household settlement and the keeping of a limited number of productive animals around centers and along milk roads. Agropastoralists, in largest numbers in SSA, are sedentary; they combine highly interacting crop and livestock management and eventually practice short or local transhumance during the wet season to avoid field crop damage. All these systems are generally very low input and/or highly extensive; they use very few chemicals, medicines, or imported concentrate feeds and are by nature quite "agroecologic" or "organic agriculture" relevant in their management. Intensified industrial livestock systems are rare and mainly limited to poultry meat and egg production. Faced with the increase in world population demand for food, priority should be given for the intensification of these low-input agricultural systems. As mentioned by Alvarez et al. (2014) and Thornton and Herrero (2001), modeling of mixed crop-livestock systems is rarely considered compared to the large variety of models for crop or livestock alone. In Africa, this mixed systems management is a main feature of large

Agroecosystem Diversity
https://doi.org/10.1016/B978-0-12-811050-8.00017-0

areas of the arid, semiarid and humid, subhumid zones extending from Senegal in the West to Ethiopia in the East (Robinson and Pozzi, 2011; Seré and Steinfeld, 1996). Livestock provides draft power to cultivate the land and manure to fertilize the soil, and crop residues are a key feed resource for the animals.

A mixed crop-livestock system can be defined (Sumberg, 2003) as "... a farming system integrating crop and livestock production activities so as to gain benefits from the resulting crop-livestock interactions." These interactions are mediated through exchanges of resources (feed and manure) between the two components of the system "... leading to greater farm efficiency, productivity and sustainability." Lemaire et al. (2014) emphasized the role of mixed crop-livestock systems for reconciling high productivity farming systems and environment quality in industrialized countries. In the nonindustrialized world, McDermott et al. (2010) argued that public and private investments in smallholder livestock systems intensification would clearly help people use their livestock enterprises as pathways out of poverty. In the Sahel region in Africa, Ickowicz et al. (2012) showed that development of this mixed farming system is the way for both increasing resilience of agriculture to climate change and improving food security.

About 320 million livestock are kept in family households in SSA.

Livestock systems in the Sahel have been for several centuries a traditional activity based on grazing of common resources of rangeland vegetation and water through pastoralism and agro-pastoralism (Dongmo et al., 2012). These systems fulfil various services for local populations through the use of livestock: capital, manure and nutrient transfer, collective use of resources, and food security risk management to cope with uncertainties (Descheemaker et al., 2016; Ickowicz et al., 2012). Nowadays, these traditional systems are strongly impacted by the rapid demographic growth and climate constraints, leading to land use changes and competition on depleted natural vegetation. This brings increased uncertainties on forage and water resource availability or accessibility that finally challenges the resilience of these traditional systems. Interactions between pastoral systems and arable cropping systems are numerous in West Africa, consisting mainly in the use of postharvest weed regrowth and crop residues as grazed resources for livestock during periods of natural forage scarcity, and of use of agricultural by-products (brans, cakes) for the supplementation of productive animals, and in counterpart exchange of direct field manuring. Tensions are, however, recurrent as population increases and as these two main land use practices are managed by different social or ethnic groups with very different strategies (Vall et al., 2006; Dugue et al., 2004).

In these generally dry or arid regions beside rain-fed systems the large irrigated perimeters, such as the one progressively developed along Senegal River or Niger River, became highly attractive opportunities for intensified cropping systems.

The objective of the chapter is to analyze the opportunities arising in the region of North Senegal for better synergies between the dynamics of development of irrigated arable cropping and livestock systems and to suggest how agronomic and environmental benefits could be provided through a more integrated system. As an illustration the chapter details a project for developing dairy production from smallholder farms through the introduction of alfalfa as a pivotal forage resource within an irrigated cropping system in the region of the Senegal River.

CROP AND DAIRY PRODUCTION IN NORTH SENEGAL: CONSTRAINTS AND OPPORTUNITIES

Regarding agricultural intensification dynamics in sub-Saharan West Africa, milk is a relevant object for considering the varying stakes in play and the innovative pathways needed for adapted intensification. Like in many other countries in West Africa, demand for milk in Senegal is rising exponentially; imports of milk powder and butter oil constitute a heavy burden for the state. Small- or medium-scale dairy producers could become important suppliers of fresh milk for the local small or large milk industries and milk chains that are emerging around many SSA cities (Corniaux et al., 2014). The general idea of investing in "conventional" milk development pathways, based on the development of specialized farms, housing high genetic merit imported cows, fed with irrigated local forages and supplemented at high energy costs with large amounts of imported corn and soya, would be, on a sole technical and economic points of view, the most effective solution. Technical aspects are well mastered and promoted as a highly specialized archetype in the usual worldwide intensifying milk agribusiness models. It is, however, highly demanding in investment, financially fragile, and quite poorly sustainable due to the complexity handling externalities and mistakes that generally reduce environmental and social sustainability of this development scheme.

In parallel in these large irrigated perimeters, intensive cropping systems have progressively emerged in the last decades, taking advantage of the highly accessible water and most of all the large area of available "commons" made originally of poor soils classically devoted to seasonal transhumant grazing of pastoral herds. They have been largely supported by land use access and investment policies (Seck, 2009). These irrigated cropping systems are managed by agro-industrial enterprises and/or large farmer cooperatives and are developing rapidly for cash crop production, highly stimulated by the growing urban and export market demand for rice, vegetables (mainly onion and tomatoes and niche export market such as baby corn, sweet potato), and for industrial crops such as sugar cane. While these systems perform highly in terms of yields, they are generally closed and interact very marginally with surrounding livestock keepers for the use of postharvest residues and agro-industrial by-products. The irrigated crops are managed within specialized monoculture systems that require medium to high fertilizer application on the generally poor and low-carbon, sandy soils. For rice and legumes, there are growing difficulties in controlling weeds, pest, and diseases, which require significant use of chemicals, impairing water quality and the sustainability of the landscape development. Market price uncertainties associated with high costs for water and other inputs constrain the profitability and the access to such systems for smallholders (Bourgoin et al., 2016). Sugar cane cultivation, which is partly mechanized, is totally integrated within the sugar industry and employs large numbers of people. Beside their main production, all these irrigated systems generate a huge biomass of by-products that are poorly recycled, generally burned, and only partly grazed.

In such a landscape, combining large, open rangeland areas and irrigated zones with small- or medium-scale integrated or associated crop and livestock farming systems based on genetic, ecologic, and social intensification would contribute to a more sustainable development (Corniaux et al., 2012). Capitalizing on the large existing biomass and livestock numbers, local knowledge, and ecologic services (e.g., soil organic matter, nutrient cycling, and soil fertility) should be more explicitly considered while considering the use of external inputs, biotechnology, irrigation, and mechanization to increase the productivity of land, animals, and labor, as well as the income of local households.

In such a pastoral Sahelian context, an emblematic example is the Laiterie du Berger (LDB) established in 2004 by Richard Toll, along the banks of the Senegal River near large rice paddies and cane irrigated crops. Along main tracks circulating among the traditional pastoral settlements the LDB motorcycles collect milk twice daily from more than 500 producers.

The major constraint is the seasonality. During the rainy season, production is high and collected milk can reach up to 6000 L per day. During the hot, dry season, milk volumes drop nearly 50%. To cope with these variations and satisfy the customer steady demand, the dairy has to use imported milk powder. Therefore, increasing the volumes collected and in particular during the dry season is a main concern for LDB. In the rangelands, its extension services promote the sedentarization in the pastoral camps of a dairy nucleus, gradually intensified (feeds, bulked forage, and genetics); they also promote small specialized dairy farm development near the cultivated zones.

The intensification scheme also involves contracting with the sugar cane industry for access to the harvested plots and baling the cane straw, which is further dispatched in the dry season to the camps and producers delivering milk to the dairy processor. This agreement between agro-industries boosted the development of the dairy business, with growing numbers of small producers becoming partners of the value chain.

So, facing the agronomic and environmental limitations for a sustainable intensification of irrigated cropping systems based mainly on monoculture management and the necessity for extending smallholder dairy producers, a more integrated crop-livestock system should be conceived and developed at the local level.

TOWARD AN INTEGRATED CROP-LIVESTOCK SYSTEM

As stated by Robinson and Pozzi (2011), a global crop-livestock system at the regional level can be composed through association of different farm types:

- livestock-only grazing systems
- rain-fed mixed crop-livestock systems
- irrigated mixed crop-livestock systems

These three types of systems are important in term of food production and livelihoods for the whole of West Africa, especially for Senegal, Niger, Mali, and Burkina Faso (Ickowicz et al., 2012). So development of an integrated crop-livestock system at the regional level in these countries should lead to several benefits:

1. an increase in forage resources for livestock-only grazing farms in terms of quantity, quality, and security through more intense interactions with rain-fed or irrigated cropping systems, to increase milk and meat production in these traditional pastoralism systems,
2. capacity to develop more intensive livestock systems based on forage produced within rain-fed and irrigated cropping systems,
3. an agronomic improvement of cash crop production and its profitability through the benefits of more diversified crop rotations.

The question is to determine which should be the "pivotal" forage crop to be inserted within arable cropping systems, to provide high-quality feed for dairy livestock and all the related agronomic benefits for improving arable cropping systems?

This "pivotal" forage species should have several characteristics:

1. to be able to produce high-quality forage, well balanced in energy and protein content, throughout the year, and capable of being easily stored, transported, and distributed during periods of feed scarcity,
2. to be a perennial crop with the potential to produce at least during three consecutive years without reseeding and soil tillage, thus considerably reducing production costs,
3. to be a legume N_2-fixing species, to be totally independent from N fertilization applications for forage production, and also provide important N residues in the soil for the follow-on crops (rice and/or vegetable), thus lowering the overall need for external and very costly fertilizers.

Alfalfa as a Source of High-Quality Forage for Local Livestock

Alfalfa (*Medicago sativa*, L.) appears to possess all the required characteristics for the pivotal forage species of integrated crop-livestock systems in irrigated conditions. This forage species is the more important cultivated forage in the world. It has a high potential productivity in South Europe under irrigation of about 20 t/ha dry matter (Gosse et al., 1982; Lemaire et Allirand, 1993) from April to October. Under subtropical and tropical conditions (e.g., South Morocco, Saudi Arabia), where the use of nondormant cultivars allows production all along the year, the annual forage production could reach higher levels. As shown by Gosse et al. (1984) and Thiébeau et al. (2011) the potential production of alfalfa is directly proportional to the quantity of incident radiation intercepted by the crop. So the high level of incident radiation all through the year in Senegal associated with the absence of low temperature period explains the higher annual production in this region compared to temperate or Mediterranean conditions, as long as irrigation is not restricted. In most of the world, alfalfa is cultivated as a perennial forage for at least 3 years and sometime more. Under dry, tropical climate, as in North Senegal, the perennial character of alfalfa could be impaired by flooding or water logging during the rainy season, but crops appeared to survive well in well-drained soil conditions. About 75 days after sowing a first harvest of forage can be obtained, and thereafter, successive harvests can be made every 45 days during the 3 years. In this way, a relatively constant forage quality can be obtained with approximately 65%–75% digestibility and 150–180 g of crude protein per kg of dry matter (Lemaire and Allirand, 1994), corresponding to an excellent nutritional quality for lactating cows despite a small excess in protein/energy ratio (INRA, 1989). Alfalfa can be used directly as green forage with daily distribution at the barn or harvested as hay and stored. A milking cow should be able to ingest approximately 20–22 kg dry matter of alfalfa forage, which should allow a daily production of about 20–25 liters of milk, depending of the genetic potential of the animal (INRA, 1989). So, it is clear that this forage resource has the potential to greatly boost the dairying capacity of the livestock in this region.

Alfalfa as a Source of Agronomic Improvement for Cash Crop Production

As alfalfa is a perennial legume species, plants are able to fix atmospheric N_2, so it does not need any N fertilizer application for reaching its potential production. N_2 fixation capacity of alfalfa is directly linked to its capacity in aboveground biomass production (Lemaire et al., 1985). For an annual production of 18–20 t/ha of dry matter in Europe, the N_2 fixation capacity of alfalfa is estimated at 400–600 kg N/ha (Justes et al., 2001). So this quantity of N fixed annually by alfalfa in Senegal conditions should be higher according to its higher dry matter production. So the first important advantage of alfalfa is to provide a high quantity of N to the agroecosystem without any N fertilizer input, which economizes both energy and greenhouse gas associated with fertilizer manufacturing and financial costs to farmers. Alfalfa cannot be cultivated as a monoculture because, after several years, development of nematodes and fungi would impair the persistence of the plant and then the productivity of the crop. It is highly recommended to have at least 3 years without alfalfa after a 3-year cultivation of alfalfa to avoid disease problems. Alfalfa has to be included within a more diversified crop rotation. Crops that could be associated with alfalfa in irrigated cropping systems are the two main cash crops actually well developed in the region: either rice in clay soils or vegetables, i.e., onion and tomato, in sandy soils. These two types of crop require a relatively high quantity of N fertilizers when they are managed in monoculture. As shown by Justes et al. (2001), and Lemaire et al. (2015), N released in soil after 3 years of alfalfa cultivation through mineralization of alfalfa plant residues and available for the following crop can be estimated at 100–150 kg N/ha. This N release from alfalfa residues declines progressively with years, but it still provides significant N to subsequent crops for 3–4 years (Angus et al., 2015). The greatest proportion of N fixed by alfalfa is exported from the field by harvested forage and ingested by animals. Generally, according to N metabolic processes, 60%–70% of this N is lost through animal excretion, but if a relevant system for manure recycling is performed, this N and the other macronutrients such as P, K, Ca, and M and the majority of micronutrients can be returned within the system for maintaining soil fertility and crop productivity without any mineral fertilizer input. In this way, introduction of a 3-year sequence of alfalfa cropping within a cash crop system should allow farmers to become totally independent from fertilizer inputs, when the cost for access to these fertilizers for small-holding farmers represents the main constraint for profitability for these producers, especially for rice (Kebbeh and Mezan, 2003; Kebbeh et al., 2004). In addition to these agronomic and financial benefits for cash crop production, introduction of alfalfa as a break in arable crop monocultures should have positive effects on weed control (Meiss et al., 2010a,b), especially on rice where weed management represents one of the more important factors limiting rice yields (Haefele et al., 2000). In the same way, a 3-year period of alfalfa could break propagation of soil pathogenesis by fungi and nematodes that are more and more frequent in pure monocultures, especially in the double rice crop system where two successive rice productions are managed each year and for several successive years (Ratnadass et al., 2013). As shown by Lemaire et al. (2014), inclusion of a perennial forage cropping sequence, such as alfalfa, within arable cropping system such as rice and/or vegetables should lead to a more sustainable farming system through an increase in the diversity of land use management (Fig. 17.1).

FIGURE 17.1 Harvest of alfalfa in the region of Diagana near the Senegal River. *Photo: Safe Nutrition.*

Alfalfa as a Source of Nutritional Supplement for the Local Population

Alfalfa is the forage crop producing the highest quantity of crude protein per ha and per year under temperate, Mediterranean, and subtropical climates (Voisin et al., 2013). A part of these proteins (about 20%) can be extracted from freshly harvested forage through a relatively simple process of pressing, juice extraction, coagulation, centrifugation, and drying to obtain a high-protein concentrated product (>50%) according to Gatineau and de Mathan (1981) and Ream et al. (1983). This process allows two products from the same fresh alfalfa forage: (1) a nutritional supplement with high protein concentration for people and (2) forage residue of high quality for livestock feeding. The value of leaf protein extract as a nutritional supplement for malnourished people has been demonstrated by several studies (Davys, 2011; FAO, 1991; EU Commission, 2009; Vyas, 2009;

Shat et al., 1981; Olatunbasun et al., 1972). Apart from its high protein content and excellent amino acid profile, this extract brings a high proportion of daily requirements for beta-carotene, iron, folic acid, Ca, Mg, and vitamins E, B_2, B_6, and B_{12}. With ingestion of only 10g per day, it provides 20% of the daily requirement in omega-3 fatty acid and a better omega-3/omega-6 fatty acid equilibrium of the global diet (Zanin, 1998). So, distribution of this type of protein extract is highly recommended for young children, pregnant women, and nursing mothers in Africa. Today, these protein extracts are industrially produced, commercialized, and distributed in West Africa through spirulina produced in an artificial system. The direct production of foliar extract from alfalfa as a coproduct of forage production should lead to a two- or three-time lower cost. The company Safe Nutrition based in Senegal is developing a pilot unit for processing alfalfa leaf extract production at the level of a small group of farms in the region of Richard Toll in association with the milking company LDB, which has the objective of developing alfalfa cropping for improving the local dairy production by small-holding farmers. So, there is a good opportunity to connect more closely the development of alfalfa cropping for both improving dairy production and developing leaf extract production. Moreover, the extraction of 15%–20% of protein would lead to a better energy/protein ratio. This forage residue can be easily dried and conditioned into pellets for distribution at an attractive price as feed resources for small-holding farmers that have no possibility for cropping alfalfa by themselves. So the association between alfalfa forage production and alfalfa leaf extract production should profit from the economy of scope generated by this coproduction system to (1) reduce the cost for access to alfalfa forage for small-holding farmers and (2) reduce the cost for alfalfa leaf extract production to offer a very accessible product to the population.

TOWARD AN INTEGRATED CROP-LIVESTOCK SYSTEM AT THE TERRITORY LEVEL

As stated by Lemaire (2014), integration of livestock production with arable cropping systems has to be conceived not only at a single farm level, where interactions between animals and crops can be managed, but also at a landscape or regional level, where these interactions between different farm entities, which are specialized either in arable crop or livestock farming, could be associated within integrated agroecosystems (see also Lemaire et al., 2017; Moraine et al., 2014). Poccard-Chapuis and Lecomte (2011) analyzed the interest in development of integrated crop-livestock systems at the regional level in tropical countries. As illustrated in Fig. 17.2, alfalfa cropping could be considered the "keystone" of a whole agroecosystem at the regional level, enhancing interactions and synergies among several agriculture sectors such as traditional livestock farms, irrigated rice and vegetable farms, dairy upstream and downstream industry, and human nutrition and health aspects.

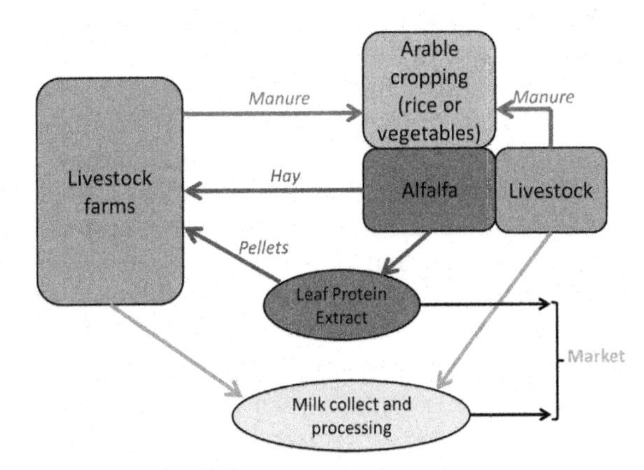

FIGURE 17.2 Schematic representation of the "keystone" role of alfalfa for enhancing interactions and synergies among different farm units at the territory level.

So, different farm unit types could be considered and related to each other:

1. Livestock-only farm units, representing either smallholder traditional farmers or more modern farms specialized in dairy production. These farms could then benefit from high-value alfalfa forage provided by an irrigated farming system and provide them in return organic matter and nutrients from manure.
2. Irrigated arable cropping farms, specialized either in rice or/and vegetable production willing to improve their cropping system through incorporation of alfalfa. These farms without their own livestock could sell alfalfa hay to livestock-only farms.
3. Integrated crop-livestock farms producing both milk and cash crops could sell or buy alfalfa forage resources according to the equilibrium of their own systems.

Moreover, we could also imagine that a sugar production system would also benefit from incorporating 3 years of alfalfa cropping within their sugar cane mono-cropping to promote a more sustainable production system. In this way, the sugar cane perimeter could be a partner with this whole agroecosystem as provider of alfalfa forage resources.

The alfalfa leaf protein extract processing can be then considered a partner of this whole system by providing an added value to alfalfa production that could facilitate and regulate exchanges of alfalfa forage resources among the different farm units. These exchanges of forage resources are the "key" to such a system: the price of alfalfa hay or pellets must not be too high for accessibility to poor small-holding livestock farmers, but it must be high enough to encourage a cash crop producer to incorporate alfalfa in their cropping systems. So the alfalfa forage market has to be regulated and should benefit from governmental subsidies within the frame of a regional development strategy. It is clear that such regional development requires not only technical and agronomic studies for evaluating the different benefits that different types of farms could benefit financially by these different interactions, but also socioeconomic studies are necessary for analyzing all the transactional costs and opportunities such an integrated system could require, as well as challenges of local governance and management for implementing these systems. The risk, if this aspect of integration at a regional level is not sufficiently taken into account, is to have a development of large, intensive dairy systems based on a large, irrigated alfalfa perimeter whose environmental and social sustainability is very questionable.

CONCLUSIONS

In West Africa, agriculture has to face two main concerns: (1) a rapid degradation of natural forage resources due to excess and uncontrolled stocking density associated with demography pressure and (2) development of modern agriculture sectors based on irrigated perimeters with large mono-cropping systems whose environmental and social sustainability is strongly questioned. Introduction of a pivotal forage crop, such as alfalfa, within integrated crop-livestock systems designed for smallholder producers at both the farm level and regional level should be an opportunity for an integrated agriculture production system, allowing development of a local dairy production based on a large network of local producers, resorption of most negative environmental impacts of crop intensification, and improvement of nutrition quality for the local people. Sustainability of such an integrated system would be based on synergies among different agriculture sectors (dairy, sugar cane, rice, and vegetable) whose technical and economic development have been managed totally separately from each other as a consequence of the paradigm of "economy of scale." So even if the project of a more integrated system such as developed earlier appears technically, agronomically, and environmentally relevant and performing, its implementation implies a strong change in socioeconomic context with a shift from "economy of scale" to "economy of scope."

References

Alvarez, S., Rufino, M.C., Vayssières, J., Salgado, P., Tittonell, P., Tillard, E., Bocquier, F., 2014. Whole-farm nitrogen cycling and intensification of crop-livestock systems in the highlands of Madagascar: an application of network analysis. Agricultural Systems 126, 25–37.

Angus, J.F., Kirkegaard, J.A., Hunt, J.R., Ryan, M.H., Ohlander, L., Peoples, M.B., 2015. Break crop and rotations for wheat. Crop & Pasture Science 66, 523–552.

Bourgoin, J.F., Jankowski, F., Camara, A.D., Diop, D., 2016. Pathways Towards Agricultural Intensification Case Study: The Delta of the Senegal River (Senegal). Coord.: C. Corniaux & C. Sall. Synth. report September 2016. 41 p. http://www.intensafrica.org/?wpdmdl=859.

Corniaux, C., Alary, V., Gautier, D., Duteurtre, G., 2012. Producteur laitier en Afrique de l'Ouest : une modernité rêvée par les techniciens à l'épreuve du terrain. Autrepart 2012 (62), 17–36.

Corniaux, C., Duteurtre, G., Broutin, C. (Eds.), 2014. Filières laitières et développement de l'élevage en Afrique de l'Ouest. L'essor des minilaiteries. Karthala, (Hommes et sociétés), Paris, pp. VI–242.

Davys, M.N.G., 2011. Leaf concentrate and other benefits of leaf fractionation. In: Thompson, B., Amoroso, L. (Eds.), Combating Micronutrient Deficiencies: Foodbased Approach. FAO, Roma.

Descheemaeker, K., et al., 2016. Climate change adaptation and mitigation in smallholder crop–livestock systems in sub-Saharan Africa: a call for integrated impact assessments. Regional Environmental Change 16 (8), 2331–2343.

Dongmo, A., Djamen-Nana, P., Vall, E., Mian-Oudanang, K., Coulibaly, D., Lossouarn, J., 2012. Du nomadisme à la sédentarisation : l'élevage d'Afrique de l'Ouest et du Centre en quête d'innovation et de durabilité. Revue d'Ethnoécologie (1), 147–161.

Dugué, P., Vall, E., Lecomte, P., Klein, H.D., Rollin, D., 2004. Evolution des relations entre l'agriculture et l'élevage dans les savanes d'Afrique de l'Ouest et du Centre. Un nouveau cadre d'analyse pour améliorer les modes d'intervention pour favoriser les processus d'innovation. OCL 11 (4–5), 268–276.

EU Commission, 2009. Décision de la commission Européenne du 13 octobre 2009 autorisant la mise sur le marché d'un extrait foliaire de luzerne (Medicago sativa) en tant que nouvel aliment ou nouvel ingrédient alimentaire en application du règlement (CE) n° 258/97 du Parlement européen et du Conseil [notifiée sous le numéro C(2009) 7641] (2009/826/CE).

FAO, Food and Agriculture Organization and World Health Organization, 1991. Besoins énergétiques et besoins en protéines - rapport d'une consultation d'experts FAO OMS ONU. Genève 1991 (série de rapports techniques).

Gastineau, I., de Mathan, O., 1981. La préparation industrielle de la protéine verte de luzerne. In: Costes, C. (Ed.), Protéines foliaires et alimentation. Gauthier-Villars, Paris, pp. 159–182.

Gosse, G., Chartier, M., Lemaire, G., Guy, P., 1982. Influence des facteurs climatiques sur la production de la luzerne. Fourrages 90, 113–135.

Gosse, G., Chartier, M., Lemaire, G., 1984. Mise au point d'un modèle de prévision de production pour une culture de luzerne. Comptes Rendus de l'Académie des Sciences 298 (III), 541–544.

Haefele, S., Johnson, D.E., Diallo, S., Wopereis, M.C.S., Janin, I., 2000. Improved soil fertility and weed management is profitable for irrigated rice farmer in Sahelian Africa. Field Crop Research 66, 101–113.

Herrero, M., Thornton, P.K., Notenbaert, A.N., Wood, S., Msangi, S., Freeman,HA, Bossio, D., Dixon, J., Peters, M., van de Steeg, J., Lynam, J., Parthasarathy Rao, P., Macmillan, S., Gerard, B., McDermott, J., Seré, C., Rosegrant, M., 2012. Smart investments in sustainable food production: revisiting mixed crop-livestock systems. Science 327, 822–825.

INRA, 1989. Ruminant Nutrition. Recommended Allowances and Feed Tables. In: Jarrige, R. (Ed.). INRA, Paris, France, p. 373.

Ickowicz, A., Ancey, V., Corniaux, C., Duteurtre, G., Poccard-Chappuis, R., Touré, I., Vall, E., Wane, A., 2012. Crop–livestock production systems in the Sahel—increasing resilience for adaptation to climate change and preserving food security. In: Meybeck, A., Lankoski, J., Redfern, S., Azzu, N., Gitz, V. (Eds.), Building Resilience for Adaptation to Climate Change in the Agriculture Sector. Proceeding of a Joint FAO/OEDC Workshop 23–24 April 2012. FAO, Rome, ISBN 978-92-5-107373-5.

Justes, E., Thiébeau, P., Cattin, G., Larbre, D., Nicolardot, B., 2001. Libération d'azote après retournement de luzerne. Un effet sur deux campagnes. Perspectives agricoles 264, 22–28.

Kebbeh, M., Miezan, K., 2003. Ex-ante evaluation of integrated crop management options for irrigated rice production in Senegal River Valley. Field Crop Research 81, 87–94.

Kebbeh, M., Miezan, K., Camara, M., 2004. Developing technology options for rice integrated crop management in the Sahel Zone of West Africa: case of irrigated rice production in Senegal River Valley. Uganda Journal of Agricultural Sciences 9, 425–431.

Lemaire, G., Cruz, P., Gosse, G., Chartier, M., 1985. Etude des relations entre la dynamique de prélèvement d'azote et la dynamique de croissance en matière sèche d'un peuplement de luzerne (*Medicago sativa* L.). Agronomie 5 (8), 685–692.

Lemaire, G., Allirand, J.M., 1993. Relation entre croissance et qualité de la luzerne : interaction génotype - mode d'exploitation. Fourrages 134, 183–198.

Lemaire, G., 2014. L'intégration Agriculture-Elevage, un enjeu mondial pour concilier production agricole et environnement. Innovations Agronomiques 39, 181–189.

Lemaire, G., Franzluebbers, A., Carvalho, P.C., Dedieu, B., 2014. Integrated Crop-Livestock Systems: strategies to achieve synergy between agricultural production and environmental quality. Agriculture, Ecosystem & Environment 190, 4–8.

Lemaire, G., Gastal, F., Franzluebbers, A.J., Chabbi, A., 2015. Grassland-cropping rotations: an avenue for agricultural diversification to reconcile high production with environmental quality. Environmental Management. https://doi.org/10.1007/s00267-015-0561-6.

Lemaire, G., Ryschawy, J., de Facio Carvalho, P.C., Gastal, F., 2017. Agricultural intensification and diversity for reconciling production and environment: role of integrated crop-livestock systems. In: Gordon, I., Squire, G., Prins, H. (Eds.), "Food Security and Nature Conservation: Conflicts and Solutions", Earthscan Book. Routledge, London, pp. 113–132.

McDermott, J.J., Staal, S.J., Freeman, H.A., Herrero, M., van de Steeg, J., 2010. Sustaining intensification of smallholder livestock systems in the tropics. Livestock Science 130, 65–109. https://doi.org/10.1016/j.livsci.2010.02.014.

Meiss, H., Médiène, S., Waldhardt, R., Caneil, J., Bretagnolle, V., Reboud, X., Munier-Jolain, N., 2010a. Perennial alfalfa affects weed community trajectories in grain crop rotations. Weed Research 50, 331–340.

Meiss, H., Mediene, S., Waldhardt, R., Caneill, J., Munier-Jolain, N., 2010b. Contrasting weed species composition in perennial alfalfas and six annual crops: implications for integrated weed management. Agronomy Sustainable Development 30, 657–666.

Moraine, M., Duru, M., Nicholas, P., Leterme, P., Therond, O., 2014. Farmins system design for innovative crop-livestock integration in Europe. Animal 8, 1204–1217.

Olatunbosun, D.A., Adadevoh, B.K., et Oke, O.L., 1972. Leaf protein A new source for the management of protein calorie malnutrition in Nigeria. Nigeria Medical Journal 2, 195–199.

Poccard-Chapuis, R., Lecomte, P., 2011. A way for developing an integrated system at the landscape level in tropical areas. In: Lemaire, G., Hodgson, J., Chabbi, A. (Eds.), Grassland Productivity and Ecosystem Services. CAB International, Wallingford, UK, pp. 271–280.

Ratnadass, A., Blanchart, E., Lecomte, P., 2013. Ecological interactions within the biodiversity of cultivated systems. In: Hainzelin, E. (Ed.), Cultivating Biodiversity to Transform Agriculture. Springer [Allemagne], Heidelberg, pp. 141–179.

Ream, H.W., Jorgensen, N.A., Koegel, R.G., Bruhn, H.D., 1983. On-farm forage harvesting: plant juice protein production system in a humid temperate climate. In: Telek, L., Graham, H.D. (Eds.), Leaf Protein Concentrates. AVI Publishing Company, Connecticut, USA, pp. 467–479.

Robinson, T.P., Pozzi, F., 2011. Mapping Supply and Demand for Animal-source Foods to 2030. Animal Production and Health Working Paper No. 2. Food and Agriculture Organisation (FAO) of the United Nations, Rome, p. 164.

Seck, S.M., 2009. Changements institutionnels et difficultés de développement hydroagricole dans le Delta du fleuve Sénégal : nouvelles dynamiques et recompositions autour de l'irrigation. In: Dansero, E., Luzzati, E., Seck, S.M. (Eds.), Organisations paysannes et développement local : leçons à partir du cas du Delta du fleuve Sénégal. L'Harmattan Italia, Torino, Italia.

Seré, C., Steinfeld, H., 1996. World Livestock Production Systems: Current Status, Issues and Trends. FAO Animal Production and Health. Paper 127. FAO (Food and Agriculture Organization of the United Nations), Rome, Italy.

Shat, F.H., Salam Sheikh, A., Farruckh, N., Rasool, A., 1981. A comparison of leaf protein concentrate fortified dishes and milk as supplements for children with nutritionally inadequate diets. Plants Foods for Human Nutrition 30, 245–258.

Sumberg, J., 2003. Toward a dis-aggregated view of crop–livestock integration in Western Africa. Land Use Policy 20, 253–264.

Thiébeau, P., Beaudoin, N., Justes, E., Allirand, J.M., Lemaire, G., 2011. Shoot:root dry matter partitioning in primary growth and regrowth of lucerne (*Medicago sativa* L.) in different seasons. European Journal of Agronomy 35, 255–268.

Thornton, P.K., Herrero, M., 2001. Integrated crop–livestock simulation models for scenario analysis and impact assessment. Agricultural System 70, 581–602.

Thornton, P.K., Kruska, R.L., Henninger, N., Kristjanson, P.M., Reid, R.S., Atieno, F., Odero, A.,

Ndegwa, T., 2002. Mapping Poverty and Livestock in the Developing World. International Livestock Research Institute, Nairobi, Kenya, p. 124.

Vall, E., Dugué, P., Blanchard, M., 2006. Le tissage des relations agriculture-élevage au fil du coton. Cahiers Agriculture 15, 72–79.

Voisin, A.-S., Guéguen, J., Huyghes, C., Jeuffroy, M.-H., Magrini, M.-B., Meynard, J.-M., Mouget, C., Pellerin, S., Pelzer, E., 2013. Legumes for feed, food, biomaterials and energy in Europe. A review. Agronomy for Sustainable Development. https://doi.org/10.1007/s13593-013-0189-y.

Vyas, S., 2009. Leaf concentrate as an alternative to iron and folic acid supplements for anaemic adolescent girls: a randomized controlled trial in India. Public Health Nutrition 2009, 1–6.

Zanin, V., 1998. Un nouveau concept nutritionnel pour l'homme: l'extrait foliaire de luzerne. Association pour la Promotion des Extraits Foliaires (APEF), Paris, France, p. 40.

CHAPTER

18

Silvopastoral Systems in Latin America for Biodiversity, Environmental, and Socioeconomic Improvements

Rogerio Martins Mauricio[1], *Rafael Sandin Ribeiro*[1], *Domingos Sávio Campos Paciullo*[2], *Mauroni Alves Cangussú*[3], *Enrique Murgueitio*[4], *Julian Chará*[4], *Martha Xochitl Flores Estrada*[5]

[1]Bioengineering Department, Universidade Federal de São João del-Rei (UFSJ), São João del-Rei, Brazil; [2]Embrapa Dairy Cattle, Rua Eugênio do Nascimento, Juiz de Fora, Brazil; [3]Centro Brasileiro de Pecuária Sustentável, Imperatriz, Brazil; [4]CIPAV – Centro para la Investigación en Sistemas Sostenibles de Producción Agropecuaria, Cali, Colombia; [5]Fundación Produce Michoacán, Morelia, Mexico

INTRODUCTION

The intensification of agriculture and livestock production systems based on *Green Revolution* methods has not succeeded in providing sufficient food for the world population in a sustainable way and has contributed to an increase in the negative effects of climate change. Livestock activities are considered responsible for a high negative impact on the environment due to their contribution to land degradation, biodiversity losses, and water pollution (Steinfeld et al., 2006).

In Latin America, livestock activity began with the colonization process during the 16th century. Cattle production systems have been historically considered the main drivers of deforestation, with around 70% of the cleared land devoted to this activity. In Brazil, for example, forests were usually cut down for wood or charcoal production and then replaced with pasture monocultures, mainly of *Brachiaria* species. In these grazing lands, biomass production is usually high during the 5–10 years following deforestation due to the remaining organic soil fertility, but after this period, livestock production falls due to low forage volume and quality. The replacement of complex forest ecosystem by pasture monoculture systems affected negatively the biodiversity and caused the degradation of large amounts of land in the region. Pasture production could be recovered using traditional methods such as plowing and chemical fertilization, but the amount of soil

Agroecosystem Diversity
https://doi.org/10.1016/B978-0-12-811050-8.00018-2

amendments and fertilizers required makes this practice unsustainable over time.

Silvopastoral systems (SPSs), which include trees (leguminous or not), are considered a sustainable alternative to increase forage quantity and quality, promote animal welfare (Souza et al., 2010, 2015; Broom et al., 2013), and diversify farm income (Yamamoto et al., 2007; Sousa et al., 2010; Paciullo et al., 2011; Rozados-Lorenzo et al., 2007; Mauricio, 2012). SPSs also have positive impacts for the recovery and conservation of biodiversity at farm and landscape level since the more heterogeneous structure provides habitat for fauna and flora and serves as wildlife corridors (Murgueitio et al., 2011). According to Sánchez and Rosales (1999), SPS can be classified in seven different types:

- SPS using natural forests: this type of system can be found in Europe and some regions of North America, among others. In Spain, the *Quercus ilex* trees are utilized for acorn production to feed pigs and consociated with the forage consumed by cattle (Izquierdo et al., 2006). In the United States, cattle and sheep graze on federal forested lands.
- Grazing in planted forests: this system is applied during forest growing, where cattle grazing is used to increase income as well as to reduce the risk of fire.
- SPS using orchards: in this system, fruit trees (e.g., citrus or olive) are associated with forage species upon which cattle or sheep graze.
- SPS using tree plantations for industrial purposes: rubber or palm trees are planted on a large scale in Asia and Brazil as source of latex, oil, and coconuts. Forage species are planted under the trees and used for cattle or sheep grazing.
- SPS using introduced trees and shrubs: in this system, trees or shrubs are planted in pasturelands to increase shade and biomass production and quality. Trees and shrubs are also used as live fences or protein fodder banks (e.g., *Leucaena leucocephala, Morus alba and Tithonia diversifolia*).
- SPS using multipurpose trees: in this mixed system, forage trees or shrubs are planted and cut for cattle, fish, or pig feeding (e.g., *Xanthosoma sagittifolium*). These animals produce an important source of nutrients (feces) to increase soil fertility and forage biomass.
- SPS for intensive cattle production (iSPS): in this system, a high density of superior nutritional-quality forage species (e.g., *L. leucocephala*) is planted and used for direct cattle browsing (beef or dairy) and to improve soil fertility.

As described before, several types of SPSs could provide options for different cropping and livestock scenarios or ecosystems. This chapter describes several research results focused on SPSs as sustainable alternatives for livestock production in the tropics. SPSs can be applied to restore soil fertility and biodiversity; to increase forage and wood biomass (including graminaceous and legume forage, shrubs, and trees); to promote income diversification; and to meet food security needs associated with environmental conservation in agriculture and livestock ecosystems. The case studies described in this chapter were developed mainly in three tropical countries—Brazil, Colombia, and México—where livestock plays not only an important role in the economy, but also in society and the environment.

Brazilian Case Studies: Natural Regeneration of Native Trees and Use of Leguminous Trees to Implement Silvopastoral Systems

Deforestation (slash and burn) is part of the process to implement monoculture pastures in large areas and different biomes of tropical countries, including Brazil. In the short-term, this practice has created a positive economic output;

however, after many years, the soil fertility and stocking rate capacity is reduced, and consequently so is farmers' income. In addition, several negative effects on the environment, such as biodiversity losses, erosion, water contamination, and forest fragmentation, also occur (Viana et al., 2002). Due to these and other factors, pasture degradation took place in approximately 100 million hectares all over Brazil (Dias-Filho et al., 2001). Most of the pasture degradation is due to the lack of appropriate cattle stocking rate control (Boddey et al., 2004) and the reduction in soil fertility (Paciullo et al., 2014) as a consequence of uneconomical use of chemical fertilizer in large areas.

In the tropics, conventional pasture management involves the elimination of trees and shrubs every year by cutting them or using herbicides. Since 1980, in a dairy farm located in Minas Gerais state in the southeast area of the Brazilian Cerrado Biome, instead of this drastic control, selective weeding was performed, protecting native tree saplings. This effort was undertaken to reduce the cost of weed control; improve animal comfort (normally during the summer, the temperature reaches 40°C); enhance forage quality/quantity; improve an extra source of income for farmers via wood production; and promote biodiversity, carbon sequestration, and erosion reduction (Viana et al., 2002). On this farm, two tree species were selected (*Zeyhera tuberculosa* and *Myracrodruon urundeuva*) based on their high natural regeneration density, fast growth, and high wood value in the market. For both tree species, young trees were selected according to the stem shape that would be more appropriate for wood production; a minimum of 4 m distance was kept between trees; and just two or three sprouts were preserved per young tree plant. The nonselected sprouts were cut at 15 cm above the soil in a bezel shape. *Braquiaria brizantha* was seeded and only moderate doses of limestone and natural phosphate rock were applied. After 20 years of growth, several experiments were performed in this area (Fig. 18.1).

The influence of trees on soil nutrient dynamics was studied (Reis et al., 2009) to understand the effects of litter fall from trees (leaves, fruits, and stems) on the soil profile contents (nitrogen, N; potassium, K; phosphorus, P; calcium, Ca; and organic matter, OM) in comparison to a monoculture area. Originally, both soils, under SPS and monoculture, were acid (pH level 5.1) and very poor in terms of P (lower than 1 mg/dm^3). Although the effects of alley crops on soil fertility are well known, the aim of the experiment was to evaluate the long-term influence of trees on this type of Cerrado soil. The results indicated that, despite higher N and K content in the litter, no

FIGURE 18.1 Natural regeneration of native trees in the Cerrado Biome, Minas Gerais (left) and Maranhao state (right), Brazil.

benefits to soil were found in the SPS. In addition, the N content (1.79%) was the minimal concentration required for a positive net mineralization of litter (Andrade et al., 2002). Consequently, the high C/N and lignin/N ratios promoted slow litter decomposition, which will take several years to release nutrients into the soil. In the SPS, litter promoted benefits in terms of top soil acidity (0–2 cm) due to Ca^{+2} and Mg^{+2}, but no effects on OM content were verified. Therefore, the complexity of the SPS, including the interactions among tree, forage, and animal, provides N, P (low concentrations), Ca, and Mg (high concentrations) that will be released at a slow pace to the soil and utilized, in the long term, for biomass production.

Apart from the soil effects, using the same system described before, the forage intake, feeding behavior, and bioclimatologic indices of pasture grass under trees in the SPS were evaluated (Souza et al., 2015). Even with the tree competition and the lower radiation values, SPSs were able to provide more forage (8%) throughout the year compared to the monoculture. Normally, it has been stated that shading has a negative effect on forage biomass, and consequently on animal production. However, forage production could benefit from low to medium levels of shade, since temperature, soil moisture, and fertility together could increase forage biomass and lessen negative shade effects on pasture (Barros et al., 2012). The shade provided by trees contributed a more comfortable environment to animals in the tropical grassland ecosystem. The daily maximum temperature in the monoculture system was 33.5°C, above the thermal comfort zone (TCZ) for adult sheep (20–28°C), while under the trees, it was 5°C lower and within the TCZ. The thermal comfort for the animals, which affects the metabolic and dietary thermogenesis, could probably explain the higher forage intake in the SPS than in the monoculture (37.3 and 34.6 g OM/kg LW/d, respectively) (Forbes, 1995), which consequently implies higher beef and milk production (Baumer, 1991). The water intake by animals from drinking spouts and forage moisture was 11% higher in the monoculture than in the SPS, which reflects the effect of heating stress suffered by the animals under full sun. The high moisture content of forage in the SPS could also have benefited the animals in fulfilling their water requirements, which suggests that animals grazing on an SPS could meet a great deal of their water needs from the green forage biomass. In addition, a lesser volume of water was consumed from spouts by animals; this is an important advantage for SPSs, as water supply is decreasing globally.

In Northeast Brazil's Cerrado Biome, after deforestation, the most appropriate trees are selected for the wood market, and the residual biomass (stems or roots) is used for charcoal or burned before forage seeding for pasture growth. During approximately 10 or 20 years of pasturing, soil fertility based on residual OM and mineral ashes left from the deforestation (as the available soil phosphorus (P) content is very low) is reduced because of inappropriate stocking rates (overgrazing) and reduced nitrogen (N) levels (Dias-Filho et al., 2001). Therefore, the pasture degradation process has negative effects on forage biomass, including a marked reduction in cattle production and a high level of invasion of nonpalatable species. Usually, there are three options for farmers in this scenario: (1) if farmers have financial support, they switch from livestock to agriculture, seeding soybean or maize after chemical fertilization on part of the farm; or (2) plantation of eucalyptus trees for cellulose, and finally, (3) farmers sell the land. As an alternative to this process, the positive results obtained by the natural regeneration method to develop the SPS were applied and expanded in Maranhao state, Northeast Brazil. The project was implemented on a beef farm located in the Cerrado Biome, where approximately 700 ha have been managed for the last 10 years by mechanical selective cutting of young tropical bushes and trees regenerated from the seedbank after deforestation, or even through seed dispersal via wind or animals. Technically, it was recommended that the canopy of the remaining trees should not intercept more than 50% of the sunlight, or the light restriction could negatively affect the growing grass. Therefore, the density of trees from the natural regeneration process was stabilized to around 200–300 trees/ha, according to the canopy shape. The yield of the system (SPS) has steadily increased when compared to the monoculture system due to a high stocking rate (two and less than one animal unit per hectare, respectively), higher animal fertility, and a lesser effect of spittlebugs (*Deois flavopicta*) on forage biomass production (Campagnani et al., 2017). The higher biomass available for cattle nutrition and consequently better income in an SPS provides financial stability and social security. In addition, an SPS can provide opportunities for economic diversification in terms of wood, carbon, and tourism activities. It is important to mention the environmental services that are potentiated by the higher biodiversity of the SPS that generate biologic control of insects (e.g., spittlebug), increase carbon sequestration, and benefit animal welfare due to the temperature reduction and consequent cattle thermal comfort. Hopefully soon, all these economic, social, and environmental services could also be quantified to implement policies and payment for economic services. The positive results of the natural regeneration method were used by The Global Agenda for Sustainable Livestock (http://www.livestockdialogue.org/) to build "The Good Practice Guideline," in which practical examples were presented that focused on maintaining, restoring, and enhancing the environmental and economic values of grasslands (http://www.livestockdialogue.org).

Animal production in SPSs with legume trees has been evaluated in Southeast Brazil (Fig. 18.2). In the first study (Paciullo et al., 2011), in which SPSs were implemented with *Brachiaria decumbens*, *Eucalyptus grandis*, and leguminous trees (*Acacia mangium* and *Mimosa artemisiana*), the daily body weight gain was higher for SPSs (665 g/heifer per day) than for *B. decumbens* monoculture (567 g/heifer per day). The higher crude protein content in the SPS may have contributed to the improved quality in the diet of heifers in the pasture with trees, thus favoring animal performance. Considering the average dry matter intake during the rainy season of 2.3% of body weight and the protein content of the pasture in each system during the rainy season, it was estimated that the heifers in the SPS were able to consume, on average, 607 g/day of Crude protein (CP), while in the monoculture, it was only 538 g/day. The

FIGURE 18.2 SPS with leguminous tree species, in Southeast Brazil.

difference in CP intake by the heifers in the two systems (69 g/day) explains in part the better performance in the SPS. The second study (Paciullo et al., 2014) compared the performance of dairy cows in two contrasting grazing systems: (1) the SPS composed of *B. decumbens* intercropped with several herbaceous and leguminous tree species; and (2) an open *B. decumbens* pasture. Milk yield was higher in the SPS (10.4 kg/cow per day) than in open pasture (9.5 kg/cow per day). The high percentage of herbaceous legumes with high crude protein content most likely had a positive impact on the nutritional quality of the diet in the SPS. The evidence indicates that the contribution of legumes to the ruminant diet results in higher cattle performance on mixed forages compared with those grazing on grass only. Both experiments (with heifers and cows) demonstrated the possibility to obtain high animal production in areas previously degraded, but that had been recovered utilizing SPS practices.

Colombia Case Studies: Silvopastoral Systems, Livestock Production, Carbon Sequestration, and Ecologic Regeneration

In this part of the chapter, we focus on the Colombia studies developed mainly by the Centre for Research on Sustainable Agriculture (CIPAV), a nongovernmental organization founded in 1986 that concentrates on the use of SPSs as a tool to promote (1) sustainable livestock; (2) ecologic restoration; (3) ecosystem services; and (4) protection of aquatic systems. This supported a strong technical capacity to design and implement SPSs that were applied in Colombia, México, Brazil, Argentina, and Panama, among other countries.

In Colombia, through several initiatives including projects using SPS approaches to ecosystem management and mainstreaming biodiversity into sustainable cattle ranching, CIPAV and other organizations have promoted the establishment of SPSs in several regions of the country. The systems include live fences, scattered trees in pastures, managed plant succession, fodder tree banks and iSPSs (Murgueitio et al., 2011). The implementation of SPSs has contributed to reduce deforestation, use of fire and pesticides, improved forage and animal productivity, biodiversity, and carbon captures in the participating farms (Murgueitio et al., 2011, 2015). Unlike conventional intensification (concentrate, fertilizer, machinery, etc.), sustainable intensification (Godfray and Garnet, 2014) promoted with iSPS is based on agroecologic principles in which shrubs (leguminous species or not) are planted in high density (Fig. 18.3). These aspects differentiate this system from

FIGURE 18.3 El Hatico farm, iSPS including *Leucaena* (shrub), *Cynodon* (grass), *Prosopis juliflora* (tree), and Lucerne cattle, Colombia.

other SPSs. In iSPS, the herbs or graminaceous stratum are composed of native (e.g., genus *Axonopus, Paspalum,* etc.) or exotic (genus *Urochloa, Megathyrsus, Pennisetum, Cenchrus,* etc.) grasses consociated with leguminous forage (e.g., genus *Stylosanthes, Arachis, Centrosema, Calopogonium,* etc.). The second stratum consists primarily of high-density (10,000 to 80,000 plants per ha) forage shrub legumes (e.g., *L. leucocephala,* Mimosoidae subfamily), but also other plant families (e.g., *Tithonia diversifolia,* Asteracea; *Guazuma ulmifolia,* Malvaceae) (Rivera et al., 2015). In a third stratum, trees are planted in rows in an east–west direction or in contour lines to help control erosion, provide shade, and other resources and services. According to the future utilization, tree species are selected for wood (e.g., *Swietenia macrophylla* King, *Cedrela odorata*), cellulose (e.g., *Pinus* or *Eucalyptus* spp), or shade and fruits (e.g., *Prosopis* ssp, *Samanea samam* or *Gliricidia sepium, Psidium guajaba, Mangifera indica*) among many others.

Regarding animal production parameters in an iSPS, research results show that due to the higher availability of forage biomass during the entire year, especially during the dry season, meat and milk production per hectare is increased compared to grass monocultures (Mahecha et al., 2011; Broom et al., 2013). The efficiency of an iSPS in producing beef is 12 times higher compared to extensive monoculture pastures, and therefore less grazing area is required. In addition, the enteric methane emission is lower when compared to traditional monoculture systems thanks to the better quality of the diet and higher productive efficiency. These aspects are key elements to mitigate and adapt animal production to the effects of global climate change (Murgueitio et al., 2014). The most common iSPS design in Colombia has *L. leucocephala* consociated with star grass (*Cynodon plectostachyus*). An experimental trial was conducted to test the effects of a diet containing 100% star grass (SG) with one composed of 76% SG and 24% *Leucaena* (SGL), a typical design of an iSPS, in terms of methane emission. The results demonstrated higher protein content and gross energy for SGL, but lower NDF content compared to SG. The associative effects of SGL and SG improved digestibility and intake, and at the same time reduced the enteric methane emissions by 3% compared to the SG diet; this could explain the better ruminant performance in an iSPS (Molina et al., 2016). Similar results were also obtained in another study in which an iSPS with *L. leucocephala, C. plectostachus,* and *Megathyrsus maximus* reduced methane emission in heifers. Although there was higher protein content, intake, and production (liveweight gain) verified in the iSPS, the enteric methane was lower than in the traditional system and could be one option to reduce GHG emissions from livestock production (Molina et al., 2015).

The shortage of fauna and flora biodiversity in the monoculture grassland, along with inappropriate stocking rates, leads to a severe pasture degradation process that is widespread all over Latin America. The association of different sources of forage biomass, shrubs, and trees presented in an SPS is being used to promote the ecologic restoration of unproductive pasture and to release land not appropriated for cattle feeding for other uses, like forest planting (Calle et al., 2014). The high biomass production from an iSPS allows the sustainable intensification of the land for animal production and consequently leaves sloping areas, riparian forests, and nonfertile land soils for biodiversity recovery or even as ecologic corridors. The increased biodiversity, natural control of insects, high biomass production and carbon sequestration, biologic fixation of nitrogen, soil organic matter improvements, low fertilizer input, hydrologic regulation, and reduction of pasture dehydration caused by wind and heating provided by an iSPS are important environmental, economic, and social services (Chara et al., 2015). One example of the impact of an iSPS on biodiversity and animal welfare is the reduction of ticks in cattle. Although the negative effects of *Rhipicephalus* (*Boophilus*) *microplus* ticks on animal production are correlated to breed, body condition, nutritional status, and environmental aspects, the study compared iSPS farms and traditional farms in Colombia and demonstrated the benefits of an iSPS on tick load. The results suggested that the life cycle of ticks is complex and is dependent on the climate as well as on the animal itself. However, the animals under effects of an iSPS showed less parasitical load in relation to the other traditional pasturing monoculture grass systems (Raquel Salazar et al., 2015), and this generates improvements to cattle welfare (Broom et al., 2013).

México case studies: Participative Approach for Implementation of Intensive SPS in México

The *Leucaena* genus was part of the culture of the indigenous people in México, utilized and managed for food and firewood. Large areas of *Leucaena* forest were destroyed during the Spanish conquests because of the intensification of agriculture (burning and plowing) and due to the high palatability of *Leucaena* for domestic animals including cattle, goats, and horses. Nevertheless, the persistence of the *Leucaena* tree and cultural resilience of the indigenous people allowed the conservation of its use (Zarate, 1997). The tropical regions of México are responsible for 50% of national livestock production. Cattle, sheep, and goats are generally produced in extensive systems that are highly dependent on the climate season (rainy and dry). Therefore, the production rates during the dry season are decreased due to low forage biomass production and arid climate. As a result, the demand for livestock products (which does not respect the appropriate livestock stocking rates according to pasture seasonality) generates negative impacts on the soil, pasture, and environment, which also affects farmer income and food security in México.

Since 1996, the Fundación Produce, a civil association of farmers, has worked on technological innovation in the livestock and agriculture sector in México. In 2006, they started the adoption of iSPS as the main tool to regain the competitiveness of the livestock sector through the sustainable use of local natural resources and low dependence on external inputs (e.g., fertilizer or seeds). Currently, there are 12,000 ha of iSPSs established (Fig. 18.4) in three regions of México (Huasteca, 3200 ha; Sur Sureste, 3000 ha; and Tierra Caliente, 6800 ha), involving 1800 farms (beef and milk) under technical supervision of the Fundación Produce (Sanches et al., 2006). The strategy applied by the foundation to expand the iSPS technology was based on a participatory approach in which producers, technicians, and researchers from national and international organizations jointly

FIGURE 18.4 iSPS composed of *Leucaena lecocephala* (tree and shrubs) and *Panincum maximum* (left) and natural regeneration of *Prosopis juliflora* and *Panicum maximum* (Michoacan state, México).

discussed the social, economic, and environmental objectives according to the agroecologic conditions. The system is denominated *silvopastoral* due to the inclusion of trees, shrubs, and pastures. It is *intensive* as it aims high biomass production per square metre, in which the animals are rotated every 3–24 h using electric fences (generated by solar panels) and mobile water drinkers.

SPS using *Leucaena* as the main shrub forage is growing fast in México. Therefore, several research projects were developed, including forage yield and quality of *L. leucocephala* and *Guazuma ulmifolia* as a pure or mixed fodder bank planted in the Yucatan peninsula, México (Casanova-Lugo et al., 2014). The results demonstrated the improvement, in terms of forage biomass and quality, when both species are consociated. However, the *Leucaena* shrub had higher protein content (228 g kg DM) than *Guazuma* (145 g kg DM) and even in the mixed fodder bank (180 g kg DM), but lower dry matter production (3.4, 4.5, and 5.1 t DM ha per season, respectively). Apart from forage production and quality, there is a growing interest in the quality of dairy products in iSPS, especially cheese, which uses 15% of the total fluid milk produced in México. The study performed by Mohammed et al. (2016) demonstrated that milk and cheese produced from iSPSs had characteristics similar (appearance, texture, color, flavor, and overall acceptability) to those produced from a monoculture grass system. The influence of *Leucaena*, as the main component of iSPS, on the intramuscular meat fat quality and fatty acid profile in bovine fattened during the drought and rainy season were evaluated. The positive relationship between concentrations of oleic acid and the reduction of cholesterol and cardiovascular diseases, as well as the benefits of omega 3 and 6 on human health, are well known (Muchenje et al., 2009). The experimental results indicated higher levels of oleic acids during the drought season and lower concentrations of linoleic, linolenic, and eicosanoic acids compared to the rainy season. iSPS was able to fatten cattle eating tropical forage and to promote a high concentration of polyunsaturated fatty acids; therefore, it was a technology to conciliate sustainable animal production and healthy food for human nutrition (Echevarría et al., 2013).

FINAL REMARKS

Cattle ranching is part of Latin American culture; it occupies 27% of the landscape and has a deep influence in the rural economy and society. The large extension of degraded land in Brazil creates the opportunity for scaling up the natural regeneration or even the use of planted leguminous trees approach, which could be a positive contribution by forestry to livestock, food security, and the environment.

There is scientific evidence showing that SPS is one sustainable alternative to change the actual poor image of the livestock activity, not only by increasing production (milk, beef, and goods), but also recovering the landscape, producing environment/ecosystems services, and generating income for rural society.

SPSs, based on different forage species, shrubs, and trees, are enhancing the most important capacity of the ruminants: turn cellulose into protein and milk without competition with human feeding. In addition, SPSs promote the sustainable intensification of land including low inputs of fossil energy, high biodiversity and biomass production, respect to animal welfare, and rational use of water. Therefore, SPS could be one option to conciliate ruminant production and environmental conservation in Latin America and other parts of the world.

Acknowledgments

PPM-FAPEMIG, CNPq, CAPES-PVE, CIPAV, EMBRAPA, CBPS, Fundación Produce and tropical farmers.

References

Andrade, C.M.S., Valetim, J.F., Carneiro, J.C., 2002. Árvores de Baginha (Stryphnodendron guianense (Aubl.) Benth.) em ecossistemas de pastagens cultivadas na Amazônia Ocidental. Brazilian Journal of Animal Science 31, 574–582.

Barros, F. de. V., Goulart, M.F., SáTelles, S.B., Lovato, M.B., Valladares, F., de Lemos-Filho, J.P., 2012. Plasticity to light of congeneric trees from contrasting habitats. Plant Biology 14, 208–215.

Baumer, M., 1991. Animal production, agroforestry and similar techniques. Agroforestry Abstracts 4, 179–198.

Boddey, R.M., Macedo, R., Tarre, R.M., Ferreira, E., Oliveira, O.C., Rezende, C.P., Cantarutti, R.B., Pereira, J.M., Alves, B.J.R., Urquiaga, S., 2004. Nitrogen cycling in Brachiaria pastures: the key to understanding the process of pasture decline? Agriculture Ecosystem Environment 103, 389–403.

Broom, D.M., Galindo, F.M., Murgueitio, E., 2013. Sustainable, efficient livestock production with high biodiversity and good welfare for animals. Proceedings of the Royal Society Biological Sciences 280, 2013–2025.

Calle, Z., Chará, J., 2014. Intensive silvopastoral systems: integration of sustainable cattle ranching, silviculture and restoration at the landscape scale. In: Calle, A., Calle, Z., Garen, E., Del Cid-Liccardi, A. (Eds.), Ecological Restoration and Sustainable Agricultural Landscapes. Environmental Leadership and Training Initiative. Yale University, New Haven, CT, pp. 27–33. Smithsonian Tropical Research Institute (Panama City). Available from: http://elti.fesprojects.net/Proceedings/2013_agrilandscapes.pdf.

Campagnani, M.O., Garcia, W.C., Rosa, L.H., Amorim, S.S., Cangussú, M.A., Mauricio, R.M., 2017. Prospection and fungal virulence associated with Mahanarva spectabilis (Hemiptera: Cercopidae) in an Amazon silvopastoral system. Florida Entomologist 100, 426–432.

Casanova-Lugo, F., Petit-Aldana, J., Solorio-Sanches, F.J., Parsons, D., Ramires-Aviles, L., 2014. Forage yield and quality of Leucaena leucocephala and Guazuma ulmifolia in mixed and pure fodder banks systems in Yucatan, Mexico. Agroforestry systems 88 (1), 29–39.

Chara, J., Camargo, J.C., Calle, Z., Bueno, L., Murgueitio, E., Arias, L., Dossman, M., Molina, E.J., 2015. In Sistemas Agroforestales. Funciones Productivas, Socioeconomicas Y Ambientales.

Dias-Filho, M.B., Davidson, E.A., de Carvalho, C.J.R., 2001. Linking biochemical cycles to cattle pasture management and sustainability in the Amazon basin. In: McClain, M.E., Victoris, R.L., Richery, J.E. (Eds.), The Biogeochemistry of the Amazon Basin. Oxford University Press, New York, pp. 84–105.

Echevarría, M.E.R., Corral-Flores, G., Sánchez, B.S., Rojo, A.D.A., Grado-Ahuir, J.A., Rodríguez-Muela, C., Palacios, L.C., Beltrán, V.E.S., Sánchez, F.F.S., 2013. Tropical and Subtropical Agroecosystems 16 (2013), 235–241.

Forbes, J.M., 1995. Voluntary food intake and diet selection in farm animals. 1.ed. Leeds: CAB International 532p.

Gea-Izquierdo, G., Cañellas, I., Montero, G., 2006. Accorn production in Spanish holm oak woodlands. Investigación agraria. Sistemas y recursos forestales 15, 339–354.

Godfray, H.C.J., Garnett, T., 2014. Food security and sustainable intensification.Phil. Trans. R. Soc. B 369 (20120273). https://doi.org/10.1098/rstb.2012.0273.

Mahecha, L., Murgueitio, M., Angulo, J., Olivera, M., Zapata, A., Cuartas, C.A., Naranjo, J.F., Murgueitio, E., 2011. Desempeño animal y características de la canal de dos grupos raciales de bovinos doble propósito pastoreando en sistemas silvopastoriles intensivos. Revista Colombiana de Ciencias Pecuarias 24, 470.

Mauricio, R.M., 2012. Comment to "Pasture shade and farm management effects on cow productivity in the tropics" by Justin A.W. Ainsworth, Stein R. Moe, C. Skarpe [Agric. Ecosyst. Environ. 155 (2012) 105–110]. Agriculture Ecossystem and Environment 161, 78–79.

Mohammed, A.H.M., Aguilar-Perez, A.F., Ayala-Burgos, A.J., Bottini-Luzardo, M.B., Solorio-Sanchez, F.J., Ku-Vera, J.C., 2016. Evaluation of milk composition and fresh soft cheese from na intensive silvopastoral system in the tropics. Dairy Science & Technology 96, 159–172.

Molina, I.C., Donneys, G., Montoya, S., Rivera, J.E., Villegas, G., Chara, J., Barahona, 2015. The inclusion of *Leucaena leucocephala* reduces the metahne productio in Lucerne heifers receiving a *Cynodon plectostachyus* and *Megathyrsus maximus* diet. Livestock Research for Rural Development 27 (5).

Molina, I.C., Angarita, E.A., Mayorga, O.L., Chara, J., Barahona-Rosales, R., 2016. Effect of *Leucaena leucocephala* on methane production of Lucerna heifers fed a diet based on *Cynodon plectostacyus*. Livestock Science 185.

Muchenje, V., Dzama, K., Chimonyo, M., Strydom, P.E., Hugo, A., Raats, J.G., 2009. Some biochemical aspects pertaining to beef eating quality. Food Chemistry 112, 279–289.

Murgueitio, E., Calle, Z., Uribe, F., Calle, A., Solorio, B., 2011. Native trees and shrubs for the productive rehabilitation of tropical cattle ranching lands. Forest Ecology and Management 261, 1654–1663.

Murgueitio, E., Flores, M., Calle, Z., Chará, J., Barahona, R., Molina, C., Uribe, F., 2015. Productividad en sistemas silvopastoriles intensivos en América Latina. In: Montagnini, F., Somarriba, E., Murgueitio, E.,

Fassola, H., Eibl, B. (Eds.), Sistemas Agroforestales. Funciones Productiva, Socioeconómica Y Ambientales. Serie Técnica Informe Técnico 402. CATIE, Turrialba, Costa Rica. Fundación CIPAV, Cali, Colombia, pp. 59–101.

Murgueitio Restrepo, E., Chará Orozco, J.D., Barahona Rosales, R., Cuartas Cardona, C.A., Naranjo Ramirez, J.F., 2014. Intensive silvopastoral systems (ISPS), miti-gation and adaptation tool to climate change. Tropical and Subtropical Agroecosystems 17 (3), 501–507.

Paciullo, D.S.C., de Castro, C.R.T., Gomide, C.A.M., Maurício, R.M., Pires, M.F.Á., Müller, M.D., Xavier, D.F., 2011. Performance of dairy heifers in a silvopastoral system. Livestock Science 141, 166–172.

Paciullo, D.S.C., Pires, M.F.A., Aroeira, L.J.M., Morenz, M.J.F., Maurício, R.M., Gomide, C.A.M., Silveira, S.R., 2014. Sward characteristics and performance of dairy cows in organic grass–legume pastures shaded by tropical trees. Animal 8 (8), 1264–1271.

Raquel Salazar, B., Rolando Barahona, R., Julian Chara, O., Maria Solange Sanchez, P., 2015. Productivity and tick load in Boss indicus x B. taurus cattle in tropical dry forest silvopastoral system. Tropical and Subtropical Agroecosystems 18, 103–112.

Reis, G.L., Lana, Â.M.Q., Maurício, R.M., Lana, R.M.Q., Machado, R.M., Borges, I., Neto, T.Q., 2009. Influence of trees on soil nutrient pools in a silvopastoral system in the Brazilian Savannah. Plant and Soil 111–117.

Rivera, J.E., Cuartas, C.A., Naranjo, J.F., Tafur, O., Hurtado, E.A., Arenas, F.A., Chara, J., Murgueitio, E., 2015. Effect of an intensive silvopastoral system (iSPS) with Tithonia diversifolia on the production and quality of milk in the Amazon foothills, Colombia. Livestock Research for Rural Development 27 (10).

Rozados-Lorenzo, M.J., Gonzalez-Hernandez, M.P., Silva-Pando, F.J., 2007. Pasture production under different tree species and densities in an Atlantic silvopastoral system. Agroforestry Systems 70, 53–62.

Sanches, M.D., Rosales, M., 1999. Agroforesteria para la produccion animal en América Latina. FAO.

Sanches, B.S., Gomez, J.C.G., Restrepo, E.M., 2006. Modelo de consenso silvipastoril, http://www.fpm.org.mx/.

Souza, L.F., Maurício, R.M., Moreira, G.R., Gonçalves, L.C., Borges, I., Pereira, L.G.R., 2010. Nutritional evaluation of Braquiarão grass in association with Aroeira trees in a silvopastoral system. Agroforestry Systems.

Souza, L.F., Mauricio, R.M., Pasciullo, D.C., Silveira, S.R., Ribeiro, R.S., Calsavara, L.H., Moreira, G.R., 2015. Forage intake, feeding behavior and bio-climatological indices of pasture grass, under the influence of trees, in a silvopastoral system. Tropical Grasslands 3, 129–141.

Steinfeld, H., Gerber, P., Wassenaar, T., Castel, V., Rosales, M., de Haan, C., 2006. Livestock's Long Shadow-Shade: Environmental Issues and Options. FAO, Rome, Italy.

Viana, V.M., Mauricio, R.M., Matta-Machado, R., Pimenta, I.A.S., 2002. Manejo de la regeneracion natural de especies arboreas nativas para la formacion de sistemas silvopastoriles en la zonas de bosques secos del surestes de Brasil. Agroforestería en las Américas 9, 48–52.

Yamamoto, W., Dewi, I.A., Ibahim, M., 2007. Effects of silvopastoral areas on milk production at dual-purpose cattle farms at the semi-humid old agricultural frontier in central Nicaragua. Agricultural Systems 94, 368–375.

Zarate, S., 1997. Domestication of cultivated Leucaena (Leguminosae) in Mexico: the sixteenth century documents. Economy Botany 51 (3), 238–250.

Further Reading

Casanova-Lugo, F., Petit-Aldana, J., Solorio-Sanches, F.J., Parsons, D., Ramires-Aviles, L., 2014. Ag0000roforestry System 88, 29–39.

Ribeiro, R.S., Terry, S.A., Sacramento, J.P., Silveira, S.Re., Bento, C.B.P., da Silva, E.F., et al., 2016. Tithonia diversifolia as a Supplementary Feed for Dairy Cows. PLoS ONE 11 (12).

Salazar, R., Barahona, R., Chará, J., Sánchez, M.S., 2015. Productividad y carga parasitaria de bovinos *Bos indicus x Bos taurus* en un sistema silvopastoril intensivo en bosque seco tropical. Tropical and Subtropical Agroecosystems 18, 103–112.

SECTION V

SOCIO-ECONOMIC OPPORTUNITIES FOR AND LOCKING EFFECTS AGAINST DIVERSIFICATION OF AGROECOSYSTEMS AT FARM AND BEYOND FARM LEVELS

CHAPTER

19

The Economic Drivers and Consequences of Agricultural Specialization

David J. Abson

Leuphana University, Lüneburg, Germany

INTRODUCTION

Three tightly related processes within the agricultural sector—intensification, specialization, and concentration (Ilbery and Bowler, 1998)—led to unprecedented changes in agroecologic systems during the 20th century (Blaxter and Robertson, 1995; Bowler, 1985), where agricultural intensification is defined here as a process whereby an increase in agricultural production per unit area (yield) is achieved via increased application of labor, external inputs (such as fertilizers and pesticides), machinery, or technology. Agricultural specialization is defined as the process of concentrating resources (labor, capital,[1] and land) on producing a limited variety of goods, and concentration is the process of consolidation of specialized production across time, space, and socioeconomic dimensions. All three processes (intensification, specialization, and concentration) led to profound changes in both the socioeconomic and ecologic functioning of agroecologic systems (Stoate, 1995b; Firbank, 2005).

Here the focus is on agricultural specialization, including temporal and spatial specialization (concentration) of agricultural production, with intensification framed as an indirect driver of specialization at multiple temporal and spatial scales. In the following sections, economic drivers such as economies of scale, competitive advantage, resource substitution, efficiency (e.g., Lambin et al., 2001; Hazell and Wood, 2008) in combination with rapid technological change (e.g., Ernoult et al., 2003) and a dominant productionist policy paradigm (e.g., Coleman et al., 1996) are used to explain the global shift toward more specialized agricultural landscapes during the 20th century (Pretty, 1998; Giampietro, 2004). Having described the process of agricultural specialization, the relations between that specialization and agroecologic simplification at multiple scales (from the genetic diversity of crops to landscape homogenization) are

[1] Where capital in this context is defined as nonfinancial assets or durable goods that can be used to produce goods or services. A distinction is made between land and capital (as factors of production) because capital can be increased by human labor.

Agroecosystem Diversity
https://doi.org/10.1016/B978-0-12-811050-8.00019-4

examined. Finally, the potential consequences of agricultural specialization are discussed in relation to four key properties of agroecosystems: productivity, stability, persistence, and justice.

Here, it should be noted that the following section takes a largely Western European/North American perspective on agricultural change and specialization. Nevertheless, many of the underlying drivers of agricultural specialization addressed here are relevant elsewhere, even if the speed, strength, and timing of those drivers may differ substantially across geographic locations and socioeconomic and ecologic systems.

DRIVERS OF AGRICULTURAL SPECIALIZATION

Historically, labor was a crucial limiting factor in agricultural production. For example, in 1790, some 90% of the American workforce was engaged in farming; this figure dropped to 41% by 1900 and less than 2% of the workforce by the year 2000 (Dimitri et al., 2005). During the late 19th and early 20th century, improved productivity and mechanization was able to overcome labor as a limiting factor (Grantham, 1993), with mechanization, in turn, driving increasing labor and production specialization (e.g., Alston et al., 2009). For example, mechanization allows larger areas of crops to be harvested within a given time frame, reducing the need for farmers to diversify cropping choices with staggered harvest dates to match available labor. This, in turn, increased the opportunities for farmers to move away from subsistence toward commercial farming because farmers could concentrate on the most profitable crops (rather than those that met dietary requirements of the farmer's family) and because of the increasing proportion of the population purchasing rather than growing their own food. The resulting shift from agriculture as a labor intensive, diversified, and largely subsistence activity to a highly specialized, capital intensive commercial (market) enterprise drove specialization in several ways.

Increased agricultural yield—for, example, the world's food production doubled from 1961 to 1996 with only a 10% increase in arable land area (Tilman, 1999)—lowered food prices and freed a large labor force to work in the factories of the industrial revolution. This, in turn, created new consumers of both agricultural and nonagricultural products (Dorward et al., 2004). Profits from agriculture were reinvested in capital, leading to further productivity gains. As subsistence food production became less important, agricultural land was increasingly consolidated (concentrated) in larger commercial landholdings (e.g., Burton and Walford, 2005). This historic concentration of landholding continues, with the average farm size in the United States doubling between 1982 and 2007 (Macdonald et al., 2013). To some extent, increasingly large farms are a response to new technologies that cannot be used efficiently on small landholdings, or for the diverse agricultural production that typified many agricultural systems in the 19th and early 20th centuries.

This "economy of size"—the ability of a farm to lower costs of production by increasing production over larger and larger areas of land—was driven not solely by mechanization. The spreading of fixed costs (largely associated with farm infrastructure costs, such as grain storage, etc.), the bulk purchase of agricultural inputs, and increased marketing power all contributed to consolidation within the agricultural sector during the 20th century (Cornia, 1985; Duffy, 2009). Taking advantage of these economies of size required concurrent agricultural specialization, with "simplified" landscapes and mono-cropping increasingly prevalent throughout the world's agricultural systems (Clay, 2013).

Economies of size and specialization are, in turn, related to other more specific problems such as the workloads associated with mixed crop-livestock farms. Integrated crop-livestock systems typically have greater workload—that is, higher labor input requirements per unit output—compared with specialized agricultural systems (Lemaire et al., 2014). To address this, farmers often choose to specialize in either crop or livestock production. For example, the higher labor requirements for dairy and livestock farming can be reduced by heavy investment in on-farm infrastructure (such as mechanized milking stalls) that are only feasible with larger herd sizes. Labor requirements for arable crops can be reduced by increased field sizes and using larger harvesting machinery. The specialized production of fodder crops further reduced the need for mixed cropping strategies for livestock farmers (De Raymond, 2013; Lemaire et al., 2014). The choice between livestock and crop specialization is determined, in part, by land suitability and availability, with larger, flatter landholdings more suited to large-scale arable specialization and smaller farms more suitable for intensified livestock and dairy production. This has led to increasing territorial separation of arable and livestock systems based on large-scale topographic patterns (Peyraud et al., 2014).

As agricultural products increasingly became market goods, comparative advantage—the ability of an economic agent to carry out a particular economic activity more efficiently (at a lower relative opportunity cost) than another agent (Ricardo, 1817)—increasingly shaped agricultural specialization. Topography, water resources, climate, growing season length, soil type, labor costs, and distance from markets all confer potential comparative advantages for certain agricultural goods in certain regions. In combination with economies of size, comparative advantage provided a strong economic rationale for both individual farmers and regional and national governments to favor larger, more specialized agricultural production.

Initially the geographic scale of such specialization in agriculture was restricted by the perishable nature of many agricultural goods, the costs of transportation, and a lack of developed markets. Regions could only specialize to the extent that they still produced the diversity of perishable goods demanded within their local markets. Similarly, inputs to cropping systems, particularly fertilizers, were largely restricted to on-farm processes (animal manures, nitrogen fixing crops, etc.), with some notable exceptions such as guano extraction and niter (potassium nitrate) mining. In this context, maintaining soil fertility, and therefore agricultural productivity, tended to require more varied and mixed farming systems (Pretty, 1999; Rackham, 2000). However, with the advent of the Harber-Bosch process (1908) for nitrogen fertilizer production, cheap mass transportation, and industrial refrigeration—the Dunedin, the first ship fitted with a compression refrigeration unit for the commercial shipping of meat, sailed from New Zealand to England in 1882 (Williscroft, 2007)—such geographic constraints to specialization began to dissolve, with specialization occurring at the regional level, not just the farm level (Chisholm, 1962; Leaman and Conkling, 1975).

For Western nations, there was a rapid increase in specialization beginning in the 1930s (Dimitri et al., 2005; Rackham, 2000; Stoate, 1995a; Winsberg, 1980). In the late 1950s the "Green Revolution"—largely premised on the development and use of high-yielding crop varieties and concurrent control of environmental variables such as nutrient loads and water availability (Foster and Rosenzweig, 1996)—led to similar transformative changes in large parts of the developing world's agricultural systems (Frankel, 2015). Here, it should be noted that it has been estimated that only approximately 20% of the increases in yield from 1961 to 1980 in developing countries are attributed to newly introduced improved crop varieties, with the additional productivity gains coming from increased usage of external inputs such as fertilizers, pesticides, irrigation, etc. (Evenson and Gollin, 2003). Despite yield gains, a lack of transportation infrastructure, market access, and investment capital in some developing nations may have limited, or at least slowed, the specialization process and the role of comparative advantage at broad spatial extents.

The need to use economies of scale to fully utilize expensive capital equipment associated with technological changes led to what Cochrane (1979) called the "technological treadmill." New technologies enabled innovators to capture windfall profits, increasing production and driving down prices. Those who do not adopt the new technologies suffer from both lower yields than adopters and shrinking profits due to the decreasing prices (Giampietro, 2003; Howkins, 2003). This, in turn, leads to small, nonadopting farms being driven out of business or being bought up by larger enterprises (Cochrane, 1979). Market forces also drive specialization due to land scarcity. Land scarcity raises land values and drives farmers to specialize on higher value market-oriented products (Lambin et al., 2001). This driver is particularly relevant for nations not yet fully integrated into global markets or with weak land tenure regimes (Ostrom et al., 1999).

In addition to market forces and concurrent technological change, an explicit productivism paradigm that arose as a response to food shortages following the Second World War (Stoate, 1995b, Ilbery and Bowler, 1998, Hazell and Wood, 2008) also increased the drive toward specialized agricultural production. Lowe et al. (1993 p. 221) defined productivism as:

> [A] commitment to an intensive, industrially driven and expansionist agriculture with state support based primarily on output and increased productivity. The concern was for 'modernization' of the 'national farm', as seen through the lens of increased production. By the 'productivist regime' we mean the network of institutions oriented to boosting food production from domestic sources.

Such productivist regimes were dominant in the West until at least the early 1990s (Wilson, 2001) and were typified by rapidly expanding globalization of food supply and industrialized agricultural production (Cloke and Goodwin, 1992; Goodman and Watts, 1997). The productivist approach can be seen in the (pre-1990s) Common Agricultural policy (Firbank, 2005), the "Atlanticist Food Order" (Le Heron, 1993), and in many polices related to the Green revolution in the developing world (Wilson and Rigg, 2003). The outcome of these productivist regimes was regional-scale agricultural specialization and increased concentration of agricultural production into large agri-business (Ilbery and Bowler, 1998; Wilson, 2001). Productivist approaches remain dominant in recent academic debates such as land sparing versus land sharing (Fischer et al., 2014), agricultural yield gaps (Mueller et al., 2012), and "sustainable intensification" (Loos et al., 2014), although increasingly these discourses frame productivity with a broader set of societal goals. Productionist approaches are now often linked to issues such as diversity conservation, food security, or sustainability. However, productionist approaches retain a clear focus on high yields, rather than better distribution of agricultural goods. These productionist approaches tend to be premised on a relatively simplistic, reductionist understanding of agroecosystems, where the problem is often seen as a technical challenge regarding how to optimize (ceteris paribus) the delivery of multiple goods from a given system. This contrasts with more holistic and system approaches to agroecosystem research that acknowledge the complexity and interdependence of agroecosystems properties and associated societal goals across multiple spatial, temporal, and socioeconomic scales (Glamann et al., 2017). The 2003 Common Agricultural Policy (CAP) reforms cut the direct link between subsidy regimes and production, leading to what has often been described as a "post-productionist" policy regime (Evans et al., 2002). However, it has been argued that those, and subsequent, reforms of CAP have done little to fundamentally shift it away from a productionist perspective (Candel et al., 2014), or to address the strong trend toward specialization of agricultural production.

The benefits to individuals or societies producing goods for which they have a comparative advantage and trading those goods for other (similarly, relatively efficiently produced) goods produced is well documented in the economic literature (e.g., Matsuyama, 1992; Hunt and Morgan, 1995). Similarly, the increase in food productivity that accompanied agricultural intensification and specialization is undeniable (Tilman, 1999; Godfray et al., 2010). Nevertheless, questions remain as to what extent such yield increases require specialized, rather than diversified, production systems (Ponisio et al., 2015). Regardless of whether or not increased specialization is a necessity for feeding a growing human population, agricultural specialization has led to serious and persistent ecologic and socioeconomic impacts.

AGRICULTURAL SPECIALIZATION, ECOLOGIC SIMPLIFICATION, AND FARMING CHARACTERISTICS

Agricultural specialization led to increasingly simplified agroecologic systems during the 20th century (e.g., Altieri and Nicholls, 2002, Table 19.1), across multiple spatial and temporal scales (Matson et al., 1997). This simplification included declines in wildlife, genetic resources, and biodiversity (Tscharntke et al., 2005) and was driven, in part, through the loss of on-farm wildlife habitats, such as field margins (Vickery et al., 2009), and the temporal and spatial synchronization and homogenization of farming activities reducing resources, particularly food, for on-farm biodiversity (Brickle and Harper, 2002; Benton et al., 2003).

Agricultural specialization has led to ecologic simplification across multiple spatial scales (from field to region), with the effects at broader scales most apparent in the most globalized

TABLE 19.1 Ecologic Simplification and Farming Characteristics Driven by Agricultural Specialization

Field to Farm Scale	Landscape Scale	Regional Scale
Increased agricultural inputs (artificial fertilizers, herbicides, and insecticides)	Increasingly capital intensive and lower labor farming	Increasing geographic separation between producers and consumers
High-yielding monocultures	Reduction in mixed farming and relatively "closed" farming systems	Increased separation of arable and pastoral farming systems
Decreased crop diversity within fields (minimizing undersowings, intercropping, and polycultures)	Increasing dependence on global commodity markets for both inputs and outputs	Lengthening supply chains
Increased field sizes, with subsequent loss of field margins, hedgerows, and other ecologically important habitats	Decreased crop diversity and decreased genetic diversity within specific crops	Increasing fragmentation of nonfarming habitats
Decreased crop rotation, use of on farm inputs	Increased synchronization and homogenization of farming activities (across both space and time)	Declines in wildlife, genetic resources, and biodiversity
Increased mechanization		Decreased land use heterogeneity

nations. Globalization increases interconnectedness of places and people through trade, information, and capital flows, leading to an increased separation between the location of production and consumption (Lambin and Meyfroidt, 2011) and allowing for greater regional-scale comparative advantages to be sought by both national government and individual farmers. For example, Brazil's institutional focus on agricultural export markets combined with a strong comparative advantage in soy production lead to more than 60% (approximately 6.4 million hectares) of the increase in cultivated cropped area in Brazil between 1996 and 2006 being dedicated to soy production (Bustos et al., 2013).

THE CONSEQUENCES OF AGRICULTURAL SPECIALIZATION

Judging the consequences of agricultural specialization requires an understanding of the goals associated with managing agroecologic systems. Gordon Conway suggests the goals of agroecosystems can, broadly, be defined as "the amounts of goods and services produced by an agroecosystem, the degree to which they satisfy human needs and their allocation among the human population. It also has a time dimension since we seek not only increased benefits in the immediate future, but also a degree of security of the longer term" (Conway, 1993, p. 50). Based on this definition, four key criteria (productivity, stability, persistence, and justice (see Box 19.1)) can be used to assess the consequences of agricultural specialization. In the following sections, each of these criteria of assessing agroecologic systems is discussed in relation to the process of agricultural specialization.

Productivity and specialization. In much of the agricultural literature, productivity changes are conceptualized only in terms of yield (production per unit area). Such measures of productivity can be conceptualized as "single factor" productivity gains (i.e., productivity gains per unit of land) (Trueblood and Ruttan, 1995). Yield change is a useful, but extremely incomplete way

BOX 19.1

FOUR AGROECOSYSTEM PROPERTIES FOR JUDGING THE CONSEQUENCES OF AGRICULTURAL SPECIALIZATION

Productivity: the (average) output of valued good and services per unit of resource input (including land, labor, capital, fertilizers, pesticides, etc.), where output can be assessed across different dimensions, including yield, calorific content, market price, or profits.

Stability: the constancy of productivity in the face of perturbations resulting from fluctuations and cycles in the surrounding ecologic and socioeconomic systems.

Persistence: the ability of the agroecosystem to maintain productivity over long periods of time, or when subject to major system perturbations.

Justice: the fair/equitable distribution of the costs and benefits of productivity among human (and potentially nonhuman) cost bearers and beneficiaries.

Adapted from Conway, G.R., 1993. Sustainable agriculture: the trade-offs with productivity, stability and equitability. In: Barbier, E.B. (Ed.), Economics and Ecology: New Frontiers and Sustanable Development. Springer.

of conceptualizing and quantifying agricultural productivity changes. Productivity is generally defined as the ratio of output to inputs (i.e., output per unit of input), and there are many inputs other than land that are used in the production of agricultural outputs.

Therefore, a more useful assessment of agricultural productivity encompasses production changes in relation to all productive inputs (including labor, capital, external inputs, etc., see Box 19.1). Such measures consider the efficient use of all resources, not just land, required for food production. However, the assessment of changes in "multifactor productivity" (Singh et al., 2000) or "total factor productivity" (Comin, 2010) is problematic due to the potential incommensurability of different productive inputs to agriculture. How, for example, do you quantify total factor productivity changes in systems where labor is substituted for capital (in the form of external inputs in the production process)?

Efforts to assess total factor productivity tend to assume no differences in the quality of inputs (e.g., all labor is treated as being equally productive), and they rely on normalization and (often relatively arbitrary) weighting to make all productive factors commensurable, so they should be interpreted with some caution. Nevertheless, it has been estimated that the total factor productivity in the UK agricultural sector increased by approximately 65% between 1973 and 2015 based on the "volume" of all input to production (DEFRA, 2016). Capalbo and Vo (1988) estimated that between 1950 and 1982 the average annual rate of growth in agricultural outputs in the United States was 1.76, while the average annual rates of growth in inputs (including labor, land, equipment, fertilizers, pesticides, and energy) was only 0.17. Trueblood and Ruttan (1995) in a review of 14 estimates of multifactor agricultural productivity in the United States found annual growth rates from the late 1970s to the mid-1990s of 1.15%–1.94%, with no clear explanation for the differences in those estimates. Multifactor agricultural productivity growth in the United States was estimated to have fallen from 2.02% per annum from 1949 to 1990 to just 1.18% from 1990 to 2007 (Alston et al., 2015). The extent to which these productivity gains results from agricultural specialization remains unclear.

Here, it should be noted that these assessments of multifactor productivity are problematic because they do not acknowledge, or account for, the externalities (the consequences of economic activity that affect third parties without this being reflected in the cost of the goods or services resulting from that economic activity) associated with agricultural production (Archibald, 1988; Capalbo and Vo, 2015). For example, the increased use of fertilizers in specialized agricultural systems, can lead to nitrogen and phosphorous runoff, and in turn, water pollution and eutrophication (e.g., Withers et al., 2014), and tillage of agricultural soils can lead to significant greenhouse gas emissions (e.g., Abson et al., 2014). The depreciation of natural capital resulting from such externalities should ideally be considered inputs into agricultural production. As such, accounts of multifactor productivity increases in the agricultural sector are likely to be considerable overestimates, particularly when dealing with specialized, industrial agricultural systems acknowledged for their considerable negative externalities (e.g., Pretty et al., 2001; Tegtmeier and Duffy, 2004). Furthermore, specialization disconnects agricultural practices from local nutrient cycling (e.g., Dorninger et al., 2017), leading to the overuse of external nitrogen and phosphorous inputs. This overuse of external inputs reduces total factor productivity while increasing environmental damage, as the flow of external inputs exceeds the ability of the ecosystem to absorb and render those flows harmless.

Despite the uncertainty and variability in estimates outlined earlier, multifactor productivity gains in increasingly specialized agrosystems appear to be significant. This is perhaps unsurprising given that increased efficiency has been a primary driving force behind agricultural specialization. Moreover, it remains unclear as to how productivity has changed for less specialized agricultural systems, or even if specialized, industrial agriculture can be considered more efficient (in terms of total factor productivity) than more traditional, diversified farming systems.

Moreover, given that some of the key factors of production (such as nitrogen) are naturally cycled within agroecologic systems, the loss of such inputs (e.g., via nutrient runoff) suggests inefficient usage. These resources are not being recovered and reused within the agroecologic system itself but continually replenished from outside the system. Diversified farming systems may allow for more efficient nutrient cycling (for example, via mixed livestock arable farming), reducing the need for external inputs and aligning production with local nutrient cycles.

Nevertheless, even if we accept that agricultural specialization has led to considerable increases in productivity, it is salient to ask to what extent this increased productivity has been at the expense of other key agroecologic properties, such as stability, persistence, and the just allocation of resources (as discussed subsequently).

Stability and specialization. The stability—the constancy of productivity in the face of relatively small perturbations resulting from fluctuations and cycles in the surrounding ecologic and socioeconomic systems—of agricultural production is potentially threatened by both ecologic (e.g., variations in rainfall, pest and disease outbreaks, extreme temperatures, etc.) and socioeconomic perturbations, largely fluctuations in market prices of either the external inputs to agricultural production or of the agricultural goods themselves (O'brien and Leichenko, 2000; Sumner, 2007). Specialization of agriculture has potential consequences for agricultural systems' responses to both ecologic and socioeconomic perturbations.

Genetic diversity within crops and cropping diversity (the number of different crops grown) have been shown to reduce agricultural systems' sensitivity to a number of ecologic/biophysical perturbations, including climate variability (Di Falco and Chavas, 2006); droughts and rainfall shocks (Di Falco and Chavas, 2008); pests (Gardiner et al., 2009), and disease outbreaks (see Box 19.2; Fraser, 2003). Similarly, by selecting a variety (portfolio) of crops with uncorrelated economic returns, it is possible to reduce the risk of income volatility due to fluctuating market prices for individual crops (Mckillop, 1989; Abson et al., 2013), in layman's terms, "the advantage of not putting all your eggs in one basket." For both ecologic and economic perturbations, having a diverse portfolio of livestock and crops (or crop genomes) spreads the risk of poor performance of a single agricultural good (e.g., wheat, milk, hay, etc.) Although, it should be noted that increased cropping/land use diversity may come as a trade-off against average levels of production or average expected returns on investment (Abson et al., 2013), particularly if this diversification occurs across relatively small spatial extents, which limit economies of scale. In addition to the risk-reducing nature of diversified farming systems, the broader scale (regional and global) homogenization of agricultural systems reduces the availability of cultivars or crops suitable for different environmental conditions. For example, it has been estimated that agricultural seed diversity in the continental United States declined by 93% between 1903 and 1983 as traditional varieties, often bred for specific environmental conditions, were replaced by a very small set of commercial cultivars (Tomaino, 2011).

This reduction in crop diversity, in turn, reduces the adaptive capacity of such systems (Kotschi, 2007), particularly when faced with new pest or disease outbreaks or environmental conditions outside those for which commercial cultivars have been optimized.

BOX 19.2

AGRODIVERSITY, STABILITY, AND THE GREAT FAMINE

Between 1845 and 1852, Ireland suffered the Great Famine (Gorta Mór), a period of unprecedented social upheaval and death in Ireland's history. Ireland's population dropped by some 1.5 million between 1841 and 1851 (approximately 18% of the population), largely due to famine-related death and emigration. As with most such events, the causes and consequences of the famine were multidimensional and complex; however, the initial causal factor was a series of outbreaks of potato blight (*Phytophthora infestans*) (Fraser, 2003; O'neill, 2009).

While large-scale potato blight outbreaks were common and widespread in Europe during the 1840s, the Irish agricultural system was particularly vulnerable because nearly one-third of the population was dependent on potatoes as their staple crop, and potato production was almost entirely limited to a single cultivar (the Irish Lumper). This dependence on a single cultivar of a single crop was largely the result of unjust property rights, a rapidly growing population, and small average landholdings, where only the Lumper potato could provide sufficient yield to feed poor tenant farmers and their families.

As a result of a lack of genetic diversity and dependency on a single crop, the disease spread more widely and rapidly and with a more devastating impact compared to outbreaks elsewhere in Europe.

Given that agricultural specialization has reduced agrobiodiversity across multiple scales (Jacobsen et al., 2013), it is likely that, regardless of changes in the exogenous perturbations in those systems, they are likely to be less stable than the more diverse systems that they have replaced (Johnson et al., 1996; Tilman and Downing, 1996), although unequivocal evidence remains sparse for the strength of the relations between changes in biologic diversity and system stability. However, it should be noted that stability of agricultural production in the face of ecologic perturbations can be maintained by the use of external inputs (irrigation, pesticides, etc.), and the increased productivity of specialized agricultural systems may provide additional capital assets to provide such stabilizing inputs (Abson et al., 2018).

Agricultural specialization leaves agroecologic systems vulnerable to both ecologic and economic shocks, largely due to the loss of functional redundancy and a narrowing of possible response in the face of exogenous perturbations. While specialized agriculture has developed a range of responses to deal with instability from lack of diversity, these responses are often themselves potentially problematic in terms of the long-term persistence of such systems.

Specialization and persistence: Persistence has a focus on maintaining systems over long terms and in the face of more severe perturbations than considered for stability. It is important to note that specialized farming seeks to reduce instability caused by ecologic volatility/perturbations by drawing on capital stocks via the use of external inputs (herbicides, pesticides, fertilizers, irrigation) to stabilize environmental conditions and ensure optimal production (Abson et al., 2018). In more traditional, diverse agroecosystems, almost all the resources used were drawn from within the farm systems themselves. As such, part of the reason for diverse farming systems was because such farming systems were largely "closed systems," and diverse cropping/farming systems help maintain the natural capital—the stocks of natural assets including, soil, air, water, and all living things that can be utilized in human's livelihood strategies—on which food production depends (Francis et al., 2003). Moreover, resource use in these traditional agroecosystems was largely limited to the ability of the system to provide those resources (including nutrient recycling rates), and diverse cropping practices evolved in part to ensure efficient resource use and cycling. Such approaches to maintaining production were also well suited to the capital assets available to most traditional/small-scale farmers (i.e., relatively low financial and infrastructure capital, but high levels of human capital, particularly labor and accessible natural capital). The reliance on within-farm capital stocks meant that considerable focus was required to maintain natural capital (even in traditional shift cultivation agricultural systems) to maintain long-term productivity. In contrast, specialization allow natural capital to be liquidated within the farm and substituted for natural capital stocks (often, as with fertilizers, heavily dependent on the use of nonrenewable natural capital stocks of fossil fuels) drawn from outside the system (Ekins, 2003; Pfeiffer, 2013).

The dependence on external input to specialized farming systems is problematic in terms of system persistence for three reasons. Firstly, the resources drawn into the system are often based on nonrenewable natural capital, and farming systems predicated on the continued use of limited stocks of nonrenewable natural capital cannot persist once those stocks are exhausted (Pfeiffer, 2013). Moreover, as the stocks of nonrenewable inputs are depleted, their price will increase, in turn increasing the cost of production, with potentially devastating impacts on food security long before the nonrenewable capital stocks are exhausted. Secondly, the use of such external inputs have a direct, negative impact on climate stability—approximately 10% of all of Europe's greenhouse gas emissions are from the agricultural sector (Eurostat, 2015)—which in turn threatens the long-term persistence of agricultural production (Schmidhuber and Tubiello, 2007; Lobell et al., 2008). Finally, extensive fertilizer use has decreased the richness of plant species—particularly nitrogen-sensitive plants—(Tilman et al., 2002; Hautier et al., 2009), diminished water quality (Turner and Rabalais, 2003; Hedley et al., 2005), and increased the occurrence of plant diseases (Tilman et al., 2002). Together with the broader declines in biodiversity associated with specialized agricultural systems (Kleijn et al., 2009), these environmental impacts potentially undermine the supporting ecosystem services on which agricultural production ultimately relies (Pretty, 1995; Tilman et al., 2002). While it is true that nitrogen-related pollution can be caused by the application of both natural and artificial fertilizers—and within both specialized and diversified farming systems—the natural nutrient cycles within more traditional mixed farming systems limited the availability of nitrogen in such systems.

The ability of highly specialized agricultural systems, predicated on the use of nonrenewable resources, to persist for the long term remains in question, and this in turn has potentially serious consequences regarding the justice of such systems.

Specialization and justice: Two main types of justice are important for judging the consequences of agricultural specialization: distributional justice—the just allocations of resources within and between generations—and procedural justice—the participatory governance by and empowerment of individuals, communities, and societies to decide how such resources are distributed (Loos et al., 2014).

It is clear that our food systems, while increasingly productive, have not solved the issue of food security and access to food (intragenerational justice). However, the extent agricultural specialization has contributed or ameliorated this failure is at best ambiguous, given that issues such as power and poverty (rather than a particular form of production) are generally considered of primary importance in access to food (e.g., Drèze and Sen, 1991). From the production side, there is evidence from the Global South that specialization and the resulting commercialization of agricultural production (in the form of the Green Revolution) led to a growing disparity between increasingly displaced or marginalized tenant farmers and land owners (e.g., Shiva et al., 1994). The costs of purchasing fertilizers, herbicides, and other yield-raising external inputs associated with specialized agriculture may exclude poor farmers from benefiting from specialization, while increasing commercialization leaves them exposed to fluctuations in market prices (Niragira et al., 2013). Here, it should be noted that the extent to which marginalization of poor farmers relates directly to specialization rather than how it is implemented (e.g., questions of procedural justice) remains an open question (Binswanger and Von Braun, 1991).

Intergenerational justice and agricultural specialization is particularly difficult to assess because judgements on this issue are in part determined by the subjective nature of assessing resource allocation between generations. The subjective nature of such assessments is typified by the notions of weak and strong sustainably. Weak and strong sustainability relate to the substitutability between different forms of capital (e.g., natural capital, financial capital, physical capital, etc.) that can be drawn on to maintain or increase human well-being. Proponents of weak sustainability argue that losses in natural capital are compensated by increases in other capital stock types. Whereas, proponents of strong sustainability argue that there are critical forms of natural capital for which no meaningful substitute exists and that nondiminishing life opportunities can only be ensured by "conserving the stock of human capital, technological capability, natural resources and environmental quality" (Brekke, 1997, p. 91). There are currently no meaningful substitutes for agricultural food production as a source of human well-being, and the loss of natural capital (biodiversity, soil fertility, etc.) in specialized agroecologic systems is usually compensated for via the use of nonrenewable natural capital inputs (largely fossil fuels). The notion that the loss of this natural capital can be compensated for via additional investment in other forms of capital stock (Stoneham et al., 2003) in the long run (despite productivity gains in the short run) therefore seems at least questionable. This, in turn, raises the question as to the appropriateness of considering weak sustainability as a useful frame for considering intergenerational justice in specialized agricultural systems. In terms of a strong sustainability perspective the association of agricultural specialization with declines in critical (nonsubstitutable) biodiversity, including agricultural biodiversity and soil fertility (Pretty, 1995), suggests that specialization may indeed be limiting the life opportunities of future generations.

CONCLUSIONS

Agricultural specialization has been driven by political, technological, and economic forces for at least the last 100 years, with profound ecologic and socioeconomic effects. Agricultural specialization has coincided with increased agricultural yields, increased multifactor agricultural productivity, and declining food prices. However, the extent to which yields, productivity, and food prices are dependent on specialization itself, rather than technological change and increased use of external inputs, remains unclear. Nor is it clear whether there are limits beyond which additional specialization does not increase productivity. Moreover, there are genuine concerns that the stability and persistence of such increasingly specialized agricultural systems is potentially eroded by the consequential loss of natural capital and the reliance on nonrenewable external inputs. Such concerns have potential far-reaching consequences for the long-term viability of our current agricultural systems. Consequently the role of increased agricultural specialization in the quest for sustainable and just food systems that meet the needs of both current and future generations is increasingly in doubt. New agricultural paradigms may be necessary that better address issues of agricultural stability, persistence, and justice that at least partially move away from specialization and embrace alternative visions of food production.

References

Abson, D.J., Fraser, E.D., Benton, T.G., 2013. Landscape diversity and the resilience of agricultural returns: a portfolio analysis of land-use patterns and economic returns from lowland agriculture. Agriculture and Food Security 2, 1.

Abson, D.J., Termansen, M., Pascual, U., Aslam, U., Fezzi, C., Bateman, I., 2014. Valuing climate change effects upon UK agricultural GHG emissions: spatial analysis of a regulating ecosystem service. Environmental and Resource Economics 57, 215–231.

Abson, D.J., Sherren, K., Fischer, J., 2018. The resilience of agricultural landscapes characterized by land sparing versus land sharing. In: Gardener, S., Ramsden, S., Hails, R. (Eds.), Agriculture Resilience: perspectives from ecology and economics. Cambridge University Press, Cambridge. In press.

Alston, J.M., Andersen, M.A., James, J.S., Pardey, P.G., 2009. Persistence Pays: US Agricultural Productivity Growth and the Benefits from Public R&D Spending. Springer Science & Business Media.

Alston, J.M., Andersen, M.A., Pardey, P.G., 2015. The Rise and Fall of US Farm Productivity Growth, 1910–2007. University of Minnesota, Department of Applied Economics.

Altieri, M., Nicholls, C., 2002. The simplification of traditional vineyard based agroforests in northwestern Portugal: some ecological implications. Agroforestry Systems 56, 185–191.

Archibald, S.O., 2016. Incorporating externalities into agricultural productivity analysis. In: Capalbo, S.M., Antle, J.M. (Eds.), Agroforestry Systems. Agricultural productivity: measurement and explanation, Abingdon; Routledge.

Benton, T.G., Vickery, J.A., Wilson, J.D., 2003. Farmland biodiversity: is habitat heterogeneity the key? Trends in Ecology and Evolution 18, 182–188.

Binswanger, H.P., Von Braun, J., 1991. Technological change and commercialization in agriculture: the effect on the poor. The World Bank Research Observer 6, 57–80.

Blaxter, K., Robertson, N., 1995. From Dearth to Plenty: The Second Agricultural Revolution. Cambridge University Press, Cambridge.

Bowler, I., 1985. Some consequences of the industrialization of agriculture in the European Community. In: Healey, M.J., Iibery, B.W. (Eds.), The Industrialization of the Countryside. Norwich: Geo Books.

Brekke, K.A., 1997. Economic Growth and the Environment: On the Measurement of Income and Welfare. Edward Elgar Publishing Ltd.

Brickle, N.W., Harper, D.G., 2002. Agricultural intensification and the timing of breeding of Corn Buntings Miliaria calandra: in an intensively managed agricultural landscape, few females attempted a second brood. Bird Study 49, 219–228.

Burton, R.J.F., Walford, N., 2005. Multiple succession and land division on family farms in the South East of England: a counterbalance to agricultural concentration? Journal of Rural Studies 21, 335–347.

Bustos, P., Caprettini, B., Ponticelli, J., 2013. Agricultural Productivity and Structural Transformation. Evidence from Brazil. Evidence from Brazil (August 13, 2013). Chicago Booth Research Paper.

Candel, J.J., Breeman, G.E., Stiller, S.J., Termeer, C.J., 2014. Disentangling the consensus frame of food security: the case of the EU Common Agricultural Policy reform debate. Food Policy 44, 47–58.

Capalbo, S.M., Vo, T.T., 1988. A review of the evidence on agricultural productivity. In: Cabalbo, S.M., Antle, J.M. (Eds.), Agricultural Productivity: Measurement and Explanation. RFF Press, Washington, DC.

Capalbo, S.M., Vo, T.T., 2015. A review of the evidence on agricultural productivity. Agricultural Productivity: Measurement and Explanation 96.

Chisholm, M., 1962. Tendencies in agricultural specialization and regional concentration in industry. Papers in Regional Science 10, 157–162.

Clay, J., 2013. World Agriculture and the Environment: A Commodity-by-commodity Guide to Impacts and Practices. Island Press.

Cloke, P., Goodwin, M., 1992. Conceptualizing countryside change: from post-Fordism to rural structured coherence. Transactions of the Institute of British Geographers 321–336.

Cochrane, W.W., 1979. The Development of American Agriculture: A Historical Analysis. U of Minnesota Press.

Coleman, W.D., Skogstad, G.D., Atkinson, M.M., 1996. Paradigm shifts and policy networks: cumulative change in agriculture. Journal of Public Policy 16, 273–301.

Comin, D., 2010. Total Factor Productivity. Economic Growth. Springer.

Conway, G.R., 1993. Sustainable agriculture: the trade-offs with productivity, stability and equitability. In: Barbier, E.B. (Ed.), Economics and Ecology: New Frontiers and Sustanable Development. Springer.

Cornia, G.A., 1985. Farm size, land yields and the agricultural production function: an analysis for fifteen developing countries. World Development 13, 513–534.

De Raymond, A.B., 2013. Detaching from agriculture? Field-crop specialization as a challenge to family farming in northern Côte d'Or, France. Journal of Rural Studies 32, 283–294.

Defra, 2016. Total factor productivity of the UK agriculture industry. In: DEFRA. Department for Environment Food and Rural Affairs, London.

Di Falco, S., Chavas, J.-P., 2006. Crop genetic diversity, farm productivity and the management of environmental risk in rainfed agriculture. European Review of Agricultural Economics 33, 289–314.

Di Falco, S., Chavas, J.-P., 2008. Rainfall shocks, resilience, and the effects of crop biodiversity on agroecosystem productivity. Land Economics 84, 83–96.

Dimitri, C., Effland, A.B., Conklin, N.C., 2005. The 20th Century Transformation of US Agriculture and Farm Policy. US Department of Agriculture, Economic Research Service Washington, DC, USA.

Dorninger, C., Abson, D.J., Fischer, J., von wehrden, H., 2017. Assessing sustainable biophysical human–nature connectedness at regional scales. Environmental Research Letters 12 (5), 055001.

Dorward, A., Fan, S., Kydd, J., Lofgren, H., Morrison, J., Poulton, C., Rao, N., Smith, L., Tchale, H., Thorat, S., 2004. Institutions and policies for pro-poor agricultural growth. Development Policy Review 22, 611–622.

Drèze, J., Sen, A., 1991. The Political Economy of Hunger: Volume 1: Entitlement and Well-being. Clarendon Press.

Duffy, M., 2009. Economies of size in production agriculture. Journal of Hunger and Environmental Nutrition 4, 375–392.

Ekins, P., 2003. Identifying critical natural capital: conclusions about critical natural capital. Ecological Economics 44, 277–292.

Ernoult, A., Bureau, F., Poudevigne, I., 2003. Patterns of organisation in changing landscapes: implications for the management of biodiversity. Landscape Ecology 18, 239–251.

Eurostat, 2015. Greenhouse Gas Emissions, Analysis by Source Sector, EU-28, 1990 and 2013. Eurostat.

Evans, N., Morris, C., Winter, M., 2002. Conceptualizing agriculture: a critique of post-productivism as the new orthodoxy. Progress in Human Geography 26, 313–332.

Evenson, R.E., Gollin, D., 2003. Assessing the impact of the green revolution, 1960 to 2000. Science 300, 758–762.

Firbank, L., 2005. Striking a new balance between agricultural production and biodiversity. Annals of Applied Biology 146, 163–175.

Fischer, J., Abson, D.J., Butsic, V., Chappell, M.J., Ekroos, J., Hanspach, J., Kuemmerle, T., Smith, H.G., Wehrden, H., 2014. Land sparing versus land sharing: moving forward. Conservation Letters 7, 149–157.

Foster, A.D., Rosenzweig, M.R., 1996. Technical change and human-capital returns and investments: evidence from the green revolution. The American Economic Review 931–953.

Francis, C., Lieblein, G., Gliessman, S., Breland, T., Creamer, N., Harwood, R., Salomonsson, L., Helenius, J., Rickerl, D., Salvador, R., 2003. Agroecology: the ecology of food systems. Journal of Sustainable Agriculture 22, 99–118.

Frankel, F.R., 2015. India's Green Revolution: Economic Gains and Political Costs. Princeton University Press.

Fraser, E.D., 2003. Social vulnerability and ecological fragility: building bridges between social and natural sciences using the Irish Potato Famine as a case study. Conservation Ecology 7, 9.

Gardiner, M., Landis, D., Gratton, C., Difonzo, C., O'neal, M., Chacon, J., Wayo, M., Schmidt, N., Mueller, E., Heimpel, G., 2009. Landscape diversity enhances biological control of an introduced crop pest in the north-central USA. Ecological Applications 19, 143–154.

Giampietro, M., 2003. Multi-scale Integrated Analysis of Agroecosystems. CRC press.

Giampietro, M., 2004. Multi-scale Integrated Analysis of Agroecosystems. CRC Press, Boca Raton.

Glamann, J., Hanspach, J., Abson, D.J., Collier, N., Fischer, J., 2017. The intersection of food security and biodiversity conservation: a review. Regional Environmental Change 17 (5), 1303–1313.

Godfray, H.C.J., Beddington, J.R., Crute, I.R., Haddad, L., Lawrence, D., Muir, J.F., Pretty, J., Robinson, S., Thomas, S.M., Toulmin, C., 2010. Food security: the challenge of feeding 9 billion people. Science 327, 812–818.

Goodman, D., Watts, M., 1997. Globalising Food: Agrarian Questions and Global Restructuring. Psychology Press.

Grantham, G.W., 1993. Divisions of labour: agricultural productivity and occupational specialization in pre-industrial France. The Economic History Review 46, 478–502.

Hautier, Y., Niklaus, P.A., Hector, A., 2009. Competition for light causes plant biodiversity loss after eutrophication. Science 324, 636–638.

Hazell, P., Wood, S., 2008. Drivers of change in global agriculture. Philosophical Transactions of the Royal Society of London B Biological Sciences 363, 495–515.

Hedley, M., Mclaughlin, M., Sims, J., Sharpley, A., 2005. Reactions of phosphate fertilizers and by-products in soils. Phosphorus: Agriculture and the Environment 181–252.

Howkins, A., 2003. The Death of Rural England: A Social History of the Countryside since 1900. Routledge.

Hunt, S.D., Morgan, R.M., 1995. The comparative advantage theory of competition. The Journal of Marketing 1–15.

Ilbery, B., Bowler, I., 1998. From agricultural productivism to post-productivism. In: Ilberry, B. (Ed.), The Geography of Rural Change. Routledge, London.

Jacobsen, S.-E., Sørensen, M., Pedersen, S.M., Weiner, J., 2013. Feeding the world: genetically modified crops versus agricultural biodiversity. Agronomy for Sustainable Development 33, 651–662.

Johnson, K.H., Vogt, K.A., Clark, H.J., Schmitz, O.J., Vogt, D.J., 1996. Biodiversity and the productivity and stability of ecosystems. Trends in Ecology and Evolution 11, 372–377.

Kleijn, D., Kohler, F., Báldi, A., Batáry, P., Concepción, E., Clough, Y., Diaz, M., Gabriel, D., Holzschuh, A., Knop, E., 2009. On the relationship between farmland biodiversity and land-use intensity in Europe. Proceedings of the Royal Society of London B Biological Sciences 276, 903–909.

Kotschi, J., 2007. Agricultural biodiversity is essential for adapting to climate change. GAIA-Ecological Perspectives for Science and Society 16, 98–101.

Lambin, E.F., Meyfroidt, P., 2011. Global land use change, economic globalization, and the looming land scarcity. Proceedings of the National Academy of Sciences 108, 3465–3472.

Lambin, E.F., Turner, B.L., Geist, H.J., Agbola, S.B., Angelsen, A., Bruce, J.W., Coomes, O.T., Dirzo, R., Fischer, G., Folke, C., George, P.S., Homewood, K., Imbernon, J., Leemans, R., Li, X., Moran, E.F., Mortimore, M., Ramakrishnan, P.S., Richards, J.F., Skånes, H., Steffen, W., Stone, G.D., Svedin, U., Veldkamp, T.A., Vogel, C., Xu, J., 2001. The causes of land-use and land-cover change: moving beyond the myths. Global Environmental Change 11, 261–269.

Le Heron, R.B., 1993. Globalized Agriculture: Political Choice. Pergamon Press, Books Division.

Leaman, J.H., Conkling, E., 1975. Transpor change and agricultural specialization. Annals of the Association of American Geographers 65, 425–432.

Lemaire, G., Franzluebbers, A., De Faccio Carvalho, P.C., Dedieu, B., 2014. Integrated crop–livestock systems: strategies to achieve synergy between agricultural production and environmental quality. Agriculture, Ecosystems and Environment 190, 4–8.

Lobell, D.B., Burke, M.B., Tebaldi, C., Mastrandrea, M.D., Falcon, W.P., Naylor, R.L., 2008. Prioritizing climate change adaptation needs for food security in 2030. Science 319, 607–610.

Loos, J., Abson, D.J., Chappell, M.J., Hanspach, J., Mikulcak, F., Tichit, M., Fischer, J., 2014. Putting meaning back into "sustainable intensification". Frontiers in Ecology and the Environment 12, 356–361.

Lowe, P., Murdoch, J., Marsden, T., Munton, R., Flynn, A., 1993. Regulating the new rural spaces: the uneven development of land. Journal of Rural Studies 9, 205–222.

Macdonald, J.M., Korb, P., Hoppe, R.A., 2013. Farm Size and the Organization of US Crop Farming. US Department of Agriculture, Economic Research Service.

Matson, P.A., Parton, W.J., Power, A., Swift, M., 1997. Agricultural intensification and ecosystem properties. Science 277, 504–509.

Matsuyama, K., 1992. Agricultural productivity, comparative advantage, and economic growth. Journal of Economic Theory 58, 317–334.

Mckillop, D., 1989. The return-risk structure of lowland agriculture in Northern Ireland. European Review of Agricultural Economics 16, 217–228.

Mueller, N.D., Gerber, J.S., Johnston, M., Ray, D.K., Ramankutty, N., Foley, J.A., 2012. Closing yield gaps through nutrient and water management. Nature 490, 254–257.

Niragira, S., D'haese, M., Buysse, J., Desiere, S., Ndimubandi, J., D'haese, L., 2013. 217-Options and impact of crop production specialization on small-scale farms. In: The Noth of Burundi. 4th International Conference of the African Association of Agricultural Economists, September 22-25.

O'brien, K.L., Leichenko, R.M., 2000. Double exposure: assessing the impacts of climate change within the context of economic globalization. Global Environmental Change 10, 221–232.

O'neill, J.R., 2009. The Irish Potato Famine. ABDO Publishing Company, Edina.

Ostrom, E., Burger, J., Field, C.B., Norgaard, R.B., Policansky, D., 1999. Revisiting the commons: local lessons, global challenges. Science 284, 278–282.

Peyraud, J.-L., Taboada, M., Delaby, L., 2014. Integrated crop and livestock systems in Western Europe and South America: a review. European Journal of Agronomy 57, 31–42.

Pfeiffer, D.A., 2013. Eating Fossil Fuels: Oil, Food and the Coming Crisis in Agriculture. New Society Publishers.

Ponisio, L.C., M'gonigle, L.K., Mace, K.C., Palomino, J., De Valpine, P., Kremen, C., 2015. Diversification practices reduce organic to conventional yield gap. Proceedings of the Royal Society of London B Biological Sciences 282, 20141396.

Pretty, J., 1999. The Living Land: Agriculture, Food and Community Regeneration in the 21st Century. Earthscan.

Pretty, J., Brett, C., Gee, D., Hine, R., Mason, C., Morison, J., Rayment, M., Van Der Bijl, G., Dobbs, T., 2001. Policy challenges and priorities for internalizing the externalities of modern agriculture. Journal of Environmental Planning and Management 44, 263–283.

Pretty, J.N., 1998. The Living Land. Earthscan, London.

Pretty, J.N., 1995. Regenerating Agriculture: Policies and Practice for Sustainability and Self-reliance. Joseph Henry Press.

Rackham, O., 2000. The History of the Countryside: The Classic History of Britain's Landscape, Flora and Fauna. Phoenix Press.

Ricardo, D., 1817. On the Principles of Political Economy and Taxation: London. publisher not identified.

Schmidhuber, J., Tubiello, F.N., 2007. Global food security under climate change. Proceedings of the National Academy of Sciences 104, 19703–19708.

Shiva, V., Littlefield, A., Gates, H., 1994. The Violence of the Green Revolution.

Singh, H., Motwani, J., Kumar, A., 2000. A review and analysis of the state-of-the-art research on productivity measurement. Industrial Management and Data Systems 100, 234–241.

Stoate, C., 1995a. The changing face of lowland farming and wildlife Part 1 1845-1945. British Wildlife 6, 341.

Stoate, C., 1995b. The changing face of lowland farming and wildlife, part 2 1945-1995. British Wildlife 7, 162–172.

Stoneham, G., Eigenraam, M., Ridley, A., Barr, N., 2003. The application of sustainability concepts to Australian agriculture: an overview. Animal Production Science 43, 195–203.

Sumner, D.A., 2007. Farm Subsidy Tradition and Modern Agricultural Realities. American Enterprise Institute, Washington, DC.

Tegtmeier, E.M., Duffy, M.D., 2004. External costs of agricultural production in the United States. International Journal of Agricultural Sustainability 2, 1–20.

Tilman, D., 1999. Global environmental impacts of agricultural expansion: the need for sustainable and efficient practices. Proceedings of the National Academy of Sciences 96, 5995–6000.

Tilman, D., Cassman, K.G., Matson, P.A., Naylor, R., Polasky, S., 2002. Agricultural sustainability and intensive production practices. Nature 418, 671–677.

Tilman, D., Downing, J.A., 1996. Biodiversity and Stability in Grasslands. Ecosystem Management. Springer.

Tomaino, J., 2011. Food Ark. National Geographic (July).

Trueblood, M.A., Ruttan, V.W., 1995. A comparison of multifactor productivity calculations of the US agricultural sector. Journal of Productivity Analysis 6, 321–331.

Tscharntke, T., Klein, A.M., Kruess, A., Steffan-Dewenter, I., Thies, C., 2005. Landscape perspectives on agricultural intensification and biodiversity–ecosystem service management. Ecology Letters 8, 857–874.

Turner, R.E., Rabalais, N.N., 2003. Linking landscape and water quality in the Mississippi River basin for 200 years. BioScience 53, 563–572.

Vickery, J.A., Feber, R.E., Fuller, R.J., 2009. Arable field margins managed for biodiversity conservation: a review of food resource provision for farmland birds. Agriculture, Ecosystems and Environment 133, 1–13.

Williscroft, C., 2007. A Lasting Legacy: A 125 Year History of New Zealand Farming since the First Frozen Meat Shipment. NZ Rural Press.

Wilson, G.A., 2001. From productivism to post-productivism... and back again? Exploring the (un)changed natural and mental landscapes of European agriculture. Transactions of the Institute of British Geographers 26, 77–102.

Wilson, G.A., Rigg, J., 2003. 'Post-productivist'agricultural regimes and the South: discordant concepts? Progress in Human Geography 27, 681–707.

Winsberg, M.D., 1980. Concentration and specialization in United States agriculture, 1939-1978. Economic Geography 183–189.

Withers, P.J., Neal, C., Jarvie, H.P., Doody, D.G., 2014. Agriculture and eutrophication: where do we go from here? Sustainability 6, 5853–5875.

CHAPTER

20

Practices of Sustainable Intensification Farming Models: An Analysis of the Factors Conditioning Their Functioning, Expansion, and Transformative Potential

Gianluca Brunori[1], *Simona D'Amico*[1,2], *Adanella Rossi*[1]

[1]**University of Pisa - Department of Agriculture, Food and Environment (DAFE) - ITALY;** [2]**Union for Ethical BioTrade, Amsterdam - The NETHERLANDS**

INTRODUCTION

The intention to balance food production, environmental preservation, satisfaction of food consumers, and adequate income generation for farmers has been showing in the debate at the both political and civil society levels. The concept of sustainable intensification has come to the fore in this debate (Levidow et al., 2014; RISE, 2014). It refers to strategies to simultaneously improve productivity in agriculture and reduce its environmental impact (Levidow et al., 2014; RISE, 2014). Some authors identify these agri-food strategies as cases of multifunctional farming or scope economy because of their ability to generate multiple, productive, and environmental outputs using the same inputs (Garbach et al., 2016).

There are many different interpretations of sustainable intensification. Some of them foster strategies that stay within mainstream, industrial agri-food systems while promoting farming techniques that seek to reduce the environmental impact of conventional farming. These techniques follow the principle of producing more while using fewer chemical inputs and have been criticized for backing a neo-productivist paradigm that locates the environmental sustainability of national food systems within current globalization patterns (Levidow, 2016). On the other hand, there are views on sustainable intensification that imply the redesign of the whole agri-food system, starting from promoting farming practices that seek to imitate natural processes (Elliott et al., 2013; TP Organics and

Agroecosystem Diversity
https://doi.org/10.1016/B978-0-12-811050-8.00020-0

IFOAM EU, 2014). They rely on resource conservation, recycling, and other aspects of the natural functioning of the farm ecosystems (Levidow et al., 2014).

Many different agri-food models are identified as expressions of sustainable intensification. They include agroecology, conservation agriculture, biodynamic, organic, integrated, and precision farming, and other models that are known for sustainability attributes (RISE, 2014). Their performances in terms of balancing productivity and environmental preservation change because of the practices used as well as in relation to the contexts where these models operate (Garbach et al., 2016; Meynard et al., 2013).

Scholars suggest avoiding normative interpretations of sustainable intensification and giving reason of all its possible expressions to learn from this variety about opportunities, challenges, and ways to realize it. They invite exploring the interconnections between production and consumption by including in the analysis different stages of the food chain such as the processing and retailing (Lamine, 2014; Levidow et al., 2014). This comprehensive approach allows reconnecting agriculture, food, and environment, which are crucial when thinking of the balance between agri-food production, market and income satisfaction, and environment preservation. Moreover, scholars recommend focusing on the ways practices of sustainable intensification are implemented in different contexts to ascertain contextual dynamics that influence this balance (Lamine, 2014; Levidow et al., 2014; Meynard et al., 2013). Finally, another perspective fosters focusing on the way practices of sustainable intensification, which start as pioneer initiatives, may spread and influence the context where they operate (Maye, 2016).

In line with these recommendations, this study explores three cases of food farms in Tuscany that practice sustainable intensification and have developed a whole chain for their products. Looking at both the production and the commercialization, it seeks to reply the following research questions:

1. How are practices configured in food chains adopting sustainable intensification methods?
2. To what extent can these practices be replicated, scaled up, and translated into the broader context where the food chains operate?
3. To what extent are these practices able to balance environment preservation, productivity, and market and income satisfaction?

To reply to these research questions, a theoretical and methodological framework has been used that combines practice theory (PT) (Schatzki, 2002) and strategic niche management theory (SNM) (Kemp et al., 1998; Loorbach, 2007). Principles and concepts of PT are instrumental to disclose how certain practices are configured in the studied food chains. SNM complements with the analysis of how practices in the studied food chains act over space and time, through the processes of replication, scaling up, and translation. PT stems from an articulated combination of studies, inspired by the work of Bourdieu (see, for example, Bourdieu, 1977) and the structuration theory of Giddens (1984). These studies share a focus on explaining how some human activities become routinized into familiar practices (Ropke, 2009). SNM is part of transition studies whose focus is on explaining transition toward sustainability, not least in the food system. SNM is particularly used to understand how innovative practices, consolidating in a specific context, might spread and contribute to change practices that are more broadly followed and, in turn, stimulate transition (Maye, 2016).

The contribution of the study is twofold. At an empirical level, it will overcome normative interpretations by showing how sustainable intensification is operated in the three cases and its results in terms of balancing environmental preservation, productivity, and market and income satisfaction in food chains. At a conceptual level, the study represents an attempt to further explore the combination between PT and SNM, with the aim of providing a temporal and contextual dimension to the analysis of practices of sustainable intensification.

The next section provides a review of the literature that has approached the study of farming models following sustainable intensification. It also describes the conceptual framework used in this study. The third section introduces the case studies, the way the framework is adapted to their investigation, and the methodology used in the analysis. The fourth section presents the results of this analysis, while the fifth section discusses them.

SUSTAINABLE INTENSIFICATION FARMING MODELS: A FRAMEWORK TO STUDY THEIR PRACTICES AND THEIR TRANSFORMATIVE POTENTIAL

The diversity in the forms and results of the agri-food models used to implement sustainable intensification depends on the practices those models build on, as well as on the relations with the contexts where they operate (Garbach et al., 2016). The many studies on these initiatives have dealt differently with these issues.

Many studies in the literature on alternative and local food systems have mostly looked at the practices. They have been exploring initiatives that seek to satisfy production and market requests while being guided by principles of environmental, social, and economic sustainability (Lamine, 2014). Their approach looks at the combinations of actors, motivations, knowledge, and understanding, means, actions, and circumstances that shape the daily functioning of the initiatives under investigation (Fonte, 2013; Veen, 2015). On the other hand, there are studies that, starting from the practices promoted by these initiatives, have explored also the relations they develop with the context where they operate. Some of these works belong to the transition studies and explore both the micro and the meso processes that are behind the different sustainable intensification initiatives and seek to understand to which extent they promote radical or incremental changes within the consolidated, industrial food system (Wiskerke and Ploeg van der, 2004; Brunori et al., 2010, 2012; Maye, 2016).

To study changes in the food system, some recent studies have combined the frame of SNM—in the interpretation of Seyfang and Haxeltine (2012)—with approaches for the study of practices (Maye, 2016). SNM is one approach within the transition studies that seek to understand sustainability transitions, looking especially at the mechanisms that can steer them (Kemp et al., 1998; Loorbach, 2007); they have been applied also in the food system (Hinrichs, 2014; Darnhofer, 2015).

Transitions are nonlinear processes resulting from the interplay of developments at the three levels of landscape, regimes, and niches (Geels and Schot, 2007). In relation to agriculture, the regime is the locus of established practices and rules, namely those that characterize the mainstream, industrial agri-food system. The niche is the locus of radical innovations, for instance, new forms of food provisioning and acquisition compared to those followed in the mainstream system. The landscape represents pressures that are exogenous to niches and regimes, such as climate change and macroeconomic processes.

Thus, a transition might take place when landscape factors destabilize regimes and niches gather momentum and increase in importance within the system. Niches are the main locus for change; however, there is no guarantee they will develop sufficiently to influence the dominant regime significantly. Understanding the relationship between niches and regimes is therefore key to understanding the nature of transitions. SNM is particularly useful to better understand niche since it focuses on how niches set up, function, and translate into the regime. SNM seeks to understand how niches can emerge through collective engagement and practice. Niches evolve according to the three modes:

- *replication*, which refers to the ability of the configured practices to replicate over time and space;
- *scaling up*, which means the opportunity for the configured practices to expand, for instance, by attracting new participants;
- *translation*, which stands for the ability of the configurations to translate into, or influence, the broader system where the initiative operates; in other words, it refers to their transformative role (Kemp et al., 1998; Loorbach, 2007; Seyfang and Haxeltine, 2012).

This study starts from recognizing the possible contribution of SNM in studying the relation between agri-food initiatives for sustainable intensification, as niche of acquisition of innovative practices around food, and their broader context, namely the regime of practices and rules that characterize the mainstream, industrial food system. Moreover, given SNM focus on understanding the emergence of niches through collective engagement and practices, this study combines this approach with PT.

PT emerges from studies interested in exploring everyday life, not least the common practices of producing and buying food. These studies stem from the work of Bourdieu (see for example Bourdieu, 1977) and the structuration theory of Giddens (1984). Recent theories of practice are heterogeneous (Røpke, 2009; Warde, 2005; Schatzki, 1996), but they share the curiosity to understand how new practices are taken up and transitioned into routinized behaviors (Shove and Walker, 2010).

For this study, we propose the interpretation of PT that emerges from the work of the philosopher Theodore Schatzki (1996, 2002), who has contributed to the formulation of a coherent approach to the analysis of practice (Røpke, 2009). In this perspective, forming a practice is about gluing activities together and enacting them. This is materially mediated as people use artifacts to shape the connections that make up a practice and to enact them (Schatzki et al., 2001). In sum, a practice is configured when there are activities, and these activities are interconnected and implemented through the following:

- actors, all people involved in the activities;
- artifacts, all the technical and social means as well as the strategies employed in the activity;
- rules, all the formal and informal codes that are relevant for the unfolding of the activity;
- understanding, the know-how that is instrumental for the activity to be implemented;
- teleo-affective structure, the motivations, objectives, beliefs, emotions, moods that move people involved in the activity (Schatzki, 1996; Schatzki et al., 2001) (hereafter indicated as motivations).

The way practices evolve, how they stabilize, and expand or fail to do so is not investigated

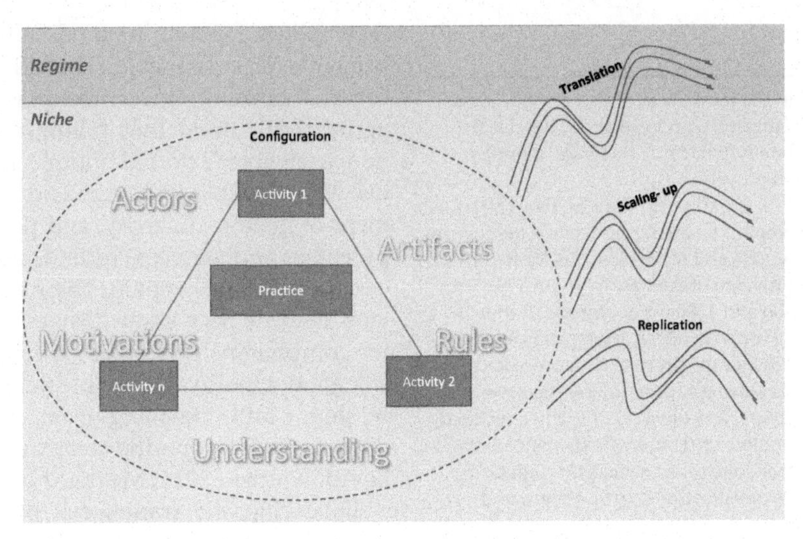

FIGURE 20.1 Practices configuration, replication, scaling up, translation.

enough in PT-based studies (Røpke, 2009). The combination of PT with SNM is here proposed to contribute to closing this gap. Reinterpreted in the light of practices, the regime in SNM can be read as the context of established practices and rules, the ways of doing typical of the mainstream agri-food system. On the other hand, niches can be interpreted as experimental practices, new ways of provisioning and acquiring food aiming at being socially, economically, and environmentally sustainable. They may stabilize and influence, even substitute, established ones. Alternatively, they may sink. The dynamics of niche evolution is here used to interpret the evolution of practices. Fig. 20.1 schematizes the way practices configure, replicate, scale up, and translate, as emerging from the combination of PT and SNM. The next section will explain how this framework is used to represent and unveil the constitution and evolution of practices of sustainable intensification.

THREE CASES OF SUSTAINABLE INTENSIFICATION IN TUSCANY: THEIR PRACTICES AND THEIR TRANSFORMATIVE POTENTIAL

Bio Colombini, *Radici*, and *Poggio di Camporbiano* are the three cases selected for this study. They are three initiatives implementing farming, processing and both direct and indirect selling of food products (see the following table for details). They operated in three different provinces, within the Tuscany region (Italy).

Bio Colombini	It is a farm of 18 ha in the province of Pisa; it is one of the biggest horticultural farms in Tuscany. The family has been running it for three generations; the last one has converted it to organic farming (in 1998) and has greatly diversified the species cultivated. It mostly grows vegetables (15 ha) and fruits (1 ha), as well as produces juices, sauces, and olive oil (2 ha). Since 2001, it has done direct selling to solidarity purchase groups (GAS), reaching about 1000 families, and

Continued

—cont'd

	in the farm shop; it also provide school canteens with vegetables. Besides the family members, there are about ten employees.
Radici	It is a family-run farm of about 30 ha, of which 5 ha are farmed, while the rest are woods and fruit trees. It is located in the province of Arezzo, in an area between 300 and 1300 mt of altitude. It mostly grows vegetables, fruits, and cereals. Most of the products are processed, producing sauces, juices, preserves, soups, and olive oil. It follows certified organic farming methods. Since 1985, it has done direct selling through GAS, farmers markets, farm outlets, and catering. Part of the selling goes through indirect channels, also abroad.
Poggio di Camporbiano	It is a 200-ha farm located in a hilly area of the province of Florence. Since its setting up, in 1988, it has been following principles of biodynamic farming,[1] which adds a stronger holistic approach to the techniques of organic farming. It is certified for both farming methods. It grows fodders, cereals, vegetables, legumes, and fruits and breeds dairy cows (about 60), goats (about 100), and horses. It produces soups, pasta, husked cereals, sauces, juices, preserves, and cheeses. It is a cooperative, run by a community of people living and working on the farm. It implemented direct selling in 1995 and sells through GAS, a farm outlet, catering, and an online shop. It uses indirect selling channels too.

We have already analyzed these cases in previous studies, and this guarantees us the availability of substantial data. There are two other important reasons why we selected them. First, they follow farming methods aligned with the principle of sustainable intensification. Besides organic farming, the three cases show some agroecologic traits: they minimize or do not use nonrenewable external inputs, commit to the use of renewable resources and to the enhancement of genetic diversity, and promote ecologic processes and services. Second, they have developed a full supply chain. These aspects are relevant because they allow answering those claims for comprehensive studies on sustainable intensification initiatives that include farming, processing, and retailing, as well as their development in specific contexts (Lamine, 2014; Levidow et al., 2014; Meynard et al., 2013).

Following the framework presented in the previous section, each of the case studies is interpreted as an initiative where two activities take place: (1) providing the genetic resources that are grown; (2) managing pre-/postharvest.

Per each activity, the configuration of the practices is analyzed by looking at the interlinkages among the following aspects: actors, artifacts, rules, understanding, and motivations. Moreover, the relation of reciprocal influence between these configurations and the context where the initiatives operate are analyzed by taking into consideration replication, scaling up, and translation of the configurations.

This analysis of the configurations of practices and their potential evolution is based on the data collected through unstructured interviews and participant observation. The analysis will show the opportunities and challenges for the mentioned initiatives and the implications for their ability to simultaneously guarantee productivity, satisfying market and incomes expectations, and fulfil environmental protection.

[1] Biodynamics has much in common with organic approaches: it emphasizes the use of manures and composts and excludes the use of artificial chemicals on soil and plants. It is unique to the biodynamic approach of the treatment of animals, crops, and soil as a single ecologic system. Various herbal and mineral additives for compost additives and field sprays are used; these are sometimes prepared through methods that are said to harvest cosmic forces.

SUSTAINABLE INTENSIFICATION FARMING MODELS: CONFIGURATION, REPLICATION, SCALING UP, AND TRANSLATION OF PRACTICES

Configuration

Providing Genetic Resources

This activity exists in providing the seeds that are necessary for farming. Participatory plant breeding (PPB) is the most common configuration for the provision of genetic resources among the studied cases. The motivations to undertake it are several. Farmers in *Radici* and *Poggio di Camporbiano* are primarily moved by rediscovering old varieties of seeds and by the willingness to produce multiple crops with ecologic farming techniques. They use PPB to fulfil these motivations since it allows the exchange of knowledge on the seeds and plant varieties that are typical of an area, and on how they can be reproduced through natural techniques.

Besides farmers, PPB may involve professional breeders, research centers, civil society organizations, policy makers, and institutions. Their involvement depends on which artifacts are created and used to carry on the activities. The artifacts used in the studied cases imply collaborations between farmers, professional breeders from research centers, and civil society organizations. They are instrumental to mobilize the genetic resources that will be used, and to monitor, evaluate, and possibly, improve their performances.

The case of *Radici* is particularly explanatory for this. They begin by producing their own seeds and researching about old varieties. Later, they interconnected with other farmers to spread this practice of self-organized breeding. An association of farmers and other people interested (including professional breeders) developed from this commitment. Moreover, the regional government set up a germplasm bank as well as promulgated a law instituting the role of "seed saver" and envisaging funds for it. Both actions facilitated the work of self-organized breeding and, afterwards, PPB.

In terms of rules, PPB refers to the international and national regulations on plant genetic resources that on one hand promote agrobiodiversity and on the other discipline the use of plant genetic resources.[2] Understandings—in the forms of traditional knowledge about old varieties and on how to breed and propagate them—play a crucial role too in PPB. The challenge is to fit traditional techniques into current, prevailing environmental, social, institutional-legislative, and economic conditions that are different compared to those where the traditional knowledge developed first. Professional breeders in research centers, universities, and civil society organizations committed to the rediscovering and revaluation of old seeds are working on this challenge by bringing out farmers' traditional knowledge and integrating it with scientific expertise as well as with lay knowledge. This work is creating a renewed knowledge fitting traditions with new conditions. This cooperation is important for farmers to face the difficulties they encounter in selecting varieties, utilizing them and assessing their multiple properties, all this from the farming-processing stages (where these varieties show

[2] The management of agrobiodiversity is influenced by the legal-institutional framework defined by the international and European policy acts that influence also at the national level. They include acts aimed at preserving and promoting biodiversity, but also at regulating the use of genetic resources, mainly on the basis of intellectual property rights, but also including special regulations for genetic material at risk of erosion. Due to the different aims, these policies are often in contradiction with each other. Moreover, this legislation is received and implemented by the Member States, which in some cases (as in Italy) envisage also the regional jurisdiction on the matter.

specific characteristics, requirements, and performances) to the quality attributes of the final products (relating to their nutritional, health, and organoleptic properties).

Managing Pre-/Postharvest

In the studied cases, multicrop farming is combined with activities of processing and direct selling. Both economic and nondirectly economic motivations are behind the choice of this type of farm organization.

On one hand, it is seen as an economically viable solution for the management of small to medium farms that seek to practice sustainable forms of farming and enter niche markets of high-quality products. *Bio Colombini* and *Radici* chose multiple cropping in consequence to their decision to enter direct selling. Managing multiple cropping was instrumental to tackle the demand for product variety, the risk of bad harvests, the continuity of supply, and the reduction of costs (warehousing, immobilized capital). Moreover, this organization was consistent with the decision of these farms to implement products processing for high value-added generation. Availability of enough variety of raw materials is crucial for processing and producing a wide range of final products; at the same time, processing is instrumental to cope with raw materials perishability, so offering a twofold benefit.

On the other hand, multicrop farming is motivated by the interest in protecting the environment and human health, as well as in protecting farming and food culture and traditions. *Poggio di Camporbiano* presents this farming strategy as a decision to follow farming methods that are as much as possible adapted to the environmental and cultural conditions where they operate.

Growing multiple crops can be the original way of running the farm, as is the case of *Poggio di Camporbiano*. For *Bio Colombini* and *Radici*, it is the result of a transition. *Bio Colombini* started with farming three to four crops, while now it farms 30 crops and multiple varieties per each crop. *Radici* doubled the number of grown crops, for a total of 30, and farms more than 100 varieties.

The studied farms employ different artifacts to implement pre- and postharvest activities and they may involve different actors such as: farmers, farm workers, agronomists-extensionists, universities and research centers, civil society organizations, consumers, policy makers, and institutions.

Managing multiple crops requires it to split the farmed land into small plots, each of them sown with a different crop, and to implement several rounds of sowing and harvesting within the same season. This sort of farm management is labor intensive. The use of machinery is limited, not least because of the small dimensions of the plots, and workers are asked to implement the tasks of sowing and harvesting as well as to look after the plot several times in the same season. This reduces the possibility of using seasonal workers. Stable workers, specialized in the management of several crops at a time with natural techniques, are crucial in multicrop farming. Working extra hours is very common too, especially for the farm owner and family farm workers. Family work is highly used together with some new, or rediscovered, types of workers, such as interns and volunteers, as well as traditional farmers. Interns might be students who are interested in gaining some experience in what could be their field of work. There are also cases of disadvantaged people covering an intern position. For them the internship serves as a process of socio-economic reintegration, and it is stimulated and funded through a legal framework set up by the regional government.[3] Volunteers are people

[3] This activity is included in the social farming model, a type of farming regulated by a national law (N.L.141/2015); the Tuscany region had, however, regulated this farming model even before, since 2010 (R.L.24/2010).

who are interested in getting in touch with where food is produced and rural lifestyles. Traditional farmers—who are not fully operational in their profession—may be hired to recover old farming knowledge.

Besides the practical aspects, employing the mentioned categories of workers has a symbolic component too. It contributes to define the farm identity (in terms of social embeddedness and ethical feature) and in the opportunity to fully valorize the family work (i.e., women or young people).

Those farms that can afford it combine intensive labor with some innovative technology. *Radici* has gone through a process of complete renovation of its farm technology to introduce a biodegradable, corn-based system of mulching to control weeds. The technology has low environmental impact and uses renewable energy rather than fossil fuel. However, it is expensive.

When moving into the postharvest activities, there are two main, sometimes combined, artifacts: processing and commercialization. Processing may either be a principal activity aimed at increasing product value or used as a secondary activity to reduce waste from unsold harvest. However, processing requires an intensive use of technology, hence it is strongly dependent on the ability to cover costs through the selling of final products and entail significant financial concerns for the farms. Moreover, newer more sustainable technology that is often preferred (such as in the case of a photovoltaic plant set up by *Poggio di Camporbiano*) can be more expensive than conventional technology.

Commercialization is mostly done directly and goes through many different channels, which include markets, in- and out-farm shops, GAS, public procurement, and private catering or restaurant services. It may concern both raw and processed products. Much labor is employed to manage logistics around the commercialization, as well as to build and maintain personal relationships with customers and clients since market channels rely on trust and the active involvement of consumers. *Bio Colombini* hired three people, working full time, to manage relations with GAS. Information and communication technology is also highly used for the commercialization. All case study farms now use at least a computer station to manage orders, especially those coming from e-commerce and GAS.

Setting up and running the described artifacts rely on a combination of traditional and modern understandings. Traditional farming knowledge is greatly used in the management of multiple crops, processing, and direct selling of products. However, it is combined with refined, or new, knowledge that allows traditional knowledge to evolve and adapt to the new conditions. The combination of knowledge results in farming, processing, and commercial practices that have low dependence on external inputs (i.e., fuel and chemicals), as well as being based on a sustainable use of human and natural resources.

As far as rules are concerned, processing internalization implies the adoption of specific procedures to obtain certain features of final products, complying with the quality standards set by public bodies to guarantee food safety and consumer health and in line with the organoleptic features that are traditionally attached to the products. The balance between cultural expectations and legal prescriptions is reached through innovation processes. Cheese production from raw milk is an example of this. Raw milk is a central ingredient to produce traditional cheese in the study area, the taste of which taste is strongly dependent on the flavoring given by herbs of grassland. However, using it has required some changes in the equipment traditionally used as well as in some procedures (e.g., utmost cleanliness in milking; substitution of wooden tools with iron tools) to guarantee a bacterial load within the ranges set by law for use of raw milk.

Replication, Scaling up, and Translation

Providing Genetic Resources

The possibility for farmers to connect with the network of professional breeders, civil society organizations, and researchers that are committed to the rediscovery of old varieties constitutes a scaling-up opportunity for PPB practices at the same time, interaction with other famers allow to share knowledge about traditional breeds and varieties. *Radici* and *Poggio di Camporbiano* stress the importance for knowledge sharing and coordination that stem from these interactions. Having a market where to valorize the products obtained from these recovered genetic resources, communicating their specific value, represents another crucial factor, which creates the conditions for the reproduction of the resources themselves and the replication of farm model built around them.

On the other hand, the risk of irregular harvests, due to the genetically variable and unstable seed varieties and the need of these varieties to adapt to the ecosystems where they are grown, represents a significant challenge that may discourage farmers from engaging in PPB. In addition, legislation that forces to produce and register seed varieties that conform to certain standards of quality, homogeneity, and stability hampers the possibility of self-breeding and exchanging genetic material. These aspects represent hindrances to replication and scaling-up of these farming models.

The hampering action of the seed regulatory system is also significant if considered in the perspective of the capacity of these farm models to influence the broader food system (translation). Initially, international and national laws regulating the management of plant genetic resources have been defined to protect breeders' rights. These regulations allow using just varieties that are registered in national or international registers and reduce the space for other forms of breeding and production of seed material under the claim of protecting consumer health. Recently, however, the legislation is moving toward the possibility of recognizing farmers' breeding too. As an effect, some regulations allow self-organized breeding and the exchange of the resulting genetic material among farmers, operating in the same territory, when the local varieties or landraces are recognized to contribute to the protection of biodiversity. This opportunity has favored the growth of interest and engagement in the concrete efforts put in place to enhance agrobiodiversity, also by actors not initially involved.

Part of this process of translation of PPB into more consolidated breeding practices is also connected with the work of advocacy and awareness raising done by some civil society organizations about the importance of rediscovering and valorizing old varieties. Slow Food, Foro Contadino,[4] and AIAB[5] are some of the organizations mentioned by *Radici* and *Poggio di Camporbiano* with this respect.

Managing Pre-/Postharvest

Opportunities and challenges for replication, scaling-up, and translation exist for pre and post-harvest activities too.

Among the factors hindering the replication and scaling-up of sustainable multicrop farming, farmers are discouraged because it takes a long time before this farming management model returns satisfactory economic results. The case of *Radici* shows that sufficient productivity levels have been reached only after a long process of varieties selection and adaptation. This kind of farming is also challenged by a generational clash. The case of *Bio Colombini* indicates that while new generations are more in favor of

[4] An organization engaged in defending and promoting peasantry.

[5] A national association for organic farming.

experimenting with sustainable multiple crop farming techniques, old generations are more reluctant, being more oriented to specialized farming.

All cases support the idea that higher level of education and awareness may play a key role in farmers' commitment in undertaking sustainable multicrop farming. In this regard, civil society organizations, farmers' associations, and public institutions may play a huge role. However, education and training can be implemented through privately hired advisory services too, but as the case of *Bio Colombini* teaches, this requires economic investments by the farmers.

Management of most of the production stages by farmers and through sustainable and culturally appropriate practices proves to have a positive impact on the relationship established with consumers. They trust the quality of the products more, as well as appreciate the story behind its production, and are more willing to consume the products frequently. Besides that, all cases show that processing does not just generate higher incomes, but it contributes to stabilize income over the years and generate employment opportunities. These dynamics are fundamental to the start-up and expansion of processing activities in farms practicing multiple cropping, and so for creating favorable conditions for replication and scaling-up. However, there are some factors that hinder them. As an overarching factor, this farming model is labor and technology intensive and, therefore, requires substantial economic investments that are not always affordable by farmers. Moreover, other challenges come from the environment where those farms operate. Often, they are in remote rural areas with scarce access to infrastructure, such as water provisioning systems, roads, and other both physical and immaterial facilities that are instrumental to the production and commercialization of products. These shortcomings increase the costs of production and commercialization, hampering the success of the production activities and thus reducing also opportunities for replication and scaling-up.

In all the case studies, it emerges that public institutions are attached with a crucial role in improving public services and infrastructure in rural areas. Further public support could be available, in the form of subsidies and loans, for improvement of on-farm infrastructure and services. Nevertheless, farmers are not always willing or able to access this kind of support. High bureaucratic obstacles, strict eligibility conditions,[6] and the predominance of a saving culture over an indebted one are among the factors limiting the farmers' willingness and ability to use loans and subsidies. On the other hand, the existence of ethical banks and ad-hoc public funds for this kind of initiative is broadening the possibility for farmers to access loans and subsidies, respectively.

Other positive inputs for replication and scaling up may arise from farmers' cooperative initiatives. Setting up these initiatives is a way to share costs and efforts, together with benefits. In particular, cooperative management models are used to share inputs, such as technological tools and other means that are used in production processes. In this context, cooperative models are also used to set up infrastructures for farming activities (e.g., water pipes for irrigation) or to direct selling (e.g., farmers market facilities), and to share or exchange products when farms fall short of one of more products to supply to consumers. They are also used to activate fair financing systems, with low or no interests

[6] Both Radici and Poggio di Camporbiano pointed out that subsidies are often committed to support just a few areas with peculiar features (e.g., areas of naturalistic relevance) and few activities (e.g., mostly processing and not farming). Moreover, the procedures look more adequate for medium to large farms managing few crops, while they become too complicated for farms managing multiple crops.

on loans, to cut costs for quality check and certification, and to set prices that are satisfactory for the both producers and consumers.

All three cases are in agreement that the commercial opportunities opened by direct selling in the last years have had a huge increase. Both *Bio Colombini* and *Poggio di Camporbiano* have been mostly working with GAS. While, for the case of *Radici*, direct selling overseas has played a crucial role. GAS proved to be a complex selling channel. Their demand sometimes fluctuates over the year, and this may generate instability. Moreover, they introduce a completely new logic in food purchasing, which builds on the active participation of consumers in the production and other decisions concerning their food provisioning. Mediating with consumers' requests is sometimes not easy for farmers and, in any case, is highly time consuming. On the other hand, GAS members are highly concerned consumers, more able and willing to share new development pathways with farmers.

It is common among multicrop farmers to think that there is a limit to scale up their businesses. For the three studied cases, growing more would mean going through huge organizational restructuring, with significant technological investments. Moreover, it would require expanding labor on the farms and face the consequent difficulties to find people who have the rights skills and willingness to work in farming. Maintaining a satisfactory direct relationship with consumers would be as such challenging.

The described multicrop farm models work in two main interrelated areas when seeking to translate in the context where they operate: civil society and institutions.

First, they open the farms to educational experiences. Consumers may join for field visits and work. Universities and other training centers may send their students for practical on field experience. Moreover, institutions and civil society organizations may set up awareness raising and training projects in collaboration with the farms. The case of *Radici* highlights how such initiatives are becoming more valuable than joining fairs in terms of opportunities they generate to spread appreciation of these farming models among farmers, consumers, and other stakeholders. However, the case of *Bio Colombini* raises concern on the feasibility of combining farming activities with educational and networking activities due to time constraints.

Other strategies used to bridge with institutions and civil society exist in developing cooperation with other actors in the food system, not least farmers' trade unions. Bridging with actors who share interest in alternative ways of farming may strengthen the social power of farmers at different levels of the food system, not least their capacity to advocating at a political level for policies and programs that are more in line with multicrop, sustainable farming.

Despite that multicrop farming still holds a minority position in the Italian agricultural landscape, there are clear signs of the translation of this farming model into broader social, economic, and political contexts. Among them are the increasing societal interest in food quality, direct food provisioning, sustainable food practices, and ecosystem services. Other indications are the development of policies, programs, and regulations in support of sustainable farming and consumption (e.g., introduction of agro-environmental measures in agricultural policies; support of forms of direct selling, such as farm shops and farmers markets, and of the introduction of local, traditional, high-quality products in public food procurement); the growing number of research projects focusing on issues related to sustainable farming and food practices, including various ecosystem services; and the openness of the seed regulatory system toward the possibility of alternative pathways in the management of genetic material.

However, there are some limitations to full translation. In addition to the general ban on the free exchange of seeds, there are still some regulatory restrictions that hinder the marketing of products from multicrop farming, since their traditional ways of production are not fully aligned with some formal hygiene and quality standards. Furthermore, agricultural and food policies are sometimes incoherent with one another. Not least, along those that support multicrops, sustainable farming, there are policies and programs that support other rural economic activities or other farming models that are competing with multicrop sustainable farming. Finally, labor regulations set some restrictions to hiring new categories of workers, such as volunteers and interns, that are highly used in the discussed farm models.

At a societal level, concerned consumption practices are limited by some habits and logistic issues that do not favor the spread of consumption of products derived from traditional, multicrop farming. These last include products that are not common in conventional diets and require consumers some learning before they can be successfully (re-)introduced. Furthermore, the provisioning of these products often occurs through unconventional channels that take higher logistic and organizational effort compared to mainstream provisioning channels. A broader translation toward these farming/food provision models indeed implies a deeper change in attitude and practices associated with food.

DISCUSSION AND CONCLUSION: OPPORTUNITIES AND CHALLENGES FOR THE SETUP AND EXPANSION OF SUSTAINABLE INTENSIFICATION FARMING MODELS

Discussions about agriculture are increasingly focusing on how to combine productivity with environmental protection. In the frame of these discussions, the concept of sustainable intensification has gained importance. There are many of these models, which include agroecological, organic, and other sorts of sustainable agro-food production and commercialization (RISE, 2014). They achieve different results, in terms of balancing productivity and environmental protection. In general, some of them remain within the productivist model, while some others imply the redesign of the whole agri-food system starting from promoting farming practices that seek to imitate natural processes (Levidow, 2016). Some scholars speak about incremental and radical changes in this regard. Incremental changes maintain the status quo; they slightly revise existing rules when the dynamics internal to the productivist, agri-food paradigm ask for this. Whereas, radical changes create new rules that seek to change this paradigm (Geels and Schot, 2007; Brunori et al., 2013; Hinrichs, 2014). The diversity in the forms and results of the changes promoted by sustainable intensification agri-food strategies depends on the practices they build on as well as on the relations with the contexts where they are operated (Garbach et al., 2016).

In this study, we have conducted a qualitative analysis of three examples of sustainable intensification in Tuscany (Italy) with three farms that have been structuring and managing whole food chains following principles of agroecology.

The framework we have used, combining PT (Schatzki, 2002) and SNM (Seyfang and Haxeltine, 2012), has been helpful to explore the aforementioned aspects. Our interest was to contribute to overcome normative understanding of sustainable intensification by unveiling under which conditions practices of sustainable intensification are configured, replicated, scaled up, and translated into the context where they operate. Moreover, we sought to reveal to which extent they manage to keep a balance among environmental protection, adequate production, and market and income satisfaction.

This work reveals that the studied cases show both a radical and a gradual interpretation of sustainable intensification. They are radical as far as the management of the food chain is concerned. With this respect, they have been going through a process of thinking and managing the whole chain in a way that is environmentally sustainable while guaranteeing productivity, income generation, and satisfaction of market requests. The result of this process so far exists in the following: introduction of PPB for the production and propagation of old and new seeds varieties; implementation of environmentally sustainable multicrop farm management practices, including organic and biodynamic techniques; internalization of processing according to environmentally sustainable practices (e.g., the use of renewable sources for processing and packing); and choice of market channels—mainly direct selling—that are interested in products that are produced respectfully for the environment, human, and animal health, and are attached with cultural value as they refer to local food traditions.

There are factors that have been facilitating the configuration of these practices and that may support their replication, scaling-up, and translation into the broader system where they operate.

Firstly, there are actors (i.e., practitioners of the food system, civil society organizations, universities, and research centers) interested or already engaged in combining environmental sustainability and productivity in the agri-food system. This has generated ethical reasons and intellectual and cultural curiosity, as well as economic and health interests.

Secondly, the development and spread of knowledge on how these models of food chain management can be set up and run is crucial too. It comes from a combination of traditional and modern knowledge, as well as from the combined action of farmers, practitioners, civil society organizations, universities, and research centers. Coordinated actions of these actors are also instrumental to raise further interest on the development of models for environmental sustainability and productivity in the agri-food sector more in general.

Thirdly, there are adequate resources available to set up and run these agri-food models. Old (e.g., traditional farmers and family labor) and new labor forces (e.g., interns, volunteers, disadvantaged people) to take on highly labor-intensive jobs. There are modern and traditional techniques and technologies (e.g., low input, traditional, and natural farming methods; machinery based on renewable energy) that are adequate for the defined strategies of food chain management. Economic resources are sufficient to undertake the strategies for sustainable and productive management of the food chains. Finally, there are policies, programs, and legislation (e.g., publicly funded research programs; public support for sustainable businesses and direct selling; institutional commitment in promoting direct selling, local products, and sustainable diets) that allow some room to set up and run sustainable food chains.

However, there are some challenges to the full exploitation of the potentialities of these initiatives. They introduce some inefficiencies in the way the established configurations function as well as hamper the opportunities for replication, scaling-up, and translating into the broader system.

The development and spread of adequate knowledge about the meanings and potentials of these production-consumption models constitute a first crucial factor. In this regard, the case studies showed that they may be impeded by generational clashes between new and old farmers, with old farmers being, sometimes, reluctant to embrace food chain restructuring. To overcome cultural conditioning, higher commitment of organizations and institutions in organizing trainings for farmers and occasions for collective learning and interactions is so essential. Knowledge and awareness raising and sharing should, however, also involve other stakeholders besides farmers, such as policy makers and consumers. On the one hand, regulations on food safety and labor are not adequate to the peculiarities of the production processes of the studied chains. This creates some impediments to product commercialization or requires some investments to make the productions meet required safety standards. Also, they reduce some opportunities to hire certain types of workers that would be valuable for these labor-intensive food chains. On the other hand, the degree of awareness of the value of these production systems/products is low in mainstream market channels. This constitutes a further limitation for these systems to scale up and translate into the established food system and shows the importance of a broader societal change in attitude and practices about food.

Another limitation arises with respect to the material resources that are needed to set up and run the studied chains. It takes a long time before they generate adequate production and income (e.g., because of irregular harvests in the early stages of the production and the time it takes to develop a market for products and a processing line). This, in turn, discourages farmers from starting them. Public institutions might play a huge role in this regard. They might increase incentives for farming activities, especially in the start-up phase, while now their attention is mostly on the processing of already active agri-food initiatives. On the other hand, procedures for accessing subsidies committed to the farming sector might be simplified. Right now, they are often not adequate for the type of sustainable multicrop farms investigated in this study, which are so discouraged from applying for certain subsidies. Together with the support given to improve organizational or material aspects of the management of production processes, as just said, the importance emerges to provide help to foster processes aimed at creating new common pools of knowledge, through interaction among the various actors involved. Signs in this direction are also coming from the new European policies for innovation strongly oriented to encourage collective action.

Finally, all the studied cases claim that they cannot grow over a certain threshold. There would not be enough workers, with adequate skills, to be employed in farming, and there would not be sufficient products to meet higher demand. However, this limitation might be overcome through replication of these initiatives according to a strategy aimed at maintaining their small scale. Developing forms of horizontal coordination among small farms might be an additional strategy.

Among the just discussed challenges, a clash between the guiding principles and practices of the studied initiatives, on one side, and the priorities and ways of operating of agri-food policies, programs, and regulations, on the other side, emerges. This represents the main reason why the cases investigated in this study might be considered examples of gradual changes. They are still operating in an institutional and political environment that is set up for conventional agri-food models. They are trying to influence the context mostly by networking with civil society, universities, and research centers to raise awareness about the importance and peculiarities of their model. However, they claim that much more can be done in this respect, but economic and time resources are insufficient.

In summary, this study shows that simultaneously improving productivity in agriculture, reducing its environmental impact, and satisfying income expectations and consumer demand, or—said with Garbach et al. (2016)—realizing scope economy in farming, is feasible. However, there are some challenges that emerge at a supply chain level and in the interaction between the supply chain and the context where it operates. These challenges hinder opportunities for these initiatives to configure and function in their daily activities as well as replicate, scale up, and translate.

This study confirms what other scientific and grey literature says about the conditions for full expression of sustainable intensification, which is to say that actions need to be taken on several aspects both directly and indirectly connected with production activities as well as environmental preservation (Elliott et al., 2013; FoEE, 2014; Lamine, 2014; Levidow et al., 2014; TP Organics and IFOAM EU, 2014). In particular, it confirms the importance to act at the level of knowledge formation involving all the stakeholders that play a role in the production-consumption practices or that influence the environment where these processes take place and may develop. To that end, the adoption of participatory methods and adequate tools of facilitation may be crucial to catch the opportunities that can stem from interaction among different sources of knowledge and to promote new learning. In addition to that, taking into consideration and combining, through a comprehensive approach, all the various aspects involved in the diffusion and growth of these systems—cultural, economic, technical, organizational, institutional, and legal ones—and reshaping them according to the sustainable and ethical claims characterizing these systems seems the most promising way to create an enabling environment to their development.

References

Bourdieu, P., 1977. Outline of a theory of Practice. Cambridge: Cambridge University Press.

Brunori, G., Rossi, A., Cerruti, R., Guidi, F., 2010. Nicchie produttive e innovazione di sistema: un'analisi secondo l'approccio delle transizioni tecnologiche attraverso il caso dei farmers' markets in Toscana. Economia Agroalimentare 3.

Brunori, G., Rossi, A., Guidi, F., 2012. On the New Social Relations around and beyond Food. Analysing Consumers' Role and Action in Gruppi di Acquisto Solidale (Solidarity Purchasing Groups). Sociologia Ruralis 52 (1), 1–30. https://doi.org/10.1111/j.1467-9523.2011.00552.x.

Brunori, G., Barjolle, D., Dockes, A.C., Helmle, S., Ingram, J., Klerkx, J., Moschitz, H., Nemes, G., Tisenkopfs, T., 2013. CAP reform and innovation: the role of learning and innovation networks. EuroChoices 12 (2).

Darnhofer, I., 2015. Socio-technical transitions in farming: key concepts. In: Sutherland, L., Darnhofer, I., Wilson, G.A., Zagata, L. (Eds.), Transition pathways towards sustainability in agriculture, Case studies from Europe. CABI, pp. 17–31, 2015.

Elliott, J., Firbank, L.G., Drake, B., Cao, Y., Gooday, R. 2013. Exploring the Concept of Sustainable Intensification, LUPG Commissioned Report, ADAS/Firbank.

Friends of the Earth Europe (FoEE), 2014. Agro-Ecology: Building a New Food System for Europe. Available from: https://www.foeeurope.org/sites/default/files/news/foee_agroecology_local_economies_280314.pdf.

Fonte, M., 2013. Food consumption as social practice: solidarity purchasing groups in Rome, Italy. Journal of Rural Studies 32, 230–239. https://doi.org/10.1016/j.jrurstud.2013.07.003.

Garbach, K., Milder, J.C., DeClerck, F.A.J., Montenegro de Wit, M., Driscoll, L., Gemmill-Herren, B., 2016. Examining multi-functionality for crop yield and ecosystem services in five systems of agroecological intensification. International Journal of Agricultural Sustainability. https://doi.org/10.1080/14735903.2016.1174810.

Geels, F., Schot, J., 2007. Typology of sociotechnical transition pathways. Research Policy 36, 399–417.

Giddens, A., 1984. The constitution of society. Berkeley, University of California Press.

Hinrichs, C.C., 2014. Transitions to sustainability: a change in thinking about food systems change? Agriculture and Human Values 31, 143–155. https://doi.org/10.1007/s10460-014-9479-5.

Kemp, R., Schot, J., Hoogma, R., 1998. Regime shifts to sustainability through processes of niche formation: the approach of Strategic Niche Management. Technology Analysis & Strategic Management 10 (2), 175–195.

Lamine, C., 2014. Sustainability and resilience in agrifood systems: reconnecting agriculture, food and the environment. Sociologia Ruralis 55 (1), 41–61. https://doi.org/10.1111/soru.12061.

Levidow, L., 2016. Sustainable intensification: agroecological appropriation or contestation?. In: Presentation at Agroecology Workshop, Brussels, on 25 May 2016.

Levidow, L., Pimbert, M., Vanloqueren, G., 2014. Agroecological research: conforming—or transforming the dominant agro-food regime? Agroecology and Sustainable Food Systems 38 (10), 1127–1155. https://doi.org/10.1080/21683565.2014.951459.

Loorbach, D., 2007. Transition Management: New Mode of Governance for Sustainable Development. International Books, Utrecht.

Maye, D., 2016. Examining innovation for sustainability from the bottom up: an analysis of the permaculture community in England. Sociologia Ruralis. https://doi.org/10.1111/soru.12141.

Meynard, J.M., 2013. Innovating in cropping and farming systems. In: Renewing innovation systems in agriculture and food. Wageningen Academic Publishers, Wageningen, pp. 89–108.

Rural Investment Support for Europe (RISE), 2014. The Sustainable Intensification of European Agriculture. Available from: http://www.risefoundation.eu/publications.

Røpke, I., 2009. Theories of practice—new inspiration for ecological economic studies on consumption. Ecological Economics 68, 2490–2497.

Schatzki, T.R., 1996. Social Practices. A Wittgensteinian Approach to Human Activity and the Social. Cambridge University Press, Cambridge.

Schatzki, T.R., 2002. The Site of the Social. A Philosophical Account of the Constitution of Social Life and Change. The Pennsylvania State University Press, University Park, Pennsylvania.

Schatzki, T.R., Knorr-Cetina, K., von Savigny, E. (Eds.), 2001. The Practice Turn in Contemporary Theory. Routledge, London.

Seyfang, G., Haxeltine, T., 2012. Growing grassroots innovations: exploring the role of community-based initiatives in governing sustainable energy transitions. Environment and Planning C: Government and Policy 30, 381–400.

Shove, E., Walker, G., 2010. Governing transitions in the sustainability of everyday life. Research Policy 39 (4), 471–476.

TP Organics & IFOAM-EU, 2014. The European Innovation Partnership: Opportunities for Innovation in Organic Farming and Agroecology. Available from: www.tporganics.eu/upload/EIP_dossier_EN.pdf.

Veen, E.J., 2015. Community Gardens in Urban Areas: A Critical Reflection on the Extent to Which They Strengthen Social Cohesion and Provide Alternative Food (Ph.D. thesis). Wageningen University, The Netherlands.

Warde, A., 2005. Consumption and theories of practice. Journal of Consumer Culture 5 (2), 131–153.

Wiskerke, J.S.C., Van der Ploeg, J.D., 2004. Seeds of Transition: Essays on Novelty Production. Niches and Regimes in Agriculture. Assen: Royal Van Gorcum.

CHAPTER

21

Environmental Benefits of Farm- and District-Scale Crop-Livestock Integration: A European Perspective

Philippe Leterme[1,2], *Thomas Nesme*[3,4], *John Regan*[4], *Hein Korevaar*[5]

[1]Agrocampus Ouest, Rennes, France; [2]UMR 1069 SAS INRA/Agrocampus Ouest, Rennes, France; [3]Bordeaux Sciences Agro, University of Bordeaux, Gradignan, France; [4]UMR 1391 ISPA, Villenave-d'Ornon, France; [5]Plant Research International, Wageningen, The Netherlands

INTRODUCTION: CONTEXT

An Increasing Global Agricultural Demand and an Ardent Need to Preserve the Environment

Global demand for food is expected to increase by 70% by 2050 (FAO, 2009a). The increase in demand for animal products driven by growing populations, incomes, and dietary preferences is stronger than that for most other food items. Global production of meat is projected to more than double from 229 million tons in 1999/2001 to 470 million tons in 2050, and that of milk to increase from 580 to 1043 million tons (FAO, 2009b). The bulk of the growth in meat and milk production will occur in developing countries, with China, India, and Brazil already representing two-thirds of current meat production, but the European Union (EU) can and must endeavor to respond to this issue, thus signifying that its agriculture must continue to be productive and competitive.

This demand must be addressed while conserving natural resources (water, soil), nonrenewable resources (phosphorus, fossil fuels), and ecosystem services (pollination, natural pest control, soil fertility), in short, the environment in its larger sense (water, soil, and air quality). It will require designing agricultural systems that use fossil fuels and natural resources in a more efficient and parsimonious way, as well as potentially becoming producers

Agroecosystem Diversity
https://doi.org/10.1016/B978-0-12-811050-8.00021-2

of raw materials for industrial and energy purposes. In addition, these systems will need to be adapted to climate change and help to mitigate it.

The Diversity of Systems: An Asset to Meet These Challenges

If appropriate technologies and policies are applied, agriculture can attenuate its environmental impacts (e.g., eutrophication) while contributing to mitigation of climate change. For example, careful recycling of organic waste as fertilizer by integrating crop and animal production makes it possible to minimize losses of nitrogen and other plant nutrients within agroecosystems, leading to mitigation of eutrophication and climate change through respectively reductions of NO_3^- leaching and N_2O emissions. A diversity of agricultural systems within the same territory is generally an essential asset, as it contributes to the diversity of land and landscape use and offers greater flexibility by spatial and temporal interactions among different components of a land use system. By this way, diversified land use can open new possibilities for combining food production with biomass production and on-farm production of renewable energy from livestock manure, perennial crops, and particular biotopes as seminatural noncultivated areas. As such, one of the most promising paths to respond to this challenge is to promote new systems that mix livestock and crops (Herrero et al., 2010).

However, Specialization Is Gaining Ground (Rieu et al., 2015)

In 2010, 47% of farms in the EU-28 were specialized in cropping, 27% in livestock, while 24% were remaining in mixed farming. Spatial specialization in animal production is important: at the NUTS3 level[1], 35 of the total 975 territorial units concentrate 25% of European animal numbers (expressed in animal units) but contain only 13% of its agricultural area. Half of European animal units are located in 114 regions, which contain 31% of its agricultural area.

Concentration of agricultural production in EU countries has been driven by the economic advantages it brings: gains in productivity, economies of scale, or economies of agglomeration. Moreover, impacts of these forces on EU agriculture have been enhanced by public policies such as the Common Agricultural Policy (CAP), environmental policy, and World Trade Organization agreements. The "common market" aimed to specialize productive agricultural areas according to their comparative advantages, which has resulted in high concentrations of animal production in parts of the EU. Regarding impacts of the current CAP reform (2015–20), the trend toward more concentration and specialization is expected to continue. A decoupling policy tends to increase agricultural specialization by renunciation of less profitable activities. The incidence of decoupling and impacts of single farm payment (with the different implementation schemes available to member states) on the degree of farm specialization was analyzed (Agrosynergie, 2013): at the EU level, between 2004 and 2009, 39% of "mixed cropping" moved to more specialized sectors (specialist cropping), 33% of "mixed livestock" moved to more specialized sectors (specialist livestock), and 34% of "mixed crop-livestock" moved to more specialized sectors, expanding cropping or breeding.

The end of milk quotas should foster concentration of milk production in basins endowed with the main "assets" to produce.

[1] The NUTS classification (nomenclature of territorial units for statistics) is a hierarchic system for dividing up the economic territory of the EU (http://ec.europa.eu/eurostat/web/nuts).

HOW TO PROMOTE NEW SYSTEMS THAT MIX LIVESTOCK AND CROPS?

To reverse this tendency toward increased specialization, the EU needs to generate new knowledge about the sustainability of mixed farming systems, establish pertinent indicators of sustainability to be able to evaluate alternate strategies and negotiate trade-offs, and develop these systems through appropriate implementation processes and public policies.

It is within this context that the CANTOGETHER program was funded by the EU and launched in 2012. CANTOGETHER designed, evaluated, and promoted new mixed crop-livestock systems at farm, district, and landscape levels to optimize energy, carbon, and nutrient flows that conserve natural resources and maximize production. Design and assessment procedures combined scientific and expert-based knowledge to allow comparison of a wide range of potential solutions. CANTOGETHER brought together data from a collection of 24 case studies based on existing networks of farms or individual farms (field-research platforms and commercial farms) across a wide variety of agricultural regions of Europe (Atlantic plain, humid mountains, continental Europe, Nordic countries, and Mediterranean zones) in which some innovative mixed farming practices and systems were implemented and monitored at farm and/or district levels (Fig. 21.1).

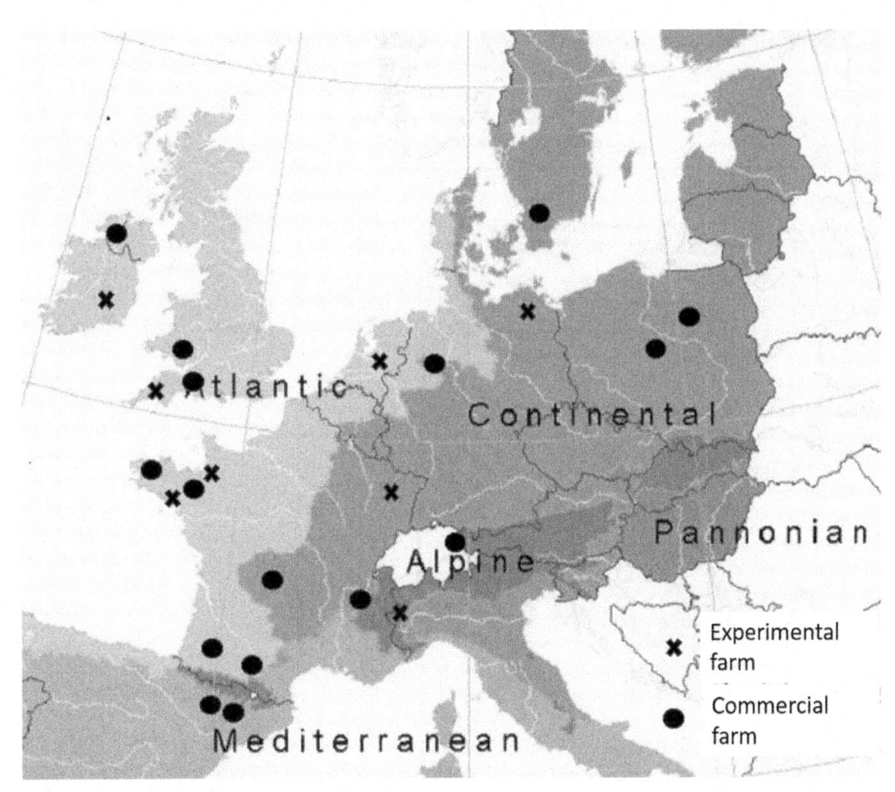

FIGURE 21.1 Localization of the CANTOGETHER case studies in the European biogeographic regions.

Integrated Crop and Livestock Systems (ICLS) at Farm and District Level

A three-sphere model (corresponding to crops, grasslands, and animals) was developed to represent mixed farms. The size of each sphere and of the overlaps and arrows between these spheres allowed the diversity of the mixed farms to be represented. Fig. 21.2 (Moraine et al., 2014) gives an example of this three-sphere representation for cooperation between lowland and upland farms in Italy.

Based on the level of diversity and synergies between constitutive elements, four types of crop-livestock integration have been identified (Moraine et al., 2013; Fig. 21.3):

- Type 1: exchange of materials (e.g., grain, forage, straw, waste as organic fertilizer) between specialized farms, regulated by the market, in a rationale of "coexistence."
- Type 2: exchange of materials between crops, grasslands, and animals (3 spheres) in a rationale of "complementarity" at the farm level. Crop systems are designed to meet the needs of livestock enterprises (need for concentrates, raw forages, and straw), and livestock waste is utilized to fertilize arable plots.
- Type 3: increased temporal and spatial interactions among the three spheres in a rationale of "local synergy": stubble grazing, temporary grasslands in rotations, and intercropped forages. A high level of diversity in farm components is targeted to enhance regulating services.
- Type 4: increased temporal and spatial interactions among the three spheres in a rationale of "territory-level synergy": organization optimizes resource allocations, knowledge sharing, and cooperation, including work.

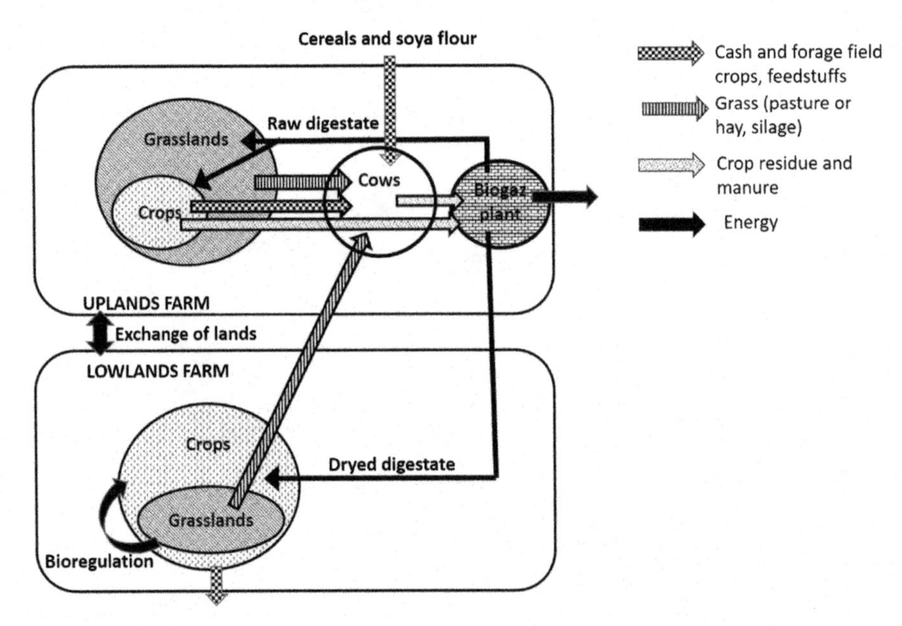

FIGURE 21.2 One mixed farm (uplands) interacting with a crop farm (lowlands). *Ovals* represent ecologic and social components. *Overlapping areas* represent spatial and temporal interactions between these compoents. *Straight arrows* represent material flows. *Curved arrows* represent ecosystem services. (Moraine et al., 2014).

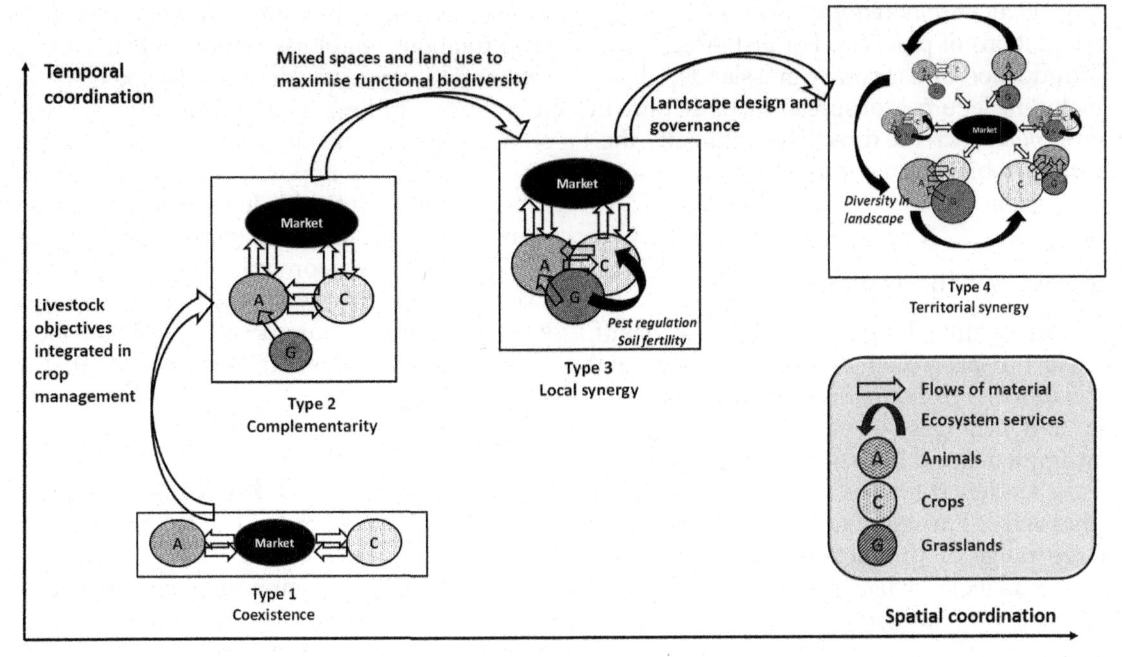

FIGURE 21.3 A generic typology of forms of crop-livestock integration according to their levels of temporal and spatial interaction. Each type has an illustrative name, and key drivers of integration necessary to pass from one type to another are noted (Moraine et al., 2013).

Types 1 and 2 focus on improving "metabolic" properties of farming systems, while types 3 and 4 focus on using ecosystem services to regulate pests and increase soil fertility.

- The "metabolic" approach is based on the principles of the industrial ecology (Ehrenfeld, 2000) and multilevel analysis (Giampietro, 2003), which aims at increasing the autonomy of the systems and the efficiency of use of the resources by reducing entering flows of natural resources (water and energy) and chemical inputs (fertilizers and pesticides). In such an approach, feeds and concentrates have to be produced and consumed within the system, and effluents have to be recycled through the organization of flows of products between the spheres. These flows can be categorized according to their "degree of circularity," i.e., the level of recycling of the material and energy resources, which characterize the level of efficiency of resource use.
- Crop-livestock integration allows one to substitute synthetic inputs by ecosystem services, through action on the biotechnical chain: practices, system properties (e.g., diversity of species, habitats, landscapes), ecologic processes (e.g., trophic chain), and services (e.g., bioregulation) (Koohafkan et al., 2011; Power, 2010). Enhancing ecosystem services through agroecosystem

management may require various adaptations of practices. For instance, introduction of temporary grasslands, including legumes, or spreading of manure in crop fields increase directly or indirectly the soil microbial populations and enhance soil biologic fertility.

Implemented Innovations

A participative design method of ICLS (with farmers but also cooperatives, environmental associations, representatives of local authorities, and managers of natural resources) has been implemented in one part of the network of case studies (Moraine et al., 2014): 15 case studies across Europe were selected to represent a large range of production systems (bovine, ovine, porcine), challenges, constraints, and resources for innovation and representing different types of crop-livestock integration (four case studies in complementarity type, four case studies in local synergy type, and seven case studies in territorial-level synergy).

For each mixed farming system selected, agronomic practices (crop rotations, soil management, etc.) and livestock practices (species selection, feeding, management, etc.) were combined into novel mixed farming systems ranging from easy-to-adopt combinations of methods to more ambitious solutions involving strategic changes at farm and district levels. These systems ensure high resource use efficiency (notably of nutrients), reduction in dependence on external inputs (fertilizers, pesticides, concentrated feeds), and acceptable environmental and economic performances.

Among the diversity of adaptation options designed in case studies (Table 21.1), few are purely "technological," except biogas production, which is cited in six case studies. The stakeholders identified mainly sets of practices based on agroecologic principles: diversification of crop rotations (eight case studies), better use of seminatural spaces such as landscape elements (five case studies) and grasslands (six case studies), and optimization of cover crops (three case studies). The importance of organizational aspects is also outlined: local market development appears in four case studies, as do forage banks or structures for exchanging products between farmers directly. Public policies appeared to be a source of support for development of ICLS but less frequently than knowledge sharing initiatives.

ASSESSMENT OF IMPLEMENTED INNOVATIONS AT THE FARM LEVEL

In a first step, an environmental assessment was performed at a large scale ("macroscale" assessment) by using a network of farms in France and Switzerland (87 Swiss and 622 French farms) to compare mixed farming systems and specialized farms. In a second step, we studied nine particular farms ("microscale assessment") where evaluations of C, N, and P fluxes and life cycle analysis calculations have been carried out.

"Macro-scale assessment" (Alig and Michler, 2015): the comparison of mixed and specialized farms showed that there are no clear relationships between the national economic (european typology OTEX) and structural (typology of Swiss farms FAT99[2]) farm classes and the environmental performance of farms. In contrast, the level of integration between crops and livestock (appreciated by considering both the diversity of productions (crops and livestock) and the use of home-grown feed) showed a statistical link between the environmental performance and the degree of mixity of farms (Fig. 21.4). In conclusion, we show that what is decisive for

[2] http://www.blw.admin.ch/old/agrarbericht0/f/anhang/procede_methodes.pdf.

TABLE 21.1 Presentation of Options of Crop-Livestock Integration in all Case Studies (Moraine et al., 2014 in Supplementary Material)

		E2	E6	E7	E8	E9	C2	C4	C5	C6	C10	C13	C14	C15	C16	C18
Adaptation options relative to a metabolic approach	Methanization/biomass boiler	X	X							X	X	X			X	
	Manure spreading optimization			X								X	X	X	X	
	Adaptation of manure production with litter type				X										X	
	Use of industrial by-products as feed or litter														X	
	Forage/grain versus manure exchanges between farms		X					X								
Adaptation options relative to an ecosystemic approach	Diversification of crop rotation soya, alfalfa, grasslands, cereal-legumes associations	X		X	X	X	X					X			X	X
	Management of cover crops to increase fodder				X			X				X				
	Exchange of lands		X									X				
	Optimization of grassland management			X	X	X			X			X		X		
	Adaptation of animal type and management				X					X						X
	Landscape management						X	X				X		X		X
Adaptation options relative to an organizational approach	Development of local markets				X	X			X			X				
	Alfalfa dehydration factory						X					X				
	Forage banks/exchange of products						X				X	X	X			
	Networks for collective learning						X					X	X	X		X
	Public support to practices change						X	X								
	Tourism							X					X	X		X

Adaptation options are presented in lines, their presence in each case study is signaled by crosses.

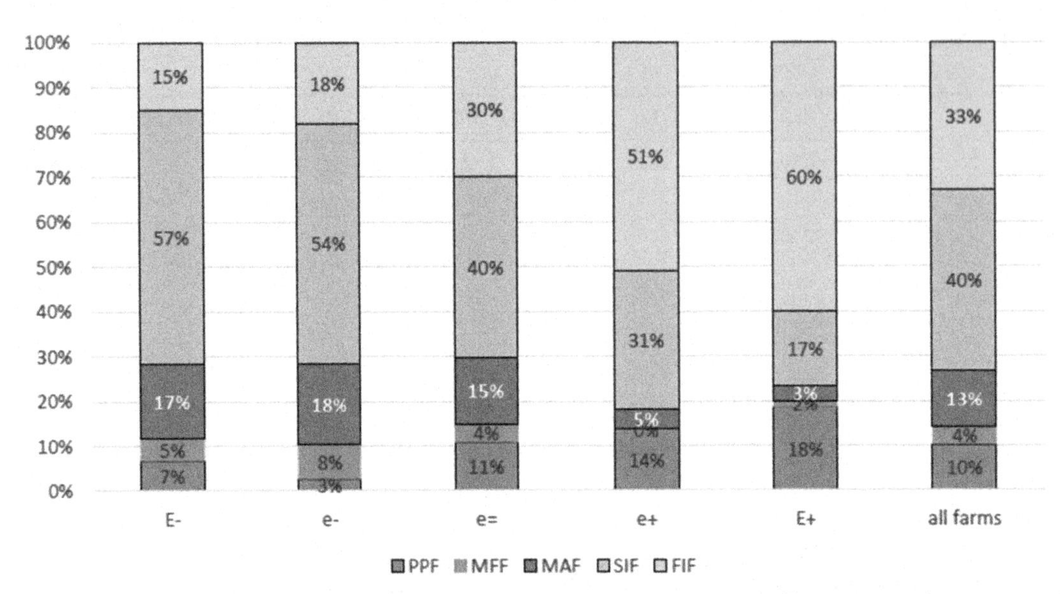

FIGURE 21.4 Distribution of the different levels of crop-livestock integration in the different classes of environmental impacts (estimated from FADN data). PPF are specialized dairy farms based on grasslands selling only animal products; MFF, MAF, SIF, and FIF are mixed farms (selling both crops and animal products) with an increasing integration between crops and animals from MFF to FIF estimated based on the use of home-grown feed. E− to E+ represent five classes of environmental impacts defined by evaluation of energy demand and risks to contribute to global warming, eutrophication, and acidification (E− gathers the farm with the worst environmental performances expressed per unit of area and per unit of product; E+ the farms performing the best both per unit of area and unit of product) (Alig and Michler, 2015).

the environmental performance of mixed farms is not the "mixity" in economic or structural terms but the physical integration between crops and animals and the effectiveness of the farm metabolism.

"Microscale" assessment was performed on a sample of nine innovative farms belonging to the collection of case studies (Chambaut et al., 2015). Implemented innovative techniques were (1) the management of farmland with low levels of inputs through original crop choices (types of rotation, variety selection, pure legume or in combination, catch crops), (2) a local animal feed (production of grain legumes or ensiled mixtures, using alfalfa as hay or dehydrated concentrate, cereals home consumed), and (3) the valuation of manure as fertilizer (separation, dehydration, composting, spot application) or as a source of energy (biogas). For each case study, the farm performance is described with and without the innovation implemented.

Table 21.2 sums up the main results of the analysis of carbon and nitrogen fluxes in terms of environmental impacts. For N losses to ground and surface water, expressed by unit of area, the implemented innovations have improved the performance in six farms of the nine; results are not as good for N losses to water expressed per unit of output (also expressed in kg of N) with an improvement in only four farms of the nine. We find the same trend for climate warming potential: an improvement in eight farms out of nine when this potential is expressed by unit of area and an overall neutral effect when expressed per unit of product. Unfortunately, the C storage capacity is generally

TABLE 21.2 Evolution of Environmental Performances of the Farms After Implementation of Innovations Toward a Better Crop-Livestock Integration (Chambaut et al., 2015)

		N water losses by unit of area (kgN-NO3/ha UAA)	N water losses by unit of product (kgN-NO3/kg N output)	N air losses by unit of area (kg N-NH3+N2O+NO /ha UAA)	N air losses by unit of product (kgN-(NH3+N2O+N O)/kgN output)	Warming potentialby unit of area (CH4+N2O+CO 2ener in kg eqCO2/ha UAA)	Warming potential/unit of product (CH4+N2O+CO 2ener in kg eqCO2/kg N output)	C storage in soil in permanent grassland (kg C stored/T C photosynth)	Total outputs by unit of area (kgN/haUAA)
	C1	-3%	-50%	-11%	-54%	-19%	-57%	-40%	91%
	E9	1%	-32%	-4%	-36%	-10%	-43%	-57%	43%
	E8	-59%	11%	-64%	-2%	-30%	74%	-14%	-40%
	E5	-25%	2%	-24%	3%	-22%	6%	5%	-26%
	E6	-12%	-11%	10%	12%	-20%	-20%	-31%	-1%
	E2	-8%	-37%	38%	-7%	21%	-22%	-1%	33%
	C16	-37%	-2%	2%	59%	-20%	40%	-5%	-35%
	E3	18%	48%	-2%	24%	-17%	11%	-52%	-20%
	E4	-5%	-1%	31%	37%	-18%	-12%	14%	-7%
Number of farms where innovation	improves	6	4	3	3	8	5	2	3
	is neutral	2	3	3	2	0	0	1	1
	degrades	1	2	3	4	1	4	6	5

Bright green indicates an improvement >25%; *light green* an improvement between 5% and 25%; *red* indicates a degradation >25%; *orange* a degradation between 5% and 25%; *white* corresponds to no significant evolution.

decreased because of the destruction of permanent grasslands. For the N losses to air, we can see that the implemented innovations have very different effects: some are highly positive (two farms on the area basis or on the product basis), but many are heavily negative (two on the area basis and three on the product basis). So, we can conclude that the implemented innovations are generally positive for water quality and global warming potential on an area basis but have various effects for protecting air quality. At the same time, the outputs per unit of area, that is to say the overall productivity, decreased in five farms out of nine. Even though this is accompanied by a reduction in inputs, this decrease in the productivity of the farm has a negative impact on economic performance. This result observed on a few farms has been confirmed by an overall economic analysis of FADN data at the European level (3 years, 2008–10), which has shown that in many cases the specialized and less integrated mixed farms achieved higher output through greater use of external inputs and ultimately were more profitable, as shown in Fig. 21.5, which presents relative median farm net income according to the main productions and the level of crop-livestock integration (Moakes et al., 2014).

Nevertheless, we have to be cautious about such comparisons because biases probably exist. For example, the technical optimization may not be at the same level between specialized systems and more complex mixed farming systems and/or the agronomic potential may not be the same for all systems, the most integrated mixed farming systems being more frequently implemented in challenging environments.

In conclusion and despite the risks of bias mentioned earlier, we can observe the same trends whatever the scale of approach or methods. Integration between crops and livestock often leads to a decrease in productivity and profitability of the farm. Environmental performances have improved overall but not necessarily for all impact categories and with a very significant effect of the functional unit employed.

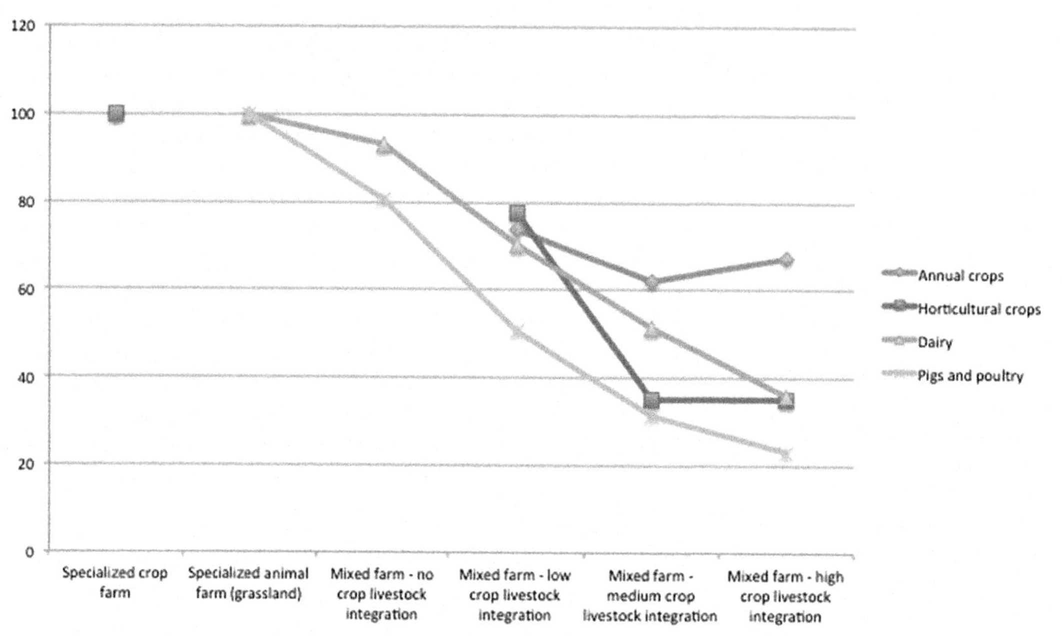

FIGURE 21.5 Farm net income by unit of area for specialized crop farm (index 100 for farms with crops as main enterprise), for specialized animal farms (index 100 for farms with animal products as main enterprise) and for mixed farms depending on their main enterprise (annual crops, horticultural crops, dairy, pig, and poultry) and their level of crop-livestock integration (Moakes et al., 2014).

Indeed, if we favor the function of land cultivation by expressing the results per ha of farmland, then the environmental impacts (water quality and global warming) are frequently reduced in the integrated systems. However, if we favor the production function by expressing results per unit of product, then the results are much less clear and vary from case to case. It must be noted that in this mode of expression of results, the action of dividing the overall impacts by the quantities of product leaving the farm inevitably puts farms that self-produced feed for their herds at a disadvantage. This artifact does not affect the general results but can particularly affect the most integrated systems in which it is precisely the self-production of feed that is a major part of the integration between crops and livestock.

ASSESSMENT OF IMPLEMENTED INNOVATIONS AT THE DISTRICT LEVEL

As shown previously, it is generally not attractive for specialized farms to become a mixed farm, because of lower income and in many cases extra labor pressure and high complexity to manage. A better option could be to collaborate with neighboring farms to exchange materials (manure, straw, and feed), land, or animals. In this way, farmers can benefit from the advantages of specialization and from the cooperation that allows them to make better use of local resources.

We tested this assumption and assessed overall consequences of collaboration between farms at the district level. A farming system approach was used to describe, analyze, and assess the following strategies to recouple crop and livestock production in five case studies:

- Two with local exchange of materials among farms: Ireland (Cavan) and Spain (Ebro Basin);
- One with provision of high-quality forages through a cooperative dehydration facility (Coopedom, Domagné in France);
- One with land sharing between dairy and arable farms (Winterswijk in the Netherlands);
- One with animal exchanges between lowland and highland regions (Thurgau and Grisons in Switzerland).

In each case, a selection of noncooperating farms (specialized and, where available, mixed) were compared with cooperating farms (specialized or mixed) using data on farming practices and organization, input use, feeding strategies, fertilizing strategies, land use, nutrient recycling, and agronomic and economic performance. The data were collected via farmer interviews (Regan et al., 2016a).

Cooperation between farms in a district via each of the crop-livestock integration strategies assessed generally led to increased access to local resources, such as land, labor, feed, or nutrients compared to specialized noncooperating farms in the same district. Table 21.3 presents these new local resources accessed through among-farm cooperation and their deployment by farmers.

The farmer's decision about how to manage these extra local resources largely determined the resulting benefits of cooperation. In the case of the land made accessible as a result of cooperation, it was used to manage excess manure/slurry, increase milk production per hectare, or broaden crop rotations. In the case of newly accessed labor, it was utilized to increase cropping intensity and increase income from outside the farm. In the case of newly accessed feed, it was utilized to increase milk production per hectare. Lastly, in the case of newly accessed nutrients, they were utilized to replace mineral fertilizer inputs. Table 21.4 lists the main benefits (+) and drawbacks (−) of these decisions (Regan et al., 2016a). In three of the four district-level crop-livestock integration strategies assessed (namely, material exchange in the Ebro Basin, forage dehydration, and animal exchange), there was a marked increase in farming intensity on cooperating farms relative to specialized farms, as indicated by farmers opting to increase (1) the number of milking cows per hectare on dairy farms and (2) the cropping intensity on arable farms.

As a result of farmers opting to use the local resources made available via cooperation to increase farming intensity as opposed to diversifying their operations, some of the expected environmental benefits of recoupling crop and livestock production at the district scale were not realized, such as lower external input use, greater farm diversification, and improved district-level nutrient autonomy. In these cases a *rebound effect*[3] was observed, which offsets the expected beneficial environmental effects of cooperation (Regan et al., 2016b). Fortunately, cooperation still had some beneficial effects for the environment and for economy by improving nutrient use efficiency. For example, in the French and Swiss case studies, the benefits of cooperation included improved productivity and lower N surplus per unit of agricultural output and so allowed farms to increase production without an increase in N surplus per hectare.

[3] In conservation and energy economics, the rebound effect (or take-back effect, RE) is the reduction in expected gains from new technologies that increase the efficiency of resource use, because of behavioral or other systemic responses. These responses usually tend to offset the beneficial effects of the new technology or other measures taken (Wikipedia: https://en.wikipedia.org/wiki/Rebound_effect_%28conservation%29).

TABLE 21.3 New Local Resources Accessed Through Among-Farm Cooperation and Their Deployment by Farmers (Regan et al., 2016a)

		Local Resources Accessed		
District	**Crop-Livestock Integration Strategy**	**Dairy Farm**	**Arable Farm**	**Deployment of Resources by Farmer**
Ebro Basin (Spain) and Cavan (Ireland)	Material exchange	Land (outlet for manure)		Export excess of manure to land located off farm
			Manure	Incorporated in soil to supply crop nutrients
		Straw input	Outlet for straw	Animal bedding
Domagne, Coopedom (France)	Industrially mediated transfers of dehydrated fodder	High-quality dehydrated alfalfa and other crops		Increase in milking herd size
Winterswijk (the Netherlands)	Land exchange		Land	Increased potato production
			Slurry and legacy nutrients	Enhance supply of nutrients to potato crop
		Labor and machinery		Plowing of grasslands
Thurgau and Grisons (Switzerland)	Animal exchange	Land (lowland)		Increase in milking herd size
		Labor and time		Increase milking herd size and production of intensive cash crops
		Land (heifer farm, highland)		Stocking of heifers on previously inaccessible steep slopes
		Time (heifer farm highland)		Take up work outside the farm

TABLE 21.4 Main Benefits (+) and Drawbacks (−) for Cooperating Farms Compared to Noncooperating Specialized Farms (Unless Indicated as Mixed) (Regan et al., 2016a)

Ebro basin	**Cooperating dairy farms**	**Cooperating arable farms**
Local exchange of manure for straw	(−) higher labor input per ha (−) lower district N autonomy (−) higher N surplus per hectare (−) higher external input (roughages; concentrates)	(−) higher mineral fertilizer use (−) higher cropping intensity (−) lower land use diversity (relative to mixed farm) (−) simpler crop rotations (relative to mixed farms)
Cavan	**Cooperating pig farms**	**Cooperating arable farms**
Local exchange of grain for slurry	(+) higher district N autonomy	(+) lower mineral fertilizer use (+) higher district N autonomy (−) higher N surplus per hectare/unit product
Coopédom	**Dairy farms that cooperate with Coopédom (A = alfalfa; B = other crops; C = alfalfa/miscanthus)**	
Provision of high quality forages through a cooperative dehydration facility	(+) higher milk yield per cow; groups A, B, and C (+) higher land use diversity; groups A and C (+) lower N surplus per unit product; groups A and B (−) higher input use (pesticide, fertilizer, and soybean); groups A, B, and C (−) lower concentrate feed autonomy; groups A, B, and C (−) higher labor per hectare and per LU; groups A, B, and C (−) lower area under permanent grassland; groups A, B, and C	
Winterswijk	**Cooperating dairy farms**	**Cooperating arable farms**
Land exchange between dairy and arable farms	(+) longer crop rotations (+) lower herbicide use when renewing grassland (+) lower fuel use (−) higher N surplus per hectare	(+) lower frequency of potato in crop rotation (+) lower mineral fertilizer use (+) lower N surplus per hectare
Switzerland	**Cooperating lowland dairy farms**	**Cooperating mountain heifer rearing farms**
Animal exchanges between lowland and highland regions	(+) higher net income per ha (+) higher feed autonomy (+) lower N surplus per ha/unit product (+) lower external input of concentrate feed (+) lower fuel use (−) higher % UAA receiving one or more pesticide applications (−) lower % UAA under permanent grassland	(+) lower on-farm labor (+) higher off-farm income (+) higher forage autonomy (+) higher district N autonomy (+) lower N surplus per hectare (+) lower external input of concentrate feed (+) lower fuel use

CONCLUSION: COMPLEMENTARITY OF INNOVATIONS AT FARM AND DISTRICT LEVELS

At the farm level, integration of crops and livestock generally leads to better environmental performances but often with lower productivity. Indeed, in many cases, specialized and less integrated farms produce higher levels of output through greater use of external inputs and ultimately make more profit. Integrated systems have lower productivity and profitability. The findings are comparable to those of Roguet et al. (2015), who found that the concentration of animal production in the EU

TABLE 21.5 Schematic Representation of the Complementary Effects of Integration of Crops and Livestock at Farm and District Level on Performances

Increasing Integration of Crops and Livestock	Environmental Performances	Economic Performances
At farm level: coexistence (type 1) → complementarity (type 2) or local synergy (type 3)	+	(0) −
At district level: types 1, 2, or 3 → territory-level synergy (type 4)	0 (+)	+

Positive, negative, and neutral effects are indicated by +, −, and 0, respectively; minority effects are indicated by ().

territory has been driven by the economic advantages brought by gains in productivity, economies of scale, or economies of agglomeration.

At the district level, cooperation between farms leads to a better valorization of resources, which in turn increases overall productivity without improving nor degrading the environmental performances. The higher productivity of farms that work together allows them to obtain better economic performances.

In summary, we can conclude that environmental quality mainly improves at the individual farm level but frequently at the cost of economic performance. Fortunately, the coordination and synergy between farms across a district is frequently associated with a rebound effect that improves economic performance, without resulting in negative ecologic impacts. This trend is summarized in Table 21.5.

Acknowledgments

This work has been funded under the EU seventh Framework Programme by the CANTOGETHER project N°289328: Crops and ANimals TOGETHER. The views expressed in this work are the sole responsibility of the authors and do not necessary reflect the views of the European Commission.

References

Agrosynergie, 2013. Evaluation of the Structural Effects of Direct Support. Financed by the European Commission. Final Report Available from: http://ec.europa.eu/agriculture/evaluation/market-and-income-reports/structural-effects-direct-support-2013_en.htm.

Alig, M., Mischler, P., 2015. Synthesis of Environmental Impacts of Mixed Versus Specialized Farms Cantogether Deliverable D2.6, 115p. Available from: https://goo.gl/IJoIDp.

Chambaut, H., Espagnol, S., Fiorelli, J.L., Foray, S., Maignan, S., Moakes, S., Mantovi, P., Ruane, E., Verloop, K., 2015. Primary Production, NPC Flows and Emissions of Innovate Mixed Farms: Synthesis of Evolutions. Cantogether Deliverable D2.8, 38p. Available from: https://goo.gl/IJoIDp.

Ehrenfeld, J.R., 2000. Industrial ecology : paradigm shift or normal science. American Behavioral Scientist 44 (2), 229–244.

FAO, 2009a. The State of Food Insecurity in the World 2009. Economic Crisis-impacts and Lessons Learned. FAO, Rome.

FAO, 2009b. The State of Food and Agriculture 2009. Livestock in the Balance. FAO, Rome.

Giampietro, M., 2003. Multi-scale Integrated Analysis of Agro-Ecosystems. CRC Press, Boca Raton, 472 p.

Herrero, M., Thornton, P.K., Notenbaert, A.M., Wood, S., Msangi, S., Freeman, H.A., Bossio, D., Dixon, J., Peters, M., van de Steeg, J., Lynam, J., Parthasarathy Rao, P., Macmillan, S., Gerard, B., McDermott, J., Seré, C., Rosegrant, M., 2010. Smart investments in sustainable food production: revisiting mixed crop-livestock systems. Science 327, 822–825.

Koohafkan, P., Altieri, M.A., Gimenez, E.H., 2011. Green Agriculture: foundations for biodiverse, resilient and productive agricultural systems. International Journal of Agricultural Sustainability 10, 61–75.

Moakes, S., Nicholas, P., Schmidt, D., 2014. Economic Analysis of Mixed and Specialised Farms in Europe Cantogether Deliverable D5.1, 46p. Available from. https://goo.gl/IJoIDp.

Moraine, M., Duru, M., Nicholas, P., Leterme, P., Therond, O., 2014. Farming system design for innovative crop-livestock integration in Europe. Animal 8 (8), 1204–1217.

Moraine, M., Therond, O., Duru, M., 2013. Design Methodology in Cantogether Project. Cantogether Deliverable D1.3, 31p. Available from: https://goo.gl/IJoIDp.

Power, A.G., 2010. Ecosystem services and agriculture: tradeoffs and synergies. Philosophical Transactions of the Royal Society B 365, 2959–2971.

Regan, J.T., Marton, J.S., Barrantes, O., Ruane, E., Hanegraaf, M., Berland, J., Korevaar, H., Pellerin, S., Nesme, T., 2016a. Does the recoupling of dairy and crop production via cooperation between farms generate environmental benefits? A case-study approach in Europe. European Journal of Agronomy.

Regan, J.T., Godinot, O., Nesme, T., 2016b. Evidence of rebound effect in agriculture: recoupling crops and livestock at the district scale does not always reduce nitrogen losses. In: Fourteenth ESA Congress, 5–9 September 2016, Edinburgh, Scotland.

Rieu, M., Perrot, C., Mann, S., Marouby, H., Roguet, C., 2015. Policy Recommendations to Support Mixed Farming Development. Cantogether Deliverables D5.5, 41p. Available from: https://goo.gl/IJoIDp.

Roguet, C., Gaigné, C., Chatellier, V., Cariou, S., Carlier, M., Chenut, R., Daniel, K., Perrot, C., 2015. Spécialisation territoriale et concentration des productions animales européennes : état des lieux et facteurs explicatifs. INRA Productions animales 28 (1), 5–22.

CHAPTER

22

Payment for Unmarketed Agroecosystem Services as a Means to Promote Agricultural Diversity: An Examination of Agricultural Policies and Issues

Clement A. Tisdell[1], *Clevo Wilson*[2]

[1]School of Economics, The University of Queensland, Brisbane, Australia; [2]QUT Business School, Economics and Finance, Queensland University of Technology, Brisbane, Australia

INTRODUCTION

At first sight, payment for unmarketed agroecosystem services seems to be a promising way of encouraging farmers to adopt or sustain agroecosystems (or modified ones) that are more environmentally friendly than otherwise and which are socially more acceptable than their alternatives. In what respects these ecosystems might be more diverse than at present is unclear because agricultural diversity has many different dimensions. Nevertheless, such payments are likely to result in increased agricultural heterogeneity in various dimensions and on different scales. In modern economies, farmers are strongly motivated to specialize in the products they produce because this is usually the most profitable course of action for them. The per-unit costs of production are usually lower when they specialize. This is because specialized equipment and production methods tailored to the culture of specific crops and different types of animal husbandry normally give higher returns than can be obtained by product diversification. Consequently, farmers must be paid subsidies or provided with some other incentive to diversify their production and land use generally.

This chapter begins with a brief discussion of the various possible dimensions of agricultural diversity. It then considers the economics of paying for unmarketed agroecosystem services (supplied at the farm level) as a means for

Agroecosystem Diversity
https://doi.org/10.1016/B978-0-12-811050-8.00022-4

encouraging on-farm diversity. Conversely, economic issues involved in policies to promote on-farm product diversity as a means to enhance the supply of agroecosystem services are also examined. Subsequently, it examines the scope for doing this to promote regional differentiation of agroecosystems and other dimensions of agricultural diversity. Issues raised in the discussion are whether rewards or penalties are more appropriate for controlling agroecosystems to ensure greater social desirability of their streams of environmental services; and whether increased on-farm product diversity is likely to reduce the aggregate supply of agricultural products.

At the outset, it might be noted that most (but not all) public policies intended to alter agroecosystems for environmental reasons do not specifically target increasing their diversity but focus on changing particular environmental spillovers or externalities generated by such systems. Consequently, they may fail to increase agricultural diversity. This type of economic analysis is usually of a partial nature and, consequently, neglects holistic or system-related effects associated with increased agricultural diversity.

THE MANY DIMENSIONS OF AGRICULTURAL DIVERSITY

Diversity is a multidimensional concept and needs to be related specifically to an object. Furthermore, it can be measured in a variety of ways. Consequently, the analysis of agroecosystem diversity is a complex subject. Nevertheless, it is possible to tell whether some systems exhibit more diversity than others; for example, whether at the farm level a limited number of crops are grown or many.

One can consider the diversity of agroecosystems at the following levels:

- at the field level
- at the farm level
- within a region
- between regions at different geographic scales, for instance, nationally or globally

There are increasing concerns that agroecosystem diversity has declined in recent times at all these levels, primarily as a result of economic drivers (Tisdell, 2015, Chapters 5–8). The types of agricultural diversity that can be considered may be related to the following nonexhaustive attributes:

- genetic diversity
- landscape variations, that is, spatial patterns of agricultural diversity
- crop and livestock mixtures
- methods of cultivation and husbandry
- combinations of cultivated and uncultivated land, as well as natural areas on agricultural land afforded protection to conserve wild species, that is, variations in landscape architecture

Another dimension is the diversity of agricultural land use and variations in the intensity of its use with the passage of time (Zermeño-Hernández et al., 2016). For instance, following the Green Revolution, the frequency of annual multiple cropping increased considerably with the same type of crop being grown more times per year (Alauddin and Tisdell, 1991). This reduces agricultural diversity of land use as a function of time and can add to agricultural sustainability problems. It is made possible both by advances in plant breeding and the increased availability and affordability of off-farm inputs (Tisdell, 2015, Chapter 5).

It seems clear that the multifunctional nature of agriculture has declined in recent times as a result of all these developments and continues to do so (Kohler et al., 2014). The possibility of paying farmers for their unmarketed environmental services has been proposed (and to some extent, adopted) as a means to slow down or reverse this trend (Baylis et al., 2008). For example, the European Union's Common Agricultural Policy rewards farmers for the multifunctionality of their activities (see Tisdell and Hartley, 2008, pp. 76–80; Potter and Burney, 2002), and Japanese agricultural policy is influenced by this approach (Iizaka and Suda, 2010). Let us, therefore, consider payments to individual farmers for the supply of unmarketed environmental services.

AN ASSESSMENT OF POLICIES FOR PAYING FARMERS FOR UNMARKETED ENVIRONMENTAL SERVICES OR REWARDING THEM FOR ON-FARM DIVERSITY

Policies of paying farmers for unmarketed environmental services or rewarding them for on-farm diversity need to be examined carefully. First, one needs to consider the reasons for such payments. These reasons can include the following:

- Such payments will improve social economic efficiency, that is, ensure socially more desirable economic outcomes than otherwise as a result of improvements in the use or allocation of resources used in agricultural production.
- The payments will improve the distribution of income, if it is agreed that farmers are deserving of higher incomes than otherwise.

Payments for these services may be made directly to farmers for these services, or they may be paid indirectly. For example, if it is judged (as it is in Japan) that the growing of rice generates significant unmarketed environmental services, the supply of rice may be subsidized as a measure to sustain rice production to obtain the unmarketed environmental benefits associated with it (Iizaka and Suda, 2010). The desirability of this approach would, however, depend upon the techniques for rice growing not substantially changing so they become less environmentally friendly. In principle, the payment of economic rewards directly to farmers based on the value of unmarketed environmental services generated by each of them are more desirable than payments of an indirect nature. However, the cost of tailoring a payment system of this kind to each individual farm may be too high to make this policy practical from an economics point of view.

Payments Should Depend on the Net Value of Unmarketed Environmental Services Supplied

There is a danger that policies to reward farmers may only take account of the unmarketed environmental services they generate and ignore the external environmental disservices they are responsible for, or put too much weight on the former to the neglect of the latter for political reasons. If so, this approach will not result in a socially beneficial allocation of resources. For example, growers of palm oil trees point out that these trees sequester carbon, but to reward them purely on this basis would be to ignore the loss of alternative vegetation cover that would sequester more carbon and fails to take account of the loss of other environmental services (such as wild biodiversity) as a result of the growing of oil palm (see, for example, Swarna Nantha and Tisdell, 2009).

Payments May Not Promote On-Farm Diversity

Furthermore, payments for environmental services may be based on too narrow a range of services and may do little or nothing to encourage land use diversity by bio-industries. However, such practices are being advocated (see, for example, Kongsager et al., 2013). Payment for carbon sequestration, for example, may extend to monoculture plantations of exotic trees. In many cases, these have negative effects on local biodiversity and are fertilized with artificial fertilizers. This can elevate levels of nitrogen and phosphorous in runoff water and groundwater with negative environmental consequences. Too often, agricultural environmental policies are based on piecemeal approaches rather than system-based ones. Consequently, they can do more harm than good environmentally.

ECONOMIC REALITY, THEORY, AND PAYMENTS FOR ON-FARM PRODUCT DIVERSITY

Where it can be demonstrated that product diversity on farms adds significantly to unmarketed off-site economic benefits (because of its favorable economic consequences), paying farmers to maintain or increase this diversity may be warranted. However, as pointed out in the chapter 21 of this book, increased on-farm product diversity does not in itself improve environmental performance, and in higher-income market-dominated economies, it can result in a substantial fall in farm profits. Let us consider this matter further.

Increasing On-Farm Product Diversification Does Not Necessarily Increase Off-Farm Environmental Benefits, and in Higher-Income Countries, It May Reduce Farming Profitability

Leterme et al. (see Chapter 21 in this book) undertook a study of a sample of farms in different EU nations to determine whether mixed crop-livestock farming resulted in environmental benefits compared to specialized farming. They found that:

> It was the physical integration and complementarity between crops and animals (e.g. home-grown feed, recycling of wastes as fertilizers), not just their coexistence which was decisive for improving environmental performance at the farm level. Unfortunately, it increased also workload, degraded productivity and economic results.

Less Developed Economies, Transitional Ones, and the Effect of Market Systems on Integrated Mixed Farming Systems

In situations, however, where the market system is less developed, the availability of agricultural labor supply is greater and the effective real wage of labor is low. Therefore, mixed farming is more economic and tends to persist in several economies (e.g., China and Vietnam) undergoing market transition (Van Asten et al., 2011). Tisdell (2010) was involved in a joint study of the nature of pig farming in Vietnam that found that greater product diversification and integration of production was practiced more by smallholders of pigs than those with large holdings. The latter depended entirely, or almost so, on purchased inputs. These purchased inputs included a lot of grains imported from the Americas, grown on broadacre specialized farms most likely having significant adverse environmental impacts. Moreover, the transport costs would have added to the carbon footprint involved.

Despite the economic and environmental case for adopting economic policies favoring smallholders of livestock compared to large holders, the Vietnamese government (as well as the Chinese) has adopted policies to favor the latter. A more neutral stance appears to be called for. As economic growth proceeds and the market system becomes more pervasive, increased specialization by farmers and reduced integration of farming activities is to be expected (Tisdell, 2010). It is well known that growing market extension results in greater specialization in economic production. Economists have traditionally argued that market extension (and increased market competition) improves economic efficiency, enhances economic growth, and reduces economic scarcity. However, this ignores their likely negative environmental and ecologic impacts. These aspects are documented in Tisdell (2015).

ECONOMIC THEORY AND PAYMENTS FOR ENVIRONMENTAL SERVICES

Consider a policy of paying farmers to engage in an integrated on-farm product diversification as a means of increasing the supply of wanted off-farm environmental services. A simple cost-benefit model will highlight some of the economic issues involved.

Payments for the provision of such services may be judged from different points of view. Consider initially whether such payments might increase collective benefits, W, from purely a resource allocation viewpoint. Where Δ represents the change in the relevant variable, this policy might be desirable if for an individual farmer,

$$\Delta W = \Delta G - \Delta L - K > 0 \qquad (22.1)$$

where ΔL represents the loss in profit of the farmer, ΔG is the increased benefit to others of this intervention, and K is the administrative cost of implementing and overseeing the policy. The latter costs can be considerable, particularly if controls have to be tailored to the exact farm property involved. To make these controls economically viable, the administration of such schemes is likely devolved to local authorities. However, this can give rise to so-called principal and agent problems: local authorities may fail to institute policies in the way wanted by a central government. There is, for example, evidence of some slippage in management control in implementing China's "Grain for Green" program that is designed to increase the supply of off-farm environmental services in selected regions of China (Zheng et al., 2011; Tisdell, 2015, Chapter 16).

One should also consider who is to fund payments for an increase in the supply of desired off-farm environmental services. It is sometimes argued that the beneficiaries should pay. If so, they will need to be identified and the amount they ought to pay will need to be determined. Again, public administration costs will be incurred, for example, in collecting payments. A policy of paying for the supply of environmental services is likely to be less attractive the higher these costs are. Legal restrictions in some countries in making direct payments to landholders are also considerable (for a detailed study, see Pagiola et al., 2013). In many cases, it is the general taxpayer who pays rather than only the beneficiaries from schemes intended to reward farmers for maintaining or increasing their supply of environmental services.

A More Specific Economic Model

The model illustrated in Fig. 22.1 highlights some of the policy issues raised by deciding to pay farmers for on-farm product diversification. There, the variable x is an indicator of the extent of the integrated diversification of farm production, and 0 corresponds to no diversification at all, and 1 indicates its maximum. The extra (marginal) off-site benefits of on-firm diversification are indicated by the line DBE, and the rate of reduction in the profit of the focal farmer is

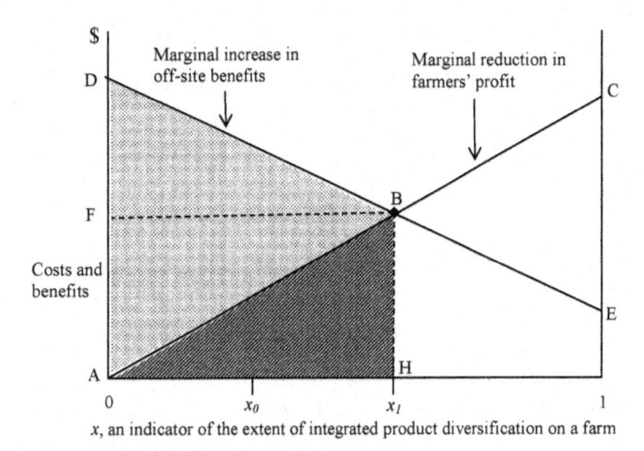

FIGURE 22.1 A diagram to illustrate the possible collective economic benefits of increased on-farm product diversification. It also highlights issues about how much compensation should be paid to farmers for increased product diversification.

indicated by line ABC. Taking into account these benefits and costs, this traditional economic approach implies that the level of on-farm product diversification equivalent to x_1 is socially optimal. At this point, the extra benefit to those living off-farm from the extra supply of environmental services by the focal farm as a result of greater product diversification equals the extra cost incurred (in terms of profit forgone) by the farmer. If initially farm production is completely reliant on one product, moving to the socially optimal position (from an allocative view) reduces the farmers' profit by an amount equivalent to the hatched area but results in an increase in off-farm benefits by an amount equivalent to this area plus the dotted area. Thus, ignoring distributional issues, the net increase in social economic benefit is equivalent to the dotted area.

How should the gains from increased product diversification be shared between the affected farmer and the off-site beneficiaries? In the case illustrated, the minimum amount that would need to be paid to the farmer to leave him/her no worse off than before the change would be an amount equivalent to the hatched triangle. This would distribute all the gains to off-site beneficiaries. If an amount equivalent to the hatched plus the dotted triangle were paid to the farmer, the farmer would secure all the collective gains of increased product diversification. Intermediate amounts of payment would result in varying degrees of gain for both the farmer and off-site beneficiaries.

The mechanics of the payment also needs to be considered. For example, the rate of payment for each unit of diversification may begin at a high rate and then decline, or start at a low rate and rise, or be at a constant rate. A payment at the constant rate AF in the case illustrated would result in a collective optimum. However, other types of payment can achieve the same result. The same result can even be achieved by a lump sum payment of not less than an amount equivalent to the hatched triangle, this payment being conditional on the farmer achieving a level of product diversification of x_1. The choice is likely to be influenced by the costs of administering the alternative forms of payment.

Note that if the focal farmer already is engaged in some integrated product diversity (for example, is at x_0 in Fig. 22.1) a smaller payment than otherwise would be needed (from an allocative perspective) to achieve a collective optimal degree of integrated on-farm diversification. If the focal farm is already at diversification level, x_1 (the social allocative optimum in Fig. 22.1), no payment would be justified on allocation grounds. Any payment would need to be justified on whether income transfers are warranted as a redistribution measure or to ensure economic justice (Seidl et al., 2002). There are many type of situations in which externalities exist, but they are irrelevant from an allocative point of view (see Tisdell, 1970, 2005, Section 3.2).

The type of traditional economic model shown in Fig. 22.1 is from a policy perspective "a magic wand" model because it does not take into account the public administration costs involved in arranging payments from agroecosystem services. When these are taken into account, such schemes become less attractive from an economic point of view. Furthermore, attention needs to be given to ensuring that the public administration of such a scheme is efficient to reduce the cost burden involved. In some cases, public administrative costs can be so high as to negate any collective benefits that would otherwise be obtained by intervention.

It may also be noted that a farmer may need to be paid more than the minimum amount indicated in Fig. 22.1 because the duration and level of payments are uncertain. This is particularly likely to be so if changing the farm product mix results in a degree of lock-in. Furthermore, the off-site benefit from integrated on-farm diversification is likely to show regional interdependence (see the previous chapter). This further complicates the analysis.

OTHER TYPES OF DIVERSITY OF AGROECOSYSTEMS

Agricultural diversity can be considered in several dimensions additional to on-farm product diversification. For example, it may be desired to have greater regional differentiation of agriculture or to sustain agricultural activities in some regions as a means of sustaining regional diversity of landscapes. The grazing of sheep on the Yorkshire Moors has resulted in a rather unique landscape that has tourist and cultural value. An economic case can exist for paying farmers in areas such as this to retain their agroecosystems (if they are in danger of disappearing) because of their value to tourists and their existence (cultural) value (Tisdell, 2009). In some regions in Europe, a bed tax has been levied on tourist accommodation to subsidize the maintenance of local agroecosystems that have touristic values. Otherwise, little or none of this value would be appropriated by local farmers.

Subsidies may also be paid to farmers for retaining or providing variations in on-farm landscapes that add to off-farm environmental services. For example, in upland areas, the building of contour embankments and the planting of hedges might be subsidized to reduce soil erosion and increase water retention resulting in off-site benefits. The retention of tree cover or the planting of trees specifically in selected positions such as along watercourses and on sloping land can yield significant off-site benefits. For instance, this may have hydrologic benefits, reduce soil erosion, and have favorable impacts on wildlife conservation. In areas subject to dryland salting (such as some parts of Australia), the retention or planting of trees can help to reduce this type of salting and result in off-site benefits. In other cases, farmers have been paid to retain many of the original vegetation and ecosystem characteristics. The biodiversity of such systems is also usually very high. Some of the examples of such landscapes include the Montado and mosaic systems from Portugal and the Spanish dehesa (Merckx and Pereira, 2015).

Another important consideration in relation to diversity of agroecosystems is their potential role in helping to conserve agricultural genetic diversity, as well as wild biodiversity. Some varieties of crops have the best chance of survival if they are grown in the area where they were developed rather than *ex situ* or preserved in seed banks (Tisdell, 2016). It may be desirable to subsidize the retention of some local agroecosystems with the purpose in mind of supporting the conservation of agricultural genetic diversity.

From an economic point of view, all of these possibilities should be subjected to social cost-benefit analysis. However, this analysis can be a costly exercise and does not usually yield categorical results. It also involves particular value judgements but usually makes no judgements about the desirability or otherwise of income distribution changes resulting from its use. It should be regarded as a supplement to policy-making rather than a final arbiter of this.

DISCUSSION

A possible alternative to rewarding farmers for undertaking socially desirable diversification of agroecosystems would be to penalize them for not doing so. In some cases, penalties (taxes or fines) can have the same allocative results as rewards (payments) (Tisdell, 2005, Section 3.3). However, these are not always politically acceptable or not practical alternatives. For instance, penalties are not a political or practical alternative if farmers are already poor. Penalties will merely add to their poverty and could lead in cases where farmers who are already near the poverty line to being placed in a desperate situation. Furthermore, views about economic justice, particularly property rights, are likely to affect decisions about whether a carrot or stick policy approach to agricultural diversification should be adopted. Some farmers in Australia, for example, believe that it is an infringement on their property rights not to be able to remove trees on their properties without a permit. They resent the possibility of not being allowed to remove trees and also being penalized for their illegal removal.

One of the consequences of adopting policies to increase greater on-farm product diversification could be to increase the market price of agricultural products. The outcome depends on how the aggregate supply schedules of the affected products are shifted. If, overall, these schedules are negatively affected because agricultural productivity as a whole declines, the real price of agricultural products will rise due to their reduced supply. On the other hand, if it can be shown that the effects on aggregate agricultural supply will be positive, the opposite outcome is to be expected. It is a matter for agricultural science to determine the productivity shifts that are likely to occur. If the first mentioned consequence prevails, it is necessary to convince consumers that the unmarketed environmental advantages of on-farm product diversification exceed the extra cost to them of agricultural products.

It is probable that measures to increase on-farm agricultural diversification yield a net social benefit in some environments but not in all. Therefore, the applicability of these policies tends to be situational. Furthermore, the time dimension needs consideration. While policies to encourage agricultural diversity could have a negative effect on agricultural supplies in the short run, they may play a positive role in sustaining agricultural productivity in the long run. As a result, the real prices of agricultural products in the long term might be lower than they otherwise would be. The aggregate productivity effects and the time pattern of these depend on biophysical relationships and need to be addressed by agricultural scientists. Economic methods are unable to provide guidance on these matters, although the predicted biophysical changes and their consequences for economic performance can be utilized in economic analysis.

CONCLUSIONS

While paying farmers for the supply of unmarketed agroecosystem services to promote agricultural diversity or conversely paying farmers to increase agricultural diversity to improve the supply of unmarketed agroecosystem services seems appealing, doing so raises many issues that may be overlooked at first sight. Furthermore, assessing such policies is difficult because agricultural diversity is a multidimensional concept. The relevant dimension(s) of agricultural diversity needs to be specified when examining policies for increasing agricultural diversity as a means to enhance the supply of unmarketed agroecosystem services. The possible role of on-farm product diversification as promoting the supply of these services was given particular attention, relying on a simple social cost-benefit formula and a specific model. This analysis enabled matters of allocative efficiency to be distinguished from those involving the distribution of income. It was emphasized that an important consideration that needs to be taken into account in assessing agricultural policies is the public administration costs and legal barriers involved in implementing them and the cost effectiveness of the interventions. This is often given no consideration in economic modelling. For a detailed discussion of some of these issues, see Pagiola et al. (2013). Consequently, "a magic wand" approach is adopted.

In the discussion, it has been pointed out that agricultural policies designed to improve environmental outcomes can be based on the provision of rewards or the imposition of penalties on agriculturalists. The choice between these alternatives is liable to be determined by income distribution considerations, views about justice (for example farmers' property rights), and political considerations.

The aggregate effect on the supply of agricultural products of policies designed to increase on-farm product diversification seems not to be well known. It is possible that policies implemented to increase greater on-farm product diversity (even if their on-farm supply is integrated) could reduce the aggregate supply of agricultural products in the short run. On the other hand, if the on-farm supply of these products is appropriately integrated, it could help sustain their supply in the long run. There is a need for more scientific evidence about the biophysical effects involved, taking into account holistic considerations.

References

Alauddin, M., Tisdell, C.A., 1991. The 'Green Revolution' and Economic Development: The Process and Its Impact in Bangladesh. Macmillan, London.

Baylis, K., Peplow, S., Rausser, G., Simon, L., 2008. Agri-environmental policies in the EU and United States: a comparison. Ecological Economics 65 (4), 753–764. https://doi.org/10.1016/j.ecolecon.2007.07.034.

Iizaka, T., Suda, F., 2010. Making device for sustainable agricultural systems: a case study of Japanese farmers' markets. In: Bonanno, A., Bakker, H., Jussaume, R., Kawamura, Y., Shucksmith, M. (Eds.), From Community to Consumption: New and Classical Themes in Rural Sociological Research (Research in Rural Sociology and Development, Volume 16), Bingley, UK. Emerald Publishing Group Limited, pp. 171–184.

Kohler, F., Thierry, C., Marchand, G., 2014. Multifunctional agriculture and farmers' attitudes: two case studies in rural France. Human Ecology 42 (6), 929–949. https://doi.org/10.1007/s10745-014-9702-4.

Kongsager, R., Napier, J., Mertz, O., 2013. The carbon sequestration potential of tree crop plantations. Mitigation and Adaptation Strategies for Global Change 18 (8), 1197–1213. https://doi.org/10.1007/s11027-012-9417-z.

Merckx, T., Pereira, H.M., 2015. Reshaping agri-environmental subsidies: from marginal farming to large-scale rewilding. Basic and Applied Ecology 16 (2), 95–103. https://doi.org/10.1016/j.baae.2014.12.003.

Pagiola, S., von Glehn, H.C., Taffarello, D., 2013. Brazil's Experience with Payments for Environmental Services, Payments for Environmental Services (PES) Learning Paper No. 2013-1. World Bank Group, Washington D.C. Available at: http://documents.worldbank.org/curated/en/554361468020374079/Brazils-experience-with-payments-for-environmental-services.

Potter, C., Burney, J., 2002. Agricultural multifunctionality in the WTO—legitimate non-trade concern or disguised protectionism? Journal of Rural Studies 18 (1), 35–47. https://doi.org/10.1016/S0743-0167(01)00031-6.

Seidl, I., Tisdell, C., Harrison, S., 2002. Environmental regulation of land use and public compensation: principles, and Swiss and Australian examples. Environment and Planning C: Government and Policy 20 (5), 699–716.

Swarna Nantha, H., Tisdell, C.A., 2009. The orangutan-oil palm conflict: economic constraints and opportunities for conservation. Biodiversity and Conservation 18, 487–502. https://doi.org/10.1007/s10531-008-9512-3.

Tisdell, C.A., Hartley, K., 2008. Microeconomic Policy: A New Perspective. Edward Elgar, Cheltenham, UK and Northampton, MA, USA.

Tisdell, C.A., 1970. On the theory of externalities. The Economic Record 46 (113), 14–25.

Tisdell, C.A., 2005. Economics of Environmental Conservation, second ed. Edward Elgar, Cheltenham, UK and Northampton, MA, USA.

Tisdell, C.A., 2009. Complex policy choices regarding agricultural externalities: efficiency, equity and acceptability. In: Beckmann, V., Padmanabhan, M. (Eds.), Institutions and Sustainability. Political Economy of Agriculture and the Environment - Essays in Honour of Konrad Hagedorn. Springer, pp. 83–106.

Tisdell, C.A., 2010. The survival of small-scale agricultural producers in Asia particularly Vietnam: general issues illustrated by Vietnam's agricultural sector, especially its pig production. In: Salazaar, A., Riva, I. (Eds.), Sustainable Agriculture: Technology, Planning and Management, New York. Nova Science Publishers, pp. 315–328.

Tisdell, C.A., 2015. Sustaining Biodiversity and Ecosystem Functions: Economic Issues. Edward Elgar, Cheltenham, UK and Northampton, MA, USA.

Tisdell, C.A., 2016. Genetic loss in food crops in the Pacific: socio-economic causes and policy issues. The Journal of Pacific Studies 36 (2), 23–40.

Van Asten, P.J.A., Wairegi, L.W.I., Mukasa, D., Uringi, N.O., 2011. Agronomic and economic benefits of coffee–banana intercropping in Uganda's smallholder farming systems. Agricultural Systems 104 (4), 326–334. https://doi.org/10.1016/j.agsy.2010.12.004.

Zermeño-Hernández, I., Pingarroni, A., Martínez-Ramos, M., 2016. Agricultural land-use diversity and forest regeneration potential in human- modified tropical landscapes. Agriculture, Ecosystems and Environment 230, 210–220. https://doi.org/10.1016/j.agee.2016.06.007.

Zheng, H., Glenwe, P., Polasky, S., Xu, J., 2011. Reputation, Policy Risk, and Land Use: A Study of China's 'Grain for Green' Programme, UNU-WIDER Working Paper, WP 2011/39. World Institute for Development Economics Research, United Nations University, Helsinki.

CHAPTER

23

Agricultural Policies and the Reduction of Uncertainties in Promoting Diversification of Agricultural Productions: Insights From Europe

Aude Ridier[1], *Pierre Labarthe*[2]

[1]Agrocampus Ouest, UMR SMART LERECO, Rennes Cedex, France; [2]INRA-SAD, UMR AGIR, Castanet-Tolosan Cedex, France

Product diversification is often presented as an evolution of farm production systems that could contribute to a more sustainable development of agriculture (Kremen and Miles, 2012). Diversifying products at farm level may come with benefits for farmers, and beyond, for society at large, by enabling a more efficient use of natural resources. But diversification also raises complex issues at the farm level, with multiple dimensions: technological, process, marketing, and organizational. It is thus a complex change for farmers, not only in terms of internal management of resources within the farms, but also in terms of relations with their sociotechnical environment, both with actors of the supply chains and with actors of agricultural knowledge and innovation systems (AKIS). Such deep changes can induce situations of strong uncertainty for farmers regarding diversification options: uncertainty about the costs and benefits of diversification, but also about the paths toward diversification. Thus, reducing uncertainty might be a key step toward diversification.

The question of the role of uncertainty in technological choices such as product diversification has gained an increasing attention among scholars in economics and social sciences (see, for instance, Marra et al., 2003; Knowler and Bradshaw, 2007; Chavas and Di Falco, 2012). This is true both in the field of microlevel analyses within standard economics and of mesolevel analyses within institutional economics. Our contention in this paper is twofold: (1) to show that these two fields of economic studies can be associated with different forms of public intervention for the support of diversification and (2) to open a debate on the need to better integrate and combine these different forms of public interventions.

Agroecosystem Diversity
https://doi.org/10.1016/B978-0-12-811050-8.00023-6

The chapter is organized as follows. In a first section, we present insights from a microeconomic perspective on the costs and benefits of product diversification. In a second section, we present insights from institutional economics on technological transitions toward diversification. In both sections, we present the type of policy instruments associated with these theories. In a third section, we propose some illustrations of these policies based on a description of instruments of the Common Agricultural Policy (CAP) of the European Union (EU). It highlights the need for more pluralistic and open debates about the instruments derived from various economic theories. A better dialogue could help prioritizing the instruments that would have the most impact according to the contexts of agricultural production.

THE COSTS AND BENEFITS OF PRODUCT DIVERSIFICATION: THE MICROECONOMIC PERSPECTIVE

In this section, we browse advances in standard microeconomics about the benefits or supplementary costs that farmers can derive from product diversification. We also want to show how elements of a standard microeconomics approach can draw the link between farm diversification and the management of uncertainties.

The Farm Multioutputs Production

The farm microeconomics underline many specificities that differentiate a farm from a classical firm. One of them is the multiproduct character of farms "by nature." Farms produce many marketable and nonmarketable goods and services in the same production process. The joint production of multiple outputs is responsible for complementarities that can enhance the productivity of farming activities and cause economies of scope (Baumol et al., 1982; Chavas and Di-Falco, 2012; Bowman and Zilberman, 2013). For instance, manure from livestock increases soil fertility and then land productivity. The crop diversification on farms that results from crop successions has some benefits on soil structure, nutrient stocks, seed stocks, crop pest population, and weed development. Also, when it concerns legumes sown as precedent or as intercrop, crop diversification can provide a natural source of nitrogen fertilization. Legume crops fix atmospheric nitrogen in soil for intercrops or for successive crops. This type of crop sequencing can be described as the production of ecosystemic services. The recognition and precise quantification of these has been investigated as a means of adequately calibrating agri-environmental support policies (Shumway et al., 1984; Peerlings and Polman, 2004; Sauer and Wossink, 2012). The promotion of appropriate support remains highly discussed. While agri-environmental policies are based on a cost-compensation principle, all the farms do not bear the same costs and benefits when diversifying. Thus, diversification is not a homogeneous process among farms. Empirical evidences in developed countries have, for instance, shown that economies of scope declined with farm size: farms have become larger and more specialized on average, so economies of scale and economies of scope are, in many cases, antagonists (Chavas and Aliber, 1993).

Diversification and the Issue of Labor Management

Farm diversification can be a response to problems of managing labor peaks and may facilitate the planning of equipment use. Diversification in products falling under different sowing and harvesting seasons is a means of decreasing labor peaks, especially when the labor resource is very limited. However, the simplification and rationalization of farming practices though the removing of some tasks (like preferring superficial or no-tillage to deep plowing) is also a serious ground for farm specialization, especially when the family labor resource is scarce or when the farm size per labor unit gets high. Farms with little family labor or part-time farmers tend to adopt less time-consuming technologies. But the labor resource might be variable from one farm to the other, and the incentive to diversify is deeply linked to the scarcity of labor resource. The marginal cost of labor is also variable among farms because of many factors: field pattern (with different plot accessibility and transport costs), productivity of machines, family labor endowment, manager skills, etc. This heterogeneity is source of different levels of motivation to diversify among farms.

Concerning labor management, diversification in products often means a diversification in tasks and the need for diversified skills. The microeconomic theory points that firms specializing in a single product, for which they have the greatest relative advantage or the least relative disadvantage given the physical and biologic factors, become more productive than diversified firms that manage several products (Castle et al., 1987). Indeed, a diversified farm also has to manage more complexity and a larger number of technologies and skills. Other studies show, for instance in developing countries, that diversification can favor labor complementarity inside groups and family. It increases the portfolio of activities, which globally enhances the livelihood. The concept of livelihood includes noneconomic attributes, such as social relationships and institutions that facilitate the access to economic growth (Ellis, 2000).

Diversification and Risk Management

The distinction made by economists between risk and uncertainty in the early 2000s is based on the predictable or unpredictable nature of outcomes. While risky situations are calculable and characterized by several possible outcomes associated with several levels of probabilities, uncertain situations are unknowable and unpredictable. Since then, many tools have been developed in two directions to support decision makers even in unpredictable situations. Firstly, building on research on cognitive psychology, a large share of the economic literature has focused on the elicitation of agents' behavior and on the characterization of their level of acceptance of risk (risk aversion) (Binswanger, 1980 for a beginning; Reynaud and Couture, 2012 for a survey). Secondly, a much less developed research stream on decision theory has sought to develop the theories and methodologies of risk assessment, to get a better knowledge of individual probabilities and individual judgments (Norris and Kramer, 1990; Nelson and Bessler, 1989). The assessment of risk and probabilities remains a complex subject matter. Historical data on past events may bring information on frequencies (like the frequency of floods in some regions, or the price fluctuation on some markets). But the challenge remains to better integrate in microeconomic models the fact that different decision makers may have different beliefs about the occurrence of this event in the future. Departing from a purely frequentist view on risk, the probability can be seen, like Savage suggested, as means of measuring subjective statements and the degree of beliefs about given events (Savage, 1972). In such approaches, risk is often represented as variations (upward and downward) around an average value. The most widespread conception of risk inherent to farming activities is linked to downside risk, i.e., the deviation from a "normal" value that would lead farmers to worse incomes. The uncertain events linked to farming activity in general are weather and market conditions that can diminish products volume or quality, inducing income losses. On the opposite, the uncertainty due to market rising prices might also give rise to improvements of farmers' situation.

In this case, uncertainty can also represent an opportunity for farmers.

Farmers have to deal with these situations. Following a Markowitz portfolio strategy, the diversification of farms can be seen as a risk management strategy that a risk-adverse farmer can implement to reduce the impact of fluctuating incomes (Markowitz, 1991). Firm diversification is a method of reducing income variability but only under certain conditions. First, the correlation of stochastic returns in the different products has to be low (Bromley and Chavas, 1989). Second, the risk-reducing effect of diversification is diminished as the number of assets is increased. Third, there is a trade-off between seeking to diminish mean income variability and seeking to increase mean income, as economies of scale, which reduce the mean cost of production and increase mean income, are favored by specialization (Robison and Barry, 1987).

The policy regulations of the farming sector in developed countries have been built upon the different elements of the microeconomics framework presented earlier. The allocative income support to farmers, implemented in the EU in 1992, was conceived to compensate for sector income losses. The counter-cyclical payments in the US Farm Bill aim at buffering market uncertainties. The decoupling and the targeting of farmers' direct payments in the EU, since 2003, was conceived to better reach the horizontal (social and environmental) objectives of the CAP while minimizing market distortions. Some of these policy instruments will be presented in Agricultural Policies and Diversification: Illustration from Europe section.

TRANSITIONS TOWARD PRODUCT DIVERSIFICATION: THE MESOECONOMIC PERSPECTIVE OF INSTITUTIONAL ECONOMICS

Overcoming the uncertainty and asymmetries of knowledge associated with diversification is thus an emerging area of research in standard microeconomic analyses. Such issues are at the very center of other economic approaches (Magrini et al., 2016), mainly institutional economics including researches embedded in evolutionary theory (Cowan and Gunby, 1996). This theory focuses on relations between economic actors involved in production (here, farmers) and other actors of knowledge and innovation systems (e.g., advisory services suppliers, applied research organizations, knowledge brokers, etc.). It offers an alternative view of the mechanisms explaining how diversification might be locked in or locked out from the technological evolution of the agricultural sector.

Product Diversification as a Radical Innovation?

In an evolutionary perspective, diversification can be represented as an innovation at farm level. It can even be considered that diversification combines different dimensions of innovation. As highlighted in the previous section, diversification implies major changes at the farm level in the use of resources and technologies by farmers. But it cannot be restricted to a product and/or process innovation. Diversification may also come with the need of marketing innovations, to articulate the needs of consumers for new products, and to develop market niches. The diversification toward the introduction of legume crops within cropping systems is a typical example (Magrini et al., 2016). Moreover, diversification might induce strong organizational changes for farmers. These changes might be internal, with new patterns of farm labor management, but also external, with new networks of relations with their sociotechnical environment. Growing new crops might, for instance, require finding new input providers (for seeds, etc.) or retailers giving access to niche or emerging markets. It may also require seeking new sources of knowledge, including advisory service organizations, but also new networks of farmers that can support the diversification process through shared knowledge and joint experiments or learning.

In short, product diversification can in many cases be pictured as a systemic or radical innovation in either one or different dimensions of innovations (McDermott and O'Connor, 2002). There is a growing agreement among scholars that such radical innovations are associated with high levels of uncertainty for users (Popadiuk and Choo, 2006). In other words, the transition toward diversification is partly determined by the dynamics of knowledge production that may reduce the uncertainty associated with its adoption. This relates to the old debate on the relations between the dynamics of innovation on the one hand and of knowledge production on the other hand. A typical illustration of these debates stems in the analysis of lock-in situations in various economic sectors.

Diversification, Uncertainty, and Knowledge Asymmetries

The debate on lock-in originates from the seminal work of David (1985), Arthur (1989) and Cowan (1990), but also from the critical and constructive analysis addressed by Liebowitz and Margolis (1990) to this concept. Technological lock-in can be defined as situations where a technology A can be adopted in the long term, or even irreversibly, to the detriment of a technology B, even though the technology B appears, ex-post, to be the most effective. Some authors have described the low level of diversification as a type of lock-in situation, for instance in the French context (Magrini et al., 2016). There are two key ingredients that lead to lock-in situations: uncertainty in the initial choices between technologies, and cumulative effects and increasing returns in the adoption of one technology. Increasing returns means that the value of adopting a technology increases with the number of users of the technology. This situation is partly explained by the material dimension of technologies, i.e., by economies of scale in their production. But increasing returns are also associated with a very strong cognitive dimension.

Indeed, more users adopting a technology may result in more knowledge being available on it. This results in growing knowledge asymmetries to the detriment of alternatives to the dominant technology. Cowan (1990) distinguishes two types of knowledge in these situations of asymmetries. The first dimension relates to *learning by doing*: the more people use a technology, the more we know how to use it. In other words, growing networks of users can reduce the uncertainty of using a technology. The second dimension relates to *learning about payoffs* that is the production of evidence about the outcomes of adopting a technology. This dimension is linked to practice, but also to R&D investments. The existence of such asymmetries of knowledge has been demonstrated in the case of agriculture (Cowan and Gunby, 1996).

As a result, evolutionary economists often plead for public investments in knowledge production on promising alternatives to support their development and overcome lock-in situations on technologies that may have adverse effects on human health or the environment (Cowan and Hultén, 1996). There is also a growing agreement on the collective dimension of this knowledge production. But there are vivid debates about how to best support this collective effort for producing knowledge on technological alternatives.

Conceptions of the Public Support to the Collective Production of Knowledge: Process Versus Infrastructure View

The mechanisms of the collective production of knowledge are central to evolutionary economics. In this theory, users select technologies and implement innovation on the basis of knowledge bases (Dosi, 1988), shared by different actors of a given sector and/or a given country. In agriculture, the cumulative construction of these knowledge bases implies farmers, advisors, engineers, researchers, etc.

If the collective dimension of knowledge production is agreed upon, there are now debates about the scale at which they can be supported. Historically, the focus was put on the identification of formal collective organizations for knowledge production. This was embodied by the settlement of the concepts of national innovation systems (Lundvall, 1992) or sectoral innovation systems (Malerba, 2002), describing the investments and systemic relations between organizations set to produce knowledge from academia to practice, respectively in various countries or sectors. In the agricultural sector, this idea was formalized with the concept of AKIS (Röling and Engel, 1990).

More recently, other researchers put an emphasis on other forms of collective organization for knowledge production, more local and/or bottom-up, using different terminologies: communities of practices, epistemic communities (Cohendet et al., 2014), niches (Geels, 2002), or learning innovation networks (Moschitz et al., 2015). Such discussions are embedded in a multidisciplinary perspective, even though the question of collective learning is also a matter of an ongoing debate among economists (Capello, 1999).

More globally, two main conceptions of agricultural innovation systems compete in the academic community (Klerkx et al., 2012): (1) an infrastructural view on the system, focusing the analysis on the presence and investments of actors (e.g., research institutes, advisory organizations, financing organizations) and the institutions that govern their behavior and interactions (Sorensen, 2011), and (2) a process view of the systems, which implies seeing innovation systems as self-organizing growing networks of actors connected to the development of a certain novelty (Ekboir, 2003). We can observe a real change in the agricultural sector, with an advocacy to switch from funding knowledge infrastructures toward supporting innovation processes (Brunori et al., 2013). Section Agricultural Policies and Diversification: Illustration from Europe will illustrate how EU's CAP is a typical example of this change in policy.

AGRICULTURAL POLICIES AND DIVERSIFICATION: ILLUSTRATION FROM EUROPE

Considering the different issues exposed in the preceding sections, we want to show how different policy instruments, including both sectoral and horizontal instruments such as agricultural sectoral policies and R&D structural policies, aiming at different objectives, can impact farmers' decisions to diversify their agroecosystems. Examples and illustrations are taken in the UE policies experiences (first and second pillars of CAP, European Innovation Partnerships, Farm Advisory Systems). The incentives will be considered at two different scales: first at the farm (micro) level and secondly at the mesolevel of the different institutions involved in agricultural knowledge and innovation networks or systems.

Incentives at a Microlevel: Supporting the "Cost" of Diversification

Uncertainty can cause a severe break in diversification adoption. The analytical framework of the public economy has neglected the issue of uncertainty and risk aversion of economic actors for a long time. The European CAP has deeply evolved since the beginning of the 21st Century to incorporate various and sometimes conflicting objectives. The successive reforms of CAP since 1992 have been driven by the willingness to increase market support orientations of agriculture in a historically highly regulated sector. The "first pillar" of the CAP represents on average 71% of the overall CAP budget 2014–20, but the ratio is variable in the different member

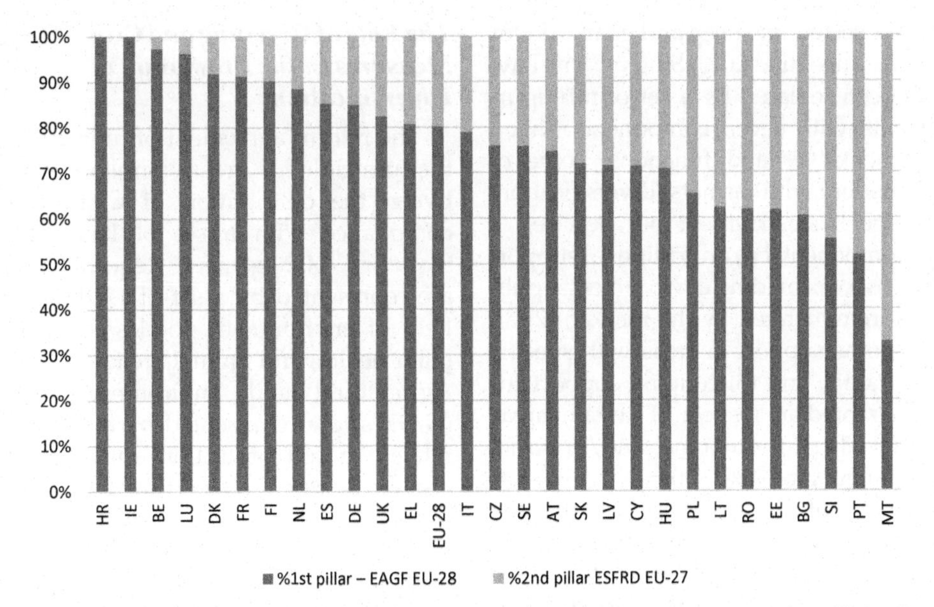

FIGURE 23.1 CAP expenditure by Member State 2014. *EU Parliament (2016). Based on European Commission, Eurostat (2016), Statistical Factsheet, Agriculture and Rural Development (updated: January 2014).*

states (Fig. 23.1). This CAP first pillar is now composed of a series of subsidies decoupled and coupled to products. They target different aims. The decoupling of direct payments from products has followed the main objective of making farmers' decision process more market oriented, whereas the convergence mechanism implemented by the last reform (2014–20) aims at raising equity in the distribution of agricultural subsidies across and within the 28 Member States. Meanwhile, some policies have remained coupled, notably, specific support to different sectors (livestock, protein crops, etc.). This decision of coupling or decoupling subsidies from products has some strong implications for the support of diversification decisions.

The Role of Coupled and Decoupled Payments in Diversification and in the Management of Uncertainties

Even if the reduction of risk and uncertainties has not always been directly targeted, the support brought by the CAP in Europe has been demonstrated to have both insurance and wealth effects on farmers' decisions. Thus, the role played by farm subsidies on hedging can benefit farmers' income in risky situations and push them to take riskier decisions (Hennessy, 1998). This is an indirect impact of CAP on risk exposure and risk behaviors, which can directly impact farmers' decisions on the level and the combination of productions in their farms.

In the EU, since the 2003 reform and following the 2009 CAP Health Check, the main part of direct aid has been decoupled and transferred to a new single payment scheme. This ensured that direct payments are compliant to WTO agreement. A Single Farm Payment has been distributed to eligible areas subject to certain conditions (notably cross-compliance), and CAP first pillar subsidies have become more and more targeted toward environmental objectives. Additional coupled payments have also been maintained for strategic purposes (through article 68): support to organic farming, to tobacco, to the insurance system; direct support to certain field crops like durum wheat, protein crops, rice, nuts and energy crops; specific coupled payments to the livestock sector, for sheep and goats and for beef and veal. Since June 2013, Single Farm Payments have been replaced by a set of multipurpose payments with seven components linked to different objectives and functions. Some of them may interfere with farmers' diversification: (1) next to the basic payment per hectare,[1] a "greening" component, i.e., an additional support to offset the cost of providing environmental public goods not remunerated by the market; (2) an additional income support in areas with specific natural constraints; and (3) coupled support for production, granted in respect of certain areas or types of farming for economic and/or social reasons (EC, Regulation 1307/2013).

The pursuance of coupled support with large allocative effects shows the willingness to support specific sectors and industries. It can be interpreted as a possibility given to member states to drive their own agricultural policy toward specific goals to build their own competitive advantage and, as a consequence, to shape their farming sector with production systems that are more or less diversified. The share of each Member State's envelope for direct support that can be used for coupled support is limited to 10%–15% of the annual national ceiling for direct payments (Fig. 23.2). Between 2015 and 2020, from the € 41–42 billion per year available to direct payments in EU-28, about 10% have been earmarked for recoupling (more than 60% for the beef and veal and milk and milk products sectors).

The Role of Greening and Cross-Compliance Measures in the Promotion of Diversification

The reform agreement of June 2013 extended the principle of cross-compliance system between the distribution of CAP support and compliance with a set of basic rules. First, Statutory Management Requirements (SMRs) are represented by a set of 12 directives in the field of environment, food safety, animal and plant health, and animal welfare. Second, Good Agricultural and Environmental Conditions (GAECs) are represented by a set of standards related to soil protection, maintenance of soil organic matter and structure, avoiding the deterioration of habitats, and water management. The possibility of receiving a green complementary payment in addition to each basic payment per hectare (up to 30% of EU's countries direct payment budget) is subject to the implementation of practices that benefit the environment, such as (1) diversifying crops[2]; (2) maintaining permanent grassland[3]; or (3) dedicating 5% of arable land to "ecologically beneficial elements." These ecologic focus areas include fallow land, field margins, hedges and trees, buffer strips, and even areas covered by catch crops or nitrogen-fixing crops.

The huge share of support potentially concerned by cross-compliance since 2006 has made this policy often considered a more powerful instrument to orient farmers' practices than the agro-environmental second pillar subsidies (Desjeux et al., 2011). If the implementation of

[1] The level is to be harmonized according to national or regional economic or administrative criteria and subject to an "internal" convergence process.

[2] Up to 30 ha: farmers have to grow at least two crops, and the main crop cannot cover more than 75% of the land; over 30 ha: farmers have to grow at least three crops, the main crop covering at most 75% of the land and the two main crops at most 95%.

[3] National or regional governments must also maintain the ratio of permanent grassland to the total agricultural area. This must not fall by more than 5% compared to the reference year.

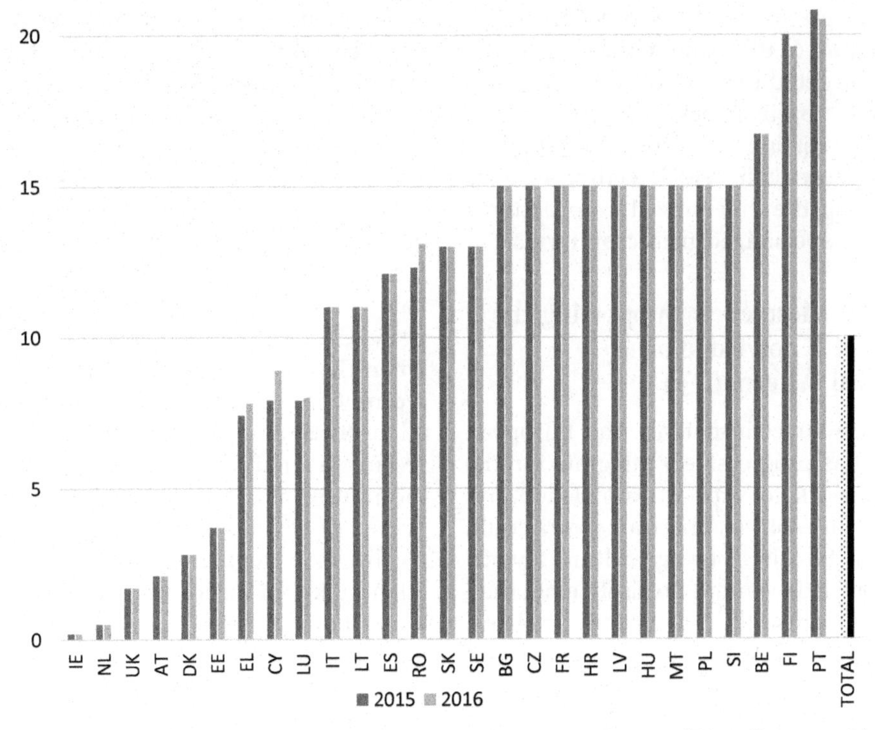

FIGURE 23.2 Financial allocations 2015–16 of voluntary coupled support in Percent of Direct Payments Scheme per Member State. *Source: European Commission, Information Note June 2016.*

the cross-compliance system is truly binding, it presents a powerful lever for diversification of crop rotation and maintenance of ecologic structural elements favorable to landscape biodiversity and to long-term resilience of production systems. A high variety of minimum requirements have been adopted in the EU, and the issue is to appreciate their degree of bindingness, considering the current practices. Some authors point out that since the minimum requirements are not binding enough, the bulk of CAP direct payments is still not targeted enough toward high nature value farming. They plead for more appropriate policy instruments, notably, to better support the maintenance of basic habitat like pastures (Oppermann et al., 2012).

The Coordination of First and Second Pillar CAP Subsidies to Support Farm Diversification

Next to specific coupled subsidies and to the greening complementary payment, agri-environment measures provide support, on a voluntary basis, to farmers who engage in long-term environmental commitments linked to the preservation of natural resources and landscape. The measures are cofinanced by member states and represent up to 22% of the CAP budget for rural development. The disposals can be seen as a reinforcement of the first pillar scheme, both aiming at the same objectives but with different means (mandatory vs. voluntary). For instance, in France, the climatic agri-environmental schemes in favor of farm extensification go in the same direction as the coupled program of the protein plan. It consists in measures targeted to practices on field crops and to mixed crop-livestock farms. It aims at improving the existing practices by reinforcing the share of pasture in livestock feeding systems and by favoring the diversity of crop patterns over 3 years in specialized field crop farms.[4]

Regulation at Mesolevel: Supporting the Production of Knowledge on Technological Alternatives

Besides microlevel incentives, the European Commission has progressively integrated in the CAP a framework to support knowledge and innovation at a mesolevel. This framework could play a role in transition of agriculture toward diversification. This is more precisely embodied in two instruments: the regulation on Farm Advisory Systems (FAS) on the one hand, and the European Innovation Partnership for Agricultural productivity and Sustainability (EIP-AGRI) on the other hand. These policy instruments are the expression of a change toward the support of innovation process rather than of the infrastructures of AKIS.

The FAS Regulation: Supporting Farmer Demand for Advisory Systems

The FAS regulation is linked to the cross-compliance of the CAP, which is the obligation for farmers to comply with a set of standards (SMRs, GAECs, etc.), regarding environment or food and safety security. If farmers do not respect these standards, they will receive penalties in the forms of reduced levels of direct aid received from the first pillar of the CAP. The main objective of the FAS regulation is to guarantee that farmers in every Member State will benefit from all the knowledge and information needed to comply with the set of standards composing cross-compliance. Concretely, each Member State of the EU had to certify some advisory organizations entitled to deliver advisory services to farmers about cross-compliance issues. In turn, farmers using these services can be refunded their expenses thanks to EU budget. Indeed, Member States have the choice to activate measure 114 of the EAFRD (second pillar of the CAP) to fund farmers' use of FAS services when charged to them. The scope of the FAS has recently been enlarged to integrate more globally all aspects of the CAP regulation, including recent diversification incentives presented in Incentives at a Microlevel: Supporting the "Cost" of Diversification section (EC, 2010).

The FAS regulation is an example of an attempt for more integrative policy at the EU level. It was the first policy instrument connecting the first and second pillar of the CAP. Moreover, it is also a policy that potentially targets the issue of uncertainty. It focuses explicitly on the production of knowledge to help farmers in complying with standards. As for the direct payments described in the former paragraph, the question of diversification is rather integrated in an indirect way.

The FAS regulation also embodies a change in supporting AKIS, with a focus on refunding farmers' demand rather than funding AKIS infrastructures (e.g., advisory organizations). But the effects of this regulation were limited (ADE, 2009), as the actual number of farmers effectively using the services has remained rather low, and as there are some very strong divergences in the levels of implementation of this policy across countries. An assessment of this measure (ADE, 2009) showed that less than 5% of farmers receiving direct payments

[4] Up to four different crops the second year and five different crops the third year, with minimum 5% of legume crops the second year. Obligation to implement crop rotations on the farm plots engaged. Limitation of pest treatments. The French climatic agri-environmental measures range from €50 to €900 per hectare.

have benefited from FAS, with variation between 0% (e.g., in France) and 30% (e.g., in United Kingdom) across countries. Most of the countries have decided not to use second pillar money to support the FAS regulation, and three countries only (Spain, Italy, and Poland) represented altogether more than 70% of the EU spending on FAS. An explanation often advanced is that the regulation failed to really connect the production of knowledge on cross-compliance with the more global needs of knowledge for farmers in their decision-making process (Koutsouris, 2014).

The EIP Regulation: Supporting Bottom-Up Initiatives for Knowledge Production

The EIP-AGRI has been launched in 2012. It brings together innovation actors (farmers, advisors, researchers, businesses, NGOs, and others). There are two key instruments within this policy: the operational groups (OGs) and the thematic networks. OGs are groups at the local level with people from different backgrounds (such as farmers, researchers, advisors, businesses, NGOs). They are set up to find an innovative solution to a shared problem or to test an innovative idea in practice. Both the size and composition of the OG depend on the project. Therefore, one OG can look completely different from another. Thematic networks are innovative projects, funded through Horizon 2020 (EU research policy), which aim to find solutions for the most urgent needs of agriculture and forestry production.

In September 2017, there were 17 Thematic Networks and 444 OGs launched in Europe. Some of them tackle explicitly diversification issues (such as OGs about the development of minor crops—blackcurrant or mustard—in Burgundy, or as OGs about the development of grazing techniques within orchards, etc.). Nevertheless, it should be noted that, in most of cases, diversification seems rather integrated in an indirect way. Moreover, the distribution of OG is at this stage really uneven across countries. There are OGs in only eight countries; Germany and Spain weigh in for more than 50% of the number of OGs, and there are not OGs yet in Eastern Europe. Nevertheless, some further developments are expected soon, with more than 3000 planned at the horizon 2020 according to the European Commission.

EIP's OGs and thematic networks are embedded in the same conception: supporting the collective production of knowledge within innovation process rather than funding the infrastructure or knowledge systems. In that respect, it is interesting to note that the money allocated to EIP was taken from the budget initially planned to fund research projects. More globally, as FAS, EIP aims to be an integrative instrument between EU's CAP and research policy (H2020 programs). Both policies can be seen as aiming to reduce innovation network failures at regional level (Crespo et al. 2014).

A Lack of Pluralistic Debates and of Integrative Policies

The case of the CAP shows that the question of uncertainty about diversification is dealt with by various policy instruments, in direct or, more often, indirect ways. On the one hand, microeconomic incentives, derived from standard economic theory, aim at compensating for losses due to the progressive decrease of market measures and due to the adoption of new productions and practices. The CAP incentives also play an indirect insurance role for farmers engaged in diversification. In that respect, the choice of maintaining coupled direct payments for certain productions appears as a key policy instrument to support diversification. On the contrary, the level of national flexibility kept on policy choices such as greening choices results in too minimal requirements on the bulk of CAP direct payments to really favor farm diversification and high nature value farming in general. On the other hand, mesoeconomic instruments, derived from evolutionary economic theory, may compensate for technological uncertainty, by supporting multiactor networks producing knowledge about diversification. Nevertheless, diversification seems to be again integrated in a rather indirect way in these networks, and there are some strong inequalities in the development of these networks across countries and regions.

Overall, diversification seems to be addressed rather indirectly by various policy instruments, with a lack of integration between these instruments. Such an integration could help to better think about which policy instrument would better support product diversification in a given context. In certain contexts, a low diversification might be first explained by price and market uncertainty that microeconomics incentives might solve the best. In other contexts, a low diversification might first be related to technological uncertainty, and a structural support to knowledge networks and production might be the most appropriate. In some cases, there is probably a need to combine both types of instruments to reduce radical uncertainty about diversification. In any case, there is a clear need to have open and pluralistic debates about the instruments that can support diversification. In other fields of academic research, some authors advocate the need to support to a pluralistic conception of science. Thus, different theories (even competing ones) could be associated with complementary solutions and policy instruments in practice (Mitchell, 2002), as they shed light on different economic mechanisms, and have different blind spots.

This can be illustrated with the case of policies supporting the diversification toward protein crops in Europe (Häusling, 2011). The history of support for this sector shows that dedicated areas have fluctuated with the amounts of aid coupled per hectare: the decoupling of aid for protein crops in 2003 has resulted in reduced areas dedicated to these crops. At the same time, there is also a clear lock-in situation; for instance, animal breeders lack knowledge about how to develop feed that is not based upon imported soybean meal (Magrini et al., 2016). It is now probably inefficient to design a public intervention that attacks one problem (the profitability gap per hectare) without thinking of the other (dissemination of information and knowledge, etc.).

CONCLUSION

Reducing uncertainty about the cost and benefits of product diversification, but also about the paths that could lead farmers toward diversification, is now a key issue debated among scholars from various theoretical standpoints. Based on a nonexhaustive literature review, our chapter demonstrates that standard microeconomic analysis on the one hand and institutional mesoeconomic analysis on the other can be associated with various policy instruments. Microeconomic analysis of diversification puts an emphasis on measuring the costs and benefits of diversification. Current development in this field highlights the difficulties to assess the diversity of farmers' behavior regarding risks and uncertainty of diversification adoption and the necessity to implement adapted allocative policies to compensate for the costs, encourage for benefits, and decrease uncertainties. Mesoeconomics analysis of diversification puts an emphasis on the collective dimension of knowledge production on specialization versus diversification. A key idea in this chapter was to show that these alternative (or even competing) economic theories might be associated with policy instruments that could be complementary in practice, but which might need to be more discussed jointly and potentially integrated, as illustrated with the CAP instruments. To achieve a better integration, there is real need to open constructive debates between policy makers and scholars in a pluralistic perspective. These debates could integrate a diversity of theoretical standpoints about how to reduce uncertainty and about the potential synergies or contradictions between policy instruments derived from alternative theories.

References

ADE, 2009. Evaluation of the implementation of the Farm Advisory System. Final report - Evaluation part.

Arthur, W.B., 1989. Competing technologies, increasing returns, and lock-in by historical events. The Economic Journal 99, 116–131.

Baumol, W.J., Panzar, J.C., Willig, R.D., 1982. Contestable Markets and the Theory of Industry Structure. Harcourt Brace Jovanovich, Orlando, Florida.

Binswanger, H.P., 1980. Attitudes toward risk: experimental measurement in rural India. American Journal of Agricultural Economics 62 (3), 395–407.

Bowman, M.S., Zilberman, D., 2013. Economic factors affecting diversified farming systems. Ecology and Society 18 (1).

Bromley, D.W., Chavas, J.P., 1989. On risk, transactions, and economic development in the semiarid tropics. Economic Development and Cultural Change 37, 719–736.

Brunori, G., Barjolle, D., Dockes, A.-C., Helmle, S., Ingram, J., Klerkx, L., Moschitz, H., Nemes, G., Tisenkopfs, T., 2013. CAP reform and innovation: the role of learning and innovation networks. EuroChoices 12, 27–33.

Capello, R., 1999. Spatial transfer of knowledge in high technology milieux: learning versus collective learning processes. Regional Studies 33, 353–365.

Castle, E.N., Becker, M.H., Nelson, A.G., 1987. Farm Business Management: the decision making process, third ed. Macmillan Publishing Company, New York.

Chavas, J.-P., Aliber, M., 1993. An analysis of economic efficiency in agriculture: a nonparametric approach. Journal of Agricultural and Resource Economics 1–16.

Chavas, J.-P., Di Falco, S., 2012. On the productive value of crop biodiversity: evidence from the highlands of Ethiopia. Land Economics 88, 58–74.

Cohendet, P., Grandadam, D., Simon, L., Capdevila, I., 2014. Epistemic communities, localization and the dynamics of knowledge creation. Journal of Economic Geography 14, 929–954.

Cowan, R., 1990. Nuclear power reactors: a study in technological lock-in. The Journal of Economic History 50, 541–567.

Cowan, R., Gunby, P., 1996. Sprayed to death: path dependence, lock-in and pest control strategies. The Economic Journal 521–542.

Cowan, R., Hultén, S., 1996. Escaping lock-in: the case of the electric vehicle. Technological Forecasting and Social Change 53, 61–79.

Crespo, J., Suire, R., Vicente, J., 2014. Lock-in or lock-out? How structural properties of knowledge networks affect regional resilience. Journal of Economic Geography 14, 199–219.

David, P.A., 1985. Clio and the economics of QWERTY. The American Economic Review 75, 332–337.

Desjeux, Y., Dupraz, P., Thomas, A., 2011. PAC et environnement: les biens publics en agriculture. INRA Sciences Sociales, Paris.

Dosi, G., 1988. Sources, procedures, and microeconomic effects of innovation. Journal of Economic Literature 26, 1120–1171.

Ellis, F., 2000. The determinants of rural livelihood diversification in developing countries. Journal of Agricultural Economics 51 (2), 289–302.

Ekboir, J.M., 2003. Research and technology policies in innovation systems: zero tillage in Brazil. Research Policy 32, 573–586.

European Commission, 2010. Report from the Commission to the European Parliament and the Council on the application of the Farm Advisory System as defined in Article 12 and 13 of Council Regulation (EC) No 73/2009, COM(2010) 665 final, 15 November 2010.

Geels, F.W., 2002. Technological transitions as evolutionary reconfiguration processes: a multi-level perspective and a case-study. Research policy 31 (8-9), 1257–1274.

Häusling, M., 2011. The EU Protein Deficit: What Solution for a Long-standing Problem? European Parliament Report [A7–0026/2011].

Hennessy, D.A., 1998. The production effects of agricultural income support policies under uncertainty. American Journal of Agricultural Economics 80, 46–57.

Klerkx, L., van Mierlo Barbara, Leeuwis C., 2012. Evolution of systems approaches to agricultural innovation: concepts, analysis and interventions. In: Darnhofer, I., Gibbon, D., Dedieu, B. (Eds.). Farming Systems Research into the 21st Century: the New Dynamic. Springer, The Netherlands, pp. 457–483.

Kremen, C., Miles, A., 2012. Ecosystem services in biologically diversified versus conventional farming systems: benefits, externalities, and trade-offs. Ecology and Society 17 (4), 40. https://doi.org/10.5751/ES-05035-170440.

Koutsouris, A., 2014. "Failing" to implement FAS under diverse extension contexts: a comparative account of Greece and Cyprus. In: Presented at the 11th European International Farming System Association (IFSA) Symposium, Berlin, 1–4 April 2014.

Knowler, D., Bradshaw, B., 2007. Farmers' adoption of conservation agriculture: a review and synthesis of recent research. Food Policy 32 (1), 25–48.

Liebowitz, S.J., Margolis, S.E., 1990. The fable of the keys. The Journal of Law and Economics 33, 1–25.

Lundvall, B.-Å., 1992. National Systems of Innovation: Towards a Theory of Innovation and Interactive Learning. Pinter, London.

Magrini, M.-B., Anton, M., Cholez, C., Corre-Hellou, G., Duc, G., Jeuffroy, M.-H., Meynard, J.-M., Pelzer, E., Voisin, A.-S., Walrand, S., 2016. Why are grain-legumes rarely present in cropping systems despite their environmental and nutritional benefits? Analyzing lock-in in the French agrifood system. Ecological Economics 126, 152–162. https://doi.org/10.1016/j.ecolecon.2016.03.024.

Malerba, F., 2002. Sectoral systems of innovation and production. Research Policy 31, 247–264.

Marra, M., Pannell, D.J., Abadi Ghadim, A., 2003. The economics of risk, uncertainty and learning in the adoption of new agricultural technologies: where are we on the learning curve? Agricultural Systems 75 (2–3), 215–234.

Markowitz, H.M., 1991. Foundations of portfolio theory. The Journal of Finance 46 (2), 469–477.

McDermott, C.M., O'Connor, G.C., 2002. Managing radical innovation: an overview of emergent strategy issues. Journal of Product Innovation Management 19, 424–438.

Mitchell, S.D., 2002. Integrative pluralism. Biology and Philosophy 17 (1), 55–70.

Moschitz, H., Roep, D., Brunori, G., Tisenkopfs, T., 2015. Learning and innovation networks for sustainable agriculture: processes of co-evolution, joint reflection and facilitation. The Journal of Agricultural Education and Extension 21, 1–11.

Nelson, R.G., Bessler, D.A., 1989. Subjective probabilities and scoring rules: experimental evidence. American Journal of Agricultural Economics 71 (2), 363–369.

Norris, P.E., Kramer, R.A., 1990. The elicitation of subjective probabilities with applications in agricultural economics. Review of Marketing and Agricultural Economics 58 (2–3), 127–147.

Opperman, R., Beaufoy, G., Jones, G., 2012. High Nature Value Farming in Europe. Verlag regionalkultur, Ubstadt-Weiher.

EU Parliament, 2016. The Common Agricultural Policy in Figures, Fact Sheets in the European Union. http://www.europarl.europa.eu/factsheets.

Peerlings, J., Polman, N., 2004. Wildlife and landscape services production in Dutch dairy farming; jointness and transaction costs. European Review of Agricultural Economics 31, 427–449.

Popadiuk, S., Choo, C.W., 2006. Innovation and knowledge creation: how are these concepts related? International Journal of Information Management 26, 302–312. https://doi.org/10.1016/j.ijinfomgt.2006.03.011.

Reynaud, A., Couture, S., 2012. Stability of risk preference measures: results from a field experiment on French farmers. Theory and Decision 73 (2), 203–221.

Röling, N.G., Engel, P.G., 1990. Information technology from a knowledge system perspective: concepts and issues. Knowledge. Technology and Policy 3, 6–18.

Robison, L.J., Barry, P.J., 1987. Competitive firm's response to risk. Macmillan.

Sauer, J., Wossink, A., 2012. Marketed outputs and non-marketed ecosystem services: the evaluation of marginal costs. European Review of Agricultural Economics 1–31.

Savage, L.J., 1972. The Foundations of Statistics. Courier Corporation.

Shumway, C.R., Pope, R.D., Nash, E.K., 1984. Allocatable fixed inputs and jointness in agricultural production: implications for economic modeling. American Journal of Agricultural Economics 66, 72–78.

Sorensen, T., 2011. Australian agricultural R&D and innovation systems. International Journal of Foresight and Innovation Policy 7, 192–212.

CHAPTER

24

Technological Lock-In and Pathways for Crop Diversification in the Bio-Economy

Marie-Benoit Magrini[1], *Nicolas Béfort*[2], *Martino Nieddu*[3,†]

[1]AGIR, Université de Toulouse, INRA, Castanet-Tolosan, France; [2]Chair in Industrial Bioeconomy, Neoma Business School; European Center in Biotechnology and Bioeconomy, Reims, France; [3]REGARDS, Université de Reims Champagne-Ardenne, Reims, France

INTRODUCTION

Cropping diversity enhances important ecosystem services, thereby reducing synthetic inputs such as pesticides and fertilizers (Stoate et al., 2009; Altieri, 1999); globally, however, harvested crops have become highly specialized. As regards global land use, Jahn et al. (2015) report that in 2012, pulse crops occupied less than 1 million square kilometers, whereas cereals 7 million, and oil crops near 3 million. More precisely, major crops such as wheat, corn, and soy occupy respectively 2.5, 2, and 1 million square kilometers (FAO). These authors also point out that there would be dramatic environmental benefits if farmlands were more diversified.

Field crop specialization is a major feature of Western countries, inherited from the agricultural revolution of the postwar period that gave preference to genetic engineering and fossil-based inputs (Vanloqueren and Baret, 2009, 2008). From the 1950s to the 1980s, the intensification of agriculture through increased mechanization, synthetic inputs, and adapted genetic selection led to crop specialization, which has continued ever since with the increasing markets for major crops (Magrini et al., 2016; Meynard et al., 2013). In addition, because of cereal and oilseed overproduction, this trend has been reinforced since 1980 by the growth of nonfood uses for major crops (biofuels, chemicals, and materials). This development continued the paradigm of specialization and has today crystallized into a new model of growth: the biorefinery-based bio-economy (Nieddu et al., 2012, 2014; Nieddu and Vivien, 2013).

Therefore today, any attempt to diversify cropping systems to reduce chemical use is faced with highly organized industrial supply chains and R&D in the agriculture and agrifood sectors (Gliessman, 2015). Feed and food markets have developed conjointly with major crops, with the result that minor crops have been gradually relegated to the background along with the

† deceased

Agroecosystem Diversity
https://doi.org/10.1016/B978-0-12-811050-8.00024-8

ecosystem services they provide, in addition to food and feed, such as grain-legumes supplying nitrogen. Indeed, grain-legumes (covering a wide variety of species as pea, fava bean, lupin, soybean, lentils, and beans) fix atmospheric nitrogen through symbiosis with soil bacteria to produce protein-rich grains (average of 22%–38% protein in dry matter), and so do not need nitrogen fertilizer. In Europe, however, the grain-legume harvested crop area has been declining for several decades, today accounting for less than 4% of field crop acreage, much less than in North America and Asia. Yet in the United States, soybean acreage accounts for nearly 33% of the harvested crop area, whereas pulses are only around 1%. European pulse crop acreage is around 0.2%, whereas dry peas, fava beans, and lupins (the three main grain-legume crops in Europe) account for 2% and soybeans, 1%. Meanwhile, cereal crops occupy more than two-thirds of the harvested crop area both in France and in the United States (Eurostat and USDA data). However, Europe has granted considerable subsidies to increase grain-legume cultivation. Faced with the continuing decline in pulses and the failure of its previous policies, the European Union has begun to question the reasons for this failure and to seek new ways to promote more diversified cropping systems using grain-legumes (Schreuder and de Visser, 2014).

The purpose of this chapter is to show that if we want to understand why public policies aimed at agriculture alone are not sufficient to reverse this situation, we must understand how the coevolution of agriculture and agrifood resulted in lock-in (Magrini et al., 2016). Institutions must take into account the design of the entire supply chain and food systems that are jointly developed with crops. Today, public authorities are promoting the bio-economy as a new driver of agricultural development (for instance, the European Commission, 2012). Yet, these public institutions must examine which bio-economy paradigm is actually promoted by the economics of the current technological pathway, to assess the impact that the bio-economy may have on crop diversity.

Lock-In: A Growing Competitiveness Gap Between Major and Minor Crops section discusses the economics of technological pathways. Using the theory of path dependency, we sketch out the major self-reinforcing mechanisms of lock-in, which explains how the joint evolution of agriculture and agrifood sectors has hindered crop diversity. Lock-In Processes That Marginalize Grain-Legumes: A French Case Study section puts these factors into perspective through a study of the history of grain-legume diversification in France. Finally, Bio-Economy: Between Agroecology and Industrializing Crop Diversity section discusses new scientific paradigms of Western agriculture based on the KBBE (knowledge-based bio-economy) as regards the challenges of cropping diversity.

LOCK-IN: A GROWING COMPETITIVENESS GAP BETWEEN MAJOR AND MINOR CROPS

Coevolutionary frameworks explain how the relationships between values, knowledge, organizations, technologies, and environments can lead to lock-in by creating strong interdependencies between actors within a production system (Kallis and Norgaard, 2010). The incumbent production system, built over time, has stabilized through multiple technical and social relationships that "lock" the actors in their choices. As explained by Geels (2011, p. 25), the dominant system is "stabilized through various lock-in mechanisms, such as scale economies, sunk investments in machines, in infrastructures and competencies. Also, institutional commitments, shared beliefs and discourses, power relations, and political lobbying by incumbents stabilize existing systems … consumer lifestyles and preferences may have become adjusted to existing technology systems. These lock-in mechanisms create path dependence and make it difficult to dislodge existing systems." Compared with traditional economics, lock-in analysis takes a different approach to explain choice in production systems, which integrates the complexity of the interdynamics of several mechanisms, such as scale economies (for a synthesis, see Dosi and Nelson, 2010). The evolutionary economics approach foregrounds historical choices and interactions across systems to understand the barriers to change. The self-reinforcing mechanisms that create lock-in and thereby discourage stakeholders from adopting alternative production systems have been identified in comprehensive and historical research. Studies of technological lock-in have focused primarily on manufacturing, transportation (David, 1985; Arthur, 1988, 1994; Liebowitz and Margolis, 1995), and energy (Cowan, 1990). Studies have also been done on the agricultural sector (Wolff and Recke, 2000; Wilson and Tisdell, 2001; Chhetri et al., 2010). In particular, studies by Cowan and Gunby (1996), Vanloqueren and Baret (2008, 2009), and Magrini et al. (2016) show how academic agricultural research has focused primarily on one type of research paradigm, oriented toward agrochemicals and genetic engineering, to the detriment of an agroecology paradigm oriented toward crop diversity. Nowadays, it is recognized that crop diversity would provide a broader range of ecologic services through their functional properties and would, therefore, enable a significant reduction in synthetic inputs (Therond et al., 2017). Yet, it is difficult to escape from the previous path that led to crop specialization by relying on a considerable use of chemicals and, thus, greater environmental problems. This trajectory can be explained by evolutionary economics, which highlights several self-reinforcing mechanisms that created (and still create) increasing returns to adoption for major crops in the agriculture and food industries.

Increasing returns to adoption (IRA) is a key concept in the theory of technological lock-in as developed by evolutionary theorists. This concept explains how a technology gradually comes to dominate other alternative technologies because it is increasingly chosen. The founding assumption of this theory is that technology is not necessarily chosen because it is the best, but it is the best because it is initially chosen, and that choice is reinforced over time. This reinforcement is primarily explained by five mechanisms (called "self-reinforcing," Arthur, 1988, 1994) that highlight the role of social action in these processes:

1. Learning by using: a technology's productive performance increases with users' experience.
2. Network externalities: the more adopters there are, the better it is for other users to adopt that technology to take advantage of additional products and services that are developed to be compatible with the dominant technology.
3. Scale economies and economies from learning by doing: the unit cost of production decreases over time as a result of volume and improved technology, making the technology even more attractive.
4. Informational increasing returns: the more a technology is used, the more it is known and understood, thus encouraging other users to adopt it.
5. Technological interrelatedness: other technologies and production standards are established in line with the dominant technology.

These returns of adoption are termed "increasing": the greater the number of users, the greater the value for these users at the expense of alternatives; or rather, the adoption of the technology increases, but it remains to be proven whether, economically, the returns (or utility derived) also increase. This is more complex to verify empirically because of the multiplicity of interrelated industries that are connected through the dominant technology. Proof can be provided by a contrary argument: businesses and investments related to this technology are economically attractive to some actors; otherwise, adoption would not increase. However, the construction of collective narratives may bias an accurate assessment of the actual economic benefit of a chosen technological pathway (in this sense, see the work of Les Levidow). Moreover, uncertainties about the alternatives—which have received less investment and less learning—and the inherent cost of change strengthen the initial choice even more over time. This progressive reinforcement may even lead to irreversible situations (Liebowitz and Margolis, 1995). In other words, the more a technology is adopted in an industry, the more that industry's performance improves and other compatible technologies develop around common standards. Thus, gradually a sociotechnical regime is created that becomes increasingly difficult to change or reverse, or to allow other alternative technologies to develop even though they may be more effective. Geels (2011) also points out that, whatever the industry, the dominant sociotechnical regime eventually ends up locked-in, in that it is unable to significantly shift its pathway to change the paradigm on which it is based.

In light of these evolutionary approaches, we can infer that in the agricultural sector, the greater the space given to major crops in rotations, the more their technical and economic performance has improved. Table 24.1 summarizes some of these mechanisms for wheat, the major global crop (Table 24.1). The adoption of major crops has been reinforced at the expense of other species, whose yields are lower and/or more variable, primarily because their farming practices and plant breeding are less advanced. One of the major self-reinforcing mechanisms privileging dominant species is the search for economies of scale both upstream and downstream in the agro-industrial system. Regarding upstream activities, since R&D is specific for each plant species in terms of varieties (selection and distribution of seeds) and in terms of synthetic inputs (licensing and distribution), investments are made depending on the likelihood of substantial production volumes to recover that investment cost. As for farms, specialization occurs because it is easier to acquire technical mastery, amortize farm equipment, and organize the work for only a few crops in which the farm specializes. Downstream, organizations for storage, processing, and selling the crops follow the same rationale: specializing the business in a few dominant species leads to economies of scale, thereby reducing the marginal cost of use for a given species. The search for economies of scale thus simplified crop systems, resulting in shorter rotations to increase the volumes of major crops and reduce the unit cost of production, collection, storage, and transport. This simplification of systems also became reinforced over time through its close dependence on the organization of the food industry and other industries dealing with farmers, which also favored economies of scale by offering few markets for diversification crops (Meynard et al., 2013).

Other self-reinforcing mechanisms should be considered in relation to actors' adaptive exceptions and those discussed by Arthur (1988). Actors bet on the future that the dominant technology will continue to grow, which encourages other actors to adopt it. Thus, predicting the future choices of other adopters reinforces any initial choices or actions taken, whether in the agroecologic or the agrochemical paradigm (for more information, refer to authors with an institutional reading of IRA, such as North). Moreover, above and beyond creating economic profits, the mechanisms for constructing communities of identity may also be self-reinforcing mechanisms; see Barthelemy and Nieddu (2004) in the case of the CAP and cereal crops.

The technological and industry complementarity between agriculture and the food industry has also had a particularly strong effect. The industrialization of food supply chains has been accompanied by a growing number of agricultural production standards that further contribute to lock-in (Busch, 2011). For example, the requirements for high wheat protein levels, which facilitates processing and cooking (soft wheat for bread and durum wheat in pasta), are high doses of nitrogen on the crops, which

TABLE 24.1 Sources of Increasing Returns to Adoption in Favor of Wheat

Source	Explanation	Example
Economies of scale and economies from learning by doing	Reduces unit production cost and improves the technology	Greater amortization of investment costs in R&D (plant chemicals, seed breeding), logistics, and processing infrastructures
		Improve mineral fertilization (better efficiency of nitrogen absorption, precision agriculture)
		Improve the technico-functional properties of wheat for food
Learning by doing	Improve yield through experience	Farmers have more experience with growing cereals and using mineral fertilizers
		Improved yields
		Loss of know-how about other cultures whose surface areas are reduced
Network externalities	Products are developed that are compatible with the dominant technology	Genetic adaptation of wheat to be able to increase nitrogen fertilization doses (semidwarf gene to avoid lodging)
		Develop pesticides/chemicals for cereals
		Develop food products based on cereals
		Absence of approved pesticides for minor crops
		Few market outlets for minor crops
Technological interrelatedness	Other technologies and production standards are established in relation to the dominant technology	Food standards about the high protein content of wheat (bread, pastries)
		Wheat varieties not suitable for intercropping systems with pulses
		Farm equipment and logistics organization are adapted to major crops
Informational increasing returns	Increased knowledge about properties and uses	More farm advising about synthetic inputs and growing cereals (by farming institutes, cooperatives, and the trade press)
		Little farm advising on managing the nitrogen cycle between crops
		Increased consumption of cereal products (greater number of processed foods)
		Increased knowledge about the technico-functional properties of cereals with the increase in food products
Institutional support	Rules and subsidies	Historical support of wheat prices
		Different tariff barriers for different species; creation of stable, clear and legible collective rules for major species

hinders the development of innovative agricultural systems that use less nitrogen fertilizer. The logistics organization of collection and storage agencies focused on major species that supply the food industry is also not suited to innovative systems that reduce inputs, such as cereal-legume intercropping (Magrini et al., 2013).

Last but not least, institutional factors have reinforced this gap between major and minor crops. In Europe, for example, CAP subsidies have historically given preference to wheat (Magrini et al., 2016). French farmers then logically followed suit and adapted to this policy.

LOCK-IN PROCESSES THAT MARGINALIZE GRAIN-LEGUMES: A FRENCH CASE STUDY

This section illustrates some of the mechanisms of IRA introduced in the previous section, such as the competition between wheat and grain-legumes (see Magrini et al., 2016, for a more detailed analysis).

France is the top producer of grain-legumes in Europe. Nevertheless, France (and Europe) has been experiencing a continuous drop in grain-legume surface area since the early 1990s. This fall is basically explained by the widening gap in farmers' annual margins compared to cereal crops. The annual gross margin for grain-legumes is now two to six times less than that of nonlegume crops in Europe (Zander et al., 2016; Dequiedt and Moran, 2015). From a broader approach, this competitiveness gap can be understood as a result of the IRA mechanisms that have shaped the competitiveness between major and minor crops.

First, there has been an even greater increase in nitrogen fertilization of cereal crops since the dramatic increase in wheat yields has focused actors' attention on developing more food products using cereals, primarily wheat (Hesser, 2006). With a growing volume of wheat, businesses have identified considerable opportunities for that commodity in food markets. Thus, between 1970 and 1980, consumption of wheat doubled in France and continues to grow: from 87 kg/year/person in the middle of the 1990s, it rose to 105 kg/year/person by the late 2000s. Conversely, grain-legume consumption has fallen sharply: from 7.3 kg/year/person in 1920 to less than 1.4 kg since the 1980s (Agreste, 2015). Grain-legume surface area for domestic human consumption has also shrunk from nearly 200,000 ha in 1960 to less than 30,000 ha in the early 21st century (Voisin et al., 2014).

Second, yields for legumes for human consumption (lentils, dry beans, etc.) have barely progressed. For protein-rich plants (including protein-rich peas, lupins, and fava beans) mainly used for feed, the wheat/pea differential also reveals demand-side preferences. Between 1960 and 1980, the yields of both wheat and protein-rich pea more than doubled in France. However, from the late 1980s, the pea yield differential with wheat grew. This happened despite public research efforts on peas, which were used as a model plant by INRA (French National Institute for Agricultural Research) for genetic research on protein-rich legumes. One reason for this gap is that the higher yield in wheat varieties with semidwarf genes enabled nitrogen doses to be increased without affecting the lodging problem. Although genetic progress has also been made for protein-rich legumes as regards lodging, their yields capped at a significantly lower level than that of cereals, and the curve of the national average yield in the last 10 years shows a clear downward trend. The pea yield was near 80% of wheat in 1990, falling to 50% in the late 2000s. Farmers tend to move minor crops to less fertile land, as they prefer to reserve the best land for species with higher yield potential. In addition, efforts at varietal selection for peas are more recent, and business investment in selection is much lower than that for major species, even in public research. Bonneuil and Thomas (2009) showed that INRA has gradually withdrawn from creating varieties (more than 100 species in 1975 to less than 10 in 2005), with a low point in 2003 when the decision was made to focus on few species in overall plant breeding research. Of these, only one grain-legume target was chosen: the winter pea. By greatly reducing genetic and breeding research on minor species, INRA thus ceased to counteract private companies' concentration of plant breeding on the dominant species linked to their market-based thinking (Meynard et al., 2013). Currently, in France, there are only six companies with a breeding program on peas, whereas there are more than 23 for wheat. The number of wheat varieties published in the European Union's Common Catalogue of Plant Varieties is also much greater than for field peas or field beans (Table 24.2).

Third, due to the key role of pesticides in the rationale of the dominant cropping system (Wilson and Tisdell, 2001), companies that market these inputs have become the main source of advice to farmers. To fight against pests, their advice usually focuses on chemical solutions, which are simple and offer spectacular efficacy, rather than preventive agronomic practices such as lengthening the crop rotation. Advising about long crop rotations is more complex. Furthermore, the subsystem of knowledge underlying agriculture is based on a logical system that prevents even considering the benefits of crop diversification, including the ecosystem services of grain-legumes. A bibliometric analysis of the French agricultural press at the end of the 2000s found that articles offering farmers reference data and advice on crop diversification, including legumes, were rare (Meynard et al., 2013). Moreover, agricultural institutes providing advice remain focused on major crops (wheat, corn, rapeseed, and sunflower). In France, a compulsory tax on major crops partially funds those institutes, whereas some grain-legumes species do not yet have such a tax. The greater the production of major crops upstream, the more funding these institutes have for R&D on major crops. Until 2015, moreover, these agricultural institutes were specialized by crop; in other words, cereals, oilseeds, and protein crops are studied in separate institutions, which does not favor crop diversification research. In that year, agricultural institutes on oilseed and protein crops were joined together (in the French Terres Inovia Institute), but they are still separate from cereal institutes.

TABLE 24.2 Number of Currently Registered Varieties of Wheat, Pea and Bean Grain-legumes in France and in Europe in 2015

	Wheat (Durum Wheat Excluded)	Pea	Bean
France	304	61	20
European Union	2234	392	134

http://ec.europa.eu/food/plant/plant_propagation_material/plant_variety_catalogues_databases/search/public/index.cfm.

Fourth, the competitiveness gap between cereal and grain-legume species has grown even wider because, historically, European grain-legumes were considered almost exclusively to be animal feed, placing them in direct competition with soybean meal, which provides cheap protein in large quantities. This orientation driven by public institutions meant that research and private stakeholders have neglected progress for grain-legumes, such as better promotion in high-value outlets such as for human food. Yet today, faced with competition from other raw materials including other co-products (rapeseed meal, or dried distillers grains primarily for ethanol production), the use of grain-legumes in animal feed has fallen sharply. From 3 million metric tons of protein-rich peas produced for feed in France in the late 1980s, fewer than 500,000 metric tons are destined for livestock today. Facing stiff competition in the market for concentrated feed, and with feed systems being built on the promotion of cereals, protein-rich grain-legumes represent less than 2% of feed formulas; they are considered a simple adjustment variable (Meynard et al., 2013). Their inclusion in formulations has fallen, particularly in recent years.

As explained before, whether for producers or for users, the IRA for major species created a competitiveness gap with grain-legumes, which price and production support can no longer compensate for in France. This public support, which has also been unstable over time, has not resulted in lasting incentives for growing pulses, unlike the market dynamics that have created significant markets for cereals and more recently for oilseeds used for biofuels and food oil. The development of biofuels and of rapeseed and sunflower oils has contributed to the growth of these crops since the late 1990s, with their meals providing new sources of protein-rich feed and in a sense being positioned as flex crops like soybeans. In addition, the reduced area of grain-legumes sends a signal of decreasing supply, prompting formulators to exclude their use in feed for fear of a lack of stock. They then become an occasional substitute depending on the ratio of market price competition between raw materials, which for farmers means that it is risky to grow them compared to other species.

As with human consumption, the low investment in research and development on these species has not created market opportunities with greater added value that would encourage their wider cultivation. Consumption of pulses declined in France during the 20th century, and half of current consumption comes from imports, while consumption of both cereals and animal products has increased significantly in recent decades (Combris and Soler, 2011). The drop in eating pulses occurred together with the growth in mass consumption of meat, which itself has followed an increase in household incomes, as observed throughout the world (Tilman and Clark, 2015). Strong sociocultural factors affect pulse consumption. According to a study on consumption on a representative group in France, pulses suffer from the image of being "old-fashioned" and have been traditionally called "the poor person's meat." In addition, this study also found that changing lifestyles in favor of fast-cooking foods have made pulses less attractive (Champ et al., 2015).

Some niche markets have, however, been created for functional ingredients, such as starch and pea proteins mostly used in biorefineries. The more traditional market for lentils or beans still exists with quality labels such as Protected Designation of Origin (see Voisin et al. (2014) for maps of pulse production and the various quality labels in France), as well as export markets such as fava beans for Egypt or green peas for India. These niches provide producers with higher prices than for feed outlets, providing greater incentives to farmers. Yet to date, these species are generally poorly represented in processed foods. Studies by the *Groupe d'Etude sur les Protéines Végétales* [Study Group on Plant Proteins] show that today, the majority of plant proteins used in processed food products in France comes from wheat, while in the rest of the world, from soybeans. These results highlight the coevolution of the supply and demand side of the agrifood sector.

This agro-industrial system (both upstream and downstream of the supply chain including research institutes) has not contributed to crop diversity, related to agriculture's great dependence on chemicals. Meanwhile, environmental concerns are growing, and dietary habits are now being questioned as people have become aware of the excess of animal-based food in human diets, calling into question the sustainability of Western diets (Clonan et al., 2015). Can these two related considerations, environment and nutrition, help to build a new pathway toward a superior technology based on more diversified crops? Which knowledge systems would support this transition? Does the new KBBE promoted by Western countries positively influence crop diversity?

BIO-ECONOMY: BETWEEN AGROECOLOGY AND INDUSTRIALIZING CROP DIVERSITY

The progressive reinforcing of a technological paradigm toward a state of lock-in, as discussed earlier, makes it very difficult for the system to unlock itself, that is to say, to begin a transition toward a new paradigm judged more desirable. The sustainability transition literature has grown considerably in the early 21st century. This new body of work, which has its roots in various schools of thought such as evolutionary economics, pragmatic sociology, and science studies, has led to the creation of a new scholarly community working on the factors that promote or hinder the transition toward greater sustainability. Within this community, the Sustainability Transition Research Network plays a key role, with a leading proposal to study the transitions of sociotechnical systems based on multilevel perspective (Geels, 2004, 2011; Smith et al., 2010), whose canonical diagram is shown in Fig. 24.1. Other approaches, more focused on transition management and governance, complete this community of thought on sustainability transitions (Elzen et al., 2004; Smith et al., 2005; Borras and Edler, 2014). We conclude this chapter by examining the driving factors for a transition of agro-food systems toward more plant diversity.

Fig. 24.1 shows three interrelated levels that shape innovation dynamics among actors driving societal functions such as agricultural and food production. The top-level landscape represents factors external to firms such as demographic or environmental pressures. The middle level is the sociotechnical regime, that is, the system composed of incumbent actors locked-in a technological trajectory shaped by scientific knowledge, consumer preferences, institutions (rules), etc. The stability of the regime means that innovations tend to be incremental and actors are unlikely to change radically. Niche innovations at the lower level are like "incubation rooms" outside the regime allowing radical innovation. Niches "provide locations for learning processes about the technology, user preferences, regulations, infrastructure, symbolic meaning etc. Niches also provide space to build social networks that support innovations." When niches manage to be diffused into the regime, then coevolution may lead to a new regime in a transition process lasting many decades. Landscape pressure on the regime would favor this diffusion. "The key point of the multilevel perspective is that system innovations occur as the outcome of linkages between developments at the multiple levels" (Hofman et al., 2004, p. 347–348); as well as the crucial role of landscape and institutional rules in orienting the regime.

Whereas the previous section focuses on how mechanisms of increasing returns of adoption reinforce an initial choice, at this point we need to think about the factors involved in choosing a new path that will trigger a new transition. Drawing on the comprehensive Multi-Level Perspective (MLP) approach, we consider that an evolution in a sociotechnical regime will combine two main dynamics of change. The first comes from niche innovation (existing outside the dominant regime) and proposes radical innovations, whose diffusion forces the dominant regime to change. This approach is part of the Schumpeterian tradition of entrepreneurs who propose new technical, product, or marketing innovations. The second dynamic of change comes from public authorities outlining a "vision of the future" that shapes new regulation, such as what the European Commission is doing today with its European Technology Platforms and proposals for the bio-economy. In this sense, change is driven by backcasting (Quist et al., 2011). For example, backcasting in the 1960s led to a vision for the future that can be summarized thus: a model of development based on scientific

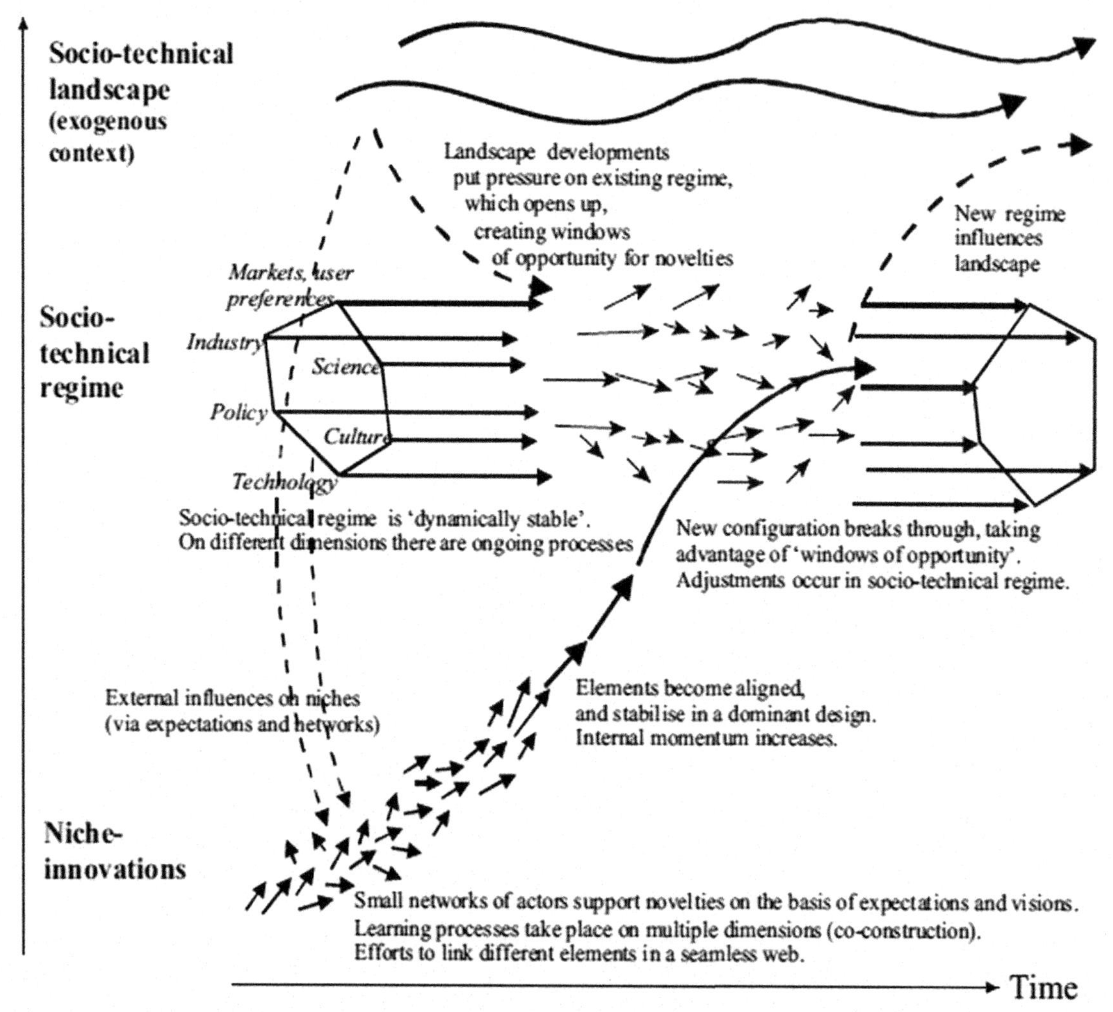

FIGURE 24.1 The multilevel perspective on sociotechnical transition dynamics (Geels, 2011).

reductionism in agronomy, the widespread use of fossil-fuel-based inputs, and crop specialization by region-enabling economies of scale. This was the initial choice that triggered the path of IRA described earlier.

What is the situation today? While there are several niche innovations for new food products or bio-based products that promote greater plant diversity, the agriculture pathway is still heavily influenced by institutional positions on the bio-economy. Thus, tensions around protein demands have led entrepreneurs to redesign food using grain-legumes, such as pasta combining wheat with lentils or chickpeas (Lascialfari and Magrini, 2016). Yet, these markets remain very marginal. Other initiatives use fractions from these species (vegetable protein materials or VPMs) in plant-based food or mixed foods (combining animal and vegetable proteins). Research by Gueguen et al. (2016) suggests strong growth in this market, estimated at over 40% between 2013 and 2018: "Representing € 7.1 billion in 2013, this market is expected to reach € 10 billion in 2018." These authors also emphasize the force of lock-in, as "wheat and/or soy proteins are components of more than 90% of new products launched on the market [of VPMs] in 2013." The growth in these plant proteins is strongly linked to new industrial capabilities currently developing throughout the world, particularly plant biorefineries, which are part of the bio-economy paradigm defined by the European Commission (2012). Here, these considerations are strongly linked to markets and only weakly to the concern for crop diversity at farm or regional levels.

These considerations lead us to different visions of greater plant diversity depending on the chosen development model for the bio-economy. While bio-economy can be understood as a reflection on the limits of human intervention on the natural environment following Gergescu-Roegen (Vivien, 1998, 1999), two other visions have largely taken precedence in bio-economy design in recent years. First, recent OECD reports focus on biotechnology applications from the genetic engineering revolution, where the living cell will become the factory of the future (OECD, 2009). In the second view, bio-economy as defined by the European Commission (2012) refers primarily to the use of biomass for food and nonfood uses, regardless of the technologies used (biotech, thermochemistry, and so on). This knowledge-based bio-economy plans to increase the industrialization of food and nonfood uses. Notwithstanding the focus on food, the dominant discourse is on nonfood uses (which masks the production of food additives resulting from the fractionation process, identical to the process for food). Moreover, despite the discourse on second-generation biorefineries (using nonfood parts of plants and organic and urban wastes), cereals, sugar, and oilseeds remain the primary raw materials of the bio-economy along with the thermochemical processes of forest industries.

Levidow et al. (2013) found that biorefinery innovation pathways have the same drivers as current production-consumption patterns that are expanding global demand for food, feed, fuel, and other uses. This kind of "techno-fix" depends on finding cheap resource supplies without paying for their societal and environmental costs. Levidow et al. (2012) discuss two rival visions for the future of the bio-economy. The first follows the path of agroecology and is more linked to concerns for crop diversity at regional or farm levels. As we see it, this vision involves changes in both farming and dietary practices, without excluding reasonable nonfood uses. The second vision is rooted in the life sciences (green and white biotech) and is promoted as a solution for continuing to increase the economies of agricultural and industrial returns. The US National Research Council "Industrialization of Biology" (2015) report sets out a roadmap in this direction.

This discussion shows two ideal types for the future of the bio-economy and therefore the future of agriculture. In the agroecology, sustainable food and nonfood view, this means adapting to the specifics of different regions and their soil and climatic conditions (a doubly green revolution) and the dual ecologic and economical intensification proposed by Griffon (2006) and others (Pretty et al., 2010). This would mean feeding more people by changing their diet toward more vegetable protein and providing ecologic services and nonfood products as a substitution for fossil fuels. This view would also require diversifying agro-industries using different business models at several regional levels and mobilizing a variety of technologies (such as an energy mix to substitute for the cheap oil era). For example, in the French research program for rural development

(PSDR4), the BIOCA project (2016–19) on the bio-economy (in the region of Champagne-Ardenne) explores this "technological mix" pathway in an industrial agriculture region.

Yet in the second, industrialization of biology vision, agronomy would shift toward the artificialization of agriculture (with biotech such as GMOs or CRIPS) and industrial processes (with white biotechnology using these biotechs). One example is the Bio-Based Industries Joint Undertaking, a new € 3.7 billion public-private partnership between http://ec.europa.eu/programmes/horizon2020/Horizon 2020 and the *Bio-based Industries Consortium*. Operating as a Horizon 2020 project, it is driven by the Vision and Strategic Innovation and Research Agenda developed by private industry, and not by the need to diversify agriculture. In this view, deeper questions about transforming diets or the link between food, feed, and nonfood are not raised. This vision remains techno-centered and does not take into account the regional identities of diets, for example. In turn, this leads to an anonymization of the plant materials used, which is likely to favor the dominant species over a variety of plant species from diverse regions.

Of course, these are two extreme ideal types; actual experiments navigate between these two poles. Both try to make their path more "green" and economically efficient, as these are required for legitimacy and credibility. Nevertheless, it is important to note that with both views, there is resistance to change related to past trajectories and to problems of social or economic acceptability. The "unlocking" of current systems requires experimenting with practices to promote new dynamics of increasing returns of adoption. Public policy frameworks are decisive in developing new practices, as shown in the agricultural modernization of the 1960s. However, the support given to different experimental pathways must be balanced; otherwise, our ability to explore the benefits of all options is greatly reduced and may condemn useful alternatives to extinction. Analyzing the different tasks in the large-scale partnership between the European Union and bio-based industries (formerly the Bridge public-private partnership) shows that today some options fostering greater crop bio-diversity are being neglected.

Acknowledgments

Translated from the French by Cynthia J. Johnson. Research for this article received support from the French Agence Nationale de la Recherche [French National Research Agency] as part of the ANR-13-AGRO-0004 project LEGITIMES (LEGume Insertion in Territories to Induce Main Ecosystem Services).

References

Agreste, 2015. Memento Alimentation. www.agreste.fr.

Altieri, M., 1999. The ecological role of biodiversity in agroecosystems. Agriculture, Ecosystems & Environment 74, 19–31.

Arthur, W.B., 1994. Increasing Returns and Path Dependence in the Economy. University of Michigan Press.

Arthur, B., 1988. Competing technologies: an overview. In: Dosi, G., Ch, F., Nelson, R., Silverberg, G., Soete, L. (Eds.), Technical Change and Economic Theory. London Pinter.

Barthélemy, D., Nieddu, M., 2004. Multifunctionality as a concept of duality in economics: an institutionalist approach. Rennes 90, 28–29. https://www.researchgate.net/profile/Martino_Nieddu/publication/242077049_MULTIFUNCTIONALITY_AS_A_CONCEPT_OF_DUALITY_IN_ECONOMICS_AN_INSTITUTIONALIST_APPROACH/links/0c96052c7f39b49af6000000.pdf.

Bonneuil, C., Thomas, F., 2009. Gènes, pouvoirs et profits: recherche publique et régimes de production des savoirs, de Mendel aux OGM. Éditions Quae.

Borrás, S., Edler, J., 2014. The Governance of Change in Socio-technical Systems: Three Pillars for a Conceptual Framework. The Governance of Socio-Technical Systems: Explaining Change. Edward Elgar, Cheltenham, UK and Northampton, MA, USA.

Busch, L., 2011. Food standards: the cacophony of governance. Journal of Experimental Botany 62 (10), 3247–3250.

Champ, M., Magrini, M.-B., Simon, N., Le Guillou, C., 2015. Les légumineuses pour l'alimentation humaine: conséquences nutritionnelles et effets santé, usages et perspectives. In: Schneider, A., Ch, H. (Eds.), Les légumineuses pour des systèmes agricoles et alimentaires durables. QUAE Editions, France.

Chhetri, N.B., Easterling, W.E., Terando, A., Mearns, L., 2010. Modeling path dependence in agricultural adaptation to climate variability and change. Annals of the Association of American Geographers 100 (4), 894–907.

Clonan, A., Wilson, P., Swift, J.A., Leibovici, D.G., Holdsworth, M., 2015. Red and processed meat consumption and purchasing behaviours and attitudes: impacts for human health, animal welfare and environmental sustainability. Public Health Nutrition 18 (13), 2446–2456.

Combris, P., Soler, L.G., 2011. Consommation alimentaire: tendances de long terme et questions sur leur durabilité. Innovations agronomiques 13, 149–160.

Cowan, R., Gunby, P., 1996. Sprayed to death: path dependence, lock-in and pest control strategies. The Economic Journal 106, 521–542.

Cowan, R., 1990. Nuclear power reactors: a study in technological lock-in. The Journal of Economic History 50 (03), 541–567.

David, P.A., 1985. Clio and the economics of QWERTY. The American Economic Review 332–337.

Dequiedt, B., Moran, D., 2015. The cost of emission mitigation by legume crops in French agriculture. Ecological Economics 110, 51–60.

Dosi, G., Nelson, R., 2010. Technical change and industrial dynamics as evolutionary processes. Handbook of the Economics of Innovation 1, 51–127.

Elzen, B., Geels, F.W., Green, K. (Eds.), 2004. System Innovation and the Transition to Sustainability: Theory, Evidence and Policy. Edward Elgar Publishing.

European Commission, 2012. Communication From the Commission to The European Parliament, The Council, The European Economic and Social Committee and The Committee of the Regions, Innovating for Sustainable Growth: A Bioeconomy for Europe {SWD(2012) 11 Final}. http://ec.europa.eu/research/bioeconomy/pdf/201202_innovating_sustainable_growth_en.pdf.

Geels, F., 2011. The multi-level perspective on sustainability transitions: responses to seven criticisms. Environmental Innovation and Societal Transitions 1 (1), 24–40.

Geels, F., 2004. From sectoral systems of innovation to socio-technical systems: insights about dynamics and change from sociology and institutional theory. Research Policy 33 (6), 897–920.

Gliessman, 2015. Agroecology: The Ecology of Sustainable Food Systems, third ed. CRC Press.

Guéguen, J., Walrand, S., Bourgeois, O., 2016. Les protéines végétales: contexte et potentiels en alimentation humaine. Cahiers de Nutrition et de Diététique 51 (4), 177–185.

Griffon, M., 2006. Pour des agricultures écologiquement intensives. Editions de l'Aube.

Hesser, L.F., 2006. The Man Who Fed the World: Nobel Peace Prize Laureate Norman Borlaug and His Battle to End World Hunger: An Authorized Biography. Park East Press, 276 pp.

Hofman, P.S., Boelie, E.E., Frank, W.G., 2004. Sociotechnical scenarios as a new policy tool to explore system innovations: Co-evolution of technology and society in the Netherland's electricity domain. Innovation 6.2, 344–360.

Jahn, J.L., Stampfer, M.J., Willett, W.C., 2015. Food, health & the environment: a global grand challenge & some solutions. Dædalus 144 (4), 31–44.

Kallis, G., Norgaard, R.B., 2010. Coevolutionary ecological economics. Ecological Economics 69 (4), 690–699.

Lascialfari, M., Magrini, M.-B., 2016. Towards more sustainable diets: insights from firms' innovation dynamics on new legumes-based food products. In: 10ème journée des recherches en sciences sociales, 8–9 décembre 2016, Paris.

Levidow, L., Birch, K., Papaioannou, T., 2012. EU agri-innovation policy: two contending visions of the bio-economy. Critical Policy Studies 6 (1), 40–65.

Levidow, L., Birch, K., Papaioannou, T., 2013. Divergent paradigms of European agro-food innovation: the Knowledge-Based Bio-Economy (KBBE) as an R&D agenda. Science, Technology & Human Values 38 (1), 94–125.

Liebowitz, S., Margolis, S.E., 1995. Path dependence, lock-in, and history. Journal of Law, Economics & Organization 11 (1), 205–226.

Magrini, M.-B., Anton, M., Cholez, C., Corre-Hellou, G., Duc, G., Jeuffroy, M.-H., Meynard, J.M., Pelzer, E., Voisin, A.-S., Walrand, S., 2016. Why are grain-legumes rarely present in cropping systems despite their environmental and nutritional benefits? Analyzing lock-in in the French agrifood system. Ecological Economics 126, 152–162.

Magrini, M.-B., Triboulet, P., Bedoussac, L., 2013. Pratiques agricoles innovantes et logistique des coopératives agricoles. Une étude ex-ante sur l'acceptabilité de cultures associées blé dur-légumineuses. Économie Rurale 338, 25–45.

Meynard, J.M., Messéan, A., Charlier, A., Charrier, F., Fares, M., Le Bail, M., Magrini, M.B., Savini, I., 2013. Crop Diversification: Obstacles and Levers. Study of Farms and Supply Chains, Synopsis of the Study Carried Out by INRA at the Request of the Ministries in Charge of Agriculture and Ecology. INRA, 62pp. https://www6.paris.inra.fr/depe/Media/Fichier/Etudes/Diversification-des-cultures/synthese-anglais.

Nieddu, M., Garnier, E., Bliard, C., 2012. The emergence of doubly green chemistry, a narrative approach. European Review of Industrial Economics and Policy (2012/1). http://revel.unice.fr/eriep/index.html?id=3455 (réédition à la demande de la revue de l'article - L'émergence d'une chimie doublement verte, Revue d'Economie Industrielle, n°132, 4ème trimestre 2010, pp.53–84.).

Nieddu, M., Vivien, F.D., 2013. Transitions Towards Bio-economy: The Case of the Biorefinery, in Bussels, Matthias, Happaerts, Sander, Bruyninckx, Hans, Evaluating Sustainability Transition Initiatives. Theorizing the Evaluation of Success in a Complex Setting 131–150. https://lirias.kuleuven.be/bitstream/123456789/379253/1/theme-4-couleur.pdf.

Nieddu, M., Garnier, E., Bliard, C., 2014. Patrimoines collectives versus exploration-Exploitation, le cas de bioraffinerie. Revue Économique 67.

OECD, 2009. The Bioeconomy to 2030: Designing a Policy Agenda. OECD Publishing, 323pp. http://www.oecd.org/futures/long-termtechnologicalsocietalchallenges/thebioeconomyto2030designingapolicyagenda.htm.

Pretty, J., Sutherland, W.J., Ashby, J., Auburn, J., Baulcombe, D., Bell, M., Bentley, J., Bickersteth, S., Brown, K., Burke, J., Campbell, H., 2010. The top 100 questions of importance to the future of global agriculture. International Journal of Agricultural Sustainability 8 (4), 219–236.

Quist, J., Thissen, W., Vergragt, P.J., 2011. The impact and spin-off of participatory backcasting: from vision to niche. Technological Forecasting and Social Change 78 (5), 883–897.

Schreuder, R., de Visser, C.H., 2014. Report EIP-AGRI Focus Group Protein Crops. Final Report, 49 pp. http://ec.europa.eu/eip/agriculture/sites/agri-eip/files/fg2_protein_crops_final_report_2014_en.pdf.

Stoate, C., Báldi, A., Beja, P., Boatman, N.D., Herzon, I., Van Doorn, A., Ramwell, C., 2009. Ecological impacts of early 21st century agricultural change in Europe—a review. Journal of Environmental Management 91 (1), 22–46.

Smith, A., Voß, J.P., Grin, J., 2010. Innovation studies and sustainability transitions: the allure of the multi-level perspective and its challenges. Research Policy 39 (4), 435–448.

Smith, A., Stirling, A., Berkhout, F., 2005. The governance of sustainable socio-technical transitions. Research Policy 34 (10), 1491–1510.

Therond, O., Duru, M., Roger-Estrade, J., Richard, G., 2017. A new analytical framework of farming system and agriculture model diversities. A review. Agronomy for Sustainable Development 37, 21.

Tilman, D., Clark, M., 2015. Food, agriculture & the environment: can we feed the world & save the earth? Dædalus 144 (4), 8–23.

Vanloqueren, G., Baret, P.V., 2008. Why are ecological, low-input, multi-resistant wheat cultivars slow to develop commercially? A Belgian agricultural 'lock-in'case study. Ecological Economics 66 (2), 436–446.

Vanloqueren, G., Baret, P.V., 2009. How agricultural research systems shape a technological regime that develops genetic engineering but locks out agroecological innovations. Research Policy 38 (6), 971–983.

Vivien, F.-D., 1999. From agrarianism to entropy : Georgescu-Roegen's bioeconomics from a Malthusian viewpoint. In: Mayumi, K., Gowdy, J.M. (Eds.), Bioeconomics and Sustainability : Essays in Honour of NIcholas Georgescu-Roegen. Edward Elgar, Cheltenham, pp. 155–172.

Vivien, F.-D., 1998. Bioeconomic conceptions and the concept of sustainable development. In: Faucheux, S., O'Connor, M., van der Straaten, J. (Eds.), Sustainable Development: Concepts, Rationalities, Strategies. Kluwer Academic Publishers, Dordrecht, pp. 57–68.

Voisin, A.S., Guéguen, J., Huyghe, C., Jeuffroy, M.H., Magrini, M.B., Meynard, J.M., Mougel, C., Pellerin, S., Pelzer, E., 2014. Legumes for feed, food, biomaterials and bioenergy in Europe: a review. Agronomy for Sustainable Development 343, 361–380.

Wilson, C., Tisdell, C., 2001. Why farmers continue to use pesticides despite environmental, health and sustainability costs. Ecological Economics 39 (3), 449–462.

Wolff, H., Recke, G., 2000. Path dependence and implementation strategies for integrated pest management. Quarterly Journal Of International Agriculture 39 (2), 149–172.

Zander, P., Amjath-Babu, T.S., Preissel, S., Reckling, M., Bues, A., Schläfke, N., Murphy-Bokern, D., 2016. Grain legume decline and potential recovery in European agriculture: a review. Agronomy for Sustainable Development 36 (2), 1–20.

SECTION VI

GLOBAL ASPECTS

CHAPTER

25

Opening to Distant Markets or Local Reconnection of Agro-Food Systems? Environmental Consequences at Regional and Global Scales

Gilles Billen[1], *Luis Lassaletta*[2], *Josette Garnier*[1], *Julia Le Noë*[1], *Eduardo Aguilera*[3], *Alberto Sanz-Cobeña*[4]

[1]Sorbonne Université, CNRS, EPHE, UMR 7619 METIS, Paris, France; [2]CEIGRAM, Agricultural Production Universidad Politécnica de Madrid, Madrid, Spain; [3]Universidad Pablo de Olavide, Sevilla, Spain; [4]ETSI Agronómica, Alimentaria y Biosistemas. Universidad Politécnica de Madrid, Madrid, Spain

INTRODUCTION

Formulated in the early 19th century by David Ricardo (1817), the theory of comparative advantages predicts the economic benefits brought by specialization of production and free trade between different regions or countries compared with autarky. Roughly, when applied to agriculture, this theory states that producing each type of agricultural product on the most suitable land and relying on international trade for redistributing food throughout the world is economically more efficient at the global scale than targeting food self-sufficiency and autonomy in each region or country. According to Ricardo, comparative advantage in agriculture comes from land's natural properties, and results in a higher rent, defined as "*that portion of the produce of the earth, which is paid to the landlord for the use of the original and indestructible power of the soil.*" These views contradict those developed earlier by the agronomist and economist James Anderson (1739–77), for whom differential rents were attributed to soils because of historical changes in their fertility rather than of "natural properties" of soils (Foster, 2000). In the latter vision, fertility was considered a historically build feature that is likely to evolve depending on the interactions between agricultural systems and soils: agricultural specialization not only exploits soil fertility, but modifies it and creates imbalances across the world (Penuelas et al., 2009; 2012;

Agroecosystem Diversity
https://doi.org/10.1016/B978-0-12-811050-8.00025-X

Bouwman et al., 2017). Ricardo's theory was subjected to several other criticisms by later economists, particularly regarding its application to food commodities (Clement, 2000; Caballero et al., 2000). However, the theory of comparative advantages still implicitly inspires many regional and global development policies.

Trade of agricultural products and regional specialization of agriculture have existed since antiquity, but their quantitative importance was long limited by the very nature of most traditional agricultural systems, in which the restitution to the soil of the nutrient exported with the harvest was restricted by limited local resources of new nutrients (Mazoyer and Roudart, 1998; McNeill, 2010; Garrabou and González de Molina, 2010). For these traditional agricultural systems, fertilization mainly depended on a close coupling between crop and livestock farming, the latter allowing transfer of fertility from semi-natural areas such as grasslands and forests to arable lands. Also, the high costs and the problems involved in long-distance transportation of food and feed implied at that time the necessity of a regional diversity of agricultural production to satisfy local needs.

The Green Revolution, with the generalization of industrial fertilizers and pesticides, as well as the development of modern transportation infrastructures, discarded these constraints. Today, there are no longer any agronomic obstacles to growing cereals or soybeans as a monoculture. Moreover, the price of shipping crop or even meat and dairy products in containers across the oceans is rather low, both in monetary and environmental terms (Weber and Matthews, 2008; Aguilera et al., 2015). This has made possible a tremendous increase of long-distance shipping of food commodities in the last 50 years. Lassaletta et al. (2014b) estimated that one-third of all proteins (a proxy for the nutritive potential of foodstuffs) produced globally are now redistributed through international trade. A study at the regional scale in France shows that the cumulated volume of gross commercial exchanges of food commodities between regions and with foreign countries amounts to more than twice the national agricultural production (Le Noë et al., 2016). Also, a growing proportion of animal products in the human diet has promoted the development of industrial animal production systems with no or only limited dependence on local feed production (Naylor et al., 2005).

There have been a number of analyses questioning this recent trend toward worldwide crop-livestock disconnection, regional specialization, and globalization of agricultural systems, from economic, societal, environmental and agronomic points of view (Schipanski and Bennett, 2012; Lassaletta et al., 2014a; Suweis et al., 2015; Oita et al., 2016; Strokal et al., 2016). In this chapter, we wish to contribute to the assessment of the consequences of this recent, unprecedented specialization and opening of agricultural systems from a purely biogeochemical perspective. Our analysis is based on the GRAFS approach (Generalized Representation of Agro-Food Systems, Billen et al., 2014; Le Noë et al., 2017), which offers a useful framework for analyzing the connections (and disconnections) between cropping and livestock farming systems, as well as between local consumption needs and agricultural production, within territorial systems, based on the description of N fluxes. A brief description of the GRAFS methodology is provided in Box 25.1.

We will apply this methodology to three main structural issues of current agro-food systems, namely, (1) crop-livestock farming connection; (2) adjustment of food production to local consumption; and (3) spatial allocation of resources. We will show that these represent powerful levers that can be activated for elaborating alternative scenarios of agro-food systems at different scales, from the farm to the globe, allowing one to compare the merits of different policy measures in terms of environmental burden. Our examples at the regional scale will be taken from contrasting systems, focusing in parallel on the

BOX 25.1

THE GRAFS APPROACH

The GRAFS approach (Billen et al., 2014) describes the agro-food system of a given regional entity by accounting for the fluxes of nutrients across and between four main compartments: cropland, grassland, livestock biomass, and the local human population (Fig. 25.1). It can be applied to several nutrient flows such as nitrogen (N), phosphorus (P), or carbon (C) (Le Noë et al., 2017), although in this chapter the focus is kept on N flows. Tracing the N flows across the agro-food system enables one to point out the N losses occurring at each stage of the chain and therefore to relate the agro-food system's structure to the environmental load in term of N pollution risks. Environmental load is defined here as (1) new N resources consumed by agriculture and people, including synthetic fertilizers, N-symbiotic fixation by legume crops, atmospheric deposition, and imported feed and food and (2) N losses resulting from N soil surplus (considered a good indicator for N losses to the hydro-system; see Garnier et al., 2016) and N volatilization from livestock excreta and synthetic N fertilizers (Sanz-Cobena et al., 2014).

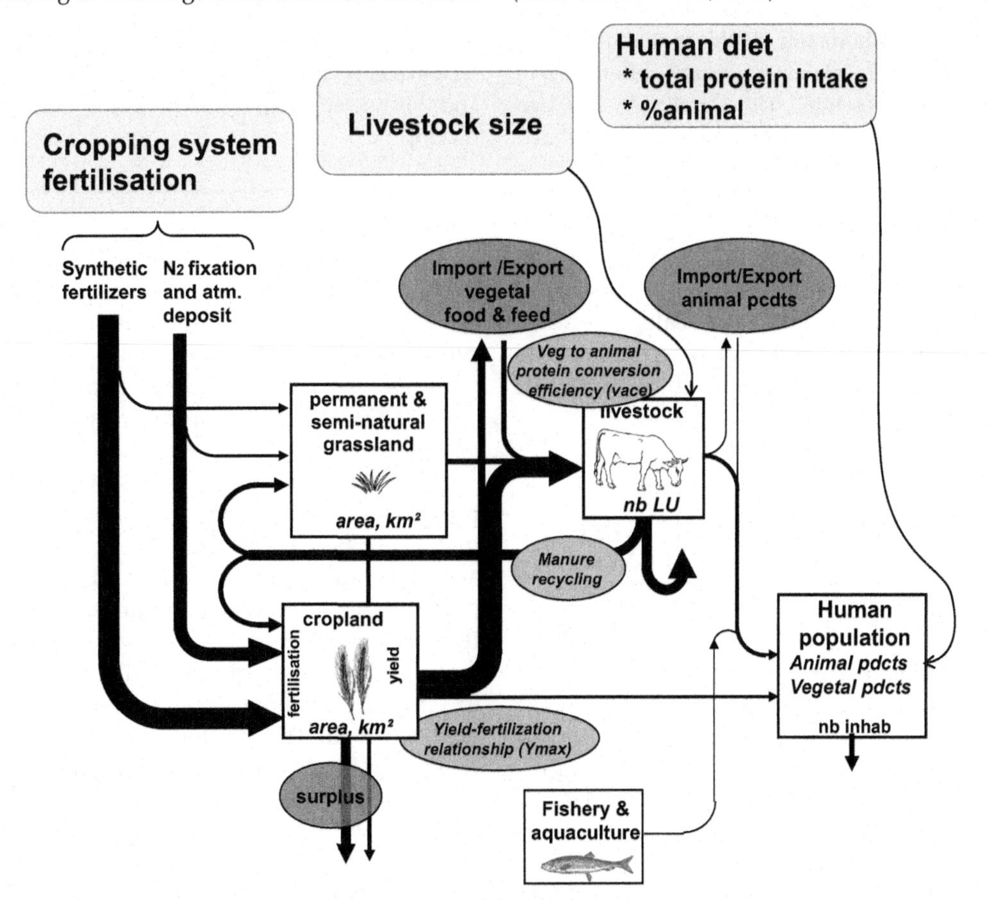

FIGURE 25.1 Conceptual model of the structure of a territorial agro-food system, according to the GRAFS approach (Billen et al., 2014).

Continued

BOX 25.1 (*cont'd*)

THE GRAFS APPROACH

The major processes involved in the agronomic function of the agro-food system can be characterized by (1) the relationship between yields and N input to the soil in cropping systems; (2) the conversion efficiency of vegetal into animal protein in livestock systems; and (3) the composition of the human diet including food waste (Billen et al., 2013). Besides, three major structural characteristics affect the environmental performance of the system: (1) the connection between crops and livestock production; (2) the balance between food production and consumption; and (3) the spatial allocation of resources in relation to spatial differences in production potential. The connection between crop and livestock production is closely related to the potential for recycling N within the land area, thus limiting the recourse to new sources of N such as synthetic fertilizers or symbiotic N fixation producing manageable amounts of manure. The balance between food production and consumption reflects the net trade pattern associated with the agro-food system. Finally, the spatial allocation of resources highlights how the specialization or diversification of the agro-food system shapes the N cycle within a territory. All these features make the GRAFS approach a well-suited methodology to explore the benefits of alternative scenarios of the agro-food systems with a biogeochemical perspective, and it is a useful tool for helping with decision-making.

regional (subnational) scale with France and the Iberian Peninsula (Le Noë et al., 2017; Lassaletta et al., 2014a), and on the global scale, where we will consider 12 major world macroregions as defined in Lassaletta et al. (2014b,c, 2016) and Billen et al. (2015).

STRUCTURAL CHARACTERISTICS OF AGRO-FOOD SYSTEMS

Crop-Livestock Farming (Dis)connection

The growing disconnection between crop and livestock farming results from the development of specialized cropping systems mainly dependent on industrial fertilizers and other agrochemicals on the one hand, and of specialized landless livestock systems based on imported feed (such as soybean, maize, and barley) on the other hand. A small proportion of manure in total N inputs to cropland soil or a large proportion of imported feed to total livestock ingestion are, respectively, good indicators of the prevalence of these two opposite specialized systems, as shown in Fig. 25.2A–D at the national scale of France, Spain, and Portugal and at the global scale.

Specialized cropping systems, with a low proportion of manure inputs to arable soils, are mainly located in the Paris basin area, as well as in the central areas of Spain and East Portugal (Fig. 25.2A). Specialized livestock farming systems are present in France mainly in Brittany and Pays de Loire, and in the Iberian Peninsula in Galicia, Catalonia, Valencia, and Murcia, as well as in all the northwest part of Portugal (Fig. 25.1C). At the global scale, the role of manure as a source of N to cropland soil is particularly low in north and southeast Americas and in India (Fig. 25.2B). In China, the percentage of manure is lower because of the very

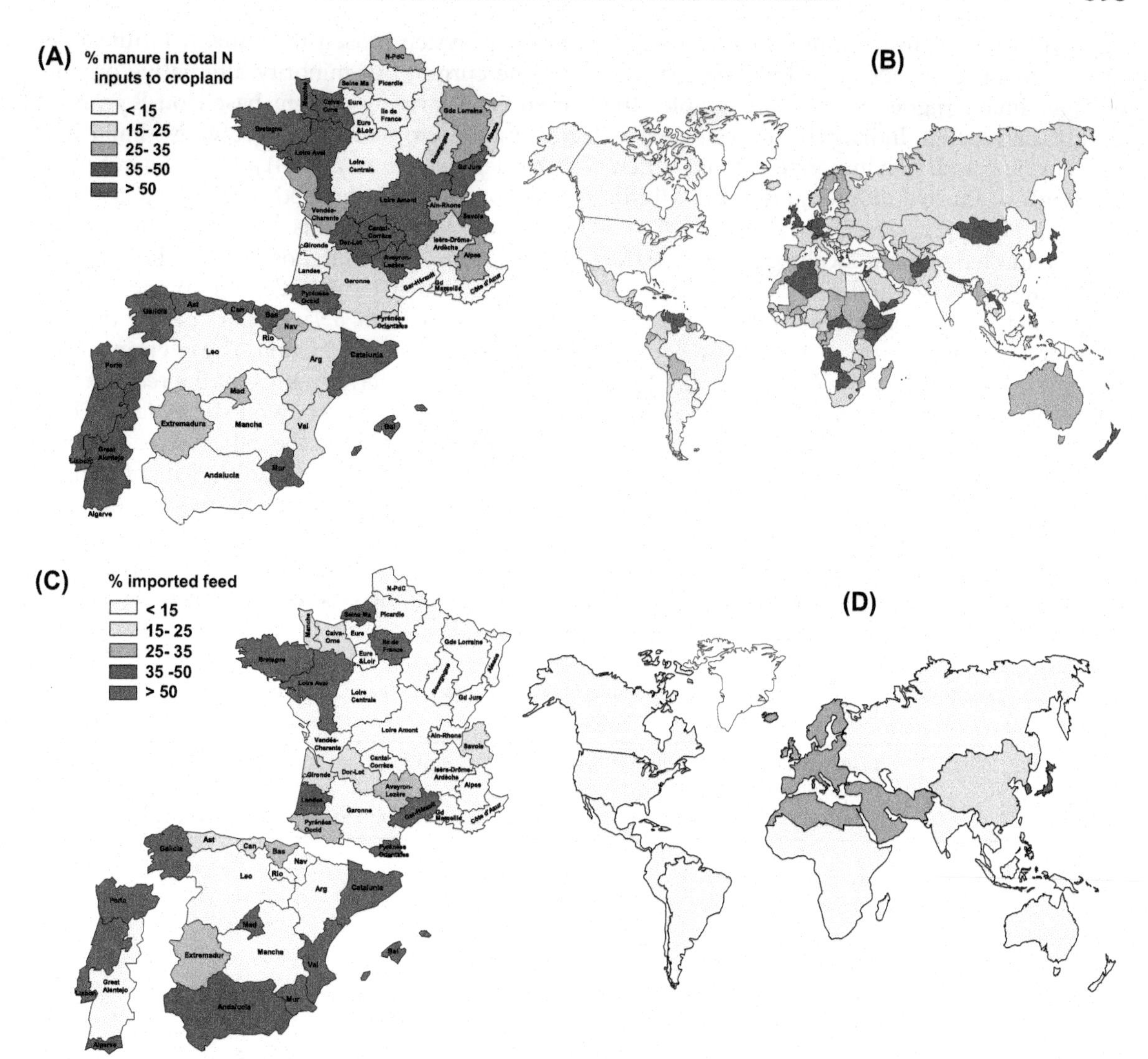

FIGURE 25.2 Geographic distribution of specialized cropping and livestock systems in France, Spain, and Portugal, and in 12 regions of the world, based on the share of manure in total N fertilization of cropland (A and B), and on the share of imported feed in total livestock nutrition (including grazing) (C and D). In the top panels, specialized cropping systems, disconnected from livestock, appear in *pale yellow* (light gray in print version). In the bottom panel, specialized livestock systems, disconnected from local cropping systems, appear in *red* (dark gray in print version) and *purple* (gray in print version). *Data from Billen, G., Lassaletta, L., Garnier, J., 2014. A biogeochemical view of the global agro-food system: nitrogen flows associated with protein production, consumption and trade. Global Food Security 3, 209–219; Lassaletta, L., Billen, G., Garnier, J., Bouwman, L., Velazquez, E., Mueller, N.D., Gerber, J.S., 2016. Nitrogen use in the global food system: past trends and future trajectories of agronomic performance, pollution, trade, and dietary demand. Environmental Research Letters 11, 095007. https://doi.org/10.1088/1748-9326/11/9/095007 for the world; from Le Noë, J., Billen, G., Lassaletta, L., et al., 2016. La place du transport de denrées agricoles dans le cycle biogéochimique de l'azote en France: un aspect de la spécialisation des territoires. Cahiers Agricultures 25, 15004; Le Noë, J., Billen, G., Garnier, J., 2017. Nitrogen, phosphorus, and carbon fluxes through the French agro-food system: an application of the GRAFS approach at the regional scale. Science of the Total Environment 586, 42–45 for France and Spain.*

high synthetic fertilizer inputs, so the absolute rate of manure is about 50 kgN/ha year on average, much higher than, for example, the 15 kgN/ha year in India. The dependence of livestock husbandry on imported feed is particularly high in Europe, the Maghreb, the Middle East, and Japan (Fig. 25.2D).

Disconnected systems locally generate greater environmental losses and use more external resources per hectare of agricultural surface than connected systems do. A comprehensive example is provided by the Garnier et al. (2016) study of the farming systems in the "Brie Laitière" region in the Paris Basin (Northern France). The authors compared four farming systems: (1) the 1950s system, when the region was still a mixed crop and livestock farming area famous for its production of Brie cheese; (2) the currently dominant livestock-free cash crop farming system based on mineral fertilization; (3) the current but minority alternative organic cash crop farming system based on 8 years of diversified crop rotation involving alfalfa and other legume crops; and (4) a fictitious but plausible scenario of reintroducing dairy cows in organic farming systems (not necessarily at the single farm scale but possibly at the regional scale) to find a local outlet for forage legume crops, which form a large part of the organic cropping system production. The environmental and agronomic performances of the reconnected organic systems were shown to be by far the best when compared with the other three systems (Table 25.1). Indeed, in addition to offering a much more diversified landscape, it relies on very few external resources for a much higher net production of vegetal and animal commodities than was the case for the 1950s system. Its

TABLE 25.1 Agronomic and Environmental Performance of Historical, Current, and Hypothesized Farming Systems in the Brie Region (Paris Basin, France)

Indicator	1950s Mixed Crop-Livestock Farming System	2010 Conventional Specialized Cash Crop System	2010 Organic Specialized Cropping System	Fictitious Organic Mixed Crop-Livestock System
Percent grassland in total agricultural area	20	0	0	20
Cash crop production (kgN/ha yr)	38	169	133	46
Animal production (kgN/ha yr)	3.5	0	0	14
Animal protein in percent of total production	8	0	0	23
Exogenous N fertilizer use (kgN/ha yr)	14	150	16	0
Cropland N soil balance surplus (kgN/ha yr)	18	30	15	14
Efficient orgC inputs to soils (kgC/ha yr)[a]	645	691	649	882

[a] *Sum of the yearly efficient organic carbon inputs to arable and grassland soils, taking into account the humification coefficient of each type of input (see Le Noë et al., 2017).*

From Garnier, J., Anglade, J., Benoit, M., Billen, G., Puech, T., Ramarson, A., Passy, P., Silvestre, M., Lassaletta, L., Trommenschlager, J.M., Schott, C., Tallec, G., 2016. Reconnecting crop and cattle farming for a reduction of nitrogen losses in an intensive agricultural watershed. Environmental Science and Policy 63, 76–90.

cropland N soil balance shows the lowest surplus, thus minimizing environmental losses. Moreover, its total input of humified efficient organic carbon (Ceff, in the sense of Vleeshouwer and Verhaegen, 2002) is by far the highest of the four systems, indicating the ability of this system to sequester organic C in the soil.

These improved performance results can be explained by better N use efficiency dueto a better balance between N inputs, crop requirements, crop production, and livestock consumption. Note that the lack of a local outlet for forage legumes is often a severe problem for the viability of the organic stockless specialized cropping systems (Nowak et al., 2013). The reintroduction of livestock breeding activity would therefore remedy this situation and ensure better economic durability (Bonaudo et al., 2014; Lemaire et al., 2014).

Another example, from extensive Andalusian rainfed agricultural systems (south of Spain) (Table 25.2), shows that connected crop and livestock systems are much less dependent on external fertilizer resources and, at least in the case of organic systems, give rise to lower environmental losses than specialized systems.

Generalizing the results of these examples, recent studies on crop and livestock integration demonstrated the ability of mixed crop and livestock systems to enhance soil fertility, nutrient use efficiency, and pest regulation compared to specialized systems (Lemaire et al., 2014; Soussana and Lemaire, 2014; Martin et al., 2016). Besides, the simultaneous presence of livestock breeding and cropping systems within a restricted area favors the fertilization of arable land by organic amendments through manure application. This type of fertilization, common in the so-called agroecologic farming systems frequent in South and Central American regions (Holt-Giménez and Altieri, 2012), enhances the quality, the quantity, and the mean residence

TABLE 25.2 Agronomic and Environmental Performance of Three Typical Rainfed Farming Systems in Andalusia (Spain)

Indicator	Conventional Stockless Cropping System	Conventional Mixed Crop and Livestock System	Organic Mixed Crop and Livestock System
Percent grassland (in total agricultural area)	0	37	54
Cash crop production (kgN/ha yr)	56	23	3
Animal production (kgN/ha yr)	0	4.5	5.9
Animal protein (in percent of total production)	0	16	66
Exogenous N fertilizer use (kgN/ha yr)	106	73	4
Cropland N soil balance (kgN/hacropland yr)	54	66	−2
Efficient orgC inputs to cropland (kgC/ha yr)[a]	159	120	194

[a] *Sum of the yearly efficient organic carbon inputs to arable soils, taking into account the humification coefficient of each type of input (see Le Noë et al., 2017).*

Data are from interviews with farmers (Aguilera, unpublished data). Unless specified, figures are per ha total utilized agricultural area.

time of organic matter in soils, which is beneficial for soil carbon sequestration, soil resilience, and its ability to provide nutrients to the crops while limiting nutrient losses (Maillard and Angers, 2014; Martin et al., 2016; Diacono and Montemurro, 2010; Kopittke et al., 2016).

At the regional and global scale, Fig. 25.3 shows that there is a general link between the degree of specialization of farming systems and the environmental burden, as defined, for example, by the potential N losses to the environment (here, indicated by the average cropland N surplus). In Spain, the highest potential losses are found in both the temperate rainfed northwest area and the Mediterranean areas when there is livestock specialization (Sanz-Cobena et al., 2014; see also Fig. 25.2) or when land is irrigated and devoted to intensive crop production.

Regional and Macroregional Mismatch Between Human Food Demand and Production

The spatial distribution of the human population and its dietary composition, hence of food demand, is quite uneven, particularly with the rapid trend of urbanization, and often does not match productivity differentials between geographic locations. Water scarcity (which might become even more severe in the future) often limits agricultural production in some areas of the world with high demographic increase, as is the case of the Mediterranean basin countries (Iglesias et al., 2011). This is a second cause, besides agricultural specialization, for disconnection between production and consumption.

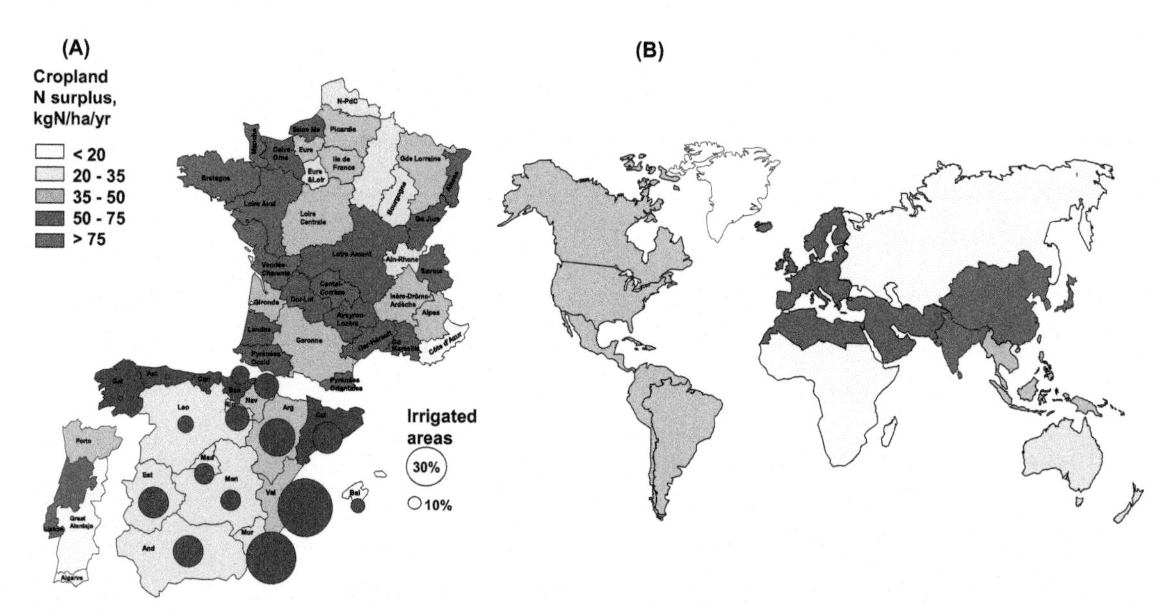

FIGURE 25.3 Geographic distribution of the cropland soil N balance surplus, taken as an indicator of environmental N losses from agriculture. For Spain, irrigated areas are treated separately from rainfed areas and represented as circles, the size of which refers to their proportion within total agricultural area. *Data from Le Noë J., Billen G., Garnier J., 2017. How the structure of agro-food systems shapes nitrogen, phosphorus, and carbon fluxes: the Generalized Representation of Agro-Food System applied at the regional scale in France. Science of the Total Environment 586: 42–55. https://doi.org/10.1016/j.scitotenv.2017.02.040 for France; Lassaletta, L., Billen, G., Garnier, J., Bouwman, L., Velazquez, E., Mueller, N.D., Gerber, J.S., 2016. Nitrogen use in the global food system: past trends and future trajectories of agronomic performance, pollution, trade, and dietary demand. Environmental Research Letters 11, 095007. https://doi.org/10.1088/1748-9326/11/9/095007 for the world and original calculations, distinguishing rainfed and irrigated surfaces, for Spain.*

The disparity of food self-sufficiency between regions in Europe, or of macroregions in the world, has greatly increased since the 1950s (Porkka et al., 2013; Lassaletta et al., 2016). At both scales, a large number of deficient regions are dependent on the import of either vegetal or animal products from a small number of highly productive regions (Fig. 25.4).

There exists an impassioned controversy on the localization of the food supply (Cowell and Parkinson, 2003). In line with the claim for food sovereignty, many citizens' movements and governments in developing countries consider that regional self-sufficiency of food production and consumption is more likely to increase food security than a globalized food system that also generates geopolitical dependency (Windfuhr and Jonsén, 2005; Holt-Giménez and Altieri, 2012). In many places, a strong wish to reconnect people with food, neighboring producers, and seasonality is currently growing (e.g., O'Kane, 2016). The opponents to these movements criticize its logic for its negation of productivity differentials between geographic locations and affirm that feeding a rapidly growing world population in a sustainable manner requires long-distance trade to ensure that food is produced most efficiently in the most suitable locations (Desrochers and Shimizu, 2008), without considering any environmental and socioeconomic implications.

An important aspect of this issue lies in the composition of the human diet. One obvious example is the incorporation in the ordinary diet of temperate countries of often luxury products from tropical areas (such as coffee, tea, bananas, etc.), which created a certain global market specialization since historical periods. But another more important and recent aspect is the share of animal products in the total protein diet, as this is a key factor determining whether it is possible or not to meet human food requirements locally (Erb et al., 2016). Billen et al. (2015) showed that a ratio of 35%–40%

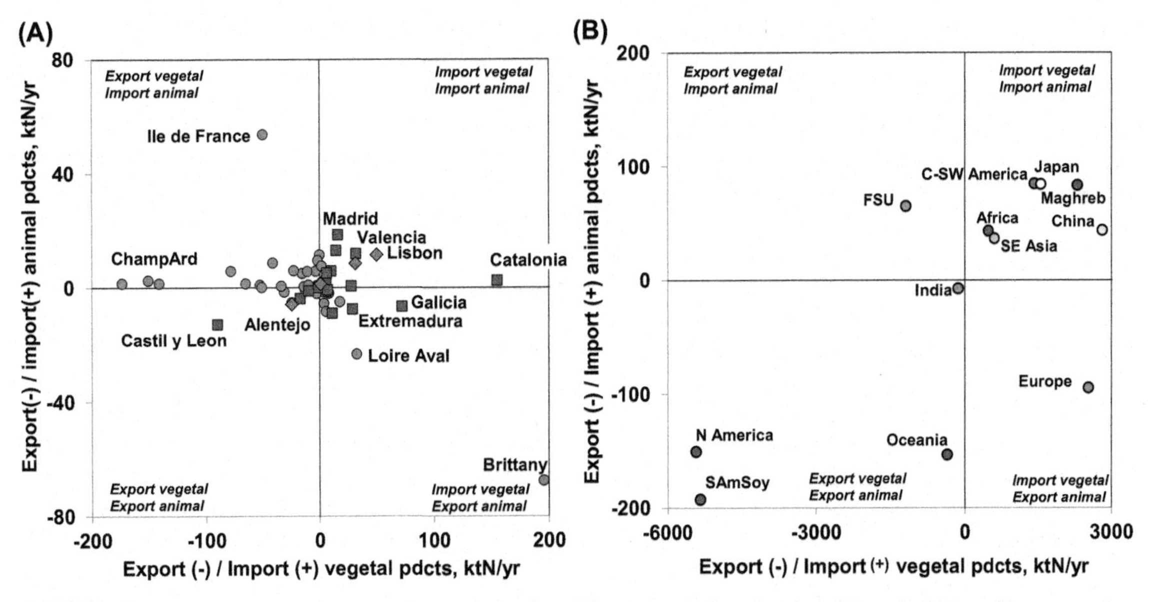

FIGURE 25.4 Degree of external dependence of French and Iberian agricultural regions (A), and of 12 world macroregions (B), as revealed by the magnitude of their net import or export of vegetal and animal products. Import is represented by positive values and export by negative ones. In panel (A), French regions are represented by *blue (gray in print version) circles*, Spanish ones by *red (dark gray in print version) squares*, and Portuguese ones as *green (light gray in print version) diamonds.*

animal protein in the diet represents both what can be shared by all inhabitants of the planet at the 2050 horizon with the current agronomic performance of the different world regions and what is desirable from a dietetic point of view. This "equitable diet" roughly corresponds to the traditional healthy Mediterranean diet, but this proportion of animal products in the total protein ingestion is largely exceeded in most western countries (Lassaletta et al., 2016); France and Spain respectively consume 69% and 60% animal proteins in the diet, excluding fish and seafood products.

Nitrogen Use Efficiency and Spatial Allocation of Fertilizing Resources

The different world megaregions, or the regions inside Spain and France, differ greatly from each other in terms of agronomic performance. Regarding cropping systems, as suggested by Lassaletta et al. (2014c), performance can be conveniently characterized by the value of the Ymax parameter of the relationship between yield (Y, expressed in N contained in the harvest, averaged over the whole crop rotation cycle) and N inputs to the soil (F, "total fertilization," the sum of all inputs of N, including fertilizers, manure, biologic N fixation, and atmospheric deposition), both expressed in kg N per ha of cropland and per year:

$$Y = Ymax \times F/(F + Ymax) \qquad (25.1)$$

This relationship has been proven a robust indicator of agronomic performance because it remains the same, under a given pedo-climatic context, whatever the crop rotations are, and in particular for organic versus conventional rotations, as shown by Anglade et al. (2015a) and Billen et al. (2018) from the thorough analysis of the French cropping systems. However, the Ymax parameter varies over more than one order of magnitude between the most fertile regions of Western Europe and North America and the less productive average cropping systems of India or Africa. Even within a country like France or Spain, Ymax values in their different regions range over more than a factor of 7 (Fig. 25.5A and B).

From relationship (1), the nitrogen use efficiency (NUE, i.e., the ratio of N output through harvest to total N input, as defined by EUNEP, 2015) in each i region can be calculated as a function of N input rate (F_i) as shown:

$$NUE_i = Y_i/F_i = Ymax_i/(F_i + Ymax_i) \qquad (25.2)$$

The marginal yield increase with respect to an increase of N input (dY_i/dF_i) has a simple relationship with NUE_i:

$$dYi/dF_i = Ymax_i^2 \Big/ (F_i + Ymax_i)^2 = NUE_i^2 \qquad (25.3)$$

Mueller et al. (2014) have developed the idea that the optimum allocation of fertilization resources among subregions of a geographic entity, i.e., the distribution of fertilization maximizing the total nitrogen use efficiency (NUE_g) for a given total crop production, is obtained when dY_i/dF_i is the same in each subregion. A formal proof of this statement is provided in Lassaletta et al. (2016). This means that the optimum fertilization allocation between regions is that corresponding to the situation when NUE and total NUE have the same value in all regions, irrespective of their differences in Ymax.

This concept has been applied at the world scale by Mueller et al. (2017) to show that there exists significant room to reduce global N pollution (up to 41%), at constant total agricultural production, just by reallocating new fertilizer resources between regions (Fig. 25.5D). They also showed that the trend observed during the last 50 years in the allocation of fertilizer resources between the different regions of the world leans toward increasing the distance from optimal distribution.

The same approach was used for France, Spain, and Portugal (each separately) (Fig. 25.5C). In all cases, the distance between

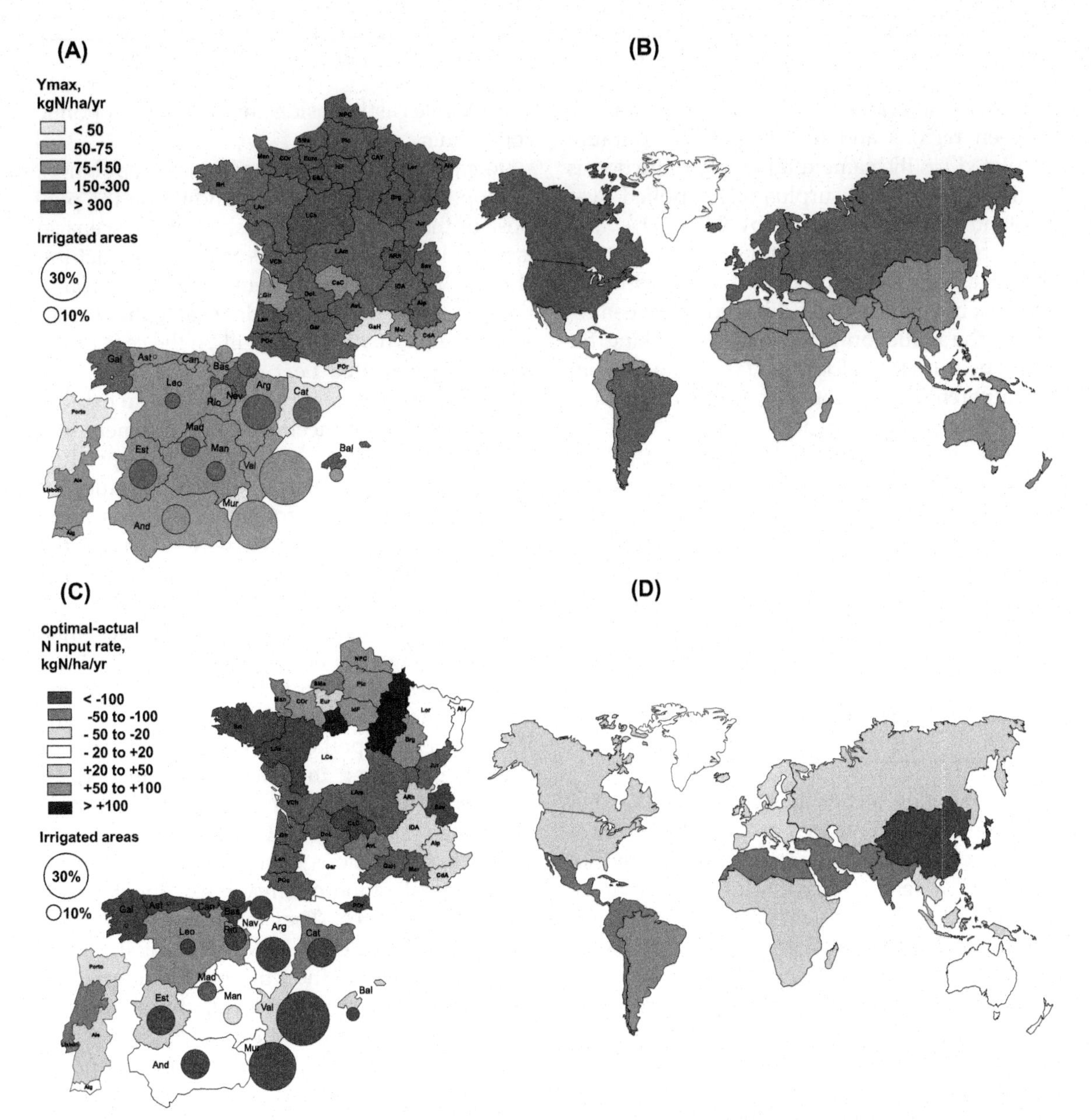

FIGURE 25.5 (A and B) Maximum yield at saturating N fertilization (Ymax) of the arable cropping systems of the different agricultural regions of France, Spain, and Portugal (A) and the world (B). For Spain, rainfed and irrigated arable cropping systems (excluding permanent crops) were distinguished; the size of the *circle* indicates the percent irrigated areas in total cropland. (C and D) Distance (in kgN/ha/year) of the current N fertilization rate of cropland with respect to the theoretically optimal allocation for the same total crop production. Red (gray in print version) indicates the actual fertilization rate above optimum allocation, blue (light gray in print version) indicates lower fertilization than optimum. France, Spain, and Portugal, treated individually (C); 12 megaregions in the world (D). *Calculations based on Le Noë, J., Billen, G., Garnier, J., 2017. Nitrogen, phosphorus, and carbon fluxes through the French agro-food system: an application of the GRAFS approach at the regional scale. Science of the Total Environment 586, 42–45 for France; Lassaletta, L., Billen, G., Grizzetti, B., et al., 2014c. 50 year trends in nitrogen use efficiency of world cropping systems: the relationship between yield and nitrogen input to cropland. Environmental Research Letters 9. https://doi.org/10.1088/1748-9326/9/10/105011; Le Noë et al. (2016) and new calculations for Spain and Portugal; Lassaletta, L., Billen, G., Garnier, J., Bouwman, L., Velazquez, E., Mueller, N.D., Gerber, J.S., 2016. Nitrogen use in the global food system: past trends and future trajectories of agronomic performance, pollution, trade, and dietary demand. Environmental Research Letters 11, 095007. https://doi.org/10.1088/1748-9326/11/9/095007; Mueller, N.D., Lassaletta, L., Runck, B., Billen, G., Garnier, J., Gerber, J.S., 2017. Declining spatial efficiency of global cropland nitrogen allocation. Global Biogeochemical Cycles 31, 245–257. https://doi.org/10.1002/2016GB005515 for the world.*

the current allocation of fertilization resources between regions and the theoretical optimum calculated for the same total crop production is significant. The total surplus of N applied over cropland could be reduced by 24%, 39%, and 37%, respectively, for France, Spain, and Portugal, with the same total national production, just by reallocating fertilization between regions. The reduction is particularly striking for Spain, because the calculation takes into account the difference in agronomic performance between rainfed and irrigated cropping systems; the latter indeed have a quite important role in the agricultural production of this country but use a disproportionately high amount of fertilizer resources. The optimal allocation of N inputs in Spain would imply reversing the current trend toward overfertilization in irrigated systems.

ELABORATION AND ASSESSMENT OF SCENARIOS

The previous section has indicated that besides agronomic practices at the plot or farm scale, three structural characteristics of territorial agro-food systems are crucial to their environmental performance. The issues at stake are (1) the composition of the human diet, (2) the local connections between food production and consumption, as well as the coupling of crop and livestock farming at the regional scale, and (3) the spatial allocation of fertilizing resources between regions of differing agronomic propensities. Changing these characteristics is the main lever for designing alternative, more sustainable systems.

Starting from the preceding description of the current structure of the agro-food systems of France, Spain, and the world, we here present alternative scenarios, elaborated in several studies and contexts, by acting on one or several of these levers from a biogeochemical point of view, while casting aside any social or economic constraints.

Regarding the size of the human population, two options are possible depending on the purpose of the scenarios established. If the scenario is intended to show the potential of structural changes in agro-food systems, then the population should be considered identical to the current reference population; this will be the case of the scenarios presented next at the scale of France and Spain. If the scenario is intended to explore the possibilities of meeting the requirement of a changing population with alternative production patterns, then the projected population of the system at the horizon concerned has to be taken into account; this will be the case for the scenarios described subsequently at the global scale.

These scenarios offer hypothetical cases illustrating the effect to be expected from structural changes brought about in the agro-food systems at the regional or global scale. By no means should these hypothetical scenarios be considered prescriptive: they are not intended to show what the future of the agro-food system should be, but simply to explore the consequences of audacious choices implying change, or of generalization of already existing, but still marginal practices. They also do not address the difficult question of how unlocking the resistance to changing the current social-technical "state of affairs," a major sociopolitical and scientific question that must be viewed in a transdisciplinary perspective, beyond the scope of the present chapter.

At the Regional Scales in France and Spain

For France, Spain, and Portugal, a scenario called Local, Organic, and Demitarian (Billen et al., 2012, 2013) has been established to explore the effects of three major structural changes in the agro-food system. This scenario was described in detail for France in Billen et al. (2016, 2018). A slightly modified and updated version is presented here, called Autonomous, Organic, and Demitarian.

The scenario assumes the generalization of a healthy human diet, in line with the "equitable diet" mentioned earlier, and also in line with the Barsac declaration (http://www.nine-esf.org/Barsac-Declaration) to halve the animal protein consumption in developed countries ("demitarian diet"). Livestock is sized to the capacity of local grassland and cropping systems to meet its feed requirements, without recourse to import from outside and also saving crop production to direct human consumption. Organic farming practices, based on manure crop fertilization and symbiotic N fixation by legumes, are generalized (see details of the scenario hypotheses in Box 25.2).

The results of this scenario are summarized in Table 25.4. In France, because of the strong reduction of meat consumption, livestock density and production are decreased by about half as a national average, but they are redistributed much more evenly among the regions (Fig. 25.6), in line with the objective of the scenario to reconnect crop and animal farming in overspecialized regions.

Overall, the French Auto-Org-Dem scenario shows an excess of animal production over human requirements, available for international export (about 26 ktN/year), contrasting with the slight net current import. The possible distribution of the interregional exchanges of animal proteins is illustrated in Fig. 25.6 and compared with the current exchanges of meat and milk revealed by the analysis of the commodity transport statistical database, SITRAM (Le Noë et al., 2016).

In the current situation, Spain is a net exporter of meat and milk for a total of 38 ktN/year, but it has to import 514 ktN/year, mostly as soybean, to feed its livestock, in good agreement with the figures of Lassaletta et al. (2014a). In the Auto-Org-Dem scenario, livestock is nearly halved and entirely sustained by local grassland and cropland production. At the national level, this is enough to meet the requirements of the population with the demitarian regime. Regions such as Madrid, Catalonia, and Andalucía are dependent on imports of animal products from other regions such as Castilla y Leon, Gallicia, and others (Fig. 25.6).

Note that the total area of grassland in France and Spain has slightly increased in the scenario, but the proportion of grassland in animal feeding has increased quite significantly from 41% to 70% in France and from 22% to 69% in Spain due to lower animal production. Soto et al. (2016) underlined the fact that current grassland production in Spain actually has the potential to be much higher than it is now, as much of it has been underused or abandoned since the 1960s.

Crop production considerably decreases in the scenario, with respect to the current observed values, as a consequence of the lack of synthetic fertilization. In France, crop production decreases by 30%. Surprisingly, however, the amount of crop production in excess of national consumption by humans and livestock is more than twice as high in the scenario than in the current situation, as a result of the decrease of forage crop demand caused by livestock reduction: the excess production increases from 300 ktN/year in 2006 to 750 ktN/year in the scenario. This excess is, however, of a different nature, from currently mostly cereals to principally forage and grain legumes in the scenario. This material, available for export, might also be suitable for nonfood uses, including methanization for energy production. (Couturier et al., 2017)

In the current situation, Spain is not self-sufficient in vegetal production and has a net balance of vegetal protein import of about 385 ktonN/year. The imports are mainly soybean from South America to feed livestock (Lassaletta et al., 2014a), while Spain exports fruits and vegetables, as well as animal products to the rest of Europe. In the Auto-Org-Dem

BOX 25.2

CONSTRUCTION OF AUTONOMOUS, ORGANIC, AND DEMITARIAN SCENARIOS FOR FRANCE AND SPAIN

The construction of the scenario involves four steps:

1. Human consumption in each region is reduced by 15% (corresponding to a reduction of food waste by 50% at the consumption stage), while the proportion of animal proteins is reduced to 40% of the total protein intake.
2. In an attempt to reconnect cropping with livestock systems and to avoid specialized landless livestock systems as well as stockless cropping systems, the number of livestock in each region is sized according to the capacity of each region to feed it locally, without importation of feed. Grassland is the preferred source of feed for ruminants and is imposed to represent at least 60% of total animal nutrition (ruminants and monogastrics), while a maximum of 50% of cropland production can be devoted to animal feeding. This implicitly imposes a rather high ratio of ruminant to monogastric livestock.
3. The permanent grassland area in each region is adjusted to a minimum of 0.5 ha/LU, but it is not allowed to decrease with respect to the current value. The arable land surface is calculated as the difference between total agricultural land (which is maintained constant at the current value) and permanent grassland areas. The overall productive capacity of the cropping systems (Ymax) is kept unchanged in the scenario, but a different cropping rotation is assumed, with more N_2-fixing legume crops and no synthetic fertilization. The basic assumption made is that, in each region, the relationship between yield and N input is the same for current and alternative agricultural practices. The overall production of the new cropping system can therefore be calculated from the total rate of N input to cropland.
   As this N input in the alternative system is very much dependent on symbiotic N_2 fixation, in addition to manure inputs, a hypothesis on the productivity of legume crops has to be made for each region. The calculation is based on a few current organic crop rotations observed in different regions of France and Spain (Table 25.3). The total N input to cropland integrated over the whole rotation cycle is then calculated as the sum of the legume N_2 fixation (calculated according to Anglade et al., 2015b) and manure application (estimated from livestock density taking into account that livestock spend 50% of the time in housing, and considering 30% loss of N during storage and handling, especially as ammonia, Sanz-Cobena et al., 2014). Total crop production is then calculated from Ymax and total N input, according to relationship (1) shown earlier. A quite similar approach is applied to calculate the N inputs and production of grassland.
4. Finally, excesses or deficits of vegetal and animal products are calculated in each region from the balance between production and requirements of humans and livestock. With these constraints, a number of highly populated regions cannot meet their own needs in animal or vegetal proteins, even taking into account the demitarian diet; in that case, the supply is ensured by possible long-distance trade, preferably from neighboring regions or from abroad.

TABLE 25.3 Current Crop Rotations in the Different Agricultural Regions and Reference Yield for the Legume Crops Involved

Crop Succession	Nb Year	Ref Yield for Legumes	kgN/ha yr Rainfed	Irrigated
PARIS BASIN (NORD AND PAS-DE-CALAIS, SEINE BASIN, ETC.)				
Alf, Alf, WW, Cer2, Sunflower, Grain legume, WW, Cer3	8	Alfalfa	250–300	–
		Grain legume (pea, lentil, faba bean, etc.)	50–100	–
SOUTHEAST FRANCE (ARDÈCHE, DRÔME, ETC.)				
Alf, Alf, WW, Cer2, Sunfl	5	Alfalfa	50–90	–
CANTABRIC				
Vetch as cover crop, maize	1	Vetch (green manure)	50	50
BASQUE COUNTRY				
Vetch as cover crop, maize	1	Vetch (green manure)	110	110
NAVARRA-RIOJA				
Chick peas, cereal	2	Chick peas	40	64
ARAGON				
Chick peas, cereal	2	Chick peas	40	64
GREAT EXTREMADURA				
Vetch, cereal	2	Vetch	45	124
MADRID				
Vetch, cereal	2	Vetch	40	130
GREAT MANCHA				
Vetch, cereal	2	Vetch	30	88
N-CASTILLA LEON				
Chick peas-cereal	2	Chick peas	26	46
Alf, root crop, potatoes, maize	4	Alfalfa	135	298
CATALONIA				
Vetch, cereal	2	Vetch	70	122
VALENCIA, MURCIA				
Vetch, cereal	2	Vetch	35	85

(*Continued*)

TABLE 25.3 Current Crop Rotations in the Different Agricultural Regions and Reference Yield for the Legume Crops Involved—cont'd

Crop Succession	Nb Year	Ref Yield for Legumes	kgN/ha yr Rainfed	Irrigated
BALEARES				
Chick peas, cereal	2	Chick peas	29	64
ANDALUCÍA				
Vetch, cereal	2	Vetch	45	100

Data from Anglade, J., Billen, G., Makridis, T., Garnier, J., Puech, T., Tittel C., 2015a. Nitrogen soil surface balance of organic vs conventional cash crop farming in the Seine watershed. Agricultural Systems 139, 82–92; Anglade, J., Billen, G., Garnier, J., 2015b. Relationships for estimating N2 fixation in legumes: incidence for N balance of legume-based cropping systems in Europe. Ecosphere 6, art37; Le Maitre, pers. comm.; Guzmán, G., Aguilera, E., Soto, D., et al., 2014. Methodology and Conversion Factors to Estimate the Net Primary Productivity of Historical and Contemporary Agroecosystems. Sociedad Espnola de Historia Agraria. Documentos de Trabajo DT-SEHA n 1407. Mayo 2014. www.seha.info; Doltra, J., 2015. Modeling nitrous oxide emissions from organic and conventional cereal-based cropping systems under different management, soil and climate factors. European Journal of Agronomy 66, 8–20. https://doi.org/10.1016/j.eja.2015.02.002; European Commission (DG ENV), 2010. Environmental Impacts of Different Crop Rotations in the European Union. Contract n 07.0307/2009/SI2.541589/ETU/B1. Final Report, 6 sept 2010.

scenario, crop production is reduced by 40% due to the lack of synthetic fertilization, but the country has a surplus of vegetal proteins of 71 ktonN/year available for export or other uses.

In terms of environmental contamination, the N surplus of cropland is reduced in France by a factor of two and in Spain by a factor of six. Losses of N to the atmosphere in the form of ammonia linked to manure management decreases by more than a factor of two in both countries.

In Fig. 25.7, the Auto-Org-Dem scenarios for France and Spain are compared with the respective current situations in terms of total N losses to the environment and of total long-distance trade required to feed the regional population and livestock. In both cases, the Auto-Org-Dem scenario meets the food requirements of the population with much less N loss and long-distance transport of food and feed.

At the Global Scale

Four scenarios of the global agro-food system at the 2050 horizon have been described by Lassaletta et al. (2016). In all scenarios, the population in each region is that projected by the United Nations (2011), while the area of cropland and grassland is kept constant at the 2009 level. The agronomic potentials (Ymax of cropland vegetal to animal protein conversion efficiency) of the crop and livestock production systems of each region are those extrapolated from past trends (thus incorporating agronomic progress) and are the same across all scenarios.

In the standard business as usual (BAU) scenario, it is assumed that production trends and regional specialization, as well as disparities in human diet among regions, continue as predicted by economic models linking them to GDP (Lotze-Campen et al., 2008, 2010, Schmitz et al., 2012, Bodirsky et al., 2012, 2014). In an alternative scenario, called the self-sufficiency/equitable diet (SSED), it is assumed that regions attempt to meet domestic demand for both animal and vegetal proteins by local, diverse production and aim at limiting their dependence on imports, while converging to the healthy, equitable human diet; interregional trade occurs only when local needs cannot be met by local production given the other constraints considered. One variant of each of these scenarios (BAU opt. fert. and SSED opt. fert.) is constructed by assuming optimized allocation of fertilization

TABLE 25.4 Summary of the Results of the Autonomous-Organic-Demitarian (Auto-Org-Dem) Scenario for the Whole of France and Spain, compared with the values for the current situation

		France		Spain	
		2006	**Auto-Org-Dem**	**2009**	**Auto-Org-Dem**
Population	Minhab	64	64	44	44
Agricultural land area	M ha	26	26	23	23
Percent grassland	%	31	33	39	40
Cropland production	ktN/yr	2098	1434	665	378
Grassland production	ktN/yr	851	931	270	432
Livestock	M LU	20	11	11	5.8
Percent ingestion on grassland	%	41	70	22	70
Percent ingestion on local cropland	%	36	30	36	30
Percent ingestion on imported feed	%	23	0	42	0
Production meat and milk	ktN/yr	277	149	188	81
Net import (+) or export (−)					
Meat and milk	ktN/yr	11	−26	−38	−1
Crop product	ktN/yr	−303	−849	−128	−71
Feedstuff	ktN/yr	284	0	514	0
Total interregional trade[a]	ktN/yr	785	62	573	84
N surplus cropland	ktN/yr	995	458	1023	161
NH_3 volatilization	ktN/yr	443	183	272	149

[a] *Excluding export to foreign countries.*

resources between the world regions. This is obtained by adjusting fertilization (either as synthetic N fertilizer application or as symbiotic N fixation through inclusion of legumes in the crop rotations) in such a way that the NUE in each region is equal to the value defined at the global scale for meeting the requirements of crop protein consumption by humans and livestock under each scenario's assumptions.

The different options taken regarding dietary choices and allocation of resources between regions result in large differences between scenarios in terms of intensity of interregional trade and environmental N contamination. These two indicators for the four scenarios are compared with the historical trajectory of the world agro-food system since 1961 (Fig. 25.8). The total global N environmental losses between 1961 and 2009 (estimated as the sum of total N surplus from arable soils and ammonia emission associated with manure management) increased by a factor of 4.5. Trade dependency (the total amount of agricultural products traded between regions) grew by a factor of seven in the same period. In the BAU standard scenario, the total N environmental losses would continue to rise by 15% in the 2009–50 period, while trade of agricultural products would more than double

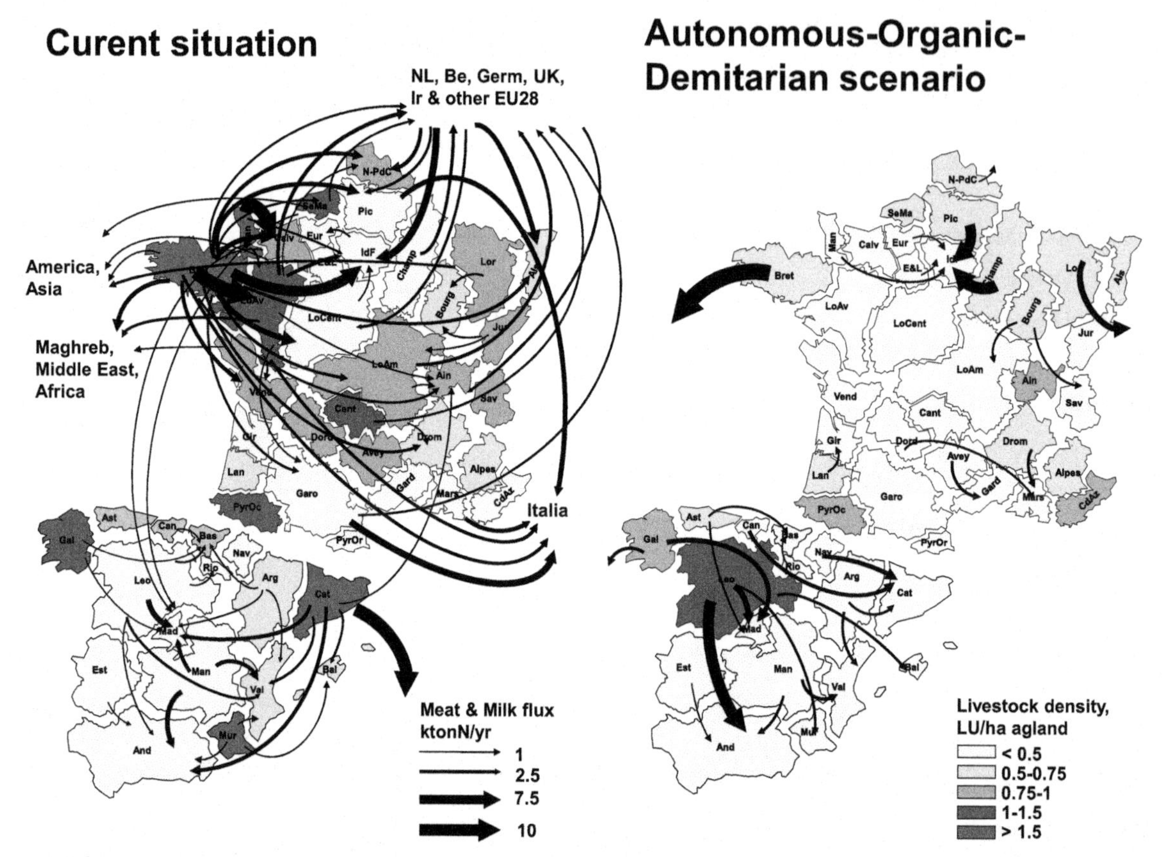

FIGURE 25.6 Distribution of livestock density and trade of animal products in the current situation (left) and in the Auto-Org-Dem scenario (right). For the current situation in France, trade fluxes are drawn according to the SITRAM database compilation (Le Noë et al., 2016). As such data are not available for Spain, the fluxes shown are just one possible distribution, taking into account the internal budget of production and consumption of animal products within each region. Similar calculations/assumptions have been made for the transport fluxes shown in the scenario for France and Spain.

(from 13 to 30 Tg N/year) compared to current levels (Fig. 25.8). In contrast, the SSED local scenario in 2050 would succeed in reducing international trade of agricultural products by 40%, and at the same time reduce environmental N losses by 33% between 2009 and 2050. The SSED scenario with optimum allocation of fertilization, which implies a complete redistribution of fertilization resources between world regions, would allow a quite significant further reduction of environmental losses. However, the optimal allocation would also lead to a large increase in the volume of trade between regions to compensate for a loss of regional self-sufficiency. As expected, this increase in N traded globally is larger in the BAU than in the SSED scenario.

The difference between the two latter scenarios illustrates the strong influence of human diet on both environmental N losses and dependence on international trade. The SSED scenarios, with a tendency toward self-sufficiency and equitable diets in all world regions, result in much less environmental N loss and trade. However, even in the SSED local scenario, the N surplus in croplands remains about twice what it could

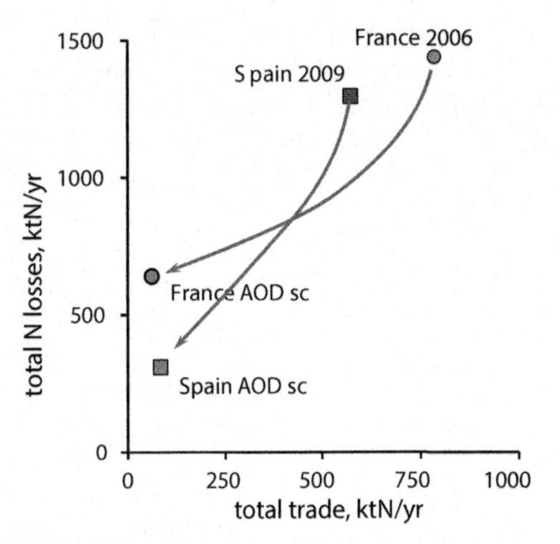

FIGURE 25.7 Comparison of the Auto-Org-Dem scenarios (*green circles [gray in print version]* and *squares* for France and Spain, respectively) with respect to the current situations in terms of total N losses to the environment and total long-distance trade required to feed the regional population and livestock.

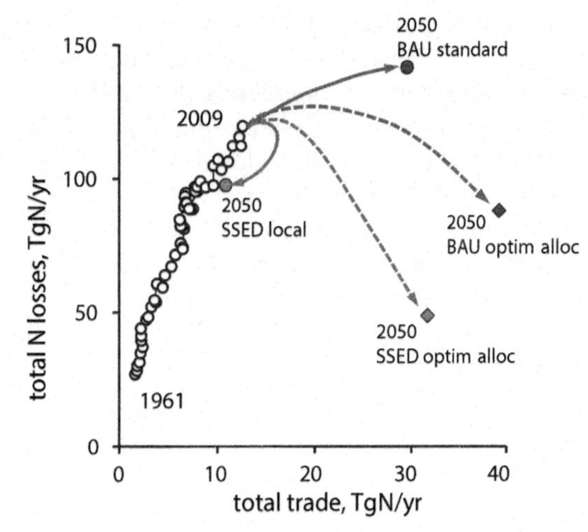

FIGURE 25.8 Volume of agricultural commodity trade (here expressed in Tg of N protein traded between 12 mega-regions) and environmental contamination (here estimated as the total losses of N from arable soils and manure management) characterizing the global agro-food system. From 1961 to 2009, the trajectory followed shows a rapid increase of both trade and contamination. The BAU standard scenario would pursue these increases. The SSED scenario, maximizing regional self-sufficiency and convergence toward an equitable diet, would reverse these trends. Optimal allocation of fertilizing resources would in both cases further reduce N contamination, but at the expense of an increased volume of food and feed traded. *Redrawn from Lassaletta, L., Billen, G., Garnier, J., Bouwman, L., Velazquez, E., Mueller, N.D., Gerber, J.S., 2016. Nitrogen use in the global food system: past trends and future trajectories of agronomic performance, pollution, trade, and dietary demand. Environmental Research Letters 11, 095007. https://doi.org/10.1088/1748-9326/11/9/095007.*

be in case of an optimal allocation of fertilization among regions (Fig. 25.8). This indicates a trade-off between the search for local self-sufficiency and that of minimizing global environmental pollution: given the distribution of the population in the world, self-sufficiency would have a cost in terms of environmental quality.

CONCLUSION

The GRAFS approach, based on the description of nutrient fluxes from production to consumption, offers a useful framework for analyzing the connections and disconnections between cropping and livestock farming systems, as well as between local consumption needs and their matches with agricultural production, within different geographic scales, from the small rural region to the continent. It also offers a tool for conceiving alternative systems with better connections and recoupling of nutrient cycles and assessing its environmental and biogeochemical imprints.

We have here illustrated this approach at the national and regional scales of France and Spain, as well as at the global scale. We have shown that projecting the current trends of specialization and disconnection of agro-food systems at both scales exacerbates negative impacts on the environment in terms of consumption of resources and N losses. We have also shown that reconnection scenarios can be designed in such a way that the environmental imprint of agro-food systems is considerably reduced, while making it possible to ensure food security.

Optimum spatial allocation of fertilization resources between geographic entities with differing agronomic propensities also has the potential to improve overall N use efficiency, at the national as well as at the global scales. At the large scale, this optimum spatial allocation of fertilization resources would require strong and efficient international governance and planning (e.g., Share Socioeconomic Pathway Scenario SSP1 for 2050, van Vuuren et al., 2017). In the current situation, local considerations, such as the respect of the pollution ceiling, are instead the basis of most environmental policies. The concept of optimum allocation might be useful, however, to indicate the margins of progress offered by the simple spatial distribution of agricultural practices at a constant production level. In the case of Spain, where our analysis distinguishes between rainfed and irrigated systems, it clearly demonstrates the excessive fertilization of irrigated systems (Fig. 25.5C).

The question addressed throughout this chapter is that of the trade-off between self-sufficiency and efficiency, which places the role of long-distance trade in the spotlight. The data presented first confirm that a better connection at the local and regional scale between production and consumption (including between crop and livestock farming) offers more opportunities to increase nutrient recycling, thus minimizing environmental losses. Scenarios privileging reconnection and food and feed self-sufficiency are most often much more environmentally virtuous than current open systems or BAU scenarios.

A historical correlation exists between population densities and the capacity of regional agricultural systems to feed them: at both regional and world scales, areas with high population density historically developed in pace with the development of locally specific agricultural systems able to meet the local human food demand (e.g., Mazoyer and Roudart, 1998; Billen et al., 2009). However, this past determinism has been surpassed, and the current distortion in the geographic distribution of the human population with respect to the agronomic potentialities of agricultural land makes long-distance trade of agricultural commodities unavoidable. Moreover, we have shown (see Fig. 25.8) that the strict application of the principles of optimum allocation of fertilizing resources to cropland among an ensemble of regions often contradicts the search for maximizing regional self-sufficiency. Long-distance trade of food and feed thus remains a necessity, as long as it can correct the distortions between human population needs and agronomic capabilities of regional systems.

This view of trade, as an adjustment variable in the search for the optimum allocation of resources, bears, however, a certain degree of naive optimism, not to say utopianism: to a large extent, the rather recent and extremely rapid trend of specialization of agricultural activities has indeed been driven, rather than followed, by a strong voluntarism of long-distance trade development, with great damage to the environment. Many authors have indeed shown how the development of international commercial exchanges of food commodities has been driven by the quest for high profits made possible by unequal terms of trade and speculation (Boris, 2010; Ziegler, 2011; Rulli and D'Odorico, 2014.). In addition, to the sometimes dramatic socioeconomic consequences on fragile populations, the current development of long-distance trade of agricultural commodities should be further questioned because of its deleterious effects on the environment.

References

Aguilera, E., Guzmán, G.I., Infante-Amate, J., Soto, D., García-Ruiz, R., Herrera, A., Villa, I., Torremocha, E., Carranza, G., González de Molina, M., 2015. Embodied Energy in Agricultural Inputs. Incorporating a Historical Perspective. Sociedad Española de Historia Agraria. DT-SEHA 1507.

Anglade, J., Billen, G., Makridis, T., Garnier, J., Puech, T., Tittel, C., 2015a. Nitrogen soil surface balance of organic vs conventional cash crop farming in the Seine watershed. Agricultural Systems 139, 82–92.

Anglade, J., Billen, G., Garnier, J., 2015b. Relationships for estimating N2 fixation in legumes: incidence for N balance of legume-based cropping systems in Europe. Ecosphere 6 art37.

Billen, G., Barles, S., Garnier, J., Rouillard, J., Benoit, P., 2009. The food-print of Paris: long term reconstruction of the nitrogen flows imported to the city from its rural Hinterland. Regional Environmental Change 9, 13–24. https://doi.org/10.1007/s10113-008-0051-y.

Billen, G., Garnier, J., Thieu, V., Silvestre, M., Barles, M., Chatzimpiros, P., 2012. Localising the nitrogen imprint of the Paris food supply: the potential of organic farming and changes in human diet. Biogeosciences 9, 607–616.

Billen, G., Garnier, J., Lassaletta, L., 2013. The nitrogen cascade from agricultural soils to the sea: modelling nitrogen transfers at regional watershed and global scales. Philosophical transactions of the Royal Society of London. Series B 368, 20130123.

Billen, G., Lassaletta, L., Garnier, J., 2014. A biogeochemical view of the global agro-food system: nitrogen flows associated with protein production, consumption and trade. Global Food Security 3, 209–219.

Billen, G., Lassaletta, L., Garnier, J., 2015. A vast range of opportunities for feeding the world in 2050: trade-off between diet, N contamination and international trade. Environmental Research Letters 10, 025001.

Billen, G., Le Noë, J., Lassaletta, L., Thieu, V., Anglade, J., Petit, L., Garnier, J., 2016. Et si la France passait au régime «Bio, Local et Demitarien» ? Un scénario radical d'autonomie protéique et azotée de l'agriculture et de l'élevage, et de sobriété alimentaire. DEMETER 2017. Club-Demeter, Paris.

Billen, G., Le Noë, J., Garnier, J., 2018. Two contrasted future scenarios for the French agro-food system. Science of the Total Environment 637–638, 695–705. https://doi.org/10.1016/j.scitotenv.2018.05.043.

Bodirsky, B.L., Popp, A., Weindl, I., Dietrich, J.P., Rolinski, S., Scheiffele, L., Schmitz, C., Lotze-Campen, H., 2012. N_2O emissions from the global agricultural nitrogen cycle – current state and future scenarios. Biogeosciences 9, 4169–4197.

Bodirsky, B.L., Popp, A., Lotze-Campen, H., Dietrich, J.P., Rolinski, S., Weindl, I., Schmitz, C., Müller, C., Bonsch, M., Humpenöder, F., Biewald, A., Stevanovic, M., 2014. Reactive nitrogen requirements to feed the world in 2050 and potential to mitigate nitrogen pollution. Nature Communications 5, 3858. https://doi.org/10.1038/ncomms4858.

Bonaudo, T., Bendahan, A.B., Sabatier, R., Ryschawy, J., Bellon, S., Leger, F., Magda, D., Tichit, M., 2014. Agroecological principles for the redesign of integrated crop–livestock systems. European Journal of Agronomy 57, 43–51.

Boris, J.P., 2010. Main basse sur le riz. Fayard, 224 p.

Bouwman, A.F., Beusen, A.H.W., Lassaletta, L., van Apeldoorn, D.F., van Grinsven, H.J.M., Zhang, J., Ittersum van, M.K., 2017. Lessons from temporal and spatial patterns in global use of N and P fertilizer on cropland. Scientific Reports 7, 40366.

Caballero, J.M., Quieti, M.G., Maetz, M., 2000. International Trade: Some Basic Theories and Concepts. FAO. http://www.fao.org/docrep/003/X7352E/X7352E02.htm.

Clément, A., 2000. La spécificité du fait alimentaire dans la théorie économique. Les fondements historiques et les enjeux. Ruralia 7, 2–12.

Couturier, C., Charru, M., Doublet, S., Pointereau, P., 2017. Le scénario Afterres 2050, Solagro. www.afterres2050.solagro.org.

Cowell, S.J., Parkinson, S., 2003. Localisation of UK food production: an analysis using land area and energy as indicators. Agriculture, Ecosystems and Environment 221–236.

Desrochers, P., Shimizu, H., 2008. Yes We Have No Bananas: A Critique of the Food Mile Perspective. Mercatus Policy Series, Policy Primer No. 8, October 2008. http://mercatus.org/PublicationDetails.aspx?id=24612.

Diacono, M., Montemurro, F., 2010. Long-term effects of organic amendments on soil fertility. A review. Agronomy for Sustainable Development 30, 401–422. https://doi.org/10.1051/agro/2009040.

Doltra, J., 2015. Modeling nitrous oxide emissions from organic and conventional cereal-based cropping systems under different management, soil and climate factors. European Journal of Agronomy 66, 8–20. https://doi.org/10.1016/j.eja.2015.02.002.

Erb, K.H., Lauk, C., Kastner, T., Mayer, A., Theurl, M.C., Haberl, H., 2016. Exploring the biophysical option space for feeding the world without deforestation. Nature Communications 7.

European Commission (DG ENV), 2010. Environmental Impacts of Different Crop Rotations in the European Union. Contract n 07.0307/2009/SI2.541589/ETU/B1. Final Report, 6 sept 2010.

EU Nitrogen Expert Panel, 2015. Nitrogen Use Efficiency (NUE) an Indicator for the Utilization of Nitrogen in Agriculture and Food Systems. Wageningen University. Alterra, PO Box 47, NL6700 Wageningen, Netherlands. Available at: http://www.eunep.com/wp-content/uploads/2017/03/Report-NUE-Indicator-Nitrogen-Expert-Panel-18-12-2015.pdf.

Foster, J.B., 2000. Marx's Ecology: Materialism and Nature. NYU Press.

Garnier, J., Anglade, J., Benoit, M., Billen, G., Puech, T., Ramarson, A., Passy, P., Silvestre, M., Lassaletta, L., Trommenschlager, J.M., Schott, C., Tallec, G., 2016. Reconnecting crop and cattle farming for a reduction of nitrogen losses in an intensive agricultural watershed. Environmental Science and Policy 63, 76–90.

Garrabou, R., Gonzales de Molina, M., 2010. La reposición de la fertilidad en los sistemas agrarios tradicionales. Icaria, 319 p.

Guzmán, G., Aguilera, E., Soto, D., et al., 2014. Methodology and Conversion Factors to Estimate the Net Primary Productivity of Historical and Contemporary Agroeco systems. Sociedad Espnola de Historia Agraria. Documentos de Trabajo DT-SEHA n 1407. Mayo 2014. www.seha.info.

Holt-Giménez, E., Altieri, M.A., 2012. Agroecology, food sovereignty, and the new green revolution. Agroecology and Sustainable Food Systems 37 (1), 90–102.

Iglesias, A., Garrote, L., Diz, A., Schlickenrieder, J., Martin-Carrasco, F., 2011. Re-thinking water policy priorities in the Mediterranean region in view of climate change. Environmental Science and Policy 14, 744–757.

Kopittke, P.M., Dalal, R.C., Finn, D., Menzies, N.W., 2016. Global changes in soil stocks of carbon, nitrogen, phosphorus, and sulfur as influenced by long-term agricultural production. Global Change Biology (n/a-n/a).

Lassaletta, L., Billen, G., Romero, E., et al., 2014a. How changes in diet and trade patterns have shaped the N cycle at the national scale: Spain (1961-2009). Regional Environmental Change 14, 784–797. https://doi.org/10.1007/s10113-013-0536-1.

Lassaletta, L., Billen, G., Grizzetti, B., Garnier, J., Leach, A.M., Galloway, J.N., 2014b. Food and feed trade as a driver in the global nitrogen cycle: 50-year trends. Biogeochemistry 118 (1–3), 225–241.

Lassaletta, L., Billen, G., Grizzetti, B., et al., 2014c. 50 year trends in nitrogen use efficiency of world cropping systems: the relationship between yield and nitrogen input to cropland. Environmental Research Letters 9. https://doi.org/10.1088/1748-9326/9/10/105011.

Lassaletta, L., Billen, G., Garnier, J., Bouwman, L., Velazquez, E., Mueller, N.D., Gerber, J.S., 2016. Nitrogen use in the global food system: past trends and future trajectories of agronomic performance, pollution, trade, and dietary demand. Environmental Research Letters 11, 095007. https://doi.org/10.1088/1748-9326/11/9/095007.

Lemaire, G., Franzluebbers, A., Carvalho, P.C.d.F., Dedieu, B., 2014. Integrated crop–livestock systems: strategies to achieve synergy between agricultural production and environmental quality. Agriculture, Ecosystems and Environment 190, 4–8.

Le Noë, J., Billen, G., Lassaletta, L., et al., 2016. La place du transport de denrées agricoles dans le cycle biogéochimique de l'azote en France: un aspect de la spécialisation des territoires. Cahiers Agricultures 25, 15004.

Le Noë, J., Billen, G., Garnier, J., 2017. Nitrogen, phosphorus, and carbon fluxes through the French agro-food system: an application of the GRAFS approach at the regional scale. Science of the Total Environment 586, 42–45.

Lotze-Campen, H., Muller, C., Bondeau, A., Rost, S., Popp, A., Lucht, W., 2008. Global food demand, productivity growth, and the scarcity of land and water resources: a spatially explicit mathematical programming approach. Agricultural Economics 39, 325–338.

Lotze-Campen, H., Popp, A., Beringer, T., Muller, C., Bondeau, A., Rost, S., Lucht, W., 2010. Scenarios of global bioenergy production: the trade-offs between agricultural expansion, intensification and trade. Ecological Modelling 221, 2188–2196.

McNeill, J.R., 2010. Du nouveau sous le soleil. Nous 517, 26.

Maillard, É., Angers, D.A., 2014. Animal manure application and soil organic carbon stocks: a meta-analysis. Global Change Biology 20 (2), 666–679.

Martin, G., Moraine, M., Ryschawy, J., Magne, M.A., Asai, M., Sarthou, J.P., Therond, O., 2016. Crop–livestock integration beyond the farm level: a review. Agronomy for Sustainable Development 36 (3), 53.

Mazoyer, M., Roudart, L., 1998. Histoire des agricultures du monde. Du Néolithique à la crise contemporaine. Seuil, Paris, 531 p.

Mueller, N.D., West, P.C., Gerber, J.S., MacDonald, G.K., Polasky, S., Foley, J.A., 2014. A tradeoff frontier for global nitrogen use and cereal production. Environmental Research Letters 9, 054002.

Mueller, N.D., Lassaletta, L., Runck, B., Billen, G., Garnier, J., Gerber, J.S., 2017. Declining spatial efficiency of global cropland nitrogen allocation. Global Biogeochemical Cycles 31, 245–257. https://doi.org/10.1002/2016GB005515.

Naylor, R., Steinfeld, H., Falcon, W., Galloway, J., Smil, V., Bradford, E., Alder, J., Mooney, H., 2005. Losing the links between livestock and land. Science 310, 1621–1622.

Nowak, B., Nesme, T., David, C., Pellerin, S., 2013. To what extent does organic farming rely on nutrient inflows from conventional farming? Environmental Research Letters 8, 044045.

Oita, A., Malik, A., Kanemoto, K., Geschke, A., Nishijima, S., Lenzen, M., 2016. Substantial nitrogen pollution embedded in international trade. Nature Geoscience 9, 111–115.

O'Kane, G., 2016. A moveable feast: contemporary relational food cultures emerging from local food networks. Appetite 105, 218–231.

Penuelas, J., Sardans, J., Alcaniz, J.M., Poch, J.M., 2009. Increased eutrophication and nutrient imbalances in the agricultural soil of NE Catalonia, Spain. Journal of Environmental Biology 30, 841–846.

Peñuelas, J., Sardans, J., Rivas-ubach, A., Janssens, I.A., 2012. The human-induced imbalance between C, N and P in Earth's life system. Global Change Biology 18, 3–6.

Porkka, M., Kummu, M., Siebert, S., Varis, O., 2013. From food insufficiency towards trade dependency: a historical analysis of global food availability. PLoS One 8.

Ricardo, D., 1817. In: Murray, J. (Ed.), On the Principles of Political Economy and Taxation (London).

Rulli, M., D'Odorico, P., 2014. Food appropriation through large scale land acquisitions. Environmental Research Letters 9, 064030.

Sanz-Cobena, A., Lassaletta, L., Estellés, F., Prado, A.D., Guardia, G., Abalos, D., Aguilera, E., Pardo, G., Vallejo, A., Sutton, M.A., Garnier, J., Billen, G., 2014. Yield-scaled mitigation of ammonia emission from N fertilization: the Spanish case. Environmental Research Letters 9, 125005.

Soussana, J.-F., Lemaire, G., 2014. Coupling carbon and nitrogen cycles for environmentally sustainable intensification of grasslands and crop-livestock systems. Agriculture, Ecosystems and Environment 190, 9–17.

Schipanski, M.E., Bennett, E.M., 2012. The influence of agricultural trade and livestock production on the global phosphorus cycle. Ecosystems 15 (2), 256–268.

Schmitz, C., Biewald, A., Lotze-Campen, H., Popp, A., Dietrich, J.P., Bodirsky, B., Krause, M., Weindl, I., 2012. Trading more food: implications for land use, greenhouse gas emissions, and the food system. Global Environmental Change 22, 189–209.

Soto, D., Infante-Amate, J., Guzmán, G., González de Molina, M., Cid, A., Aguilera, E., García-Ruiz, R., 2016. The biomass metabolism of Spain 1900-2008. Transformations in the agroecosystems during the socio-ecological transition. Ecological Economics 128, 130–138.

Strokal, M., Ma, L., Bai, Z., Luan, S., Kroeze, C., Oenema, O., Velthof, G., Zhang, F., 2016. Alarming nutrient pollution of Chinese rivers as a result of agricultural transitions. Environmental Research Letters 11, 024014.

Suweis, S., Carr, J.A., Maritan, A., Rinaldo, A., D'Odorico, P., 2015. Resilience and reactivity of global food security. Proceedings of the National Academy of Sciences 112, 6902–6907.

United Nations, 2011. World Population Prospects: The 2010 Revision.

van Vuuren, D.P., Stehfest, E., Gernaat, D.E.H.J., Doelman, J.C., van den Berg, M., Harmsen, M., de Boer, H.S., Bouwman, L.F., Daioglou, V., Edelenbosch, O.Y., Girod, B., Kram, T., Lassaletta, L., Lucas, P.L., van Meijl, H., Müller, C., van Ruijven, B.J., van der Sluis, S., Tabeau, A., 2017. Energy, land-use and greenhouse gas emissions trajectories under a green growth paradigm. Global Environmental Change 42, 237–250. https://doi.org/10.1016/j.gloenvcha.2016.05.008.

Vleeshouwers, L.M., Verhagen, A., 2002. Carbon emission and sequestration by agricultural land use: a model study for Europe. Global Change Biology 8, 519–530.

Weber, C., Matthews, H., 2008. Food-miles and the relative climate impacts of food choices in the United States. Environmental Science and Technology 42, 3508–3513.

Windfuhr, M., Jonsén, J., 2005. Food Sovereignty: Towards Democracy in Localized Food Systems. FIAN-International. ITDG Publishing, Warwickshire, UK.

Ziegler, J., 2011. Destruction massive. Géopolitique de la faim. Le Seuil, Paris, 347 p.

CHAPTER

26

Scaling From Local to Global for Environmental Impacts From Agriculture

Marty D. Matlock

University of Arkansas, Fayetteville, AR, United States

INTRODUCTION

Environmental impacts from agriculture have been explored in depth in the previous chapters. They include soil erosion and health, water quality and quantity, habitat degradation and loss, energy use, greenhouse gas emissions, and air pollution (Table 26.1). Sustainability is the process of continual improvement, where these impacts are reduced across multiple scales over time through innovations in technologies and practices. The producers (farmers) can control some drivers of negative environmental impacts, but their decisions are constrained by geography, external policies, processes, technologies, and legacy conditions. The process of improving those impacts requires complex decision-making across the entire agricultural supply chain.

Agricultural supply chains are complex, ranging from locally produced fruits and vegetables to globally distributed commodity grains (Ahumada and Vilalobos, 2009). The agri-food supply chain includes the entire life cycle of food products, including production, processing, storage, wholesale transportation and distribution, retail sales, consumption, and disposal. The agri-food supply chain is shaped by market forces (profitability, competition, supply/demand, consumer preferences), agri-food production and trade policies (tariffs, price supports, subsidies, and risk management instruments), and external forces (weather, political instability, disease) (Fig. 26.1). The effectiveness of this complex, global exchange of agri-food products is usually measured by efficiency, flexibility, responsiveness, and food quality (Aryamyan et al., 2007). Communication and automation technologies are driving efficiency even further throughout the agri-food supply chains by reducing spoilage and other waste streams.

Producers usually control only the production phase of this life cycle, which happens on the farm. The farm is a system with inputs and outputs, each determined by the decisions made by the farmer to maximize profits while reducing risks (Fig. 26.2). This process of risk optimization is at the heart of sustainability decision-making. Producers manage risks differently, depending

Agroecosystem Diversity
https://doi.org/10.1016/B978-0-12-811050-8.00026-1

TABLE 26.1 Environmental Impacts From the Global Agri-Food Supply Chain

Environmental Impact	Metric	Scale of Management
Soil erosion	Kg/ha-yr	Local
Soil health	Soil organic carbon, g/cm^3	Local
Air pollution	Pollutant concentrations, μg/cm^3, pm10	Local to regional
Water quality	Pollutant concentrations, mg/L	Regional
Water quantity	Scarcity, L/kg crop	Regional
Habitat degradation	Critical species impacts or loss/yr	Regional/national
Habitat loss	Ha/ecoregion, Ha/yr	Regional/national
Energy use	MJ/kg crop, MJ/yr	Local/global
Greenhouse gas emissions	kg/kg crop, MT/yr	Local/global

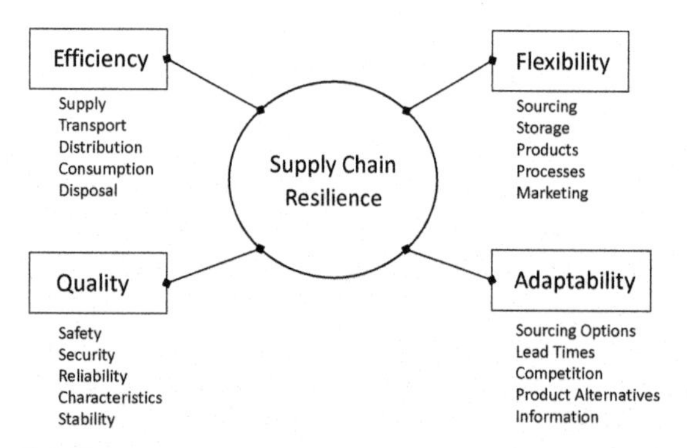

FIGURE 26.1 Criteria for a sustainable, effective agri-food supply chain. *Modified from Aramyan, L.H., Oude Lansink, A.G., Van Der Vorst, J.G., Van Kooten, O., 2007. Performance measurement in agri-food supply chains: a case study. Supply Chain Management: International Journal 12 (4), 304–315.*

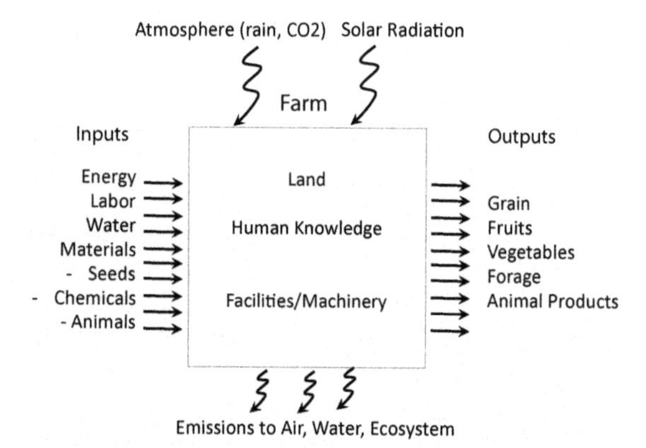

FIGURE 26.2 Systems diagram of a simplified farm operation, with inputs, outputs, and emissions.

on their histories, levels of capitalization, categories of production, and scales of operation. Each input to production is a cost of operation, thus each loss to the system through emissions is a loss of value to the farm. Producers have an inherent self-interest in reducing environmental impacts; however, the practices that can lead to those reductions are also costs. Producers exercise a wide range of options in how they manage these impacts, referred to as freedom to operate. Emissions from farm operations have potential negative impacts, as discussed in previous chapters. If farms emit too many pollutants the environmental impacts can lead to increased regulations or even prohibitions on production choices. Producers must manage costs of reductions of impacts with value of those reductions in saved products and the freedom to operate that comes from responsible production.

The decisions available to an agricultural producer to reduce impacts while maximizing profits vary by region and production category.

Production categories are very generally divided into animal agriculture (beef, swine, poultry, eggs, dairy, sheep, goats, other), row crop agriculture (rice, corn, soy, wheat, sorghum, rye, oats, barley, cotton, and other grains), and specialty crops (fruits, nuts, fresh greens, and vegetables). Each category and combination of categories contained within a production system defines the range of sustainability practices available to the producer. The scale of consideration affects the array of production systems within the landscape, resulting in increasingly complex decision criteria. This chapter will explore the impact of scale on environmental indicators of sustainability, the decisions that producers can make to influence those indicators, and the tradeoffs that commonly occur in this decision-making process.

SCALES OF ENVIRONMENTAL IMPACTS

The biosphere is composed of coupled cycles of energy, mass, and living organisms, each interacting in specific manners at different scales of consideration. The agri-food supply chain is a subcomponent of this global dynamic system. The continuum of scales for agricultural production environmental impacts begins at the field and extends to the global biosphere. Very few of the environmental impacts from agricultural production are globally fungible; greenhouse gas emissions and energy use may be the only two. Most are bound by conditions of the location in which they occur.

Local Scale

The highest resolution (smallest) scale for decision-making that affects environmental impacts from agricultural production is local, which for this discussion means the farm. Farms are the smallest unit of enterprise and thus the coherent scale to influence operational and management decisions. Farms are composed of a number of interconnected elements depending on the categories of production. The basic components of the farm include facilities (housing, storage, and animal production barns), machinery (tractors, combines, implements, etc.), land (fields, orchards, forests, prairies, pasture, and streams), labor, and human knowledge (Fig. 26.2). The largest area of impacts derives from human-dominated activities on the land, at scales of the field. Fields are discrete, homogenously managed parcels of land. This is the scale over which the producer has the highest influence.

Environmental impacts from local activities are those that directly derive from field activities at the farm level (Table 26.1). These include soil erosion (wind and water), soil acidification, fertility loss, water quality from agrichemicals, nutrient, and soils, atmospheric emissions (particulates, ammonia, nitrous oxide, and other greenhouse gasses), and water scarcity. Local impacts can be aggregated to regional and even global levels, but effective decision-making to reduce these impacts occurs at the local scale.

Regional Scale

The regional scale is where most impacts from agricultural activities are manifested. The environmental impacts on water and habitat at the regional scale are the product of the activities across all land use classes within the region. Regional scale is defined by the processes being assessed or managed. Landscape processes such as ecosystem services and biodiversity should be evaluated at the ecoregional level (Omernik and Bailey, 1997). For water resource management, regional scale is defined by the watershed. The watershed is an area of land that drains surface runoff to a discrete outlet (stream, river, lake, reservoir, or estuary) (Omernik and Bailey, 1997). Watersheds can be as large as 60,000 ha in size (United States Geological Survey Hydrologic Unit Code 8);

they often transcend municipal, provincial, and national geopolitical boundaries. The regional scale often includes multiple land use classes, including agriculture, residential, industrial, urban, forest, pasture, prairie, river/lake, and coastal. Global watershed delineation was formalized in 2003, providing a standardized geospatial reference point for analyzing and managing land and water resources (WRI, 2003).

Regional impacts on water from agriculture include water quality degradation and water scarcity. Water quality degradation is largely the result of cumulative nutrient, sediment, and pesticide loading to waterbodies within a watershed. Reducing water quality impacts is complicated by the complex land use distribution within a watershed. Agricultural, municipal, and industrial discharges often contribute to the loss of water quality integrity at the watershed scale. Soil erosion is the most pernicious water quality impact at the regional scale.

The aggregate impact of erosion from multiple farms in a watershed converge within the relatively narrow river channel, concentrating the impacts. Erosion-derived sediment degrades productive potential of the land; destroys habitat in streams, rivers, and reservoirs; reduces clarity of the water column with suspended materials; and transports pesticides and nutrients associated with the soil particles downstream. Water quality degradation due to nutrient loads is the next highest impact at the regional level. Desirable water quality is defined generally based upon the uses humans have for the water (drinking, irrigation, fisheries, etc.). When nutrients (predominantly nitrogen and phosphorus) occur at too high levels, water quality is degraded through increased algal productivity. This process, called eutrophication, can result in dissolved oxygen extremes in the water, as photosynthesizing algal biomass produces oxygen during the day, and as the biomass respires at night, it consumes oxygen. These rapid and dramatic swings in oxygen concentration can be lethal to fish, benthic macroinvertebrates, and other life forms that depend on the water for gas exchange. In addition, eutrophication can decrease water pH to lethal levels for many living organisms by increasing the rate of carbon uptake from the water column (Gypens et al., 2011).

Another common water quality impact from agriculture at the regional scale is the accumulation of pesticides (insecticides, fungicides, and herbicides) from field runoff and overspray. In the United States the occurrences and concentrations of these contaminants has been in decline for over 30 years (Rasmussen et al., 2015; Orlando, 2013). Urban sources of pesticides increased in the United States over the last 20 years, while human health risks from agricultural sources decreased (Stone et al., 2014). This decrease is attributed to the broad-scale adoption of modern agricultural production technologies such as conservation tillage, seed biotechnology, pesticide application and use regulations, and precision agriculture (Gustafson et al., 2014). Globally, pesticide occurrences above human health thresholds are increasing, suggesting a need for improved technologies and regulatory management of these chemicals (Stehle and Schulz, 2015).

Water scarcity is very simple in concept but harder to manage. Scarcity is the result of water demand and withdrawal being greater than supply. Water supplies in a given region or watershed come from surface flows (rivers, lakes, and reservoirs) and groundwater (riparian and geologic). The potential for disruption of agriculture production due to prolonged drought is very high. Competing demands for water from municipal and industrial sectors compounds the complexity of mitigating scarcity. Reservoirs are manmade containment pools of water created by damming rivers. They are the dominant mechanism humans use to capture and expand availability of water over an annual period. However, reservoirs can only capture water that flows through the watershed; if there is no snow or rain, there is no water to feed the reservoir. Changing climate conditions are resulting in disruptions of precipitation frequency and intensity, especially in midcontinent areas where much of Earth's agricultural production occurs (Meehl et al., 2007).

When biodiversity and habitat are the major impacts of concern, regional scales are defined by ecoregions rather than watersheds. Ecoregions are areas of similar soils, geology, land surface forms, climate, and natural vegetation (Omernik, 1987). The ecoregion classification scheme developed by Omernik (1987) uses watershed and basin boundaries to define ecoregion boundaries, so it provides a coherent geospatial framework for regional management of natural resources (Omernik and Bailey, 1997). Biodiversity is affected at the local scale by land use decisions that disrupt critical habitat, but the cumulative impacts on biodiversity from habitat loss are realized at the ecoregion scale. A species can persist in an area where habitat loss occurs so long as critical areas of refugia persist and are interconnected by functional corridors. Biodiversity loss occurs when these areas of refugia within an ecoregion are reduced below a critical threshold for a species, or when connections between those areas are disrupted, so organisms can no longer complete their life cycle requirements (pollination, reproduction, distribution, brooding, etc.).

National Scale

The most identifiable scale of human activities is the national scale. Economic data are aggregated and reported at the national scale. National scales range from just larger than regional scales for island nations and some European nations to subcontinental and continental scales. Agricultural policies on supports, risk management, trade, practices, technologies, and liabilities are made at the national scale. National authority generally supersedes regional and local authority, so transboundary conflicts are often resolved through national policies. Environmental impacts at the national scale are aggregated from regional impacts (soil erosion, water quality, air quality, greenhouse gas emissions). Habitat degradation and loss are often aggregated and reported at the national level (Table 26.1).

Global Scale

Global scale includes the global atmosphere, biosphere, geosphere, and technosphere (Fig. 26.3). Human impacts at the global scale are largely associated with acceleration of global geochemical cycling. Global carbon, nitrogen, and phosphorus cycles have been addressed in previous chapters; their impacts are often assessed at global levels. The approach is most often a mass balance assessment over time, showing the movement of each geochemical across global compartments.

Greenhouse gas movement from the geosphere to the atmosphere is most commonly assessed at the global level, as discussed in previous chapters. Carbon (C) is extracted from the geosphere as fossil fuel, transformed to greenhouse gas emissions in the technosphere, and accumulates in the atmosphere, where it contributes to the greenhouse effect, driving global climate change. In addition, anoxic metabolism of carbon molecules in the biosphere results in emissions of methane gas, a potent greenhouse gas, into the atmosphere. The global nitrogen cycle is driven by the extraction of nitrogen (N) gas from the atmosphere into the technosphere using the Haber-Bosch method to create nitrogen fertilizers. This process requires energy (usually from natural gas, a fossil fuel). Fertilizer application on the biosphere increases carbon uptake from the atmosphere, but it also results in release of nitrous oxide, a very potent greenhouse gas, to the atmosphere. Inefficient nitrogen fertilization also increases nitrogen loading to the hydrosphere, which contributes to eutrophication of bays and estuaries in the marine environment, thus creating oxygen

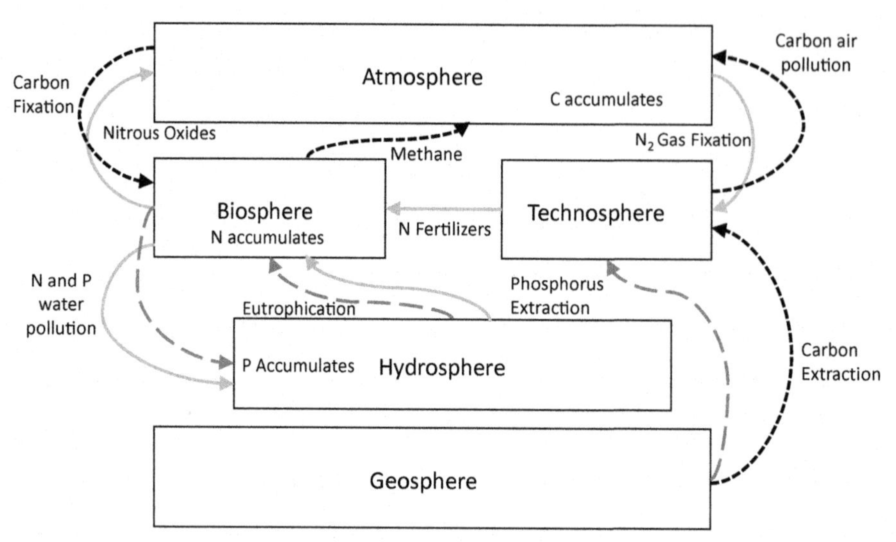

FIGURE 26.3 Exchange of carbon (C), nitrogen (N), and phosphorus (P) between global storage pools.

deficiencies and estuarine dead zones around the world. The global phosphorus cycle begins when phosphorus (P) is extracted from the geosphere as mineral rock, processed through the technosphere and applied to the biosphere as fertilizer, where it increases crop productivity. Phosphorus movement to the hydrosphere causes eutrophication of freshwater systems, resulting in water quality degradation. Phosphorus ultimately is transported to the ocean, where it cycles through the biosphere, accumulates in the mesotrophic zone, and is ultimately redeposited into the geosphere on the ocean bed. The decision processes that affect global impacts most often occur at the smallest scale: the field.

DECISION-MAKING ACROSS SCALES

Reducing impacts from agricultural production systems requires making better decisions at each scale of activity, from local to global. The impacts of scale on the decision-making process include systems knowledge, technological access, economic frameworks, and regulatory constraints. At the local level, decisions made by agricultural producers have the greatest potential for reducing environmental impacts. Global impacts such as greenhouse gas emissions, energy use, and nutrient cycling are also managed by local-scale decisions.

Many of the decisions farmers make require sophisticated tradeoffs between many variables, including economics, risk management, and environmental protection. Increasing water use efficiency may decrease energy efficiency, for example. Diversifying production strategies on the land may increase land area required to produce an economically viable crop, which can result in loss of biodiversity as marginal lands are moved into production. Choosing not to use biotechnology may increase energy use, soil erosion, and pesticide pollution. These tradeoffs are not simple, and they are influenced by many factors, including short- and long-term economic risk, market demands, available technology, public policies and regulations, governmental incentive programs, and producer culture. Farmers make local decisions in most cases without information regarding the global impacts of those decisions. Even if they knew the global impacts, the motivations for decisions are often financial, logistical, and cultural. These motivations are not easily altered by simple information.

There are very few tools available for farmers to support more effective complex decision-making. Farmers need real-time geospatial modeling of crop yields with economic inputs and environmental impacts to make daily choices on management strategies to manage risks and optimize profits. These types of models are becoming more accessible, but mostly occur in disparate elements across the academic community. Geospatial crop modeling decision support systems for agricultural producers would give farmers better knowledge of the complex systems they are operating in and would allow them to make more sustainable decisions.

Regional decision-making requires collective action, since no single land owner or authority has jurisdiction over most regional areas. The decision tools at the regional scale are limited to incentive programs that provide cost supports or subsidies for desirable actions, and regulations that penalize undesirable actions. Most nations do not have watershed or ecoregion authority structures. Where they do exist, they are predominantly resource management authorities designed to allocate resources and collect fees, not to manage for reduced environmental impacts. The few watershed-level authorities that do have regulatory jurisdictions have proven to be very effective at conserving water resources and protecting water quality. Similar ecoregion-scale authorities could have success in protecting and preserving critical habitats and thus preserving biodiversity.

National-scale decision-making for environmental impacts is also a mixture of regulations and incentives. The United States Department of Agriculture and the EU Common Agriculture Policy both provide incentives for agricultural producers to implement practices that preserve and protect the environment (Husand and Johansson, 2016). Moving from practice-based incentive payment structures to outcomes-based payment systems would likely provide more impact for agricultural protection of environmental resources at the national scale. Under an outcomes-based (also called results-based) scheme the agricultural producer could receive remuneration for ecosystem services provided by his/her lands at greater value than the costs of implementation. Thus, these ecosystem services could be economic drivers for producer decision-making, rather than costs to be recovered or managed (Daugbjerg and Swinbank, 2015).

Global-scale decision-making to reduce environmental impacts from agricultural activities is limited to international agreements and treaties. The Paris Climate Agreement (PCA) required participating countries to submit Intended Nationally Determined Contributions to collectively reduce greenhouse gas emissions to prevent a greater than 2-degree Celsius increase in global temperatures (Rogelj et al., 2016). There is still much disagreement within the PCA on how to account for the impacts of land use change; some argue for a national annual accounting based upon emissions associated with land use changes, while others prefer a longer-term averaging of trends (Grassi and Dentener, 2015).

The World Trade Organization (WTO) represents another global-scale effort to create common rules and policies for trade practices that are equitable and promote responsible business decisions. However, WTO policies have been antithetical to the protection of environmental resources and services (Tisdell, 2001). However, increasing conflicts over national agriculture support policies, tariffs, and international resource conflicts will likely result in WTO becoming more engaged in arbitration of conflicts. This quasi-regulatory role will be enhanced through the WTO's authority to levy penalties and restrictions in trade (Mathiason and Cabral, 2015). The complexity and uncertainty associated with global decision-making processes makes it highly unlikely that this scale of process will promote practices at the local scale to protect environmental services and resources.

CONCLUSIONS

Agricultural production occupies more than 40% of Earth's surface and is the anchor for economic prosperity of every community, yet we do not have coherent strategies to manage the natural resources upon which agricultural production depends (Tanentzap et al., 2015). Agricultural expansion is driving habitat loss across the tropics and water resource depletion in almost every ecoregion on Earth. The challenge of managing environmental resources and services requires scale-specific strategies. Almost all decisions that directly affect the impacts of agriculture on the environment are made locally, while the processes that influence those decisions are made nationally. The gap between national policy and local decisions must be filled with regional frameworks that create watershed- and ecoregion-specific policies for incentivizing desirable outcomes and penalizing undesirable outcomes. These policies must be promulgated at national scales and supported at the global level. Transnational interests as manifested in WTO policies must yield jurisdiction to regional decision-making with regard to environmental resources and services.

References

Ahumada, O., Villalobos, J.R., 2009. Application of planning models in the agri-food supply chain: a review. European Journal of Operational Research 196 (1), 1–20.

Aramyan, L.H., Oude Lansink, A.G., Van Der Vorst, J.G., Van Kooten, O., 2007. Performance measurement in agri-food supply chains: a case study. Supply Chain Management: An International Journal 12 (4), 304–315.

Daugbjerg, C., Swinbank, A., 2015. Three decades of policy layering and politically sustainable reform in the European Union's agricultural policy. Governance 29 (2), 1468–1491.

Grassi, G., Dentener, F., 2015. Quantifying the Contribution of the Land Use Sector to the Paris Climate Agreement. Report No. EUR 27561, European Union, JRC Science Hub. https://doi.org/10.2788/096422.

Gustafson, D.I., Collins, M., Fry, J., Smith, S., Matlock, M., Zilberman, D., Shryock, J., Doane, M., Ramsey, N., 2014. Climate adaptation imperatives: global sustainability trends and eco-efficiency metrics in four major crops–canola, cotton, maize, and soybeans. International Journal of Agricultural Sustainability 12 (2), 146–163.

Gypens, N., Lacroix, G., Lancelot, C., Borges, A.V., 2011. Seasonal and inter-annual variability of air–sea CO_2 fluxes and seawater carbonate chemistry in the Southern North Sea. Progress in Oceanography 88 (1), 59–77.

Hasund, K.P., Johansson, M., 2016. Paying for environmental results is WTO compliant. EuroChoices 15 (3), 33–38.

Mathiason, T., Cabral, A., 2015. Symposium: managing the global environment through trade: WTO, TPP, and TTIP negotiations, and bilateral investment treaties versus regional trade agreements: introduction. American University International Law Review 30 (3), 1.

Meehl, G.A., Stocker, T.F., Collins, W.D., Friedlingstein, P., Gaye, A.T., Gregory, J.M., Kitoh, A., Knutti, R., Murphy, J.M., Noda, A., Raper, S.C., 2007. Global climate projections. Climate Change 3495, 747–845.

Omernik, J.M., 1987. Ecoregions of the conterminous United States. Annals of the Association of American Geographers 77 (1), 118–125.

Omernik, J.M., Bailey, R.G., 1997. Distinguishing between watersheds and ecoregions. JAWRA 33 (5).

Orlando, J.L., 2013. A Compilation of US Geological Survey Pesticide Concentration Data for Water and Sediment in the Sacramento-San Joaquin Delta Region, 1990-2010. US Department of the Interior, US Geological Survey.

Rasmussen, J.J., Wiberg-Larsen, P., Baattrup-Pedersen, A., Cedergreen, N., McKnight, U.S., Kreuger, J., Jacobsen, D., Kristensen, E.A., Friberg, N., 2015. The legacy of pesticide pollution: an overlooked factor in current risk assessments of freshwater systems. Water Research 84, 25–32.

Rogelj, J., Den Elzen, M., Höhne, N., Fransen, T., Fekete, H., Winkler, H., Schaeffer, R., Sha, F., Riahi, K., Meinshausen, M., 2016. Paris Agreement climate proposals need a boost to keep warming well below 2 C. Nature 534 (7609), 631–639. PAri.

Stehle, S., Schulz, R., 2015. Agricultural insecticides threaten surface waters at the global scale. Proceedings of the National Academy of Sciences of the United States of America 112 (18), 5750–5755.

Stone, W.W., Gilliom, R.J., Ryberg, K.R., 2014. Pesticides in US streams and rivers: occurrence and trends during

1992–2011. Environmental Science and Technology 48 (19), 11025–11030.

Tanentzap, A.J., Lamb, A., Walker, S., Farmer, A., 2015. Resolving conflicts between agriculture and the natural environment. PLoS Biology 13 (9), e1002242.

Tisdell, C., 2001. Globalisation and sustainability: environmental Kuznets curve and the WTO. Ecological Economics 39 (2), 185–196.

WRI, 2003. Watersheds of the World. IUCN-The World Conservation Union, the International Water Management Institute (IWMI), the Ramsar Convention Bureau and the World Resources Institute (WRI), Washington, DC, ISBN 1-56973-548-4.

Further Reading

Lobell, D.B., Burke, M.B., Tebaldi, C., Mastrandrea, M.D., Falcon, W.P., Naylor, R.L., 2008. Prioritizing climate change adaptation needs for food security in 2030. Science 319 (5863), 607–610.

Matlock, M.D., Morgan, R.A., 2011. Ecological Engineering Design: Restoring and Conserving Ecosystem Services. John Wiley & Sons.

Nemecek, T., Heil, A., Huguenin, O., Meier, S., Erzinger, S., Blaser, S., Dux, D., Zimmermann, A., 2007. Life Cycle Inventories of Agricultural Production Systems. Final report ecoinvent v2. 0 No, 15.

CHAPTER

27

An Evolutionary Perspective on Industrial and Sustainable Agriculture

John Gowdy[1], *Philippe Baveye*[2]

[1]Rensselaer Polytechnic Institute, Troy, NY, United States; [2]UMR ECOSYS, AgroParisTech, Université Paris-Saclay, Thiverval-Grignon, France

INTRODUCTION

The achievements of industrial agriculture are impressive. Per capita yields for the world's most important crops have increased dramatically in the decades following WWII, although there is some evidence that growth rates in yields for some major crops have declined in recent years (Funk and Brown, 2009). New food production technologies are truly amazing and seem to be keeping pace with population growth. But what are the costs and what are the prospects for maintaining this high-tech, high-input, high-cost food producing system? Industrial agriculture has come with immense direct and indirect environmental costs including groundwater depletion, soil loss, the degradation of freshwater ecosystems, antibiotic resistance, greenhouse gas emissions into atmosphere, and biodiversity loss. There are also considerable social costs from the economic "rationalization" of food production. Agricultural mechanization and rural unemployment have led to an unprecedented worldwide population movement from rural to urban areas. Agricultural expansion and deforestation have also contributed to climate change, and these effects are likely to impose disproportionate costs on the world's poorest countries. It seems prudent to step back and take stock of where we are and evaluate the likely consequences of business as usual policies. If current practices are unsustainable, how can we move to a path that is more environmentally friendly and socially desirable?

Among economists, the term "sustainability" has come to mean maintaining the ability of the economy to produce a nondeclining output of goods and services (Pearce and Atkinson, 1993). Applied to agriculture, this simply means continuing to produce enough food to meet the needs of a growing population. Even within this narrow conception of sustainability, economists recognize that sustaining economic output depends on maintaining the stock of capital needed to produce it, including so-called natural capital, the inputs from nature necessary to support economic activity (Dasgupta and Mäler, 2000). But the focus is still only on increasing productivity and efficiency, that is, increasing

Agroecosystem Diversity
https://doi.org/10.1016/B978-0-12-811050-8.00027-3

yields per hectare through technology and capital intensification. The term "sustainable agriculture" is thus framed within the prevailing narrative of modernization and globalization and the neoliberal worldview supporting it. The focus on increasing agriculture output ignores both the "invisible" environmental and social costs and also the invisible subsidies supporting industrial agriculture (Wise, 2004).

There exists a long history of criticism of the economic notion of sustainability and the differences between economic, social, and ecologic sustainability (Baveye et al., 2013; Gowdy, 1997; Kallis et al., 2013; Muradian et al., 2013; Westman, 1977). The environmental destructiveness of industrial agriculture has been well documented (Ponting, 2007, Union of Concerned Scientists, Wise, 2004). But the conflict between the requirements of the industrial economy and the social good has taken on a new urgency with the acceleration of environmental degradation and social instability (Barnosky et al., 2012, WWF, 2014). The loss of nonhuman nature in recent decades has been dubbed "biological annihilation" (Ceballos et al., 2017). The buildup of greenhouse gases is also accelerating, and its predicted consequences seem to be more severe than those estimated just a few years ago. A Yale University study found that global climate models have significantly underestimated the rise in global temperatures from a doubling of CO_2: between 5 and 5.3°C rather than the 2 to 4.7°C predicted by the 2013 IPCC report (Tan et al., 2016.) A recent EPA report concluded that without rapid reductions in greenhouse gas emissions "the increase in annual average global temperatures relative to preindustrial levels could reach 9°F (5°C) by the end of this century" (EPA, 2017). Unsustainable agricultural practices have contributed substantially both to the loss of the nonhuman natural world and to global warming. The agricultural sector is also likely to be the first to feel the negative effects of these phenomena.

The global economy, including the agricultural sector, is a highly evolved interlocking system of technologies, belief systems, and institutions that work together to reinforce resource exploitation and growth (Gowdy and Krall, 2016). As an ultrasocial evolutionary system, the economy is driven by changes in external environmental conditions and the basic requirements of sustaining the inputs necessary for producing food. But evolution cannot see ahead. The economic system can adapt to future detrimental changes only if it is presently receiving information relevant to those coming changes. Climate change and the loss of the nonhuman world have not, so far, seriously affected the global economy, so there are no "market signals" for the system to respond to. If our agricultural system is unlikely to adapt "naturally" in time to avoid serious disruption, how can we devise public policies to begin to move toward a sustainable evolutionary path, one not based on growth, greed, and exploitation? We do not have to change everything at once, just start down another evolutionary path.

ROADBLOCKS TO SUSTAINABLE AGRICULTURE

Lock-in and path dependency: Brian Arthur (1994) and Paul David (1985) pioneered the concept of lock-in and path dependency in economic systems. Some policies lock us in to specific technologies and power relationships (industrial agriculture, for example) and others leave open future possibilities (preserving or enhancing soil quality, for example). A seemingly minor change can either open new possibilities or restrict future options. Agricultural policies can either open or close pathways for future development. An evolutionary approach is particularly important in assessing the biologic effects of industrial agriculture—biodiversity loss and pesticide resistance, for example—and in assessing the ability of the agricultural sector to adapt to climate change (Carroll et al., 2014).

Magrini, Befort, and Nieddu (Chapter 24) point out that food production and distribution systems have coevolved with a few major crops. Focusing food production on a few crops has allowed the capture of efficiencies from economies of scale in production, research and development, and distribution systems. But it has also diminished the acreage of some minor crops like grain-legumes that provide important ecosystem services like nitrogen fixation (Magrini, Befort, and Nieddu, Chapter 24). In terms of sustainability and equity, there are apparently much better systems for growing food than the current industrial model. But it is costly to switch since we are locked into technologies that are interconnected and difficult to change (Johnson, 2014).

Enlightened government policies can help overcome these difficulties. An example of overcoming technological lock-in through public policy is given by Cowan and Gunby (1996). In the 1940s and 1950s, Israeli farmers used native insect predators, like ladybugs, to control pests in citrus orchards. But this changed when new insect pests appeared in the 1960s and aerial spraying of insecticides also killed the beneficial predators. To kill the new pests, farmers had to spray, but this looked them into the new technology because the beneficial predators were killed. Government regulation got them out of this dilemma by prohibiting spraying of nonselective pesticides within 200 m of a citrus grove. The government provided assistance to farmers to switch back to biologic insect control.

Private interest versus the social good: Market economies are driven by the quest for surplus value and profit. The starting point for standard economic theory is that of a rational individual acting at a point in time (the immediate present). In economic terms, this means maximizing the discounted flow of income or output, given the available resources. Economists use this individualistic framework to judge public investment decisions. Just as a rational individual should maximize the returns to her investment portfolio, so too should governments make investments to maximize total economic output measured by gross domestic project (GDP). How does this relate to agricultural policies that affect environmental quality? If we take maximizing GDP as our starting point, then the question of how to value the environment is narrowed to "How much do particular features of the natural world (ecosystem services) contribute to the discounted flow of economic output?" If ecosystem services are degraded, will overall income diminish or increase? There are benefits from increased production, but there are also costs from environmental degradation. The question of distribution is ignored, and public and private interests are conflated.

Economic theory recognizes that markets work properly only if price signals are "correct." *Market failures* (monopoly power, public goods, and externalities) can occur when prices fail to register true value. Ecosystem services do not traditionally have prices, so there is no way that their social value is reflected in market outcomes. Correcting externalities is the justification for putting prices on ecosystem services that contribute to economic well-being but are outside the realm of market exchange. Monetizing ecosystem services dominates the economic approach to environmental protection (Baveye et al., 2013). Once the correct prices are calculated, externalities can be "internalized" by subsidizing those who "own" ecosystem services through direct payments by governments (PES).

There are several objections to the externality approach to environmental policy. First, there are other kinds of market failure including the concept of public goods. Public goods have the characteristics of being nonexcludable and nonrival. Once provided, anyone can use them, whether or not they pay, and one person's use of a public good does not preclude its use by someone else. National parks and public open space are examples. Many kinds of ecosystem services—watershed protection, flood control by wetlands—could be better managed as public goods rather than as externalities. There is some evidence that doing so is more cost effective than payments for ecosystem services. Curran et al. (2016) compared the cost effectiveness of payments for ecosystem services to farmers in Kenya versus land purchases. They found that outright purchases led to larger reserves, better species protection, and lower unit costs of conservation.

It is increasingly recognized that traditional measures of economic activity (GDP) do not fully capture the full social and environmental costs and benefits of production and consumption (Stiglitz et al., 2010). A large and growing literature in behavioral economics suggests that beyond a certain income level, income and happiness are not strongly correlated (Easterlin, 1995; Frank, 1999). There exists considerable evidence that interacting with the natural world has strong positive effects on human well-being. These nonmonetary values of the natural world are usually ignored even though they are critical to a comprehensive policy analysis.

Government subsidies to agricultural technology and energy: Market-based agriculture policies encouraging speculation and large capital investments, combined with the long time periods between investment and crop production, can result in disruptive price and output fluctuations. This has had the indirect effect of public funding of massive government payments to farmers and landowners to cover the difference between cost of production and commodity prices (Jordan et al., 2017; Wise, 2004). Between 1997 and 2006, agricultural producers received 30% of their net income from direct government subsidies (ERS, 2007). Also important are the massive subsidies to fossil fuels—amounting to over $5 trillion in 1995—that encourage mechanization and global trade in agricultural products (Coady et al., 2017). According to a study by the World Watch Institute (2017), agricultural subsidies by the world's top 21 countries totaled $486 billion in 2012. In addition to these subsidies, often disguised by terms like "crop insurance," governments provide massive indirect funding through research and development for industrial agriculture and energy technology.

THE CAUSES AND CONSEQUENCES OF AGRICULTURAL MODERNIZATION: THE ROLE OF GOVERNMENT POLICY

Industrial agriculture did not naturally evolve because it is more efficient and out-competed small farms. The path to intensification was paved with rules, regulations, and subsidies that favored corporate agriculture (Wise, 2004; Union of Concerned Sciences, 2017). The relative lack of attention given to agricultural subsidies and their effects is due in large part to the prevailing false dichotomy between "government" and "the market." Karl Polanyi (1944) emphasized the way that markets are deeply shaped and created by policy: "The road to free markets was opened and kept open by an enormous increase in continuous, centrally organized and controlled interventionism ... Administrators had to be constantly on the watch to ensure the free working of the system."

In every developed economy, government spending dominates, representing between 40% and 50% of GDP. Markets have always been shaped and directed by active government policies favoring one sector of the economy or another. It is not an exaggeration to say that every major technological change in recent decades can trace most of its funding and development back to the state (Gowdy et al., 2016; Mazzucato, 2013). The only question is in whose interest does the government act? Neoliberalism, the political and intellectual driver of modernization, does not advocate laissez-faire but rather subsidies and regulations favoring markets, often at the expense of individual well-being (Büscher, 2008; Helmreich, 2007; Mirowski, 2013;

Rulli et al., 2013). Industrial agriculture is the result of conscious, although sometime haphazard, intervention by the public sector. Industrial agriculture is not the "natural" result of efficiency and competition in a competitive free market economy. Furthermore, even in instances where industrial agriculture is more efficient (monoculture capturing economies of scale, for example), this does not automatically justify it. As Bromley (1990) points out, efficiency is only one of many policy criteria. There is no scientific reason why efficiency should trump other goals like equity, diversity, and resilience.

The question for global agriculture is not whether or not to "interfere" with the market but rather which system, sustainable agriculture or industrial agriculture, should public money and policy support? There is no scientific reason why policies cannot be designed and implemented to support a more environmentally friendly and socially responsible system. Reinforcing the argument is the fact that, for a substantial portion of the world's population, the promise of science and technological progress has fallen short (Sarewitz, 2016). The rise of populist movements around the world has focused attention on differential effects of public science and technology policies. It is time for policy makers to consider the secondary and distributional effects of agricultural research, subsidies, and trade policies.

Agriculture and Ecosystems

Diversity and redundancy is an important feature of stable evolutionary systems (Duffy et al., 2017). Redundancy is an adaptive feature in nature to deal with sudden change as a perturbation due to external factors. Controlled experiments indicate that more biodiverse ecosystems are more resilient to environmental stresses such as drought (Naeem, 1998, Tilman and Downing, 1994). Yet the economic logic of surplus value maximization forces simplification. With state subsidized industrial agriculture, it is easier and cheaper to simplify production through monoculture and the uniform application of pesticides and fossil fuel-based fertilizer. Subsidies like crop insurance shift the risks associated with monoculture and the elimination of redundancy from the producers to the general public in the form of taxes. Profit maximization also drives simplification. By contrast, sustainable agricultural systems like agroecology are consciously designed to be more diverse and more resilient. The relationship between mechanized agriculture and biodiversity loss, and the benefits of small-scale multicrop systems are beginning to be documented, but much more work needs to be done on the system-wide and indirect effects on biodiversity of industrial agriculture. Büscher (2008, p. 231) writes: "... conservation biology must be extended into the social sciences, but this should then be based on rigorous empirical research that shows that reality is about inequality, gray zones, and winners and losers, rather than mere neoliberal win-win ideas of consensus around competition and the market."

Industrial Agriculture and Urbanization

One of the most serious and still underappreciated effects of industrial agriculture is the massive redistribution of the world's population from rural to urban areas. From 1960 to 2016 the proportion of the world's population living in cities increased from 33% to 54% (Liu and Li, 2017). Since 1960, rural populations have declined by 73% in Brazil, 47% in China, 18% in India, and 44% in the Russian Federation (Liu and Li, 2017). This has resulted in massive environmental stress and massive social upheaval. Lui and Li (2017, p. 276) write, "In developing countries, millions of subsistence farmers seek work in cities such as Delhi and Lagos in Nigeria each year, as climate change and agribusiness drive livelihoods to the brink.

When rural populations wither, villages see labor shortages, recession and social degradation." Again, the massive exodus from rural areas to cities is not a "natural" phenomenon. It is the partly result of government policies favoring mechanization and standardization over labor-intensive, complex, and small-scale farming.

ALTERNATIVE PATHS

Agroecology can be considered the overarching scientific framework for developing specific alternatives to industrial agriculture. According to Altieri and Nicholls (2015), "Agroecology is a science that draws on social, biological and agricultural sciences and integrates these with traditional knowledge and farmers' knowledge. ... At the heart of the agroecology strategy is the idea that an agricultural system should mimic the functioning of local ecosystems, thus exhibiting tight nutrient cycling, complex structure and enhanced biodiversity."

Agroecology applies ecologic principles to food production systems. It does not reject all technology but attempts to minimizes or eliminate pesticides, commercial fertilizers, and monoculture. It is multidisciplinary in the sense that it focuses on both the biologic and social aspects of agriculture. It focuses on four properties of agriculture systems: productivity, stability, sustainability, and equity. Especially important to this field is the incorporation of indigenous farming practices (Garí, 2001). Perhaps the most important feature of agroecology is its insistence on the integration of ecologic and social science principles in the study of agricultural systems (Dalgaard et al., 2003).

Agroecology is not a "solution" but rather a generic framework for evaluating specific local and diverse approaches (Wezel et al., 2009). Within this framework, several promising developments are taking hold. A few examples are the following.

The local food movement: One of the most successful social movements in recent decades is community supported agriculture (CSA). CSAs have many different forms, but the focus is on high quality, low environmental impact, and local farming. They depend on direct contracts between consumers and local farmers. The basic feature of a CSA is consumers who are willing to commit to buying a whole season's worth of food and farmers who are willing to work with consumers to produce the kinds of crops they want (Kleppel, 2014). According to the US Department of Agriculture, there were 7398 CSAs in the United States in 2015, accounting for 7% of direct to consumer sales by farmers (USDA, 2016).

The ugly food movement: The objective of the ugly food movement is to prevent the waste of perfectly good fruits and vegetables that do not meet the standards of appearance that consumers are used to. The move has been particularly successful in Europe. In 2014 the European Union eliminated regulations prohibiting the sale of odd or misshaped fruits and vegetables. Public information campaigned have been successful by introducing terms like "wonky," "inglorious," or "naturally imperfect" to promote ugly food.

http://time.com/3,761,942/why-people-are-falling-in-love-with-ugly-food/.

The healthy food movement: Has more food made people less healthy? A related question is "Is food too cheap?" According to the World Health Organization (WHO), 57 of 129 countries are struggling both with obesity and malnutrition. Eight hundred million people are malnourished worldwide, but over twice that number (2 billion) are overweight or obese. The consumption of healthy food (fish, nuts, and vegetables) is declining, and people are eating too much fat, salt, and sugar (Haddad et al., 2016). Government subsidies for sugar and corn have been blamed for much of the decline in food quality. Agricultural research money is disproportionately spent on crops like maize (accounting for 45% of private sector agricultural

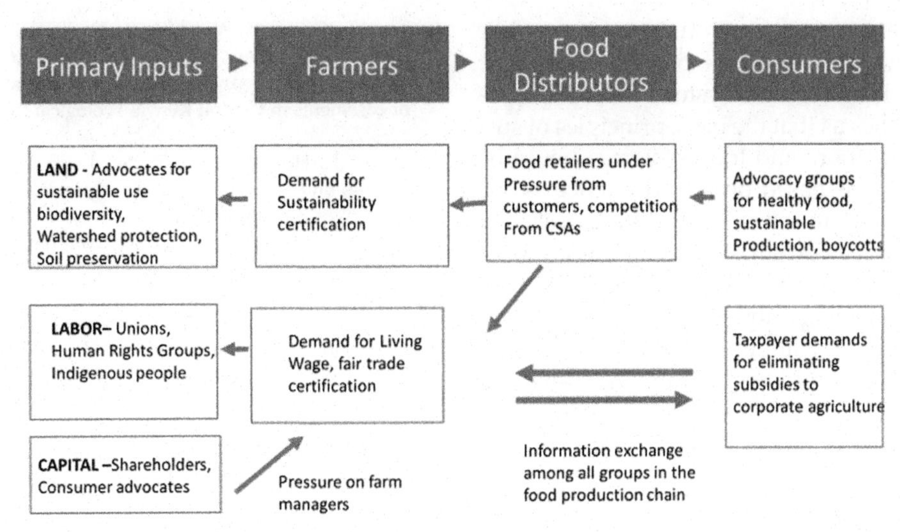

FIGURE 27.1 Points of entry in the food production chain to influence behavior and public policy.

research spending) compared to research on more nutritious fruits and vegetables. But consumers are having a tremendous effect at the retail level, and this is putting pressure on food distributors and producers to fundamentally change the way food is produced.

ENTRY POINTS FOR CHANGE

Some points for entry to change the way food is produced can be seen in Fig. 27.1. Food production is a value chain starting with the primary inputs land, labor, and capital and ending with the food consumer. At various points in the value chain, pressure can be brought to bear to change the way food is produced and consumed.

Consumer advocacy groups can use boycotts and advertising campaigns to pressure food outlets like supermarkets to sell organic and fair wage products. Supermarkets, in turn, force producers to change food production practices. An important point to keep in mind is that the agricultural system is constantly evolving, and as with any evolutionary process, a small nudge can sometimes move the system to a radically different development path.

CONCLUSION

The industrial high-tech, high-input agricultural system did not evolve naturally. It is the result of accidents of history, technological lock-in, and path dependence and massive government subsidies. This does not mean that it will be easy to change, but it moves the question of what is the best agricultural policy from "how do we encourage more technology and intensification?" to "how do we redirect public funding to create a food system that is environmentally sustainable and socially beneficial?" Given all the constraints and all the interlocking and self-reinforcing elements of the industrial agricultural, how can we begin to move the system? A good start is to keep two simple insights in mind. First is to recognize that industrial agriculture is the result of deliberate and lavishly funded public initiatives; it did not naturally come into existence because it is the best system that won out in the competitive struggle for survival.

An equivalent commitment on the part of public agencies could put us on the path to sustainable and equitable agricultural systems. Second, agroecology teaches us that the basic principles of sustainable agriculture and food systems should be universal, that is based on sound science, while specific solutions to achieve such a system should be local. Science can give us some basic guidelines about such things as maintaining soil fertility, protecting aquifers, and enhancing biodiversity. But within these parameters, specific initiatives should begin at the local level.

References

Altiery, M., Nicholls, C., 2015. Agroecology: Key Concepts, Principles and Practices. Third World Network and SOCLA. Available at: http://agroeco.org/wp content/uploads/2015/11/Agroecology-training-manual-TWN-SOCLA.pdf1.

Arthur, B., 1994. Increasing Returns and Path Dependence in the Economy. University of Michigan Press, Ann Arbor.

Barnosky, A., Hadley, D., Bascompte, J., Berlow, E., Brown, J., Fortlius, M., Getz, W., Harte, J., Hastings, A., Marquet, P., Martinez, N., Mooers, A., Roopnarine, P., Vermeij, G., Williams, J., Gillespie, R., Kitzes, J., Marshall, C., Matzke, N., Mindell, D., Revilla, E., Smith, A., 2012. Approaching a state shift in earth's biosphere. Nature 486, 52–58.

Baveye, P., Baveye, J., Gowdy, J., 2013. Monetary evaluation of ecosystem services: getting the timeline right. Ecological Economics 95, 231–235.

Bromley, D., 1990. The ideology of efficiency. Journal of Environmental Economics and Management 19, 86–107.

Büscher, B., 2008. Conservation, neoliberalism, and social science: a critical reflection on the SCB 2007 Annual Meeting in South Africa. Conservation Biology 22, 229–231.

Carroll, S., et al., 2014. Applying evolutionary biology to address global challenges. Science 346, 313–323 (8 other authors).

Ceballos, G., Ehrlich, P., Dirzo, R., 2017. Biological Annihilation via the Ongoing Sixth Mass Extinction Signaled by Vertebrate Population Losses and Declines, PNAS Early Edition http://www.pnas.org/content/114/30/E6089.

Coady, D., Parry, I., Sears, L., Shang, B., 2017. How large are global fossil fuel subsidies? World Development 91, 11–27.

Cowan, R., Gunby, P., 1996. Sprayed to death: path dependence, lock-in and pest control strategies. Economic Journal 106, 521–542.

Curran, M., Kiteme, B., Wünscher, T., Koellner, T., Hellweg, S., 2016. Pay the farmer, or buy the land?—Cost effectiveness of payments for ecosystem services versus land purchases or easements in Central Kenya. Ecological Economics 127, 59–67.

Dalgaard, T., Hutchings, N., Porter, J., 2003. Agroecology, scaling, and interdisciplinarity. Agricultural Ecosystems and Environment 100, 39–51.

Dasgupta, P., Mäler, K.-G., 2000. Net national product, wealth, and social well-being. Environment and Development Economics 5, 69–93.

David, P., 1985. Clio and the economics of QWERTY. The American Economic Review 75, 332–337.

Duffy, E., Godwin, C., Cardinale, B., 2017. Biodiversity effects in the wild are common and as strong as key drivers of productivity. Nature 549, 261–264.

Easterlin, R., 1995. Will raising the incomes of all increase the happiness of all? Journal of Economic Behavior and Globalization 27, 35–47.

EPA, 2017. U.S. Global Change Research Program Climate Science Special Report. https://assets.documentcloud.org/documents/3920195/Final-Draft-of-the-Climate-Science-Special-Report.pdf.

ERS, 2007. Farm Income Data Files. ERS, Washington, DC. www.ers.usda.gov/data/FarmIncome/finfidmu.htm. Cited in Jordan et al. 2014.

Frank, R., 1999. Luxury Fever: Why Money Fails to Satisfy in an Age of Excess. Free Press, New York.

Funk, C., Brown, M., 2009. Declining Global per Capita Agricultural Production and Warming Oceans Threaten Food Security. NASA, Washington, DC. http://digitalcommons.unl.edu/cgi/viewcontent.cgi?article=1128&context=nasapub.

Garí, J., 2001. Biodiversity and indigenous agroecology in Amazonia. Ethnoecologica 5, 21–37.

Gowdy, J., 1997. The value of biodiversity: markets, society, and ecosystems. Land Economics 73, 25–41.

Gowdy, J., Krall, L., 2016. The economic origins of ultrasociality. Behavioral and Brain Sciences 39, 1–60.

Gowdy, J., Mazzucato, M., Page, S., van den Bergh, J., van der Leeuw, S., Wilson, D.S., 2016. Shaping the evolution of complex societies. In: Wilson, D.S., Kirman, A. (Eds.), Complexity and Evolution: A New Synthesis for Economics. In: Lupp, J. (Ed.), Strüngmann Forum Reports, vol. 19. MIT Press, Cambridge, MA.

Haddad, L., Hawkes, C., Webb, P., Thomas, S., Beddington, J., Waage, J., Flynn, D., 2016. A new global research agenda for food. Nature 540, 30–32.

Helmreich, S., 2007. Blue-green capital, biotechnological circulation and an oceanic imaginary: a critique of biopolitical economy. BioSocieties 2, 287–302.

Johnson, N., 2014. For farmers, using pesticides is a lot like picking the wrong smartphone. Grist. http://grist.org/

food/for-farmers-using-pesticides-is-a-lot-like-picking-the-wrong-smartphone/.

Jordan, N., et al., 2017. Sustainable development of the agricultural bio-economy. Science 316, 1570–1571 (13 other authors).

Kallis, G., Gómez-Baggethun, E., Zografos, C., 2013. To value or not to value? That is not the question. Ecological Economics 94, 97–105.

Kleppel, G., 2014. The Emergent Agriculture: Farming, Sustainability and the Return of the Local Economy. New Society Publishers.

Liu, Y., Li, Y., 2017. Revitalize the world's countryside. Science 548, 275–277.

Mazzucato, M., 2013. The Entrepreneurial State. Anthem Press, London.

Mirowski, P., 2013. Never Let a Serious Crisis Go to Waste: How Neoliberalism Survived the Financial Crisis. Verso, London.

Muradian, R., et al., 2013. Payments for ecosystem services and the fatal attraction of win-win solutions. Conservation Letters 6, 274–279 (23 other authors).

Naeem, S., 1998. Species redundancy and ecosystem reliability. Conservation Biology 12, 39–45.

Pearce, D.W., Atkinson, G., 1993. Capital theory and the measurement of sustainable development: an indicator of 'weak' sustainability. Ecological Economics 8 (2), 103–108.

Polanyi, K., 1944. The Great Transformation. Beacon Press, Boston.

Ponting, C., 2007. A New Green History of the World. Penguin, New York.

Rulli, M.C., Saviori, A., D'Odorico, P., 2013. Global land and water grabbing. Proceedings of the National Academy of Sciences 110 (3), 892–897.

Sarewitz, D., 2016. Donald Trump's appeal should be a call to arms. Nature 536, 17.

Stiglitz, J., Sen, A., Fitoussie, J.-P., 2010. Mismeasuring Our Lives: Why GDP Doesn't Add Up: The Report. New Press, New York.

Tan, I., Storelvmo, T., Zelinka, M., 2016. Observational constraints on mixed-phase clouds imply higher climate sensitivity. Science 352, 224–227.

Tilman, D., Downing, J., 1994. Biodiversity and stability in grasslands. Nature 367, 363–365.

Union of Concerned Scientists, 2017. Hidden Costs of Industrial Agriculture. http://www.ucsusa.org/food_and_agriculture/our-failing-food-system/industrial-agriculture/hidden-costs-of-industrial.html#.WblIHuTXvDc.

U.S. Department of Agriculture, 2016. Direct Farm Sales of Food: Results from the 2015 Local Food Marketing Practices Survey. https://www.agcensus.usda.gov/Publications/2012/Online_Resources/Highlights/Local_Food/LocalFoodsMarketingPractices_Highlights.pdf.

Westman, W., 1977. How much are nature's services worth? Science 197, 960–964.

Wezel, A., Bellon, S., Doré, T., Francis, C., Vallod, D., David, C., 2009. Agroecology as a science, a movement and a practice. Annual Review of Agronomy and Sustainable Development 29, 503–515.

Wise, T., 2004. The paradox of agricultural subsidies: measurement issues, agricultural dumping, and policy reform. In: Global Development and Environment Institute Working Paper No. 04-02. http://ase.tufts.edu.gdae.

World Watch Institute, 2017. http://www.worldwatch.org/agricultural-subsidies-remain-staple-industrial-world-0.

World Wide Fund, 2014. Living Planet Report 2014. Gland, Switzerland. http://wwf.panda.org/about_our_earth/all_publications/living_planet_report/.

Further Reading

Magrini, M.-B., Befort, N., Nieddu, M., 2018. Technological Lock-in and Pathways for Crop Diversification in the Bio-economy (Chapter 24), THIS VOLUME.

McCauley, D., Pinsky, M., Palumbi, S., Estes, J., Joyce, F., Warner, R., 2015. Marine defaunation: animal loss in the global ocean. Science 347 (248), 254.

Pretty, J., 2012. Agriculture and food systems: our current challenge. Chapter 2. In: Rosin, C., Stock, P., Campbell, H. (Eds.), Food Systems Failure: The Global Food Crisis and the Future of Agriculture. Earthscan, New York.

Waring, T., Brooks, J., Kline, M., Goff, S., Gowdy, J., Jacquet, J., Janssen, M., Smaldino, P., 2015. A multi-level evolutionary framework for sustainability analysis. Ecology and Society 20 (2). https://doi.org/10.5751/ES-07634-200234.

Wilson, D.S., Gowdy, J., 2013. Evolution as a general theoretical framework for economics and public policy. Journal of Economic Behavior and Organization 90S, S3–S10.

CHAPTER

28

Current and Potential Contributions of Organic Agriculture to Diversification of the Food Production System

Verena Seufert[1,2], *Zia Mehrabi*[1], *Doreen Gabriel*[3], *Tim G. Benton*[4]

[1]Institute for Resources, Environment and Sustainability (IRES), University of British Columbia, Vancouver, BC, Canada; [2]Institute of Meteorology and Climate Research - Atmospheric Environmental Research (IMK-IFU), Karlsruhe Institute of Technology (KIT), Garmisch-Partenkirchen, Germany; [3]Institute for Crop and Soil Science, Julius Kühn-Institut, Braunschweig, Germany; [4]School of Biology, University of Leeds, Leeds, United Kingdom

INTRODUCTION

Diversity is a key organizing principle of natural ecosystems, and as previous chapters of this book have laid out in detail, diversification of agroecosystems can provide many benefits for the sustainability of food production and represents an important path for moving toward a more sustainable food system. But even though the promises of more diversified agriculture at the local and farm scale have often been emphasized (Benton et al., 2003; Frison et al., 2011; Kremen et al., 2012; Lin, 2011), the feasibility and potential impacts of diversified agriculture at a larger scale are yet poorly understood. Given current trends toward homogenization of agriculture through, for example, decreases in crop diversity (Aguilar et al., 2015; Khoury et al., 2014), decreased integration of crop and livestock systems (Naylor et al., 2005), and increases in farm and field sizes in many regions of the world (Lowder et al., 2016), it is important to (1) understand both the extent and impacts of more diversified farming systems currently already in place, as well as (2) identify ways for reversing current homogenization trends and identify potential pathways toward more diversified agriculture. Organic agriculture represents an alternative farming system that is built upon agroecologic principles that naturally encourage diversity. Being the most prominent alternative farming system currently in place, it offers a great case study to examine these questions and assess the current and future opportunities for diversification of food production. In the following, we will examine both our understanding of the current extent of diversification implemented in and fostered by organic systems

Agroecosystem Diversity
https://doi.org/10.1016/B978-0-12-811050-8.00028-5

(Current Contribution of Organic Agriculture to Diversification section), as well as the opportunities offered by organic agriculture for moving toward a more diversified food system, including a discussion of current food system trends that oppose or foster more diversified organic agriculture or organic practices (Potential Contribution of Organic Agriculture to Diversification of the Food Production System section).

CURRENT CONTRIBUTION OF ORGANIC AGRICULTURE TO DIVERSIFICATION

Diversity is a key component of organic systems resulting from the agroecologic design principles of organic management. Even though the term and concept of diversity did not feature prominently in the work of organic pioneers like Albert Howard, Rudolf Steiner, or Eve Balfour—as the concept of biodiversity was developed much later than the original organic movement—diversity is a key component of modern conceptualizations of organic agriculture. A commonly used definition of organic agriculture defines it as "a holistic production management system which promotes and enhances agroecosystem health, including biodiversity, biological cycles, and soil biological activity" (FAO and WHO, 2001). This definition recognizes the importance of biologic processes for agricultural production and thus quite closely matches the definition of diversified farming systems proposed by Kremen et al. (2012), who define these as systems that intentionally include functional biodiversity to generate ecosystem services that are critical to agriculture. While organic systems, as they manifest themselves today on the ground, do not necessarily *have* to implement diversification practices or actually lead to diversification outcomes (as highlighted by Kremen et al. (2012), and as we will discuss throughout this book chapter), we argue that diversification is key to the original concept of organic farming (Van Bueren et al., 2002; van Elsen, 2000). Farmer surveys have also shown that organic farmers are often motivated by values of nature conservation and environmental stewardship (Stobbelaar et al., 2009). Based on the hypothesis that organic agriculture in its theoretical conceptualization represents a diversified farming system, we will examine the evidence for how diversification practices are in practice realized in today's organic farms.

Diversification Practices Implemented by Organic Farmers

Farm-Level Diversification

Farm-level diversification practices range from the genetic diversity of crop and livestock breeds used, over the temporal and spatial diversity of crop and livestock associations, to the structural components of the farm, like field sizes and amount and configuration of noncrop habitat. While we attempt to separate our discussion of diversification practices into farm- and landscape-level practices, we recognize that this represents a continuum rather than a clear binary distinction (Kremen et al., 2012) (Fig. 28.1), and that some practices span from the field to the landscape scale.

CROP AND LIVESTOCK GENETIC DIVERSITY

Crops in organic systems experience different and typically more variable environments than crops under conventional management, as they, for example, need to access different forms of nutrients and are exposed to different pest and weed pressures (Van Bueren et al., 2002; Wolfe et al., 2008). In theory, we might expect organic farms to use different crop varieties than conventional farms, and also use varieties with higher genetic diversity, as this would provide better adaptation to the more variable environmental conditions experienced under organic management (Wolfe et al., 2008). Some organic regulations emphasize that "plant varieties should be selected to maintain genetic diversity"

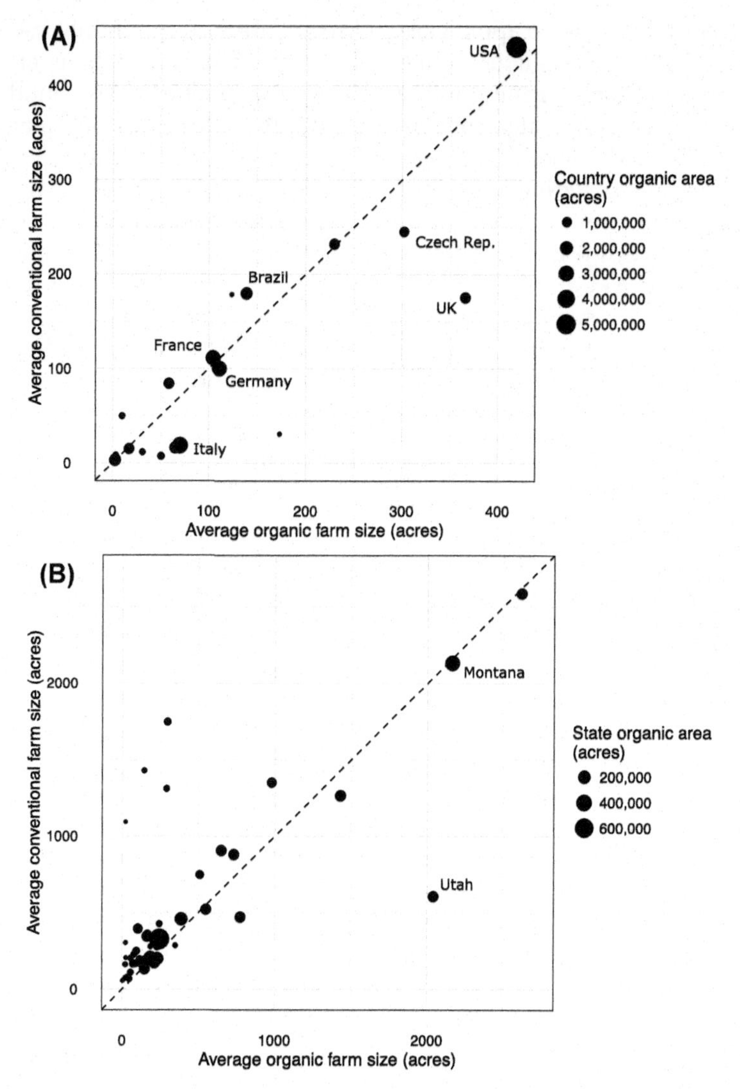

FIGURE 28.1 Average organic versus conventional farm size in the top 20 countries with the highest organic area (A), as well as different US states (B). The *dotted line* indicates equal organic and conventional farm sizes. Size of dots shows the total organic area per country or state. Note that panel a does not include Argentina and Uruguay, as in these countries average size of organic farms is >5 times larger than the average size of conventional farms. National farm size data for organic farms is from FiBL and IFOAM (2016), usually for the year 2014. National farm size data for conventional farms is from FAO (2000), usually for the year 2000. US state level farm size data is from USDA (2014a) for organic farms for the year 2014 and from USDA (2014b) for conventional farms for the year 2012. Data sources: FiBL and IFOAM (2016), FAO (2000), USDA (2014a, b)

(IFOAM, 2006, p. 20). In practice, however, some studies have estimated that more than 95% of organic crop production still relies on crop varieties bred for conventional agriculture, due to the lack of targeted breeding programs for organic agriculture (van Bueren et al., 2011). Census data from apple production in the United States shows that organic farmers grow some different, but not more diverse apple varieties (Slattery et al., 2011). This suggests that crop genetic diversity in organic farming today might not be much different from that in conventional agriculture. Increased genetic diversity and more targeted organic breeding programs are strongly needed and would most likely reduce the yield gap between organic and conventional agriculture (Wolfe et al., 2008). However, progress is made in some areas. For example, German wheat varieties can be tested under and registered for organic production methods. This allows to better identify additional traits important for organic production such as competitive ability against weeds or ability to deal with low fertilization levels (as are common in organic farming). Hence incorporating organic farming into variety registration trials may enhance the diversity of varieties and help to diversify breeding goals and strategies of commercial breeders.

Similarly, it is also to be expected that organic livestock production would use different livestock breeds and have different breeding goals than conventional livestock production, as organically raised livestock experiences different production environments due to the requirements imposed by organic farming regulations, for example, regarding the use of antibiotics, feed composition, or outdoor housing (Ahlman et al., 2014). Organic livestock is generally exposed to more variable environmental conditions and require more robust breeds, which often is dependent on higher genetic diversity at the population level (Boelling et al., 2003). Some organic regulations emphasize that the choice of organic livestock breeds should "take account of their capacity to adapt to local conditions, their vitality and their resistance to disease and a wide biological diversity should be encouraged" (EU, 2008, p. 2). Even though data on livestock breeds used in organic agriculture are scarce, some literature shows that organic dairy farms are more likely to use crossbreeds or other local breeds rather than the Holstein breed, which is commonly used in conventional dairy production, in Sweden and the United States (Sorge et al., 2016; Sundberg et al., 2009). In contrast to crops, in many regions, there have been targeted efforts to avoid the loss of livestock genetic diversity, as well as considerable conservation efforts to save traditional livestock breeds (Hall and Ruane, 1993). It is thus likely that organic farming uses more local and genetically more diverse breeds that are more suitable to the different livestock production methods in organic systems.

CROP AND LIVESTOCK ASSOCIATIONS

Farm-level crop diversity can be achieved both in space (e.g., through multicropping) as well as in time (e.g., through crop rotation). Crop rotations typically form an integral part of organic management, as sequential cultivation of different types of crops is used to suppress pests and diseases, to maintain soil fertility, and to prevent soil erosion and nutrient losses. Most organic regulations require or recommend multiannual crop rotations that include green manures and sometimes cover crops (Seufert et al., 2018). Census data on the actual use and types of crop rotations and crop associations used by organic farmers is not available to date. Ponisio et al. (2015) showed that in 13% of studies examining yields of organic agriculture for annual crops organic systems had longer crop rotations than conventional systems (while the remaining 85% of studies had similar rotation lengths), while 24% of the 92 studies on annual crops included in their analysis did not use any crop rotation in the organic system. Similarly, in 15% of the yield studies, organic systems used multicropping (i.e., the cultivation of more than one crop species on the same field within a growing season), while conventional systems did not (while the use of multicropping was the same in the remaining 85% of studies). Given the importance of crop rotations for organic management and given the crop rotation requirements in many organic regulations, this appears like a small difference. This data comes, however, both from experimental field trials as well as from farmer's fields, and it is thus not necessarily representative of the entire organic farm population. Some primary studies have observed slightly higher diversity of crops in organic than in conventional farms (Belfrage et al., 2005; Kragten and de Snoo, 2008; Weibull et al., 2003). Organic farms appear thus to often but not always have higher crop diversity than conventional farms.

For livestock, some studies show that organic dairy farms often include more other species like chicken, dairy animals for slaughter, and pigs than conventional dairy farms (Sorge et al., 2016). But census data on livestock associations of organic farms are not available.

FARM-SCALE CROP-LIVESTOCK INTEGRATION

Organic regulations do not allow landless livestock systems and typically require a minimum amount of grazing (Von Borell and Sørensen, 2004), and the European regulation, for example, also requires at least 50% of animal feed to come from the same farm unit or, if this is not feasible, from farms in the region (EU, 2008). This means that organic livestock farms have to include pasture, but they also often include cropland for feed production. But organic livestock farms often still represent specialized livestock farms, even if they also cultivate crops for feed production. Farms that cultivate multiple types of crops (e.g., arable and permanent crops), multiple types of livestock, or both crops and livestock are as rare amongst organic farms as amongst all farms in Germany (i.e., totaling up to 14% of both the organic and total farm population, Table 28.1). In Italy and France, such organic polyculture farms are also still rare (14% of all organic farms in Italy and 16% in France) but slightly more common than amongst all farms (Table 28.1). While crop-animal integration is thus enhanced in organic agriculture, most organic farms are still rather specialized.

Landscape-Level Diversification

It is important to include a landscape perspective in any assessment of diversification impacts of agriculture (Benton et al., 2003; Tscharntke et al., 2005). Landscape diversification practices include compositional changes like the inclusion

TABLE 28.1 Mixed Farming in Organic Agriculture in the Three European Countries With the Highest Organic Area

	France[a]		Italy[b]		Germany[c]	
Farm Type	**Organic Farms (%)**	**All Farms (%)**	**Organic Farms (%)**	**All Farms (%)**	**Organic Farms (%)**	**All Farms (%)**
Mixed crops[d]			8.0	6.5	2.2	1.2
Mixed livestock[d]	16.4	12.8	0.7	0.3	1.7	2.7
Mixed crop-livestock[d]			5.6	2.2	10.0	10.4

Data shows proportion of all farms practicing mixed farming in 2010 for France and Italy and for 2013 for Germany.

[a] *Source: for organic and Agreste, 2010 for conventional farms.*

[b] *Source: Calculated based on Bioreport 2013 (RRN, 2013) for organic and Istituto Nazionale di Statistica, 2010, for all farms.*

[c] *Source: Calculated based on Statistisches Bundesamt, 2014a for organic and Statistisches Bundesamt, 2014b for all farms.*

[d] *Mixed crop farms are defined as farms practicing a mixture of arable, horticultural and/or permanent crops; mixed livestock farms are farms raising both grassfed and grainfed livestock; mixed crop-livestock farms are farms practicing a combination of cropping and livestock activities (and where neither of these makes up >2/3 of the operation).*

of noncrop habitats like field boundaries, hedgerows, and ditches, as well as configurational changes like smaller field or farm sizes, but also integration (rather than specialization) of crop and livestock systems (Benton et al., 2003). These changes can be implemented at the field scale (leading to within-farm heterogeneity), at the farm scale (leading to between-farm heterogeneity), or at the landscape scale (leading to between-region heterogeneity). Given that organic agriculture represents a farming system and set of management practices that are typically limited to single fields or farms so does typically not include landscape-scale interventions, we focus here on farm-level management practices that lead to landscape-level diversification.

NONCROP HABITATS.

Some organic regulations require farmers to set aside land in forest or grasslands that are not cultivated or intensively grazed (e.g., AUS, 2009; NPOP, 2005). And census data shows that in the United States, 66% of organic farms maintain buffer strips on their farms, and 34% of all organic farmers (USDA, 2014a) and 73% of organic apple producers (compared to only 49% of conventional apple producers, Slattery et al., 2011) purposefully maintain habitats for beneficial insects or vertebrates. Organic farmers also often participate in other agri-environmental schemes that aim to increase the amount of habitat for wildlife on farms (Stobbelaar et al., 2009). Some scientific studies have, in fact, observed higher habitat diversity and higher amounts of noncropped habitat on organic than on conventional farms (Fuller et al., 2005; Norton et al., 2009; Weibull et al., 2003). Yet other studies have observed no such differences (Chamberlain et al., 1999; Kragten and de Snoo, 2008; Schneider et al., 2014). While there thus appears to be a trend toward higher amounts of noncrop habitats on organic farms, this does not seem to be the case for all farms.

FARM HETEROGENEITY.

Data on field sizes on organic farms is rare, but some studies suggest that organic farms have smaller fields than conventional farms (Belfrage et al., 2005; Chamberlain et al., 1999; Stobbelaar et al., 2009). Whether this also holds true in those regions where organic farms are larger than conventional farms (Fig. 28.1) is not sure, as field size is often correlated with farm size (Belfrage et al., 2005). Other types of configurational heterogeneity also can be higher on organic farms, including field boundaries and hedgerow structure (Chamberlain et al., 1999; Fuller et al., 2005). While not all organic farms thus appear to have smaller fields than their conventional counterparts, some of them show increased configurational heterogeneity.

FARM SIZE

Organic agriculture is often thought to be associated with smaller farms. Organic farms might be smaller in size due to the demands of organic management, or smaller farms might be more likely to be organic for economic (e.g., to access premium prices) or structural reasons. A study from the United Kingdom showed, for example, that organic farming was often more prevalent in areas less suited to intensive arable farming, characterized by high altitude and slope where farm size is small and farms are more likely to be mixed or dairy (Gabriel et al., 2009).

Census data shows, however, a more mixed picture (Fig. 28.1). While in the United States organic farms are typically, on average, slightly smaller than conventional farms (Fig. 28.1B), in numerous other countries (including e.g., Argentina, Uruguay, Italy, United Kingdom, Czech Republic), organic farms are, on average, larger than conventional ones (Fig. 28.1).

It is important to note that the larger or smaller farm size of organic farms could also be related to farm type, as organic agriculture is more common in livestock system (which are typically larger in size, e.g., in Argentina), as well as in horticultural systems (which are typically smaller in size, e.g., in California). But data from Italy suggests that organic farms can be larger than their conventional counterparts even within the same farm type (RRN, 2013).

REGIONAL CROP-LIVESTOCK INTEGRATION.

Integration of crop and livestock systems can take place both at the farm scale (as discussed before) and also at a larger landscape scale (e.g., through the use of animal manure on neighboring arable or horticultural farms, or through the use of animal feed from neighboring arable farms). As mentioned earlier, the European organic regulation requires animal feed to come from the same farm or from farmers in the region (EU, 2008). A study from France showed that organic farms, particularly in mixed agricultural districts, engage in intensive exchanges of nutrients (in the form of animal manure, organic fertilizers, and animal feed), typically over short distances (from 10 to 50 km, Nowak et al., 2015). Even though the study by Nowak et al. (2015) did not include a comparison with conventional farms, the high reliance of organic farms on animal manure and regional livestock feed suggests that organic agriculture enhances the regional-scale integration of crop and livestock systems.

Influence of Organic Agriculture on Diversification Outcomes

After having reviewed the evidence on the implementation of *diversification practices* by organic farms (i.e., components that are directly managed by farmers), here, we will review evidence on the impacts of organic management on *diversification outcomes* (i.e., positive consequences of more diversified farming systems proposed in the literature).

Proposed positive outcomes of diversification include biodiversity and biodiversity-mediated regulating and supporting ecosystem services like pest control, pollination, and soil fertility, but also yield stability and resilience, as well as nutritional quality and diversity of agricultural products (Frison et al., 2011; Kremen et al., 2012; Lin, 2011). The benefits of organic management for biodiversity at the field and farm scale have been demonstrated across many studies and many regions (Bengtsson et al., 2005; Kennedy et al., 2013; Schneider et al., 2014; Tuck et al., 2014), but whether these benefits extend beyond the farm is more contested. Some studies have shown positive landscape-scale impacts of organic management on biodiversity (Clough et al., 2007; Gabriel et al., 2010; Rundlöf et al., 2008), even though these benefits were not always ubiquitous across all organism groups (Clough et al., 2007; Gabriel et al., 2010), while other studies have shown no landscape-level effects of organic management (Schneider et al., 2014).

This increased biodiversity in organic fields is expected to lead to increased provision of biodiversity-mediated ecosystem services. Organic agriculture has, for example, been shown to enhance the evenness of arthropod communities (Crowder et al., 2010), as well as the abundance of natural enemies but not of pest species (Bengtsson et al., 2005) and thus to provide enhanced biologic pest control (Birkhofer et al., 2016; Östman et al., 2003). Pollination success (Andersson et al., 2012; Kremen et al., 2002), as well as the incidence of wild pollinators (Kennedy et al., 2013; Klein et al., 2012; Kremen et al., 2002), is also typically increased with organic management, although some pollinator species seem to require a seminatural habitat in the surroundings to benefit from organic management (Brittain et al., 2010; Klein et al., 2012). Similarly, organic agriculture has been shown to lead to increased soil organic matter content (Gattinger et al., 2012; Leifeld and Fuhrer, 2010) and generally increased soil fertility (Mäder et al., 2002; Watson et al., 2002), due to higher organic matter inputs (Gattinger et al., 2012; Leifeld and Fuhrer, 2010) but also due to higher soil biologic activity (Mäder et al., 2002).

The higher provision of biodiversity-mediated ecosystem services like pest control, pollination, and soil fertility under organic agriculture would suggest a positive impact on the productivity of organic systems (Bommarco et al., 2013). But organic agriculture typically shows lower yields than conventional agriculture (de Ponti et al., 2012; Ponisio et al., 2015; Seufert et al., 2012). This lower yield is associated both with ecosystem disservices to agriculture like high weed pressure and pest damage (Clark et al., 1998; Penfold et al., 1995), but also with nutrient limitation, particularly low crop nitrogen availability (Berry et al., 2002; Seufert et al., 2012). The inclusion of diversification measures in organic management (and not in conventional management) reduces this yield gap (Ponisio et al., 2015). This suggests that the inclusion of diversification practices in organic agriculture leads to yield benefits but still is not able to entirely close the yield gap to intensive conventional systems.

But yield studies typically only examine yields of a single crop in a single year and over a limited period of time. Long-term studies have often shown yield declines in continuously cultivated monoculture systems (Bennett et al., 2012), as well as due to soil erosion from intensive cultivation (Crosson, 2016), while more diversified systems are proposed to provide more stable yields over longer time periods (Lin, 2011). Evidence on the influence of organic management on crop yield variability is mixed (Clark et al., 1999; Delmotte et al., 2011; Lotter et al., 2003; Smith et al., 2007). The trade-offs between higher soil organic matter under organic management leading to more stable yields under drought conditions (Lotter et al., 2003) and higher and more variable weed pressures and more variable nitrogen supply leading to increases in yield variability (Clark et al., 1999; Delmotte et al., 2011) are not sufficiently understood yet.

Some studies have suggested that more diversified cropping systems enhance nutritional security due to the provision of a more diversified diet (Frison et al., 2011). Studies examining this claim for organic agriculture have, to our knowledge, not been conducted to date. While studies on the nutritional diversity provided by organic farms are thus very rare, studies examining the nutritional quality of organically produced foods have been conducted in large numbers (Barański et al., 2014; Smith-Spangler et al., 2012; Średnicka-Tober et al., 2016). But despite this wide evidence base, it is still highly debated whether organically produced food actually has higher amounts of nutrients that provide health benefits to consumers (Barański et al., 2014; Brandt et al., 2011; Dangour et al., 2010; Smith-Spangler et al., 2012).

Overall, our understanding of diversification outcomes of organic agriculture is more advanced than our knowledge of the diversification practices actually implemented on organic farms (Fig. 28.2). On the one hand, this indicates that diversification practices, which lead to the observed positive diversification outcomes, are probably being implemented on organic farms (even if we do not have data on them). On the other hand, this also suggests that further research is warranted to characterize the actual management practices used by organic farmers better.

POTENTIAL CONTRIBUTION OF ORGANIC AGRICULTURE TO DIVERSIFICATION OF THE FOOD PRODUCTION SYSTEM

In the previous sections of this chapter, we have laid out how diversification underpins the key agroecologic principle of organic agriculture, and we reviewed the often rather limited knowledge on the diversification practices implemented by organic farmers, as well as the

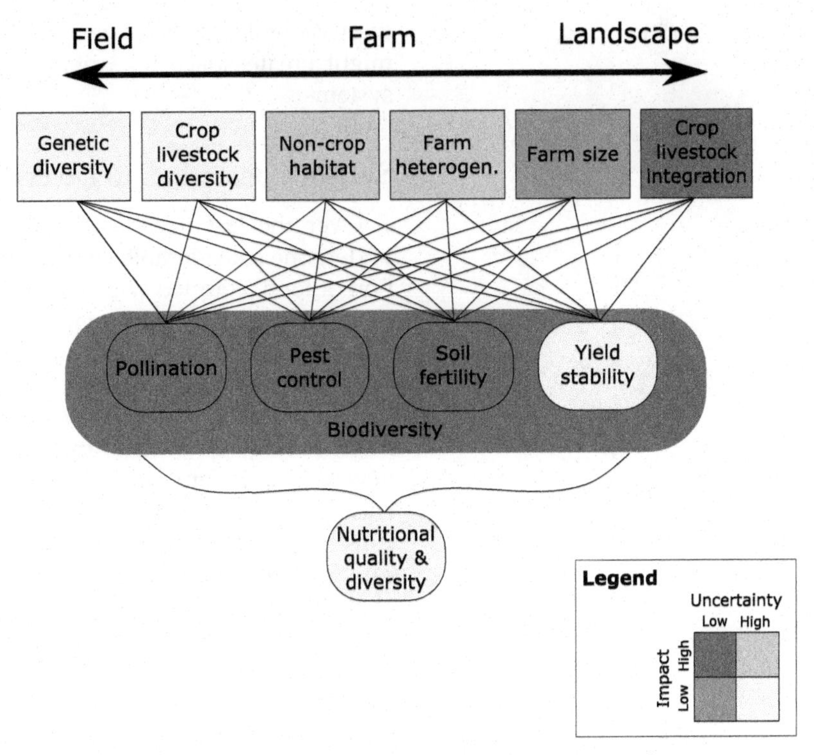

FIGURE 28.2 Diversification practices (*square boxes*) and diversification outcomes (*rounded boxes*) of organic farming from field over farm to landscape scale. Diversification outcomes include biodiversity, and biodiversity-mediated ecosystem services like pollination, pest control, soil fertility and yield stability, as well as nutritional quality and diversity of the agricultural output. *Light shades* indicate a low impact of organic (relative to conventional), as well as high uncertainty in our understanding of these impacts. *Dark shades* indicate a high impact on diversification practices and outcomes, as well as low uncertainty (see legend for details). Note that crop-livestock integration is only depicted at the landscape-scale, even though it takes place at the farm-scale as well (see text for further discussion). *Adapted from Kremen, C., Iles, A., Bacon, C.M., 2012. Diversified farming systems: an agroecological, systems-based alternative to modern industrial agriculture. Ecology and Society 17 (4), 44.*

empirical evidence on the diversification outcomes of organic agriculture. This review did not only highlight many knowledge gaps but also a wide range of levels of diversification amongst organic farmers (see Fig. 28.3B). Despite this uncertainty and the high variability, it is, however, safe to assume that, on average, organic systems represent more highly diversified systems than their conventional counterparts (Fig. 28.3A and B). This does not mean that every organic system is more diversified than every conventional system. Conventional farms also fall along a diversification spectrum (Fig. 28.3A and B). Diversification can thus also take place in conventional systems, while some organic systems can be characterized by a lack of diversity (Kennedy et al., 2013; Kremen et al., 2012).

We will now turn toward a discussion of future diversification trajectories of the food production system and the role organic agriculture might be able to play in this. For this purpose, we will first (1) discuss current system constraints toward diversification, (2) the different types of diversification opportunities of organic agriculture, and we will conclude with a

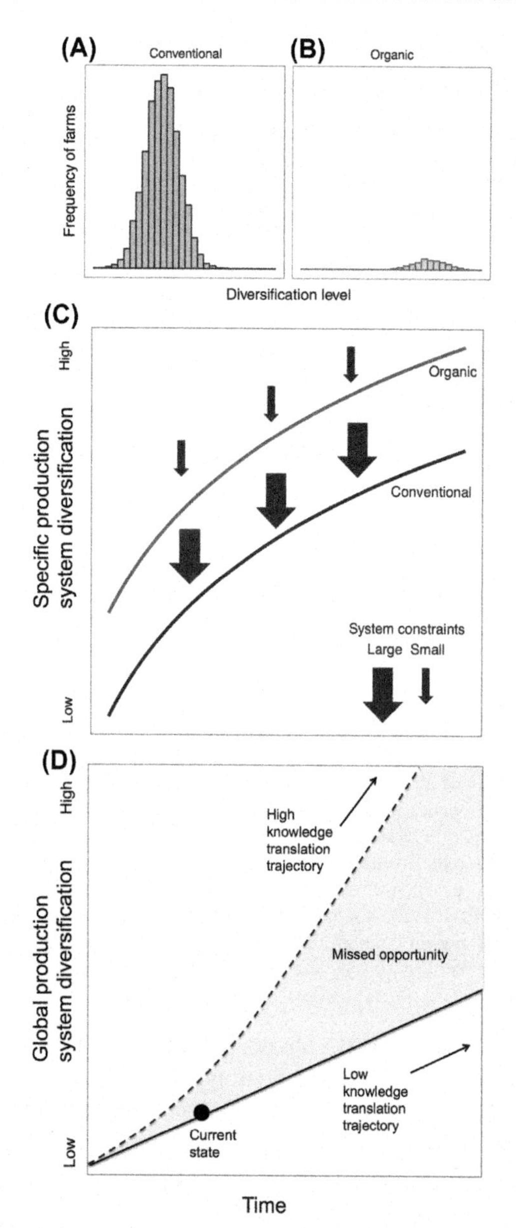

FIGURE 28.3 Current (panels A, B) and potential contribution (panels C, D) of organic agriculture to diversification of the food production system through diversification within organic systems (panel C) and through knowledge transfer to conventional systems (panel D). Panels (A) and (B) show today's distribution of organic and conventional farms along a diversification spectrum. Organic farms are, on average, more diversified but considerably less common than conventional farms. Panel (C) shows the potential diversification trajectories within organic and conventional farming systems into the future and the different system constraints for achieving this diversification potential. Panel (D) shows the potential diversification trajectories of the whole food production system if knowledge translation from organic to conventional systems takes place (upper *dotted line*), or if the opportunity for knowledge translation is missed (lower *solid line*).

discussion of (3) broader food system trends that might create space for organic or organic-like systems.

System Constraints to Diversification

Over the last decades the food system has undergone considerable changes. Commodification of food, coupled with market liberalization, has resulted in an increasingly globalized food system (Puma et al., 2015) focused on intensive production. This globalization and industrialization of the food system enabled us to produce food in large quantities and at low prices as never seen before, thus providing access to cheap food to a growing population of urban consumers and reducing the relative number of calorie-deficient people by about half (FAO et al., 2015).

These developments in the food system have been driven by market liberalization and the pursuit of economic growth. The maximization of demand growth is viewed as a prerequisite of economic growth (Lebow, 1955), and the need to not only meet domestic demand but also supply into international markets has entrenched the interests of postwar economic powers (McKeon, 2014). Market liberalization postulates that wherever people have money to pay, a market without barriers and distorting subsidies will stimulate supply, and competition will drive down prices, increase efficiencies, and maximize the "comparative advantage" of specific locations. This ideology has led to both oversupply and specialization, resulting in intensive production of a few commodity crops (Cassidy et al., 2013) concentrated to those areas that allow large-scale, highly efficient and high-yielding production (Foley et al., 2011, see Fig S5a). This rhetoric of producing more food ever more cheaply is further supported by more recent calls for the need to increase food production into the future to meet demands of a growing and wealthier population: that "food production must increase by 60%–100% by 2050" (Tilman et al., 2011).

All of these food system trends combined represent strong system constraints to the diversification of food production systems in conventional agriculture (Fig. 28.3C). If past trends are continued, further specialization, intensification, and homogenization of food production will ensue in the coming decades.

The Organic Diversification Opportunity

As we have argued before, organic agriculture is based upon agroecologic principles that *require* diversity for the production of agricultural commodities. Due to the prohibition of synthetic inputs, organic farmers are faced with more limited options to manage weeds, pests, and soil fertility. This forces organic farmers to rely more strongly on biodiversity and biodiversity-mediated supporting and regulating ecosystem services for the production of crops and livestock. This, in turn, encourages organic farmers to incorporate diversification practices into their farm management. In addition, organic farmers can access premium prices and thus offset the sometimes higher costs of using diversification practices rather than inputs to manage pests, weeds, and soil fertility.

This does not mean, however, that diversification in organic agriculture does not also experience system constraints. As our review of diversification practices and outcomes in organic agriculture in Current Contribution of Organic Agriculture to Diversification section has highlighted, not all organic farms actually represent diversified production systems. The "conventionalization" hypothesis argues that when organic is scaled up, it increasingly incorporates aspects of intensive conventional systems like consolidation of farm units, the use of external inputs, and monocropping (Buck et al., 1997; Darnhofer et al., 2010). While this hypothesis has been supported by trends observed in California (Buck et al., 1997) and Australia (Lockie et al., 2004), the conventionalization of organic farms appears to be limited to date in most parts of Europe (Darnhofer et al., 2010), New Zealand (Campbell and Liepins, 2001), and Canada (Hall and Mogyorody, 2001). This suggests that some of the constraints driving the food production system toward homogenization and specialization (see System constraints to diversification) are also acting in organic agriculture and leading to undiversified organic systems. But, overall, organic systems still experience fewer system constraints toward diversification than conventional systems (Fig. 28.3C).

These smaller system constraints for diversification have led to organic systems becoming important experimental grounds for the development of diversified agriculture in the modern context. Organic agriculture has, for example, led the way in the use of intercropping, trap crops, and mulches to manage pests (Zehnder et al., 2007), the development of botanical insecticides (Isman, 2008), the use of cover crops, mulches, and crop rotations for weed management (Bàrberi, 2002), the use of microorganisms as biofertilizers (Berg, 2009), or the use of different types of composts and cover crops for nutrient management (Maltais-Landry et al., 2015).

The continued expansion of organic agriculture could thus provide the means for more innovation in diversity research and the building of a future knowledge base on diversified agricultural systems. And this knowledge can be translated and incorporated into conventional agriculture to enhance the diversification of the whole food production system (Fig. 28.3D). Given the currently small extent of organic agriculture (Fig. 28.3B), such cross-fertilization of diversification knowledge is likely to be critical for scaling up the contributions of organic agriculture to the diversification of the food production system beyond the relatively small niche that it is currently occupying. Through such knowledge transfer, organic agriculture could have an impact on the food production system that goes far beyond the c. 1% of global agricultural land it is currently occupying.

Creating Space for Organics

We have laid out why we believe that organic agriculture provides important opportunities toward enhanced diversification of the food production system. But organic agriculture, or the inclusion of organic practices in conventional management, still appears to be at risk of being relegated to the sidelines, given current strong system constraints toward diversification in the broader food system. The neoliberal ideology underpinning our global food system at first sight trumps anything other than a niche role for organic agriculture. This is primarily because of the yield penalty associated with organic production (de Ponti et al., 2012; Ponisio et al., 2015; Seufert et al., 2012): if organic agriculture produces less, when the world demands more, it will require more land (which may not be available) to meet the demand (Connor and Mínguez, 2012). But there are several reasons why the future food production system could and should create more space for organics, which we will lay out in turn subsequently.

Firstly, the continued calls for increased food production are based on the (questionable) assumption that demand growth is inevitable. Analyses of how per capita demand changes with income are based on historical records, and particularly dietary transitions associated with the development of the "Westernized diet" (Tilman et al., 2011). But despite ample supply of food the current food system and the "Westernized diet" have not resulted in adequate nutrition to everybody. A relatively small handful of commodity crops (Cassidy et al., 2013), grown at scale, underpin the production of calorically dense, but nutritionally poor, diets, especially for the poor (Darmon and Drewnowski, 2015). This reliance on few crops has resulted in increasing global homogenization of diets (Khoury et al., 2014). Although there are some 800 million people who are calorically deficient, there are now more obese people in the world than underweight, and obesity is becoming the predominant form of malnutrition (NCD-RisC, 2016), with its associated noncommunicable diseases. This applies across both the rich and poor worlds. Given that there is increasing recognition of the health externalities associated with nutrient-poor, calorically rich diets based on animal products, processed grains, sugar, and oil, and the escalating global healthcare costs associated with noncommunicable, dietary-related diseases, it is increasingly a strong assumption that in the future, dietary transitions will follow similar paths to the past. Due to the inefficiencies associated with animal production, current and future populations could easily be fed on a less meat-intensive diet (Cassidy et al., 2013). In addition, the production of food at levels that are relatively cheaper than at any time in history has made food into a cheap commodity, which in turn incentives the wasteful use of food, resulting in up to a third of foods grown being lost or wasted (Foley et al., 2011). The scale of food waste and the potential for relatively "painless" interventions makes reducing it politically tractable, and reductions in food waste could provide another important contribution toward making more with less (Foley et al., 2011). Finally, the 2015 Paris Climate Agreement requires increasingly stringent action on emissions. The global agri-food system accounts for about 30% of greenhouse gas emissions, more than any other sector. Analyses suggest that changing diets, rather than changing production, is the most potent route to decarbonize the economy (Bajzelj et al., 2014). Changing diets for health, climate change mitigation, or any other reason, coupled with waste reduction, has clear scope to overcome the need to produce more food.

Secondly, current intensive production is built upon the assumption that externalizing environmental costs to create cheap food is necessary. This assumption is wrong for two reasons. Firstly, it emphasizes efficiency over resilience. While a rich literature is suggesting that there exist trade-offs between intensive production and different aspects of environmental impact (German et al., 2016), the relationships between sustainability, resilience, and yields are much less understood. However, as the weather changes, resilience in production may become a stronger driver of agricultural production, and one might expect highly efficient, intensive systems to be inherently less resilient due to their specialization than more diversified production systems.[1] Conversion to more resilient organic or organic-like methods, to counteract climate risk, may become a more rational strategy with time. Secondly, the externalization of environmental costs assumes that agriculture is separate from land or water. Though, in fact, the environmental degradation caused by agriculture is threatening the very natural resources upon which food production is based. As populations and their demands grow, the interconnectivity of food with water, energy, and land systems is becoming more apparent. And managing the "food, water, energy, land nexus" requires optimization rather than maximization of production.

Thirdly, the ideology of market liberalization, which is currently driving oversupply, homogenization, and specialization in the food system, might not be suitable to a world under climate change. In a stable world, a country concentrating production on crops where it has a comparative advantage, exporting its excess, and using that income to import what it does not grow is a rational strategy. However, increasing reliance on trade for local food security carries with it a systemic risk (Puma et al., 2015), particularly in a world characterized by a changing climate, where weather risks (e.g., from extreme events like droughts) can exasperate other existing risks. In such a less stable world an (over-)reliance on trade for food security may become increasingly problematic. A "bet-hedging" strategy to encourage local diversification of agriculture may, instead, become more attractive. Furthermore, as climate change influences weather patterns in unprecedented ways, local diversification of agriculture may reduce the variability in incomes, even if it is at the expense of mean income (Abson et al., 2013).

Another driver for diversification of local production systems, away from global trade, is rural development. Large-scale agriculture with its reliance on capital inputs typically reduces rural employment relative to small-scale enterprises (White, 2012). With urban infrastructure, especially of megacities, of increasing policy concern (Satterthwaite et al., 2010), maintaining rural employment through incentivizing small-scale producers (as is already done in many rural regions in Europe) may become more common in the future.

Finally, consumer demand for different modes of production is becoming an increasingly important driver in the food system, and food politics are increasingly consumer- rather than producer-led (Magnan, 2012). For example, with increasingly prominent focus on food authenticity and safety, short supply chains often are preferred by consumers (Kneafsey et al., 2013). While consumers are traditionally

[1] But see Influence of organic agriculture on diversification outcomes for a discussion of knowledge gaps on resilience in organic agriculture.

assumed to select foods primarily on price, other attributes do matter: quality, provenance, trust, and environmental and social attributes (Zander and Hamm, 2010). And as consumers become more informed, they typically express a greater willingness to pay for added values of food (Zander and Hamm, 2010). If there is demand for a diversity of locally produced food, it creates demand for a diversity of production, which may promote organic farming systems. Food sovereignty is the ability of a people to determine attributes of their food system, and while it is typically represented by smallholder systems (such as La Via Campesina), the European Union's rejection of genetically modified food or increasing pesticide restrictions is arguably an expression of the citizen consumer's desire not to have technologies imposed. Given the increasing awareness of consumers of environmental and health externalities of food production, it is not impossible to imagine changing patterns of purchasing behavior in a pro-health and pro-environmental way increasingly driving food production systems.

Given the health and environmental externalities associated with the current direction of travel in the food system, policy support to reduce demand and thus reduce the externalities of food production may grow. In addition, the systemic risks associated with relying on trade for local food security may be growing. And consumers, who are increasingly driving the food system, may demand more localized, healthier, and environmentally friendlier food choices. These factors, collectively, may drive the system in a different direction: different things being grown, in different ways.

Conclusion

Organic agriculture, which is based on diversification practices, can provide important contributions toward designing more diverse agroecosystems. The full potential of organic agriculture for the diversification of food systems will only come through the effective collaboration and knowledge translation across the organic sector to the vast majority of conventional farms. In a business-as-usual, liberalized, globalized, food system, organic farming's yield penalty (and thus greater price per unit output) will relegate it to a niche area, given the economic pressures to grow ever more, ever more cheaply. However, as explored earlier, there are a variety of drivers that could stimulate a tipping point toward increasing adoption of organic farming or a range of organic practices being becoming more integrated into conventional agriculture. These drivers may include building climate resilience, building diversified local food systems, greater public/private incentives to manage the food-water-land nexus, and compromise productivity for sustainability, as well as consumer focus on sustainable, healthy diets. The need for climate change mitigation could drive significant incentives around dietary change or farming practice change (e.g., a carbon tax on food emissions or a carbon premium to farmers for storing carbon). Social change can be rapid, given the right conditions.

References

Abson, D.J., Fraser, E.D., Benton, T.G., 2013. Landscape diversity and the resilience of agricultural returns: a portfolio analysis of land-use patterns and economic returns from lowland agriculture. Agriculture and Food Security 2, 1.

Aguilar, J., Gramig, G.G., Hendrickson, J.R., Archer, D.W., Forcella, F., Liebig, M.A., 2015. Crop species diversity changes in the United States: 1978–2012. PLoS One 10, e0136580.

Ahlman, T., Ljung, M., Rydhmer, L., Rocklinsberg, H., Strandberg, E., Wallenbeck, A., 2014. Differences in preferences for breeding traits between organic and conventional dairy producers in Sweden. Livestock Science 162, 5–14.

Andersson, G.K., Rundlöf, M., Smith, H.G., 2012. Organic farming improves pollination success in strawberries. PLoS One 7, e31599.

AUS, 2009. National standard for organic and biodynamic produce. In: A.Q.a.I.S. (Ed.), Organic Industry Export Consultative Committee, Edition 3.4.

Bajzelj, B., Richards, K.S., Allwood, J.M., Smith, P., Dennis, J.S., Curmi, E., Gilligan, C.A., 2014. Importance of food-demand management for climate mitigation. Nature Climate Change 4, 924–929.

Barański, M., Średnicka-Tober, D., Volakakis, N., Seal, C., Sanderson, R., Stewart, G.B., Benbrook, C., Biavati, B., Markellou, E., Giotis, C., 2014. Higher antioxidant and lower cadmium concentrations and lower incidence of pesticide residues in organically grown crops: a systematic literature review and meta-analyses. British Journal of Nutrition 112, 794–811.

Bàrberi, P., 2002. Weed management in organic agriculture: are we addressing the right issues? Weed Research 42, 177–193.

Belfrage, K., Björklund, J., Salomonsson, L., 2005. The effects of farm size and organic farming on diversity of birds, pollinators, and plants in a Swedish landscape. Ambio 34, 582–588.

Bengtsson, J., Ahnström, J., Weibull, A.-C., 2005. The effects of organic agriculture on biodiversity and abundance: a meta-analysis. Journal of Applied Ecology 42, 261–269.

Bennett, A.J., Bending, G.D., Chandler, D., Hilton, S., Mills, P., 2012. Meeting the demand for crop production: the challenge of yield decline in crops grown in short rotations. Biological Reviews 87, 52–71.

Benton, T.G., Vickery, J.A., Wilson, J.D., 2003. Farmland biodiversity: is habitat heterogeneity the key? Trends in Ecology and Evolution 18, 182–188.

Berg, G., 2009. Plant–microbe interactions promoting plant growth and health: perspectives for controlled use of microorganisms in agriculture. Applied Microbiology and Biotechnology 84, 11–18.

Berry, P., Sylvester Bradley, R., Philipps, L., Hatch, D.J., Cuttle, S.P., Rayns, F., Gosling, P., 2002. Is the productivity of organic farms restricted by the supply of available nitrogen? Soil Use and Management 18, 248–255.

Birkhofer, K., Arvidsson, F., Ehlers, D., Mader, V.L., Bengtsson, J., Smith, H.G., 2016. Organic farming affects the biological control of hemipteran pests and yields in spring barley independent of landscape complexity. Landscape Ecology 31, 567–579.

Boelling, D., Groen, A.F., Sørensen, P., Madsen, P., Jensen, J., 2003. Genetic improvement of livestock for organic farming systems. Livestock Production Science 80, 79–88.

Bommarco, R., Kleijn, D., Potts, S.G., 2013. Ecological intensification: harnessing ecosystem services for food security. Trends in Ecology and Evolution 28, 230–238.

Brandt, K., Leifert, C., Sanderson, R., Seal, C., 2011. Agroecosystem management and nutritional quality of plant foods: the case of organic fruits and vegetables. Critical Reviews in Plant Sciences 30, 177–197.

Brittain, C., Bommarco, R., Vighi, M., Settele, J., Potts, S.G., 2010. Organic farming in isolated landscapes does not benefit flower-visiting insects and pollination. Biological Conservation 143, 1860–1867.

Buck, D., Getz, C., Guthman, J., 1997. From farm to table: the organic vegetable commodity chain of Northern California. Sociologia Ruralis 37, 3–20.

Campbell, H., Liepins, R., 2001. Naming organics: understanding organic standards in New Zealand as a discursive field. Sociologia Ruralis 41, 22–39.

Cassidy, E.S., West, P.C., Gerber, J.S., Foley, J.A., 2013. Redefining agricultural yields: from tonnes to people nourished per hectare. Environmental Research Letters 8, 034015.

Chamberlain, D., Wilson, J., Fuller, R., 1999. A comparison of bird populations on organic and conventional farm systems in southern Britain. Biological Conservation 88, 307–320.

Clark, M., Ferris, H., Klonsky, K., Lanini, W., Van Bruggen, A., Zalom, F., 1998. Agronomic, economic, and environmental comparison of pest management in conventional and alternative tomato and corn systems in northern California. Agriculture, Ecosystems and Environment 68, 51–71.

Clark, S., Klonsky, K., Livingston, P., Temple, S., 1999. Crop-yield and economic comparisons of organic, low-input, and conventional farming systems in California's Sacramento Valley. American Journal of Alternative Agriculture 14, 109–121.

Clough, Y., Holzschuh, A., Gabriel, D., Purtauf, T., Kleijn, D., Kruess, A., Steffan-Dewenter, I., Tscharntke, T., 2007. Alpha and beta diversity of arthropods and plants in organically and conventionally managed wheat fields. Journal of Applied Ecology 44, 804–812.

Connor, D.J., Mínguez, M.I., 2012. Evolution not revolution of farming systems will best feed and green the world. Global Food Security 1, 106–113.

Crosson, P.R., 2016. Productivity Effects of Cropland Erosion in the United States. Routledge.

Crowder, D.W., Northfield, T.D., Strand, M.R., Snyder, W.E., 2010. Organic agriculture promotes evenness and natural pest control. Nature 466, 109–112.

Dangour, A.D., Lock, K., Hayter, A., Aikenhead, A., Allen, E., Uauy, R., 2010. Nutrition-related health effects of organic foods: a systematic review. The American Journal of Clinical Nutrition 92, 203–210.

Darmon, N., Drewnowski, A., 2015. Contribution of food prices and diet cost to socioeconomic disparities in diet quality and health: a systematic review and analysis. Nutrition Reviews 73 (10), 643–660.

Darnhofer, I., Lindenthal, T., Bartel-Kratochvil, R., Zollitsch, W., 2010. Conventionalisation of organic farming practices: from structural criteria towards an assessment based on organic principles. A review. Agronomy for Sustainable Development 30, 67–81.

de Ponti, T., Rijk, B., Van Ittersum, M.K., 2012. The crop yield gap between organic and conventional agriculture. Agricultural Systems 108, 1–9.

Delmotte, S., Tittonell, P., Mouret, J.-C., Hammond, R., Lopez-Ridaura, S., 2011. On farm assessment of rice yield variability and productivity gaps between organic and conventional cropping systems under Mediterranean climate. European Journal of Agronomy 35, 223–236.

EU, 2008. Commission Regulation (EC) No 889/2008 laying down detailed rules for the implementation of Council Regulation (EC) No 834/2007 on organic production and labelling of organic products with regard to organic production, labelling and control. In: Union, E. (Ed.), L 250/1, Official Journal of the European Union.

FAO, IFAD, WFP, 2015. The state of food insecurity in the world 2015. In: FAO (Ed.), Meeting the 2015 International Hunger Targets: Taking Stock of Uneven Progress. FAO, Rome.

FAO, 2000. 2000 World Census of Agriculture. FAO Statistical Development Series 12, Rome, p. 238.

FAO, WHO, 2001. Codex Alimentarius - Organically Produced Foods. World Health Organization (WHO) & Food and Agriculture Organization of the United Nations (FAO), Rome, p. 65.

FiBL, IFOAM, 2016. The World of Organic Agriculture. Statistics and Emerging Trends 2016, Frick, Switzerland.

Foley, J.A., Ramankutty, N., Brauman, K.A., Cassidy, E.S., Gerber, J., Johnston, M., Mueller, N.D., O'Connell, C., Ray, D., West, P.C., Balzer, C., Bennett, E.M., Carpenter, S.R., Hill, J., Monfreda, C., Polasky, S., Rockstrom, J., Sheehan, J., Siebert, S., Tilman, D., Zaks, D., 2011. Solutions for a cultivated planet. Nature 478, 337–342.

Frison, E.A., Cherfas, J., Hodgkin, T., 2011. Agricultural biodiversity is essential for a sustainable improvement in food and nutrition security. Sustainability 3, 238–253.

Fuller, R., Norton, L., Feber, R., Johnson, P., Chamberlain, D., Joys, A., Mathews, F., Stuart, R., Townsend, M., Manley, W., 2005. Benefits of organic farming to biodiversity vary among taxa. Biology Letters 1, 431–434.

Gabriel, D., Carver, S.J., Durham, H., Kunin, W.E., Palmer, R.C., Sait, S.M., Stagl, S., Benton, T.G., 2009. The spatial aggregation of organic farming in England and its underlying environmental correlates. Journal of Applied Ecology 46, 323–333.

Gabriel, D., Sait, S.M., Hodgson, J.A., Schmutz, U., Kunin, W.E., Benton, T.G., 2010. Scale matters: the impact of organic farming on biodiversity at different spatial scales. Ecology Letters 13, 858–869.

Gattinger, A., Muller, A., Haeni, M., Skinner, C., Fliessbach, A., Buchmann, N., Mäder, P., Stolze, M., Smith, P., Scialabba, N.E.-H., 2012. Enhanced top soil carbon stocks under organic farming. Proceedings of the National Academy of Sciences 109, 18226–18231.

German, R.N., Thompson, C.E., Benton, T.G., 2016. Relationships among multiple aspects of agriculture's environmental impact and productivity: a meta-analysis to guide sustainable agriculture. Biological Reviews 92 (2), 716–738.

Hall, A., Mogyorody, V., 2001. Organic farmers in Ontario: an examination of the conventionalization argument. Sociologia Ruralis 41, 399–422.

Hall, S.J., Ruane, J., 1993. Livestock breeds and their conservation: a global overview. Conservation Biology 7, 815–825.

IFOAM, 2006. The IFOAM Norms for Organic Production and Processing. Version 2005. International Federation of Organic Agriculture Movements (IFOAM).

Isman, M.B., 2008. Botanical insecticides: for richer, for poorer. Pest Management Science 64, 8–11.

Istituto Nazionale di Statistica, 2010. Sesto censimento general dell'agricoltura - Caratteristiche strutturale delle aziende agricole. Istituto Nazionale di Statistica, Rome 221. https://www.istat.it/it/files/2011/03/142512_Vol_VI_Cens_Agricoltura_INT_CD_1_Trimboxes_ipp.pdf.

Kennedy, C.M., Lonsdorf, E., Neel, M.C., Williams, N.M., Ricketts, T.H., Winfree, R., Bommarco, R., Brittain, C., Burley, A.L., Cariveau, D., 2013. A global quantitative synthesis of local and landscape effects on wild bee pollinators in agroecosystems. Ecology Letters 16, 584–599.

Khoury, C.K., Bjorkman, A.D., Dempewolf, H., Ramirez-Villegas, J., Guarino, L., Jarvis, A., Rieseberg, L.H., Struik, P.C., 2014. Increasing homogeneity in global food supplies and the implications for food security. Proceedings of the National Academy of Sciences 111, 4001–4006.

Klein, A.M., Brittain, C., Hendrix, S.D., Thorp, R., Williams, N., Kremen, C., 2012. Wild pollination services to California almond rely on semi-natural habitat. Journal of Applied Ecology 49, 723–732.

Kneafsey, M., Venn, L., Schmutz, U., Balázs, B., Trenchard, L., Eyden-Wood, T., Bos, E., Sutton, G., Blackett, M., 2013. Short Food Supply Chains and Local Food Systems in the EU. A State of Play of Their Socio-economic Characteristics. JRC Scientific and Policy Reports. Joint Research Centre Institute for Prospective Technological Studies, European Commission.

Kragten, S., de Snoo, G.R., 2008. Field-breeding birds on organic and conventional arable farms in The Netherlands. Agriculture, Ecosystems and Environment 126, 270–274.

Kremen, C., Iles, A., Bacon, C., 2012. Diversified farming systems: an agroecological, systems-based alternative to modern industrial agriculture. Ecology and Society 17 (4), 44.

Kremen, C., Williams, N.M., Thorp, R.W., 2002. Crop pollination from native bees at risk from agricultural intensification. Proceedings of the National Academy of Sciences 99, 16812–16816.

Lebow, V., 1955. Price competition in 1955. Journal of Retailing 31, 5–10.

Leifeld, J., Fuhrer, J., 2010. Organic farming and soil carbon sequestration: what do we really know about the benefits? AMBIO: A Journal of the Human Environment 1–15.

Lin, B.B., 2011. Resilience in agriculture through crop diversification: adaptive management for environmental change. BioScience 61, 183–193.

Lockie, S., Lyons, K., Lawrence, G., Grice, J., 2004. Choosing organics: a path analysis of factors underlying the selection of organic food among Australian consumers. Appetite 43, 135–146.

Lotter, D., Seidel, R., Liebhardt, W., 2003. The performance of organic and conventional cropping systems in an extreme climate year. American Journal of Alternative Agriculture 18, 146–154.

Lowder, S.K., Skoet, J., Raney, T., 2016. The number, size, and distribution of farms, smallholder farms, and family farms worldwide. World Development 87, 16–29.

Mäder, P., Fliessbach, A., Dubois, D., Gunst, L., Fried, P., Niggli, U., 2002. Soil fertility and biodiversity in organic farming. Science 296, 1694–1697.

Magnan, A., 2012. Food Regimes. Oxford University Press, Oxford.

Maltais-Landry, G., Scow, K., Brennan, E., Vitousek, P., 2015. Long-term effects of compost and cover crops on soil phosphorus in two California agroecosystems. Soil Science Society of America Journal 79, 688–697.

McKeon, N., 2014. Food Security Governance: Empowering Communities, Regulating Corporations. Taylor & Francis.

Naylor, R., Steinfeld, H., Falcon, W., Galloway, J., Smil, V., Bradford, E., Alder, J., Mooney, H., 2005. Losing the links between livestock and land. Science 310, 1621–1622.

NCD-RisC, 2016. Trends in adult body-mass index in 200 countries from 1975 to 2014: a pooled analysis of 1698 population-based measurement studies with 19·2 million participants. The Lancet 387, 1377–1396.

Norton, L., Johnson, P., Joys, A., Stuart, R., Chamberlain, D., Feber, R., Firbank, L., Manley, W., Wolfe, M., Hart, B., Mathews, F., MacDonald, D., Fuller, R.J., 2009. Consequences of organic and non-organic farming practices for field, farm and landscape complexity. Agriculture, Ecosystems and Environment 129, 221–227.

Nowak, B., Nesme, T., David, C., Pellerin, S., 2015. Nutrient recycling in organic farming is related to diversity in farm types at the local level. Agriculture, Ecosystems and Environment 204, 17–26.

NPOP, 2005. In: M.o.C.a.I (Ed.), National Programme for Organic Production. Department of Commerce.

Östman, Ö., Ekbom, B., Bengtsson, J., 2003. Yield increase attributable to aphid predation by ground-living polyphagous natural enemies in spring barley in Sweden. Ecological Economics 45, 149–158.

Penfold, C.M., Miyan, M., Reeves, T., Grierson, I.T., 1995. Biological farming for sustainable agricultural production. Animal Production Science 35, 849–856.

Ponisio, L.C., M'Gonigle, L.K., Mace, K.C., Palomino, J., de Valpine, P., Kremen, C., 2015. Diversification practices reduce organic to conventional yield gap. Proceedings of the Royal Society of London B Biological Sciences 282, 20141396.

Puma, M.J., Bose, S., Chon, S.Y., Cook, B.I., 2015. Assessing the evolving fragility of the global food system. Environmental Research Letters 10, 024007.

Rete Rurale Nazionale, 2013. Bioreport 2013 - L'agricoltura biologica in Italia. Ministero delle Politiche Agricole, Alimentari e Forestali, Rome 156. http://www.sinab.it/sites/default/files/share/BIOREPORT_2013_WEB%5B1%5D.pdf.

RRN, 2013. Bioreport 2013-L'agricoltura biologica in Italia, 2007-2013. In: R.R.N. (Ed.), Ministero Delle Politiche Agricole. Alimentari e Forestali, Rome, p. 156.

Rundlöf, M., Bengtsson, J., Smith, H.G., 2008. Local and landscape effects of organic farming on butterfly species richness and abundance. Journal of Applied Ecology 45, 813–820.

Satterthwaite, D., McGranahan, G., Tacoli, C., 2010. Urbanization and its implications for food and farming. Philosophical Transactions of the Royal Society B: Biological Sciences 365, 2809–2820.

Schneider, M.K., Lüscher, G., Jeanneret, P., Arndorfer, M., Ammari, Y., Bailey, D., Balázs, K., Báldi, A., Choisis, J.-P., Dennis, P., 2014. Gains to species diversity in organically farmed fields are not propagated at the farm level. Nature Communications 5, 4151.

Seufert, V., Ramankutty, N., Foley, J., 2012. Comparing the yields of organic and conventional agriculture. Nature 485, 229–232.

Seufert, V., Ramankutty, N., Mayerhofer, T., 2017. What is this thing called organic? - How organic farming is codified in regulations. Food Policy 68, 10–20.

Slattery, E., Livingston, M., Greene, C., Klonsky, K., 2011. Characteristics of conventional and organic apple production in the United States. In: Service/USDA, E.R. (Ed.), A Report from the Economic Research Service FTS-347-0 USDA. United States Department of Agriculture, pp. 1–27.

Smith, R.G., Menalled, F.D., Robertson, G.P., 2007. Temporal yield variability under conventional and alternative management systems. Agronomy Journal 99, 1629.

Smith-Spangler, C., Brandeau, M.L., Hunter, G.E., Bavinger, J.C., Pearson, M., Eschbach, P.J., Sundaram, V., Liu, H., Schirmer, P., Stave, C., 2012. Are organic foods safer or healthier than conventional alternatives?: a systematic review. Annals of Internal Medicine 157, 348–366.

Sorge, U.S., Moon, R., Wolff, L.J., Michels, L., Schroth, S., Kelton, D.F., Heins, B., 2016. Management practices on organic and conventional dairy herds in Minnesota. Journal of Dairy Science 99, 3183–3192.

Średnicka-Tober, D., Barański, M., Seal, C., Sanderson, R., Benbrook, C., Steinshamn, H., Gromadzka-Ostrowska, J., Rembiałkowska, E., Skwarło-Sonta, K., Eyre, M., 2016. Composition differences between organic and conventional meat: a systematic literature review and meta-analysis. British Journal of Nutrition 115, 994–1011.

Statistisches Bundesamt, 2014a. Land- und Forstwirtschaft, Fischerei - Betriebe mit ökologischem Landbau Agrarstrukturerhebung. Fachserie 3, Reihe 2.2.1. Statistisches Bundesamt, Wiesbaden 105. https://www.destatis.de/DE/Publikationen/Thematisch/LandForstwirtschaft/Betriebe/OekologischerLandbau2030221139004.pdf?__blob=publicationFile.

Statistisches Bundesamt, 2014b. Land- und Forstwirtschaft, Fischerei - Betriebswirtschaftliche Ausrichtung und Standardoutput Agrarstrukturerhebung. Fachserie 3, Reihe 2.1.4. Statistisches Bundesamt, Wiesbaden 397. https://www.destatis.de/DE/Publikationen/Thematisch/LandForstwirtschaft/Betriebe/BetriebswirtschaftlicheAusrichtungStandardoutput2030214139004.pdf?__blob=publicationFile.

Stobbelaar, D.J., Groot, J.C., Bishop, C., Hall, J., Pretty, J., 2009. Internalization of agri-environmental policies and the role of institutions. Journal of Environmental Management 90, S175–S184.

Sundberg, T., Berglund, B., Rydhmer, L., Strandberg, E., 2009. Fertility, somatic cell count and milk production in Swedish organic and conventional dairy herds. Livestock Science 126, 176–182.

Tilman, D., Balzer, C., Hill, J., Befort, B.L., 2011. Global food demand and the sustainable intensification of agriculture. Proceedings of the National Academy of Sciences of the United States of America 108, 20260–20264.

Tscharntke, T., Klein, A.M., Kruess, A., Steffan-Dewenter, I., Thies, C., 2005. Landscape perspectives on agricultural intensification and biodiversity–ecosystem service management. Ecology Letters 8, 857–874.

Tuck, S.L., Winqvist, C., Mota, F., Ahnström, J., Turnbull, L.A., Bengtsson, J., 2014. Land-use intensity and the effects of organic farming on biodiversity: a hierarchical meta-analysis. Journal of Applied Ecology 51, 746–755.

USDA, 2014a. 2014 Organic Survey.

USDA, 2014b. 2012 Census of Agriculture - United States, Summary and State Data. USDA, National Agricultural Statistics Service.

van Bueren, E.L., Jones, S.S., Tamm, L., Murphy, K.M., Myers, J.R., Leifert, C., Messmer, M., 2011. The need to breed crop varieties suitable for organic farming, using wheat, tomato and broccoli as examples: a review. NJAS - Wageningen Journal of Life Sciences 58, 193–205.

Van Bueren, E.L., Struik, P., Jacobsen, E., 2002. Ecological concepts in organic farming and their consequences for an organic crop ideotype. NJAS - Wageningen Journal of Life Sciences 50, 1–26.

van Elsen, T., 2000. Species diversity as a task for organic agriculture in Europe. Agriculture, Ecosystems and Environment 77, 101–109.

Von Borell, E., Sørensen, J.T., 2004. Organic livestock production in Europe: aims, rules and trends with special emphasis on animal health and welfare. Livestock Production Science 90, 3–9.

Watson, C., Atkinson, D., Gosling, P., Jackson, L., Rayns, F., 2002. Managing soil fertility in organic farming systems. Soil Use and Management 18, 239–247.

Weibull, A.-C., Östman, Ö., Granqvist, Å., 2003. Species richness in agroecosystems: the effect of landscape, habitat and farm management. Biodiversity and Conservation 12, 1335–1355.

White, B., 2012. Agriculture and the generation problem: rural youth, employment and the future of farming. IDS Bulletin 43, 9–19.

Wolfe, M., Baresel, J., Desclaux, D., Goldringer, I., Hoad, S., Kovacs, G., Löschenberger, F., Miedaner, T., Østergård, H., Van Bueren, E.L., 2008. Developments in breeding cereals for organic agriculture. Euphytica 163, 323–346.

Zander, K., Hamm, U., 2010. Consumer preferences for additional ethical attributes of organic food. Food Quality and Preference 21, 495–503.

Zehnder, G., Gurr, G.M., Kühne, S., Wade, M.R., Wratten, S.D., Wyss, E., 2007. Arthropod pest management in organic crops. Annual Review of Entomology 52, 57–80.

Index

'*Note:* Page numbers followed by "f" indicate figures, "t" indicate tables and "b" indicate boxes.'

B

C

H

I

T

Printed in the United States
By Bookmasters